Methods in Enzymology

Volume 84
IMMUNOCHEMICAL TECHNIQUES
Part D
Selected Immunoassays

METHODS IN ENZYMOLOGY

EDITORS-IN-CHIEF

Sidney P. Colowick   Nathan O. Kaplan

Methods in Enzymology

Volume 84

Immunochemical Techniques

Part D
Selected Immunoassays

EDITED BY

John J. Langone

LABORATORY OF IMMUNOBIOLOGY
NATIONAL CANCER INSTITUTE
NATIONAL INSTITUTES OF HEALTH
BETHESDA, MARYLAND

Helen Van Vunakis

DEPARTMENT OF BIOCHEMISTRY
BRANDEIS UNIVERSITY
WALTHAM, MASSACHUSETTS

1982

ACADEMIC PRESS

A Subsidiary of Harcourt Brace Jovanovich, Publishers

New York   London
Paris   San Diego   San Francisco   São Paulo   Sydney   Tokyo   Toronto

ACADEMIC PRESS, INC.
111 Fifth Avenue, New York, New York 10003

United Kingdom Edition published by
ACADEMIC PRESS, INC. (LONDON) LTD.
24/28 Oval Road, London NW1 7DX

Library of Congress Cataloging in Publication Data
Main entry under title:

Immunochemical techniques.

(Methods in enzymology; v. 84)
Includes bibliographical references and index.
1. Radioimmunoassay. 2. Immunoassay. I. Langone,
John J. (John Joseph), Date. II. Van Vunakis,
Helen, Date. III. Series.
QP601.M49 vol. 84 [QP519.9.R3] 574.19'25s 82-1678
ISBN 0-12-181984-1 [616.07'57] AACR2

PRINTED IN THE UNITED STATES OF AMERICA

82 83 84 85   9 8 7 6 5 4 3 2 1

Table of Contents

Section I. Oncofetal Proteins

Section II. Proteins and Peptides of the Blood Clotting System

Section III. Metal and Heme Binding Proteins

Section IV. Nucleic Acids and Their Antibodies

Section V. Toxins

Section VI. Endogenous Compounds of Low Molecular Weight

Section VII. Drugs

A. Antineoplastic Agents

B. Drugs Active on the Nervous System and Neurotransmitters

Contributors to Volume 84

Article numbers are in parentheses following the names of contributors.
Affiliations listed are current.

M. AKHTAR (20), *Microbiology Research Division, Health Protection Branch, Health and Welfare Canada, Ottawa K1A OL2, Ontario, Canada*

PHILLIP W. ALBRO (47), *Laboratory of Environmental Chemistry, National Institute of Environmental Health Sciences, Research Triangle Park, North Carolina 27709*

NAOMI ARIEL (3), *Department of Microbiology, Israel Institute for Biological Research, P.O. Box 19, Ness Ziona, Israel*

LUC BÉLANGER (2), *Molecular Oncology Laboratory, L'Hôtel-Dieu de Quebec Research Center, Laval University, Quebec G1R 2J6, Canada*

RITA A. BLANCHARD (5), *Department of Medicine, Tufts–New England Medical Center, Boston, Massachusetts 02111*

ALAN BROUGHTON (32), *Nichols Institute, San Juan Capistrano, California 92675*

JOHN C. BROWN (26), *Department of Physiology, Faculty of Medicine, University of British Columbia, Vancouver, British Columbia V6T 1W5, Canada*

VINCENT P. BUTLER, JR. (8, 42), *Department of Medicine, College of Physicians and Surgeons, Columbia University, New York, New York 10032*

ROBERT E. CANFIELD (8), *Department of Medicine, College of Physicians and Surgeons, Columbia University, New York, New York 10032*

ANDRE CASTONGUAY (49), *Division of Chemical Carcinogenesis, American Health Foundation, Valhalla, New York 10595*

MIROSLAV CESKA (19), *Department of Immunology, Sandoz Forschungsinstitut, A-1235 Wien, Austria*

KUN CHAE (47), *Laboratory of Environmental Chemistry, National Institute of Environmental Health Sciences, Research Triangle Park, North Carolina 27709*

JAMES G. CHAFOULEAS (10), *Department of Cell Biology, Baylor College of Medicine, Houston, Texas 77030*

CHIN C. CHANG (9), *Merrell Research Center, Merrell Dow Pharmaceuticals Inc., Cincinnati, Ohio 45215*

INDER J. CHOPRA (22), *Division of Endocrinology, Department of Medicine, Center for the Health Sciences, University of California School of Medicine, Los Angeles, California 90024*

GEORGE CLARK (47), *Department of Immunology, University of California, Berkeley, California 94720*

ROBERT J. CONNOR (46), *Division of Biological Effects, Bureau of Radiological Health, Food and Drug Administration, Washington, D.C. 20201*

MARIA DA COSTA (31), *Department of Medicine, New York Medical College, New York, New York 10029*

JOHN R. DEDMAN (10), *Department of Medicine, Division of Endocrinology, University of Texas Health Science Center at Houston, Houston, Texas 77030*

M. A. DELAAGE (25), *Centre d'Immunologie, INSERM-CNRS, Case 906, 13288 Marseille Cedex 9, France*

M. L. DE REGGI (25), *Centre d'Immunologie, INSERM-CNRS, Case 906, 13288 Marseille Cedex 9, France*

N. DICKIE (20), *Microbiology Research Division, Health Protection Branch, Health and Welfare Canada, Ottawa K1A OL2, Ontario, Canada*

ROSS DIXON (35), *Research Division, Hoffmann-La Roche Inc., Nutley, New Jersey 07110*

ROSS C. DONEHOWER (30), *Johns Hopkins Oncology Center, Baltimore, Maryland 21205*

ix

JILL R. DRYBURGH (26), *Department of Biochemistry, University of Surrey, Guildford, Surrey, England*

EVA ENGVALL (1), *La Jolla Cancer Research Foundation, La Jolla, California 92037*

CHARLES ERLICHMAN (30), *Department of Medicine, Ontario Cancer Institute, Toronto, Ontario M4X 1K9, Canada*

DAVID S. FREEMAN (36), *Department of Pharmaceutical Sciences, College of Pharmacy, North Dakota State University, Fargo, North Dakota 58105*

BARBARA C. FURIE (5), *Department of Medicine and Department of Biochemistry and Pharmacology, Tufts–New England Medical Center, Boston, Massachusetts 02111*

BRUCE FURIE (5), *Department of Medicine and Department of Biochemistry and Pharmacology, Tufts–New England Medical Center, Boston, Massachusetts 02111*

JUSTINE S. GARVEY (9), *Department of Biology, Syracuse University, Syracuse, New York 13210*

HILDA B. GJIKA (36), *Department of Biochemistry, Brandeis University, Waltham, Massachusetts 02154*

JAN A. GUTOWSKI (21), *Connaught Research Institute, Willowdale, Ontario M2N 5T8, Canada*

A. M. HAGENAARS (18), *National Institute of Public Health, Laboratory for Immunology, P. O. Box 1, 3720 BA Bilthoven, The Netherlands*

JUNE W. HALLIDAY (11), *Department of Medicine, University of Queensland, Royal Brisbane Hospital, Herton 4029, Australia*

SALLY E. HAYS (38), *Stanford Research Institute, Menlo Park, California 94025*

R. HENDRIKS (39), *Department of Drug Metabolism and Pharmacokinetics, Janssen Pharmaceutica N.V., B-2340 Beerse, Belgium*

J. HEYKANTS (39), *Department of Drug Metabolism and Pharmacokinetics, Janssen Pharmaceutica N.V., B-2340 Beerse, Belgium*

M. H. HIRN (25), *Centre d'Immunologie,*

INSERM-CNRS, *Case 906, 13288 Marseille Cedex 9, France*

LEON W. HOYER (4), *Department of Medicine, University of Connecticut Health Center, Farmington, Connecticut 06032*

DIANE M. JACOBS (21), *Department of Microbiology, State University of New York at Buffalo, Buffalo, New York 14214*

R. S. KAMEL (29), *Department of Chemical Pathology, St. Bartholomew's Hospital, London EC1A 7HL, England*

KAREN L. KAPLAN (6, 7), *Department of Medicine, College of Physicians and Surgeons, Columbia University, New York, New York 10032*

JOAN L. KLOTZ (15), *Division of Cytogenetics and Cytology, City of Hope National Medical Center, 1500 East Duarte Road, Duarte, California 91010*

S. KOZAKI (18), *University of Osaka Prefecture, College of Agriculture, Department of Veterinary Science, Sakai-Shi, Osaka 591, Japan*

MARK J. KRANTZ (3), *Department of Medicine, Cross Cancer Institute, Edmonton, Alberta T6G 1Z2, Canada*

SUZANNE LAFERTÉ (3), *McGill Cancer Centre, McGill University, Montreal, Quebec H3G 1Y6, Canada*

J. LANDON (29), *Department of Chemical Pathology, St. Bartholomew's Hospital, London EC1A 7HL, England*

JOHN J. LANGONE (28, 48, 51), *Laboratory of Immunobiology, National Cancer Institute, National Institutes of Health, Bethesda, Maryland 20205*

MICHAEL I. LUSTER (47), *National Toxicology Program, National Institute of Environmental Health Sciences, Research Triangle Park, North Carolina 27709*

RENÉ MASSEYEFF (2), *Laboratoire d'Immunologie, Faculté de Médicine, Université de Nice, 06034 Nice-Cedex, France*

JAMES D. MCKINNEY (47), *Laboratory of Environmental Chemistry, National Institute of Environmental Health Sciences, Research Triangle Park, North Carolina 27709*

ANTHONY R. MEANS (10), *Department of*

Cell Biology, Baylor College of Medicine, Houston, Texas 77030

A. WAYNE MEIKLE (44), Department of Internal Medicine, University of Utah College of Medicine, Salt Lake City, Utah 84132

M. MICHIELS (39), Department of Drug Metabolism and Pharmacokinetics, Janssen Pharmaceutica N.V., B-2340 Beerse, Belgium

JOHN G. MOFFATT (33), Institute of Bio-Organic Chemistry, Syntex Research, 3401 Hillview Avenue, Palo Alto, California 94304

CHARLES E. MYERS (30), Clinical Pharmacology Branch, National Cancer Institute, National Institutes of Health, Bethesda, Maryland 20205

HIROSHI NAKAZATO (17), Laboratory of Biotechnology, Suntory Institute for Biomedical Research, Shimamoto-Cho, Osaka 618, Japan

HYMIE L. NOSSEL (8), Department of Medicine, College of Physicians and Surgeons, Columbia University, New York, New York 10032

S. NOTERMANS (18), National Institute of Public Health, Laboratory for Zoonoses and Food Microbiology, P. O. Box 1, 3720 BA Bilthoven, The Netherlands

TADASHI OKABAYASHI (33), Shionogi Research Laboratories, Shionogi and Co. Ltd., Fukushima-ku, Osaka 553, Japan

F. ÖTTING (43), Department of Proteinchemistry, Center of Research, Grünenthal GmbH, D-5100 Aachen, Federal Republic of Germany

JOHN OWEN (6, 7), Department of Medicine, College of Physicians and Surgeons, Columbia University, New York, New York 10032

MIRIAM C. POIRIER (46), In Vitro Pathogenesis Section, National Cancer Institute, National Institutes of Health, Bethesda, Maryland 20205

RUSSELL E. POLAND (38), Division of Biological Psychiatry, Department of Psychiatry, Harbor-UCLA Medical Center, Torrance, California 90509

J. J. PRATT (27), Isotopenlaboratorium, Academisch Ziekenhuis, Oostersingel 59, 9713 EZ Groningen, The Netherlands

VIC RASO (34), Division of Biochemical Pharmacology, Sidney Farber Cancer Institute, and the Department of Pathology, Harvard Medical School, Boston, Massachusetts 02115

DAVID J. ROBISON (5), Department of Biochemistry and Pharmacology, Tufts University School of Medicine, Boston, Massachusetts 02111

P. E. ROSS (24), Department of Medicine, Ninewells Hospital and Medical School, Dundee DD1 9SY, Scotland

SHELDON P. ROTHENBERG (31), Department of Medicine, New York Medical College, New York, New York 10029

ROBERT T. RUBIN (38), Division of Biological Psychiatry, Department of Psychiatry, Harbor-UCLA Medical Center, Torrance, California 90509

ERKKI RUOSLAHTI (1), La Jolla Cancer Research Foundation, La Jolla, California 92037

TAKESHI SASAKI (16), The Second Department of Internal Medicine, Tohoku University School of Medicine, Sendai, Miyagi 980, Japan

ROBERT P. SCHLEIMER (37), Division of Clinical Immunology, Johns Hopkins University School of Medicine, Baltimore, Maryland 21239

HARTMUT R. SCHROEDER (23), Immunochemistry, Ames Division of Miles Laboratories, Inc., P. O. Box 70, Elkhart, Indiana 46515

WEI-CHIANG SHEN (50), Department of Pathology, Boston University School of Medicine, Boston, Massachusetts 02118

SYDNEY SPECTOR (40, 41), Roche Institute of Molecular Biology, Nutley, New Jersey 07110

CHARLES R. STEINMAN (13, 14), Department of Medicine, Mount Sinai School of Medicine, New York, New York 10029

MARVIN J. STONE (12), The Charles A. Sammons Cancer Center, Baylor University Medical Center, Dallas, Texas 75246

MINDY M. TAI (5), *Immunology Program, Tufts University School of Medicine, Boston, Massachusetts 02111*

TAKEHIKO TAKATORI (45), *Department of Legal Medicine, Hokkaido University School of Medicine, Sapporo 060, Japan*

BARBARA B. TOWER (38), *Equine Research Center, California Polytechnic University, Pomona, California 91768*

NORMA C. TRABOLD (4), *Department of Medicine, University of Connecticut Health Center, Farmington, Connecticut 06032*

DORIS TSE-ENG (42), *Department of Medicine, College of Physicians and Surgeons, Columbia University, New York, New York 10032*

MARJATTA UOTILA (1), *La Jolla Cancer Research Foundation, La Jolla, California 92037*

RONALD J. VANDER MALLIE (9), *New England Nuclear Corporation, North Billerica, Massachusetts 01862*

HELEN VAN VUNAKIS (48, 49, 51), *Department of Biochemistry, Brandeis University, Waltham, Massachusetts 02254*

MICHAEL R. WATERMAN (12), *Department of Biochemistry, The University of Texas Health Science Center at Dallas, Dallas, Texas 75235*

WILMAR M. WIERSINGA (22), *Klinick Voor Inwendige Ziekten, Academisch Ziekenhuis, Wilhelmina Gasthuis, Amsterdam, The Netherlands*

JAMES T. WILLERSON (12), *Department of Internal Medicine, The University of Texas Health Science Center at Dallas, Dallas, Texas 75235*

M. G. WOLDRING (27), *Isotopenlaboratorium, Academisch Ziekenhuis, Oostersingel 59, 9713 EZ Groningen, The Netherlands*

Preface

Previous volumes (70, 73, 74) of Immunochemical Techniques deal with the properties of antigen–antibody interactions, procedures for the preparation of immunochemical reagents, and various techniques for the detection and estimation of antigens, antibodies, and circulating immune complexes. This volume contains descriptions of immunoassays for some biologically important molecules, including macromolecules as well as low-molecular-weight nonantigenic compounds that require chemical modification before they can be used in the production of specific antibodies. One major goal is to illustrate the variety of immunological procedures involved in the development of a successful assay for substances of diverse structural and biochemical properties.

Immunoassays for many other molecules of biochemical and clinical interest appear throughout the Series. They are indexed under specific categories in this volume [51], and those articles that contain comprehensive lists of references to original work are noted.

In addition to methodologies, these contributions provide information and examples that can be used to properly assess the suitability of employing immunochemical techniques to answer specific scientific queries. Only a mere fraction of molecules that are of interest to the scientist has been dealt with in the vast literature devoted to immunoassays of proteins, peptides, nucleic acids, carbohydrates, lipids, cell surface antigens, viruses, bacteria, and other low-molecular-weight compounds of endogenous and exogenous origin. Immunochemical procedures will continue to proliferate since highly sensitive and specific analytical methods are of fundamental importance to many areas of research.

We are grateful to the authors for their contributions, advice, and cooperation. A few of our contributors also deserve special thanks for their patience. Publication of their articles was originally scheduled for an earlier volume, but modifications in the format rendered them more suitable for this volume.

JOHN J. LANGONE
HELEN VAN VUNAKIS

METHODS IN ENZYMOLOGY

EDITED BY

Sidney P. Colowick and Nathan O. Kaplan

VANDERBILT UNIVERSITY
SCHOOL OF MEDICINE
NASHVILLE, TENNESSEE

DEPARTMENT OF CHEMISTRY
UNIVERSITY OF CALIFORNIA
AT SAN DIEGO
LA JOLLA, CALIFORNIA

I. Preparation and Assay of Enzymes
II. Preparation and Assay of Enzymes
III. Preparation and Assay of Substrates
IV. Special Techniques for the Enzymologist
V. Preparation and Assay of Enzymes
VI. Preparation and Assay of Enzymes (*Continued*)
Preparation and Assay of Substrates
Special Techniques
VII. Cumulative Subject Index

METHODS IN ENZYMOLOGY

EDITORS-IN-CHIEF

Sidney P. Colowick   Nathan O. Kaplan

xxii   METHODS IN ENZYMOLOGY

VOLUME 80. Proteolytic Enzymes (Part C)
Edited by LASZLO LORAND

VOLUME 81. Biomembranes (Part H: Visual Pigments and Purple Membranes, I)
Edited by LESTER PACKER

VOLUME 82. Structural and Contractile Proteins (Part A: Extracellular Matrix)
Edited by LEON W. CUNNINGHAM AND DIXIE W. FREDERIKSEN

VOLUME 83. Complex Carbohydrates (Part D)
Edited by VICTOR GINSBURG

VOLUME 84. Immunochemical Techniques (Part D: Selected Immunoassays)
Edited by JOHN J. LANGONE AND HELEN VAN VUNAKIS

VOLUME 85. Structural and Contractile Proteins (Part B: The Contractile Apparatus and the Cytoskeleton) (in preparation)
Edited by DIXIE W. FREDERIKSEN AND LEON W. CUNNINGHAM

VOLUME 86. Prostaglandins and Arachidonate Metabolites (in preparation)
Edited by WILLIAM E. M. LANDS AND WILLIAM L. SMITH

VOLUME 87. Enzyme Kinetics and Mechanism (Part C: Intermediates, Stereochemistry, and Rate Studies) (in preparation)
Edited by DANIEL L. PURICH

VOLUME 88. Biomembranes (Part I: Visual Pigments and Purple Membranes, II) (in preparation)
Edited by LESTER PACKER

VOLUME 89. Carbohydrate Metabolism (Part D) (in preparation)
Edited by WILLIS A. WOOD

VOLUME 90. Carbohydrate Metabolism (Part E) (in preparation)
Edited by WILLIS A. WOOD

Section I

Oncofetal Proteins

[1] Radioimmunoassay of α-Fetoprotein with Polyclonal and Monoclonal Antibodies

By Erkki Ruoslahti, Marjatta Uotila, and Eva Engvall

Introduction

αFetoprotein (AFP) is a major plasma protein of the early human fetus. In addition to the embryonal and fetal periods, elevated AFP levels have been found in the serum of mice and humans with primary liver cancer and teratocarcinomas (see Ref. 1). In fact, the AFP assay is widely used as a diagnostic test in these malignancies. The production of AFP by germ cell tumors and primary liver cancer is utilized to monitor cancer patients. AFP is a model marker in the studies concerning the expression of fetal antigens in cancer. Determination of AFP in maternal serum and in amniotic fluid during pregnancy has also proved to be an important tool in prenatal diagnosis of spina bifida and congenital nephrosis. The recent observation that human AFP exhibits microheterogeneity, which can be exploited diagnostically, may further expand the indications for assay of AFP.[2–4]

Measurement of AFP is generally carried out using radioimmunoassay capable of measuring a few nanograms of AFP per milliliter in biological fluids.[5] Standard procedures for such assays are described in this chapter. Since the availability of purified AFP is a factor that limits the wide application of the AFP assay, attention is also given to the methods available for the preparation of standard AFP for the assays. A recent promising development in the realm of immunoassays is the hybridoma technique.[6] It allows production of large quantities of homogeneous antibody from malignant cell hybrids of antibody-producing lymphocytes and myeloma cells. It can be expected that such antibodies will by virtue of their monospecificity replace conventional antisera used in various immunochemical tests. Production of monoclonal antibodies to human AFP and the partial characterization of highly sensitive radioimmunoassays utilizing such antibodies will also be described in this chapter.

[1] E. Ruoslahti, and M. Seppälä, Adv. Cancer Res. **29**, 275 (1979).
[2] E. Ruoslahti, E. Engvall, A. Pekkala, and M. Seppälä, Int. J. Cancer **22**, 515 (1978).
[3] M. Seppälä, T. Ranta, P. Aula, and E. Ruoslahti, in "Carcino-Embryonic Proteins. Chemistry, Biology, Clinical Applications." (F.-G. Lehmann, ed.), Vol. I. Elsevier/North Holland, Amsterdam, 1979.
[4] C. J. Smith, P. C. Kelleher, L. Bélanger, and L. Dallaire, Br. Med. J. **1**, 920 (1979).
[5] E. Ruoslahti and M. Seppälä Int. J. Cancer **7**, 218 (1971).
[6] G. Köhler and C. Milstein, Nature (London) **256**, 495 (1975).

Purification of AFP

Immunoadsorbents are commonly utilized to purify AFP in quantities permitting establishment of a radioimmunoassay.

The protein solution is incubated with an anti-AFP immunoadsorbent which results in binding of AFP to the adsorbent while the other proteins remain in solution. The adsorbent is washed and the AFP eluted with a chaotropic solution. The AFP preparations obtained contain small amounts of contaminating proteins, mainly γ-globulin released from the adsorbent. This can be removed by incubation of the AFP solution with anti-γ-globulin conjugated to Sepharose. Normal serum components can similarly be removed by treatment with anti-normal serum conjugated to Sepharose. Immunological purification methods require the use of antibodies with comparatively low affinity,[7] which facilitate the elution of the AFP complexed with antibody.

Production of Anti-AFP

Three principal ways of producing antibodies with low affinities are available. When conventional antisera are prepared, high doses of antigen and a short course of immunization favor the production of low-affinity antibodies in high titers.

We produce anti-AFP sera for immunochemical purification by immunizing goats, sheep, or rabbits with 0.5 to 1 mg of purified AFP. The protein is dissolved in phosphate-buffered saline (PBS) (0.5 ml) and mixed with 1.0 ml of Freund's complete adjuvant (Difco, Detroit, Mi.) by repeated passages through an 18-gauge injection needle. The antigen paste is injected subcutaneously to the buttocks of the animal. Three injections are given at 2-week intervals. To obtain large amounts of early antiserum, the animals are exsanguinated 10 days after the last injection (about 5 weeks after the beginning of immunization).

A recent development in this area is the use of immunochromatography on cross-reacting antibodies with very low affinity.[8] This method makes use of the fact that AFP is present in all mammalian species (and in chicken) as immunologically and chemically related, but not identical, molecules (see Ref. 1). Antibodies against AFP of one species have a lower affinity to AFP from another species than they have to the AFP used for immunization. Such cross-reactive antibodies can be utilized to prepare immunoadsorbents possessing a very low affinity. AFP bound to such columns can be eluted under relatively mild conditions.

Monoclonal antibodies represent a promising solution to the prepara-

---

[7] E. Ruoslahti, *Scand. J. Immunol. Suppl.* **3**, 39 (1976).
[8] E. Ruoslahti, *J. Immunol* **121**, 1687 (1978).

tion of antibodies suitable for immunological purification of AFP. We have prepared a large set of monoclonal antibodies to human AFP.[9,9a] Some of these bind AFP with a low affinity. When used to prepare immunoadsorbents, these antibodies allow elution of AFP at lower concentrations of the eluting agent (urea) than is the case when monoclonal antibodies with high affinity or conventional antibodies are used (Fig. 1). Details of the production of monoclonal anti-AFP for immunoadsorbents and for radioimmunoassay are given below.

Activated Sepharose

Sepharose activated with cyanogen bromide treatment[10] can be purchased commercially (Pharmacia Fine Chemical, Piscataway, N.J., or Sigma Chemical Co., St. Louis, Mo.), and its preparation will not be described here. In case large quantities of activated Sepharse are needed, it may be economical to perform the activation, in which case the cross-linked type of Sepharose should be used. Detailed descriptions for the precedures required are available.[7,11,12]

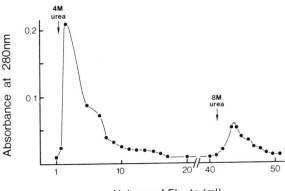

Volume of Eluate (ml)

FIG. 1. Isolation of human AFP at an analytical scale on anti-AFP conjugated to Sepharose. Four milligrams of monoclonal antibodies with low affinity to AFP were coupled to 1 ml of cyanogen bromide-activated Sepharose. AFP from 200 ml of second trimester amniotic fluid was bound to the column and eluted with 4 and 8 M urea. A γ-globulin fraction from a conventional antiserum is used in a similar way. However, a larger column would be needed to obtain the same binding capacity because of the lower concentration of antibodies relative to other γ-globulin, and the elution requires larger amounts of urea at higher concentrations (see text for details).

[9] M. Uotila, E. Ruoslahti, and E. Engvall, *Mol. Immunol.* **17**, 791 (1980).
[9a] M. Uotila, E. Ruoslahti, and E. Engvall, *J. Immunol Meth.* **42**, 11 (1981).
[10] R. Axén, J. Porath, and S. Ernback, *Nature (London)* **214**, 1302 (1967).
[11] R. Axén and J. Porath, *Eur. J. Biochem.* **18**, 351 (1971).
[12] P. Cuatrecasas, *J. Biol. Chem.* **245**, 3059 (1970).

Preparation of γ-Globulin for the Conjugation

The IgG fraction from the antiserum can be prepared using any of several methods available. A practical procedure is to precipitate a γ-globulin fraction from serum using 18% Na_2SO_4. The detailed procedure is as given below.

A few comments on the γ-globulin fraction to couple are in order. Immunoadsorbents leak some of the coupled protein when the bound antigen is eluted. The more nonantibody protein is coupled to the adsorbent, the less favorable is the ratio of eluted antigen to proteins leaking from the adsorbent. In this sense, a crude γ-globulin fraction such as that obtained after sodium sulfate precipitation is not the ideal preparation for immunoadsorbents. Such adsorbents will leak not only IgG, but also other contaminating serum proteins. However, the sodium sulfate precipitation method is a practical one, and the losses associated with other methods that could be employed to isolate pure IgG or specific antibody are avoided. Ammonium sulfate cannot be used directly for coupling because the ammonium ions in the solution would interfere with the coupling of the amino groups of the protein to the activated Sepharose.

It is obvious that a high ratio of specific to nonspecific antibody in the antiserum used to prepare an adsorbent will be an advantage. With monoclonal antibodies, very high concentrations of specific antibody in the ascites of hybridoma-bearing mice can be achieved. Adsorbents with high proportions of specific antibody can be prepared from such ascites making the monoclonal antibodies particularly suitable for the preparation of immunoadsorbents.

Starting Material for AFP Isolation

Amniotic fluid from the second trimester and cord serum (AFP content about 15 and 50 $\mu g/ml$, respectively) are suitable sources of AFP for purification. Serum or ascites from patients with an AFP-producing liver cancer or teratocarcinoma can also be used when available. A common mistake in immunoadsorbent purification is to start with too small quantities of AFP-containing material. The yield of the entire procedure rarely exeeds 20%. Accordingly, there is no point in attempting the purification from samples containing less than 10 mg of total AFP.

Human serum contains immunoglobulins that bind to Sepharose and will elute with AFP from the anti-AFP adsorbent but they can be removed by a prior treatment of the starting material with plain Sepharose. Such treatment will reduce the contamination in the subsequent isolation of AFP.

Detailed Procedure

Preparation of γ-Globulin

To 1 ml of antiserum (rabbit or goat) add 180 mg of sodium sulfate (anhydrous). The solution is stirred gently at room temperature (vigorous stirring will result in a finely dispensed precipitate which is difficult to spin down in a clinical centrifuge). After 1 hr the precipitate is collected by centrifugation, washed using 180 mg/ml of sodium sulfate in PBS, and dissolved in 0.1 M NaHCO$_3$, 0.5 M NaCl for conjugation at a concentration of 10–20 mg protein per ml.

Coupling of Anti-AFP γ-Globulin to Activated Sepharose

The protein solution is added to activated Sepharose in bicarbonate (10–20 mg protein per ml activated Sepharose) and the mixture is stirred overnight at +4° or for 4 hr at room temperature. The adsorbent is washed with the coupling buffer, incubated for 2 hr in 1 M ethanolamine, pH 8.0 (to block remaining active groups on Sepharose), washed again with the coupling buffer, and with the agent that is going to be used later for elutions (8 M urea in 0.05 M Tris–HCl buffer or 0.1 M glycine–HCl pH 2.8). The washing is done on a glass filter using suction to facilitate the process. About 10 adsorbent volumes of each washing solution are used. The adsorbent is stored in PBS containing 0.02% sodium azide.

Use of Immunoadsorbent

1. The serum, amniotic fluid or ascites sample is passed through a column of plain Sepharose of the same size as the serum sample (one-tenth of the volume of amniotic fluid). The Sepharose can be regenerated by washing with 2 column volumes of 8 M urea in 0.05 M Tris–HCl, pH 7.5 followed by PBS.
2. The Sepharose-absorbed serum (amniotic fluid) is next passed through an anti-AFP Sepharose column. The column is washed with large quantities of PBS (20 column volumes).
3. The column is then eluted with 8 M urea in 0.05 M Tris–HCl buffer, pH 7.5. The elution of AFP is monitored by UV absorbance and by immunodiffusion against anti-AFP. The amount of AFP present in the original sample may have exceeded the capacity of the column. The unbound fraction is therefore also tested against anti-AFP in immunodiffusion, and the isolation procedure is repeated until the sample no longer contains AFP.
4. The AFP-containing fractions are pooled, dialyzed against distilled water, and lyophilized. The AFP is tested at 5 mg/ml against anti-

normal human serum and against antiserum to serum proteins from the species that was used to prepare the adsorbent (e.g., goat). If impurities are found (this usually is the case), the AFP preparation is dissolved in PBS and passed through a column of Sepharose-coupled anti-normal human serum and/or anti-normal goat serum. AFP is detected in the fractions by UV absorbance and its purity tested by immunodiffusion as described above. It may be necessary to repeat the column procedures a number of times to obtain several milligrams of AFP free of immunologically detectable contaminants. The columns can be regenerated by washing with 2 column volumes of 8 M urea followed by PBS. Adsorbents can be reused up to 50 to 100 times.

5. We routinely add gel filtration on Sephadex G-200 as the last step to eliminate possible nonantigenic contaminants which would not be removed by the antibody columns. The purity of the final preparation should be tested by polyacrylamide gel electrophoresis with or without sodium dodecyl sulfate.

Various fractionation procedures have been employed in the purification of AFP. These include isoelectric focusing, affinity chromatography on Sepharose-linked concanavalin A (Con A), DEAE cellulose chromatography, and removal of albumin by affinity chromatography on Sepharose-Blue Dextran. They appear to result in an appreciable enrichment of AFP and should be useful in conjunction with immunochemical or other procedures (see Ref. 1 for references).

Radioimmunoassay of AFP

Production of Conventional Antisera for AFP Radioimmunoassay

Antisera to be used for RIA are obtained as was described above for the antisera to be used in immunoadsorbents, but after the initial three injections, the immunization is continued with monthly booster injections for 3 to 6 months. We have observed a 10-fold increase in the sensitivity of the radiommunoassays carried using anti-AFP from the same animal after 4 months as compared to 5 weeks immunization.

Testing and Absorption of Antisera

We test our anti-AFP sera in immunodiffusion against fetal serum, fetal extracts, amniotic fluid, and normal human serum. Several dilutions of sera and extracts (e.g., undiluted, 1:10, and 1:100) are used to avoid missing weak antibodies because of antigen excess. A single strong precipitation line against fetal serum appears when a satisfactory antiserum has been obtained. After an extended period of immunization, the antisera often contain small amounts of antibodies against normal human serum proteins, usually albumin and transferrin. Owing to its lower sensitivity, the immunodiffusion test cannot detect very small amounts of contaminating antibodies. Therefore, irrespective of whether contaminating antibodies are found in this test, we absorb our antisera with an excess of normal human serum proteins conjugated to Sepharose.

Monoclonal Antibodies to AFP

Monoclonal antibodies offer great promise as immunochemical tools, especially as reagents in immunoassays, but such antibodies have not yet been used extensively in assays of substances of clinical interest. We have recently shown that monoclonal antibodies to AFP can be utilized to establish highly sensitive radioimmunoassays for AFP[9] and enzyme immunoassays.[9a] Furthermore, assays based on monoclonal antibodies seem to have a specificity similar to conventional assays as indicated by the similarity of the AFP levels obtained using the two types of assays.[9,9a] The clinical AFP assay as applied to prenatal diagnosis depends on comparison of the level of AFP obtained in a given case to the normal AFP level at that age of gestation. Since the differences between normal and pathological AFP levels are small, accuracy of the test is of prime importance. The monoclonal antibodies could, by reducing the variables of the assay, increase its accuracy. It is not practical to describe here the entire methodology involved in the production of monoclonal antibodies. The approach we have taken is not different from the procedures generally used. An excellent detailed description of the production of hybridomas has appeared.[13] The procedures we have successfully utilized are briefly described here.

Immunization of Mice. Mice were immunized by subcutaneous injections of 50 μg of purified AFP emulsified in Freund's complete adjuvant. The injections were repeated 2–3 times at 3–4 week intervals. A final injection of AFP was given intraperitoneally or intravenously without adjuvant, and 3–5 days later the spleens of the mice were removed for hybridization.

Cell Lines. We have used the myeloma cell lines P3 X63-Ag 8 (X63), P3 NS1/1-Ag 4-1 (NS1), and SP2/0-Ag 14 (SP2) (Cell Distribu-

[13] V. T. Oi and L. A. Herzenberg *in* "Selective Methods in Cellular Immunology" (B. B. Mishell and S. M. Shiigi, eds.), Chap. 17, pp. 351–372. Freeman, San Francisco, California, 1980.

tion Center, Salk Institute, San Diego, Ca.) and X63 Ag8.653 for hybridizations. Cell line P3 X63 produces a myeloma protein consisting of a γ_1 heavy chain and κ light chain.[6] Cell line NS1 produces intracellular light chain (κ). This has been found to be secreted in hybrid cells derived from the NS1 line,[14] but we have not detected myeloma light chains in the monoclonal antibodies synthesized by hybrids derived from this cell line. The SP2[15] and X63-Ag8.653[16] lines do not produce their own immunoglobulin.

Monoclonal antibodies derived from these different myeloma lines seem to have different properties. Radioimmunoassays have shown that unlabeled AFP is a poor inhibitor of the binding of the labeled AFP in all cases where antibodies from P3 X63-derived clones were used. All clones derived from the other cell lines have produced antibodies which give highly sensitive assays.

Hybridomas obtained from the P3 X63 line produce mixed IgG molecules which represent all permutations that can be obtained from the heavy and light chains of the myeloma protein and the antibody molecule. It is possible that such scrambled molecules have low affinities to the antigen. Based on these findings, it may not be advisable to use the P3 X63 line for hybridizations if the aim is to obtain antibody for radioimmunoassays. However, the low affinity antibodies from P3 X63 hybridomas are excellent for the preparation of immunoadsorbents.

Hybridization. We have performed the hybridizations and the subsequent culture and cloning of the hybrids essentially according to Oi and Herzenberg.[13] Briefly, 10^8 spleen cells are mixed with $1-5 \times 10^7$ myeloma cells in Dulbecco's medium (Flow Laboratories, Inglewood, Ca.). The cells are spun down, and 1 ml 50% polyethylene glycol (MW 1500, J. T. Baker Chemical Co., Phillipsburg, N.J.) is added slowly (during 45 sec) with gentle mixing of the cells. After another 45 sec, the cell suspension is diluted to 25% polyethylene glycol, left for 4 min then diluted further with gentle mixing of the cells. The cells are spun down, washed, and distributed into three 96-well plates (Linbro, Flow Laboratories) in Dulbecco's medium containing 10% fetal calf serum (Flow Laboratories), 100 IU/ml penicillin, 100 μg/ml streptomycin, and 5×10^{-5} M 2-mercaptoethanol. Half of the culture medium in each well is replaced by fresh medium containing HAT: 10^{-4} M hypoxanthine, 1.6×10^{-5} M thymidine (Calbiochem-Behring Corp., La Jolla, Ca.) and

[14] G. Köhler, S. C. Howe, and C. Milstein, *Eur. J. Immunol.* **6,** 292 (1976).
[15] M. Shulman, C. D. Wilde, and G. Köhler, *Nature (London)* **276,** 269 (1978).
[16] J. F. Kearney, A. Radbruch, B. Liesegang, and K. Rajewsky, *J. Immunol* **123,** 1548 (1979).

4×10^{-7} M aminopterin (ICN Pharmaceuticals, Inc., Cleveland, Oh.), on days 1 and 2 and every 3 days thereafter for 2 weeks. After 2 weeks in HAT, the cultures are transferred to HT medium which does not contain aminopterin and after another week to usual medium. The cultures are tested for anti-AFP activity from 2 weeks on using the enzyme-linked immunosorbent assay, ELISA.[17,18] In this, wells in nontreated polystyrene microtiter plates (Disposotray, Flow Laboratories) are coated with 0.1 or 1 μg/ml of AFP. Culture fluid samples diluted 2-fold in phosphate-buffered saline containing Tween 20 are incubated in the coated wells for 2 hr at 37°. After washing, rabbit antibodies to mouse immunoglobulins labeled with peroxidase are incubated in the wells for another 2 hr. Enzyme activity bound to the wells is measured using o-phenylenediamine as chromogen and recorded using Titertek Multiskan (Flow Laboratories). The positive cultures detected by this assay are then cloned repeatedly using the limiting dilution method and then grown in culture or as ascitic tumors in mice.

Antiimmunoglobulin Sera

Most laboratories are basing their AFP radioimmunoassay on the double antibody principle. The antiserum necessary for this procedure is raised as follows.

Anti-goat (sheep) and anti-mouse γ-globulin reagents are prepared in rabbits and anti-rabbit γ-globulin in goats. (For the purpose of anti-IgG reagents, sera from goats and sheep are interchangeable.) The animals are immunized following the schedule given for the production of anti-AFP. Rabbits receive 1 mg and goats (or sheep), 5 mg per injection of γ-globulin prepared as follows: γ-globulin is precipitated from normal goat (sheep) or rabbit serum with sodium sulfate as described above. The precipitate is dissolved using half of the original serum volume of 0.05 M Tris buffer, pH 7.0. Seven milliliters of this solution is fractionated on a 2.5 × 90 cm Sephadex G-200 column and the IgG peak is collected.

Testing of Anti-γ-Globulin Sera

We test the antisera against different dilutions of goat or rabbit serum in immunodiffusion. In addition to IgG, these antisera usually react with a number of serum proteins, but anti-IgG dominates, and the point of equivalence can be roughly estimated. A suitable carrier IgG–

[17] E. Engvall and P. Perlmann, *J. Immunol.* **109**, 129 (1972).

[18] E. Engvall, this series, Vol. 70, p. 419.

anti-γ-globulin system for the radioimmunoassay is then sought (see below).

Radiolabeling of AFP

The chloramine-T method is suitable for the radioiodination of AFP.[19] The lactoperoxidase method[20] has also given good results. The details of the former procedure are given below.

Radioiodination (Detailed Procedure)

1. Add 1 to 2 mCi of carrier-free [125]I (10–20 μl, IMS 30, Radiochemical Centre, Amercham, England) to a small test tube.
2. Add 20 μl 0.5 M phosphate buffer, pH 7.0.
3. Add 5 to 10 μg of AFP in distilled water.
4. Add 25 μl fresh chloramine-T solution (4 mg/ml in 0.05 M phosphate buffer, pH 7.0).
5. After mixing, stop the reaction as soon as possible (10–20 sec) by adding 100 μl sodium metabisulfite (2.4 mg/ml $Na_2S_2O_5$ in 0.05 M sodium phosphate, pH 7.0).
6. Add 20 μl NaI (2%).

The products are fractionated immediately on Sephadex G-25 (column volume about 10 ml). The column is equilibriated with 0.05% gelatin in PBS containing 0.02% sodium azide. We use gelatin as the protein diluent rather than, e.g., purified albumin because albumin preparations may contain small amounts of AFP. The reaction mixture is eluted with gelatin-PBS and 1 ml fractions are collected. The radioactivity in aliquots from each tube is counted. ^{125}I-labeled AFP elutes in the first peak. After the procedure has been standardized, it is no longer necessary to elute the second peak containing the iodide. This is discarded with the column. The peak AFP fraction is diluted 1:10 in gelatin-PBS and stored at +4° (or at −20° divided into aliquots). For use in radioimmunoassay, the ^{125}I-labeled AFP is diluted further in gelatin-PBS to give about 20,000 cpm in 100 μl.

In the iodination procedure 50 to 90% of the radioactivity is incorporated into the protein. The label is usable for about 6 weeks, after which the assay tends to become unstable. Fractionation of the label on Sephadex G-200 at this point will remove radioactivity associated with low-molecular-weight material and restore its usability. However, the sensitivity

[19] F. C. Greenwood, W. M. Hunter, and J. S. Glover, *Biochem. J.* **89**, 114 (1963).
[20] G. S. David and R. A. Reisfeld, *Biochemistry* **13**, 1014 (1974).

of the assay might be reduced, and we usually prefer to carry out fresh radioiodination rather than refractionation.

Properties of the Labeled AFP. About 90% of the radioactivity in freshly labeled AFP is bound to antibody in antibody excess. The iodinated AFP elutes from Sephadex G-200 as a single peak at the same position as unlabeled AFP. Gel electrophoresis in the presence of SDS results in a single radioactive band at MW 70,000. If less than 80% of the radioactivity in freshly labeled AFP is bound at antibody excess, this suggests that the AFP used for labeling contains impurities. Such label may still give a satisfactory assay since only one of the reagents, the AFP or the antibody, needs to be absolutely specific in a competitive radioimmunoassay. However, impurities introduce an uncertain factor in the assay and should be avoided.

Our experience suggests that the binding of labeled AFP to some monoclonal antibodies is particularly sensitive to iodination damage of the AFP. This and the inherent specificity of the monoclonal antibodies make them particularly suitable for the evaluation of labeled AFP.

Radioimmunoassay

The salient features of our radioimmunoassay procedure are as follows: ^{125}I-labeled AFP is incubated with a limiting dilution of anti-AFP in the presence of known amounts of standard AFP or test samples. The antibody-bound and free label are separated using precipitation of anti-AFP with the second antibody (anti-γ-globulin).

AFP Standards to RIA. The primary standard is prepared using purified AFP. The standard solutions are prepared in a protein-containing diluent. In case the assay is being used for serum samples, the diluent is normal human male serum (in PBS), diluted 1:25 (like unknown serum samples) in PBS.

From a stock solution with a known AFP content, standard solutions containing 0.1, 0.25, 0.5, 1.0, 2.5, 5.0, 10.0, and 25.0 ng per ml of AFP are prepared in the protein-containing diluent buffer. We base the concentration of our AFP standard on protein determination according to Lowry *et al.*[21] According to our experience, purified AFP at a concentration of 1 mg/ml is stable in deep-frozen solution for up to 6 months. It may be preferable to store the purified AFP lyophilized at −20° and to prepare solutions for the standard curve from this. It should be pointed out that in our hands the weight of a given purified AFP preparation and the result of the Lowry assay do not agree completely. Usually, the Lowry assay indi-

[21] O. H. Lowry, N. J. Rosebrough, A. L. Farr, and R. J. Randell, *J. Biol. Chem.* **193**, 265 (1951).

cates a protein content which is 80–85% of the dry weight. This introduces a corresponding difference in the standardization. Measurement of absorbance is not a reliable way of quantitating AFP, since AFP binds chromogenic substances such as bilirubin.[22]

International Standard for AFP. Collaborative studies have shown that there is a large variation between the AFP values obtained in different laboratories for the same sample. This is not a feature peculiar of AFP but rather seems to be an inherent characteristic of many protein radioimmunoassays. An international standard preparation consisting of pooled human cord serum has been prepared.[23] The average value of one international unit, 1.17 ng, was obtained for this standard in different laboratories. Because of the high variation, it was decided to express the international reference preparation in terms of arbitrary units.

It is recommended that all published work should include a reference to how the local standard compares with the international standard. However, the use of international standard does not justify omission by individual laboratories to estimate their own normal ranges, since differences in methodology and antisera may affect the level even when the same standard is used. Racial differences in AFP levels may exist as well.

In order to economize the use of purified AFP and the international standard, it is advisable to use a secondary standard such as cord serum. Dilutions of this standard are included in every assay, and the secondary standard is periodically checked against a primary standard.

Radioimmunoassay of AFP (Detailed Procedure)

The general layout of the assay is shown in the table. An appropriate amount of goat (or rabbit) serum should be added to each tube to provide carrier γ-globulin for the precipitation with the second antibody. The amount needed varies between 1 and 10 μl per tube depending on the IgG content and the amount of precipitate desired. If the antiserum is from rabbits, 10 μl of normal rabbit serum per tube is usually optimal. The tubes are incubated for 24 hr at $+20°$ or for 72 hr at $+4°$. The total radioactivity of each tube can be counted during the incubation period. Usually it is sufficient to use an average total radioactivity for the whole series of tubes.

Antibody-bound label is precipitated by adding an amount of anti-goat (or anti-rabbit) γ-globulin serum giving a clearly visible precipitate. The precipitate appears after 15 min and is spun down after 1 hr at 3000 rpm (15–30 min). The supernatant is removed with a Pasteur pipet connected

---

[22] E. Ruoslahti, E. Estes, and M. Seppälä, *Biochim. Biophys. Acta* **578,** 511 (1979).
[23] Report of a Collaborative Study, *Clin. Chim. Acta* **96,** 59 (1979).

PROCEDURE FOR AFP RADIOIMMUNOASSAY

Tube no.	AFP standard	Protein diluent (ml)	Antiserum (ml)	AFP label[a] (ml)	Test sample
1	—	1.1	—	0.1	—
2	—	1.0	0.1	0.1	—
3	0.1 ng/ml (1 ml)	—	0.1	0.1	—
4	0.25 (1 ml)	—	0.1	0.1	—
5	0.5 (1 ml)	—	0.1	0.1	—
6	1.0 (1 ml)	—	0.1	0.1	—
7	2.5 (1 ml)	—	0.1	0.1	—
8	5.0 (1 ml)	—	0.1	0.1	—
9	10.0 (1 ml)	—	0.1	0.1	—
10	25.0 (1 ml)	—	0.1	0.1	—
11	—	—	0.1	0.1	Sample A 1:25, 1 ml
12	—	—	0.1	0.1	Sample B 1:25, 1 ml
13	—	—	0.1	0.1	Sample C 1:25, 1 ml

[a] May include the appropriate amount of goat (rabbit) serum to provide carrier γ-globulin for precipitation.

to a water suction or a vacuum line with an intervening trap. The radioactivities of each precipitate are then counted in a gamma counter. The percentage of added radioactivity found in the tube without added antiserum (#1) indicates the amount of radioactivity which is nonspecifically bound in the precipitate. This is usually 2 to 5%.

Parameters Affecting the First and Second Immune Reactions in AFP RIA

We adjust the percentage bound label to about 40%, but have used binding percentages ranging from 10 to 60%. Low binding percentage gives a test with a higher sensitivity, but it also results in a less stable test.

The incubation time used (24 hr at $+20°$) is a compromise between the reaching of equilibrium between AFP and anti-AFP, and the duration of the test. A longer incubation (e.g., 48 hr) will give a higher binding and also a better stability of the test.

The second antibody reaction is of great importance to the final outcome of the test. We use a high concentration of the precipitating system. This allows short incubation (1 hr) with a complete precipitation of γ-globulin. The precipitates should be visible within 15 min in this system. If less concentrated reagents are used, the incubation time has to be prolonged.

Reliability Experiments

Specificity. Diluted (1:10) normal human serum inhibits in RIA. This inhibition is caused by AFP in normal human serum since it can be totally abolished by absorption with anti-AFP immunoadsorbent.[5] The lack of inhibition by absorbed normal human serum also means that normal human serum proteins do not cause nonspecific inhibition in AFP RIA.

Sensitivity. The sensitivity of the AFP RIA is 0.1 to 0.25 ng/ml.[5] However, since we dilute the samples 1:25, the final sensitivity of the assay is 5 to 10 ng/ml. This is a convenient cut-off level since all exceptional AFP levels are detected, but normal sera do not inhibit or inhibit only marginally.[5]

The *intraassay variation* of the AFP RIA is 13 to 18% at concentrations from 0.25 to 2.5 ng/ml, and 6 to 12% at concentrations from 5 to 25 ng/ml. The *between-assay variation* is 16%.[24]

Radioimmunoassay with Monoclonal Antibodies

We have also used monoclonal antibodies for radioimmunoassay of AFP.[9] An assay basically identical to the one described above for conventional antibodies can be used. However, due mainly to incidental factors, we have modified the procedure slightly. We have started using 5% horse serum as the protein diluent and precipitate the antigen–antibody complexes with 12.5% (final concentration) polyethylene glycol (MW 6000, Sigma Chemical Co., St. Louis, Mo.). Typical inhibition curves obtained with monoclonal antibodies are shown in Fig. 2.

It can be seen that highly sensitive assays can be obtained with monoclonal antibodies that have a high affinity (50/3 in Fig. 2). However, many of the monoclonal antibodies we have raised have had low affinities, giving assays of poor sensitivity. As discussed above, such antibodies were regularly obtained when the IgG-producing P3 X63 myeloma line was used for the fusions. The antibodies from hybrids derived from non-IgG producing parental cell lines had intermediate or high affinities.

When we have tested serum and amniotic fluid samples in assays that utilize monoclonal antibodies, the results have been similar to what is obtained with a conventional assay. However, we have tested only a limited number of samples and more experience is needed before a given anti-

---

[24] E. Ruoslahti and M. Seppälä, *in* "Handbook of Radioimmunoassay" (G. E. Abraham, ed.), Chap. 18, pp. 543–569. Dekker, New York, 1977.

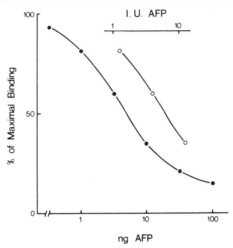

FIG. 2. Monoclonal antibody was incubated with ¹²⁵I-labeled AFP in the presence of various amounts of purified AFP (●) or units of AFP in the international AFP standard of the International Agency for Research on Cancer (○). Total volume was 1 ml, and incubation time, 20 hr. Separation between free and antibody bound antigen was accomplished by precipitation with polyethylene glycol.

body can be accepted as a reliable reagent. Monoclonal antibodies are thought to react with a single determinant only in a protein molecule. Since the size of an antigenic determinant is about 8 amino acids,[25] a single determinant could by chance be shared by other proteins. This, however, does not seem to be common, since monoclonal antibodies have been found to be quite specific. Our AFP assays, for instance, show that the proteins in normal serum do not interact with the monoclonal anti-AFPs, since normal serum does not inhibit in the assay. A specificity problem, therefore, does not seem to exist.

A more serious problem could be the occurrence of genetic variants in the population. While a change in a single determinant due to, e.g., a point mutation, may have only a slight effect or no effect at all on the result in a conventional assay, it could completely change the result of a monoclonal assay if the change happens to be in the determinant detected by this antibody. A large number of samples will have to be tested with a monoclonal antibody intended for clinical use to exclude this possibility. It should be pointed out that no genetic variants are known for AFP in any species. The third important aspect to discuss with regard to specificity of monoclonal antibodies is the existence of microheterogeneous variants.

[25] M. Z. Atassi and W. Zablocki, *J. Biol. Chem.* **252,** 8784 (1977).

AFP exists as variants that differ with respect to molecular properties. One set of such variants can be distinguished in human AFP by chromatography on the lectin concanavalin A.[2] The monoclonal antibodies we have do not differentiate between the Con A variants (or other variants of AFP) since we found that such antibodies bound as much of the radioactive AFP as conventional antibodies, and direct assays using purified Con A variants have not disclosed any differences in their reactivities with the monoclonal antibodies (our unpublished results). It may be possible to produce such antibodies specific for the variants by selecting from a large number of monoclonal antibodies. Such work is now in progress in our laboratory. It would be desirable to have antibodies that would distinguish the Con A variants because these have been shown to be of diagnostic significance as discussed below.

The immunological reactivity of AFP with monoclonal antibodies and perhaps even conventional antisera could also be influenced by ligands bound to AFP. AFP binds fatty acids[26] and bilirubin.[22] The presence of fatty acids in AFP influences the spectrum of simultaneously bound bilirubin,[22] suggesting that fatty acids may cause conformational changes in the AFP molecule. Such changes could affect the reactivity of monoclonal antibodies whose determinants are in the affected domain(s) of the molecule. To avoid these problems, one could use several monoclonal antibodies recognizing different antigenic determinants on the AFP molecule.

Assay for the Concanavalin A Variants of AFP

We recently found that human AFP can be separated into subfractions by chromatography on Con A. About 2–5% of fetal serum AFP does not bind to Con A, and this is a constant feature throughout gestation. In contrast, 15–40% of AFP in amniotic fluid during the second trimester is nonreactive with Con A.

AFP is produced by the yolk sac and the fetal liver,[27] and the yolk sac may be the source of the AFP nonreactive with Con A. This is suggested by the finding that half of the AFP in sera from patients with yolk sac tumors is Con A nonreactive.[2] In mice, yolk sac-derived AFP does not bind to Con A while AFP originating in the liver does.[28] We hypothesize that the yolk sac contributes significantly to the AFP pool in the amniotic fluid during early gestation. It has been shown recently[3,4,29] that the ratio of the Con A nonreactive to total reactive AFP is lower than normal in

---

[26] D. C. Parmelee, M. A. Evenson, and H. F. Deutsch, *J. Biol. Chem.* **253**, 2114 (1978).
[27] D. Gitlin, A. Perricelli, and G. M. Gitlin, *Cancer Res.* **32**, 979 (1972).
[28] E. Ruoslahti, and E. Adamson, *Biochem. Biophys. Res. Commun.* **85**, 1622 (1978).
[29] E. Ruoslahti, A. Pekkala, D. E. Comings, and M. Seppälä, *Br. Med. J.* **2**, 768 (1979).

amniotic fluids from pregnancies where the fetus has anencephaly, spina bifida, or congenital nephrosis. This seems to be due to an increased leakage of the Con A reactive AFP from the fetal circulation to the amniotic fluid. The test for the Con A variants is rapidly gaining acceptance as an adjunct to the regular AFP test in the diagnosis of these fetal anomalies.

Con A Affinity Chromatography

Affinity chromatography on Con A-Sepharose (Pharmacia Fine Chemicals, Piscataway, N.J., 10 mg of Con A/ml of gel) is performed using a buffer containing 0.5 M sodium chloride and 1 mM each of manganese chloride, magnesium chloride, and calcium chloride in 0.1 M sodium acetate, pH 6.5. The Con A-Sepharose columns are equilibrated with this buffer and washed with it after application of the sample. Elution is performed with 1 M α-methyl mannoside (Sigma Chemical Co., St. Louis, Mo.) in the same buffer. When necessary, fractions from an initial separation can be pooled, dialyzed against distilled water, lyophilized, dissolved in the starting buffer, and refractionated on Con A-Sepharose. Distribution of AFP in the fractions is monitored by radioimmunoassay.

For purposes of routine testing, the percentage of AFP nonreactive with Con A can be measured by fractionating 0.1 ml amniotic fluid on a 1 ml column of Con A-Sepharose. The amount of AFP in the first 2 ml eluant obtained with the initial column buffer (the Con A nonreactive fraction) is compared with the total amount of AFP applied to the column and expressed as a percentage of nonreactive AFP.

Acknowledgments

Our work on AFP is supported by grant CA 27460 from the National Cancer Institute.

[2] Enzyme Immunoassay of Human α_1-Fetoprotein

By Luc Bélanger and René Masseyeff

α_1-Fetoprotein (AFP) is one of the first molecules for which immunoenzymatic methods of assay were developed. A variety of EIA systems have been tested with this antigen, among which the two methods described here are probably the most extensively characterized and best established for routine operation in the clinical laboratory.

The double antibody assay developed by Bélanger's group is a competitive system adapted from conventional radioimmunoassays and adhering to the general principles of saturation analysis. Enzyme-labeled (tracer) and unlabeled (assay sample) antigens compete for a limited amount of antibody and the extent of tracer–antibody reaction measures the amount of unlabeled antigen reacting in the system. The sandwich enzyme immunoassay established by Masseyeff's group is a noncompetitive system adapted from two-site immunoradiometric assays. Here, the test antigen reacts successively with stoichiometric excesses of unlabeled solid phase antibody and enzyme-labeled soluble antibody (tracer). The extent of tracer–antigen reaction measures the amount of antigen reacting in the system.

These two systems are considered below in regard to their essential requirements and characteristics. Their detailed development and characterization, as well as alternative procedures for enzyme immunoassay of AFP in man and other species, are discussed in Refs. 1–9.

Double Antibody Enzyme Immunoassay

The double antibody assay operates as follows. A fixed amount of AFP/alkaline phosphatase conjugate (tracer) reacts with a fixed amount of rabbit anti-AFP serum (first antibody) and variable amounts of free AFP (assay samples). The tracer–antibody complexes are then precipitated with a sheep anti-rabbit IgG serum (second antibody). The alkaline phosphatase activity recovered in the precipitate measures the concentration of AFP in the assay samples.

The working reagents used in the assay are listed in Table I and the assay protocol is given in Table II. The system is discussed under headings of requirements (antigens, antibodies, stock and working reagents), procedure (standard and modified assays), and performance (specificity, sensitivity, precision, and practicality).

[1] L. Bélanger, C. Sylvestre, and D. Dufour, *Clin. Chim. Acta* **48**, 15 (1973).
[2] L. Bélanger and D. Dufour, in "Alpha-feto-protein" (R. Messeyeff, ed.), p. 533. INSERM, Paris, 1974.
[3] L. Bélanger, D. Hamel, D. Dufour, and M. Pouliot, *Clin. Chem.* **22**, 198 (1976).
[4] R. Daigneault and L. Bélanger, in "Protides of the Biological Fluids" (H. Peeters, ed.), p. 597. Pergamon, Oxford, 1976.
[5] L. Bélanger, *Scand. J. Immunol.* **8**(Suppl. 7), 33 (1978).
[6] R. Maiolini, B. Ferrua, and R. Masseyeff, in "Alpha-feto-protein" (R. Masseyeff, ed.), p. 581. INSERM. Paris, 1974.
[7] R. Maiolini, B. Ferrua, and R. Masseyeff, *J. Immunol. Methods* **6**, 355 (1975).
[8] R. Maiolini and R. Masseyeff, *J. Immunol. Methods* **8**, 223 (1975).
[9] R. Masseyeff, *Scand. J. Immunol.* **8**(Suppl. 7), 83 (1978).

TABLE I
WORKING REAGENTS FOR THE DOUBLE ANTIBODY ENZYME IMMUNOASSAY OF AFP

BSA-PBS	Phosphate-buffered saline (0.15 M NaCl, 0.05 M PO_4^{3-}, pH 7.2) containing 1% bovine serum albumin (Sigma, Fraction V) and 0.01% NaN_3
Blank solution	Barbital buffer 0.05 M, pH 8.6 (Fisher Electrophoretic Buffer # 1) containing 1% (w/v) BSA, 10% (v/v) normal sheep serum (decomplemented at 56° for 30 min), 0.08% (v/v) normal rabbit serum (decomplemented), 0.85% NaCl, and 0.01% NaN_3
First antibody solution	Blank solution containing a working concentration of rabbit anti-AFP serum
Second antibody solution	Sheep anti-rabbit IgG serum (decomplemented) diluted in BSA-PBS so that 0.1 ml will precipitate the antigen contained in 0.8 ml of first antibody solution
Substrate solution	Sodium carbonate buffer 0.05 M, pH 9.8, containing 1 mM $MgCl_2$ and 1 mg/ml p-nitrophenyl phosphate (Sigma tablets #104-105)
Tracer	AFP/enzyme conjugate diluted to a working concentration in BSA-PBS
Standards	BSA-PBS containing 2, 5, 10, 20, 50, 100, 200, and 500 ng AFP/ml
Assay samples	Decomplemented; diluted in BSA-PBS as required
Quality controls	Cord blood serum diluted to 20 and 120 ng AFP/ml in normal adult serum (decomplemented)
Storage	The substrate solution is prepared fresh. The tracer is kept at 4°. All other reagents are filtered on Millipore and frozen (they may also be kept at 4° for months if preservation is adequate). Repeated freeze-thaw cycles of antigen and antibody solutions are to be avoided

Requirements

Pure AFP. The preparation of tracer, antiserum and standards makes pure AFP almost an absolute requirement for setting up the system.

Several efficient physicochemical or immunological methods have now been developed for the isolation of human AFP. Initially, one can obtain highly purified material by chromatography of early fetal serum or AFP-rich amniotic fluid on Dextran-Blue agarose (to remove albumin), followed by electrophoresis in nondenaturing 7.5% polyacrylamide (we use 16 × 180-mm rod gels); AFP is eluted from gel slices by diffusion in saline and residual acrylamide is removed by filtration on a glass filter. Once monospecific anti-AFP serum is available, pure antigen can be relatively easily obtained by immunological methods. Routinely, we use the technique of Nishi[10] for isolation of cord blood serum AFP by immuno-

[10] S. Nishi, *Cancer Res.* **30**, 2507 (1970).

TABLE II
PROTOCOL FOR THE DOUBLE ANTIBODY ENZYME IMMUNOASSAY OF AFP

1. Pipet 0.1 ml of assay samples in duplicate into 12 × 75-mm glass tubes
   1–2 : BSA-PBS (blank)
   3–4 : BSA-PBS (zero dose)
   5–20: Standards
   21–   : Test and quality control samples
2. Add 0.1 ml of tracer to all tubes
3. Pipet 0.8 ml of blank solution in tubes 1–2, and 0.8 ml of first antibody solution in the other tubes. Vortex-mix and allow to stand overnight at room temperature
4. Add 0.1 ml of second antibody solution to all tubes. Vortex-mix and allow to stand at room temperature until precipitation is complete (≈3 hr)
5. Centrifuge all tubes (22°, 10 min, 3000 rpm). Aspirate supernates as completely as possible, being careful not to disturb the pellets. Add 2 ml of PBS, vortex-mix, centrifuge, and aspirate the supernates. Repeat the pellet-wash operation one more time
6. Transfer tubes into an ice-water bath. Add 1 ml of ice-cold substrate solution. Vortex-mix 2–3 sec at high speed
7. Transfer tubes into a 37° water bath and incubate for 30 min. Shake tubes regularly during incubation
8. Transfer tubes back into the ice-water bath. Let cool down and add 0.2 ml of NaOH 0.1 M. Vortex and transfer to room temperature
9. Read absorbance at 400 nm using blank tubes (#1–2) for zero calibration of the spectrophotometer
10. Plot absorbance of standards, or percentage maximum binding [(absorbance standard/absorbance zero-dose) × 100], as a function of their AFP content. Calculate AFP concentration of test and quality control samples from their absorbance or percentage maximum binding referred to the standard curve

precipitation with sheep anti-AFP serum. This technique is simple, fast, and reasonably efficient: it uses complete antiserum and crude antigen source, and milligrams of AFP can be isolated in 2 days with a yield of about 30%, through easy centrifugation and gel filtration steps.

The purity of the AFP preparations is verified by electrophoresis in nondenaturing polyacrylamide gels (single band in postalbumin position) and by double diffusion and immunoelectrophoresis against anti-human serum protein antiserum. Immunologically purified AFP is also verified with an antiserum directed against the anti-AFP source used for immunopurification (in our case, anti-sheep serum protein). In case of residual contamination, final purification is achieved by preparative polyacrylamide gel electrophoresis as described above.

Stock Tracer. Pure AFP is conjugated to alkaline phosphatase (Sigma Type VII, from calf intestinal mucosa) using glutaraldehyde as linking agent.[11] The enzyme suspension (0.3 ml, 1.5 mg) is centrifuged 10 min at 4°, 1000 rpm. The pellet is dissolved in 0.5 ml of PBS (Table I) containing

[11] S. Avrameas, *Immunochemistry* 6, 43 (1969).

0.5 mg of pure AFP. The mixture is dialyzed overnight at 4° against PBS. The solution is then adjusted to 0.75 ml with PBS, and glutaraldehyde (Fisher biological grade) is added to a final concentration of 0.4% (15 μl of a 20% solution in PBS). After 6 hr of incubation at room temperature, the mixture is chromatographed on a Sephadex G-200 column equilibrated in PBS. The excluded fraction is stabilized in 4% BSA/0.01% NaN$_3$ and stored at 4°. The conjugate is stable for at least 12 months.

Working Tracer. The working concentration of enzyme tracer must be such that a suitable substrate color reaction develops with zero-order kinetics in all assay tubes [in our conditions, the absorbance increases linearly for at least 30 min and reads an adequate 0.5 to 0.6 A_{400} unit in zerodose (maximum binding) tubes]. The proper amount of tracer is determined by assaying zero-dose tubes (Table II) with serial dilutions of stock tracer in the presence of a large excess of first antibody. The dilution selected is that which yields an immunoprecipitable enzyme activity double that required for zero-dose tubes in the regular assay (i.e., 0.5 to 0.6 A_{400} unit after 15 min). In the present system this amounts to about 10 ng of tracer AFP per assay tube, assuming that the stock tracer contains all of the 500 μg of AFP used for conjugation (considering the sensitivity of the assay, this relatively high concentration of tracer points to a substantial loss of AFP antigenic sites occurring during conjugation).

Anti-AFP Serum. A specific AFP antiserum can initially be raised by immunization with fetal serum and absorption of the antiserum with normal adult serum. When pure antigen is available, the following scheme is followed for immunization of rabbits, sheep, or goats. The animal is sensitized with 0.1 to 0.2 mg of antigen emulsified in 1 ml of complete Freud's adjuvant and injected in 10 intradermal spots in the back. Four weeks later, and every other week thereafter, the animal receives an increasing dose of antigen (0.2 to 1 mg) emulsified in 1 ml of incomplete Freund's adjuvant and administered in several subcutaneous and intramuscular locations. AFP is a strong immunogen and a satisfactory antiserum is generally obtained after two or three boosters.

The specificity of the antisera is verified by immunoelectrophoresis and double diffusion against normal human adult plasma and against other specific sources of potential contamination (in our case sheep serum as it is used for AFP immunopurification). RIA/EIA-grade rabbit anti-AFP serum is now available commercially.

Working First Antibody. The working concentration of first antibody should bind 50 of the tracer in the absence of competing antigen. This is determined by assaying zero-dose tubes (Table II) with serial dilutions of rabbit anti-AFP serum in the presence of working concentration of tracer. The proper antibody dilution is at mid-point on the resulting tracer binding curve. Rabbits almost invariably produce high titer high-affinity anti-AFP sera, typically used in the assay at a dilution of 1:50,000 or less.

AFP Standards. Preparations of pure AFP are quantitated by the Lowry method against bovine serum albumin (Sigma, Fraction V). Standards are prepared individually from stock AFP (avoid serial dilutions) in BSA-PBS. Upon storage of pure AFP, dimers may form (this is checked by electrophoresis in nondenaturing polyacrylamide gels): such preparations are not to be used for standards. Secondary standards can be prepared from cord blood serum calibrated by reference to the primary standard. Standards are stored at $-20°$ in 1–2 ml aliquots. They are kept at $4°$ for day-to-day use. One nanogram is equivalent to about 0.9 IU of the 72/225 International Reference Preparation supplied by the World Health Organization (Dr. Ph. Sizaret, International Agency for Research on Cancer, Lyon, France).

Rabbit IgG and Sheep Anti-Rabbit IgG Serum. Rabbit IgG is purified from serum by ammonium sulfate precipitation and anion exchange chromatography[3] or by affinity chromatography on Protein-A Sepharose (Pharmacia). Sheep anti-rabbit IgG is raised following the immunization protocol described for AFP except that immunogen doses are higher (0.5 to 5 mg). Rabbit IgG and anti-rabbit IgG sera are available commercially.

Procedure

A number of variations can be made on the protocol suggested in Table II without affecting the overall performance of the system. For instance, temperature conditions and incubation times can be varied according to the kinetics of association and dissociation of AFP with its antibody.[3] The results would also likely be equally satisfactory if buffers other than barbital were used, volumes of reagents were modified, or preinactivation at $56°$ (serving principally to enhance the second antibody reaction) was eliminated.

Two more significant modifications can be brought into the protocol which may suit better one's specific needs.

Speeded Assay Using PEG-6000. Advantage can be taken of the property of polyethylene glycol (PEG) to enhance antigen–antibody reactions.[12] By adding 4% w/v PEG-6000 (and 1% v/v Tween-20) to the first antibody solution and raising the temperature to $37°$, Daigneault[4] was able to reduce the first and second antibody incubations to 2 hr and 30 min, respectively. This fast (4 hr) assay has a higher background but compares with the regular assay in terms of sensitivity and precision.

[12] K. Hellsing, *in* "Protides of the Biological Fluids" (H. Peeters, ed.), p. 579. Pergamon, Oxford, 1973.

Sequential Saturation Assay. Advantage can also be taken of the high affinity of anti-AFP sera (and the consequent slow dissociation of antigen –antibody complexes) to develop a sequential saturation type of assay. The only modification in the regular protocol is that sample and tracer AFP are added sequentially to the first antibody solution (addition of tracer is delayed by 4–8 hr). The result of this is to shift the standard curve to the left (Fig. 1), with a gain in sensitivity, a lowered analytical range, and a potential gain in precision (owing to the less stringent precision required for addition of tracer).

Performance

Specificity. The specificity of the assay is dictated primarily by the purity of the tracer and by the specificity of the first antibody. The latter must be given more consideration than in an RIA as the tracing material may contain potential cross-reacting substances.

Nonspecific interferences (by endogenous alkaline phosphatase or by sample components altering antigen–antibody reactions or tracer enzyme activity) appear to be negligible in this system. Standards prepared in

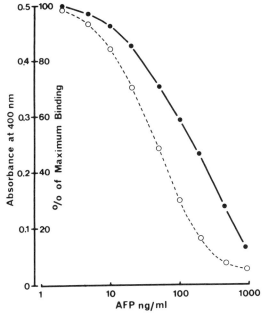

FIG. 1. Standard dose–response curves in the double antibody enzyme immunoassay of AFP using equilibrium (●) or sequential saturation (○) procedures.

buffer or in serum produce parallel displacement curves, and analytical recovery of exogenous AFP was found to be accurate in a variety of normal and pathological sera and amniotic fluids.[3] Thus there should be no need for individual sample blanks in the assay.

Sensitivity. The sensitivity of the assay depends largely on the amount of tracer required for proper signal detection. The present end-point photometric conditions place the limit of sensitivity to around 5 ng/ml, which suits all AFP levels of clinical interest. Alternative enzyme monitoring systems (kinetic analyses, more sensitive substrate) and/or tracers with higher enzyme specific activity could reduce the concentration of tracer in the assay and drive down the sensitivity threshold.

Precision. A quality control scheme must be implemented for effective routine operation of the assay. We monitor precision basically by running with each assay a "high" and a "low" reference sample (quality controls, Table I). The values serve to plot the variability limits and calculate the interassay coefficient of variation (CV). In the range 10–200 ng/ml, the CV should be within 10–12%. Within-run precision ranges between 5 to 8%.

The critical operation, with respect to precision, resides in handling the bound tracer. In particular, meticulous care must be given to pellet washing and enzyme measurement to avoid any loss of precipitate and to achieve rigorous uniformity for substrate incubation.

Practicality. Up to tracer processing, the procedure of assay is simple, straightforward, and essentially optimized with reagent dispensers and sample dilutors. After the separation step however, the use of automated equipment can improve very significantly the practicality and reliability of the assay by eliminating manually tedious, time-consuming, error-prone operations. For instance, absorbance readings and calculation of results can be reduced to minutes with the use of an automated spectrophotometer and a computer program correlating AFP concentrations and absorbance values.[4] A fast centrifugal enzyme analyzer can monitor kinetics of substrate reaction and integrate absorbance readings, calculation of results and data output.[5]

The pellet washing step remains a limitation. Measuring free instead of bound tracer clearly would simplify the assay, improve its precision, and increase its potential for complete automation. This may be possible if interferences by endogenous alkaline phosphatase and sample effects on enzyme measurement are obviated or accounted for.

Sandwich Enzyme Immunoassay

Our group published in the past two different methods of enzyme immunoassay of human and rat AFP. The first one[7] was a replica of the method described by Miles and Hales,[13] as immunoradiometric assay. This method was very sensitive but had a too limited quantitative range, especially for a protein showing exceptionally wide pathological variations. It was never optimized and we investigated a sandwich method.[8] This latter method was introduced for routine clinical use in January 1976, and was successfully used for nearly 50 months. Its main features were the use of a solid phase made of microcrystalline cellulose onto which the γ-globulin fraction of an AFP antiserum was attached by covalent bounds, and the use of glucose oxidase for conjugation to pure specific antibodies against AFP. This method gave satisfactory results but involved tedious steps of washing by centrifugation. Also, the preparation of the solid phase and conjugate was long and delicate. We then gradually changed nearly all the reagents of this sandwich method for reasons that are discussed below. The assay described here has been in daily routine use for more than a year.

Reagents

Anti-AFP serum. Monospecific and high-affinity anti-AFP serum is raised in sheep by monthly injections of 0.3 mg of human AFP (purified by immunosorption) emulsified in an equal volume of complete Freund's adjuvant. Intradermal injections are made in the back at 20 different sites. Six months later, the first samples of blood are taken. The antiserum is checked for specificity by immunodiffusion and in any case absorbed on a matrix of glutaraldehyde-insolubilized normal human serum.[14]

AFP Standard. We use as an internal standard the serum of a hepatoma patient calibrated against pure AFP and diluted in chicken serum. One nanogram is equivalent to 0.742 IU of the International Reference Preparation (WHO 72/225).

γ-Globulin and IgG Fraction. Part of the antiserum is precipitated with 18% Na_2SO_4 (final concentration) and reconstituted back to its original volume with PBS. The IgG fraction is obtained from the remaining antiserum by caprylic acid precipitation followed by batch absorption on DE 52 Whatman cellulose.[15]

Labeling Procedure. The IgG fraction is coupled to sodium *m*-periodate-activated horse radish peroxidase (HRP, grade 1, Boehringer Mannheim, West Germany) following the method of Wilson and Nakane.[16] After reduction with sodium borohydride, the conjugate is iso-

[13] L. Miles and C. Hales, *J. Biochem.* **108**, 611 (1968).
[14] S. Avrameas and T. Ternynck, *Immunochemistry* **6**, 53 (1969).
[15] M. Steinbuch, R. Audran, and L. Pejaudier, *C. R. Soc. Biol. Sci. Paris* **164**, 296 (1970).
[16] M. B. Wilson and P. K. Nakane, *in* "Immunofluorescence and Related Staining Techniques" (W. Knapp, K. Holubar, and G. Wick, eds.), p. 215. Elsevier/North Holland, Amsterdam, 1978.

TABLE III
WORKING REAGENTS USED FOR THE SANDWICH ENZYME IMMUNOASSAY OF AFP

PBSS	Phosphate buffer 0.2 M, pH 7.2 containing 10% (v/v) normal sheep serum and 0.01% Thimerosal
Solid phase antibody	Polystyrene balls coated with sheep anti-AFP γ-globulins
Tracer	PBSS containing in 0.3 ml 100 ng of the anti-AFP-HRP conjugate
Wash and dilutant	Distilled water; chicken serum
Substrate solution	0.1 M phosphate citrate buffer, pH 5.5, containing 5 mM H_2O_2 and 16 mM o-phenylenediamine dihydrochloride (Sigma)
Standards	Chicken serum containing 3 to 120 or 0.3 to 12 ng AFP/ml

lated by precipitation with an equal volume of saturated cold neutral ammonium sulfate solution. After two washings with 50% ammonium sulfate, the conjugate is dissolved in PBSS buffer (Table III) and filtered on a 0.45-μm Millipore membrane. Aliquots are kept at $-20°$ or lyophilized.

Solid Phase Antibody. Polystyrene balls (6.5 mm) (Precision Plastics Co. CHICAGO, Ill.) are washed with 0.1 M phosphate buffer pH 7.2 and

TABLE IV
PROTOCOL FOR THE SANDWICH ENZYME IMMUNOASSAY OF AFP

1. Pipet 0.1 ml of assay samples in duplicate into disposable polystyrene tubes
    1–2 : Chicken serum (zero dose)
    3–10: Standards
    11– : Test and quality control samples.
2. Add 0.4 ml of PBSS. Vortex-mix
3. Drop one antibody-coated polystyrene ball in each tube. Stopper tubes and incubate 2 hr in a water bath at 45°
4. Wash polystyrene balls three times with 5 ml of distilled water
5. Add 0.3 ml of tracer. Stopper tubes and incubate 90 min at 45°
6. Wash polystyrene balls three times with distilled water
7. Add immediately 0.3 ml of substrate solution. Incubate 30 min in the dark at room temperature
8. Stop the reaction by adding 2 ml of 1 N HCl
9. Read absorbance at 492 nm using the zero-dose tubes for blank calibration of the spectrophotometer
10. Plot absorbance of standards as a function of their AFP content. Calculate AFP concentration of test samples from their absorbance referred to the standard curve

As an alternative procedure, the different incubations are performed in disposable tray wells on which cover sealers are applied. Balls are washed with the PENTAWASH automatic washing device from ABBOTT. Just before the enzymatic reaction, the wet balls are transferred into polystyrene tubes and the substrate is added. After stopping the reaction, absorbance are measured directly through the tube in the QUANTUM 1 photometer from ABBOTT

immersed in an anti-AFP sheep γ-globulin solution at 20 mg/liter. After incubating for 1 hr at 45° in a water bath, balls are washed with the same phosphate buffer and finally stored at 4° in this buffer with 0.1, sodium azide as preservative.

Method

The working reagents used in the assay are listed in Table III and the assay protocol is given in Table IV.

Results

The assay requires 4.5 hr for completion. The use of a washing device and of a computerized photometer improves the practicality of the test. A typical standard curve, ranging from 0 to 120 ng/ml, is shown in Fig. 2. Threshold sensitivity, calculated as the lowest value which differs significantly (with an interval of confidence of 99%) from the zero standard, is 0.35 ng/ml. When using 0.3 ml of undiluted sample in the first step, the detection limit can be shifted down to 0.1 ng/ml. The within-run precision

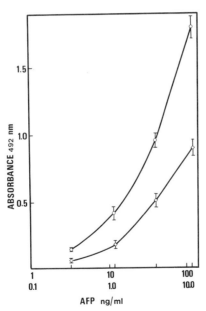

FIG. 2. Standard dose–response curves in the sandwich enzyme immunoassay of AFP. Each point is represented with two standard deviations. (O) Curve ranging from 3 to 120 ng/ml, obtained with the regular procedure (0.1 ml sample diluted with 0.4 ml of PBSS); (◇) curve ranging from 0.3 to 12 ng/ml obtained when 0.3 ml of undiluted sample is used for the first incubation (see Results).

estimated on four different samples at 3, 12, 40, and 95 ng/ml levels yielded coefficents of variation ranging from 1.3 to 5.7%. The day-to-day reproducibility was estimated with a control serum at 25 ng/ml assayed in 50 different series. The coefficient of variation was 7.4%. The mean AFP value in a population of 171 apparently normal adults was 3.5 ng/ml with a standard deviation of 2.1 ng/ml. No false positives were detected in sera containing high titers of anti-globulin factors.

Discussion

This assay has definite advantages over the first sandwich method that we developed. The choice of a relatively high temperature for the different reactions permits them to reach equilibrium faster. Since no daily variations can occur, as when using incubation at room temperature, comparison between daily series is better. Contrary to microcrystalline cellulose, polystyrene balls are easy to wash, avoiding the centrifugation steps. The coating procedure requires no chemicals, it is simple, reproducible, and allows the preparation of large batches of balls which, in addition, show an exceptional stability even at 37°. This type of surface for coating was the only one which, at least in our hands, gave really satisfactory results. With antibody-coated tubes or microtitration plates comparable reproducibility was never achieved.

With the idea of increasing the signal and decreasing the "noise" (i.e., the background color) we preferred in our first method to prepare the conjugate from pure antibodies obtained by immunoabsorption rather than from the γ-globulin fraction of the same antiserum. Further experiments showed that the γ-globulin conjugates gave negligible background absorbances. Conversely, the efficiency of the pure antibody conjugate was questionable since five times more of this conjugate was required to reach the same absorbance than with the γ-globulin conjugate. Our interpretation was that the antibodies possessing the highest affinity were not eluted from the immunosorbent. These antibodies are precisely those possessing the best efficiency at low antigen concentration. Their loss may explain the poor performance of conjugates made from pure antibodies.

The periodate technique produces high activity conjugates. Since the standard curves do not differ significantly whether or not the conjugate is further purified by gel filtration, the preparations probably contain little unconjugated material. Using the same periodate technique the conjugates made with peroxidase are more active than those prepared with glucose oxidase. An important gain in sensitivity was also obtained by the use of o-phenylenediamine dihydrochloride instead of ABTS [2,2'-azino-di(3-ethylbenzthiazoline) 6'-sulfonate] or o-toluidine as chromogen in the enzymatic reaction.

Our routine assay is submitted to an internal quality control program. It consists in following the daily variation of (a) a control serum, (b) the average of the normal values (< 15 ng/ml) on both a LEVEY JENNINGS graph and on a cumulative sums (CUSUM) diagram. No noticeable drift of our internal standard has occurred. We had the opportunity of preparing samples of pure AFP on several occasions and used them to recalibrate the assay. Very close values were obtained in all cases. The excellent stability of AFP and of the reagents may explain this fact. Especially the conjugate showed no apparent loss of activity for months when stored frozen or lyophilized. It was also quite stable in liquid form at 4°.

As it is known that rheumatoid factor is a potential cause of error in sandwich assays,[8] an excess of sheep serum is added in the incubation buffer in order to neutralize the antiglobulinic factors. For several years, a confirmatory test was applied to every abnormally elevated serum. This study was made on more than 1000 samples and did not reveal one single false positive case.

We can conclude that this type of assay is reproducible, simple to perform, and in spite of its short performance time is very sensitive. The preparation of the reagents is not a problem provided that a high avidity antiserum is available. Large batches of reagents can be prepared, owing to their stability. All in all the assay appears to meet the requirements of the clinical laboratory and can replace radioimmunoassay.

Conclusion

After 10 years of research and development, as their initial limitations are being progressively overcome, their advantages increasingly recognized, and their tremendous potential better and better exploited, enzyme immunoassays are finding diversified and expanding uses and applications in laboratory medicine. The two systems described here are illustrative of the challenge met in routine clinical work and show that EIAs can readily equal or surpass any alternative for specificity, sensitivity, precision, speed, convenience, and overall reliability.

Acknowledgments

This work was supported in part by grants from the Medical Research Council and the National Cancer Institute of Canada (L.B.). Data were reproduced from Ref. 3 with permission of the editor of *Clinical Chemistry*.

[3] Radioimmunoassay of Carcinoembryonic Antigen

By MARK J. KRANTZ, SUZANNE LAFERTÉ, and NAOMI ARIEL

Carcinoembryonic antigen (CEA) is an oncofetal glycoprotein expressed in the human fetal digestive system and in adenocarcinomas of the digestive system. It was first identified in colonic tumor tissue by means of antisera raised against human colonic tumor extracts and absorbed with normal tissue extracts in order to render the sera specific for tumor antigen(s).[1] CEA was subsequently shown to be present in fetal digestive tissues, but was not detectable, by techniques available at that time, in any other normal or neoplastic tissues.[2] The first radioimmunoassay capable of detecting CEA in the circulation of cancer patients was reported in 1969,[3] since then a number of other radioimmunoassay techniques have been developed, and extensive clinical studies have been carried out. These studies have shown the clinical value of CEA assays in assessing the prognosis of patients with colorectal cancer, in the postoperative management of cancer patients, and in monitoring antitumor therapy. The assay is not considered useful for the early detection of cancer because false-positive and false-negative results are obtained in this application.[4] Concurrent research on the chemistry and biology of CEA has demonstrated the heterogeneous nature of the antigen, and the existence of cross-reacting substances in various normal and cancerous tissues. It is not yet clear whether CEA heterogeneity and antiserum cross-reactivity are significant factors contributing to the false-positive results noted above that limit the clinical applications of the serum CEA radioimmunoassay. For further information the reader is referred to recent reviews on the clinical applications of the CEA assay,[5,6] the chemistry of CEA,[7] and the properties of some known cross-reactive antigens.[8]

[1] P. Gold and S. O. Freedman, *J. Exp. Med.* **121,** 439 (1965).
[2] P. Gold and S. O. Freedman, *J. Exp. Med.* **122,** 467 (1965).
[3] D. M. P. Thomson, J. Krupey, S. O. Freedman, and P. Gold, *Proc. Natl. Acad. Sci. U.S.A.* **64,** 161 (1969).
[4] I. R. Mackay, *in* "Immunodiagnosis of Cancer, Part 1" (R. B. Herberman and K. R. McIntire, eds.), p. 255. Dekker, New York, 1979.
[5] G. Reynoso and M. Keane, *in* "Immunodiagnosis of Cancer, Part 1" (R. B. Herberman and K. R. McIntire, eds.), p. 239. Dekker, New York, 1979.
[6] A. Fuks, J. Shuster, and P. Gold, *in* "Cancer Markers: Diagnostic and Developmental Significance" (S. Sell, ed.), p. 315. Humana Press, Clifton, New Jersey, 1980.
[7] D. G. Pritchard and C. W. Todd, *in* "Immunodiagnosis of Cancer, Part 1" (R. B. Herberman and K. R. McIntire, eds.), p. 165. Dekker, New York, 1979.
[8] S. von Kleist and P. Burtin, *in* "Immunodiagnosis of Cancer, Part 1" (R. B. Herberman and K. R. McIntire, eds.), p. 322. Dekker, New York, 1979.

METHODS IN ENZYMOLOGY, VOL. 84

The purpose of this article will be to provide methods for the purification of CEA, the preparation of antisera, and the use of these reagents in a double antibody radioimmunoassay for CEA which is suitable as a research tool. The assay described below is rapid, convenient, and reliable, and the sensitivity is adequate for most research applications. References to other procedures will be provided to help investigators adapt the assay to their individual needs. We would like to emphasize that this assay has not been designed for clinical investigations and has not been tested in this application. For clinical studies, standardized reagents and procedures are of great importance, and assays developed specifically for this purpose are commercially available. Much of the clinical data to date has been obtained by the zirconyl phosphate (Z-gel) method,[9,10] but correlations of the results obtained by other assay methods have been reported recently and some differences have been observed.[11-13] The clinical use of an assay which has not been validated by clinical trials is not recommended.

Purification of CEA

Most of the commonly used procedures for the purification of CEA are based on the method described by Krupey et al.,[14] which consists of perchloric acid (PCA) extraction, two gel filtration steps, and preparative electrophoresis. In the procedure described below, the electrophoresis step is replaced by two ion exchange steps, and a single gel filtration step is used.

Hepatic metastases from primary adenocarcinomas of the colon are obtained at autopsy and stored at $-20°$. Since tissues vary in CEA content, a small sample is first assayed by Ouchterlony double diffusion,[15] to determine whether or not purification is feasible. A small section is cut from the frozen tumor tissue and homogenized with 4 volumes of distilled water in an ice-cooled, Virtis Model 23 homogenizer for about 15 min. The crude homogenate and 2-fold serial dilutions are tested by Ouchterlony reaction against anti-CEA antiserum. If unabsorbed antiserum is em-

[9] T. M. Chu and G. Reynoso, *Clin. Chem.* **18,** 918 (1972).

[10] H. J. Hansen, J. J. Snyder, E. Miller, J. P. Vandevoorde, O. N. Miller, L. R. Hines, and J. J. Burns, *Hum. Pathol.* **5,** 139 (1974).

[11] C. Wagener and H. Breuer, *J. Clin. Chem. Clin. Biochem.* **16,** 601 (1978).

[12] Y. D. Kim, J. T. Tomita, J. R. Schenck, C. Moeller, G. F. Weber, and A. A. Hirata, *Clin. Chem.* **25,** 773 (1979).

[13] P. Sizaret and J. Estève, *J. Immunol. Methods* **34,** 79 (1980).

[14] J. Krupey, T. Wilson, S. O. Freedman, and P. Gold, *Immunochemistry* **9,** 617 (1972).

[15] Ö. Ouchterlony and L.-Å. Nilsson, *in* "Handbook of Experimental Immunology, Vol. 1: Immunochemistry" (D. M. Weir, ed.), p. 19.1. Blackwell, Oxford, 1978.

ployed (see below, preparation of antisera), more than one reactive component is found in many tissues. We have seldom observed cross-reactive antigens in tumor tissues where CEA is undetectable, and in cases where multiple antigens are present, the precipitin line nearest the antigen well can usually be assigned to CEA. If a sample known to contain CEA is included in the analysis, positive identification of CEA can be made even with an unabsorbed antiserum. Homogenates of typical tissue samples show reactivity with our antisera when diluted 2-fold, and these usually yield 20 to 40 mg of purified CEA per kg of tissue. A few homogenates are reactive when diluted 8-fold and yield proportionally more CEA. If no reactivity can be demonstrated on Ouchterlony plates, purification is not ordinarily attempted, even though small amounts of CEA might be demonstrated by more sensitive methods.

Step 1. Crude Extract When frozen tissue is employed for purification, it is diced without prethawing into 2-cm pieces, using a heavy knife or small axe and a hammer. Normal tissue is discarded, and 500 g to 1 kg of partially frozen tumor tissue is added to 4 volumes of cold distilled water (w/v) in an ice-chilled stainless-steel vessel. The sample is homogenized (Brinkman Polytron, Model PT 45 equipped with a PT45/2K generator) with periodic interruption to maintain the temperature below 10°. Disintegration is usually completed in less than 15 min, and the homogenate is clarified by centrifugation at 8000 g for 20 min at 4°. All subsequent steps are performed at 4°.

Step 2. Perchloric Acid. An equal volume of 2 M perchloric acid (PCA) is added to the crude extract, and the mixture is stirred for 10 min. The precipitate is removed by centrifugation at 8000 g for 20 min and the supernatant is neutralized with NH_4OH. The sample is desalted by ultrafiltration (see below) before assays for protein and CEA reactivity are carried out.

Step 3. CM-Cellulose. CM-cellulose chromatography (Whatman CM-52) is carried out in 15 mM sodium acetate, pH 5.0, containing 1 mM EDTA. CM-52 is suspended in column buffer and the pH is adjusted to 5.0. Several buffer changes are made batchwise until the pH and conductivity of the supernatant have stabilized. One milliliter of packed volume is used per 10 mg of protein to be fractionated. In order to achieve maximum flow rates a thick slurry of the ion exchange should be packed rapidly into the column. For typical extracts, containing 1–2 g of protein, a bed height of 5–10 cm is prepared in a column of diameter 5 cm.

The neutralized PCA supernatant is desalted and the pH and conductivity are adjusted to correspond to the column buffer, by ultrafiltration or dialysis. The sample volume is not critical for ion exchange chromatography, but the overall speed of processing is improved if the volume is re-

duced about 10-fold before loading the column. We have used a hollow
fiber system (Amicon Model DC2 with H1P10 cartridge) for high-speed
processing, or a stirred pressure cell (Amicon Model 2000 with PM 30
membrane) which is satisfactory but not as rapid. Alternatively the sample can be dialyzed, lyophilized, and redissolved in a small volume of column buffer, but this procedure requires more time and is less convenient
than ultrafiltration, especially for preparations as large as described here.
After the sample has been equilibrated with column buffer it is passed
through the CM-cellulose and washed with 2 column volumes of buffer.
The effluent is collected as a single fraction and contains virtually all of
the CEA.

   Step 4. DEAE-Cellulose. DEAE-cellulose chromatography (Whatman
DE-52) is carried out in 15 mM Tris-acetate, pH 7.0, containing 1 mM
EDTA (starting buffer). DE-52 is equilibrated with starting buffer and
packed as described above for CM-52. A column 2.5 × 20 cm (100 ml) is
used for samples containing up to 1 g of protein. Most preparations contain less protein at this step, but standardization of the column size simplifies elution and analysis procedures. The effluent from the CM-cellulose
column is adjusted to pH 7.0 by the addition of Tris base, and the conductivity is adjusted to that of the starting buffer by dilution. The sample volume may be reduced by pressure dialysis to facilitate loading of the column. The sample is loaded and the column is washed with several bed
volumes of starting buffer until the A_{280} of the effluent is less than 0.02.
CEA is eluted with a linear gradient of NaCl, using 300 ml of starting
buffer in the first chamber and 300 ml of the same buffer containing 0.3 M
NaCl in the second chamber, and 10 ml fractions are collected. A typical
elution profile is shown in Fig. 1. CEA is detected in the first broad peak,
by Ouchterlony reaction and by its characteristic mobility on SDS–polyacrylamide gel electrophoresis (SDS–PAGE), as depicted in Fig. 2.
Cross-reacting antigen(s) are often detected as well, in some or all of these
fractions, if unabsorbed anti-CEA antiserum is used. CEA containing
fractions are selected for further purification without regard to the presence of cross-reacting antigens; however, the second major protein peak
(beginning at fraction 21, Fig. 1) is carefully excluded because it contains
polypeptides which are not readily separated from CEA by gel filtration.

   Step 5. Gel Filtration. The pooled fractions from the DEAE-cellulose
step are either dialyzed, lyophilized, and redissolved in a small volume, or
the volume is reduced by pressure dialysis (Amicon Model 202 stirred cell
with PM-30 membrane). Gel filtration is carried out on a column
(2.6 × 90 cm) of Sephacryl S-300 (Pharmacia) in 0.1 M Tris–HCl, pH
7.5, containing 1 mM EDTA. A column of this size is adequate for up to
75 mg of protein in a sample volume of 5 ml or less. Cross-reacting anti-

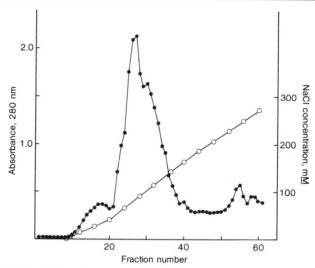

Fig. 1. DEAE-cellulose chromatography of CEA. The sample was loaded and eluted as described in the text. CEA was present in tubes 10–25. Absorbance at 280 nm, ●; NaCl concentration, ○.

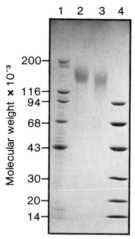

Fig. 2. Analysis of CEA by SDS–PAGE. Electrophoresis was carried out in a slab gel 0.75 mm thick (Bio-Rad) using a 5–20% linear polyacrylamide gradient and a discontinuous buffer system [U.K. Laemmli, *Nature (London)* **227**, 680 (1970)]. The top of the photograph corresponds to the interface between the stacking and running gels. The gel was fixed and stained with 0.2% Coomassie Blue G-250 in 40% methanol/10% acetic acid, and destained by diffusion in 7% methanol/7% acetic acid. Lane 1, molecular weight standards (Bio-Rad); lane 2, CEA purified from a PCA extract; lane 3, CEA purified without the use of PCA; lane 4, molecular weight standards (Pharmacia).

gens, as well as other proteins, are not completely resolved from CEA, but can be effectively removed at this step if the column fractions are analyzed by Ouchterlony reaction and SDS–PAGE before pooling. Only those fractions are selected which give a single precipitin line with unabsorbed anti-CEA antiserum, and which contain a single component of about 180,000 daltons as depicted in Fig. 2. The pooled fractions are dialyzed against distilled water and lyophilized.

Results. The results of a representative purification of CEA are summarized in the table. The tumor tissue contained typical amounts of CEA and cross-reacting antigen(s). PCA treatment of the crude extract precipitates about 95% of the protein. CEA is recovered from the supernatant in good yield and is more than 20-fold purified. The pH and ionic strength must be carefully controlled for the two ion exchange steps in order to ensure efficiency and reproducibility. CEA is an acidic macromolecule and does not bind to CM-cellulose, but at the specified pH and ionic strength about half of the total protein binds to the column and is removed. CEA binds weakly to DEAE-cellulose at pH 7.0 and is eluted at the beginning of the salt gradient as a relatively broad peak. At about 50 mM NaCl a large peak of protein begins to elute, and is not completely resolved from the CEA peak (Fig. 1). Analysis by SDS–PAGE shows that fractions from this region of the gradient contain large amounts of polypeptides of 40,000 to 60,000 daltons, as well as some CEA. These proteins are not readily separated from CEA by gel filtration, presumably because of aggregation under nondenaturing conditions; therefore, only fractions free of these proteins are pooled for the final purification step. The re-

PURIFICATION OF CEA[a]

Purification step	Total protein[b] (mg)	Total CEA[c] reactivity (mg)	Relative purification	Recovery (%)
1. Crude extract	27,000	110	1	
2. Perchloric acid	1,020	96	23	87
3. CM-cellulose	450	86	47	78
4. DEAE-cellulose	41	32	191	29
5. Gel filtration	19	17	220	15

[a] Purification from 535 g of tumor tissue.
[b] Determined by the method of Lowry *et al.* [O. H. Lowry, N. J. Rosebrough, A. L. Farr, and R. J. Randall, *J. Biol. Chem.* **193**, 265 (1951)].
[c] Apparent total weight of CEA in the fraction, determined by radioimmunoassay using unabsorbed anti-CEA antiserum. The total reactivity includes CEA and cross-reactive antigens known to be present in semipurified fractions. Some loss of CEA during purification can therefore be attributed to removal of cross-reactive substances.

covery of CEA reactivity at the DEAE step is somewhat low, about 40%, because some of the CEA is not included in the pool, and because some cross-reacting antigen(s) are removed. About 4-fold purification is achieved at this step, and the selected fractions contain mainly CEA and cross-reacting antigen(s). At the gel filtration step about 50% of the applied CEA reactivity is recovered as pure CEA. The removal of all cross-reacting material contributes to the apparently low yield at this step. Analysis of the purified product by SDS–PAGE is shown in Fig. 2.

Purification of CEA without the Use of PCA

Most workers have purified CEA from PCA extracts, because solubility in PCA was part of the original definition of this antigen, and because PCA treatment precipitates most of the proteins present in the crude extract. To obtain a product corresponding to that originally described, PCA extraction is the method of choice; however, if a product corresponding to the native molecule is desired, purification without the use of PCA may be preferrable. The PCA step can be omitted from the purification scheme described above, and pure CEA is obtained by ion exchange and molecular sieve chromatography. The procedure, modified as described below, is less convenient because of the large amount of protein and large volume that must be handled at each chromatographic step.

A crude extract containing up to 15 g of protein is equilibrated with buffer and passed through a 1 liter bed of CM-cellulose. The flow rate is improved by stirring the top portion of the column when it becomes heavily loaded with protein. The effluent is loaded onto a DEAE-cellulose column (5×20 cm), and proteins are eluted with 1.5 liters of each of the gradient components described above. The protein profile at this step is somewhat more complex than for a PCA extract, but fractions are pooled by the same criteria of excluding the second major protein peak which begins to elute at about 50 mM NaCl. We have used a 5×90 cm column of Sepharose 4B for the gel filtration step, but better resolution of the CEA peak would probably be achieved with Sepharose 6B, or with Sephacryl S-300 which we have used for PCA extracted preparations. The yield of CEA is similar to that obtained from PCA extracts.

SDS–PAGE analysis of a sample of CEA purified by this method is compared with a PCA extracted sample in Fig. 2. The differences in mobility and heterogeneity are representative of different preparations in general, and cannot be specifically attributed to the different purification procedure. No proteins other than CEA are detectable, and after radiolabeling the CEA is bound quantitatively by anti-CEA antisera. CEA prepared in this manner is virtually indistinguishable from PCA-treated CEA by radioimmunoassay (Fig. 4).

Characterization of Purified CEA

CEA is a relatively large glycoprotein with a carbohydrate content of 50% or more. Because the carbohydrate composition and structure are heterogeneous, many of the usual criteria used to demonstrate homogeneity are not satisfied. CEA subfractions have been isolated but there is no generally accepted procedure to provide immunologically distinct fractions. Because CEA is defined by its reactivity with particular antisera, immunochemical analysis with an antiserum of established CEA specificity is required. The following criteria of CEA identity and purity are suggested:

1. A single component with an apparent molecular weight of about 180,000 should be demonstrated by SDS–PAGE. The stained band is diffuse, and has low mobility unless the gel concentration is low. Because of the high carbohydrate content of CEA, molecular weight estimates by this method are probably inaccurate, but the absence of lower molecular weight cross-reacting antigens should be established. SDS–PAGE of ^{125}I-labeled CEA is a more sensitive test of purity. The gel may be analyzed by counting gel slices or by autoradiography of the dried gel. For semiquantitative evaluations, overexposure of film beyond the range of film image density must be avoided.
2. The purified antigen should give a reaction of complete identity by Ouchterlony double diffusion analysis with a known CEA standard, using a previously characterized anti-CEA antiserum.
3. The ^{125}I-labeled CEA should give a typical binding curve with an antiserum of known CEA specificity and titer. Detailed comparisons of individual reagents may reveal some differences, but the anti-CEA antiserum should quantitatively precipitate the labeled antigen.
4. The CEA should have activity in radioimmunoassay which is similar to that of other CEA preparations. Unfortunately, methods for the standardization of materials have not been established, and only limited amounts of antigen or antibody can be obtained from single preparations, so some variation can be expected. When activity is reported on a dry weight basis, errors may be introduced if samples contain salt or differ in degree of hydration.

Physical and Chemical Properties of CEA

CEA was originally defined by its immunological properties, and because of its heterogeneous physical and chemical properties, immunoreactivity is still the most definitive property of the antigen. Preparative isoelectric focusing and density gradient centrifugation were employed to isolate a more homogeneous CEA fraction which was designated CEA-S, and a number of molecular parameters were reported.[16] CEA-S has a sedimentation coefficient of 6.6 S, a diffusion constant of 3.05×10^{-7} cm^2/sec, a Stokes radius of 65 Å, and an estimated molecular weight of 181,000. The pI is 4.5 and the density is 1.41 g/ml. A radioimmunoassay based on this preparation was reported to have increased tumor specificity.[17]

The chemical properties of CEA have been summarized in review articles.[7,18] The amino acid compositions of various CEA preparations are similar, with a relatively high content of hydrophilic residues. The molecule contains six disulfide bonds, but no free cysteine thiol groups. A single N-terminal sequence has been reported. Methionine is reported to be absent, but the determination of small amounts of this residue is difficult. The carbohydrate content of CEA is usually 50% or greater, and variations among different preparations are observed. The carbohydrate is probably linked to the protein by N-glycosidic bonds between N-acetylglucosamine and asparagine, and information about other sugar linkages has been obtained by methylation analysis. Galactosamine is absent, or present only in very low amounts.

Multiple acidic species of CEA have been demonstrated by isoelectric focusing, but nevertheless, isoelectric variants have similar immunoreactivity.[19] Some of this heterogeneity can be attributed to sialic acid residues, but removal of sialic acid does not yield a completely homogeneous preparation. CEA is also heterogeneous with respect to concanavalin A binding, presumably because of differences in the detailed carbohydrate structure of individual molecules, but fractions separated by different lectin affinity appear to be immunologically similar.[20] The existence of subspecies of CEA having unique physical, chemical, or biological properties has been postulated,[17] but further work will be needed to fully evaluate this hypothesis.

Preparation of Antisera

Various anti-CEA antisera as well as sheep anti-horse IgG and goat anti-rabbit IgG are prepared by the following protocol. Antisera are raised

[16] E. F. Plow and T. S. Edgington, *Int. J. Cancer* **15,** 748 (1975).
[17] E. F. Plow and T. S. Edgington, *in* "Immunodiagnosis of Cancer, Part 1" (R. B. Herberman and K. R. McIntire, eds.), p. 181. Dekker, New York, 1979.
[18] W. D. Terry, P. A. Henkart, J. E. Coligan, and C. W. Todd, *Transplant. Rev.* **20,** 100 (1974).

in goats, sheep, and horses by intramuscular injection of 500 μg of purified antigen emulsified in complete Freund's adjuvant (Difco) followed by booster injections of 100 μg at monthly intervals. Serum is collected 1 week after the second and each subsequent boost. Antisera are raised in rabbits by injection of 50 μg of antigen in complete Freund's adjuvant, equally divided between one intramuscular site and multiple intradermal sites. Monthly booster injections of 25 μg are divided between an intramuscular site and a subcutaneous site, and serum is collected by the schedule described above. Antisera are stored in aliquots at $-20°$. It should be noted that various investigators have used different amounts of antigen and modes of injection, and that antisera have been obtained with significantly smaller amounts of CEA.[21]

Commercially available IgG preparations (Miles) are injected to prepare the anti-IgG antisera. These antisera are tested for effectiveness as second antibody by titration at a concentration of first antibody known to give quantitative reaction with the antigen. Maximum precipitation is usually obtained over a broad range of anti-IgG concentrations. The amount used in the radioassay is about 50% more than the minimum required for quantitative precipitation.

Absorption of anti-CEA antisera with normal tissue extracts or with known cross-reacting antigens is recommended by many workers, but no standardized procedure has been established. We have used dialyzed and lyophilized aqueous extracts of normal colon, normal lung, normal liver, and pooled normal serum. Ten milligrams of each extract is added per ml of antiserum, and the mixture is stirred slowly for 1 hr at 37°. It is then stirred overnight at 4°, and clarified by centrifugation at 10,000 g for 30 min. Reactivity with some of the known cross-reacting antigens[8] is greatly reduced or eliminated by this procedure, but true monospecificity for CEA cannot be assured. Individual antisera may differ in their content of cross-reacting antibodies, and require more or less extensive absorption. Elimination of all detectable cross-reactivity in Ouchterlony analysis does not ensure specificity in the radioimmunoassay, and sufficient absorption can only be assured insofar as sensitive assays for undesired antibodies are available. Antisera absorbed as described here will contain many soluble tissue components, and the possible effects of these materials in subsequent experiments is unknown. Absorption with insolubilized or immo-

[19] J. E. Coligan, P. A. Henkart, C. W. Todd, and W. D. Terry, *Immunochemistry* **10,** 591 (1973).

[20] H. S. Slayter and J. E. Coligan, *Cancer Res.* **36,** 1696 (1976).

[21] M. L. Egan, J. T. Lautenschleger, J. E. Coligan, and C. W. Todd, *Immunochemistry* **9,** 289 (1972).

bilized antigens might be preferrable if complications due to the presence of soluble antigens are observed or anticipated.

Radioiodination

The chloramine-T method[22] is convenient and satisfactory for the radioiodination of CEA. The reaction should be carried out in a well-ventilated fume hood. A suitable reaction mixture contains 25 μl of 0.5 M sodium phosphate, pH 7.5, 5 to 20 μg of CEA (usually 20 μl or less), 1 mCi of sodium [^{125}I]iodide (10 μl) and 10 μg of chloramine-T (10 μl). After 30 to 90 sec, the reaction is terminated by the addition of 50 μg of sodium metabisulfite (10 μl). The exact volumes of the various components are not critical, but the correct pH must be maintained and the total reaction volume should be minimized in order to obtain high specific radioactivity.

The reaction mixture is diluted with 50 μl of 0.1 M potassium iodide and 100 μl of 0.05 M sodium phosphate buffer, pH 7.5, containing 0.1% bovine serum albumin (phosphate-BSA buffer), and an aliquot of 5 μl is set aside for determination of specific radioactivity. The labeled protein is separated from the reaction mixture by gel filtration on a 10 ml column of Sephadex G-25 equilibrated with phosphate-BSA buffer. ^{125}I-labeled CEA is recovered in the void volume and is stored at 4°.

It is important to prepare labeled antigen of high specific activity in order to maximize the sensitivity of the assay, but the immunoreactivity of the antigen may be altered if the reaction conditions are too severe. Incorporation of 1–2 atoms of iodine per antigen molecule ensures that most of the molecules will be labeled, and damage will usually be minimized.[22] Since the radiolabeled antigen is used in the assay without any unlabeled antigen as carrier, it is convenient to determine the specific activity with an aliquot of the unfractionated reaction mixture, where the protein concentration is most accurately known. The specific activity is estimated by immunoprecipitation, since CEA is at least partially soluble in many reagents ordinarily employed to precipitate proteins. Dilutions of the unfractionated iodination reaction mixture are prepared in radioimmunoassay diluent (see below) so that 100 μl will contain about 20,000 cpm of protein bound iodine (usually 20–50% of the total counts). The protein bound counts are measured by immunoprecipitation at an antiserum dilution known to give quantitative precipitation, as determined by the titration curve. The specific radioactivity is calculated from this measurement, the dilution factor, and the known amount of CEA used in the labeling reaction. The specific activity is usually 30,000 to 60,000 cpm ^{125}I

[22] W. M. Hunter, *in* "Handbook of Experimental Immunology, Vol. 1: Immunochemistry" (D. M. Weir, ed.), p. 14.1. Blackwell, Oxford, 1978.

per ng CEA, which corresponds to about 15–30 μCi/μg at a counting efficiency of 70%.

Radioimmunoassay

Titration Curve. The binding of a fixed amount of ^{125}I-labeled CEA is measured at serial antiserum dilutions, to estimate the titer of the antiserum with respect to the particular preparation of labelled antigen, and to evaluate the immunoreactivity of the antigen. The diluent of all assay components is 0.05 M boric acid–borax buffer, pH 8.5, containing 0.5% (v/v) of normal serum of the same species as the antiserum. The buffer is prepared by mixing 0.1 M stock solutions of boric acid and sodium tetraborate in a ratio of 3 : 1 and diluting 2-fold. Each assay tube (12 × 75 mm, polypropylene) contains 500 μl of diluted antiserum (or only normal serum in the case of the blank) and about 20,000 cpm of ^{125}I-labeled CEA in 100 μl of diluent. Tubes are incubated for 2 hr with slow shaking in a 37° water bath, followed by the addition of 150 μl of undiluted anti-IgG antiserum (determined to be in excess by pretitration). After 1 additional hour at 37° and 1 hr in an ice bath (or overnight at 4°), precipitates are collected by centrifugation at 8000 g (Sorvall HS4 rotor) for 25 min. The supernatant is decanted and discarded, and the precipitates are counted.

The antibody-specific binding is calculated for each serial dilution of antiserum according to Eq. (1).[23]

$$\% \text{ Bound} = \frac{B - N}{T - N} \times 100 \tag{1}$$

Representative titration curves are shown in Fig. 3 for a horse anti-CEA antiserum before and after absorption with normal tissue extracts. Each point represents the average of duplicate determinations with agreement usually within 5%. Ordinarily, 90% or more of the radioactivity is precipitable with freshly radioiodinated CEA, and nonspecific precipitation is 4–8% of the total counts. The maximum extent of precipitation decreases upon storage, and we usually discard labeled preparations after about 3 weeks. The immunoprecipitability of labeled CEA can often be at least partially restored by chromatography on Sephadex G-150, which removes degradation products of lower molecular weight. The antiserum titer is defined as the dilution at which 50% of maximal binding occurs. The titers of unabsorbed horse antisera are typically in the range of 20,000 to 50,000, while those of unabsorbed rabbit antisera are usually 8,000 to 12,000. Absorption with normal tissue extracts by the procedure described here usually reduces the titer by a factor of 2 to 4.

[23] *B*, counts bound; *T*, total counts; *N*, counts bound in the absence of immune serum.

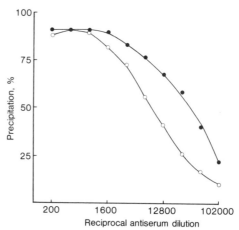

FIG. 3. Titration of anti-CEA antisera with ^{125}I-labeled CEA. Experimental procedures and calulations are described in the text. Unabsorbed antiserum, ●; absorbed antiserum, ○.

Inhibition Assay. The inhibition assay is carried out at an antiserum dilution such that 50% of the maximal binding occurs when no inhibitor is present. For the standard curve, each assay tube contains 500 μl of diluted antiserum and 50 μl of diluent containing 1 to 100 ng of unlabeled CEA. The blank tubes contain only diluent and represent the nonspecific precipitate. Tubes containing antiserum but no CEA represent the maximum precipitation in the absence of inhibitor. For samples containing an unknown amount of CEA to be assayed, 50 μl of the sample, or of an appropriate dilution, are added instead of the CEA standard. All tubes are incubated for 2 hr at 37°, followed by the addition of about 20,000 cpm of ^{125}I-labeled CEA in 100 μl of diluent. Subsequent incubations, the addition of second antibody, and the collection of precipitates are carried out as described above for the titration curve.

Results. The inhibition for each standard or unknown is calculated according to Eq. (2).[23,24]

$$\% \text{ Inhibition} = \left[1 - \frac{B - N}{B_0 - N}\right] \times 100 \qquad (2)$$

Representative standard inhibition curves are shown in Fig. 4. Each point represents the average of duplicate determinations, with agreement usually within 10%. Two CEA preparations were tested; one was extracted with PCA while the other was purified without the use of PCA. The PCA treated CEA was used as radioactive tracer for all the curves, and the antiserum was raised against PCA treated CEA. The antiserum

[24] B_0, counts bound in the absence of unlabeled CEA.

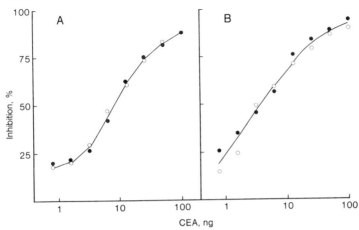

FIG. 4. Standard CEA inhibition curves. Experimental procedures and calculations are described in the text. (A) Unabsorbed antiserum, (B) absorbed antiserum. CEA purified from a PCA extract, O; CEA purified without the use of PCA, ●.

titer, before and after absorption, was determined from the binding curves shown in Fig. 3; the unabsorbed antiserum (Fig. 4A) was used at a dilution of 1:40,000, and the absorbed antiserum (Fig. 4B) at a dilution of 1:10,000. Identical aliquots from the same CEA dilutions were used as inhibition standards for the two antiserum preparations.

The two CEA preparations are indistinguishable when the unabsorbed antiserum is employed (Fig. 4A). Some discrimination of the antigens might be possible with absorbed antiserum (Fig. 4B), but more data are needed to establish the significance of the differences. The sensitivity is better with the absorbed antiserum; about 4 ng of either CEA preparation gives 50% inhibition, compared to about 8 ng with the unabsorbed antiserum. In addition, the shape of the inhibition curves is changed somewhat by absorption of the antiserum. The theoretical and practical significance of these differences is unclear. The curves illustrate the sort of differences that may generally be expected when individual reagents are compared, and the differences might not be reproduced with other reagents prepared in the same fashion.

Discussion

The double antibody radioimmunoassay procedure described here has been designed for speed and convenience, with reproducibility and sensitivity adequate for most research applications. The assay can be completed in 1 day, all materials are easily prepared, sample volumes are large enough for convenient handling, and a large number of samples can be processed at one time. The precipitants are bulky and stable, allowing easy separation from the supernatant. Considerable time is saved by not washing the precipitates, and the risk of partial loss of the precipitate is reduced. The counts in the nonspecific precipitate are higher as a result, but still acceptable because the total reaction volume is relatively large.

We have used this procedure for several years, to monitor the purification of CEA and cross-reacting antigens, to compare the reactivities of different CEA and anti-CEA preparations, to evaluate the effects of modifications of the CEA molecule, and to compare cross-reacting antigens with CEA. We have not used this assay to measure CEA levels in serum, and have not tested it in any clinical application.

The procedure can be modified if necessary to accommodate sample volumes larger than 50 μl by reducing the volume alloted to the addition of antiserum, but adequate carrier serum must then be provided in the other reaction components. The possibility of systematic interference by extraneous sample components should be evaluated under the usual assay conditions. For example, if samples contain human serum, it should be verified that human IgG does not interfere with the action of the second antibody, and the CEA standards should be assayed in the presence of normal human serum devoid of CEA.

The main disadvantage of double antibody methods in general is the requirement for large amounts of second antibody, which usually cannot be used at high dilution, and the resulting significant increase in assay cost. Carrier nonimmune serum is added to produce a bulky precipitate, but the optimal amount may vary for different second antibody preparations. Serial dilutions of carrier normal serum should be tested over a range of second antibody concentrations to find conditions that minimize consumption of second antibody while maintaining effective precipitation of the antigen.

No single radioimmunoassay procedure will satisfy the needs of all investigators, and a number of different assay systems for CEA have been reported. The antibody-bound fraction can be separated from free antigen by ammonium sulfate precipitation,[3] or by the Z-gel method,[10] procedures which do not require second antibody. The Z-gel method can accommodate larger sample volumes than most assays. A double antibody technique employing three radioisotopes has been reported which provides a means to monitor the precipitation of IgG by the second antibody and to correct for incomplete removal of supernatant.[21] A double antibody assay employing 25-μl samples for the measurement of serum CEA levels has been useful for screening large numbers of samples.[4,25] An assay using

[25] J. M. MacSween, N. L. Warner, A. D. Bankhurst, and I. R. Mackay, *Br. J. Cancer* **26**, 356 (1972).

tubes coated with anti-CEA antibody was reported in 1973,[26] and more recently several other solid-phase assay procedures have been described.[12,27,28] Solid-phase assays in general are sensitive, rapid, and more convenient than solution methods, but preparation and standardization of the reagents are more complex.

The titration and inhibition curves shown in Fig. 3 and Fig. 4 were prepared by subtracting nonspecific binding (N) for each data point. Presentation of inhibition curves in this manner is useful for the comparison of different antigen preparations, or curves from different experiments. More uniform curves are obtained because the upper and lower limits are fixed at 100 and 0%, independent of variations in nonspecific binding (N) and maximum binding (B_0).

When the standard inhibition curve is used to estimate the CEA content of test samples, the counts bound (B), the precentage bound (B/T), or the percentage bound relative to maximum binding (B/B_0) is often plotted versus log CEA. The use of a logarithmic scale for the antigen coordinate is recommended because it facilitates interpolation at low antigen values. Nonspecific binding (N) and the maximum percentage bound in the absence of inhibitor (B_0/T) should be monitored as indicators of the quality of the assay. None of these methods results in a linear standard curve over the entire working range, and the scatter is not uniform. Linearization by more complex mathematical transformations increases the nonuniformity of the variance, and an alternative approach for the statistical quality control of radioimmunoassays has been presented.[29] Ten parameters were utilized and those recommended as the most informative were B_0/T, the 50% intercept, within-assay variance, and between-assay variance.

Interpretation of Results

In spite of extensive investigations carried out with CEA radioimmunoassays during the past 10 years, interpretation of the results remains controversial in many cases, and a number of significant problems have not been resolved. Because of the heterogeneous nature of the antigen, CEA is still usually defined by its reactivity with particular antisera. Individual antiserum preparations contain heterogeneous populations of antibodies, and different antisera may have different properties even if raised

[26] T. A. McPherson, P. R. Band, M. Grace, H. A. Hyde, and V. C. Patwardhan, *Int. J. Cancer* **12**, 42 (1973).

[27] R. Wang, E. D. Sevier, R. A. Reisfeld, and G. S. David, *J. Immunol. Methods* **18**, 157 (1977).

[28] R. Zimmerman, *J. Immunol. Methods* **25**, 311 (1979).

[29] D. Rodbard, P. L. Rayford, J. A. Cooper, and G. T. Ross, *J. Clin. Endocrinol. Metab.* **28**, 1412 (1968).

against the identical antigen preparation. CEA and anti-CEA antisera are thus defined in a circular fashion, by means of heterogeneous reagents that are available in limited amounts. Furthermore, generally acceptable criteria for the comparison and standardization of reagents have not yet been defined. The problems of standardization are illustrated by a recent comparison of four CEA preparations, assayed by three different radioimmunoassay methods.[13] Different relative potencies were obtained by the different assays, and some of the results varied as a function of the CEA concentration.

The problem is further complicated by the fact that a large number of cross-reacting substances have already been identified, the probability that more will be discovered, and the difficulty in establishing identity or nonidentity solely by immunological methods. Immunological identity, or cross-reactivity, of antigen preparations can only be assigned with respect to the particular antiserum considered. Hypothetically, an antiserum could be reactive only with common sites shared by two different macromolecules, and the unique features of the antigens would be completely obscured. In many cases, purified antigens are not available at all, or the quantity is insufficient for independent comparisons by physical and chemical methods. In addition to antigens that are clearly cross-reactive, normal tissues may also express low levels of an antigen that is very similar, if not identical to CEA. Purified antigen preparations have been obtained from normal bowel washings[30,31] and from normal colonic mucosa,[32] and these could not be distinguished from tumor CEA.

It is now clear that radioimmunoassays for CEA suffer from a number of problems related to the nature of the antigen, and that results must be evaluated with awareness of the limitations of the assay. The availability of monoclonal antibodies would solve the problems of antiserum heterogeneity and lack of standardization, but selection of the most appropriate specificity may be difficult. As of this time, tumor-specific antigenic determinants of CEA have not been rigorously identified, and the preparation of "specific" anti-CEA antisera remains an empirical process.

Acknowledgments

This work was supported by grants from the National Cancer Institute of Canada and the Medical Research Council of Canada. M. J. K. is a Research Scholar of the National Cancer Institute of Canada.

[30] M. L. Egan, D. G. Pritchard, C. W. Todd, and V. L. W. Go, *Cancer Res.* **37**, 2638 (1977).
[31] M. L. Egan and V. L. W. Go, *in* "Immunodiagnosis of Cancer, Part 1" (R. B. Herberman and K. R. McIntire, eds.), p. 304. Dekker, New York, 1979.
[32] R. Fritsche and J. P. Mach, *Immunochemistry* **14**, 119 (1977).

Section II

Proteins and Peptides of the Blood Clotting System

[4] Immunoradiometric Assays for Factor VIII Antigens: Coagulant Protein (Antihemophilic Factor) and Factor VIII-Related Protein (von Willebrand Factor)

By LEON W. HOYER and NORMA C. TRABOLD

General Considerations

It is now recognized that plasma factor VIII is a complex of two proteins that have distinct immunologic and biochemical properties as well as separate genetic control. Factor VIII procoagulant activity (VIII:C) (antihemophilic factor) is the property of a small glycoprotein that is activated — and subsequently inactivated — by thrombin. Its presence in plasma is under the control of a X chromosome gene and it comprises only a small part of the protein mass of the factor VIII complex. Factor VIII-related protein (VIIIR) (von Willebrand factor) is a polymeric glycoprotein that circulates in large forms ($0.85-12 \times 10^6$ daltons) and is required for normal platelet interaction with damaged blood vessels. It is synthesized in endothelial cells under the control of an autosomal gene and can be reduced to homogeneous ca. 200,000 dalton subunits when incubated with mercaptoethanol or dithiothreitol. Severe VIIIR deficiency prolongs the bleeding time and a functional *in vitro* estimate of VIIIR content can be obtained by assays of ristocetin-induced platelet agglutination.

Both VIII:C and VIIIR can be measured by immunoassays, and immunoradiometric methods provide an excellent combination of specificity, sensitivity, and simplicity. We have used the same general approach to measure the two different proteins. The VIII:C antigens (VIII:CAg) are measured with high titer human anti-VIII:C derived from patients with autoantibodies or transfused hemophilic patients who have developed inhibitors. VIIIR antigens (VIIIR:Ag) are measured with rabbit antibodies obtained by immunization with purified factor VIII. Both assays use labeled antibody that has been purified from immune complexes. Antigen measurement in these assays is based on the differential solubility of the antigen–antibody complexes and free antibody in ammonium sulfate solutions.

Measurement of Factor VIII-Related Antigen (VIIIR:Ag)

Antibody Source

Satisfactory results have been obtained with several different antibodies obtained from rabbits immunized with purified human factor VIII.

METHODS IN ENZYMOLOGY, VOL. 84

While the factor VIII purification method is not critical, if there is trace contamination with other proteins, the antiserum must be absorbed before use. We have used sera prepared from animals immunized with factor VIII obtained by the method of Zimmerman and co-workers[1] and by the method of Bouma and co-workers.[2] A commercial antiserum (Behring Diagnostics, Woodbury, NY) has also been successfully used with this method.

The rabbit sera are heated to 56° for 30 min, centrifuged at 2000 g for 15 min at 4°, absorbed with $Ca(PO_4)_3$ (10 mg/ml serum)for 10 min at room temperature, and centrifuged at 2000 g for 15 min at 4°. When necessary, rabbit sera are absorbed with proteins that are soluble in 3% ethanol at − 3° but which are precipitated when the ethanol concentration is brought to 8%. The most efficient use of these proteins is to couple them to agarose (Sepharose 2B-CL)[3] so that they can be reused. In a typical preparation, 490 mg protein derived from 300 ml normal human plasma was incubated with 80 ml of activated agarose and 92% of the protein was coupled. Antisera are absorbed by running them over a column of agarose-protein (2 ml antiserum/ml beads) at 20 ml/hr at room temperature. An equal volume of saturated ammonium sulfate is added to the absorbed serum and the insoluble proteins removed by centrifugation at 16,000 g for 15 min at 4°. The precipitate is dissolved in a small volume of saline and dialyzed against 0.04 M phosphate, pH 8.0. The absorbed antisera form a single line on Ouchterlony immunodiffusion when tested with a factor VIII-rich concentrate of normal human plasma (cryoprecipitate dissolved in $\frac{1}{10}$ of the volume of plasma from which it was prepared) and have no lines when tested with a similar concentrate of plasma from a patient with severe von Willebrand's disease.

Antibody Purification

IgG can be isolated from the antiserum by a variety of techniques, all of which are satisfactory. We have used DEAE-ion exchange chromatography: the antiserum (or a globulin fraction obtained by precipitation of plasma proteins with an equal volume of saturated ammonium sulfate) and DE-52 (Reeve-Angel, Clifton, N.J.) are equilibrated with 0.04 M phosphate at pH 8.0. A ratio of 100 mg globulin/30 ml DEAE-cellulose has provided satisfactory purification. The IgG fractions (protein not absorbed by the column) have been characterized by immunoelectrophoresis using a potent anti-whole rabbit serum. Although small amounts

[1] T. S. Zimmerman, O. D. Ratnoff, and A. E. Powell, *J. Clin. Invest.* **50**, 244 (1971).
[2] B. N. Bouma, Y. Wiegerinck, J. J. Sixma, J. A. van Mourik, and I. A. Mochtar, *Nature (London) New Biol.* **236**, 104 (1972).
[3] J. Porath, R. Axen, and S. Ernback, *Nature (London)* **215**, 1491 (1967).

of transferrin are present in some preparations in addition to the IgG, this is easily removed in subsequent purification steps. The IgG is then dialyzed against borate-buffered saline, 7.85[4] and stored at $-20°$.

The purified rabbit IgG has been satisfactorily labeled by both the iodine monochloride and lactoperoxidase methods. The latter technique[5] is now routinely employed in our laboratory. The following reagents are added in sequence to 50 μl IgG anti-VIIIR:Ag (0.2–0.4 mg) in a 12 × 75-mm polystyrene tube: 5 μl of ^{125}I (0.5 mCi of carrier-free Na^{125}I for protein labeling), 5 μl lactoperoxidase (2 mg/ml) in 0.05 M sodium phosphate pH 7.5, and 5 μl of H_2O_2 (0.44 mM, freshly prepared by dilution in the phosphate buffer). The mixture is gently vortexed for 1 min at room temperature and the iodination terminated by adding 0.5 ml of the phosphate buffer and 50 μl of normal rabbit serum. The solution is then dialyzed against one liter of borate-buffered saline, pH 7.85. Over 80% of the added ^{125}I is bound to protein and >95% of the counts are precipitated by 10% trichloroacetic acid at 4°.

Immune complexes are prepared by incubating the labeled anti-VIIIR:Ag with 5 ml of normal human plasma (30 min at 37°) and the mixture is gel filtered using a 6% agarose column equilibrated with borate-buffered saline. The flow rate is maintained at 20 ml/hr with a peristaltic pump and 5 ml fractions are collected (Fig. 1A). The void volume fractions contain immune complexes and 3–17% of the labeled IgG. The amount of labeled IgG incorporated into immune complexes depends on the potency of the antibody preparation.

Normal rabbit serum (0.25 ml) is added to the pooled void volume fractions and the mixture dialyzed at room temperature for 3 hr against 1 liter of 0.05 M glycine, 0.1 M NaCl, pH 2.4. The mixture is then gel filtered on Sephadex G-200 equilibrated with the glycine–saline buffer. The flow rate is maintained at 10 ml/hr with a peristaltic pump with 5 ml fractions are collected in tubes that contain 3 ml of borate buffer, pH 8.4[4] (Fig. 1B). The fractions containing IgG (the second peak) are pooled for use in the immunoassay and 2 ml normal rabbit serum is added.

The dissociated purified anti-VIIIR:Ag is then concentrated by adding an equal volume of saturated ammonium sulfate, adjusting the pH to 8, mixing for 0.5 hr at room temperature, and centrifuging at 16,000 g for 20 min at room temperature. Over 99% of the radioactivity is recovered in the precipitate which is dissolved in 5 ml borate–saline. The antibody is then "precycled" to remove any material insoluble in 25% saturated ammonium sulfate by adding 2.5 ml of normal rabbit plasma and 2.2 ml satu-

[4] W. E. Vannier, W. P. Bryan, and D. H. Campbell, *Immunochemistry* **2**, 1 (1964).

[5] J. E. Thorell and B. G. Johansson, *Biochim. Biophys. Acta* **251**, 363 (1971).

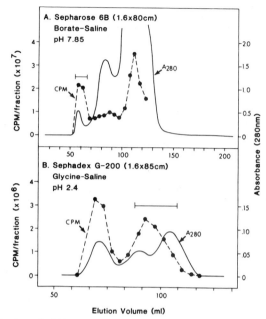

FIG. 1. Purification of rabbit anti-VIIIR:Ag for use in immunoradiometric assay. (A) Separation of nonantibody IgG from immune complexes composed of ^{125}I-labeled rabbit IgG and human VIIIR:Ag. Five milliliters of normal human plasma and 0.3 mg labeled IgG (0.4 mCi) were incubated together for 30 min prior to gel filtration on a 1.6 × 80-cm column of 6% agarose equilibrated with borate-saline. The bar indicates those fractions that were pooled and dissociated at pH 2.4. (B) Separation of dissociated ^{125}I-labeled anti-VIIIR:Ag from undissociable immune complexes and denatured IgG. The immune complexes from (A) were acidified prior to gel filtration on a 1.6 × 85-cm column of Sephadex G-200. The fractions containing free antibody (indicated by the bar) were pooled for use in the immunoassay.

rated ammonium sulfate to the dissolved precipitate, mixing for 0.5 hr at room temperature, and centrifuging at 16,000 g for 20 min at room temperature. The *precipitate* containing approximately 10% of the radioactivity is then discarded and the supernatant dialyzed against borate-buffered saline. After dialysis, the material is stored in 1 ml aliquots at −20°.

The protein concentration of the purified antibody solution is calculated from the cpm/ml and the specific activity of the initial IgG preparation. The amount of labeled antibody used in the immunoradiometric assay is chosen to permit satisfactory counting statistics within a reasonable period of time.[6]

---

[6] L. W. Hoyer, *J. Lab. Clin. Med.* **80**, 822 (1972).

The Immunoradiometric Assay

The assay is carried out in 12 × 75-mm polystyrene tubes. To undiluted normal rabbit plasma (0.1 ml) and dilutions of test material in borate-buffered saline (0.1 ml) are added 5–15 ng labeled antibody (0.2 ml containing 2–7000 cpm). The mixture is incubated for 30 min at 37° in most assays since this provides satisfactory sensitivity. Increased sensitivity may be obtained by continuing the incubation at room temperature or 37° overnight. At the end of the incubation, 0.4 ml of 50% saturated ammonium sulfate is added, the contents mixed and the tubes left at room temperature for an additional 30 min. The precipitate is then separated by centrifugation at room temperature (2600 *g* for 15–30 min) and washed once with 1 ml of 25% saturated ammonium sulfate. The precipitated radioactivity is determined by counting for 10 min using a crystal scintillation detector.

Duplicate determinations are done for each dilution of test material and for dilutions of pooled normal plasma standards that are included in each assay. When a 30-min incubation is used, the standard curve is prepared with dilutions of normal plasma between 1/20 and 1/640 (Fig. 2). Extended incubation increases the sensitivity so that normal plasma dilutions as low as 1/10,000 give values above the blank.

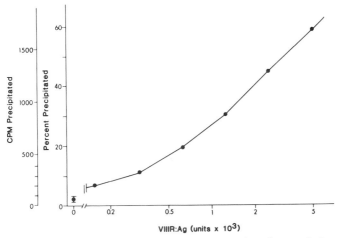

Fig. 2. Immunoradiometric assay of VIIIR:Ag. Data are given for a typical experiment. The amount of VIIIR:Ag in the 0.1 ml assay volume is indicated for dilutions of the pooled normal plasma standard between 1/20 and 1/640. The amount of ^{125}I-labeled antibody precipitated in 25% saturated ammonium sulfate given on the vertical axis: both cpm and percentage of total radioactivity are noted. This assay was carried out by the standard procedure using a 30-min incubation. Assay sensitivity can be increased 20-fold if the labeled antibody is incubated with test materials for 16 hr prior to the separation of immune complexes with ammonium sulfate.

Assay characteristics include sensitivity of 0.002 U/ml when the 30-min incubation period is used and 0.001 U/ml with prolonged incubation. The assay reproducibility is such that repeated determinations of a single plasma on different days give a coefficient of variation of 9%. The specificity of the assay has been determined with plasmas from patients with severe von Willebrand's disease; they have less than 0.001 U/ml.

There is good correlation of VIII:Ag with VIII:C activity in normal plasma ($r = 0.82$).[6] Normal values of VIIIR:Ag are found in hemophilic plasmas and reduced levels are found in plasmas from patients with von Willebrand's disease.[6]

Measurement of Factor VIII Procoagulant Antigen (VIII:CAg)

Antibody Source

Immunoradiometric assay of VIII:CAg requires a high titer human inhibitor. Both spontaneous autoantibodies and antibodies developing in multitransfused hemophiliacs are satisfactory if their titer is sufficiently high. We have used a very potent (3800 Bethesda U/ml) spontaneous inhibitor in most of our studies, but three other antibodies with titers > 1000 Bethesda units/ml have been satisfactory.[7] There is a theoretical advantage in using material obtained from a patient with an autoantibody who has not received multiple transfusions since it is less likely to be contaminated by antibodies to other human plasma proteins. For this reason, most inhibitor antibodies can be used without absorption.

Antibody Purification

IgG has been separated from other serum proteins by the caprylic acid method of Steinbuch and Audran,[8] and Fab' fragments have been prepared by sequential pepsin digestion, reduction with β-mercaptoethanol, and Sephadex G-100 gel filtration.[9] In a typical preparation, 10 ml of 0.06 M sodium acetate (pH 4) is added with vigorous stirring to 5 ml of antibody plasma. After thorough mixing, 68 mg of caprylic acid is added per ml of plasma while the solution is vigorously stirred at room temperature. After 30 min incubation, the non-IgG plasma proteins are separated by centrifugation at 12,000 g for 15 min at room temperature. The supernatant is then concentrated 3-fold by negative-pressure ultrafiltration using standard dialysis tubing and dialyzed against 0.1 M sodium acetate, pH 4.5. Pepsin dissolved in the acetate buffer is added to the IgG (0.02 mg pepsin/mg IgG) and the digestion is allowed to proceed at 37° overnight.

[7] J. Lazarchick and L. W. Hoyer, *J. Clin. Invest.* **62**, 1048 (1978).
[8] M. Steinbuch and R. Audran, *Arch. Biochem. Biophys.* **134**, 279 (1969).
[9] G. M. Edelman and J. J. Marchalonis, *in* "Methods in Immunology and Immunochemistry," (C. A. Williams and M. W. Chase, eds.), Vol. 1, p. 405. Academic Press, New York, 1967.

After the pH is adjusted to 7.4, a small amount of precipitate is removed by centrifugation (12,000 g for 10 min at room temperature). This protein is then gel filtered over a 1.6 × 80-cm column of Sephadex G-100 in borate-buffered saline. Proteins of the initial peak are pooled and concentrated to 5 ml by negative-pressure ultrafiltration at room temperature. The F(ab')$_2$ is then reduced by adding 2-mercaptoethanolamine–HCl to a final concentration of 0.02 M and incubated at 37° for 75 min. The reaction is stopped by adding iodoacetamide, pH 8, to a final concentration of 0.022 M. The solution remains at pH 7.4 while gently mixed for 15 min at room temperature. The gel filtration is then repeated using the same Sephadex G-100 column. The fractions of the second protein peak (ca. 50,000 daltons) are pooled, concentrated to 2–3 mg/ml by negative-pressure ultrafiltration, and stored at −20° in 0.5-ml aliquots. There is good separation of Fab' from any residual F(ab')$_2$ in this gel filtration since the unreduced material remains at the void volume.

Fab' is iodinated for use in the immunoassay by the method described for IgG anti-VIIIR:Ag. The protein concentration and incubation conditions are identical to those already described.

Complex formation between labeled Fab' and VIII:C is achieved by incubating 0.1–0.3 mg antibody (70 Bethesda units in 50 μl) and a Factor VIII concentrate containing 50–70 units of VIII:C. The immune complexes obtained during 4 hr incubation at 37° are separated by gel filtration using 6% agarose in borate-buffered saline (Fig. 3A). Fractions eluting before free Fab' are pooled and dissociated after bovine serum albumin is added to a final concentration of 1 mg/ml. The mixture is brought to pH 3.5 with special precautions taken to avoid any lower pH. This is accomplished by the addition of 0.1 M HCl until the pH is approximately 4 followed by a final adjustment with 0.01 M HCl. The slow addition is done with thorough mixing on a magnetic stirrer. After the mixture has been held at 37° for 30 min, it is returned to pH 7.4 with 0.1 M NaOH, and the mixture is centrifuged to remove precipitated proteins (12,000 g for 10 min at room temperature). The solution is then concentrated to 5–10 ml by dialysis against Aquacide (Calbiochem, Costa Mesa, CA), and dissociated antibody is separated from undissociated immune complexes and from denatured protein by Sephadex G-200 gel filtration. The purified antibody (late eluting fractions corresponding to free Fab') contains 30–60% of the radioactivity (Fig. 3B). It should be noted that the optimum conditions for dissocation of immune complexes may be different for different antibodies. If the antibody is stable at lower pH values, it may be possible to obtain better yields by using a lower pH. We have also used 4 M guanidine, pH 7, as a dissociating agent.[10] Although the yields have

[10] B. Furie, K. L. Provost, R. A. Blanchard, and B. C. Furie, *J. Biol. Chem.* **253**, 8980 (1978).

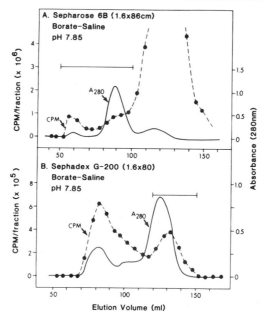

Fig. 3. Purification of human anti-VIII:CAg for use in immunoradiometric assay. (A) Separation of nonantibody Fab' from immune complexes composed of ^{125}I-labeled human Fab' and human VIII:CAg. Fifty units of a factor VIII concentrate and 0.3 mg labeled Fab' (0.4 mCi) were incubated together for 4 hr prior to 6% agarose gel filtration. The bar indicates those fractions that contained immune complexes. They were pooled and dissociated at pH 3.5. (B) Separation of dissociated ^{125}I-labeled anti-VIII:CAg from undissociable immune complexes and denatured Fab'. The immune complexes from (A) were incubated at pH 3.5 for 30 min and neutralized prior to gel filtration on Sephadex G-200. The fractions containing free Fab' (indicated by the bar) were pooled for use in the immunoassay.

been comparable, the method requires that the second gel filtration be carried out in guanidine and is, therefore, somewhat more cumbersome.

As in the VIIIR:Ag assay, the amount of purified anti-VIIICAg used in each assay tube is arbitrary. We have chosen to dilute the antibody so that it has no less than 10,000 cpm/ml,[7] and 1 mg/ml BSA (final concentration) is added.

The Immunoradiometric Assay for VIII:CAg

The assay is carried out in 12 × 75-mm polystyrene tubes to which are added sequentially 0.1 ml of human IgG (20 mg/ml in borate-buffered saline) included as a carrier protein so that the precipitate is large enough for easy separation, 0.1 ml of test material or dilution of test material in borate-buffered saline, and 0.1 ml of the purified radiolabeled Fab' anti-VIII:CAg. The mixture is incubated at 37° for 4 hr to obtain maximal complex formation, or at 37° for 0.5 hr and at 4° for 18 hr. The results obtained by the two alternative incubation periods are similar. Subsequently, the bound antibody is separated from free Fab' by adding 0.2 ml of a solution containing 95% saturated ammonium sulfate. The precipitated proteins are separated by centrifugation (2800 g for 30 min) after a 30-min incubation at room temperature and are washed twice with 1 ml of 38% saturated ammonium sulfate. The radioactivity of the washed precipitates is then determined. Duplicate determinations are done for at least two dilutions of test material, and the assay values are determined by reference to a standard curve obtained with dilutions of pooled normal plasma (Fig. 4).

VIII:C can be detected at levels as low as 0.01 U/ml in most assays, with the limiting factor being nonspecific precipitation of radiolabel in the blank tubes (8–21% depending upon the antibody preparation). The reproducibility of the assay is such that repeated assays for the same plasma on different days have a coefficient of variation between 7 and 12%. The specificity in the assay has been verified by demonstrating no detectable reactivity in plasmas from patients with severe classic hemophilia.

As described above, the method is satisfactory for quantitative assay of column fractions, other materials with low protein concentrations, and plasma samples diluted at least 1 to 4 with borate-buffered saline. The incubation mixture has been modified slightly for more concentrated samples to avoid the influence of protein concentration. Thus, when plasmas

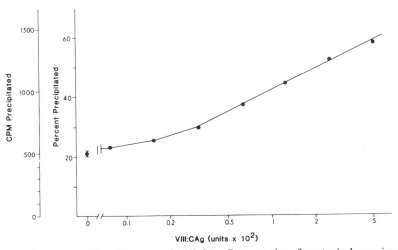

FIG. 4. Immunoradiometric assay of VIII:CAg. Data are given for a typical experiment. The amount of VIII:CAg in the 0.1 ml assay volme is indicated for dilutions of the pooled normal plasma standard between 1/2 and 1/128. The vertical axis indicates the amount of radioactivity precipitated in 38% saturated ammonium sulfate.

are assayed at concentrations less that 1 : 4, both standards and unknowns are diluted in hemophilic plasma that has no detectable VIII:CAg. Concentrations of VIII:CAg as low as 0.01 U/ml can be measured in undiluted plasma samples if this precaution is observed.

When VIII:CAg values are compared with VIII:C levels in normal human plasmas, a very good correlation is detected ($r = 0.90$).[7] Plasmas from patients with von Willebrand's disease also have similar VIII:C and VIII:CAg values. In hemophilia, most patients with severe disease have undetectable (less than 0.01 U/ml) VIII:CAg and VIII:C. Low concentrations of VIII:CAg have been found in 25% of patients with severe classic hemophilia and comparable values for VIII:C and VIII:CAg are found in patients with mild and moderate disease. Some (ca. 10%) of mild and moderate hemophiliacs have normal levels of VIII:CAg even though the VIII:C activity is less than 0.1 U/ml.[7]

The assay is sensitive and reproducible and it can detect proteins that have only one VIII:CAg antigenic determinant. This is an important advantage when compared to two-site immunoradiometric assays that require divalent antigens. The assay has considerable potential in studies of VIII:C structure and the characterization of plasma from patients with hemophilia and von Willebrand's disease. It is of demonstrated value in the prenatal diagnosis of hemophilia using small plasma samples obtained by fetoscopy.[11]

[11] S. I. Firshein, L. W. Hoyer, J. Lazarchick, B. G. Forget, J. C. Hobbins, L. Clyne, F. A. Pitlick, W. A. Muir, I. R. Merkatz, and M. J. Mahoney, *N. Eng. J. Med.* **300**, 937 (1979).

[5] Conformation-Specific Antibodies: Approach to the Study of the Vitamin K-Dependent Blood Coagulation Proteins

By BRUCE FURIE, RITA A. BLANCHARD, DAVID J. ROBISON, MINDY M. TAI, and BARBARA C. FURIE

Introduction

The antibody combining site is a sensitive probe of the structure and conformation of a macromolecule against which it is directed. Antibody populations raised against an intact native protein may be purified and antibody subpopulations directed against a particular set of structural determinants on the protein isolated. These subpopulations may serve as useful reagents for the qualitative and quantitative evaluation of the structure and the conformational states of these protein antigens.

The binding of antibody to antigen is the result of the specific interaction of the molecular surface of the antigen with the complementary features on the surface of the antibody combining site. This binding reaction requires alignment, in three dimensions, of complementary charged and hydrophobic regions. Within the context of the molecular motion of both the antibody combining site and the antigenic surface of the protein, maximal binding strength is attained when the antigenic determinant occupies the optimum number of binding subsites within the antibody combining site. The relative degree of interaction (cross-reactivity) of monospecific antibody with different antigens is quantitatively related to the degree of structural similarity between them. The relative degree of interaction of monospecific antibody with different conformational states of a flexible antigen is also quantitatively related to the degree of structural similarity between these conformers. Using these conformation-specific antibodies, it is possible to study the conformational states of a particular antigenic surface on a protein.

The degree of binding observed in the interaction of monospecific antibody with a series of antigens which share identical primary structures but which have significant differences in their time-averaged three-dimensional structure may be interpreted in several manners. We may consider the induced-fit hypothesis, originally put forth for interpretation of enzyme-substrate binding but appropriately applied to antibody–antigen interaction.[1] The antigen is conformationally motile. The antibody combining site may be considered as a fixed structure which can bind to only one of the conformational states of the antigen. As antibody and antigen come in contact, the tertiary structure of the antigen is altered by the antibody combining site (i.e., induced). The final structure of the antigen is that conformer which is most suited thermodynamically for occupancy of the antibody combining site.

Alternatively, we may consider a model in which the conformationally flexible antigen exists in solution in an array of conformational states.[2] The occupancy of each state is determined by an equilibrium constant which describes this conformational equilibrium. For purposes of simplicity, we have previously considered a two-state model in which the antigen exists in its native, folded form or in its random, unfolded form. If it is assumed that the antibody combining site can interact with only one form, a simple mathematical model can be employed to evaluate the relative concentration of molecules in each conformational state.

[1] D. E. Koshland, Jr., *In* "The Enzymes" (P. D. Boyer, *et al.*, eds.), pp. 1–305. Academic Press, New York, 1959.
[2] B. Furie, A. N. Schechter, D. H. Sachs, and C. B. Anfinsen, *J. Mol. Biol.* **92**, 497 (1975).

It should be emphasized that the fundamental experimental observation in such studies, regardless of the interpretive model employed, is that an antibody directed against an antigenic determinant in a polypeptide binds quantitatively differently to solution conformers of polypeptides with identical primary structure. This difference may be seen as a change in the apparent binding constant, K_A, describing antibody–protein complex formation and may be demonstrated by an alteration of the antibody–antigen binding curve.

One advantage of the immunologic approach to the study of the three-dimensional structure of proteins is that specific regions of the protein may be identified as participants in a conformational transition. If an antibody is isolated to a limited, discrete portion of a protein surface, change in the interaction of the antibody with the protein will reflect changes in the structure of that portion of the protein surface.

We have applied these principles to the study of the conformational states of blood coagulation proteins, including prothrombin and Factor X. Prothrombin is the plasma glycoprotein zymogen of the serine protease, thrombin. Prothrombin, like the other vitamin K-dependent coagulation factors (Factor X, Factor IX, and Factor VII), contains γ-carboxyglutamic acid.[3,4] This amino acid is formed by the posttranslational, vitamin K-dependent enzymatic carboxylation of 10 glutamic acid residues near the amino-terminal end of the protein.[5] These γ-carboxyglutamic acid residues participate in the formation of two high-affinity metal binding sites and numerous lower affinity sites.[6,7] In the presence of calcium, prothrombin undergoes a conformational change that facilitates the Factor Xa-catalyzed conversion of prothrombin to thrombin on membrane surfaces.[8-10]

Under conditions of vitamin K deficiency or in the presence of a vitamin K antagonist, carboxylation is impaired and abnormal prothrombins are formed.[11,12] The resulting des-γ-carboxyglutamyl-prothrombins bind

[3] J. Stenflo, P. Ferlund, W. Egan, and P. Roepstorff, *Proc. Natl. Acad. Sci. U.S.A.* **71**, 2730 (1974).

[4] G. L. Nelsestuen, T. H. Zytokovicz, and J. B. Howard, *J. Biol. Chem.* **249**, 6347 (1974).

[5] J. W. Suttie, *CRC Crit. Rev. Biochem.* **8**, 189 (1980).

[6] R. Sperling, B. C. Furie, M. Blumenstein, B. Keyt, and B. Furie, *J. Biol. Chem.* **253**, 2893 (1978).

[7] Y. Nemerson and B. Furie, *CRC Crit. Rev. Biochem.* **1**, 45 (1980).

[8] G. L. Nelsestuen, *J. Biol. Chem.* **251**, 5648 (1976).

[9] F. G. Prendergast and K. G. Mann, *J. Biol. Chem.* **252**, 840 (1977).

[10] B. Furie and B. C. Furie, *J. Biol. Chem.* **254**, 9766 (1979).

[11] J. E. Nilehn and P. O. Ganrot, *Scand. J. Clin. Lab. Invest.* **22**, 17 (1968).

[12] P. O. Ganrot and J. E. Nilehn, *Scand. J. Clin. Lab. Invest.* **22**, 23 (1968).

calcium weakly and exhibit little intrinsic clotting activity.[13-15] These forms of prothrombin may be unable to adopt the appropriate conformation required for conversion to the active protease during blood coagulation. The normal activation of prothrombin relies on the presence of γ-carboxyglutamic acid residues and upon the ability of these residues in the presence of metal ions to stabilize a conformation of the molecule.

Factor X is a plasma glycoprotein that is the zymogen of the protease, Factor Xa. Factor X has a molecular weight of 56,000, two polypeptide chains linked by a single disulfide bond, and twelve γ-carboxyglutamic acid residues in the light chain.[16-20] Zymogen activation of Factor X involves the proteolytic cleavage of a single peptide bond in the heavy chain. This results in a structurally subtle but functionally important conformational transition.[21]

In order to study conformational states of prothrombin, the differences in the tertiary structures that may distinguish abnormal prothrombins from normal prothrombin and the structural differences between Factor X and Factor Xa, we have employed antibody subpopulations purified by affinity chromatography as immunochemical probes of protein structure:

1. Two of the antibody subpopulations, anti-prothrombin·Ca(II) and anti-$(12-44)_N$, have been isolated from anti-prothrombin antisera. Anti-$(12-44)_N$ is an antibody preparation which contains antibodies specific for the γ-carboxyglutamic acid-rich region of prothrombin, amino acid residues 12 to 44.[22] Anti-prothrombin-Ca(II) reacts specifically with the metal ion-dependent conformation of prothrombin.[23]

2. An abnormal prothrombin-specific antibody subpopulation has been isolated from anti-abnormal prothrombin antisera which does not interact with prothrombin.[15] This antibody subpopulation has been used in a radioimmunoassay to measure levels of abnormal prothrombin in the presence of prothrombin. This antibody popula-

[13] G. L. Nelsestuen and J. W. Suttie, *J. Biol. Chem.* **247,** 8176 (1972).

[14] J. Stenflo and P. O. Ganrot, *Biochem. Biophys. Res. Commun.* **50,** 98 (1973).

[15] R. A. Blanchard, B. C. Furie, and B. Furie, *J. Biol. Chem.* **254,** 12513 (1979).

[16] C. M. Jackson and D. J. Hanahan, *Biochemistry* **7,** 4506 (1968).

[17] C. M. Jackson, *Biochemistry* **11,** 4873 (1972).

[18] K. Fujikawa, M. E. Legaz, and E. W. Davie, *Biochemistry* **11,** 4882 (1972).

[19] J. B. Howard and G. L. Nelsestuen, *Proc. Natl. Acad. Sci. U.S.A.* **72,** 1281 (1975).

[20] H. C. Thorgersen, T. E. Petersen, L. Sottrup-Hensen, S. Magnusson, and H. R. Morris, *Biochem. J.* **175,** 61 (1978).

[21] B. Furie and B. C. Furie, *J. Biol. Chem.* **251,** 6807 (1976).

[22] B. Furie, K. L. Provost, R. A. Blanchard, and B. C. Furie *J. Biol. Chem.* **253,** 8980 (1978).

[23] M. M. Tai, B. C. Furie, and B. Furie *J. Biol. Chem.* **255,** 2790 (1980).

tion probably recognizes differences in both the tertiary and primary structures of these prothrombin species.

3. Antibodies specific for the heavy chain and the light chain of Factor X have been isolated from anti-Factor X antisera.[24] These antibodies have been used to localize domains of Factor X which undergo a conformational transition during zymogen activation.

General Considerations

Antibody Preparation

Immunization

A primary consideration concerning the production of conformation-specific antibodies is the nature of the three-dimensional structure of proteins presented as immunogen to the recipient animal. Specifically, what is the effect of Freund's adjuvant on the tertiary structure of the protein? Based upon studies on antibodies directed against the native form and antibodies directed against the unfolded, random form of a domain near the COOH-terminal of staphylococcal nuclease, indirect evidence has accrued to support the idea that the solution structure of a polypeptide is not altered by adjuvant.[2,25] Electron spin resonance studies by Berzofsky et al. on hemoglobin A have indicated directly that the three-dimensional structure of the protein is not perturbed in adjuvant.[26] Further support, based upon studies of myoglobin, have also been forthcoming.[27] It is now reasonable to assume the structure of immunogen in adjuvant presented to the immunized animal and the solution structure of the immunogen are very similar, if not identical.

Immunogen Purification

When antibody subpopulations are to be isolated by sequential immunoadsorption using affinity chromatography, the requirements for homogeneity of the immunogen become less stringent. Unlike immunoassays based upon the use of whole antisera, our studies of conformation-specific antibodies have required significant purification of the antibody subpopulation. During this purification antibodies directed against impurities contaminating the immunogen are usually removed.

---

[24] B. Keyt, B. C. Furie, and B. Furie, unpublished results.

[25] D. H. Sachs, A. N. Schechter, A. Eastlake, and C. B. Anfinsen, Proc. Natl. Acad. Sci. U.S.A. **69**, 3790 (1972).

[26] J. A. Berzofsky, A. N. Schechter, and H. Kon, J. Immunol. **116**, 270 (1976).

[27] J. A. Smith, J. G. R. Hurrell, and S. J. Leach, J. Immunol. **118**, 226 (1977).

This is particularly so when multiple purification steps are employed using progressively smaller fragments of the original immunogen.

Purification of Antibody Subpopulations to Specific Regions of Proteins

Purification of Protein Fragments for Immunoadsorption. Two general methods are available to obtain the necessary polypeptide fragments to use as immunoadsorbents: (1) proteolytic digestion of the native protein and purification of peptides corresponding to the region of interest; (2) synthesis of peptides which correspond to particular regions of the protein. Sequential proteolytic digestion using enzymes (e.g., trypsin, chymotrypsin, *S. aureus* protease) or chemical cleavage methods (e.g., cyanogen bromide, *o*-iodosobenzoic acid) can yield useful fragments but at significant costs. Large quantities of purified protein must be readily available (>100 mg). The size and sequence of fragments that may be obtained are limited to those which are amenable to specific cleavage and purification techniques. Purification, and rigorous characterization of the product can consume considerable time. Yet, overall, these methods can yield good quality reagents for affinity chromatographic immunoabsorption. In our studies of the immunochemistry of the vitamin K-dependent proteins, we have employed this strategy.

Alternatively, solid phase peptide synthesis offers several attractive advantages. This approach offers considerable flexibility to design and prepare a wide variety of peptides, including those which may differ from the structure of the immunogen. These peptides can be synthesized in considerable quantity. Because of the difficulty of purifying the product from a heterogeneous mixture of similar byproducts, the size of synthetic peptides is usually limited to about 15 residues. Application of preparative high performance liquid chromatography to this problem may increase the efficiency of this synthetic approach. In addition, this technology is rapidly improving, and should allow for preparation of longer peptides in reasonable yields.

It should be mentioned that an antibody purification scheme may be designed in which fragments prepared by both proteolytic cleavage and chemical synthesis are employed. Often, proteolytic fragmentation may be employed to obtain large (molecular weight 6000–25,000) polypeptides while peptide synthesis may be used to prepare small peptides (molecular weight 1500–2000) which correspond to an antigenic determinant.

Purification of Antibody Subpopulations

Antibodies specific for limited surface structures of a protein are usually trace components of the unfractionated antisera prepared to the protein. Purification of this antibody subpopulation to functional homogeneity is a difficult but necessary requirement to enable its use as a conformation-specific reagent. Using affinity chromatography, we have employed the method of Smith *et al.* to minimize nonspecific protein binding to the affinity matrix.[28] Tween 20 (a nonionic detergent) and NaCl at high concentration are used to eliminate nonspecific ionic and hydrophobic interaction between the serum proteins and the antigen-derivatized matrix. This approach significantly decreases the amount of unrelated protein that coelutes with the specific antibodies of interest.

In addition, we usually design a purification scheme in which multiple affinity chromatographic steps are sequentially employed. Each affinity matrix contains a smaller covalently bound polypeptide than that used before it. This approach also minimizes coelution of contaminating antibodies.

In general, the quality of the product obtained by affinity purification depends in part upon the availability of an affinity matrix containing a specific ligand to which a protein may be bound and upon the availability of an elution system which specifically removes bound protein. For antibody–antigen interactions, the first criteria is easily met through utilization of the protein or a fragment of the protein. In general, it has not proven necessary to attach this fragment to the affinity matrix through an "arm" or an extension. The specific elution system is more problematic, and is likely the reason that many antibody preparations remain, to some extent, functionally heterogeneous. Such a system may require large quantities of antigen to use as an eluting agent. More often, systems which disassociate antibody from antigen nonspecifically are used. These systems include acetic acid or citric acid buffers, ammonium hydroxide, chaotropic salts, or guanidine–HCl. Each must be evaluated experimentally, since their efficacy is not predictable. A satisfactory elution system does not alter antibody structure or function, does not irreversibly alter the binding characteristics of the affinity matrix, and elutes the bound antibody quantitatively. In our work, we have found 4 M guanidine–HCl to be suitable.

Radioimmunoassay: Direct Binding Assay

The interaction of trace amounts of ^{125}I-labeled antigen with varying amounts of antibody is used to determine the antibody concentration to be used in the competitive binding assay. In addition, information obtained from an assay of this type is useful for comparing the relative amounts of antibody present in the preparation which react with differ-

---

[28] J. A. Smith, J. G. Hurrell, and S. J. Leach, *Anal. Biochem.* **87**, 299 (1978).

ent but similar antigens, in assessing the immunoreactivity of the iodi-nated antigen and in assaying the effect of a change in the environment on the interaction of the radiolabeled antigen and antibody. Further-more, the information obtained from the direct binding assay serves as an adjunct to the interpretation of the results of the competitive binding assay. For example, a protein which appears to displace radiolabeled antigen effectively in the competitive binding assay may do so either by competing with radiolabeled antigen for antibody or by altering the ra-diolabeled antigen. Specifically, proteolytic degradation of the antigen or masking of its antigenic determinants by a component in the assay system must be considered.

There are several disadvantages to the use of the direct binding assay. It is costly in terms of the amount of antibody required. At high antibody concentrations the amount of second antibody may have to be increased to obtain complete separation of bound and free radiolabeled antigen when the double antibody method is used. If different proteins or fragments are to be compared they must each be radiolabeled and assayed separately. Lastly, the actual concentration of the radiolabeled antigen may not be precisely known.

Radioimmunoassay: Competitive Binding Assay

The competitive binding radioimmunoassay may be used to measure the relative amount of a fixed concentration of radiolabeled antigen bound to a fixed concentration of antibody in the presence of different amounts of unlabeled antigen competitor. In these experiments increas-ing concentrations of unlabeled antigen compete to displace radiola-beled antigen from antibody. Comparison of the displacement curve generated by the unlabeled primary antigen with the displacement curve generated by a competing antigen allows an evaluation and com-parison of the similarity of the tertiary structure of antigenic determi-nants between the antigen and the competitor. For example, a displace-ment curve generated by competing antigen which is identical to that of the primary antigen suggests that the antibody recognizes no difference in the antigenic determinants between the antigen and the competing protein. This suggests that all determinants which exist on the primary antigen also exist on the competitor at the same concentration. Com-peting proteins of identical primary structure which effect no displace-ment of the radiolabeled antigen from the antibodies have no antigenic similarity to those determinants against which the antibodies are directed. Competitors may effect complete displacement or only partial displacement of the radiolabeled antigen from the antibody. Complete displacement suggests that the competitor and the original antigen share some similarity in all the antigenic determinants against which the population of antibodies is directed. Partial displacement suggests that the competitor and the original antigen share some antigenic determinants but that other determinants which exist in the primary antigen are missing in the competitor. (For further treatment of this subject, see this series, Vol. 70 [1].)

The results of the radioimmunoassay may be misinterpreted if the radiolabeled antigen is destroyed by proteolytic degradation or by complex formation which leads to masking of the antigenic determinants by another component of this system. If such were the case radiolabeled antigen would appear to be displaced from antibody although the competitor and radiolabeled antigen may share no antigenic determinants. Lack of displacement could also be misinterpreted to indicate that the competitor shares *no* antigenic determinants with the original antigen. However, if the association constant describing the interaction of the competitor and antibody is many orders of magnitude lower than that of the original antigen, it may be difficult to achieve satisfactory concentrations of competing antigen. However, a direct binding assay may demonstrate interaction of the radiolabeled competitor with antibody at high antibody concentration.

Specific Methods

Immunization

The proteins were prepared for immunization by exhaustive dialysis at 4° against 0.04 M Tris–HCl, 0.15 M NaCl, pH 8.1. The protein solution (1 mg/ml) was emulsified with an equal volume of complete Freund's adjuvant. Two milliliters of the CFA-protein emulsion was injected subcutaneously into a New Zealand White rabbit at multiple sites (0.4 ml/site) on the back and axillary and femoral folds. Immunization with the emulsion was repeated twice at 10–14 day intervals. Booster injections with an incomplete Freund's adjuvant–protein emulsion were repeated at monthly intervals. Five to seven days after each injection, animals were bled from the central ear artery. Approximately 50 ml of blood from each animal was collected in glass tubes and allowed to clot. After rimming the tube with a wooden applicator stick, the clots were allowed to retract for 24–72 hr at 4°. The serum was collected following centrifugation at 500–1000 g for 15 min at 4°, and stored at − 15°.

The antibody concentration was estimated by quantitative precipitin analysis. Quantitative precipitin analyses were performed in polyethylene microcentrifuge tubes. Tubes contained whole antisera (100 μl), antigen in varying concentrations (1.0 to 150 μg) 0.04 M Tris–HCl, 0.15 M NaCl, pH 8.1, and 1 mM CaCl$_2$, in a total volume of 250 μl. The tubes were mixed vigorously, incubated for 1 hr at room temperature and at 4° for 72 hr. The precipitate was centrifuged and washed twice with 200 μl of 0.04 mM Tris–HCl, 0.15 M NaCl (pH 8.1), at 4°. The washed precipitate was dissolved in 1 ml of 1 N NaOH and the absorbance at 280 nm was determined.

Purification of Protein Antigens

Prothrombin and Factor X were purified from citrated bovine plasma by barium citrate adsorption and elution and DEAE-Sephadex chromatography.[10,29,30] Abnormal prothrombin was isolated from citrated plasma obtained from a calf treated with sodium warfarin. This protein was purified from barium citrate-absorbed plasma by DEAE-Sephacel chromatography, affinity chromatography using anti-prothrombin covalently bound to agarose, and Sephacryl S200 gel filtration.[15]

Preparation of Protein Fragments

Fragment 1 and prethrombin 1 (Fig. 1) were isolated from the thrombin digest of prothrombin using SP-Sephadex.[31] Fragment 12–44 was isolated from the tryptic digest of fragment 1 by barium citrate adsorption

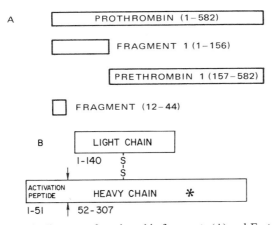

FIG. 1. Schematic diagrams of prothrombin fragments (A) and Factor X (B).

[29] M. P. Esnouf, P. H. Lloyd, and J. Jesty *Biochem. J.* **131**, 781 (1973).
[30] S. P. Bajaj, R. Byrne, T. Nowak, and F. J. Castellino, *J. Biol. Chem.* **252**, 4758 (1977).
[31] R. Lundblad, *Biochemistry* **10**, 2501 (1971).

and gel filtration on Sephadex G-75.[32] Fragments of Factor X were prepared by standard methods.[24]

Preparation of Agarose Derivatives for Affinity Chromatography

Prothrombin, prothrombin fragments, Factor X, and fragments of Factor X covalently bound to Sepharose 4B were prepared in the following manner:

Activation of Sepharose 4B. Sepharose 4B was activated by reaction with cyanogen bromide (Eastman) using standard methods.[33] An appropriate amount of Sepharose 4B was washed five times with distilled water (5 volumes). The washed agarose was resuspended in an equal volume of distilled water and the pH adjusted to 11.0 with 1 N NaOH; cyanogen bromide (200 mg/ml Sepharose) was added. The reaction mixture was maintained below 23° by the addition of ice and at pH 11.0 by the dropwise addition of 5 N NaOH for 9 min. The reaction was quenched by the addition of ice. The activated Sepharose was washed with cold phosphate buffered saline (0.2 M potassium phosphate, 0.15 M NaCl pH 7.4) over a scintered glass funnel.

Conjugation of Proteins and Protein Fragments to Agarose. The protein to be coupled to the activated Sepharose 4B was prepared by dialysis in phosphate-buffered saline for 18 hr at 4°. The protein solution was added to the activated Sepharose at a ratio of 2.0–3.0 mg protein per milliliter of Sepharose. The material was covered and stirred gently overnight at 4°. The protein-Sepharose suspension was then filtered on a scintered glass funnel and washed with 1 liter of 1% ethanolamine. The coupling efficiency was estimated by measuring the amount of protein remaining in the first filtrate. The protein-Sepharose was then washed successively with 1 liter of phosphate-buffered saline, 1 M NaCl and 0.15 M NaCl. The efficiency of coupling ranged from 85 to 95%. The protein-Sepharose was stored at 4° in 0.2% sodium azide.

Affinity Purification of Antibody Subpopulations

We have designed three separate strategies for the affinity purification of antibody subpopulations. An example of each approach is presented below. In the first example, anti-(12–44)$_N$, an antibody subpopulation directed against a γ-carboxyglutamic acid-rich region of prothrombin, was purified from anti-prothrombin antiserum by a multistep procedure.[22] Progressive enrichment of the specific antibody population is accomplished using affinity matrices containing progressively smaller fragments

[32] B. C. Furie, M. B. Blumenstein, and B. Furie, *J. Biol. Chem.* **254**, 12521 (1979).
[33] P. Cuatrecasas, M. Wilchek, and C. B. Anfinsen, *Proc. Natl. Acad. Sci. U.S.A.* **61**, 636 (1968).

which include the antigenic determinant(s) of interest. The second example illustrates the isolation of specific antibodies which bind to a protein but do not bind to an analogous protein identical except for minor changes in the primary structure. Antibodies specific for abnormal prothrombin were purified which do not interact with prothrombin.[15] These proteins differ by the presence of 10 additional carboxyl groups in amino acid residues (γ-carboxyglutamic acid) near the NH_2-terminus. A third example is the purification of a population of antibodies directed against a conformation of prothrombin stabilized by metal ions.[23] In this method an alteration in the environment of a protein (removal of ligand) may effect a significant change in the tertiary structure of some regions of the protein while other regions remain unaltered. Subpopulations of antibodies specific for a particular ligand-stabilized localized conformation of a protein or fragment may be isolated from the whole population of antibodies directed against the protein by altering the conformation of the protein during affinity chromatography.

Affinity Chromatography

Affinity chromatography was performed at room temperature. A buffer system (BBS) which included borate buffer, 1 M NaCl and 0.1% Tween 20 was used to minimize nonspecific interaction of the serum proteins with the column matrix.[28] The antibodies bound to the affinity column should be eluted by reagents which disrupt the antibody–antigen reaction without damaging the antibody or the antigen-Sepharose support. In these experiments antibodies were eluted with 4 M guanidine hydrochloride, immediately pooled, and dialyzed at 4° against 0.04 M Tris–HCl, 0.15 M NaCl, pH 8.1. All affinity columns were monitored spectrophotometrically at 280 nm. Antibodies were concentrated by ultrafiltration using an Amicon PM-30 membrane to a concentration of 0.5–2.0 mg/ml and stored at $-20°$.

Anti-(12–44)$_N$: Sequential Affinity Purification

Prothrombin fragment (12–44) is derived from a region of prothrombin which contains eight of the ten γ-carboxyglutamic acid residues of prothrombin. The antibody subpopulation directed against prothrombin fragment (12–44) was purified by sequential immunoadsorption (Fig. 2).

Rabbit antiserum (4 ml) against bovine prothrombin was applied to a column (2 × 7 cm) containing prothrombin-Sepharose. The column was equilibrated and eluted with BBS. After the large peak of unbound serum proteins was eluted and the column washed with BBS, the bound antibody population was eluted with 4 M guanidine–HCl. These antibodies

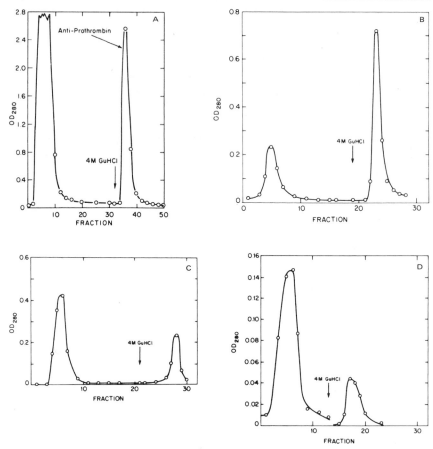

FIG. 2. Purification of anti-$(12-44)_N$ antibodies by sequential immunoadsorption using affinity chromatography. (A) Purification of anti-prothrombin on prothrombin-Sepharose. (B) Purification of anti-fragment 1 from anti-prothrombin on fragment 1-Sepharose. (C) Purification of anti-fragment 1 minus prethrombin 1 on prethrombin 1-Sepharose. (D) Purification of anti-$(12-44)_N$ on fragment $(12-44)$-Sepharose. Bound antibody fractions were eluted with 4 M guanidine–HCl. From Furie et al. (1978).[22]

were immediately dialyzed against 0.04 M Tris–HCl, 0.15 M NaCl, pH 8.1 at 4°.

The population of antibodies interacting with the NH_2-terminal third of prothrombin (fragment 1) was separated from those interacting with the prethrombin 1 portion by affinity chromatography of the purified anti-prothrombin antibodies over a fragment 1-Sepharose column. Approximately 5–6 mg of purified anti-prothrombin antibodies were applied to a column (1 × 6 cm) of fragment 1-Sepharose and washed with 1 mM CaCl$_2$ in BBS. Anti-fragment 1 antibodies were eluted with 4 M guanidine –HCl and dialyzed as described above.

Since these anti-fragment 1 antibodies still contained a subpopulation of antibody which reacted with prethrombin 1, the anti-fragment 1 antibodies were further purified by affinity chromatography over a prethrombin 1-Sepharose column. Approximately 7 mg of purified anti-fragment 1 was applied to a 1.5 × 7-cm column of prethrombin 1-Sepharose in 1 mM CaCl$_2$ in BBS. The first peak eluted contained the specific anti-fragment 1 antibodies which do not cross-react with prethrombin 1. This was the antibody population from which anti-fragment (12–44)$_N$ was purified.

The purified anti-fragment 1 antibodies minus those antibodies which cross-react with prethrombin 1 (1.6 mg) were then applied to a column (1 × 12 cm) of fragment (12–44)-Sepharose equilibrated in BBS and 1 mM CaCl$_2$. Anti-fragment (12–44)$_N$ was eluted with 4 M guanidine–HCl and dialyzed as described above.

Anti-Prothrombin–Ca(II) Antibody: Ligand-Induced Conformational Change of Protein Antigen on Affinity Matrix

Rabbit anti-bovine prothrombin antiserum (7 ml) was applied to a prothrombin-Sepharose column (2 × 10.5 cm) equilibrated with 0.04 M Tris–HCl, 1 M NaCl, 1 mM CaCl$_2$, pH 8.1. After elution of the unbound serum protein peak, the antibody was eluted with 4 M guanidine–HCl. These anti-prothrombin antibodies were dialyzed exhaustively against 0.04 M Tris–HCl, 0.15 M NaCl, pH 8.1 at 4°.

Anti-prothrombin–Ca(II) antibodies were then isolated on a prothrombin-Sepharose column (Fig. 3). The purified anti-prothrombin antibodies from the first column were applied to a prothrombin-Sepharose column (2 × 10.5 cm) equilibrated with 0.04 M Tris–HCl, 1 M NaCl, 1 mM CaCl$_2$, pH 8.1. Anti-prothrombin–Ca(II) antibodies were eluted with 0.04 M Tris–HCl, 1.0 M NaCl, 3 mM EDTA, pH 8.1. The eluted antibodies were dialyzed exhaustively in 0.04 M Tris–HCl, 0.15 M NaCl, pH 8.1 at 4°.

Antiabnormal Prothrombin-Specific Antibodies: Antibodies Which Distinguish between Proteins with Minor Primary Structure Differences

Antibodies specific for abnormal prothrombin were purified from rabbit anti-abnormal prothrombin antiserum.[15] Abnormal prothrombin, isolated from plasma of animals treated with sodium warfarin, is in limited supply. Instead, a form of abnormal prothrombin, decarboxylated prothrombin, was prepared semisynthetically by the thermal decarboxylation of γ-carboxyglutamic acid residues on prothrombin.[34] This material

---

[34] P. M. Tuhy, J. W. Bloom, and K. G. Mann, *Biochemistry* **18**, 5842 (1979).

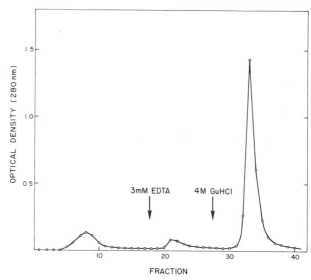

Fig. 3. Purification of anti-prothrombin–Ca(II) antibodies. Purified anti-prothrombin antibodies were applied to a prothrombin-Sepharose column equilibrated with 1 mM CaCl$_2$. Anti-prothrombin–Ca(II) antibodies were eluted with 3 mM EDTA. From Tai *et al.* (1980).[23]

is suitable for affinity immunoabsorption but is too structurally heterogeneous to be employed as immunogen. Anti-abnormal prothrombin antiserum (5 ml) was applied to des-γ-carboxy fragment 1-Sepharose column (1 × 4.5 cm) equilibrated with BBS containing 0.1% Tween 20. After elution of the unbound serum protein peak with this buffer, the bound antibody was eluted with 4 M guanidine–HCl. This antibody was dialyzed exhaustively against 0.04 M Tris–HCl, 0.15 M NaCl (pH 8.1) at 4°. The anti-des-γ-carboxy fragment 1-antibodies were applied to a prothrombin-Sepharose column (1.5 × 8 cm) equilibrated in 0.04 M Tris–HCl, 0.15 M NaCl, pH 8.1. Antibodies which failed to bind to this second column (0.6 mg) were considered anti-abnormal prothrombin-specific antibodies.

Anti-Factor X Antibodies

Antibodies to Factor X were isolated by affinity chromatography. Rabbit anti-bovine Factor X antiserum (2 ml) was applied to a 2 × 5-cm Factor X-Sepharose column equilibrated with 1 M NaCl, 50 mM Tris–HCl, 1 mM CaCl$_2$, pH 7.5 at 25°. Antibodies which bound to the column were eluted with 4 M guanidine–HCl. These anti-Factor X antibodies were dialyzed exhaustively against 40 mM Tris–HCl, 0.1 M NaCl, pH 7.5 at 4°. Anti-heavy chain antibodies were purified from anti-Factor X antibodies by affinity chromatography using Factor X heavy chain-Sepharose. The protein which did not bind to Factor X heavy chain-agarose was pooled and subsequently applied to a Factor X light chain-Sepharose column. Anti-heavy chain and anti-light chain antibodies were eluted with 4 M guanidine–HCl from heavy chain-Sepharose and light chain-Sepharose columns, respectively. Anti-des-peptide Factor Xa [anti-$(52-307)_N$] antibodies were purified from anti-heavy chain antibodies by affinity chromatography on a des-peptide Factor Xa-Sepharose column.

Radioimmunoassay

Preparation of ^{125}I-Labeled Bovine Prothrombin, Abnormal Prothrombin, and Factor X

Bovine prothrombin and abnormal prothrombin were labeled with ^{125}I by the chloramine-T method of Hunter and Greenwood.[35] Approximately 1–2 mg/ml of protein was dialyzed against 100 volumes of 0.2 M sodium phosphate, pH 7.5 at 4°. The sodium phosphate concentration of the protein sample to be iodinated was then adjusted to 0.3 M pH 7.5. Na^{125}I (2 mCi) was added to a solution of 50–100 μg protein in 180 μl of phosphate buffer followed immediately by the addition of 10 μl of the chloramine-T solution (2.5 mg/ml). After 45 sec of gentle mixing, the reaction was quenched by the addition of 20 μl sodium metabisulfite (2.5 mg/ml). An equal volume of 20% bovine serum albumin was then added to further quench the reaction and to minimize the nonspecific adsorption of ^{125}I-labeled protein to vessel surfaces. Free iodine was removed by either gel filtration or exhaustive dialysis at 4° against 0.04 M Tris–HCl, 0.15 M NaCl, pH 8.1 containing 1 mM benzamidine hydrochloride. Factor X was labeled by the Bolton–Hunter method.[36]

^{125}I-labeled prothrombin and abnormal prothrombin may be stored at −70° for up to 2 months. Either benzamidine hydrochloride (1 mM) or m-(O-(ethoxysulfoxyethylamino)phenoxypropoxybenzamidine hydrochloride covalently attached to Sepharose was added to inhibit proteolytic degradation.

The reaction conditions employed result in a specific radioactivity of approximately 120–150 Ci/mmol of protein. The molar stoichiometry of iodine to protein is approximately 1:20.

Binding Studies. The interaction of the purified antibodies with prothrombin, abnormal prothrombin, and Factor X was evaluated using a direct binding radioimmunoassay. The double antibody precipitation method was used to separate bound and free radiolabeled antigen. A 250-

[35] W. M. Hunter and F. C. Greenwood, *Nature (London)* **194**, 495 (1962).
[36] A. E. Bolton and W. M. Hunter, *Biochem. J.* **133**, 529 (1973).

μl solution containing ^{125}I-labeled antigen (10^{-9} to $10^{-8} M$), antibody (10^{-10} to $10^{-7} M$), and 100 μg of bovine albumin in 40 mM Tris–HCl, 0.15 M NaCl, pH 8.1, was adjusted to 1 mM CaCl$_2$. Solutions were incubated at 4° for 18 hr prior to the addition of 190 μg of goat anti-rabbit γ-globulin. After 15 min at 25°, 40 μg of rabbit γ-globulin was added. The solution was incubated at 4° for 4.5 hr during which time a visible precipitate was formed. The precipitate was sedimented by centrifugation in a Beckman model 152 Microfuge. A 50-μl aliquot of the supernatant was obtained and assayed for ^{125}I. The pellet was resuspended, washed twice with 40 mM Tris–HCl, 0.15 M NaCl, pH 8.1, at 25° and assayed for ^{125}I in a Beckman Gamma 8000 γ-scintillation spectrometer. From these data, the concentrations of antigen bound to antibody and free in solution were calculated. These data are presented as the percentage of antigen bound as a function of antibody concentration. In experiments to evaluate the effect of Ca(II) on prothrombin–antibody binding, all reagents were rendered as metal-free as possible with Chelex 100.

Competition Studies. The displacement of ^{125}I-labeled antigen from specific antibody subpopulations by unlabeled proteins or protein fragments was assessed using a competition radioimmunoassay. Experiments were performed in polyethylene microcentrifuge tubes containing 250 μl of a solution including ^{125}I-labeled antigen, antibody (approx. $10^{-8} M$), 100 μg of bovine albumin, and unlabeled competing antigen in 40 mM Tris–HCl, 0.15 M NaCl, pH 8.1, 1 mM CaCl$_2$. As described above, free antigen was separated from antibody-bound antigen employing the double antibody precipitation method. Data are presented as the percentage of radiolabeled antigen bound to antibody.

Characterization of the Interaction of Protein Antigens and Antibody Subpopulations

Anti-(12–44)$_N$

Anti-(12–44)$_N$, purified from anti-bovine prothrombin antiserum, is directed against a γ-carboxyglutamic acid-rich region of prothrombin.[22] This antibody forms soluble complexes with prothrombin and inhibits the conversion of prothrombin to thrombin in plasma. This antibody preparation, albeit functionally monospecific, is heterogeneous with regard to antigen affinity and isoelectric point. As shown in Fig. 4, prothrombin and fragment 1 compete equally with ^{125}I-labeled prothrombin for anti-(12–44)$_N$. This suggests that the three-dimensional antigenic structure of the region from residue 12 to residue 44 in both these polypeptides is essentially identical. A 250-fold molar excess of fragment (12–44) to prothrombin and a 100-fold molar excess of abnor-

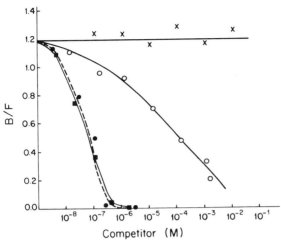

FIG. 4. Interaction of anti-$(12-44)_N$ with prothrombin and prothrombin fragments. Competition radioimmunoassay. Prothrombin (■——■), fragment 1 (●——●), fragment $(12-44)$ (○——○), and γ-carboxyglutamic acid (×——×). From Furie *et al.* (1978).[22]

mal prothrombin are required to inhibit 50% of the prothrombin from binding anti-$(12-44)_N$. These results suggest that the time-averaged structure of fragment $(12-44)$ is not similar to the region 12 to 44 on prothrombin. Using the analysis applied to staphylococcal nuclease,[2] a conformational equilibrium constant (K_{conf}) of 0.004 may be estimated for fragment $(12-44)$.

The interaction of anti-$(12-44)_N$ with prothrombin is significantly altered by metal ions.[10] About 50% of the anti-$(12-44)_N$ antibodies bind only to a metal-stabilized conformer of prothrombin. Metal ions which can facilitate recognition of antigenic determinants on prothrombin by the anti-$(12-44)_N$ subpopulation include Ca(II), Sr(II), Zn(II), Ba(II), Fe(III), Mn(II), Pr(III), La(III), Gd(III), and Mg(II). At Ca(II) concentrations of 0 to 3 mM, increasing metal concentration yielded increased antibody·prothrombin binding (Fig. 5). Half-maximal binding changes were observed at Ca(II) concentrations of 0.2 mM. Using Mn(II), Mg(II), and Gd(III), half-maximal binding changes were observed at metal concentrations of 20 μM, 0.2 mM, and 0.1 μM, respectively. These concentrations correspond closely to the binding constant, K_D, describing the interaction of each metal ion with prothrombin and the concentration of each metal ion required to effect half-maximal fluorescence and circular dichroism changes.[10] We believe that these experiments indicate that occupancy of the high-affinity metal binding sites by metal ions is associated with a structural transition that includes

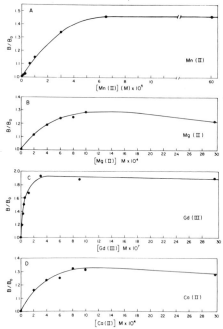

FIG. 5. Effect of metal ions on the interaction of anti-$(12-44)_N$ antibodies and bovine prothrombin. All reagents were rendered metal-free using Chelex 100. Binding is expressed as the ratio of binding of antibody–antigen in the presence of metal ions to the binding of antibody-antigen in the absence of metals (B/B_0). Metal ions studied include (A) Mn(II), (B) Mg(II), (C) Gd(III), (D) Ca(II). From Furie and Furie (1979).[10]

changes in the three-dimensional structure of the γ-carboxyglutamic acid-rich domain.

Anti-prothrombin–Ca(II)

The anti-$(12-44)_N$ antibodies were selected for the ability to recognize the region 12 to 44 in prothrombin. In contrast, the antiprothrombin–Ca(II) antibodies were isolated from anti-prothrombin antiserum on the basis of an absolute requirement for the three-dimensional structure of prothrombin present in the prothrombin·metal complex. These antibodies formed precipitating complexes with prothrombin in the presence but not in the absence of Ca(II).[23] In the radioimmunoassay, the anti-prothrombin–Ca(II) antibodies bound prothrombin tightly in the presence of Ca(II) but not in the absence of Ca(II) (Fig. 6). Like anti-$(12-44)_N$, these antibodies are heterogeneous with regard to antigen affinity.

These antibodies were used to localize the metal-dependent struc-

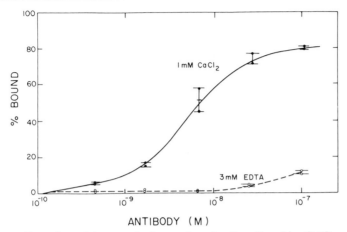

FIG. 6. Effect of metal ions on the interaction of anti-prothrombin–Ca(II) antibodies and prothrombin. Studies were performed in the presence of $CaCl_2$ (●——●) or EDTA (○——○). From Tai et al. (1980).[23]

tures on prothrombin. Anti-prothrombin–Ca(II) binds to fragment 1 and, more specifically, to the NH_2-terminal cyanogen bromide fragment of prothrombin, fragment (1–70). It does not bind to prethrombin 1, abnormal prothrombin, or fragment (71–156). These results indicate that the metal-induced structural transitions in prothrombin involve structures near the NH_2-terminus of the protein.

Significantly, these antibodies also offer a useful reagent to measure prothrombin in the presence of abnormal prothrombin. Anti-prothrombin antibodies bind equally well to prothrombin and abnormal prothrombin. However, anti-prothrombin–Ca(II) antibodies bind only to prothrombin and can be employed to assay prothrombin in complex biological fluids which also contain decarboxylated forms of prothrombin. As will be described, this may have relevance to the clinical laboratory.

Anti-abnormal Prothrombin-Specific Antibodies

The purified antibody subpopulation, representing about 3–5% of the total anti-abnormal prothrombin in the antiserum, was found by radioimmunoassay to bind to abnormal prothrombin but did not cross-react with prothrombin (Fig. 7).[15] In a competition assay, prothrombin in 1600-fold molar excess of abnormal prothrombin did not inhibit the interaction of these specific antibodies with abnormal prothrombin. Using these antibodies, abnormal prothrombin was found to be a com-

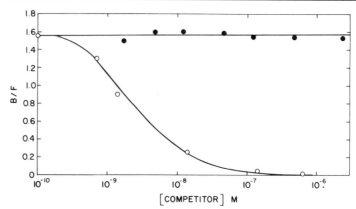

FIG. 7. Binding of anti-abnormal prothrombin antibodies to abnormal prothrombin (O——O) and prothrombin (●——●). Competition radioimmunoassay.[15]

ponent of plasma obtained from a calf treated with warfarin; however, abnormal prothrombin was not detected in normal bovine plasma.

Parallel studies have also been performed with anti-human abnormal prothrombin-specific antibodies. Abnormal prothrombin, measured in the presence of prothrombin with a concentration of about 150 μg/ml, could not be measured in normal human plasma. An upper limit of 0.030 μg/ml was established.[37] Measurement of this protein in blood may have important implications for the monitoring of oral anticoagulant therapy, detection of vitamin K deficiency, or the evaluation of liver function.

Anti-Factor X Antibodies

Conformation-specific antibodies against distinct regions of Factor X were employed to localize determinants which are altered during zymogen activation.[24] All antibodies were derived from anti-Factor X antiserum. Anti-(light chain)$_N$, anti-(heavy chain)$_N$, and anti-(des-peptide heavy chain)$_N$, directed against the active three-dimensional structure of Factor X, bound Factor X, and Factor Xa, as determined by radioimmunoassay. Anti-(light chain)$_N$ interacted with Factor X and Factor Xa equally (Fig. 8A). Anti-(heavy chain)$_N$ and anti-(des-peptide heavy chain)$_N$ bound Factor X preferentially to Factor Xa (Fig. 8B). These differences in binding between antibodies and Factor X or Factor Xa were interpreted as manifestations of changes in conformation or conformational equilibria of domains of the protein which contain antigenic determinants. These results indicate that during zymogen activation of

[37] R. A. Blanchard, B. C. Furie, M. Jorgensen, S. F. Kruger, and B. Furie, N. Engl. J. Med. **305**, 242 (1981).

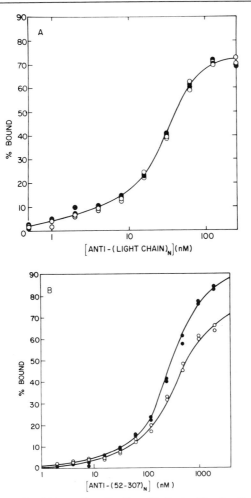

FIG. 8. Interaction of anti-Factor X antibodies with Factor X and activated Factor X. (A) Anti-(light chain)$_N$ antibodies. (B) Anti-(des-peptide heavy chain) antibodies, also known as anti-(52–307)$_N$, directed against the heavy chain. Factor X (●); Factor Xa (○). Direct binding radioimmunoassay.[24]

Factor X to Factor Xa, structural transitions are restricted to the heavy chain of the protein.

Discussion

In these studies we have isolated subpopulations of antibodies directed against limited, specific determinants on the protein surface.

These antibodies are conformation-specific, and, as such, are specific for the native structure of the protein antigen. We have used such antibody preparations as reagents for several purposes: (1) to identify specific regions of a protein which undergo a conformational transition upon ligand binding (prothrombin) or zymogen activation (Factor X); (2) to quantitate structurally similar protein antigens in plasma (prothrombin and abnormal prothrombin); (3) to monitor conformational transitions in biological fluids (prothrombin).

These approaches offer some unique advantages to the study of macromolecular structure. Indeed, we perceive these techniques as means to extend the type of information traditionally derived by physical methods, such as optical and NMR spectroscopy. With particular regard to large proteins, conformational changes may be observable by spectroscopy but localization or assignment of such a perturbation to a particular domain of the protein is often problematic. Using antibodies directed against a region or domain of a macromolecule, the quantitative description of antibody protein binding may be interpreted with regard to changes in the structure of a particular region. Furthermore, very limited quantities of protein are required for such studies. Sensitive immunochemical assays are highly reliable with antigen concentrations of the order of 10^{-6} to 10^{-10} M. Although instrument sensitivity varies, the concentrations of protein required for circular dichroism and fluorescence measurements, absorption difference spectroscopy, and NMR spectroscopy are of the order of 10^{-6}, 10^{-5}, and 10^{-3} M, respectively. Artifacts due to protein association may be avoided at a lower concentration, a particular problem in studies of proteins that function physiologically at very low concentrations. An additional advantage of immunologic approaches to the study of protein structure is that such studies can be performed in complex biological fluids or on immobilized macromolecules, as with membrane-bound proteins. Whether in plasma, cytosol, or cell homogenates, the conformation of a particular protein may be evaluated in its natural milieu.

These methods are not without their problems or their hazards. The antibodies used in these studies are derived from heterologous antisera. After purification, the subpopulations are restricted and some are operationally monospecific. However, these antibody preparations are all structurally and functionally heterogeneous. They are characterized by multiple bands upon isoelectric focusing and demonstrate varying affinities for the antigen. For these reasons it is difficult to extract quantitative information from the analysis of antibody–antigen interaction using theoretical models which assume homogeneity of the antibody and the protein antigen. It is likely that with the application of conformation-specific monoclonal antibodies obtained from murine lymphocyte hybrids to these studies,[38] more rigorous analysis of the primary data may become practical. It is becoming increasingly clear that to understand the function of biologically important macromolecules in their physiologic context, new conceptual and technological approaches must be evolved. These approaches must accommodate the detailed structural study of molecules in complex biological fluids or on membrane surfaces. The use of conformation-specific antibodies as probes of protein structure may offer at least one strategy toward this goal.

[38] R. M. Lewis, B. C. Furie, and B. Furie, *Blood* **58**, 220A (1981).

[6] Radioimmunoassay of Platelet Factor 4

By Karen L. Kaplan and John Owen

Platelets are small anucleate cells which circulate in the blood. Their major physiologic function is to provide initial hemostasis when injury to vascular endothelium occurs. This is accomplished by the formation of a platelet plug. Experimental data indicate that endothelial injury is followed by adhesion of platelets to exposed vascular subendothelial connective tissue with secretion of platelet granule contents into the surrounding plasma. One of the released materials, adenosine diphosphate, induces aggregation of platelets from the circulating blood with formation of the platelet plug. The plug is then consolidated by thrombin-induced fibrin formation. Thrombin can also stimulate further platelet aggregation and release. In pathologic states, particularly in arterial thrombosis, *in vivo* platelet aggregation and release may be excessive and contribute significantly to the disease process. Until recently, methods for detecting increased platelet activation *in vivo* were essentially limited to studies of platelet survival and turnover, with the assumption that increased activation would lead to shortened survival. Experimental evidence in animals has shown that this assumption is not always valid.[1] In the last several years, the possibility of using measurements of secreted platelet proteins as indices of *in vivo* platelet activation has been investigated and it appears that such measurements are useful in detecting *in vivo* release in thrombotic disorders. There are two secreted proteins specific to platelets for which radioimmunoassays have been developed; these are platelet

[1] H. J. Reimers, R. L. Kinlough-Rathbone, J. P. Cazenave, A. F. Senyi, J. Hirsh, M. A. Packham, and J. F. Mustard, *Thromb. Hemostas.* **35**, 151 (1976).

factor 4 and β-thromboglobulin. Radioimmunoassay of platelet factor 4 will be described in this chapter and of β-thromboglobulin in the following chapter (this volume [7]).

Platelet factor 4 is a heparin-neutralizing protein. Its presence in platelets was first noted in the 1940s because of an association between heparin sensitivity and thrombocytopenia.[2-4] A platelet fraction responsible for the anti-heparin activity was separated by Van Creveld and Paulssen,[5,6] and this material was called platelet factor 4 by Deutsch.[7] The term platelet factor 4 is now reserved for a homogeneous protein with potent heparin neutralizing activity which has been isolated from platelets in several laboratories[8-17] and has been shown to have a molecular weight of 27,000 or 30,000 by ultracentrifugal analysis[8,10] and by amino acid sequencing to be composed of four identical subunits with 70 residues each and a subunit molecular weight of 7767.[14-17] Platelet factor 4 appears to be released from platelets bound to a proteoglycan carrier of molecular weight 59,000,[8,10] but it can be dissociated from the carrier at high ionic strength ($\mu = 0.75$).[8] The fully saturated complex has a molecular weight of 350,000,[9,10] and is composed of a dimer of two carrier molecules each of which binds four molecules of platelet factor 4.

Radioimmunoassays for platelet factor 4 have been developed in a number of laboratories.[13,18-22] All of these assays have used ^{125}I-labeled

[2] T. R. Waugh and D. W. Ruddick, *Can. Med. Assoc. J.* **51**, 11 (1944).

[3] J. G. Allen, G. Bogardus, L. O. Jacobson, and C. L. Spurr, *Ann. Int. Med.* **27**, 382 (1947).

[4] C. L. Conley, R. C. Hartmann, and J. S. Lalley, *Proc. Soc. Exp. Biol. Med.* **69**, 284 (1948).

[5] S. Van Creveld and M. M. P. Paulssen, *Lancet* **2**, 242 (1951).

[6] S. Van Creveld and M. M. P. Paulssen, *Lancet* **1**, 23 (1952).

[7] E. Deutsch, S. A. Johnson, and W. H. Seegers, *Circ. Res.* **3**, 110 (1955).

[8] R. Kaser-Glanzmann, M. Jakabova, and E. F. Luscher, *Experientia* **28**, 1221 (1972).

[9] R. Kaser-Glanzmann, M. Jakabova, and E. F. Luscher, *Haemostasis* **1**, 136 (1972/73).

[10] S. Moore, D. S. Pepper, and J. D. Cash, *Biochim. Biophys. Acta* **379**, 370 (1975).

[11] R. I. Handin and H. J. Cohen, *J. Biol. Chem.* **251**, 4273 (1976).

[12] S. P. Levine and H. Wohl, *J. Biol. Chem.* **251**, 324 (1976).

[13] K. L. Kaplan, H. L. Nossel, M. Drillings, and G. Lesznik, *Br. J. Haemat.* **39**, 129 (1978).

[14] F. J. Morgan, G. S. Begg, and C. N. Chesterman, *Thromb. Haemostas.* **38**, 231 (1977).

[15] D. A. Walz, V. Y. Wu, R. de Lamo, H. Dene, and L. E. McCoy, *Thromb. Res.* **11**, 893 (1977).

[16] T. F. Deuel, P. S. Keim, M. Farmer, and R. L. Heinrikson, *Proc. Natl. Acad. Sci. U.S.A.* **74**, 2256 (1977).

[17] M. Hermodson, G. Schmer, and K. Kurachi, *J. Biol. Chem.* **252**, 6276 (1977).

[18] A. E. Bolton, C. A. Ludlam, D. S. Pepper, S. Moore, and J. D. Cash, *Thromb. Res.* **8**, 51 (1976).

[19] S. P. Levine and L. S. Krentz, *Thromb. Res.* **11**, 673 (1977).

[20] R. I. Handin, M. McDonough, and M. Lesch, *J. Lab. Clin. Med.* **91**, 340 (1978).

[21] C. N. Chesterman, J. R. McGready, D. J. Doyle, and F. J. Morgan, *Br. J. Haematol.* **40**, 489 (1978).

[22] T. T. S. Ho, D. A. Walz, and T. R. Brown, *Clin. Chim. Acta* **101**, 225 (1980).

platelet factor 4 as tracer, labeled using the chloramine-T method of Greenwood et al.,[23] the Bolton–Hunter reagent,[24] or lactoperoxidase.[22] Antisera to platelet factor 4 have been raised in goats[13] and rabbits.[18-22] Methods for separating bound tracer from free tracer have included double antibody precipitation[18,21] and ammonium[19,20] or sodium[13] sulfate precipitation. The method to be described in detail in this chapter is the method currently used in this laboratory for assaying platelet factor 4 in plasma.

Isolation of Platelet Factor 4 Antigen

Platelet factor 4 was prepared from concentrates of fresh platelets prepared from blood anticoagulated with acid citrate dextrose.[8,13] The platelets were separated from the plasma by centrifugation for 15 min at 3000 g at room temperature and then resuspended in 0.12 M NaCl, 0.1% Na_2-EDTA, 0.6% glucose. Platelets were again separated by centrifugation and then resuspended in 0.135 M NaCl, 0.6% glucose. After separation by centrifugation the platelets were resuspended in 0.12 M NaCl, 0.01 M Tris–maleate, pH 7.4. Following a final separation by centrifugation the platelets were resuspended at 10^{10} platelets/ml in 0.12 M NaCl, 0.01 M Tris–maleate, pH 7.4 which also contained 0.3 mM $MgCl_2$, 1.0 mM KCl, 0.5 mM $CaCl_2$, and 0.56 mM glucose. This platelet suspension was stirred at 37° for 3 min with 1 unit/ml human thrombin. The aggregated platelets were removed by centrifugation at 3000 g for 15 min at 3°. The supernatant solution was dialyzed against 0.00015 M NaCl with 25 μM Tris, 1 μM Na_2-EDTA, 3.8 μM sodium borate, pH 8.8 and then lyophilized. The lyophilized material was reconstituted to 1/1000 of its original volume and then chromatographed on BioGel A15M (Fig. 1) equilibrated with 0.15 M NaCl, 25 mM Tris, 1 mM Na_2-EDTA, 3.8 mM sodium borate, pH 8.8.[10] After washing the column with the equilibration buffer the PF4 was then eluted with a buffer containing 0.75 M NaCl, 25 mM Tris, 1 mM Na_2-EDTA, 3.8 mM sodium borate, pH 8.8. A sharp protein peak with heparin-neutralizing activity was obtained which was homogeneous on sodium dodecyl sulfate–polyacrylamide gel electrophoresis and had an amino acid composition corresponding to those published for platelet factor 4.[13] Platelet factor 4 was also prepared from the supernatant of platelets frozen and thawed after washing as described above.[13] Contaminating proteins were removed by 50% ammonium sulfate precipitation.[12] The supernatant was dialyzed extensively against 0.0001 M phosphate–0.00015 M NaCl, pH 8.0, lyophilized, and reconstituted in

[23] F. C. Greenwood, W. M. Hunter, and J. S. Glover, Biochem. J. 89, 114 (1963).
[24] A. E. Bolton and W. M. Hunter, Biochem. J. 133, 529 (1973).

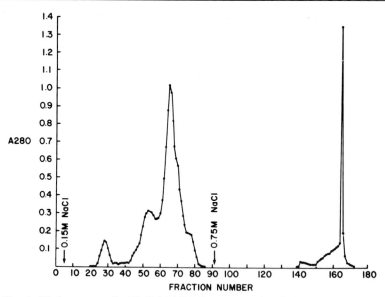

FIG. 1. Elution pattern for BioGel A15M chromatography of platelet supernatant.

0.1 M phosphate–0.15 M NaCl, pH 8 and chromatographed on heparin-Sepharose prepared by the method of Iverius.[25] Platelet factor 4, eluted with 0.1 M phosphate–1.5 M NaCl, pH 8,0 (Fig. 2) was homogenous on sodium dodecyl sulfate–polyacrylamide gel electrophoresis and had the appropriate amino acid composition.[13]

Preparation of Radiolabeled Platelet Factor 4

Heparin-sepharose purified platelet factor 4 has been used for radiolabeling with ^{125}I by the chloramine-T method of Greenwood *et al.*[13,23] The radiolabeling is carried out in siliconized glassware. All reagents are dissolved in 0.5 M phosphate buffer because platelet factor 4 is insoluble at physiologic ionic strength. 2 mCi ^{125}I is incubated with 10 μg platelet factor 4 and 50 μg chloramine-T in a volume of 50 μl for 10 sec. Sodium metabisulfite (60 μg) in 10 μl is then added, followed by 100 μl buffer containing 0.45 M NaCl, 0.05 M Tris, 2% horse serum, 0.02% sodium azide, 5 μg/ml heparin, pH 8.0, and the protein bound ^{125}I is separated from unbound iodine by chromatography on a 10 ml Sephadex G50 column equilibrated with the same buffer. Radiolabeled PF4 is stored at $-50°$ and retains its immunoreactivity for 6–10 weeks.

[25] P. H. Iverius, *Biochem. J.* **124,** 677 (1971).

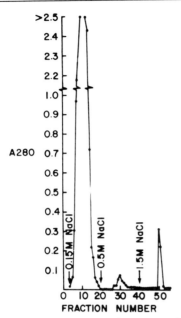

FIG. 2. Elution pattern for heparin-Sepharose chromatography of platelet supernatant.

Preparation of Antiserum

Antiserum to platelet factor 4 was prepared by immunization of a goat with platelet factor 4 prepared by BioGel A15M chromatography and mixed with complete Freund's adjuvant. The immunization was performed by Atlantic Antibodies, Westbrook, Maine. The antiserum titer was assessed by binding of radiolabeled platelet factor 4. At a 1/10000 dilution of antiserum 85% of ^{125}I-platelet factor 4 was bound by the goat antiserum.[13]

Radioimmunoassay Method

The radioimmunoassay is performed at room temperature using siliconized glassware. The buffer for this assay is 0.1 M NaCl, 0.05 M Tris, 0.02% sodium azide, 5 μg/ml heparin, pH 8.0 containing 2% horse serum. This buffer is used for dilution of plasma samples and for dilutions of radiolabeled platelet factor 4, and standards. Buffer containing a 1/1200 dilution of normal goat serum is used for diluting the goat anti-platelet factor 4. Antiserum is used at an initial dilution of 1/100,000, at which dilution 30–40% of the radiolabeled tracer is bound. Tracer is added so that when the tracer is fresh 10,000 bound cpm are obtained in the absence of unlabeled platelet factor 4. This corresponds to 200 pg ^{125}I-labeled platelet factor 4 per tube. Antibody bound ^{125}I-labeled platelet factor 4 is separated from free tracer by addition of a 1/100 dilution of donkey anti-goat serum in Tris-buffered saline with horse serum and 2% polyethylene glycol. The tubes are incubated at room temperature for 1 hr, then 2 ml of Tris-buffered saline without protein is added and the tubes are centrifuged for 20 min at room temperature at 3000 g, the supernatant is removed by aspiration, and the precipitate counted for 1–3 min. Longer counting times are used as the tracer becomes older, to maintain 10,000 bound counts in the antiserum control. The volumes of reagents and times of incubation for the assay are shown in Table I. Additions are made in the order shown in the table, beginning with the buffer.

Radioimmunoassay Results

A mean standard curve with 95% confidence limits for the platelet factor 4 radioimmunoassay is shown in Fig. 3. Fifty percent displacement of binding occurs with 3.4 ng/ml platelet factor 4 and 20% displacement is seen with 0.7 ng/ml. For binding between 20 and 80% the between assay coefficient of variation is approximately 10%. Recovery of platelet factor 4 added to platelet free plasma is shown in Table II.

TABLE I
ASSAY OF PLATELET FACTOR 4[a]

	Buffer[b]	Sample[b]	Tracer[b]	Antiserum[c,d]	DAG[e,f]	TBS[g]
Buffer control	300	0	50	0	0	0
Second antibody control	300[h]	0	50	0	500	2000
Antiserum control	200	0	50	100	500	2000
Standards[b]	0	200	50	100	500	2000
Plasma samples[i]	0	200	50	100	500	2000
Antiserum control	0	0	50	100	500	2000

[a] Volumes are given in μl. Assay is carried out in triplicate.
[b] Dilutions made in buffer (0.1 M NaCl, 0.05 M Tris, 0.02% sodium azide, 5 μg/ml heparin, pH 8.0, containing 2% horse serum).
[c] Diluted in buffer containing 1/1200 dilution of normal goat serum.
[d] Incubate for 2 hr at room temperature.
[e] Donkey antigoat antiserum (Miles, lot D 7) diluted 1/100 in buffer containing 2% polyethylene glycol.
[f] Incubate for 1 hr at room temperature.
[g] TBS: 0.1 M NaCl, 0.05 M Tris, pH 8.0.
[h] 200 μl buffer + 100 μl buffer containing 1/1200 dilution of normal goat serum.
[i] Plasma diluted 1 in 6 in Buffer for initial assay.

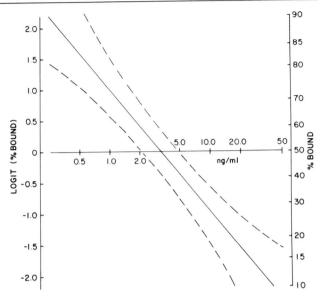

FIG. 3. Standard curve for the platelet factor 4 radioimmunoassay. The solid line represents the mean standard curve and the dashed lines represent the 5 and 95% confidence limits for the standard curve.

Blood Sample Collection and Preparation for Radioimmunoassay

Venous blood (2.5–10 ml) is collected into a chilled siliconized vacutainer tube containing $\frac{1}{10}$ volume of anticoagulant which contains 1400 U/ml heparin, 1000 U/ml Trasylol, 0.01 M adenosine, and 0.02 M theophylline in 0.13 M NaCl, 7.1 mM sodium acetate, 7.1 mM sodium barbital, pH 7.4. This anticoagulant is used in our laboratory for collec-

TABLE II
RECOVERY OF PLATELET FACTOR 4 ADDED TO
NORMAL PLASMA[a]

Expected (ng/ml)	Observed (ng/ml)	% Recovery
98.5	92.4	94
51.0	50.4	99
27.2	25.8	95
15.3	13.2	86
9.36	9.6	103
6.4	6.5	101
3.4	3.4	100

[a] Mean recovery = 97% ± 5.8% (SD).

tion of blood samples for assay for β-thromboglobulin and fibrinopeptide A as well as for platelet factor 4. Thus heparin is included to inhibit thrombin action, Trasylol to inhibit plasmin action, and adenosine and theophylline to inhibit platelet release. The blood tube is put immediately into an ice bath and maintained there for up to 1 hr before centrifugation at 3000 g for 20 min at 4°. The top 1 ml of plasma is then centrifuged at 46000 g for 10 min at 4° and the supernatant from the high-speed centrifugation is stored at −50° until assay.

Results of Platelet Factor 4 Radioimmunoassay in Clinical Blood Samples

Using the blood collection and preparation techniques and the assay method described in this chapter, the plasma concentration of platelet factor 4 in a group of 312 individuals was found to be described by a log normal distribution with a median value of 6.0 ng/ml and 5 and 95% values of 1.7 and 20.9 ng/ml (Fig. 4). It is important that if multiple blood samples are to be collected from a single venipuncture, the sample for platelet factor 4 be collected in the first tube since thrombin generation stimulated by the needle within the vein can cause dramatic increases in platelet factor 4 concentration in the plasma. This principle is illustrated in Fig. 5 which shows platelet factor 4 concentrations in sequential 10 ml tubes from a single venipuncture compared with fibrinopeptide A concentrations in the same tubes.

Results of Platelet Factor 4 Radioimmunoassay in Disease States

Clinical studies from this laboratory have included a pilot study of patients with peripheral vascular disease and a study of disseminated intra-

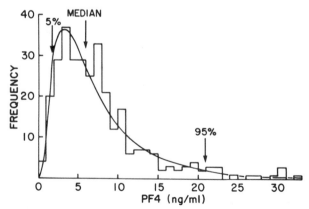

FIG. 4. Distribution of platelet factor 4 (PF4) values in normal individuals.

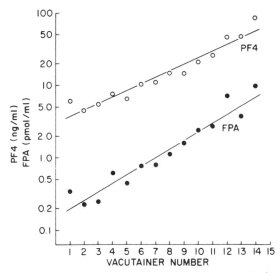

FIG. 5. Levels of platelet factor 4 (PF4) and fibrinopeptide A (FPA) in sequential vacutainer tubes collected from the same venipuncture.

vascular coagulation in patients undergoing hypertonic saline induced
abortion.

The peripheral vascular disease study[26] involved eight patients with
arteriographically demonstrated peripheral arterial disease. The mean
platelet factor 4 level in these patients was slightly but not significantly
elevated compared with a group of normals. Only two of the patients had
levels more than two standard deviations above the normal mean. Mean
β-thromboglobulin level in the patients was significantly elevated compared with a group of normals, with four patients having levels outside the
normal range (see this volume [7]).

Women undergoing hypertonic saline induced abortion were studied
as a model of disseminated intravascular coagulation.[27] Plasma samples
were collected before saline injection and 1, 2, 4, 24, and 48 hr after saline
injection. The mean plasma level of platelet factor 4 was elevated at 1 and
2 hr after saline injection (26.3 and 23.6 ng/ml) compared with the preinjection sample (15.4 ng/ml), and the changes were significant at the 5%
level by the paired t test. Changes in the mean plasma level of β-throm-
boglobulin were more highly significant (see this volume [7]).

In the studies just described plasma levels of β-thromboglobulin were

[26] K. L. Kaplan, in "Hemostasis and Thrombosis" (T. Spaet, ed.), Vol. 5. Grune & Stratton, New York, 1980.
[27] H. L. Nossel, J. Wasser, K. L. Kaplan, K. S. LaGamma, I. Yudelman, and R. E. Canfield, J. Clin. Invest. 64, 1371 (1979).

consistently abnormal in the patient groups while changes in plasma levels of platelet factor 4 were either not significant or just significant. There is good evidence that similar quantities of these proteins are released *in vitro*, thus it is likely that the more rapid disappearance of platelet factor 4 than β-thromboglobulin from the plasma[28] explains the smaller elevations of platelet factor 4 levels in these patients. This rapid disappearance time is apparently due to the binding of platelet factor 4 to endothelial cells.[28-30] Workers in other laboratories have, however, reported significant elevations in platelet factor 4 levels in patients with myocardial infarction,[20,21,31] coronary artery disease,[32,33] prosthetic heart valves,[18,34,35] and diabetes mellitus.[36] It is not clear why these latter studies have found elevated levels of platelet factor 4. It may be, particularly in those studies where changes in platelet factor were highly correlated with changes in β-thromboglobulin, that *in vitro* release contributed to the measured levels.

A review of the experience in this laboratory and in the literature suggests that with *in vivo* release there are small changes in plasma levels of platelet factor 4 with significantly greater changes in plasma levels of β-thromboglobulin. When *in vitro* release occurs both platelet factor 4 and β-thromboglobulin levels rise and platelet factor 4 levels are highly correlated with the β-thromboglobulin levels. Finally, during heparin administration levels of platelet factor 4 rise markedly,[28,30] while β-thromboglobulin levels are unchanged.

[28] J. Dawes, R. C. Smith, and D. S. Pepper, *Thromb. Res.* **12**, 851 (1978).
[29] C. Bush, J. Dawes, D. S. Pepper, and Å. Wasteson, *Thromb. Haemostas.* **42**, 43 (1979).
[30] C. W. Pumphrey, D. S. Pepper, and J. Dawes, *Thromb. Haemostas.* **42**, 43 (1979).
[31] C. A. Ludlam, J. R. O'Brien, A. E. Bolton, and M. Etherington, *Thromb. Res.* **15**, 523 (1979).
[32] S. P. Levine, J. A. Lindenfeld, N. M. Raymond, and L. S. Krentz, *Clin. Res.* **27**, 299A (1979).
[33] D. J. Doyle, C. N. Chesterman, J. F. Cade, J. R. McGready, G. C. Rennie, and F. J. Morgan, *Blood* **55**, 82 (1980).
[34] R. Dudczak, H. Niessner, E. Thaler, K. Lechner, K. Kletter, H. Frishauf, E. Domanig, and H. Aicher, *Thromb. Haemostas.* **42**, 72 (1979).
[35] G. Cella, D. A. Lane, V. V. Kakkar, A. Donato, and S. Dalla Volta, *Thromb. Haemostas.* **42**, 134 (1979).
[36] J. Zahavi, N. A. G. Jones, D. J. Betteridge, J. Leyton, D. J. Galton, S. E. Clark, and V. V. Kakkar, *Thromb. Haemostas.* **42**, 334 (1979).

[7] Radioimmunoassay of β-Thromboglobulin

By KAREN L. KAPLAN and JOHN OWEN

Introduction

In recent years radioimmunoassay of specific proteins which are secreted by platelets in response to stimuli such as thrombin, collagen, and ADP has been advocated as a method for detection of *in vivo* platelet activation in patients with thrombotic disorders. The first such protein for which a radioimmunoassay was reported was β-thromboglobulin.[1,2] This protein was isolated from the supernatant of thrombin treated washed platelets[3,4] and found to have a molecular weight by ultracentrifugation of 36,000.[3,4] It has subsequently been shown to be composed of 4 identical subunits of molecular weight 8851 and containing 81 amino acids.[5] The isoelectric point of β-thromboglobulin is 7.0.[6] β-Thromboglobulin appears to be derived from a precursor protein, low-affinity platelet factor 4, which has an isoelectric point of 8.0[6] and which has been shown to differ from β-thromboglobulin only in its four amino-terminal amino acids.[6] Antibodies produced in response to immunization with β-thromboglobulin cross-react with low-affinity platelet factor 4 and vice versa. Both β-thromboglobulin and low-affinity platelet factor 4, are, however, distinct, both immunologically and chemically, from platelet factor 4, described in this volume [6].

The methods for isolation of β-thromboglobulin, preparation of antiserum, and radiolabeling of β-thromboglobulin described in this chapter are similar to those previously reported from this laboratory.[7] The method for radioimmunoassay is a modification of the method previously described.[7] There are a number of other methods published for radioimmunoassay of β-thromboglobulin.[1,2,8,9]

[1] C. A. Ludlam, S. Moore, A. E. Bolton, and J. D. Cash, *Lancet* **2**, 259(1975).
[2] C. A. Ludlam, S. Moore, A. E. Bolton, and J. D. Cash, *Thromb. Res.* **6**, 543 (1975).
[3] S. Moore, D. S. Pepper, and J. D. Cash, *Biochim. Biophys. Acta* **379**, 360 (1975).
[4] S. Moore and D. S. Pepper, *in* "Platelets in Biology and Pathology" (J. L. Gordon, ed.), p. 293. Elsevier/North-Holland, Amsterdam, 1976.
[5] G. S. Begg, D. S. Pepper, and C. N. Chesterman, *Biochemistry* **17**, 1739 (1978).
[6] B. Rucinski, S. Niewiarowski, P. James, D. A. Walz, and A. Z. Budzynski, *Blood* **53**, 47 (1979).
[7] K. L. Kaplan, H. L. Nossel, M. Drillings, and G. Lesznik, *Br. J. Haematol.* **39**, 129 (1978).
[8] A. E. Bolton, C. A. Ludlam, S. Moore, D. S. Pepper, and J. D. Cash, *Br. J. Haematol.* **33**, 233 (1976).
[9] T. T. S. Ho, D. A. Walz, and T. R. Brown, *Clin. Chim. Acta* **101**, 225 (1980).

Isolation of Antigen

Concentrates of fresh platelets prepared from blood anticoagulated with acid-citrate-dextrose were centrifuged at 1200 g for 15 min at 20°. The sedimented platelets were resuspended in 0.12 M NaCl, 0.1% Na$_2$-EDTA, 0.6% glucose. The platelets were then sedimented by centrifugation at 1200 g for 15 min at 20° and resuspended in 0.135 M NaCl, 0.6% glucose. After an additional centrifugation the platelets were resuspended in 0.12 M NaCl, 0.01 M Tris–maleate pH 7.4. After a final centrifugation the platelets were resuspended to a concentration of 10^{10} platelets/ml in a buffer containing 0.12 M NaCl, 0.01 M Tris–maleate pH 7.4, 0.3 mM MgCl$_2$, 1 mM KCl, 0.5 mM CaCl$_2$, 0.56 mM glucose. After adding thrombin to give a final concentration of 1 unit/ml, the platelet suspension was stirred for 3 min at 37°. The aggregated platelets were sedimented by centrifugation at 3000 g for 15 min at 4°. The supernatant was decanted and then heated to 56° for 10 min to inactivate thrombin and denature fibrinogen. For preparation of semipurified β-thromboglobulin for immunization, the supernatant was dialyzed against 0.005 M phosphate buffer, pH 7.0, lyophilized, reconstituted to 0.15 M phosphate with deionized water and chromatographed on a 2.6 × 100-cm column of Sephadex G-200 (Fig. 1) equilibrated and run with 0.12 M NaCl, 0.03 M Tris–HCl, pH 7.4. Fractions 85–93 as identified by A_{280} contained β-thromboglobulin, and were used for immunization as described below.

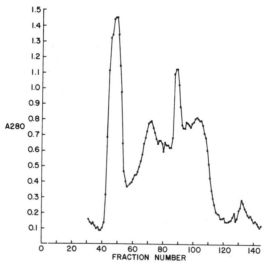

FIG. 1. Elution pattern of platelet supernatant on a 2.6 × 100-cm column of Sephadex G-200 equilibrated with 0.12 M NaCl, 0.03 M Tris–HCl, pH 7.4. Fractions were 3.2 ml.

[10] F. C. Greenwood, W. M. Hunter, and J. S. Glover, *Biochem. J.* **89,** 114 (1963)

Purified β-thromboglobulin for use as tracer and standard in the radio-immunoassay was prepared as described by Moore and Pepper.[3,4] The supernatant of the thrombin-treated washed platelets was dialyzed against a buffer containing 0.00015 M NaCl, 0.001 mM EDTA, 0.0038 mM sodium borate, and 0.025 mM Tris, pH 8.8, lyophilized and reconstituted to 1/1000 the original volume. This material was chromatographed on a 2.6 × 100 cm column of BioGel A15M (Fig. 2) equilibrated with buffer containing 0.15 M NaCl, 1 mM EDTA, 3.8 mM sodium borate, and 25 mM Tris, pH 8.8. Fractions (4.8 ml) were collected, and β-thromboglobulin was eluted in fractions 60–74. These fractions were dialyzed against buffer containing 0.00015 M NaCl, 0.001 mM EDTA, 0.0038 mM sodium borate, and 0.025 mM Tris, pH 8.8, lyophilized, reconstituted to 1/1000 the original volume, and chromatographed on a 1.6 × 100 cm column of Sephadex G-200 (Fig. 3) equilibrated with 0.15 M NaCl, 1 mM EDTA, 3.8 mM sodium borate, 25 mM Tris, pH 8.8. The third protein peak by A_{280} (fractions 38–50, 1.6 ml/fraction) was dialyzed against 0.5% pyridine, 10% acetic acid, pH 3.2, lyophilized and reconstituted to 3 ml and chromatographed on a 1.6 × 100-cm column of Sephadex G-75 (Fig. 4) equilibrated with 0.5% pyridine, 10% acetic acid pH 3.2. The second protein peak contained β-thromboglobulin which was homogeneous on sodium dodecyl sulfate–polyacrylamide gel electrophoresis and which had an amino composition comparable to that published previously.[4,7]

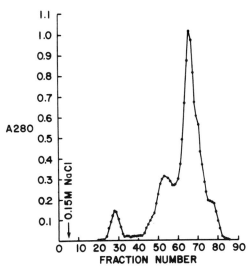

Fig. 2. Elution pattern of platelet supernatant on a 2.6 × 100-cm column of BioGel A15M equilibrated with 0.15 M NaCl, 1 mM EDTA, 3.8 mM sodium borate, 25 mM Tris, pH 8.8.

Preparation of ^{125}I-labeled β-Thromboglobulin

Purified β-thromboglobulin is radiolabeled by the chloramine-T technique.[10] All glassware is siliconized and all reagents are dissolved in 0.05 M phosphate buffer, pH 7.5. ^{125}I (2 mCi) is incubated with 10 μg β-thromboglobulin and 50 μg chloramine-T in a volume of 50 μl for 10 sec. Sodium metabisulfite (60 μg) in 10 μl is added, followed by 100 μl of Tris-buffered saline (0.1 M NaCl, 0.05 M Tris, 2% horse serum, 0.02% sodium azide, 5 μg/ml heparin, pH 8.0). Protein bound ^{125}I is separated from unbound ^{125}I by chromatography on a 10 ml column of Sephadex G-50 equilibrated with Tris-buffered saline. The specific activity of the tracer is 100,000 cpm/19 pg.

Preparation of Anti-β-Thromboglobulin Antiserum

The semipurified β-thromboglobulin described above was used for immunization. Protein (0.44 mg) was dissolved in 1 ml of 0.15 M NaCl and emulsified with an equal volume of complete Freund's adjuvant. Rabbits received weekly toe pad injections of 0.2 ml for 3 weeks, then bi-weekly

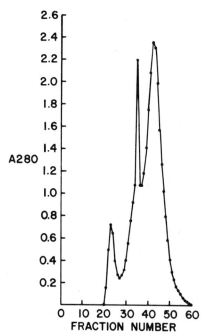

FIG. 3. Elution pattern of fractions 60–74 from BioGel A15M column on a 1.6 × 100-cm column of Sephadex G-200 equilibrated with buffer as for Fig. 2.

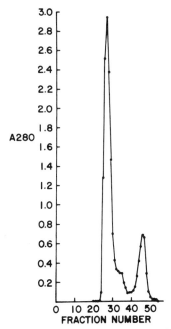

FIG. 4. Elution pattern of fractions 38–50 from Sephadex G-200 on a 1.6 × 100 cm column of Sephadex G-75 equilibrated with 0.5% pyridine, 10% acetic acid, pH 3.2.

haunch injections of 0.2 ml for 2 months, and monthly haunch injections of 0.2 ml thereafter. Titer of antiserum was assessed by the binding of [125I]-labeled β-thromboglobulin. At a dilution of 1/4000 the antiserum after 11 months of immunization could bind 92% of the [125I]-labeled β-thromboglobulin.

Radioimmunoassay of β-Thromboglobulin

For assay of plasma samples the following radioimmunoassay method is recommended. This is a modification of the method described by Kaplan *et al.*[7] All glassware for the assay is silizonized. Two buffers are used for the assay: Tris-buffered saline (0.10 M NaCl, 0.05 M Tris–HCl, 0.02% sodium azide, 5 μg/ml heparin, pH 8.0) and Tris-buffered saline containing 2% horse serum. The buffer without horse serum is used for the initial 1 in 50 dilution of plasma samples. The buffer with 2% v/v horse serum is used for dilution of standards, controls, and any secondary dilution of plasma samples. [125I]-labeled β-thromboglobulin is used at a dilution which gives 10,000 bound cpm in the antiserum control when the tracer is new, and antiserum is used at a fixed dilution of 1/32,000. antibody bound

TABLE I

	2% Buffer[a]	Sample	Tracer	Antiserum[b]	$(NH_4)_2SO_4{}^c$
Buffer control	300[f]	0	50	0	1000
Sulfate control	300	0	50	0	1000
Antiserum control	200	0	50	100	1000
Standards[d]	0	200	50	100	1000
Plasma samples[e]	0	200	50	100	1000
Antiserum control	200	0	50	100	1000

[a] Buffer containing 2% v/v horse serum.
[b] Mix, overnight incubation at 4°.
[c] Mix, 20 minutes incubation at 4°.
[d] Standards are diluted in 2% buffer.
[e] Plasma diluted 1 in 50 in buffer without horse serum. If further dilution is needed use 2% buffer.
[f] Volumes given in microliters.

tracer is separated from unbound tracer by precipitation with 35% w/v ammonium sulfate at 4°. After centrifugation at 3000 g for 20 min at 4° the supernatant is discarded and the precipitate counted for 1–3 min as needed to obtain a minimum of 10,000 counts in the antiserum control. The volumes and times of incubation for the assay are shown in the table. Reagents are added in order shown, from left to right. A mean standard curve with 5 and 95% confidence limits is shown in Fig. 5. Fifty percent displacement

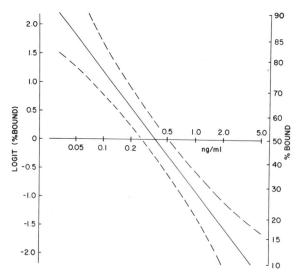

FIG. 5. Standard curve for the β-thromboglobulin radioimmunoassay. The solid line represents the mean standard curve and the dashed lines represent the 5 and 95% confidence limits for the standard curve.

of binding occurs with 0.37 ng/ml and 20% displacement with 0.075 ng/ml.

Collection and Preparation of Blood Samples for Radioimmunoassay for β-Thromboglobulin

Blood samples (2.5–10.0 ml) are collected by clean venipuncture into chilled siliconized vacutainer tubes which contain $\frac{1}{10}$ volume of anticoagulant (1400 U/ml heparin, 1000 U/ml Trasylol, 0.01 M adenosine, 0.02 M theophylline in 0.13 M NaCl, 7.1 mM sodium acetate, 7.1 mM sodium barbital, pH 7.4). The same collection tubes are used for samples for platelet factor 4 and fibrinopeptide A assay; thus heparin and Trasylol are included to inhibit thrombin and plasmin, respectively, and adenosine and theophylline are included to inhibit platelet release. The blood sample is immediately put into an ice bath and maintained there until it is processed further, within 1 hr of collection. The sample is centrifuged for 20 min at 3000 g at 4° and the top 1 ml of plasma is removed and centrifuged at 46,000 g for 10 min at 4°. The supernatant from this centrifugation is virtually free of platelets and is stored at −50° until assay.

When multiple blood samples are to be obtained from a single venipuncture, it is essential that the sample for β-thromboglobulin, platelet factor 4, and fibrinopeptide A radioimmunoassay be collected first as there is a steady increase in the concentration of these peptides when sequential tubes are collected and assayed for β-thromboglobulin and fibrinopeptide A as shown in Fig. 6. This increase is apparently due to thrombin generation stimulated by the needle within the vein.

Results of Radioimmunoassay of β-Thromboglobulin in Clinical Studies

Using the blood collection and radioimmunoassay techniques described above, there is a highly skewed distribution of normal values. The data appear to be logarithmically distributed (Fig. 7) with a median of 17.8 ng/ml and 5–95% range of 6.6 to 47.9 ng/ml. Survey of a group of patients hospitalized on a general medical ward showed elevated levels of β-thromboglobulin in patients with inflammatory, vascular, renal, or neoplastic disease, showing that elevations in β-thromboglobulin are not specific for clinically evident thrombotic disease. Study of a group of patients with angiographically demonstrated peripheral vascular disease showed a significantly elevated mean plasma level of β-thromboglobulin in the patients compared with a group of normal individuals. All patients had levels above the normal mean but only three had levels outside the normal

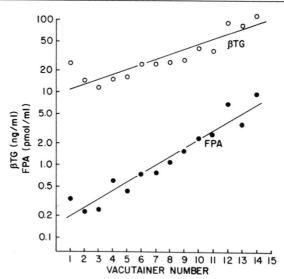

FIG. 6. Levels of β-thromboglobulin (βTG) and fibrinopeptide A (FPA) in sequential vacutainer tubes collected from a single venipuncture.

range, implying that increased platelet release is not detectable in all patients with peripheral vascular disease.[11]

In a study of disseminated intravascular coagulation in women undergoing hypertonic saline induced abortion[12] significant elevations in mean plasma levels of β-thromboglobulin (as compared with the preinjection value) were found at 1, 2, and 4 hr after saline injection. At 24 hr after saline injection the mean value was equal to the pre-injection value, and

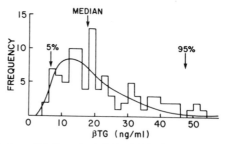

FIG. 7. Distribution of β-thromboglobulin (βTG) levels in normal individuals.

[11] K. L. Kaplan, in "Hemostasis and Thrombosis" (T. Spaet, ed.). Grune & Stratton, New York, 1980.
[12] H. L. Nossel, J. Wasser, K. L. Kaplan, K. S. LaGamma, I. Yudelman, and R. E. Canfield, J. Clin. Invest. 64, 1371 (1979).

by 48 hr after injection the mean plasma β-thromboglobulin value was significantly less than the preinjection value. This decrease suggests that early pregnancy elevates the mean plasma level of β-thromboglobulin. The mechanism for this increase is not known. The time course of elevation of β-thromboglobulin levels was parallel to that of fibrinopeptide A levels, implying that thrombin was responsible for platelet release in this model.

There are numerous reports in the literature of elevated plasma levels of β-thromboglobulin in a wide spectrum of clinical disorders. These have recently been reviewed.[11] In summary, elevated plasma levels of β-thromboglobulin have been reported in patients with venous thrombosis,[1,2,13–15] peripheral arterial disease,[11,13,16–18] diabetes mellitus,[19–22] coronary artery disease,[23,24] myocardial infarction,[15,25,26] prosthetic heart valves,[2,27–29] renal failure,[30–32] hyperlipidemia,[33] disseminated intravascular coagulation,[12,34] and malignant disease.[35,36] As indicated in the recent review[11] the results of some of these studies are open to question with regard to *in vitro* versus *in vivo* release, but overall the evidence appears to be strong that there is increased *in vivo* platelet release in many of these disorders which is reflected by the plasma level of β-thromboglobulin. The increased plasma levels of β-thromboglobulin in patients with renal failure may be due to a failure of clearance of β-thromboglobulin rather than to increased *in vivo* release.[31]

[13] G. Cella, J. Zahavi, H. A. de Haas, and V. V. Kakkar, *Br. J. Haematol.* **43**, 127 (1979).
[14] R. C. Smith, J. Duncanson, C. V. Ruckley, R. G. Webber, N. C. Allan, J. Dawes, A. E. Bolton, W. M. Hunter, D. S. Pepper, and J. D. Cash, *Thromb. Haemostas.* **39**, 338 (1978).
[15] C. A. Ludlam, J. R. O'Brien, A. E. Bolton, and M. Etherington, *Thromb. Res.* **15**, 523 (1979).
[16] D. Bevan, C. J. P. Yates, J. A. Dormandy, and P. T. Flute, *Thromb. Haemostas.* **42**, 147 (1979).
[17] J. Zahavi, W. A. P. Hamilton, J. J. G. O'Reilly, S. E. Clark, L. T. Cotton, and V. V. Kakkar, *Thromb. Haemostas.* **42**, 146 (1979).
[18] M. J. G. O'Reilly, J. Zahavi, M. Dubiel, L. T. Cotton, and V. V. Kakkar, *Thromb. Haemostas.* **42**, 338 (1979).
[19] A. W. Burrows, S. I. Chavin, and T. D. R. Hockaday, *Lancet* **1**, 235 (1978).
[20] F. E. Preston, B. H. Marcola, I. D. Ward, N. R. Porter, and W. R. Timperley, *Lancet* **1**, 238 (1978).
[21] J. Zahavi, N. A. G. Jones, D. J. Betteridge, J. Leyton, D. J. Galton, S. E. Clark, and V. V. Kakkar, *Thromb. Haemostas.* **42**, 334 (1979).
[22] G. Schernthaner, K. Silberbauer, H. Sinzinger, and M. Muller, *Thromb. Haemostas.* **42**, 334 (1979).
[23] D. J. Doyle, C. N. Chesterman, J. F. Cade, J. R. McGready, G. C. Rennie, and F. J. Morgan, *Blood* **55**, 82 (1980).
[24] P. Han, A. G. G. Turpie, E. Genton, and M. Gent, *Thromb. Haemostas.* **42**, 59 (1979).
[25] A. Hughes, S. Daunt, G. Vass, and J. Wickes, *Thromb. Haemostas.* **42**, 357 (1979).
[26] M. Cortellaro, C. Boschetti, G. Fassio, and E. E. Polli, *Thromb. Haemostas.* **42**, 486 (1979).

[27] C. A. Ludlam, N. Allen, R. B. Blandford, R. Dowdle, N. Bentley, and A. L. Bloom, *Thromb. Haemostas.* **42**, 329 (1979).
[28] R. Dudczak, H. Niessner, E. Thaler, K. Lechner, K. Kletter, H. Frischauf, E. Domanig, and H. Aicher, *Thromb. Haemostas.* **42**, 72 (1979).
[29] G. Cella, D. A. Lane, V. V. Kakkar, A. Donato, and S. Dalla Volta, *Thromb. Haemostas.* **42**, 134 (1979).
[30] C. A. Ludlam and J. L. Anderton, in "Proceedings of a Conference on Platelet Function Testing" (H. J. Day, H. Holmsen, and M. B. Zucker, eds.), p. 267. DHEW Publication No. (NIH) 78-1087, 1978.
[31] D. Depperman, K. Andrassy, H. Seelig, E. Ritz, and D. Post, *Thromb. Haemostas.* **42**, 416 (1979).
[32] D. Green, S. Santhanam, F. A. Krumlovsky, and F. del Greco, *Thromb. Haemostas.* **42**, 416 (1979).
[33] J. Zahavi, D. J. Betteridge, N. A. G. Jones, D. J. Galton, J. Leyton, and V. V. Kakkar, *Thromb. Haemostas.* **42**, 424 (1979).
[34] C. W. G. Redman, M. J. Allington, F. G. Bolton, and G. M. Stirrat, *Lancet* **2**, 248 (1977).
[35] R. J. Farrell, M. J. Duffy, and G. J. Duffy, *Thromb. Haemostas.* **42**, 143 (1979).
[36] B. J. Boughton, M. J. Allington, and A. King, *Br. J. Haematol.* **40**, 125 (1978).

[8] Immunoassay of Human Fibrinopeptides

By VINCENT P. BUTLER, JR., HYMIE L. NOSSEL, and
ROBERT E. CANFIELD

Human fibrinogen is a dimeric protein, each monomer of which contains three polypeptide chains (Aα, Bβ,γ) which are cross-linked with each other and with the other monomeric unit by disulfide bonds. The primary amino acid sequence of human fibrinogen (MW 340,000) has been determined and it has been found that the Aα-chain contains 610 amino acids, the Bβ-chain 461 amino acids, and the γ-chain 410 amino acids.[1] The conversion of fibrinogen to an insoluble fibrin clot is an event of central importance in normal hemostasis as well as in arterial and venous thrombosis. This conversion occurs as the result of three sequential alterations in the covalent structure of the fibrinogen molecule: (1) the conversion by the proteolytic enzyme, thrombin, of fibrinogen to fibrin I as the result of the release of the 16 amino acid fibrinopeptide A (FPA; Fig. 1) from the amino-terminal ends of the Aα-chains; (2) the conversion of fibrin I to fibrin II via release by thrombin of the 14 amino acid fibrinopeptide B (FPB; Fig. 1) from the amino-terminal ends of the Bβ-chains; and, (3) the cross-linking by factor XIIIa (transglutaminase) of glutamyl and

---

[1] R. F. Doolittle, K. W. K. Watt, B. A. Cottrell, D. D. Strong, and M. Riley, *Nature (London)* **280**, 464 (1979).

```
                                          TH              TH
                                          ↓               ↓
      1    2    3    4    5    6    7    8    9   10   11   12   13   14   15   16  17   18   19  20
A   ALA-ASP-SER-GLY-GLU-GLY-ASP-PHE-LEU-ALA-GLU-GLY-GLY-GLY-VAL-ARG-GLY-PRO-ARG-VAL-

                PL
                ↓
     21   22   23   24   25   26   27
    VAL-GLU-ARG-HIS-GLN-SER-ALA-

                                                      TH
                                                      ↓
      1    2    3    4    5    6    7    8    9   10   11   12   13   14  15   16   17   18   19   20
B   PYR-GLY-VAL-ASN-ASP-ASN-GLU-GLU-GLY-PHE-PHE-SER-ALA-ARG-GLY-HIS-ARG-PRO-LEU-ASP-

      PL   PL   PL
      ↓    ↓    ↓
     21   22   23   24   25   26   27   28   29   30   31   32   33   34   35   36   37   38   39   40
    LYS-LYS-ARG-GLU-GLU-ALA-PRO-SER-LEU-ARG-PRO-ALA-PRO-PRO-PRO-ILE-SER-GLY-GLY-GLY-

           PL
           ↓
     41   42   43   44   45   46
    TYR-ARG-ALA-ARG-PRO-ALA-
```

Fig. 1. Amino-terminal sequences of the (A) Aα- and (B) Bβ-chains of human fibrinogen. Arrows indicate bonds at which cleavage by thrombin (TH) or plasmin (PL) occurs.

lysyl residues in the α- and γ-chains of fibrin "monomer," resulting in the formation of insoluble, cross-linked polymeric fibrin.[2]

Since the detection of thrombin action has important clinical and investigative applications in the study of human thrombosis, radioimmunoassays (RIAs) for FPA[3-9] and for FPB[10] have been developed for use in the study of the thrombotic process *in vivo* and *in vitro*. Both assays have been utilized in the *in vitro* study of thrombin action and the FPA assay has proved to be useful in the *in vivo* study of the production of fibrin I by thrombin. However, FPB is rapidly converted by carboxypeptidase B in plasma to its desarginyl derivative with a major decrease in immunoreactivity[11]; an RIA for desarginyl–FPB has therefore been developed and

[2] B. Blombäck, B. Hessel, D. Hogg, and L. Therkildsen, *Nature (London)* **275**, 501 (1978).
[3] H. L. Nossel, L. R. Younger, G. D. Wilner, T. Procupez, R. E. Canfield, and V. P. Butler, Jr., *Proc. Natl. Acad. Sci. U.S.A.* **68**, 2350 (1971).
[4] H. L. Nossel, I. Yudelman, R. E. Canfield, V. P. Butler, Jr., K. Spanondis, G. D. Wilner, and G. D. Qureshi, *J. Clin. Invest.* **54**, 43 (1974).
[5] W. B. J. Gerrits, O. T. N. Flier, and J. van der Meer, *Thromb. Res.* **5**, 197 (1974).
[6] A. Z. Budzynski and V. J. Marder, *Thromb. Diath. Haemorrh.* **34**, 709 (1975).
[7] M. Cronlund, J. Hardin, J. Burton, L. Lee, E. Haber, and K. J. Bloch, *J. Clin. Invest.* **58**, 142 (1976).
[8] C. Kockum, *Thromb. Res.* **8**, 225 (1976).
[9] V. Hofmann and P. W. Straub, *Thromb. Res.* **11**, 171 (1977).
[10] S. B. Bilezikian, H. L. Nossel, V. P. Butler, Jr., and R. E. Canfield, *J. Clin. Invest.* **56**, 438 (1975).
[11] K. S. La Gamma and H. L. Nossel, *Thromb. Res.* **12**, 447 (1978).

employed as an index of the *in vivo* conversion of fibrin I to fibrin II by thrombin.[12]

Other proteolytic enzymes, notably plasmin, also act on the fibrinogen molecule. Plasmin action is felt to play an important pathophysiological role in disseminated intravascular coagulation (DIC). Plasmin is capable of cleaving many peptide bonds in all three polypeptide chains of fibrinogen. Among the earlier cleavage products are a series of relatively large carboxy-terminal Aα-chain fragments and a 42 residue amino-terminal peptide from the Bβ-chain (Fig. 1); later products following exhaustive *in vitro* cleavage include a 23 residue amino-terminal peptide from the Aα-chain (Fig. 1).[13,14] The amino terminal peptides cleaved by plasmin from the Aα- and Bβ-chains cross-react, to varying degrees, with antibodies to FPA and to FPB, respectively, and therefore may exhibit immunoreactivity in fibrinopeptide RIAs. However, if one employs anti-FPA or anti-FPB sera which exhibit minimal cross-reactivity with the larger plasmin-produced amino-terminal peptides, one can detect these larger molecules by the demonstration of a significant increase in FPA and/or FPB immunoreactivity following thrombin treatment of fibrinogen-free plasma extracts. Thus, by measuring thrombin-increasable fibrinopeptide immunoreactivity, the FPA and FPB RIAs have been useful in the study of plasmin action *in vivo* and *in vitro*.[15-17]

This review will deal with the development of RIAs for human fibrinopeptides (FPA, FPB, and desarginyl-FPB) and their application in the study of the action of proteolytic enzymes on the amino-terminal ends of the Aα- and Bβ-chains of human fibrinogen, both *in vivo* and *in vitro*.

Preparation of Immunogen

Although immunization of experimental animals with free fibrinopeptides may result in antibody formation,[15] titers are usually low. Accordingly, fibrinopeptides have been coupled as haptens to antigenic protein carriers such as human or bovine serum albumin, porcine thyroglobulin, or chicken egg albumin.[3-10] All of these carriers have proved to be satis-

[12] T. Eckhardt, H. L. Nossel, A. Hurlet-Jensen, K. S. La Gamma, J. Owen, and M. Auerbach, *J. Clin. Invest.* **67**, 809 (1981).

[13] P. Wallén, *Scand. J. Haematol.* [*Suppl.*] **13**, 3 (1971).

[14] E. J. Harfenist and R. E. Canfield, *Biochemistry* **14**, 4110 (1975).

[15] H. L. Nossel, V. P. Butler, Jr., G. D. Wilner, R. E. Canfield, and E. J. Harfenist, *Thromb. Haemostas.* **35**, 101 (1976).

[16] H. L. Nossel, J. Wasser, K. L. Kaplan, K. S. LaGamma, I. Yudelman, and R. E. Canfield, *J. Clin. Invest.* **64**, 1371 (1979).

[17] V. P. Butler, Jr., D. A. Weber, H. L. Nossel, D. Tse-Eng, K. S. LaGamma, and R. E. Canfield, submitted.

factory; however, there are theoretical disadvantages to human albumin because antisera produced in response to human albumin-fibrinopeptide conjugates may contain antibodies to other human peptides (present as contaminants in many commercial albumin preparations), and to egg albumin because antibodies to the albumin carrier may react with this protein when it is employed as the diluent in RIA systems. FPA, which contains one free amino and five free carboxyl groups (Fig. 1), has been coupled to protein carriers by the carbodiimide method (which forms peptide bonds between free amino and carboxyl groups of carrier and of fibrinopeptide)[18] and by the glutaraldehyde method (which cross-links free amino groups).[19] The amino-terminal group of FPB is blocked and hence FPB contains no free amino group. Since, as will be discussed in a subsequent section, it is more convenient to introduce a phenolic group for radioiodination into the amino-terminal end of fibrinopeptides, it was considered desirable to attempt to elicit anti-FPB antibodies which preferentially reacted with the carboxyl-terminal end of FPB. Accordingly, an FPB analog was synthesized in which the amino-terminal pyroglutamic residue of native FPB was replaced by glutamic acid which, of course, contains a free amino group; the FPB analog was then conjugated to bovine serum albumin by the glutaraldehyde method.[10]

Reagents

Bovine serum albumin (Miles Laboratories, Kankakee IL)

FPA, prepared by solid phase peptide synthesis or isolated from human fibrinogen according to the method of Blombäck *et al.*[20]

FPB analog, prepared by solid phase peptide synthesis

1-Ethyl-3-(3-dimethylaminopropyl)-carbodiimide HCl (CDI; Pierce Chemical Co., Rockford IL)

Glutaraldehyde (Eastman Organic Chemicals, Rochester NY)

0.1 *M* phosphate buffer, pH 7.0 (glutaraldehyde method only)

Procedure

A. FPA: carbodiimide method

1. FPA (15 mg) and BSA (30 mg) are dissolved in 1 ml distilled water.

2. CDI (300 mg) is dissolved in 1 ml distilled water and added *immediately* dropwise with stirring to the FPA-BSA mixture; stir 1 hr at room temperature.

3. Dialyze against several changes of distilled water for 24 hr.

[18] T. L. Goodfriend, L. Levine, and G. D. Fasman, *Science* **144,** 1344 (1964).

[19] M. Reichlin, this series, Vol. 70 [8].

[20] B. Blombäck, M. Blombäck, P. Edman, and B. Hessel, *Biochim. Biophys. Acta* **115,** 371 (1966).

4. Lyophilize and store in refrigerator or freezer in tightly sealed container; allow container to reach room temperature before reopening.

5. To estimate extent of hapten conjugation to protein, perform amino acid analyses on FPA, BSA, and conjugate; incorporation of glycine into BSA may be used as index of extent of conjugation. Alternatively, ability of conjugate to inhibit a well-characterized FPA–anti-FPA system can be used to estimate the degree of incorporation of immunoreactive FPA.

B. FPB: glutaraldehyde method

1. FPB analog (8 mg) and BSA (40 mg) are dissolved in 4 ml 0.1 M sodium phosphate buffer, pH 7.0.

2. Two milliliters freshly prepared glutaraldehyde (0.021 M in 0.1 M sodium phosphate buffer, pH 7.0) is added dropwise with stirring to the BSA-FPB analog mixture; stir 2 hr at room temperature.

3. Dialyze 24 hr against several changes of 0.15 M saline.

4. Lyophilize as described above for BSA-FPA conjugate.

5. Estimate extent of hapten incorporation as described above for BSA-FPA conjugate.

Comments. In preparing conjugates, we would prefer to use equal weights of fibrinopeptide and protein carrier, but have been limited by the small quantities of fibrinopeptides available. A desarginyl–FPB–protein conjugate has not been synthesized; all desarginyl-FPB assays to date have employed anti-FPB sera. The carbodiimide reaction may be carried out either in water or in pH 7.0 phosphate-buffered saline, but high concentrations of phosphate must be avoided, as they will interact with the carbodiimide reagent. Tris buffer must be avoided with the carbodiimide and glutaraldehyde conjugation procedures because of the interference caused by the amino groups of trisaminomethane. Estimation of the extent of hapten incorporation is useful in determining that the conjugation procedure has been successful, but is not an essential step as the optimal degree of hapten incorporation has not been determined.

Immunization

Immunization by other routes[21] or of other species may prove to be satisfactory but, in our experience, prolonged immunization of rabbits for several months with fibrinopeptide–albumin conjugates in complete Freund's adjuvant mixture is required to elicit antifibrinopeptide antibodies of satisfactory titer for use in RIA studies.

---

[21] J. L. Vaitukaitis, this series, Vol. 73 [2].

Procedure
1. The albumin–fibrinopeptide conjugate is dissolved in isotonic saline or phosphate buffered saline (pH 7.0–7.4) at a concentration of 0.5–2 mg/ml and mixed with a precisely equal volume of Freund's complete adjuvant mixture and carefully emulsified as described elsewhere.[22] The mixture is stored at 4° and reemulsified before each use.
2. After obtaining, if desired, a preimmunization bleeding, each rabbit is given 0.1 ml of emulsion in toepads of the front feet and 0.2 ml in the looser toepads of the rear feet. One and two weeks later the procedure is repeated (using a different toepad on each occasion). Thereafter, rabbits are given 0.4 ml intramuscular injections, 0.2 ml on each side, every 2 weeks until serum antibody titers are satisfactory in at least one animal; 3–8 months may be required to elicit satisfactory antibodies. After a rabbit has begun to produce antibodies of suitable titer, monthly immunization may suffice.
3. Animals are usually bled from an ear vein 5–7 days following an antigen injection. We have customarily obtained serum for our studies but it should be noted that, unlike plasma, serum contains thrombin which may create problems in certain studies, as outlined below. Serum is stored either in the frozen state or, at 4°, in the presence of a preservative such as thimerosal (Merthiolate, Eli Lilly & Co., Indianapolis IN; 0.1 ml of a 1% solution per 10 ml of serum).

Radiolabeled Fibrinopeptide Analogs

Neither FPA nor FPB contains a tyrosine or histidine residue suitable for radioiodination. Accordingly, it is necessary to introduce a phenol-containing residue. Synthetic fibrinopeptide analogs containing a tyrosine residue have been employed.[6,7] However, in our experience, the method of Goodfriend and Ball,[23] which involves the coupling of 3-(*p*-hydroxy-phenyl)-propionic acid (desaminotyrosine) to the amino termini of FPA and of an FPB analog has been most convenient for this purpose. Desaminotyrosyl-fibrinopeptide derivatives can readily be radioiodinated by the chloramine-T method of Hunter and Greenwood.[24] The procedure outlined below has been used for the synthesis of a radioiodinated desaminotyrosyl-FPA analog, but a similar procedure may be employed in the

[22] B. A. L. Hurn and S. M. Chantler, this series, Vol. 70 [5].
[23] T. L. Goodfriend and D. L. Ball, *J. Lab. Clin. Med.* **73**, 501 (1969).
[24] W. M. Hunter and F. C. Greenwood, *Nature (London)* **194**, 495 (1962).

preparation of radioiodinated desaminotyrosyl–FPB and desaminotyrosyl–desarginyl–FPB analogs.[10,12]

Reagents

FPA, FPB analog, or desarginyl–FPB analog
Desarginyl–FPB analog is prepared by incubation of 1 mg of the FPB
analog at 37° with 0.06 units porcine pancreatic carboxypeptidase
B (Worthington Biochemical Corp., Freehold NJ) in 1 ml 0.1 M
NH_4HCO_3 for 24 hr, followed by inactivation of carboxypeptidase
B by adding *O*-phenanthroline (Sigma Chemical Co., St. Louis
MO) to a final concentration of 0.01 M.[11,12]
3-(*p*-Hydroxyphenyl)-propionic acid, *p*-nitrophenyl ester (Vega Lab-
oratories, Inc., Tucson AZ) or *N*-hydroxysuccinimidyl ester (Vega
Laboratories or Calbiochem-Behring Corp., La Jolla CA)
^{125}I, carrier-free in 0.1 N NaOH
Dimethylformamide, chloroform, ethyl ether (analytical reagent
grade)
Acetic acid, 10%
0.1 M NH_4HCO_3
BioGel P-2 polyacrylamide beads (200–400 mesh), Bio-Rad Labora-
tories, Richmond CA
Chloramine-T, Eastman Organic Chemicals, Rochester NY
Sodium metabisulfite
0.5 M phosphate buffer, pH 7.5
Tris–NaCl buffer, pH 7.5 (0.1 M NaCl, 0.05 M Tris)
Sephadex G-10, Pharmacia Fine Chemicals, Piscataway NJ
Rabbit serum albumin, Miles Laboratories, Kankakee IL, 20 mg/ml
in 0.5 M phosphate buffer, pH 7.5

Procedure

1. To 4 mg FPA and 25 mg 3-(*p*-hydroxyphenyl)-propionic acid ester
   is added 3 ml dimethylformamide.
2. After constant stirring at 37° for 18 hr, 2 ml distilled water and
   0.2 ml 10% aqueous acetic acid are added.
3. The mixture is extracted twice with 7 ml of 1:1 chloroform–ethyl
   ether to remove unreacted reagents and byproducts.
4. After extraction, the mixture is centrifuged 5 min at 500 g at room
   temperature in a glass-stoppered tube; the aqueous phase is then
   lyophilized.
5. The lyophilized peptide is dissolved in 4 ml 0.1 M ammonium bicarbonate and subjected to gel filtration on a 20 × 1.2-cm column
   of Biogel P-2 which has been suspended in 0.1 M ammonium bicarbonate. One milliliter fractions are eluted with the same buffer
   and monitored for absorbance at 280 nm and for ninhydrin reac-

tivity (as assessed by absorbance at 570 nm). Fractions in the initial peak exhibiting maximum 280 nm (desaminotyrosyl) and 570 nm (peptide) absorbance are pooled and stored at $-70°$ in 0.1-ml aliquots; from the two absorbance values, it has been estimated that the final products contain 0.9–1 molecule of desaminotyrosine per molecule of peptide.

6. To 2–5 mCi carrier-free ^{125}I in 10 μl 0.1 M NaOH in a 0.3 ml microflex vial (Kontes Glass Co., Vineland NJ), 30 μl 0.5 M phosphate buffer, pH 7.5, is added, followed by 10 μg desaminotyrosyl–FPA in 15 μl distilled water.

7. Freshly prepared chloramine-T (50 μg in 15 μl 0.5 M phosphate buffer, pH 7.5) is promptly added with mixing, followed 30 sec later by freshly prepared sodium metabisulfite (96 μg in 20 μl 0.5 M phosphate buffer, pH 7.5).

8. Rabbit serum albumin (20 μl) and 0.5 M phosphate buffer, pH 7.5 (200 μl) are added in rapid succession.

9. The solution is immediately subjected to gel filtration in a 10 ml serological pipette on Sephadex G-10 which has been equilibrated with Tris-buffered NaCl, pH 7.5; 0.5 ml fractions are collected and their radioactivity counted.

10. The fraction (usually the 3.5–4 ml fraction) containing the initial peak of radioactivity is diluted 1:10 in Tris-buffered ovalbumin, pH 8.5 (see following section), and stored in 50-μl aliquots at $-20°$ for use in binding studies. Freshly prepared tracer should exhibit 80–90% immunoreactivity (i.e., percentage bound by an excess of antibody), but percentage immunoreactivity decreases with time. A tracer preparation can ordinarily be used until its immunoreactivity falls below 60% (usually 6–8 weeks after the labeling procedure).

Antibody Detection

The binding of radioiodinated fibrinopeptide derivatives by their corresponding antibodies can be demonstrated by any of the methods commonly employed to detect the binding of low-molecular-weight substances by antibodies. In our experience, a coated charcoal method for the separation of free and bound tracer has been found to be most convenient. We have modified the classical method of Herbert et al.[25] in two

---

[25] V. Herbert, K.-S. Lau, C. W. Gottlieb, and S. J. Bleicher, J. Clin. Endocrinol. Metab. 25, 1375 (1965).

ways. First, we have not found it necessary to use dextran. Second, we have employed ovalbumin rather than HSA, because most HSA preparations contain significant quantities of fibrinopeptides which interfere with the binding of the labeled fibrinopeptide derivative by antibody.[4,10]

Reagents
Tris-buffered ovalbumin, pH 8.5
0.1 M NaCl, 0.05 M Tris base, adjusted to pH 8.5 with concentrated HCl
0.1% ovalbumin (3X-crystallized, ICN Life Sciences Group, Cleveland OH)
Freshly prepared weekly
Norit "A" charcoal (neutral)—Amend Drug & Chemical Co., Irvington NJ
1% (w/v) suspension in Tris-buffered ovalbumin
[125]I-labeled fibrinopeptide derivative
0.2 μCi/ml in Tris-buffered ovalbumin
Antiserum and control serum from nonimmunized animal
Dilute 1:20 in Tris-buffered ovalbumin; then prepare serial 2-fold dilutions of antiserum

Procedure
1. To duplicate 400-μl aliquots of buffer, add 50 μl ^{125}I-fibrinopeptide derivative, followed by 50 μl of serial dilutions of antiserum (1:20 to 1:2560); mix well on a vortex mixer after each addition. Duplicate control tubes should be included, containing: buffer instead of antiserum (blank, or charcoal control); 1:20 normal serum instead of antiserum (serum blank; ordinarily comparable to charcoal control); and buffer instead of tracer (background control). Additional tubes containing only buffer and tracer (to which charcoal will not be added in Step 3) should be included for determination of total counts (standards).
2. Incubate for 2 hr or overnight at 4°.
3. Resuspend charcoal by vigorous shaking and maintain a uniform suspension by continuous magnetic stirring. Using a 1-ml Cornwall syringe pipet and a 4-in. 14-gauge cannula (Arthur H. Thomas Co., Philadelphia PA), add 0.5 ml to all tubes (except tubes for total count determination, to which 0.5 ml buffer should be added). Subject all tubes to immediate vortex mixing.
4. Centrifuge immediately at 4° for 10 min at 6100 g.
5. Aspirate 0.5 ml of supernatant and count in a gamma spectrometer.
6. The amount of supernatant radioactivity present in blank, or "charcoal control," specimens is subtracted from the supernatant radio-

activity present in all test and standard ("total count") tubes. The bound radioactivity in all test specimens is then expressed as the percentage of radioactivity present in the supernatant, in comparison with the radioactivity present in standard tubes.

7. The titer of each antiserum is expressed as the greatest dilution of that antiserum capable of binding 50% of immunoreactive tracer. Ordinarily, 60–90% of a tracer preparation can be bound by an excess of antifibrinopeptide antibody; the maximal amount bound under these conditions is considered to represent immunoreactive tracer. For a discussion of the significance of titers, see Butler and Tse-Eng.[26]

Antibody Specificity

The specificity of antifibrinopeptide antibodies may be established by comparing the abilities of fibrinopeptides and of fibrinogen (as well as of other fibrinopeptide-containing protein fragments or peptides, if available) to inhibit the binding, by antibody, of the radiolabeled fibrinopeptide analogs.[3,10,12,15,17,27–29]

Reagents

Tris-buffered 0.1% ovalbumin, pH 8.5
  Prepared weekly; used as diluent for all reagents
Norit "A" charcoal (neutral), 1% suspension
^{125}I-labeled fibrinopeptide derivative, 0.2 μCi/ml
Antifibrinopeptide serum
  Diluted to a concentration capable of binding 40–50% of immunoreactive tracer.
  Containing hirudin (Sigma Chemical Company, St. Louis Mo.; grade IV), 0.1–0.2 units per ml of diluted serum
Inhibitors
  Fibrinopeptides, 0.1–20 pmol/ml
  Fibrinogen (dialyzed to remove free fibrinopeptides), 0.1–500 pmol/ml
  If available, fibrinogen fragment D, the amino-terminal disulfide knot of fibrinogen, reduced and carboxymethylated Aα- or Bβ-

[26] V. P. Butler, Jr. and D. Tse-Eng, this volume [42].
[27] R. E. Canfield, J. Dean, H. L. Nossel, V. P. Butler, Jr., and G. D. Wilner, *Biochemistry* **15**, 1203 (1976).
[28] G. D. Wilner, H. L. Nossel, R. E. Canfield, and V. P. Butler, Jr., *Biochemistry* **15**, 1209 (1976).
[29] C. Kockum, *Thromb. Res.* **19**, 639 (1980).

chains, or amino-terminal peptides derived from the Aα- or Bβ-chains[16,27,29]; 0.1–500 pmol/ml

Procedure

1. To duplicate 400-μl aliquots of buffer, containing various concentrations of inhibitor, 50 μl of tracer is added, followed by vortex mixing and by the subsequent addition of 50 μl of an appropriate dilution of antifibrinopeptide serum, also followed by vortex mixing. Controls should include standard or "total count" tubes, blank or "charcoal control" tubes, and background control tubes as described in step 1 of the procedure for antibody detection. In addition, control tubes should be included in which tracer is added to an excess of antibody to determine the maximal binding of tracer (immunoreactive tracer control).

2–6. As outlined in steps 2–6 of the procedure described above for antibody detection.

7. Percentage of tracer (or of immunoreactive tracer) bound is plotted on the ordinate of semilogarithmic graph paper against the $\log_{10}$ of the concentration of each inhibitor. The concentration of each compound required to produce 50% inhibition of binding under these conditions (I_{50}) is determined and compared with the I_{50} of the fibrinopeptide being tested simultaneously.

Comments. In experiments of this type, the binding of tracer by antibody may vary in different experiments, and the I_{50} for each inhibitor may vary accordingly. It is therefore essential that the inhibitory capacity of each compound be compared with the inhibitory capacity of an appropriate reference compound (i.e., FPA or FPB), as determined in the same experiment, before it is compared with the inhibitory capacity of another compound determined in another experiment. Thus, the relative molar inhibitory capacity, or I_{50} of test compound/I_{50} of fibrinopeptide, has, in our experience, provided a convenient method of comparing the immunoreactivities of compounds assessed with one antiserum in separate experiments.

The validity of all results is dependent on the absence of any release of fibrinopeptide from fibrinogen or from the various fibrinopeptide-containing protein fragments and peptides; since such release could be caused by thrombin present in the antiserum, these experiments are carried out in the presence of hirudin. The validity of these experiments is also dependent on the presence of the anticipated molar quantities of FPA and/or FPB in the various test proteins and peptides; the fibrinopeptide content of these compounds can readily be assessed by treating each with thrombin (in the absence of hirudin) and ascertaining that the anticipated molar quantity of immunoreactive fibrinopeptide has been released.

Choice of Antiserum for Radioimmunoassay

It has been indicated above that prolonged immunization is required to elicit antifibrinopeptide antisera of sufficiently high titer to permit their use in RIA studies. Most such antisera contain antibodies of sufficiently high affinity (average intrinsic association constants of 10^9 or greater) to permit their use in the development of sensitive RIA procedures. However, antifibrinopeptide sera vary in their dissociation rate constants, and the charcoal used in the separation of free from bound tracer competes effectively with antibody for tracer molecules following their dissociation from antibody.[30,31] Accordingly, before an antiserum is employed in the development of an RIA employing a charcoal separation step, it should be determined that such dissociation, with resultant displacement of tracer from antibody to charcoal, is not sufficiently rapid to interfere with the assay procedure. For use of a given antiserum in a charcoal RIA procedure in our laboratory, the decrease in antibody-bound tracer must be less than 10% in the 10–15 min following addition of charcoal.

As has been indicated previously, antifibrinopeptide sera vary considerably in their cross-reactivity with fibrinopeptide-containing protein or peptide molecules. For example, anti-FPA sera which cross-react extensively with Aα1(Ala)-23(Arg) cannot distinguish this peptide from FPA in test specimens. However, treatment of Aα1(Ala)-23(Arg) with thrombin will cause a significant increase in immunoreactivity with other antisera which can discriminate effectively between these two peptides; hence, when such an antiserum is employed, a lack of an increase in immunoreactivity following thrombin treatment indicates the presence of FPA and the absence of the larger peptide (a product of the action of plasmin on the Aα-chain of fibrinogen), while an increase in immunoreactivity after thrombin treatment indicates the presence of this or a larger NH_2-terminal Aα-chain peptide.[15,27] Similarly, using an antiserum which discriminates effectively between FPB and Bβ1(Pyr)-42(Arg), it has been possible to identify Bβ1(Pyr)-42(Arg), a product of the action of plasmin on the NH_2-terminal end of the Bβ-chain of fibrinogen, by the detection of thrombin-increasable FPB in clinical blood samples.[16] Thus, in general, it is advisable to select, for use in RIA procedures, antifibrinopeptide sera which do not cross-react extensively with longer NH_2-terminal fibrinopeptide-containing segments of the Aα- or Bβ-chains of fibrinogen.

Another consideration in the choice of antiserum for use in FPB RIA procedures is related to the fact that carboxypeptidase B in human plasma rapidly cleaves the COOH-terminal arginine residue from FPB. If one

[30] R. C. Meade and T. J. Kleist, *J. Lab. Clin. Med.* **80**, 748 (1972).
[31] T. W. Smith and E. Haber, *Pharmacol. Rev.* **25**, 219 (1973).

wishes to discriminate between FPB and its desarginyl derivative, one should employ an antiserum whose binding of the ^{125}I-FPB derivative is not effectively inhibited by desarginyl–FPB. On the other hand, if one wishes to measure desarginyl–FPB as an index of FPB release, it is desirable to employ an antiserum which cross-reacts extensively with desarginyl–FPB; using such an antiserum and an ^{125}I-labeled desarginyl–FPB derivative, a desarginyl–FPB RIA has recently been developed and employed in clinical studies.[12]

Processing of Blood Samples for RIA

Several factors may interfere with the accurate measurement of fibrinopeptides in clinical blood samples. Blood obtained after a traumatic venipuncture may exhibit spuriously high fibrinopeptide levels, presumably as the result of activation of coagulation enzymes by tissue factor; thus, blood should be assayed only if obtained by a nontraumatic venipuncture. To minimize *in vitro* fibrinopeptide release, freshly drawn blood is mixed with heparin and, to minimize *in vitro* plasmin action on the NH_2-terminal ends of the Aα- and Bβ-chains, freshly drawn blood is also mixed with trasylol. Blood is chilled at 4° to further minimize the action of proteolytic enzymes on fibrinogen. Although the cross-reactivity of fibrinogen with the antisera used in fibrinopeptide RIAs is minimal, the high concentrations of fibrinogen in plasma (2–4 mg/ml) would interfere with the RIA procedure. An ethanol precipitation step has therefore been employed to remove fibrinogen as well as residual proteolytic activity.[4]

After removal of ethanol (and water) by evaporation, the fibrinogen-depleted plasma extract is reconstituted with water. Subsequent processing of the reconstituted plasma varies somewhat, depending upon the assay(s) to be performed. In the case of FPA, the reconstituted plasma extract may be assayed directly without additional processing.[4,32] In the cases of assays for FPB or for desarginyl–FPB, adsorption with bentonite has been used for the selective removal of Bβ1(Pyr)-42(Arg) and Bβ1(Pyr)-21(Lys), both of which are capable of producing interference in the RIA procedures for the smaller peptides.[12] The presence of Bβ1(Pyr)-42(Arg) and /or Bβ1(Pyr)-21(Lys) in reconstituted plasma can be detected if there is an increase in FPB immunoreactivity following thrombin treatment. To measure thrombin-increasable FPB (TIFPB), the bentonite adsorption step is omitted and the heparin in the plasma extract (approximately 3 U/ml) is neutralized by the addition of protamine sulfate; the

[32] H. L. Nossel, M. Ti, K. L. Kaplan, K. Spanondis, T. Soland, and V. P. Butler, Jr., *J. Clin. Invest.* **58**, 1136 (1976).

extract is then assayed for FPB immunoreactivity before and after thrombin treatment.[16]

Reagents

Heparin (Hynson, Westcott and Dunning, Inc., Baltimore MD)
Trasylol (FBA Pharmaceuticals, New York NY)
Anticoagulant solution
    Heparin (1500 U/ml) and Trasylol (1000 U/ml) in 0.15 M NaCl
Ethanol
    Anhydrous denatured ethanol, formula L982 (Matheson, Coleman
    and Bell, East Rutherford NJ)
Porcine pancreatic carboxypeptidase B (COBC; Worthington Bio-
    chemical Corp., Freehold NJ)
Bentonite (laboratory grade; Fisher Scientific Co., Springfield NJ)
Tris–ovalbumin buffer, pH 8.9(0.05 M Tris–HCl, 0.1 M NaCl, 0.1%
    ovalbumin)
Protamine sulfate (Eli Lilly & Co., Indianapolis IN)
Bovine thrombin (Parke Davis Company, Detroit MI)

Procedure

1. Blood is collected from an antecubital vein via a 19-gauge needle
    into a polyethylene syringe in the presence of a stopwatch. If
    blood flow is not free or if collection of the sample requires longer
    than 45 sec, the specimen is not assayed.
2. The blood is immediately transferred to a conical plastic tube con-
    taining $\frac{1}{10}$ volume of anticoagulant solution; after gentle mixing,
    the tube is placed on ice.
3. The anticoagulated blood is immediately centrifuged at 1700 g at
    4° for 20 min, and the plasma aspirated.
4. Within 30 min of blood collection, 2 ml plasma is added to an
    equal volume of chilled denatured ethanol in a 12 ml conical cen-
    trifuge tube in an ice-water bath.
5. After 30 min incubation, the tube is centrifuged at 1500 g at 4° for
    20 min; the supernatant is recentrifuged in an identical manner to
    eliminate any traces of precipitate.
6. The supernatant is then dried at 37° in a Brinkmann concentrator
    (Brinkmann Instruments Inc, Westbury NY) and reconstituted to
    its initial volume with deionized water.

Subsequent steps vary with the peptide being assayed, as follows:

7a. FPA: The reconstituted extract may be assayed for FPA without
    further processing.[4,32]
7b. Desarginyl–FPB and FPB: For the measurement of desarginyl–
    FPB, complete conversion of FPB to its desarginyl derivative is as-
    sured by incubating the reconstituted plasma with carboxypepti-

dase B (0.1 U/ml final concentration) at 37° for 5 min (this step is omitted in FPB assays). To remove carboxypeptidase B as well as Bβ1(Pyr)-42(Arg) and Bβ1(Pyr)-21(Lys), 2 ml of reconstituted plasma is added to 1 ml of a 2% (w/v) bentonite suspension in Tris–albumin buffer, pH 8.9. After 30 sec of vortex mixing and 15 min on an Ames aliquot mixer (Ames Co., Elkhart IN), the bentonite is sedimented by centrifugation at 4800 g at 4° for 10 min. The supernatant plasma is aspirated and assayed for desarginyl–FPB or for FPB.[12]

7c. TIFPB: The heparin in the reconstituted plasma is neutralized by the addition of protamine sulfate (final concentration 150 μg/ml). The reconstituted plasma is tested for FPB immunoreactivity before and after treatment with thrombin (10 U/ml for 1 hr at 37°).[16]

Comments Because of its instability in plasma, direct assay of FPB has not yielded satisfactory results and measurement of desarginyl–FPB has provided a more accurate reflection of fibrin II formation *in vivo*. In instances in which direct measurement of FPB in plasma is considered desirable, addition of *O*-phenanthroline (2 *M* solution in absolute ethanol) to anticoagulated blood (step 2) at a final concentration of 0.01 *M* effectively inhibits carboxypeptidase B action *in vitro*. It should be noted, however, that considerable conversion of FPB to its desarginyl derivative will already have occurred *in vivo* prior to and during venipuncture.[12]

The possible *in vitro* release by plasmin and perhaps other proteolytic enzymes of NH$_2$-terminal peptides of the Aα- and Bβ-chains such as Aα1(Ala)-23Arg, Bβ1(Pyr)-42(Arg), and Bβ1(Pyr)-21(Lys) is a matter of major concern in the processing of blood for fibrinopeptide assays. The Arg-Ala bond between positions 42 and 43 of the Bβ-chain is particularly sensitive to enzyme action and hence the ethanol precipitation step in the TIFPB procedure is particularly important. In our experience, release of Aα1(Ala)-23(Arg) has not been a problem when blood is properly processed; however, it is of sufficient concern that samples should be periodically treated with thrombin to determine whether such treatment causes an increase in FPA immunoreactivity with an antiserum which does not cross-react extensively with Aα1(Ala)-23(Arg). Although the 23-amino acid peptide is not as reactive as FPA with such antisera, it is sufficiently immunoreactive that appreciable quantities can yield spuriously high FPA values.[15]

Radioimmunoassay Procedure

Samples processed as described above are subjected to radioimmunoassay together with a set of standards containing appropriate concentrations of fibrinopeptide in Tris-buffered 0.1% ovalbumin, pH 8.5. All reagents (Tris-buffered 0.1% ovalbumin, tracer, and antiserum) are prepared as described above for the specificity studies except that the charcoal is suspended at a concentration of 1.25% (w/v) in Tris-buffered 0.35% ovalbumin. Normal and abnormal plasma samples are included in each assay, and controls are included as described above for the specificity studies. The RIA procedure described below has been used for FPA measurement, but is equally satisfactory for FPB assays.

Procedure

1. To 100 μl of diluted antiserum is added 500 μl test sample or standard; all specimens are analyzed in duplicate.
2. After incubation at 4° for 24 hr, 50 μl radioiodinated FPA derivative (approximately 10,000 cpm) is added.
3. After 1 hr at 4°, 500 μl cold (4°) charcoal suspension is added.
4. After vigorous vortex mixing, the tubes are immediately centrifuged at 1700 g at 4° for 10 min.
5. The supernatant fluid, representing antibody-bound radioactivity, is decanted and counted in a gamma spectrometer.
6. The percentage of tracer bound is calculated as described in step 6 of the above section on antibody detection and a curve is plotted from the results obtained with fibrinopeptide standards as described in step 7 of the above section on antibody specificity. Fibrinopeptide concentrations in test specimens are determined by comparison with the standard curve.

Comments Other workers have used the dioxane,[5] ammonium sulfate,[7] or double antibody[33] methods to separate free from bound radiolabeled fibrinopeptide derivative. Recently, a solid-phase enzyme immunoassay for FPA has been described.[34]

Applications

The FPA assay has been used extensively in the clinical measurement of plasma FPA levels. Normal plasma FPA concentrations are in the 0.6–2.0 pmol/ml range. Elevations are usually found in patients with active thromboembolic disease, but elevations are also commonly encountered in patients with disseminated intravascular coagulation and various inflammatory, infectious, and neoplastic disorders.[4–9,32] Despite the lack of specificity of elevated FPA concentrations, plasma FPA levels decrease rapidly following the administration of heparin to patients with thromboembolic disease, and the response of FPA levels may prove useful as an

[33] J. Harenberg, G. Hepp, and H. Schmidt-Gayk, *Thromb. Res.* **15**, 513 (1979).
[34] J. Soria, C. Soria, and J. J. Ryckewaert, *Thromb. Res.* **20**, 425 (1980).

index of the adequacy of anticoagulant therapy in such patients.[35] The generation of FPA in freshly obtained clinical blood samples has been employed as an index of thrombin activity in these specimens.[32] The FPA assay has also been used in studies of the *in vivo* kinetics of FPA and also in *in vitro* studies of the kinetics of the proteolytic action of various enzymes on the NH_2-terminal end of the Aα-chain of human fibrinogen.[10,36] The FPB assay has also been very useful in the studies of the action of proteolytic enzymes on the NH_2-terminal end of the Bβ-chain of fibrinogen[10,36] but, as noted above, the presence of carboxypeptidase B in plasma has interfered with the direct assay of FPB in human blood. Levels of desarginyl–FPB rather than of FPB appear to reflect fibrin II formation *in vivo;* normal desarginyl–FPB levels are less than 0.6 pmol/ml, and elevated levels have been detected in patients with *in vivo* fibrinogenolysis secondary to the intraamniotic administration of hypertonic saline in the induction of abortion. The desarginyl–FPB RIA has also been used in studies of the *in vivo* kinetics of desarginyl–FPB as well as in *in vitro* studies of the action of carboxypeptidase B on FPB.[12] Finally, the TIFPB RIA has been used in *in vivo* and *in vitro* studies of the action of plasmin on the NH_2-terminal end of the Bβ-chain of fibrinogen, resulting primarily in the release of Bβ1(Pyr)-42(Arg).[16]

Acknowledgment

Supported by USPHS program project grant HL 15486.

[35] I. M. Yudelman, H. L. Nossel, K. L. Kaplan, and J. Hirsh, *Blood* **51,** 1189 (1978).
[36] S. B. Bilezikian and H. L. Nossel, *Blood* **50,** 21 (1977).

Section III

Metal and Heme Binding Proteins

[9] Radioimmunoassay of Metallothioneins

By JUSTINE S. GARVEY, RONALD J. VANDER MALLIE, and
CHIN C. CHANG

Introduction

Metallothionein (Mt),[1] a cytoplasmic, low-molecular-weight, metal-binding, cysteine-rich protein was first isolated (1957) from equine kidney cortex tissue by Margoshes and Vallee.[1a] Most of the literature on the subject of Mt has been published since 1970 and an introduction to much of it is made available in the published proceedings of the first international meeting on metallothionein.[2] Also available in this general reference is a recommended nomenclature which has been followed in the preparation of this manuscript.

The vertebrate species in which Mt is found are both extensive in number and varied in type; but, recent findings of the protein in microorganisms suggest Mt may be ubiquitous. Additionally, the finding of a very high degree of homology in the amino acid sequence of the protein isolated from various species suggests a selective pressure for its preservation throughout evolutionary development. Although these findings are factual evidence, highly suggestive that Mt has an important biological function, this function is presently unknown.

In addition to Mt being present in organisms that have not been intentionally exposed to a heavy metal, increased synthesis of the protein is easily obtained by exposure of rodents to salts of heavy metals, e.g., Zn, Cd, Hg, Cu, and Ag. Laboratory animals are not only a source of Mt that can be isolated for study of the biochemical and physiochemical properties of Mt; but, animal models are useful in resolving the role of Mt in the altered health status of organisms, including humans, that experience heavy metal exposure.

For the many unresolved questions about this interesting molecule, a

---

[1] Abbreviations: BBS, borate-buffered saline; cFA, complete Freund adjuvant; GDA, glutaraldehyde; ifA, incomplete Freund adjuvant; iv, intravenous; Mt, metallothionein; the isoforms in rat hepatic tissue are designated Mt-1 and Mt-2 (these terms, like all terms referring to metallothionein, conform to nomenclature proposed for uniform adoption in Ref. 2); NRS, normal rabbit serum; RIA, radioimmunoassay.

[1a] M. Margoshes and B. L. Vallee, *J. Am. Chem. Soc.* **79**, 4813 (1957).

[2] J. H. R. Kägi and M. Nordberg (eds.), Metallothionein (Proceedings of the first international meeting on metallothionein and other low molecular weight metal-binding proteins, Zurich, July 17–22, 1978). *Experientia Suppl.* **34**, (1979).

METHODS IN ENZYMOLOGY, VOL. 84

sensitive assay method, such as the described RIA, should prove to be a useful tool.

Isolation and Purification of Mt

The isolation of rat hepatic Mt was performed as outlined in Fig. 1. Male Sprague–Dawley rats, 200–250 g, received 3–5 subcutaneous injections of $CdCl_2$, with an initial injection of 0.25 mg Cd^{2+}/kg and subsequent daily injections of 1.0 mg Cd^{2+}/kg. The last injection usually contained 1–2 μCi ^{109}Cd. Two to three days after the last injection of $CdCl_2$, the rats were sacrificed and the livers were excised and placed into prechilled (0–4°) 0.1 M ammonium formate buffer, pH 7.4. A 20% homogenate in 0.1 M ammonium formate buffer, pH 7.4, was prepared by homogenizing small portions of liver with a Potter-Elvehjem tissue

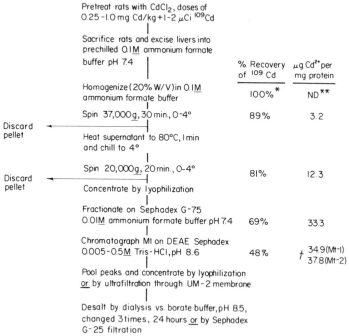

FIG. 1. Isolation of rat hepatic metallothionein (Mt). This scheme yields pure Mt-2 whereas further purification steps are required to obtain pure Mt-1. The values in the recovery and purification columns are representative of the more than 20 separate preparations performed to obtain Mt-1 and Mt-2 for use as immunogens and test antigens. *Recovery in subsequent steps is relative to assumed complete recovery at this step. **Not determined. †These ratios were determined using unadjusted Lowry values; as adjusted Lowry values they increase by a factor of 2.5 to yield 87.2 for Mt-1 and 94.5 for Mt-2. Modified from R. J. Vander Mallie and J. S. Garvey, *Immunochemistry* **15**, 857 (1978), with permission from the publisher.

grinder. The homogenate was centrifuged at 37,000 g for 30 min. The supernatant was heated in a water bath and held at 80° for 1 min. Immediately after heat treatment, the solution was cooled in an ice-water bath. The chilled, heat-treated supernatant was then centrifuged at 20,000 g for 20 min at 4°. The supernatant was concentrated by lyophilization, redissolved in 0.01 M ammonium formate buffer, pH 7.4, and chromatographed on Sephadex G-75. The Sephadex G-75 column (93 × 3 cm) was eluted with 0.01 M ammonium formate buffer, pH 7.4, 0.002% hibitane. The eluant from all fractionations was continuously monitored at 254 and 280 nm. Fractions of 4–5 ml were collected and 1-ml aliquots of these were assayed for radioactivity in a gamma counter. Fractions with radioactivity that measured greater than 10% of the maximum found in the single composite peak were pooled and lyophilized. Metallothionein consistently eluted at a V_e/V_o ratio of 1.9–2.1. The lyophilized protein was redissolved in 5 mM Tris–HCl, pH 8.6, and rechromatographed on A-25-120 DEAE-Sephadex and eluted with a linear gradient of 0.005–0.5 M Tris–HCl, pH 8.6. The DEAE-Sephadex was prepared by swelling 15 g of the dry form in 250 ml 0.005 M Tris–HCl, pH 8.6 buffer overnight, followed by washing with 18 liters of the same buffer. The washed DEAE was then used to prepare a 2.5 × 16 cm column on which the separation was performed. The two resulting peaks that contained [109]Cd were separately pooled and concentrated by either lyophilization or ultrafiltration through UM-2 membranes. Because these peaks have a characteristically strong absorbance at 254 nm (due to the Cd-mercaptide complex formed when cadmium is the predominant metal bound to thionein) they are easily identified by monitoring the absorbance at this wavelength. The concentrated protein was desalted either by dialysis (for 24 hr against borate buffer, pH 8.5, using three changes of dialyzate and dialysis tubing with a molecular weight cutoff of 6000–8000) or by Sephadex G-25 (equilibrated with borate buffer) filtration. The desalted protein was lyophilized and stored at −20° for future use.

As an alternative to injecting the rats with [109]Cd *in vivo*, [109]Cd (2–10 μCi) was added to the heated supernatant. The "spiked" supernatant was placed at 4° for 12–24 hr before it was concentrated by lyophilization. This modification eliminates the need to dispose of radioactive animal carcasses. The rest of the procedure was identical to that above.

Using the described procedure two isoforms of Mt are isolated, Mt-1 and Mt-2. The two forms are resolved by DEAE-Sephadex chromatography (Fig. 2). Although the second peak so obtained contains only Mt, it was found that the first peak contains two proteins: Mt and an unidentified nonmetalloprotein. Electrophoretic examination of successive portions (labeled a, b, c, d on Fig. 2) of the first peak showed the presence of

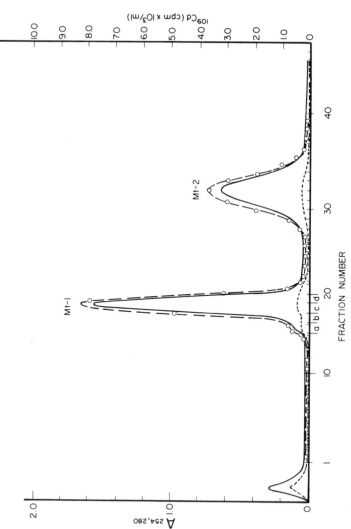

FIG. 2. DEAE-Sephadex fractionation of the main [109]Cd-containing peak from Sephadex G-75 (Fig. 1) into Mt-1 and Mt-2. The gel was 16 × 2.5 cm and elution was at a flow rate of 1 ml/min from a gradient of 0.005–0.350 M Tris–HCl. The fractions were collected using a peak separator mode (ISCO monotor UA-5) actuated to collect a maximal volume of 20 ml. Ultraviolet absorption scans were obtained by monitoring the eluant at both 254 nm (——) and 280 nm (– – –). [109]Cd (O——O) was determined by direct counting of 1-ml aliquots of the eluted fractions in a gamma counter with sodium iodide crystal. The first peak was divided into fractions a, b, c, and d which were examined by disc gel electrophoresis (see Fig. 3). Modified from R. J. Vander Mallie and J. S. Garvey, *Immunochemistry* **15**, 857 (1978), with permission from the publisher.

two distinct protein components in the ascending region of the peak, but only one present in the descending portion (Fig. 3). It was determined by electrophoresis (Fig. 3) and autoradiography (Fig. 4) that the protein with the higher electrophoretic mobility was Mt, while the other, which still remains unidentified, has the following characteristics: (1) does not bind [109]Cd injected *in vivo* as a tracer for Mt (Fig. 4); (2) contains little or no cadmium or zinc (Table I); and (3) has significantly fewer SH groups than metallothionein (Table I). Thus, the contaminating nonmetalloprotein is not thionein (the apoprotein of metallothionein); however, electrophoretically pure metallothionein can be obtained by pooling selectively the fractions corresponding to the descending region of the first 254 nm absorbing

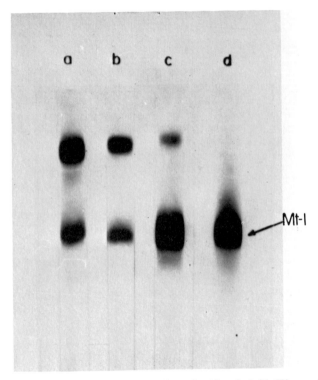

FIG. 3. Electrophoresis in cylindrical 7% polyacrylamide gels (with 5% cross-linkage) of four samples obtained separately from the peak labeled Mt-1 (see abscissa of Fig. 2). A slow migrating component is present early, i.e., is strong in fraction *a*, diminishes in fractions *b* and *c*, and is absent in the late eluting fraction *d*. In view of the results from [109]Cd detection in gels (Fig. 4), the pure component obtained in fraction *d* is designated Mt-1. The μg protein applied to each gel were: *a*, 17; *b*, 18; *c*, 65; and *d*, 77 (as determined by unadjusted Lowry assay). Modified from R. J. Vander Mallie and J. S. Garvey, *Immunochemistry* **15**, 857 (1978), with permission from the publisher.

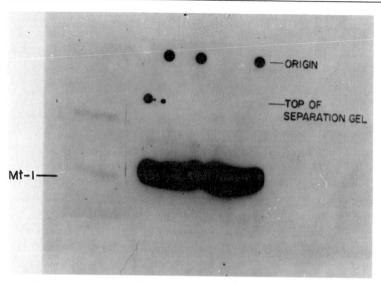

FIG. 4. Electrophoresis of an aliquot from fraction *b* of the peak designated Mt-1 in Fig. 3. A 7% polyacrylamide gel (5% cross-linkage) was cast as a slab gel that was both stained for protein and examined as an autoradiograph. The stained gel is on the *left* and the autoradiograph is on the *right*. The amount of protein used was 1 mg/well (10,000 cpm) for the autoradiograph and 300 μg for the stained gel (as determined by unadjusted Lowry assay). The origin and the top of the gel are defined by ^{32}P ink. These results confirm the identity of Mt-1 in Fig. 3. Modified from R. J. Vander Mallie and J. S. Garvey, *Immunochemistry* **15,** 857 (1978), with permission from the publisher.

protein peak eluted during DEAE-Sephadex chromatography. The two isometallothioneins, Mt-1, isolated from the descending region of the first protein peak and Mt-2, the whole second peak, have been characterized with respect to cadmium, zinc, SH content, and an absorption ratio of 254 to 280 (see Table I). The values obtained are in good agreement with those published by others.[3,4] Despite the foregoing indications that the two Mt were pure, the possibility was still considered that contaminating nonmetalloprotein might cause some interference in the RIA. A further, and final, solution to this problem was use of an affinity chromatography matrix specific for free sulfhydryl groups that permitted the nonmetalloprotein species to be purified[5]; this protein did not interfere with the described RIA of metallothionein (see Table II).

---

[3] R. H. P. Bühler and J. H. R. Kägi, *FEBS Lett.* **39,** 299 (1974).
[4] R. D. Anderson, W. P. Winter, J. J. Mager, and I. A. Bernstein, *Biochem. J.* **174,** 327 (1978).
[5] R. J. Vander Mallie and J. S. Garvey, *J. Biol. Chem.* **254,** 8416 (1979).

TABLE I

SULFHYDRYL AND METAL CONTENT OF DEAE-SEPHADEX FRACTIONS (a) AND (d) OF MT-1 AND PEAK OF MT-2[a]

Fraction or peak	Sulfhydryl		Metal				Absorption A_{254}/A_{280}[a]
	mol SH/ mol protein (no reduction)[b]	mol SH/ mol protein (with reduction)[c]	mol Cd^{2+}/ mol protein[d]	mol Zn^{2+}/ mol protein[d]	mol metal/ mol protein	mol SH/ mol metal	
Mt-1(d)	8.72	21.8	3.8	2.8	6.6	3.3	13.5
Mt-2	9.01	23.1	3.4	3.6	7.0	3.3	13.4
Mt-1(a)	0.70	4.0	0.5	0.2	0.7	5.7	N.D.[e]

[a] All calculations were based on an assumed molecular weight of 7000 for Mt.
[b] Determined by the method of G. L. Ellman, Arch. Biochem. Biophys. 82, 70 (1959).
[c] Determined by the method of D. Cavallini, M. T. Graziani, and S. Dupré, Nature (London) 212, 294 (1966).
[d] Measured by atomic absorption spectrophotometry.
[e] N.D., not determined.

Preparation of Antigen and Immunization Procedures

Although we did not pursue a systematic study of all parameters of an immunization schedule, we will outline the methods which, in our hands, were successful for antibody production against the two isolated isoforms of metallothionein. These antibodies demonstrated sufficient avidity to be useful for an RIA.

To begin, let us note that the successful response of only one or two rabbits is necessary for the production of high avidity antiserum to be used in an RIA. Thus, good antibody production in only one or two out of ten rabbits may be considered acceptable. It has been found, in general, that an increase in the immunogenicity of a protein can be achieved by increasing the effective size of the protein. There are several ways in which this has been accomplished. For example, in many instances a low-molecular-weight protein has been covalently coupled, using a bifunctional reagent such as carbodiimide or GDA (glutaraldehyde), to a larger protein thereby producing an immunogenic protein X-protein Y complex.[6,7] Another method found similarly effective in this regard has been the self-polymerization of a protein via a bifunctional reagent.[8,9]

We have obtained antibody against both isometallothioneins by immunization of New Zealand white rabbits with Mt that was cross-linked with GDA to form high-molecular-weight polymers. The GDA polymerized form of Mt proved to be a better immunogen than the monomeric form.[10] Although we found that rabbits immunized with polymerized Mt give a better response than those immunized with monomeric Mt it should be noted that not all rabbits immunized with polymerized Mt produced adequate antibody responses. Therefore, we suggest that several rabbits (>4) be immunized with the use of several booster injections in order to obtain one or more animals with a suitable antibody response.

To polymerize Mt into either soluble or insoluble complexes, the following procedures were used:

Soluble Polymers. To a solution of Mt at a concentration of 4–5 mg/ml in 0.1 M acetate buffer, pH 6.2, was added an amount of 2.5% GDA to obtain a final molar ratio of GDA to protein of 17 to 1 (molecular weight of Mt assumed to be 7000). The solution of GDA and protein was

[6] D. N. Orth, *in* "Methods in Hormone Radioimmunoassay" (B. M. Jaffe and H. R. Behrman, eds.), p. 125. Academic Press, New York, 1974.

[7] B. M. Jaffe and J. H. Walsh, *in* "Methods of Hormone Radioimmunoassay" (B. M. Jaffe and H. R. Behrman, eds.), p. 251. Academic Press, New York, 1974.

[8] H. Daughtery, J. E. Hopper, A. B. McDonald, and A. Nisonoff, *J. Exp. Med.* **130**, 1047 (1969).

[9] A. F. S. A. Habeeb and R. Hiramoto, *Arch. Biochem. Biophys.* **126**, 16 (1968).

[10] R. J. Vander Mallie and J. S. Garvey, *Immunochemistry* **15**, 857 (1978).

allowed to react spontaneously for 1 hr at room temperature. Then sufficient solid lysine was added to obtain a final concentration of 0.10 M lysine. After an additional hour, the reaction mixture was dialyzed against BBS for 24 hr at 4°. The product of this reaction has been shown to be a soluble polymer of Mt with a molecular weight greater than 60,000.[10]

Insoluble Complexes. To a solution of Mt at a concentration of 1–2 mg/ml in 0.10 M acetate buffer, pH 6.2, was added an amount of 2.5% GDA sufficient to obtain an 80 to 1 molar excess of GDA. The reaction was allowed to proceed for 1 hr at room temperature and then lysine was added so that the final concentration of lysine was 0.10 M. At this extreme excess of GDA, insoluble aggregates were formed. The solid precipitate was dispersed mechanically by passing the precipitate through a syringe and employing a series of successively higher gauge needles in order to disperse the precipitate into small particles. The reaction mixture (precipitate plus soluble products) was dialyzed against BBS for 24 hr at 4° and then was used as an immunogen.

Immunization. The following two protocols were used to produce anti-Mt in New Zealand white rabbits:

A. Insolubilized polymeric Mt (1–2 mg) in a volume of 0.75 ml was mixed with cFA in a 1 : 1 ratio (v/v) and emulsified in a double barrel syringe. The emulsion was then injected into multiple sites of a rabbit: subcutaneously into both hind foot pads and into 2–4 sites on either side of the spinal column in the lumbar region, intramuscularly into the flank of each hind leg, and intraperitoneally. Four weeks later the rabbit was reinjected with the same dose of Mt (insolubilized). For the second injection the adjuvant was prepared by mixing together iFA and cFA in a 9 to 1 ratio; otherwise the reimmunization was the same as the primary with the exception that the hind foot pads were never reinjected. The rabbit was boosted similarly at 12 and 28 weeks. Ear vein bleedings of 20–30 ml blood were taken at least once every 2 weeks to obtain serum for testing.

B. Soluble polymers of Mt were used as the immunogen for this procedure. Polymerized Mt, 1–2 mg in 0.75 ml BBS was mixed in a 1 : 1 ratio with cFA, emulsified in a double barrel syringe, and injected into multiple sites as was described above. One week after the initial injection a booster injection of the same dose was given. At the beginning of the fifth week a series of weekly iv injections of polymerized Mt was started and continued for 4 consecutive weeks. The iv injection was administered easily via a 22-gauge needle inserted into the marginal ear vein of the rabbit. The dose for the iv injections was 0.5 mg soluble GDA polymerized Mt in 1 ml buffered saline. (As a prophylactic measure for anaphylactic shock symptoms occurring at the time of intravenous injection, a preinjection of 1 ml Benadryl may be given intramuscularly 1 hr earlier. Alternatively, anaphylactic symptoms may be counteracted when they occur if a syringe is filled with epinephrine and held by an assistant for immediate injection.) The antibody titer increased rapidly after the first injection so it is suggested that the rabbit be bled 5 days after each injection. The antiserum should be tested immediately for precipitating antibody titer and likewise for ABC-20 value in order that the rabbit can be bled out if the antibody has sufficient avidity. "Bleeding out" is recommended for this procedure since the elicitation of antibody may be of short duration.

Detection of Antibody

Screening of antibody for high avidity antibody may be accomplished in several ways. We suggest the following methods.

Interfacial Ring Test. This test is carried out using the procedure as described by Garvey et al.[11] A solution of Mt at a concentration of 1 mg/ml should overlay antiserums from different bleedings. Those antiserums which show a precipitation reaction should be assayed for their antigen binding capacity (ABC).

ABC Assay. In order to obtain a relative measure of the avidity of selected antiserums the antigen ABC value for different antiserums may be obtained using a modified "Farr" ABC test.[12] This assay uses saturated ammonium sulfate to separate antigen–antibody complexes from free antigen, and it can be performed easily with several antiserums of potential usefulness. The following procedure was used:

Serums were serially diluted in borate buffer, pH 8.5, containing 10% NRS. Either Mt-1 or Mt-2 labeled with ^{125}I (see RIA section for details of iodination) was diluted to a concentration of 0.30 μg protein/ml in borate buffer, pH 8.5, containing 1% NRS. Diluted antibody and antigen, 0.5 ml each, were mixed and incubated overnight at 4°. After this incubation, 0.6 ml of chilled saturated ammonium sulfate was added to each assay tube and the precipitation of complexes was allowed to proceed for 30 min at 4°. The assay tubes were then centrifuged at 1800 g for 30 min at 4°. The supernatant was decanted and the pellets were resuspended in 0.5 ml of chilled 40% saturated ammonium sulfate. After recentrifugation at 1800 g for 20 min, the pellets were counted for a minimum of 4000 counts in a gamma counter. From plots of percentage counts precipitated vs log antibody dilution, the dilution of serum that bound 20% of the

[11] J. S. Garvey, N. E. Cremer, and D. H. Sussdorf, "Methods in Immunology," 3rd Ed., p. 313. Benjamin, Reading, Massachusetts, 1977.
[12] R. S. Farr, *J. Infect. Dis.* **103,** 239 (1958).

added antigen was interpolated and used to calculate the antigen binding capacity (ABC-20) of the serum.

Ouchterlony Analysis. An initial examination of the specificity of the antiserum was accomplished by Ouchterlony analysis.[11] Specific reactions between an antiserum and various test proteins were observed. Cross-reactions between Mt-1 and Mt-2 and selected fractions from the Sephadex G-75 chromatographic separation which either contain or do not contain Mt, were examined so that an initial appraisal of the specificity of the antiserum could be made. Ideally, the antiserum should be monospecific and only react with Mt, which is the result we obtained. However, if there are antibodies present against proteins other than Mt, then it must be demonstrated that these antibodies do not interfere with the RIA of Mt. This is possible to demonstrate if the Mt labeled with ^{125}I is pure, since the only reaction measured will be between Mt and anti-Mt.

Development of RIA for Rat Mt

Protein determinations of rat hepatic Mt were estimated by the Folin–Lowry protein assay, using bovine serum albumin as the reference protein, and also by the Nessler reaction. Because the estimated protein content of rat Mt by the Lowry method was higher by a factor of 2.5 than that determined by nitrogen content using the Nessler reagent, values obtained by using the Lowry method were divided by 2.5. (See also Isolation and Purification of Mt). Rat Mt standards, in amounts ranging from 40 to 25,000 pg/100 μl in Tris buffer (0.05 M Tris–HCl containing 0.02% sodium azide, pH 8.0) containing 0.1% gelatin, were frozen until time of use.

Rat Mt was iodinated by a method adapted from Bolton and Hunter,[13] i.e., the protein was reacted initially with N-succinimidyl-3-(4-hydroxyphenyl) propionate (NSHPP); subsequently, ^{125}I was incorporated into the protein using the chloramine-T reaction.

Rat Mt, 75–100 μg as determined with known specific activity resulting from trace-labeling with ^{109}Cd during induction (see Preparation of Antigen) was dissolved in water to a concentration of 2.5 mg protein/ml.

NSHPP (8.0 mg) was dissolved in 50 ml 50% (v/v) toluene and ethyl acetate to obtain a solution containing 160 μg NSHPP/ml. The reaction of NSHPP with Mt utilizes a NSHPP:Mt molar ratio of 20:1; thus, a volume ($\sim$320 μl) of dissolved NSHPP (appropriate in amount for this reaction ratio with the previously determined quantity of 75–100 μg Mt) was placed into a 1.5 ml polypropylene microcentrifuge tube. Thorough drying was achieved at 24–26° using a gentle stream of nitrogen gas delivered from a 9-in. disposable Pasteur pipet with a flame bent tip.

[13] A. E. Bolton and W. M. Hunter, *Biochem. J.* **133**, 529 (1973).

The following steps of the Mt iodination procedure were performed at 4°. The Mt, dissolved earlier, was added to the dried NSHPP. During the next 45 min, the reaction mixture was agitated by either constant, gentle stirring with a glass rod or intermittant low-speed vortex mixing for 2 sec at 5–10 min intervals. At the end of the 45 min reaction, 1 ml 0.05 M sodium phosphate, pH 7.5, was added to the reaction mixture. The NSHPP-Mt was then separated from the nonreacted components on a 1 × 10-cm Sephadex G-25 column and eluted with the same phosphate buffer as added earlier. Gravity flow was used to collect the eluate at 20 drops per tube in 15 tubes.

The protein peak was identified by the [109]Cd-containing fractions (void volume) and the 2–3 fractions (35–50 μg Mt) with the highest radioactivity were pooled into a dialysis bag and concentrated by pervaporation at room temperature to obtain a final concentration of ~0.5 mg Mt/ml. The final recovery of Mt from the complete procedure was ~25% of the initial Mt (based on the recovery of protein).

To 3–5 μg protein in a volume of 5–10 μl (NSHPP-Mt in 0.05 M sodium phosphate, pH 7.5 as determined by [109]Cd specific activity) was added 10 μl 0.25 M sodium phosphate buffer, pH 7.5, followed by the addition of 1 mCi ^{125}I in NaOH. (The molarity of buffer added prior to the ^{125}I should be sufficient to neutralize the base in which the iodine is commercially supplied and specified for "iodination of protein.") Chloramine-T (2.5 mg/ml in 0.05 M sodium phosphate buffer, pH 7.5, 20 μl) was added; and after 60 sec, sodium metabisulfite (1.2 mg/ml in 0.05 M sodium phosphate buffer, pH 7.5, 100 μl) was added to stop the reaction. Labeled protein was separated from other reaction products by filtration on a 1 × 20 cm Sephadex G-25 column. Tris buffer containing 0.1% gelatin was used to equilibrate and to elute the column. A peristalic pump was used to facilitate the flow rate at 1.5–2.0 ml/min; and, 20 fractions of 20 drops per fraction were collected. Those 3–4 tubes of the void volume peak that showed the highest cpm (when each fraction was counted for 0.02 min in a gamma counter) were pooled as the ^{125}I-Mt for use in the assay; this ^{125}I-Mt was diluted to 70,000 cpm per 50 μl (using Tris buffer containing 0.1% gelatin) and frozen until time of use. Maximal precipitable ^{125}I-Mt, as determined in antibody excess, will be 60–65% immediately following iodination (see titration in Fig. 5). This value will be decreased substantially with time; and, consequently the ^{125}I-Mt is usable in the RIA for only 2–3 weeks postiodination.

Rabbit anti-rat Mt (primary antibody) was titrated in order to determine the concentrations usable in the assay. ("Initial" concentration specifies that of the reagent as added to the assay and "final" as its concentration in an assay volume of 500 μl exclusive of the volume of second

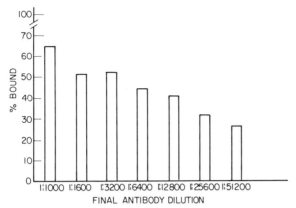

FIG. 5. Final concentrations of 1 : 1000–1 : 51,200 rabbit anti-rat Mt were assayed for capacity to precipitate ^{125}I-Mt. In antibody excess, e.g., 1 : 1000, 65% of the added ^{125}I-Mt was
routinely bound (maximal precipitable) immediately following iodination. A 1 : 25,600 dilution of primary antibody (final concentration) precipitated 50% of the maximal precipitable
^{125}I-Mt. Modified from C. C. Chang, R. J. Vander Mallie, and J. S. Garvey, *Toxicol. Appl.
Pharmacol.* **55,** 94 (1980).

antibody.) To 10 × 75 mm plastic tubes were added the following: 400 μl
Tris buffer containing 0.1% gelatin and 0.02% Na azide; 50 μl primary antibody in dilutions of 1 : 100–1 : 5120 prepared in Tris buffer containing
50 mM EDTA and 1 : 40 NRS; and 50 μl ^{125}I-Mt. After an incubation at 4°
for 6–8 hr, 100 μl second antibody (goat anti-rabbit IgG) was added and
incubation at 4° was continued for 40–48 hr. The precipitates were collected by centrifugation and their ^{125}I content was determined by gamma
counting. The primary antiserum dilution which precipitated 50% of the
maximal precipitable antigen as determined in second antibody excess,
was used in subsequent assays. From Fig. 5, this final dilution is 1 : 25,600
but 1 : 20,000 is used as a conservative final dilution of primary antibody.
NRS may be used in a final concentration range of 1 : 100–1 : 400; however, the optimal NRS concentration is determined from a titration of the
second antibody with dilutions of NRS.

Each preparation of goat anti-rabbit IgG (second antibody) was titered
in a range of dilutions, e.g., 1 : 1, 1 : 2, 1 : 4, 1 : 6, and 1 : 10 prepared in
100 μl Tris buffer and used under assay conditions optimized for 50%
maximal binding by the primary antibody. The next to lowest concentration of second antibody which precipitated completely all antigen–antibody complexes was used in future assays. (If all the above range of concentrations resulted in complete precipitation, then a range of higher
dilutions, e.g., 1 : 12, 1 : 15, and 1 : 20 should be tested.)

The following is a procedure for the optimized RIA for Mt:

Step 1. Standards and unknowns are added to 10×75 mm plastic tubes in a final volume of 400 μl of Tris buffer that contains gelatin and sodium azide. Included in the standard curve are controls for "maximal and nonspecifically bound" labeled antigen where "maximal bound" is the binding of labeled antigen by primary antibody in the absence of unlabeled antigen and "nonspecifically bound" is the binding of labeled antigen in the absence of the primary antibody. The latter is a measurement of binding by components in the assay other than the primary antibody.

Step 2. Rabbit anti-rat Mt in Tris buffer is added in a final volume of 50 μl at that concentration which precipitates 50% of maximal precipitable ^{125}I-Mt; this reagent contains EDTA at 50 mM initial and 5 mM final concentration and NRS at 1:40 initial and 1:400 final concentration. The assay tubes are stored at 4° for 18–20 hr.

Step 3. ^{125}I-Mt in Tris buffer containing 0.1% gelatin is then added in a final volume of 50 μl (approximately 70,000 cpm); and the assay tubes are again stored at 4° for 6–8 hr.

Step 4. (Beginning with this step and continuing to the counting stage, the procedure should be performed at 4° and with precooled solutions.) Goat anti-rabbit IgG in 100 μl Tris buffer is added to each tube in a concentration to completely precipitate all antigen–antibody complexes and the tubes are stored for 48 hr at 4°.

Step 5. Following the addition of 2 ml Tris buffer, the precipitates are collected by centrifugation at 4500–5000 g. The supernatants are decanted and discarded as liquid radioactive waste. The wall of each tube is wiped with a cotton tip applicator stick and the precipitates are counted in a gamma counter.

Validation of RIA for Rat Mt

The sensitivity of the RIA is primarily dependent upon the avidity of the primary antibody. A standard curve, consisting of Mt concentrations less than the lower detection limit through Mt concentrations greater than the upper detection limit, is assayed routinely. That portion of the sigmoid curve shown in Fig. 6 which demonstrates a linear dose–response relation is the usable range of Mt concentration for quantitation of unknowns. The present RIA for rat Mt has a routine lower limit of 200 pg and a linear range of 200–10,000 pg. This range can be extended at both lower and upper limits if a logit transformation is used (see section on data handling).

The specificity of the antibody (see Detection of Antibody) for antigen binding can be verified by assaying various concentrations of unrelated substances, e.g., proteins that may be found ubiquitous in the presence of

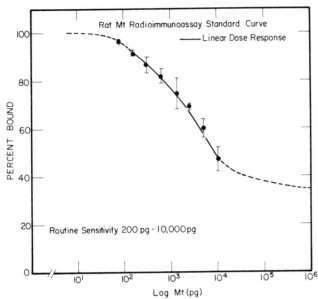

FIG. 6. A concentration range of rat Mt (78–20,000 pg) was assayed. Log Mt vs percent ^{125}I-Mt bound is shown. Unknowns were assayed in volumes such that the total Mt quantitated in the assay volume was within this range.

antigen. Nonmetalloprotein, β_2-microglobulin, C-reactive protein and bovine serum albumin were assayed in varying amounts and none of these proteins inhibited significantly the binding of labeled Mt to the primary antibody (see Table II).

The accuracy by which the RIA predicts a quantity of sampled Mt can be established by addition of a known quantity of pure Mt (e.g., 1.0 ng as determined by Lowry protein assay) to the assay for quantitation from the standard curve. Moreover, pure Mt may be added to experimental samples already measured for Mt; and the repeated assay of these samples allows assessment of possible interference by components within the unknown samples (see Utilization of Mt RIA that follows).

Utilization of Mt RIA

Binding curves obtained from assays in which unlabeled Mt of varying species of origin is present as competitor with ^{125}I rat Mt for binding with the primary antibody allows the degree of cross-reactivity with rat Mt to be assessed. Binding of the unlabeled rat Mt and likewise different species of Mt, e.g., hamster Mt (ovarian cell line), equine Mt (hepatic), and human Mt (hepatic) were shown to cross-react completely with radiola-

TABLE II
PROTEINS SHOWING NO CROSS-REACTION WITH MT

Exp	Mt	Test protein	ng[a]
I	Mt-1		2.1×10^2
		BSA	1.0×10^5
II	Mt-1		1.3×10^2
		C-reactive protein	1.3×10^4
III	Mt-1		1.3×10^1
		β_2-microglobulin	2.0×10^4
IV	Mt-2		8.0×10^0
		Nonmetalloprotein	8.0×10^2

[a] Amount of competing Mt which produced 50% inhibition of the binding of anti-Mt with ^{125}I rat Mt or the maximal amount of test protein used. The amount given for the test protein produced no inhibition and is the highest amount tested in a particular experiment. Exp I and Exp II adapted from R. J. Vander Mallie and J. S. Garvey, *Immunochemistry* **15**, 857 (1978) with permission from the publisher. Exp III and Exp IV adapted from R. J. Vander Mallie and J. S. Garvey, *J. Biol. Chem.* **254**, 8416 (1979) with permission from the publisher.

beled rat Mt (see Fig. 7). Additionally, Mt, differing in metal of induction, i.e., Cd (Mt-1 and Mt-2), Zn (Mt-A and Mt-B), as well as the apoprotein, showed complete cross-reactions.[5,10]

The complete cross-reactivity of rat and human species of Mt (see Fig. 7) suggested the clinical applicability of these reagents, i.e., rat Mt and its

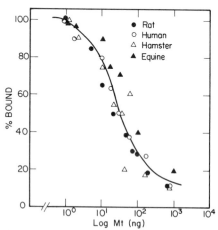

FIG. 7. Cross-reactivity of Mt from various species. Nanogram concentrations of pure Mt (equine, hamster, human, and rat) were compared for their relative binding ability to anti-rat Mt. All species of Mt tested showed complete cross-reactivity with rabbit anti-rat Mt. Modified from R. J. Vander Mallie and J. S. Garvey, *J. Biol. Chem.* **254**, 8416 (1979), with permission of the publisher.

specific antibody for the detection of human Mt in physiological fluids. In the assay for human Mt, similar volumes of control as well as experimental samples of urine and plasma are assayed in order to determine nonspecific binding attributable to these fluids. When Mt was assayed in human serum and urine it was found that sample volumes of 20 μl or less caused no significant deviation from the standard curve and required no correction for nonspecific binding. However, 50 μl or more sample volume resulted in significant deviation from the standard curve and required correction for nonspecific binding (see Fig. 8 and Ref. 14 for the detailed description of the validation of RIA for human Mt and Ref. 15 for the utilization of the assay in an assessment of an occupationally exposed population).

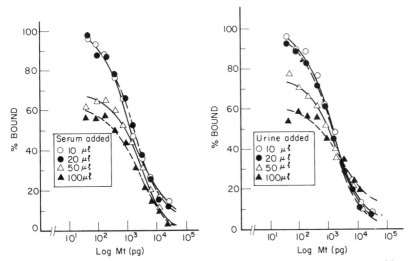

FIG. 8. The effect of human physiological fluid proteins on the RIA was assessed by assaying human Mt standard in the presence of 10, 20, 50, and 100 μl control serum or urine. In the presence of 10 or 20 μl serum or urine, the slope of the standard curve is not significantly different from that of the slope obtained in the presence of buffer only. In the presence of 50 or 100 μl serum or urine, the slopes of the curves do differ significantly from the standard curve in the presence of the buffer only ($p < 0.05$). Reproduced from C. C. Chang, R. J. Vander Mallie, and J. S. Garvey, *J. Toxicol. Appl. Pharmacol.* **55**, 94 (1980), with permission of the publisher.

[14] C. C. Chang, R. J. Vander Mallie, and J. S. Garvey, *Toxicol. Appl. Pharmacol.* **55**, 94 (1980).

[15] C. C. Chang, R. Lauwerys, A. Bernard, H. Roels, J. P. Buchet, and J. S. Garvey, *Environ. Res.* **23**, 422 (1980).

Data Handling of Unknown Samples

Unknown samples are assayed in aliquots such that the amount of Mt in the assay tube is within the linear range of the standard curve. The standard curve may be expressed as "percentage bound vs log quantity of Mt," where percentage bound is $100 \, (B/B_0)$. B is counts bound by either a known amount of Mt (standard curve) or by an unknown sample of Mt, and B_0 is "maximal bound" (labeled Mt) in the absence of unlabeled Mt. Both B and B_0 have been corrected for nonspecifically bound (other binding than by the primary antibody).

Another useful method in developing a standard curve is to plot logit B/B_0 against log quantity of antigen, where logit B/B_0 is $\log [(B/B_0)/(1 - B/B_0)]$; and, as mentioned above, B and B_0 are corrected for "nonspecifically bound" prior to the logit transformation.

Acknowledgments

Previously published material is based upon research supported by National Science Foundation Grant No. ENV 77-07896. Any findings, conclusions, or recommendations expressed in this publication are those of the authors and do not necessarily reflect the view of the National Science Foundation. Previously unpublished material is from investigations supported by Grant No. ES 01629-01A2 from the National Institute of Environmental and Health Sciences.

[10] Radioimmunoassay of Calmodulin

By James G. Chafouleas, John R. Dedman, and Anthony R. Means

Introduction

Calmodulin is a multifunctional Ca^{2+}-binding protein involved in the regulation of cyclic nucleotide and glycogen metabolism, Ca^{2+}-dependent protein phosphorylation, cell division, and cell motility (for review see 1–4). The importance of this protein is evidenced by its ubiquity in eukaryotes and the highly conserved nature of its primary sequence. The

[1] A. R. Means and J. R. Dedman, *Nature (London)* **285**, 73 (1980).
[2] W. Y. Cheung, *Science* **207**, 19 (1980).
[3] J. H. Wang and D. M. Waisman, *Curr. Top. Cell. Reg.* **15**, 47 (1979).
[4] C. B. Klee, T. H. Crouch, and P. G. Richman, *Ann. Rev. Biochem.* **49**, 487 (1980).

method most commonly used for the determination of calmodulin levels relies on its ability to activate a calmodulin-dependent form of cyclic nucleotide phosphodiesterase. Such an assay is capable of measuring only the "biologically active" calmodulin present. Furthermore since the assay requires Ca^{2+}-dependent binding, interference from other calcium binding proteins may yield artificially lower values than actually exist. Such limitations, however, are not associated with radioimmunoassays, for detection is predicated on the highly specific interactions between antibody and antigen. The following is a discussion on the development of a radioimmunoassay for calmodulin.

Purification of Calmodulin

Because of its highly conserved primary sequence, any tissue which affords good yields of calmodulin may be used as a source for the pure protein. Although calmodulin can be purified to homogeneity from rat testis by the following technique with a yield of 200 mg/kg (wet wt), use of other tissues may require modifications to this scheme.

1. Homogenization and Heat Treatment

One to two kilograms of frozen rat testes are thawed in deionized water (1:1, w/v) and homogenized at 0° with a Brinkman Polytron (3 min bursts with a PT-35 ST generator, setting 6). Aliquots (200 ml) of the homogenate are rapidly brought to a boil (approximate time required: less than 1 min) in a microwave oven, then immediately cooled (to 5°) by immersion into a methanol: dry ice bath. Once cooled, the precipitate is removed by centrifugation at 20,000 g for 20 min. The resultant supernatant solution is adjusted to a final concentration of 10 mM imidazole, 1 mM EGTA pH 6.1.

2. Ion Exchange Chromatography

The protein solution (approximately 1500 ml/kg testes) is applied to a Whatman DE-52 column (2.5 × 40 cm) previously equilibrated with 10 mM imidazole, 1 mM EGTA pH 6.1. The column is washed with the same buffer until the A_{280} < 0.050 at which time a 2000 ml linear gradient (0 to 0.4 M NaCl) is initiated. Fractions (20 ml) are then tested for calmodulin activity by the phosphodiesterase activation assay.[5] The calmodulin-containing fractions are pooled, dialyzed against 5 mM ammonium carbonate, and lyophilized.

[5] E. G. Beale, J. R. Dedman, and A. R. Means, *Endocrinology* **101**, 1621 (1977).

3. Gel Filtration

The lyophilized sample from Step 2 is resuspended in approximately 20 ml of 10 mM imidazole, 1 mM EGTA, 100 mM NaCl, 0.02% NaN$_3$ pH 6.1 and chromatographed through an Ultragel AcA44 (LKB) column (2.5 × 92 cm) previously equilibrated with the same buffer. Fractions are tested for calmodulin activity as before (Step 2) and those calmodulin-containing fractions which have an $A_{260}/A_{280} = 1$ are pooled. The overall yields and fold purifications for each step starting from 2 kg of rat testes are summarized in the table.

Radioiodination of Calmodulin

While several different methods are available for labeling a protein with ^{125}I, the conjugation method first described by Bolton and Hunter[6] is most amenable for labeling calmodulin to high specific radioactivity with essentially no loss in immunological or biological activity.[7]

One millicurie of ^{125}I-labeled Bolton–Hunter reagent, (^{125}I-labeled N-succinimidyl-3-(4-hydroxyphenol) propionate (1600 Ci/mmol supplied in benzene from Amersham) is evaporated to dryness under a stream of nitrogen in a conical tube. The walls of the tube are washed with 200 μl of benzene (desiccated with calcium hydride) and the benzene removed as before. Ten microliters of 0.5 mg/ml calmodulin in 125 mM borate buffer pH 8.4 is applied directly to the residue at the bottom of the conical tube and incubated for 30 min at 4°. The reaction is then terminated by the addition of 150 μl of 200 mM glycine in the same buffer and a subsequent 15 min incubation at 4°. The sample is quantitatively removed and applied to a Sephadex G-25 (Pharmacia) column (0.7 × 20 cm) previously equilibrated with 0.05 M phosphate, 0.05% NaN$_3$, 0.25% (w/v) gelatin, 0.1 M

PURIFICATION OF CALMODULIN FROM 2 kg OF RAT TESTES

Procedure	Calmodulin (units)	Total protein (mg)	Specific activity (units/mg)	Purification (fold)	Yield (%)
Homogenate	6.0 × 10⁶	181,400	33		100
Heat treatment	5.6 × 10⁶	2,508	2,233	68	93
DE-52	4.8 × 10⁶	213	22,535	683	80
Ultragel AcA44	4.5 × 10⁶	206	21,845	662	75

[6] A. E. Bolton and W. M. Hunter, *Biochem. J.* **133,** 529 (1973).
[7] J. G. Chafouleas, J. R. Dedman, R. P. Munjaal, and A. R. Means, *J. Biol. Chem.* **254,** 10262 (1979).

NaCl (pH 7.5). The reaction tube is washed with 50 μl of 200 mM glycine in 125 mM borate buffer pH 8.4 which is added to the column. The sample is chromatographed and 30 1-ml fractions collected. Five-microliter aliquots of each fraction and the reaction vessel are counted for ^{125}I in a Searle gamma counter. ^{125}I-labeled calmodulin eluted from the column in the V_0 is readily separated from the protein-free ^{125}I-labeled Bolton–Hunter reagent eluting in the V_I. In order to determine the total ^{125}I conjugated to the 5 μg of calmodulin, the total cpm present in the V_0 is corrected for recovery of ^{125}I-labeled calmodulin on the G-25 column (routinely 90%). Since only protein bound ^{125}I-labeled Bolton–Hunter reagent will remain attached to the reaction vessel (in the case of calmodulin $<$ 1%) the cpm present in the vessel are added to the corrected V_0 value. This total radioactivity is then divided by the total mass of 5 μg to generate the total cpm/μg calmodulin. This procedure routinely conjugates 1.5 mol of ^{125}I/mol of protein resulting in a specific activity of approximately 2400 Ci/mmol or 1.56 $\times$ 10^8 cpm/μg.

Production of Antiserum and Purification of Antibody

The specificity of any radioimmunoassay is dependent on the quality of the antiserum available. While there are various methods available to make proteins more antigenic, these procedures generally result in a loss in sensitivity to the native form of the protein. For this reason we have endeavored to generate antibodies against the native protein. While we initially reported the production of monospecific antibodies in goat to native calmodulin,[8] the goat antiserum was inappropriate for use in a radioimmunoassay. When sheep were used instead of goats, antisera were produced which were satisfactory for use in the development of a radioimmunoassay for calmodulin.[7] The following is the protocol used in sheep.

1. Immunization Schedule

Electrophoretically pure calmodulin (10 mg in 1 ml of 0.9% saline) is emulsified with an equal volume of complete Freund's adjuvant. Ten milligrams protein is injected subcutaneously at multiple sites in the lower back of an adult sheep. Subsequent booster injections of 2 mg calmodulin each are administered at 42, 56, 70 days and every 60 days thereafter. Test bleeds are obtained at 2-week intervals following the administration of each booster injection. Upon removal, the blood is allowed to clot, and the serum obtained by centrifugation at 6000 g for 20 min at 4°. The serum

---

[8] J. R. Dedman, M. J. Welsh, and A. R. Means, *J. Biol. Chem.* **253**, 7515 (1978).

from each bleed (approximately 500–700 ml) is divided into 100-ml aliquots and can be stored at −20° for at least 18 months.

2. Purification of Antibody

Serum (100 ml) is thawed at 4° and brought to 50% saturation with solid ammonium sulfate. The suspension is slowly stirred for 2 hr at 4° followed by centrifugation at 10,000 g for 30 min. The precipitate is resuspended in approximately 50 ml of 125 mM borate buffer, 75 mM NaCl pH 8.4 and dialyzed extensively against several changes of the same buffer for 18 hr at 4°. Ten milliliters of this solution is then applied to a 10 mg calmodulin-Sepharose affinity column prepared by the method of Dedman et al.[8] The column is washed with the borate buffer until the A_{280} < 0.005 at which time 200 mM glycine pH 2.7 is applied to the column. Figure 1 demonstrates a typical column profile of such an affinity run. The fractions which are eluted with the glycine buffer and exceed an A_{280} of 0.100 are pooled and dialyzed against 125 mM borate, 75 mM NaCl pH 8.4. This pooled sample consists of antibodies to calmodulin to be used in the radioimmunoassay. The specificity of the affinity-purified antibodies is presented in Fig. 2 which demonstrates that only calmodulin is immunnoprecipitated from a total in vitro translation product. Those fractions which originally passed through the calmodulin affinity column (affinity flow through) should be pooled, reapplied to the affinity column, and processed as before. This procedure should be repeated until no protein is eluted from the column by the glycine buffer (usually three cycles). The

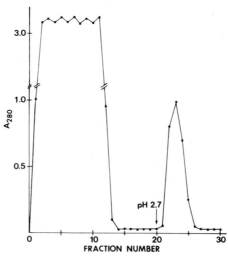

FIG. 1. Purification of sheep anticalmodulin by antigen-affinity chromatography.

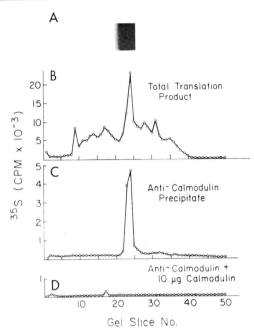

FIG. 2. Analysis of sheep anticalmodulin immunoprecipitate by SDS–polyacrylamide gel electrophoresis. Total poly(A)-containing RNA from the electroplax of *E. electricus* was translated in the presence of [^{35}S]methionine in a rabbit reticulocyte lysate cell free translation system by the procedure of Means *et al.* [A. R. Means, J. P. Comstock, G. C. Rosenfeld, and B. W. O'Malley, *Proc. Natl. Acad. Sci. U.S.A.* **69**, 1146 (1972)] and the labeled translation product immunoprecipitated as described by Chafouleas *et al.*[7] (A) Authentic rat testis calmodulin; (B) total translation product; (C) immunoprecipitate by anticalmodulin (2 μg); (D) immunoprecipitate by 2 μg anticalmodulin plus 10 μg unlabeled rat testis calmodulin.

final affinity flow through is devoid of antibodies to calmodulin and can be used as control serum.

Development of the Radioimmunoassay

1. Anticalmodulin Dilution

Since the effective antibody titer fluctuates from one preparation to the next, it is important to generate an antibody dilution curve for each preparation of calmodulin affinity-purified antibody. In so doing, the correct concentration of affinity-purified antibody can be determined for use in the radioimmunoassay.

A series of dilutions are made from the stock of calmodulin affinity-purified antibody (starting concentration is usually between 200 to 300 μg

IgG/ml). The dilutions are prepared with an appropriate amount of affinity run through as carrier to assure that a constant IgG concentration of 800 µg/ml is maintained throughout the dilution series. The assay is performed in 12 × 75 mm culture tubes by first adding 300 µl of radioimmunoassay buffer (125 mM borate, 1 mM EGTA, 75 mM NaCl, 20 µg/ml bovine serum albumin pH 8.4), followed by the addition of 100 µl of the same buffer containing 10,000 to 15,000 cpm ^{125}I-labeled calmodulin. One hundred microliters of the appropriate antibody dilution (in duplicate) is then added and the samples vortexed. Nonspecific background is determined by substituting 100 µl of the affinity flow through (800 µg/ml) for anticalmodulin. The samples are incubated for 18 hr at 25°. The length of incubation should be tested since we have found that 90% of the total binding observed at 18 hr has occurred after only 5 hr. Although the sheep anticalmodulin preparations are precipitating, separation of bound from free ^{125}I-labeled calmodulin is facilitated by the addition of protein A.[7] We use the commercial fixed cell preparation of *Staphylococcus aureus*, Pansorbin, available from Calbiochem. Ten microliters of a 10% solution of Pansorbin is added, the assay tubes are vortexed and incubated for 30 min at 25°. The tubes are centrifuged at 1000 g for 10 min in a Beckman TJ6 centrifuge. The supernatant solution is decanted and while in the inverted position, the mouth of the tube is blotted on absorbant paper. The pellet is washed twice with 1 ml of radioimmunoassay buffer and processed as before. The relative amounts of ^{125}I present in the final pellets is determined on a Searle gamma counter. Figure 3 demonstrates a

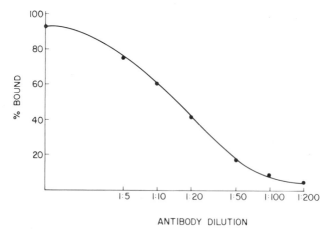

Fig. 3. Sheep anticalmodulin dilution curve. Calmodulin affinity-purified sheep anticalmodulin was diluted (starting with 300 µg of anticalmodulin/ml) and tested for its ability to bind ^{125}I-labeled calmodulin.

typical antibody dilution experiment. The data are obtained by subtracting the nonspecific background (routinely less than 100 cpm) and dividing each duplicate set by the total cpm of ^{125}I-labeled calmodulin initially added to each assay tube. The antibody dilution which precipitates between 30 to 40% of the total ^{125}I-labeled calmodulin is used for all subsequent assays.

2. Preparation of Standards

Initially standards should be prepared from homogeneous calmodulin. Since calmodulin can be readily purified to homogeneity the preparation is relatively simple. A starting stock solution of calmodulin is made in radioimmunoassay buffer devoid of bovine serum albumin at a concentration of 500 μg/ml. This final concentration can be routinely verified spectrophotometrically if rat testis calmodulin is used since $\epsilon = 0.21$ (1 mg/ml). From this stock a series of dilutions are made using the radioimmunoassay buffer. The dilution series should initially range from 10 μg/ml to 1 ng/ml. However, once the limits of the assay are determined a more limited series of concentrations is required.

3. Sample Preparation

Tissues to be assayed are collected on ice and homogenized in 125 mM borate buffer, 5 mM EGTA, 75 mM NaCl pH 8.4 (5 : 1, v/w) with a Brinkman polytron homogenizer (5 × 30 sec bursts using a PT-10 generator at a setting of 4.5). Tissue culture cells as well as other isolated cell preparations can be homogenized by sonication in the same buffer using a Brownwill Biosonik IV Sonicator. Following homogenization, appropriate aliquots are removed for total protein and DNA determinations. Homogenates are brought to a rapid boil in a microwave oven (less than 1 min) and rapidly cooled to 4° by immersion into a methanol : dry ice bath. The heat-treated samples are centrifuged at 10,000 g for 30 min and the supernatant solutions are used for the assay. Preparation of samples by this method routinely yields quantitative recovery of calmodulin. When samples cannot be heat-treated because of other considerations 0.1% Triton X-100 should be included in the homogenization buffer to assure total release of membrane bound calmodulin. Prior to assay, samples should be diluted to a calmodulin concentration of 50 to 100 ng/ml. Since calmodulin levels are routinely 0.1 to 1% of the total protein this can be accomplished to a first approximation by considering the total protein content of each sample.

4. Radioimmunoassay

Initially 200 to 300 μl of radioimmunoassay buffer is added to the entire series of 12 × 75 mm culture tubes used in the assay. Since the final reaction volume is 500 μl, the actual volume of buffer per tube will be dependent on the total volume of the other components added. Both standards and samples are added to their appropriate tubes in 100 μl volumes, followed by the addition of 100 μl of a fixed number of cpm ^{125}I-labeled calmodulin (10,000–15,000) in radioimmunoassay buffer. Finally, the reaction is initiated by the addition of 100 μl of the appropriately diluted anticalmodulin (the amount required to achieve 30–40% binding of the tracer). Nonspecific background is determined by substituting 100 μl of the affinity run through (800 μg/ml) for the anticalmodulin. Samples are processed as described above for the antibody dilution curve and total radioactivity in each tube obtained. Figure 4 depicts a typical standard curve generated by the above procedure. The highly conserved nature of this protein is clearly demonstrated in this figure, since the same competition curve is generated by calmodulins purified from such diverse phyla as higher plants (the peanut) to mammals (cow brain). In addition, the same competition curves are generated whether calmodulin is in its purified state or in a heat-treated tissue extract. Because of this fact, standards

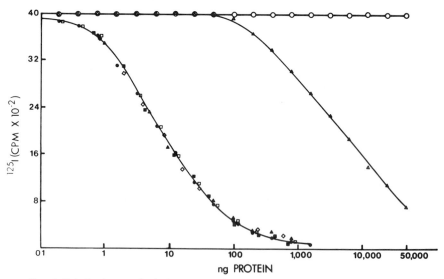

FIG. 4. Relative immunological cross-reactivity of homologous calcium binding proteins. Homogeneous calmodulins from rat testis (●——●), bovine brain (□——□), A. hypogaea (■——■); troponin C (△——△), parvalbumin (○——○); heat-treated supernatant solutions from the electroplax of E. electricus (◇——◇) and C. reinhardii (▲——▲).

made from heat-treated extracts may be substituted for the pure protein standards once their calmodulin levels have been determined with the pure protein standards. Since the determination of calmodulin levels by the radioimmunoassay is predicated on comparing the degree of competitive binding by a heat-treated sample to a standard curve, it is important to ascertain the degree to which other proteins may cross-react with the antibody. As shown in Fig. 4, only troponin C, a protein with 50% sequence homology and biological cross-reactivity to calmodulin, exhibits immunological cross-reactivity (at a 665-fold greater concentration than calmodulin). The data presented in this figure were calculated by the log–logit computer program based on the assay statistics derived by Midgley et al.[9] and Duddleson et al.[10] A typical assay has an intraassay variability of 3% and an interassay variability of 5%. The limit of detection is approximately 15 pg and the assay sensitivity is 115 pg. Nonspecific background is generally less than 100 cpm but varies depending on the gamma counter used.

Conclusion

The development of a calmodulin radioimmunoassay similar to the one described in this chapter is straightforward. Large quantitites of the pure protein are easily obtained for immunization and affinity resin production. In addition, calmodulin can be radiolabeled to high specific radioactivity. Because of the highly conserved nature of calmodulin, the major difficulty is in the development of the antibody. The procedure for antibody production outlined above has proven to be successful in sheep; however, species differences have been observed, and use of other animals may require modifications of this scheme. Once developed, the radioimmunoassay affords a simple, but sensitive means of measuring total calmodulin levels in essentially all eukaryotic cells.

Acknowledgments

We are grateful to Mr. Charles R. Mena for his excellent technical help as well as Mr. Mick Scheib for his help in the preparation of the manuscript. We would also like to thank Dr. Howard Kirchick for his invaluable aid in the use of the log–logit computer program.

[9] A. R. Midgely, Jr., G. D. Niswender, and R. W. Rebar, Acta Endocrinol. Suppl. 142, 247 (1969).
[10] W. G. Duddleson, A. R. Midgely, Jr., and G. D. Niswender, Comput. Biomed. Res. 5, 205 (1972).

[11] Immunoassay of Ferritin in Plasma

By JUNE W. HALLIDAY

The presence of the iron storage protein ferritin in human serum (ferritin may be measured in either plasma or serum by the methods described in this article and the words plasma and serum are therefore interchangeable) in certain pathological conditions, such as acute liver disease, was demonstrated by Riessman and Dietrich in 1956.[1] However the immunoprecipitation technique which they used was sufficiently sensitive only for the detection of ferritin concentrations greater than 1000 μg/liter and permitted the detection of only those ferritins which contained a high concentration of iron in the molecule. Some 10 years elapsed before Aungst[2] showed that a more sensitive immunological technique, a modified Ouchterlony technique of double diffusion in agar gel using antiserum to human ferritin, permitted the detection of ferritin in body fluids at concentrations of approximately 100 μg/liter. Beamish and his colleagues[3] used a further refinement of counterimmunoelectrophoresis against antisera to ferritin derived from human liver to detect ferritin in serum.

The development of a sensitive immunoradiometric technique by Addison et al.[4] in 1972 first permitted the measurement of nanogram quantities of ferritin. This resulted in their discovery that ferritin is a constituent of all normal human serum and that its concentration in serum is directly proportional to the total body iron stores, i.e., the concentration is low (less than 10 μg/liter) in iron deficiency and high (often greater than 1000 μg/liter) in uncomplicated iron overload. The realization that the determination of serum ferritin concentration is a useful diagnostic technique in clinical medicine has resulted in the publication of numerous methods and modifications thereof and in attempts to improve the sensitivity and specificity of these methods. In the present chapter several immunological methods for the determination of serum ferritin concentration are described in detail and the major problems associated with each type of assay are discussed. Since some of the problems associated with its measurement relate to the structure of the ferritin molecule a short discussion of the structure is included.

[1] K. R. Reissman and M. R. Dietrich, J. Clin. Invest. 35, 588 (1956).
[2] C. W. Aungst, J. Lab. Clin. Med. 71, 517 (1966).
[3] M. R. Beamish, P. Llewellin, and A. Jacobs, J. Clin. Pathol. 24, 581 (1971).
[4] G. M. Addison, M. R. Beamish, C. N. Hales, M. Hodgkin, A. Jacobs, and P. Llewellins, J. Clin. Pathol. 25, 326 (1972).

Ferritin and Isoferritins

Unlike many other proteins for the measurement of which radioimmunoassays are used, ferritin is a large protein of MW approximately 450,000. It is a spherical protein that consists of 24 subunits and is capable of binding up to 4000 atoms of iron per molecule, the iron being located within the hollow center of the molecule. Ferritin is unusally stable to heat (80°), a fact which is used to assist in its purification. The ferritin molecule is also known to form polymers. It has been demonstrated recently that some ferritins such as serum ferritin, contain carbohydrate.[5-7] It has also been demonstrated that, unlike tissue ferritins which contain a considerable proportion of iron within the molecule, circulating plasma ferritin contains relatively little iron. Even in disorders associated with gross iron overload the iron content of serum ferritin is low (0.023 to 0.067 μg Fe/μg protein).[8]

A heterogeneity in the structure of ferritin has been observed using several different techniques. After polyacrylamide gel electrophoresis differences in electrophoretic mobility are apparent in ferritins derived from different organs in the same subject as well as in neoplastic tissues. Further, proteins of differing isoelectric points have been observed using the technique of isoelectric focusing. These proteins have been termed isoferritins. It is now known that each tissue contains a number of different isoferritins and that the isoferritin profile of tissue ferritin is organ-specific.[9] It has also been demonstrated that changes in the isoferritin profile of serum and tissue ferritins occur in certain clinical conditions,[10] notably in iron overload and malignancy.

The above studies have clearly demonstrated that differences in iron content occur in tissue and serum ferritins, and also that ferritin is not a structurally homogeneous protein and is capable of existing in oligomeric forms. Immunological differences in the various isoferritins or in oligomeric forms would be expected to give rise to serious methodological problems. Thus if standard ferritins used in assays of serum ferritin failed

[5] M. Worwood, M. Wagstaff, B. M. Jones, S. Dawkins, and A. Jacobs, in "Proteins of Iron Metabolism" (E. B. Brown, P. Aisen, J. Fielding, and R. Crichton eds.), p. 79. Grune & Stratton, New York, 1977.

[6] D. J. Lavoie, D. M. Marcus, K. Ishikawa, and I. Listowsky, in "Proteins of Iron Metabolism" (E. B. Brown, P. Aisen, J. Fielding, and R. Crichton eds.), p. 71. Grune & Stratton, New York, 1977.

[7] J. W. Halliday, U. Mack, and L. W. Powell, Br. J. Haematol. 42, 535 (1979).

[8] M. Worwood, W. Aherne, S. Dawkins, and A. Jacobs, Clin. Sci. Mol. Med. 48, 441 (1975).

[9] L. W. Powell, E. Alpert, J. W. Drysdale, and K. S. Isselbacher, Br. J. Haematol. 30, 47 (1975).

[10] J. W. Halliday and L. W. Powell, in "Progress in Hematology" (E. B. Brown ed.), Vol. XI, p. 229. Grune & Stratton, New York, 1979.

to detect some of the isoferritins present in abnormal sera, considerable underestimation of the serum ferritin concentration could result.

Most ferritin assays now used routinely make use of ferritin purified from either human liver or spleen. These ferritins consist primarily of the more basic isoferritins. The antisera used are raised against these same tissue ferritins. It is now known that differences in immunological reactivity do exist in isoferritins[11-13] and a more detailed discussion of these factors and their effects on the serum ferritin assay follows later in this chapter.

Immunoassays

Four main kinds of assay have been used to determine serum ferritin concentration and will be included in this discussion. These are as follows:

1. Immunoradiometric assays (IRMA): (a) IRMA based on the method of Addison et al.[4]; (b) 2-site "sandwich" IRMA based on the methods of Miles et al.[14] and Halliday et al.[15]

2. Radioimmunoassay (RIA) based on the methods of Luxton et al.[16] and Barnett et al.[17]

3. Solid-phase enzyme-labeled immunosorbent assay (ELISA) based on the method of Anaokar et al.[18]

Methods

Methods for the preparation of reagents which are used in all assay types have been grouped for convenience at the beginning and before specific assay techniques are considered. Similarly, methods for the assessment of protein concentration and purity of standard preparations are also included here.

---

[11] M. Worwood, B. M. Jones, and A. Jacobs, *Immunochemistry* **13**, 477 (1976).

[12] P. Arosio, M. Yakota, and J. W. Drysdale, *Cancer Res.* **36**, 1735 (1976).

[13] J. T. Hazard, M. Yakota, P. Arosio, and J. W. Drysdale, *Blood* **49**, 139 (1977).

[14] L. E. M. Miles, D. A. Lipschitz, C. P. Bieber, and J. D. Cook, *Anal. Biochem.* **61**, 209 (1974).

[15] J. W. Halliday, K. L. Gera, and L. W. Powell, *Clin. Chim. Acta* **58**, 207 (1975).

[16] A. W. Luxton, W. H. C. Walker, J. Gauldie, M. A. M. Ali, and C. Pelletier, *Clin. Chem.* **23**, 683 (1977).

[17] M. D. Barnett, Y. B. Gordon, J. A. L. Amess, and D. L. Mollin, *J. Clin. Pathol.* **31**, 742 (1978).

[18] S. Anaokar, P. J. Garry, and J. C. Standefer, *Clin. Chem.* **25**, 1426 (1979).

Purification of Tissue Ferritin

a. Chromatographic Method [modification of the method of Drysdale and Munro[19] *as described by Halliday et al.*[15]*]*

Ferritin is prepared from either normal human liver or normal human spleen tissue obtained at necropsy not more than 12 hr after death. The tissue may be either processed on the same day or stored at $-20°$ until required.

The tissue is homogenized in 2 volumes of distilled water and heated to $70-75°$ for 5 min with vigorous continuous stirring, cooled rapidly to $4°$ in ice and centrifuged at 1500 g for 20 min. The ferritin is precipitated from the supernatant by 50% saturated (31% w/v) $(NH_4)_2SO_4$ and allowed to stand overnight at $4°$. After centrifugation at 1500 g for 30 min the precipitate is resuspended in 0.02 M phosphate buffer pH 7.4 containing 0.1 M sodium chloride (PBS) and dialyzed against the same buffer for 48 hr at $4°$. The dialyzed solution is further centrifuged at 30,000 g for 30 min and the pellet discarded. The supernatant containing ferritin is then further centrifuged at 100,000 g for 2 hr. The resulting pellet is resuspended in PBS and subjected to gel chromatography on Sepharose 6B (column size 2.6 × 90 cm) eluting with PBS. (The column may be previously standardized with a small amount of ^{125}I-labeled purified ferritin.) The protein concentration of the column eluate is monitored by spectrophotometric measurement of absorption at 280 nm and the ferritin-containing fractions are combined and concentrated on a stirred ultrafiltration cell using a membrane of MW exclusion approximately 30,000.

The concentrated ferritin solution is then applied to a Sephadex G-200 column (1.6 × 90 cm) and eluted with PBS.

The ferritin-containing fractions are again pooled and the ferritin further purified by preparative electrophoresis on polyacrylamide gel followed by electrophoretic elution of the monomer ferritin band into (0.065 M) Tris–borate buffer pH 8.6.

b. Crystallization Method [modification of the method of Mazur and Shorr[20]*]*

Human liver or spleen tissue is obtained and homogenized in 2 volumes distilled water as for method (a) (above).

The homogenate is centrifuged at 10,000 g for 20 min and the supernatant heated at 70 to 75° for 5 min with vigorous stirring.

The homogenate is again centrifuged at 10,000 g for 20 min.

[19] J. W. Drysdale and H. N. Munro, *Biochem. J.* **95,** 851 (1965).
[20] A. Mazur and E. Shorr, *J. Biol. Chem.* **176,** 771 (1948).

After the pH of the supernatant is adjusted to 4.6 with 50% acetic acid, it is allowed to stand 24 hr at 4° and further centrifuged at 10,000 g for 20 min.

The ferritin is crystallized by the addition of 4 g cadmium sulfate (3 $CdSO_4\cdot8H_2O$) per 100 ml of supernatant. The solution is allowed to stand at 4° for a minimum of 24 hr.

The crystals are recovered by centrifugation at 1500 g for 20 min and washed three times in 4% (w/v) cadmium sulfate.

The crystals are redissolved in a minimum volume of 2% (w/v) $(NH_4)_2SO_4$ and recrystallized 3 more times.

The recrystallized ferritin is precipitated by addition of an equal volume of saturated $(NH_4)_2SO_4$ and allowed to stand overnight at 4°. The precipitate is recovered by centrifugation at 1500 g and the ferritin redissolved in 2% $(NH_4)_2SO_4$. The ferritin is reprecipitated 3 more times to remove traces of cadmium sulfate. The dissolved ferritin is then passed over a large Sephadex G-200 column and eluted with PBS.

The ferritin may be concentrated either by ultrafiltration, e.g., on an Amicon ultrafiltration cell model S2 using an SM11730 membrane, or by ultracentrifugation for 2 hr at 105,000 g at which speed iron-rich ferritin can be recovered in the pellet.

Comments

The use of either human liver or human spleen preparations as standard ferritin produces no significant difference in the results of ferritin determination on any single serum sample.[21]

Normal human liver yields approximately 0.5 mg ferritin per gram wet weight of tissue by the crystallization method and considerably higher yields by the chromatographic method although recoveries from the final purification step of preparative polyacrylamide gel electrophoresis may be relatively poor (30–40%). Careful attention to purification procedures is essential whichever method is selected and provided this attention is given, either of the above procedures produces a final product which at a given protein concentration has comparable immunoreactivity. Simpler chromatographic techniques have been reported to produce a less pure ferritin preparation.[22] The high-speed centrifugation step (105,000 g for 2 hr) results in a ferritin of high iron concentration which may have an advantage of added stability to storage since iron has been reported to stabilize the molecule.[23] However a potential disadvantage should also be

[21] L. Gonyea-Stewart and C. H. Arnt, *Clin. Chem.* **25**, 1344 (1979).
[22] L. M. Gonyea, C. M. Lamb, R. D. Sundberg, and A. S. Deinhard, *Clin. Chem.* **22**, 513 (1976).
[23] J. W. Drysdale and H. N. Munro, *J. Biol. Chem.* **241**, 3630 (1966).

mentioned, i.e., the removal of the more acidic isoferritins of different immunological reactivity.

The Storage of Ferritin

Purified ferritin may be denatured by freezing and thawing and should be kept at 4° preferably at concentrations of 100 μg/ml to 1 mg/ml in phosphate or veronal-buffered saline pH 8.0. It may also be stored at 4° as a precipitate with half-saturated ammonium sulfate. Sodium azide should be added to prevent bacterial contamination. Immunological reactivity may change with storage. The reason for this has not been established and is currently the subject of investigation by International and National Standardization Committees. The ferritin concentration in sera stored at $-20°$ for up to 2 years has remained constant.

Criteria for Purity of Ferritin

Electrophoresis of the purified ferritin carried out on both a 6% polyacrylamide gel and on a polyacrylamide gradient gel (4 to 24% polyacrylamide) will identify contaminants of a similar molecular size or of a similar charge to size ratio. If a single band is evident and a protein stain (0.01% Coomassie Blue, 5% acetic acid, 50% ethanol) and iron stain (2% potassium ferricyanide, 1.2 M HCl, 4.5% trichloroacetic acid) are coincident using both electrophoretic techniques, then a high degree of purity can be assumed. An added criterion of purity is the demonstration that radiolabeled monospecific antiferritin antibody binds to the band. It is essential that the gels be ''overloaded'' with protein so that any small amounts of contaminants can be seen (most frequently in the position of serum albumin). Once a monospecific antibody to highly purified tissue ferritin has been prepared, affinity chromatographic techniques can be used to purify ferritin (see under purification of antiferritin antibody).

As mentioned above, final purification by preparative gel electrophoresis and elution of the monomer ferritin band is recommended.

Preparation of Antiserum to Human Tissue Ferritin

a. Immunization

Antiserum may be prepared in either rabbits or goats. If rabbits are used, at least 4 animals should be immunized as the titers may vary widely from animal to animal. Each animal is given 1 ml of an emulsion containing approximately 500 μg of ferritin protein in 0.5 ml saline and 0.5 ml

Freund's adjuvant. The mixture is repeatedly aspirated through a 21-gauge needle and injected subcutaneously in multiple sites. The injections are repeated 3 times at weekly intervals. An antiserum of usable titer (greater than 1 : 16) can usually be obtained 10 days after the last injection. Booster injections of ferritin in saline may be given at intervals of several months.

Blood is removed from ear veins (rabbits) or jugular veins (goats) and allowed to clot. To minimize nonspecificity, antisera from more than one animal should not be pooled. The antiserum may be stored in aliquots at $-20°$ with added (0.1%) sodium azide.

b. Antiserum Titration

Agar gel diffusion methods are convenient for the determination of the titer of an antiserum. If 0.1 ml antigen at a concentration of approximately 1 mg/ml is placed in a 5 mm well in the center of an agar gel plate and surrounding wells contain serial dilutions of the antiserum to be tested then the lowest concentration of antiserum to produce a precipitation line between the wells containing antigen and antibody is recorded as the titer of the antiserum. Relative antiferritin antibody titers may be then determined by radial immunodiffusion in agarose.

Procedure

One percent (w/v) agarose in PBS is heated in a boiling water bath until the agarose is completely dissolved. The solution is cooled to 60° and purified spleen or liver ferritin is added to a final concentration of 100 μg/ml. The solution is poured rapidly into a petri dish or onto a glass plate to a depth of 3 mm; 8 holes of 4 mm diameter are punched in the gel. The holes are separated by at least 1.5 cm. The antiserum or antibody preparation (10 μl) is carefully pipetted into 2 holes and a series of dilutions of a previously standardized antiserum is pipetted into the 6 remaining holes. The plate is incubated at 4° for 24 hr and the diameter of the resulting precipitation rings is measured. The square of the diameter is plotted against the antiserum titer to provide a standard curve from which the titer of the newly prepared antiserum is derived.

Measurement of Ferritin Protein Concentration

The protein concentration of ferritin samples is generally measured by either the original method of Lowry et al.[24] or some modification thereof such as that of Schacterle and Pollack,[25] or by a microKjeldahl method.

[24] O. H. Lowry, N. J. Rosebrough, A. L. Farr, and R. T. Randall, J. Biol. Chem. **193**, 265 (1951).

The protein contains a variable amount of iron and the biuret reaction is unsuitable as a measure of protein concentration.[21] Kjeldahl analysis yields somewhat higher results than Lowry determinations using bovine serum albumin as a protein standard.[21] Nevertheless, because of the greater convenience and wide acceptance of the latter method for protein determination it has been the method of choice in virtually all immunoassays for human ferritin. Final absorbance should be measured at 720 to 750 nm.

Preparation of Purified Antiferritin Antibodies

a. Diazocellulose Immunoadsorbent Method

Diazocellulose may be prepared from aminocellulose (*m*-aminobenzyloxymethyl cellulose, Miles Laboratores, Kanakee, Ill.) powder and Schweitzer's reagent (cupric ammonium hydroxide) as described by Miles and Hales.[26]

Cupric ammonium sulfate is prepared by the addition of 2.5 g sodium hydroxide dissolved in distilled water to 5 g copper sulfate ($CuSO_4 \cdot 5H_2O$) in 500 ml distilled water. The resulting precipitate is carefully filtered, washed with 3 liters of distilled water, and redissolved in 100 ml concentrated ammonium hydroxide. Aminocellulose (250 mg) is added slowly with constant stirring to 82.5 mg sucrose in 25 ml cupric ammonium sulfate. After 1 hr any undissolved material is removed by centrifugation at 1500 *g* for 15 min. Distilled water is added to the resulting clear solution to a final volume of 100 ml. Approximately 8 ml of 10% sulfuric acid is added until a flocculent precipitate appears and the color is a faint blue. The white precipitate is removed by centrifugation at 800 *g* for 10 min and washed three times with distilled water. The resulting precipitate is suspended in distilled water to give an approximate concentration of 10 mg/ml. Twenty-five milliliters of suspension containing 250 mg aminocellulose is cooled to 4° in ice (use within 24 hr).

Fifty milliliters of 15% HCl is cooled to 4° and added at 4° to the aminocellulose. Two milliliters of 1% sodium nitrite ($NaNO_2$) at 4° is added and the mixture is allowed to stand for 30 min at 4° in ice. Approximately 1 to 2 g of urea is added gradually in 50 to 100 mg quantities until no further dark blue color is formed with iodide starch paper. This ensures that excess nitrous acid groups have been removed. The diazocellulose is washed rapidly by centrifugation three times with distilled water and once with 0.05 *M* borate buffer pH 8.6.

Diazocellulose (250 mg) suspended in 10 ml 0.1 *M* borate buffer pH

---

[25] G. R. Schacterle and R. L. Pollack, *Anal. Biochem.* **51**, 654 (1973).
[26] L. E. M. Miles and C. N. Hales, *Biochem. J.* **108**, 611 (1968).

8.6 is reacted in the dark at 4° with 500 mg ferritin. Occasional slow stirring is necessary. After the addition of 1 g glycine in 10 ml borate buffer to inactivate unreacted diazo groups the suspension is left to stand at 4° for a further 24 hr and washed exhaustively with a BSA buffer (0.05 M veronal buffer pH 8.0 containing 1 g/liter bovine serum albumin). Rabbit antiferritin antiserum, diluted 1 in 5 with BSA buffer is added to the washed ferritin-immunoadsorbent in the proportion of 0.2 ml diluted antiserum per 1.0 mg of adsorbent. After incubation overnight at 4°, excess unbound antiserum is removed by centrifugation at 1500 g for 20 min. The addition of antiserum to immunoadsorbent is repeated a total of 5 times. The antiferritin–ferritin-immunoadsorbent is finally washed 5 times in 0.1 M borate buffer pH 8.6 and stored in amounts of 1 to 1.5 mg in small plastic tubes at −20° until used for radioiodination.

b. Cyanogen Bromide-Activated Sepharose Method

Cyanogen bromide-activated Sepharose 4B (CNBr-Sepharose) (Pharmacia, Sweden) is hydrated in 1 mM HCl. To 1.0 g of CNBr-Sepharose is added 30 mg purified human liver (or spleen) ferritin dissolved in 0.1 M borate buffer pH 8.0 containing 0.5 M sodium chloride (B8B) and the mixture shaken overnight at 4°. The adsorbent is washed with 0.1 M ethanolamine pH 8.4 in 1 mM sodium chloride and poured into a chromatography column. The immunoadsorbent is then washed with 0.1 M acetate pH 4 followed by B8B a total of 5 times.

The globulin fraction is precipitated from 20 ml of rabbit antiserum to human liver ferritin by 50% saturation [31% (w/v)] with ammonium sulfate. The precipitate is dialyzed against PBS. This globulin fraction is passed through the Sepharose-ferritin affinity column and washed with B8B. Purified antiferritin antibody is then eluted from the column using 3 volumes of 0.01 M phosphate buffer pH 6.0 containing 3 M sodium thiocyanate followed immediately by B8B. The thiocyanate is removed as rapidly as possible by immediate passage of the eluted antibody through a column of Sephadex G-25 and elution with borate buffer pH 8.0 followed by dialysis for 24 hr at 4° against borate buffer pH 8.0. After determination of the protein concentration and titer of the antibody (see above), aliquots of 100 μl are stored at −20°.

Comments

The diazocellulose immunoadsorbent presents more difficulties in preparation than the CNBr-Sepharose. While the former method has the theoretical advantage that iodination may be performed while the antibody is bound to the immunoadsorbent, thus protecting at least one antigen binding site, it is often difficult to achieve labeled antibodies of high specific activity. The major problem associated with the CNBr-Sepharose method relates to the need to remove all traces of thiocyanate rapidly prior to any attempt to iodinate the protein. N.B. Failure to remove thiocyanate rapidly may result in denaturation of the protein and failure to remove all traces of thiocyanate may result in failure of iodination. The preparation of purified antiferritin antibodies is probably the most difficult part of the assay.

Iodination Procedures

All protein iodination techniques carry a safety hazard because of the potential for inhalation and absorption of volatilized radioactivity. Thus all such techniques must be carried out in a well-ventilated laboratory hood which meets radiation safety requirements. Several pairs of latex gloves should be worn and disposed of at various steps during the procedure to prevent contamination of laboratory areas. The ingestion of 300 mg of potassium iodide orally immediately before iodination procedure is advocated by some to reduce thyroid uptake of any absorbed ^{125}I. A portable laboratory monitor for low energy isotopes calibrated for use with ^{125}I, e.g., Mini instruments Model 5.42 (Essex, England) should be available for frequent monitoring of laboratory working areas.

The method of Hunter and Greenwood[27] using chloramine-T as described in detail by Ryan et al.[28] may be used to iodinate antiferritin antibodies bound to immunoadsorbent, free purified antiferritin antibodies, or purified ferritin protein for use in the RIA. In each case the procedure is similar, i.e., approximately 10 to 50 μg of protein is subjected to the procedure and after the reaction is complete the unreacted ^{125}I is removed by gel chromatography.

a. Chloramine-T Iodination of Antibody Bound to Immunoadsorbent[27]

Antiferritin–ferritin-diazocellulose adsorbent complex (1.0 to 1.5 mg) is washed 5 times with 0.1 M borate buffer pH 8.0 and the washed precipitate is resuspended in approximately 50 μl of the buffer. This immunoadsorbent, together with 1 to 2 mCi ^{125}I in a volume of approximately 5 μl, is added to a small conical reaction vial containing 25 μl (50 μg) freshly prepared chloramine-T in 0.1 M borate. After approximately 30 sec the reaction is stopped by the addition of 100 μl of freshly prepared sodium meta-

[27] W. M. Hunter and F. C. Greenwood, *Nature (London)* **194**, 495 (1962).
[28] S. Ryan, L. R. Watson, M. Tavassoli, R. Green, and W. H. Crosby, *Am. J. Hematol.* **4**, 375 (1978).

bisulfite (1.2 mg/ml in 0.1 M borate) and mixing is continued for a further 30 sec. Approximately 300 μl of BSA buffer is used to transfer the mixture to a column of Biogel P2 (200 to 400 mesh) made up in BSA buffer pH 8.0. This column may be made conveniently in a 10 ml plastic syringe. The free ^{125}I is eluted with 20 to 30 ml BSA buffer and the iodinated antibody is then eluted from the immunoadsorbent by successive addition of 10 ml, 0.1 M acetate buffer pH 4.5, 10 ml 0.001 M HCl, 10 ml 0.0034 M HCl, and 10 ml 0.01 M HCl. One-milliliter fractions are collected and the radioactivity of 10 μl samples is counted in an autogamma counter. The fractions obtained from elution with 0.001 M HCl and 0.0034 M HCl (which contain the highest counts) are usable in the assay.

b. Chloramine-T Method for Iodination of Free Antiferritin Antibody or Ferritin Protein

The method is similar to that described for iodination of the ferritin–antiferritin immunoadsorbent except that approximately 50 μl of purified antiferritin antibody in borate buffer pH 8.0 and containing 50 to 100 μg of protein is iodinated with 1 to 2 mCi [125]I. The chromatographic purification step is much simpler. In a single passage through either Sephadex G-25 (1 × 10 cm), or BioGel P-10, eluting with phosphate buffer pH 7.5, the high-molecular-weight iodinated protein is eluted in a single peak, the position of which is determined by counting 10 μl samples of each 1 ml fraction. The free ^{125}I is eluted in the total volume of the column.

c. Acylating Agent Method for Iodination of Free Antiferritin Antibody or Ferritin Protein

The method of Bolton and Hunter[29] provides a simple alternative to the Chloramine-T iodination method and is less damaging to the protein. An iodinated acylating agent [^{125}I-labeled 3-(4-hydroxyphenyl)propionic acid N-hydroxysuccinimide ester (Bolton–Hunter reagent) containing approximately 1500 Ci/mmol] may be obtained from New England Nuclear (549 Albany Street, Boston, Mass.). An amount containing approximately 0.5 mCi is dried in a small tube and immediately placed in an ice bath. Purified antiferritin antibody in amount 50 to 100 μl and of protein concentration 1.0 mg/ml is added to the labeled acylating agent and allowed to stand 1 hr in the ice bath with frequent mixing. Glycine (0.5 ml, 200 mM) in 100 mM borate buffer pH 8.5 is added to prevent subsequent iodination of carrier proteins.

After 5 min the reaction mixture is subjected to chromatographic separation on Sephadex G-75 (1 × 23 cm) and elution with 50 mM phosphate

[29] A. E. Bolton and W. M. Hunter, *Biochem. J.* **133,** 529 (1973).

buffer pH 7.5. Fractions of 1 ml are collected and monitored as in the chloramine-T technique. The first radioactive peak contains the labeled antibodies. This is stored at 4° and maintains its activity for up to 4 weeks.

Comments

The Bolton and Hunter technique is rapid, convenient, and appears to be more uniformly successful than the more damaging Hunter and Greenwood technique. It is also less hazardous to personnel.

If purified ferritin is to be iodinated, the procedure is similar except that amounts of 10 μg of ferritin protein in 10 μl buffer are iodinated with 1 mCi of ^{125}I and a Sepharose 6B column is used to separate the labeled ferritin from the free iodine. Pure ferritin is eluted by Tris–HCl buffer (0.02 M, pH 7.5 containing 10 g/liter BSA) in the second peak of ^{125}I from the column.

Ferritin Assay Techniques

Immunoradiometric Assays (IRMA)

Two methods have been described using labeled antibodies to assay antigens in solution. In the first method labeled purified antibodies in solution are reacted with the antigen in solution. The unreacted radioactive antibodies are removed from solution by addition of a solid-phase antigen which can be removed by centrifugation. The radioactive antigen–antibody complex remains in solution and the radioactivity can be counted. This is the IRMA[4] (Fig. 1). In the second method the unknown antigen is bound to insolubilized antibodies and excess radiolabeled purified antibodies are added in solution. This labeled antibody binds to other antigen-binding sites on the insolubilized antigen–antibody complex. After the reaction, the excess radioactive antibody is washed away and the amount bound to insolubilized antigen is counted directly. As the amount of antigen increases more radioactive antibody is bound. The counts bound are therefore proportional to the amount of antigen present. This is the "2-site" or "sandwich" IRMA[14,15] (Fig. 1).

a. IRMA

In the IRMA as originally described by Addison *et al.*,[4] ^{125}I-labeled antibody after elution from the diazocellulose immunoadsorbent is reacted with either a known concentration of standard ferritin, or an unknown serum. Following this reaction, 50 μl of a horse spleen ferritin diazocellulose immunoadsorbent is added to the tubes. After centrif-

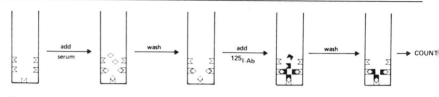

2-Site IRMA

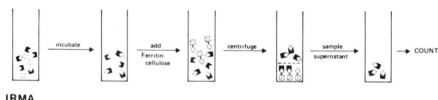

IRMA

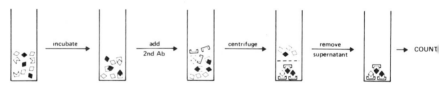

RIA

♦   125I Ferritin
◇   Ferritin (serum)
◄   125I-Ab
⊠   Ab
○⊏ Ferritin-cellulose
⊏   2nd Ab

Fig. 1. Diagrammatic illustration of the three different assay procedures described in the text: 2-site IRMA, IRMA, and RIA.

ugation to remove the adsorbed antibody the unreacted antibody in the supernatant is counted. Calculations are considered in detail after the 2-site IRMA. A standard curve is plotted as supernatant radioactivity (y axis) against ferritin concentration (x axis) (Fig. 2).

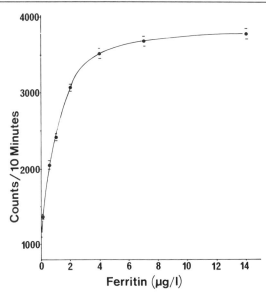

FIG. 2. A typical standard curve for the IRMA. Each symbol represents mean ± standard deviation.

Comments

The sensitivity of this method is 0.2 μg/liter ferritin protein. A more convenient and widely used method makes use of the fact that the antibodies may be bound to polystyrene and hence to the walls of tubes or of pitted beads from which the unbound radioactivity is easily removed by washing.

b. 2-Site IRMA

Flat bottomed polystyrene tubes (e.g., Falcon Products, Los Angeles, Calif.) are coated by addition of 500 μl of whole rabbit antiserum to human liver ferritin at a dilution of 1:10,000 (suitable dilution is dependent on antibody titer) in 0.015 M phosphate-buffered saline (PBS) pH 7.4 containing 0.01% azide. After incubation for 24 hr at 20°, the contents are aspirated and the tubes washed 10 times with PBS and dried and stored at 4°. These coated tubes are best left for several days before use and are stable for up to 8 weeks.

Standard curves are obtained by quadruplicate estimations on ferritin standard solutions (100 μl per tube) containing 0.5, 1, 2, 4, 6, 8, 10 μg/liter of ferritin in veronal buffered saline (0.05 M veronal, 0.1 M sodium chloride pH 8.0) containing 4% bovine serum albumin (VBSA). One part in 20 of normal rabbit serum may also be added to allow for some observed inhibitory effects of serum.[14] A series of 4 replicate blank tubes containing VBSA without added ferritin is used to determine nonspecific binding of labeled antibody to the tubes.

The Assay Procedure

Human serum (100 μl) at 2 dilutions (usually 1:10 and 1:20) in VBSA is added to duplicate or triplicate antibody-coated polystyrene tubes and allowed to stand either for 24 hr at 20° or at 37° for 4 hr. The contents of the tubes are aspirated and the tubes washed 8 times with VBSA. ^{125}I-labeled antiferritin antibody (100 μl) containing 80,000 to 100,000 counts per minute (cpm) is added carefully to the bottom of each tube. After incubation for a further 2 hr at 20° the contents are again aspirated and the tubes washed 5 times with PBS and counted.

Figure 3 shows a typical standard curve obtained in this assay.

Calculations

The simplest method for the calculation of results in an unknown sample is the preparation of a standard dose–response curve as shown in Fig. 4. The mean radioactivity in replicate standard tubes is plotted against ferritin concentration, after subtraction of mean counts of the blank tubes from the mean of replicate standard tubes.

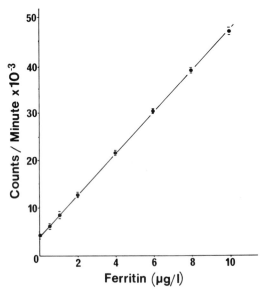

FIG. 3. A typical standard curve for the 2-site IRMA. Each symbol represents mean ± standard deviation.

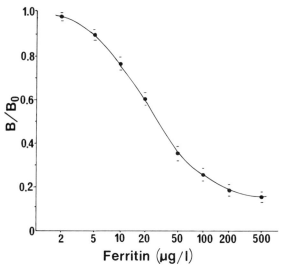

FIG. 4. A typical standard curve for the RIA. Each symbol represents mean ± standard deviation.

A mathematical method of least squares regression avoids bias in constructing the dose–response curve and a desk-top calculator may readily be used to facilitate the analysis of a large assay. A logit/log plot[30] may also be used to linearize the results. In this method the logit of Y [the tube radioactivity minus zero blank, as a fraction of the total increase in radioactivity from zero to infinite dose (M)] is plotted against the log of the ferritin concentration.

Thus if S is the net cpm for standards and U for unknowns

$$Y_s = S/M \text{ for standards}$$
$$Y_u = U/M \text{ for unknowns}$$

Then

$$\text{logit } Y_s = \log_e \left(\frac{Y_s}{1 - Y_s} \right)$$
$$\text{logit } Y_u = \log_e \left(\frac{Y_u}{1 - Y_u} \right)$$

Least squares regression is used to calculate the relationship of logit (Y_s) as ordinate against log X as abscissa (where X is the ferritin concentration of the standard in μg/liter). Thus the slope (b) and intercept (a) of the logit/log plot are determined

$$\text{logit } Y_s = a + [b \log_e (X)]$$

[30] D. Rodbard, W. Brison, and P. L. Rayford, *J. Lab. Clin. Med.* **74**, 770 (1969).

The ferritin concentration of the unknown is determined from the following expression:

$$\text{ferritin concentration} = D \exp \left(\frac{\log (Y_u) - a}{b} \right)$$

where D is the dilution.

The dose–response of the highest concentration of reference ferritin standards cannot be considered as the "infinite" dose–response because of the "high-dose hook effect" (see later in this chapter). If facilities are available a computer searching technique may be used in which the infinite dose–response is serially increased with repeated least squares regression for the standard curve being determined at each new infinite dose-response. Because of this effect it is advisable to restrict the calculation of unknowns to within the working range of the assay (0.25 to 50 μg/liter) and to repeat determinations on samples which require dilution to fall within this range.

Comments

Accuracy of the assay. The recovery of added ferritin, determined by the addition of 0.5, 1.0, 2.0, and 2.5 μg/liter ferritin, respectively, to the 5 μg/liter standard, ranges from 93 to 100%. It is important to note that if recovery studies are performed by addition of purified ferritin to concentrated serum before dilution, assay recoveries of only 60 to 80% have been reported. The reasons for this apparent inhibitory effect of undiluted serum have not been determined.[31]

Sensitivity of the Assay. The smallest value which may be distinguished from zero may be calculated from the 95% confidence units at the zero point of the standard surve. The sensitivity calculated from 20 standard curves is 0.25 μg/liter.

Analysis of Variability. Analysis of variability within the same assay (intraassay) shows that the standard deviation is directly proportional to the concentration of measured ferritin. The variability is increased for plasma ferritins below 10 μg/liter but the coefficient of variation remains constant over the standard curve for plasma ferritin concentrations from 10 to 200 μg/liter. The mean coefficient of variation for quadruplicate determinations is approximately 4%. The variation between 20 consecutive assays (interassay) of a sample, performed in the same laboratory is less than 10% for ferritin concentrations from 10 to 200 μg/liter and up to 15% for values below 10 or above 800 μg/liter.

[31] D. A. Lipschitz and J. D. Cook, *in* "Proteins of Iron Metabolism" (E. B. Brown, P. Aisen, J. Fielding, and R. Crichton eds.), p. 433. Grune & Stratton, New York, 1977.

Specificity

Most sera when assayed simultaneously at 4 dilutions produce a linear response when plotted arithmetically. Occasional sera do not dilute out in a linear fashion and serial dilutions result in an ever-increasing value for the serum ferritin. It is possible that this phenomenon may result from the existence of isoferritins of different immunological reactivity in sera from certain patients, e.g., patients studied during phlebotomy therapy for idiopathic hemochromatosis. Isoferritin profiles of serum ferritin have been shown to change during such therapy[32] and Saab *et al.*[33] have shown differences in immunological reactivity of serum ferritin from patients with other conditions such as malignancy. An assay directed more specifically toward the assay of ferritins of more acidic isoferritin profile such as placental or HeLa cell ferritin may result in a higher result. This subject will be discussed in more detail later.

Inhibitory Effects. 1. It has been shown that inhibition of the ferritin assay may occur in the presence of serum and the use of normal rabbit serum at a 1/20 dilution has been reported to decrease the dose–response curve 3- to 4-fold as compared with BSA buffer.[14] This effect may be removed either by the addition of rabbit serum 1/20 to the standard curve as mentioned above or by the use of serial dilutions of a normal human serum of known ferritin concentration as a "secondary" standard.

2. A feature of the 2-site IRMA which has been described is the inhibition of the dose–response curve at very high ferritin concentrations: the "high-dose hook effect."[34] This paradoxical fall in radioactivity in the presence of a great excess of antigen has been attributed to the binding of antigen to low affinity antibodies coating the tubes and the subsequent loss of the soluble antigen–antibody complex with washing. This effect is not apparent in our hands until well beyond the upper limit of the normal standard curve. The hook effect has recently been shown to occur only at very high concentrations ($>90,000$ μg/liter) provided an adequate number of final washes (at least 5) of the tubes are carried out. It is readily detected if samples are routinely assayed at more than one dilution. Further assay at 2 appropriate dilutions can then be performed. The use of a highly purified solid-phase

[32] J. W. Halliday, L. V. McKeering, R. Tweedale, and L. W. Powell, *Br. J. Haematol.* **36,** 395 (1977).

[33] G. A. Saab, R. Green, and W. H. Crosby, *Proc. Soc. Exp. Biol. Med.* **161,** 444 (1979).

[34] R. Green, L. R. Watson, G. A. Saab, and W. H. Crosby, *Blood* **50,** 545 (1977).

antibody with high affinity for ferritin is also recommended to avoid this problem.[35]

The 2-site IRMA appears to have some advantages over the conventional IRMA. It is more economical of purified antigen and it can be adapted to recognize only antigens with two specific sites. It is apparently free from the allosteric effects which are sometimes observed in IRMA and conventional RIA techniques. As serum is removed prior to the addition of labeled antibody for the second reaction any nonspecific inhibitory effect of serum may be minimized whereas in IRMA, serum exerts an effect throughout the assay. At high antigen dose the IRMA is subject to antibody exchange in the second reaction. In the 2-site IRMA the investigator should beware of inconstancy of solid-phase washing. A major advantage of the 2-site system is the very low zero dose–response and nonspecific binding which rarely exceeds 2% of the total radioactivity in the system. In the IRMA the zero dose–response may be as high as 40% of the radioactivity added since it is determined by the efficiency of binding of the excess labeled antibody. Thus the 2-site IRMA allows greater precision at the low end of the dose–response curve which is most important for the detection of iron-deficient states. The 2-site assay is readily automated and requires no centrifugation during the performance of the assay.

Radioimmunoassay (RIA) (from the method of Luxton et al.[16])

Several groups have now reported techniques for the radioimmunoassay of serum ferritin.[16,17,36,37] The radioimmunoassay makes use of labeled antigen rather than labeled antibody. ^{125}I-labeled ferritin is mixed with either ferritin standard or unknown serum. Antiferritin antiserum is then added in an amount which is not sufficient to bind the ^{125}I-labeled ferritin or ferritin in the unknown. There is thus a competition of the labeled and unlabeled ferritin for binding sites on the antibody. Further rabbit antiserum is added as a carrier and antiserum raised in sheep, goat, or donkey against rabbit immunoglobulins is then added. After a 24-hr incubation the soluble ferritin–antiferritin complexes which have been precipitated by this antiserum are removed by centrifugation and counted. The radioactivity of the precipitated ferritin–antiferritin complex reflects the relative proportions of labeled and unlabeled ferritin in the original mixture. As the concentration of ferritin in

[35] P. K. Li, J. R. Humbert, and C. S. Cheng, *Clin. Chem.* **24**, 1650 (1978).
[36] D. J. Goldie and M. J. Thomas, *Ann. Clin. Chem.* **15**, 102 (1978).
[37] W. M. Deppe, S. M. Joubert, and P. Naidoo, *J. Clin. Pathol.* **31**, 872 (1978).

the unknown serum increases so the percentage of labeled ferritin bound decreases (Fig. 1).

The sensitivity of the assay as first reported was much less than that of the IRMA but recent improvements have resulted in a sensitivity of approximately 2 μg/liter which is suitable for clinical use although still not quite as sensitive as the IRMA.

Purified ferritin may be iodinated by either the Chloramine-T method or by the Bolton and Hunter reagent as described above.

The radioimmunoassay requires the determination of the maximal and minimal (i.e., nonspecific) binding of ^{125}I-labeled ferritin fractions. Maximal binding of the ^{125}I-labeled ferritin is determined by the addition of excess of antiferritin antibody in the first step of the procedure and nonspecific binding to the second antibody is determined in the absence of the antiferritin antibody. Fractions selected for use in the assay are those with maximal specific and minimal nonspecific binding. The labeled ferritin should be diluted in BSA buffer containing 1 g/liter BSA to provide 30,000 cpm and approximately 1 ng of ferritin protein in 100 μl samples. This dilution is generally around 1 in 50 to 1 in 100. The dilution of antiferritin antibody to be used in the assay is selected such that approximately 50 to 60% of the labeled ferritin counts are bound. This dilution should be of the order of 1:300,000 to 1:500,000 if the original antibody used is of sufficiently high titer.

Ten ferritin standards ranging in concentration from 2.5 to 250 μg/liter are made by dilution in VBSA buffer (containing 4% BSA). All standards are measured in triplicate and in each assay, tubes are included for duplicate estimates of maximal binding, nonspecific binding, total counts together with two control sera known to fall within the ferritin concentration of the assay standard curve.

The Assay Procedure

Diluted antiferritin antibody (100 μl) is added to a 10 × 75 mm tube containing 100 μl of labeled ferritin (approximately 30,000 cpm) together with 100 μl of either buffer, standard ferritin, or unknown serum. The tubes are centrifuged at 2500 g and after incubation at 4° overnight 100 μl of antibody to rabbit immunoglobulin (at a dilution of 1 in 20) is added. The tubes are again centrifuged at 2500 g for 30 min at 4° and then incubated at 37° for 5 hr. The separation of bound and free moieties may be facilitated by the addition of 100 μl of polyethylene glycol which allows the incubation time to be shortened to as little as 5 min at room temperature. After centrifugation for 30 min at 2500 g the supernatant (free) fraction is aspirated and the precipitated (bound) fraction counted in an autogamma counter. The washing of the final precipitate, e.g., twice with 500 μl Tween 20/saline solution (0.1% Tween 20 in 0.9% NaCl) followed by centrifugation at 2500 g and aspiration before counting may improve the agreement between duplicates.

Calculations

The maximal (B_0) and nonspecific binding is calculated from the counts in the appropriate tubes. A standard curve of B/B_0, i.e., the ratio of the counts bound by each standard dilution (B) to the counts bound in the absence of added ferritin (B_0) may be plotted on the linear axis of 3-cycle semilogarithmic paper and from this line the B/B_0 for the unknown serum and hence the ferritin concentration may be calculated (Fig. 4).

Comments

The mean inter- and intraassay coefficients of variation for ferritin concentration in the normal range have been estimated at approximately 7 and 13%, respectively.[16] In the radioimmunoassay a "high-dose hook effect" has not been reported. Measurements of the ferritin concentration of iron-deficient sera will be made on the less sensitive portion of the standard curve and the coefficient of variation within a single assay may approach 20%. The decreased sensitivity at low ferritin concentrations is important because the major use of the assay in the clinical laboratory is in the diagnosis of iron deficiency.

An early report of a less sensitive radioimmunoassay for ferritin in which the range of ferritin measured in the assay was 0.2 to 1.0 μg and the antibody dilution used was 1:640 indicated an inhibitory effect of iron in the molecule.[38] However as serum ferritin has been shown to contain relatively little iron within the molecule even in iron overload, this appears to be of little practical importance.

Solid-Phase Enzyme-Labeled Immunosorbent Assay for Plasma Ferritin (ELISA) (based on the method of Anaokar et al.[18])

The development of immunoassays which avoid the use of radioactivity and yet which provide adequate sensitivity is greatly to be desired. Enzyme-linked immunoassays fulfill this criterion. The principle of the assay is the same as that of the 2-site IRMA assay except that an enzyme instead of ^{125}I is conjugated to the second antibody. After incubation of the antibody-bound enzyme with an appropriate substrate

---

[38] F. S. Porter, J. Lab. Clin. Med. **83**, 147 (1974).

such as a chromogen, the color development is directly proportional to the amount of enzyme and hence of antibody bound to the insolubilized antigen on the walls of the tube. A number of such assays for ferritin have now been published but most had a lower limit of sensitivity of approximately 5 μg/liter which is barely adequate for the accurate detection of iron deficiency. Alkaline phosphatase was used to label the second antibody[39,40] in the earlier published studies. Watanabe and his co-workers in Japan have developed an assay with a sensitivity of 0.25 μg/liter using 3-D-galactosidase as the marker.[41] This method is somewhat complex for routine use. Horseradish peroxidase has been used in the method of Zuyderhoudt et al.[42] and this method has now been adapted by Anaokar et al.[18] for use with a sensitive chromogen. This assay has a reported sensitivity of 1 μg/liter which is better than some radioimmunoassays reported and the standard curve is linear up to 25 μg/liter of ferritin.

The preparation of a peroxidase-IgG conjugate is carried out by a 2-step method.[43] Horseradish peroxidase (5.0 mg) is dissolved in 1 0 ml of 0.3 M sodium bicarbonate and mixed with 1.0 ml of a 10 g/liter solution of fluorodinitrobenzene in ethanol for 1 hr at room temperature. Ethylene glycol (1.0 ml, 0.016 M) is added, mixed 1 hr at room temperature, and the reaction mixture dialyzed against 10 mM carbonate buffer pH 9.5 at 4°. This results in a "peroxidase aldehyde" which is stable for at least 1 month at 4°. This is conjugated to antiferritin rabbit IgG which has been prepared from rabbit antiferritin antiserum by precipitation with $(NH_4)_2SO_4$.

The enzyme–IgG conjugate is concentrated to 500 μl by ultrafiltration and subjected to Sepharose 6B chromatography and elution with PBS. The dark brown conjugate is eluted in the void volume and is stable for several months when stored in 2% BSA.

The chromogenic reagent used is a 2% solution of 2,2'-azinodi(3-ethylbenzthiazoline-6-sulfonate) in water. The working substrate consists of a 1 in 20 dilution of the chromogenic reagent in 0.1 M citrate buffer pH 4.0 to which is added 0.2 ml of a 0.6% solution of hydrogen peroxide.

The assay is performed according to the method for the 2-site IRMA up to the addition of the enzyme-labeled antibody. After incuba-

[39] L. Thériault and M. Pagé, Clin. Chem. 23, 2142 (1977).
[40] R. L. Fortier, W. P. McGrath, and S. L. Twomey, Clin. Chem. 25, 1466 (1979).
[41] N. Watanabe, Y. Niitsu, S. Ohtsuka, J. Koseki, Y. Kohgo, I. Urushizaki, K. Kato, and E. Ishikawa, Clin. Chem. 25, 80 (1979).
[42] F. M. J. Zuyderhoudt, W. Boers, C. Linthorst, G. G. A. Jörning, and P. Hengeveld, Clin. Chim. Acta 88, 37 (1978).
[43] P. Nakane and J. Kawasi, J. Histochem. Cytochem. 22, 1084 (1974).

tion for 2 hr with this reagent the tubes are washed 5 times with PBS, 0.1 ml of the working substrate solution is added, and the tubes are incubated for 1 hr at room temperature. The reaction is stopped by the addition of 25 μl of sodium azide. Five hundred microliters of 0.1 M citrate buffer pH 2.8 is added and the absorbance at 410 nm is measured.

The standard curve is produced by plotting absorbance against ferritin concentration.

Comments

In the initial studies using this method[18] a "high-dose hook effect" was not observed nor was any significant inhibitory effect of serum apparent.

Greater sensitivity was achieved by coating the reaction tubes with the IgG fraction of the antiferritin antiserum at a concentration of 5 mg/liter.

The appropriate dilution of the IgG peroxidase conjugate must be determined such that a linear standard curve results.

Optimal concentrations of hydrogen peroxide and chromogen and incubation for 1 hr with these substrates must be strictly adhered to if the assay is to be successful.

The intraassay and interassay coefficients of variation are of the order of 6 and 10%, respectively, and a correlation coefficient of 0.964 was ob tained[18] in a comparison of the results obtained by the 2-site IRMA and the ELISA method.

Effect of Isoferritin Composition on Ferritin Assays

As described above, ferritins isolated from human tissues and serum have been shown to contain a number of isoferritins and organ-specific isoferritin profiles have been shown to occur.[9] The ferritins of heart and kidney contain more acidic isoferritins than those of liver and spleen while the most prominent isoferritin in serum is the most basic (pI 5.8) and probably represents apoferritin.[12] Changes in the isoferritin profile of both tissues and serum have been observed in iron storage disorders, in malignancies, and during phlebotomy therapy.[10]

The isoferritins of normal human heart and HeLa cells are primarily of the more acidic type and have been shown to differ in immunological reactivity from the ferritins of liver and spleen.[11,12] Since acidic isoferritins have been detected in some tumors, assays for the more acidic isoferritins have been developed. An assay based on HeLa ferritin has been shown to result in the detection of substantially higher ferritin levels in the sera of some patients with malignancies, than levels detected with the liver or spleen ferritin assay.[10] Assays using placental ferritin and leukemic ferritin have also been used in patients with malignancies, while Jones and

Worwood[44] have developed an assay for heart ferritin. The latter assay showed little reactivity with spleen ferritin; however circulating ferritin even from patients with myocardial infarction, leukemia, or carcinoma appeared to contain only a small proportion of molecules with the immunological characteristics of acidic heart ferritin. Thus, while the apparent level of serum ferritin may theoretically be dependent on the serum isoferritin population, the specificity of the antiferritin antibodies and the type of ferritin used as a standard, in practical terms most of the serum ferritin in normal subjects or subjects with iron overload consists of the most basic isoferritin. Hence the use of an antiferritin antiserum directed against liver or spleen will not invalidate the results obtained in untreated iron overload or iron deficiency states. It is possible that underestimation of more acidic isoferritins may occur in some patients with malignancies if standard assays are used and "tumor-specific" ferritin assays which are currently being investigated may provide useful data in such patients.

In summary, a solid-phase "sandwich-type" IRMA has advantages in sensitivity, specificity, and economy of reagents over the conventional IRMA, and it allows greater sensitivity at the low-end of the dose–response curve than does the RIA.

A disadvantage is the "high-dose hook effect" which may occur at very high ferritin concentrations and attempts to avoid the need for dilution of test sera by use of an extended standard curve should be discouraged. Iodination of purified antibody in the free form is preferable as it results in a higher specific activity of the product.

The conventional RIA is satisfactory in the normal range for serum ferritin but is less sensitive at the low-end of the dose–response curve, i.e., in the range of iron deficiency. This method does not appear to exhibit a "high-dose hook effect" but, as with the IRMA, sera with very high ferritin values will require dilution to fall within the assay range.

The advantage of the ELISA lies in the avoidance of the use of radioactive isotopes and the need for expensive counting equipment. Since the principle is that of the solid-phase IRMA then a "high-dose hook effect" might be expected to occur at very high ferritin concentrations although it has not yet been reported. If the sensitivity of this method can be maintained at 1 μg/liter or below and critical attention is paid to the conditions of the enzyme assay, this technique may well be the method of choice for the future.

Acknowledgments

This work was supported by the National Health and Medical Research Council of Australia. The author wishes to thank Mr. U. Mack for his assistance in preparing figures.

[44] B. M. Jones and M. Worwood, *Clin. Chim. Acta* **85,** 81 (1978).

[12] Radioimmunoassay of Myoglobin

By Marvin J. Stone, James T. Willerson, and
Michael R. Waterman

Myoglobin is a single polypeptide chain of 153 amino acids and molecular weight of approximately 17,000. It contains one heme prosthetic group per molecule and is found only in skeletal and cardiac (i.e., striated) muscle.[1] Immunochemical analysis of myoglobin from many species has revealed the presence of five antigenic reactive regions in the native protein.[2] Cardiac and skeletal muscle myoglobins appear to be immunochemically identical. Damage to striated muscle often is associated with myoglobin release into the circulation and, subsequently, this low-molecular-weight protein may appear in the urine. Elevated serum levels of myoglobin occur in individuals with acute myocardial infarction and in those with genetic or acquired skeletal muscle disorders.[1,3,4] A variety of immunochemical techniques have been employed to identify myoglobinemia and myoglobinuria; these include immunodiffusion, quantitative microcomplement fixation, and radioimmunoassay.[3,4] This article describes a sensitive, specific, and reproducible radioimmunoassay method for quantification of myoglobin in human serum.[4]

Preparation of Myoglobin

Myoglobin from human heart, obtained at autopsy within 12 hr of death from noncardiac conditions, is prepared by a modification of the method of Yamazaki *et al.*[5] Stepwise fractionation of the blended muscle mince is carried out with 50% and then 60% ammonium sulfate; myoglobin remains in the supernatant. After exhaustive dialysis of the supernatant fraction against 5 mM Tris–HCl buffer, pH 8.5, it is loaded onto a DEAE-cellulose column equilibrated with this buffer. Two zones of color develop on the column during elution with the same buffer; the lower zone is brown (metmyoglobin), and the upper zone is red (oxyhemoglobin). The myoglobin zone is sliced from the column, and the protein is eluted from the DEAE-cellulose with 10 mM phosphate buffer, pH 7.0.

[1] L. J. Kagen, "Myoglobin. Biochemical, Physiological, and Clinical Aspects." Columbia University Press, New York, 1973.

[2] M. Z. Atassi, *Immunochemistry* **12**, 423 (1975).

[3] L. Kagen, S. Scheidt, L. Roberts, A. Porter, and H. Paul, *Am. J. Med.* **58**, 177 (1975).

[4] M. J. Stone, J. T. Willerson, C. E. Gomez-Sanchez, and M. R. Waterman. *J. Clin. Invest.* **56**, 1334 (1975).

[5] I. Yamazaki, K. Yokata, and K. Shikama, *J. Biol. Chem.* **239**, 4151 (1964).

The final purification step involves passage of the myoglobin fraction through a BioGel P-10 column equilibrated with 10 mM phosphate buffer, pH 7.0. The isolated myoglobin is shown spectrophotometrically to be free of hemoglobin, the α-band absorbance of carbonmonoxymyoglobin being at 578 nm while that of carbonmonoxyhemoglobin is at 568 nm.[6] The absence of other proteins is demonstrated by polyacrylamide disc gel electrophoresis. The myoglobin concentration is determined using an extinction coefficient of 14.1 mM^{-1} cm^{-1} at 578 nm for the carbonmonoxy form. The final preparation is divided into aliquots and stored at $-70°$.

Immunization of Rabbits

Antisera to human heart myoglobin are prepared by immunizing adult New Zealand white rabbits. The antigen (0.5 mg) is emulsified in complete Freund's adjuvant and injected into multiple subcutaneous sites at monthly intervals. The rabbits are bled 10–14 days after each booster injection, and the serum is stored at $-20°$.

Radiolabeling of Myoglobin

Iodination of myoglobin is accomplished by the method of conjugation labeling described by Bolton and Hunter.[7] One milligram of N-succinimidyl-3-(4-hydroxyphenyl) propionate is dissolved in 50 ml benzene and a tube containing 5 μl of this solution is taken to dryness under vacuum. Ten microliters of 0.5 M phosphate buffer, pH 7.0, 1 mCi ^{125}I, and 50 μg chloramine-T in 10 μl distilled water, are added. Ten seconds later the reaction is terminated by the addition of 50 μg (in 10 μl distilled water) of sodium metabisulfite. The iodination procedure is carried out at room temperature. The labeled product is extracted into 1 ml benzene and recovered by evaporation of the solvent under nitrogen. Myoglobin (2 μg) in 10 μl of 0.25 M borate buffer, pH 8.5, is added to the dried iodinated ester, and the reaction mixture is agitated for 20 min at 0°. After addition of 0.1 ml of 0.2 M glycine, the mixture is loaded onto a 10 ml BioGel P-60 column and eluted with distilled water. Aliquots of each fraction (0.5 ml each) are assayed for radioactivity in a gamma counter. The first peak of radioactivity usually is confined to three tubes. Diluted aliquots from these tubes are tested for binding to antibody (see below) and are stored at $-20°$ for further use. Labeled myoglobin can be kept under these condi-

[6] E. Antonini and M. Brunori, "Hemoglobin and Myoglobin and Their Reactions with Ligands." North-Holland Publ., Amsterdam, 1971.
[7] A. E. Bolton and W. M. Hunter, *Biochem. J.* **133**, 529 (1973).

tions for up to 4 weeks. Myoglobin also can be radiolabeled utilizing the conventional chloramine-T method.[8]

Radioimmunoassay Procedure

All determinations (standards and unknowns) are performed in 12 × 75-mm glass tubes containing 0.05 M borate buffer, pH 7.8, in the presence of 0.1% sodium azide and 5% sterile normal rabbit serum. The optimal dilution of antiserum is determined by the addition of a constant amount of ^{125}I-labeled myoglobin to serial dilutions of antiserum obtained from a single bleeding. The antiserum dilution (e.g., 1:8000) at which maximum binding occurs is utilized in subsequent assays. Approximately 6000 cpm of ^{125}I-labeled myoglobin is present in each tube. Ten to fifty microliters of test serum (from normal subjects or patients) is added to each sample tube. Twenty microliters of each serum is routinely assayed. For those samples in which the myoglobin content is so high as to fall on the flat portion of the standard curve, the determination is repeated with 10 μl of serum. The total volume per tube is 0.5 ml. All determinations are performed in triplicate and a standard curve (Fig. 1) is included in each assay. The standard curve is prepared by the addition of increasing

[8] M. Reichlin, J. P. Visco, and F. J. Klocke, *Circulation* **57**, 52 (1978).

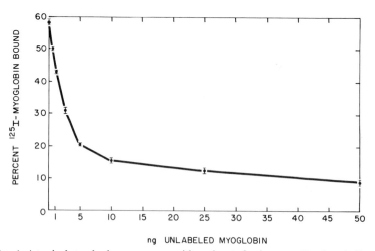

FIG. 1. A typical standard curve, as run with each set of unknowns. Brackets indicate the range of triplicate determinations. In this assay, as little as 0.5 ng of unlabeled myoglobin resulted in detectable reduction in binding of ^{125}I-labeled antigen to antibody. When the results are plotted on a semilogarithmic scale, the curve is linear in the range of 0.5–10 ng. (Reproduced from *The Journal of Clinical Investigation*.)

amounts of unlabeled myoglobin (instead of test serum) to a series of tubes containing the remaining reactants. Tubes are routinely incubated at 4° for 24 hr. Separation of free from antibody-bound labeled myoglobin is accomplished by the addition of cold, saturated ammonium sulfate to a final concentration of 50%. After mixing, the tubes are centrifuged at 2000 g for 15 min and the supernatant fractions are removed. Precipitates are counted in the original tubes in an automated gamma counter; a minimum of 10,000 counts/tube is obtained. Binding data are calculated by a modification of the log–logit method using a programmable electronic calculator or time-sharing computer.[9]

This assay can detect 0.5 ng of myoglobin and is not affected by hemolysis, lipemia, or storage of serum at − 20°. In a single assay, triplicate determinations generally agree within 5%. The coefficient of variation between assays is 3–11%.[4,10] There is no evidence of significant cross-reactivity with any of the following proteins: human hemoglobin A, lactate dehydrogenase, creatine kinase, citrate synthase, cytochrome c, or glucose-6-phosphate dehydrogenase. The mean serum myoglobin value for 92 normal adults is 28.9 ± 17.3 (SD) ng/ml with a range of 6–85 ng/ml. A value in excess of 85 ng/ml is taken as abnormal.[4,10] The same radioimmunoassay can be employed to detect myoglobinemia in dogs except that canine cardiac muscle myoglobin is used as antigen.[11] The method is not satisfactory for quantification of myoglobin in urine; possible explanations include alteration of the myoglobin molecule in urine or the presence of an inhibitor. In addition, myoglobin is known to be catabolized by the kidney and other organs so that the amount excreted in the urine is only a fraction of that present in plasma.[1]

The assay described above can be shortened appreciably by reducing the incubation interval from 24 to 4 hr.[10,12] This is accomplished by the use of siliconized tubes and by placing the tubes on a shaker. The shorter assay is as sensitive as the 24-hr procedure (Fig. 2) but the coefficient of variation is 18%.[10] Elevated serum levels of myoglobin are adequately detected utilizing the abbreviated method.

The separation of free from antibody-bound labeled myoglobin also can be performed using a second antibody (e.g., sheep or goat anti-rabbit IgG).[4] A commercial kit using this method is available and appears to

[9] D. Rodbard, W. Bridson, and P. L. Rayford, J. Lab. Clin. Med. **74**, 770 (1969).
[10] G. Gilkeson, M. J. Stone, M. Waterman, R. Ting, C. E. Gomez-Sanchez, A. Hull, and J. T. Willerson, Am. Heart J. **95**, 70 (1978).
[11] M. J. Stone, M. R. Waterman, L. R. Poliner, G. H. Templeton, L. M. Buja, and J. T. Willerson, Angiology **29**, 386 (1978).
[12] M. J. Stone, M. R. Waterman, D. Harimoto, G. Murray, N. Willson, M. R. Platt, G. Blomqvist, and J. T. Willerson, Br. Heart J. **39**, 375 (1977).

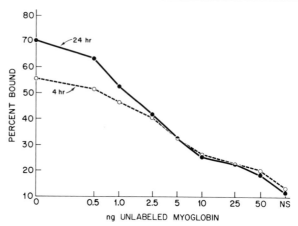

FIG. 2. Standard curves utilizing the 4- (dashed line) and 24-hr (solid line) radioimmuno-assay determinations for serum myoglobin. (Reproduced from the *American Heart Journal*.)

yield satisfactory results.[13-15] Other investigators have reported radioim-munoassay procedures which appear comparable to the one described above.[8,16,17]

Conditions Associated with Hypermyoglobinemia

Serum myoglobin levels determined by radioimmunoassay are ele-vated in more than 90% of patients with clinical and laboratory evidence of acute myocardial infarction (AMI), if serum samples are obtained on admission to the hospital.[4,8,12-15,17] In patients with chest pain or conges-tive heart failure but without evidence of AMI, serum myoglobin levels remain normal.[4,12] Several clinical studies have documented that hyper-myoglobinemia often precedes elevation of serum creatine kinase in pa-tients with AMI.[4,8,12] In dogs with experimentally produced AMI, serum myoglobin values rise within 2 hr and peak within 6 hr following coronary artery occlusion.[11] Moreover, peak serum myoglobin levels correlate with histologic infarct size.[11]

Since myoglobins of skeletal and cardiac muscle origin are immuno-

[13] A. P. Varki, D. S. Roby, H. Watts, and J. Zatuchni, *Am. Heart J.* **96**, 680 (1978).
[14] N. P. Kubasik, W. Guiney, K. Warren, J. P. D'Souza, H. E. Sine, and B. B. Brody, *Clin. Chem.* **24**, 2047 (1978).
[15] D. K. Oxley, M. R. Bolton, and C. W. Shaeffer, *Am. J. Clin. Pathol.* **72**, 137 (1979).
[16] T. G. Rosano and M. A. Kenny, *Clin. Chem.* **23**, 69 (1977).
[17] K. Miyoshi, S. Saito, H. Kawai, A. Kondo, M. Iwasa, T. Hayashi, and M. Yagita, *J. Lab. Clin. Med.* **92**, 341 (1978).

SERUM MYOGLOBIN LEVELS DETERMINED BY RADIOIMMUNOASSAY IN
VARIOUS CONDITIONS[a]

Conditions in which serum myoglobin levels are increased
Acute myocardial infarction (AMI)
Open heart surgery
Exhaustive exercise
Skeletal muscle damage
Patients and genetic carriers of progressive muscular dystrophy
Shock
Severe renal failure
Following intramuscular injections (variable)
Conditions in which serum myoglobin levels remain normal
Healthy adults
Chest pain without AMI
Congestive heart failure without AMI
Cardiac catheterization
Cardioversion
Moderate bicycle ergometer exercise

[a] See text.

chemically indistinguishable, the radioimmunoassay described is not specific for the cardiac muscle protein. Skeletal muscle damage (rhabomyolysis) also is associated with increased serum myoglobin levels.[4,12,17–19] Hypermyoglobinemia has been detected in patients and genetic carriers of progressive muscular dystrophy and thus may constitute a useful adjunct in carrier detection of this sex-linked disorder.[17,20] Raised serum myoglobin levels also have been observed during the first 24 hr after open heart surgery, and in patients in shock or with severe renal failure.[10,12,13] Serum myoglobin levels may be variably elevated following intramuscular injections; a rise in myoglobin values may be related to specific types of medication as well as to depth of intramuscular injection.[4,13] Serum myoglobin levels remain normal in patients undergoing cardiac catheterization, elective cardioversion, and bicycle ergometer exercise.[10,12] The various conditions associated with elevated or normal levels of myoglobin in serum are listed in the table.

Acknowledgments

This work was supported in part by NIH Ischemic Heart Specialized Center of Research Grant HL-17669.

[18] W. S. Ritter, M. J. Stone, and J. T. Willerson, *Arch. Intern. Med.* **139,** 644 (1979).
[19] L. J. Kagen, *J. Am. Med. Assoc.* **237,** 1448 (1977).
[20] L. J. Kagen, *Arch. Intern. Med.* **139,** 628 (1979).

Section IV

Nucleic Acids and Their Antibodies

[13] Quantitation of Submicrogram Amounts of DNA by Rocket Electrophoresis

By CHARLES R. STEINMAN

Introduction

Colorimetric or fluorimetric assays for DNA, particularly at concentrations of about 1 μg/ml, have generally been found to lack adequate sensitivity and/or specificity especially when potentially interfering cross-reactants were present as is frequently the case with crude biological mixtures.[1-5] Of the remaining nonimmunologic assays, the use of RNA–DNA hybridization has allowed quantitation of DNA in this range, but is difficult to perform and is of only limited applicability.[6] Among immunoassays for DNA, counterimmunoelectrophoresis and hemagglutination inhibition have been successfully employed as semiquantitative techniques.[5,7] For more precise quantitation assays based on complement fixation or competitive radioimmunoassay have generally been used,[8-10] although each of these approaches also has limitations.

Rationale

The method of "rocket immunoelectrophoresis," as described by Laurell,[11] offers a convenient, simple approach to this problem provided one can extend its sensitivity to concentrations of DNA in the range of 1 μg/ml and below since visible "rockets" are not formed under these conditions. In order to allow visualization of these rockets we stained the washed plates with ethidium bromide, a dye that binds to polynucleotides

[1] G. N. Abraham, C. Scaletta, and J. H. Vaughan, *Anal. Biochem.* **49**, 547 (1972).
[2] T. Yamashita and M. Yamada, *Chem. Abstr.* **46**, 10362b (1952).
[3] P. G. Rochmis, H. Palefsky, M. Becker, H. Roth, and N. J. Zvaifler, *Ann. Rheum. Dis.* **33**, 357 (1974).
[4] G. M. Richards, *Anal. Biochem.* **57**, 369 (1976).
[5] C. R. Steinman, *J. Clin. Invest.* **56**, 512 (1975).
[6] C. R. Steinman, *J. Clin. Chem.* **21**, 407 (1975).
[7] D. Koffler, R. Carr, V. Agnello, R. Thoburn, and H. G. Kunkel, *J. Exp. Med.* **134**, 294 (1971).
[8] M. Seligman and R. Arana, *in* "Nucleic Acids in Immunology" (O. J. Piescia and W. Braun, eds.), p. 98. Springer-Verlag, Berlin and New York, 1968.
[9] E. V. Barnett, *Arthritis Rheum.* **11**, 407 (1958).
[10] G. R. V. Hughes, S. A. Cohen, R. W. Lightfoot, Jr., J. I. Meltzer, and C. L. Christian, *Arthritis Rheum.* **14**, 259 (1971).
[11] C. Laurell, *Anal. Biochem.* **15**, 45 (1966).

with consequent marked enhancement of its red fluorescence upon ultraviolet irradiation.[12] After removal of unbound dye the precipitates, although still not visible even under UV illumination, could be photographed by prolonged exposures using a red filter to eliminate the background resulting from the visible blue light emitted by the UV source. In the course of these studies it was found that standard solutions of DNA added to normal human (DNA-free) plasma yielded anomalously low assay results when compared with identical concentrations of DNA in buffer. Postulating that DNA–protein aggregates prevented free migration of DNA into the gel in the case of plasma, we found that addition of 1 M guanidine hydrochloride to the test samples eliminated the discrepancy, presumably by disrupting such aggregates.

Materials

DNA was either from human placenta (Calbiochem) or calf thymus (Worthington) and was quantitated by optical density at 260 nm assuming $E_{1\,cm}^{1\,\%} = 200$. Ethidium bromide and agarose were purchased from Sigma. The latter material was described as exhibiting "medium" electroendosmosis. Guanidine hydrochloride (Ultrapure grade) was from Schwarz/Mann. Antiserums from four different patients with SLE have been successfully employed in this assay although the maximum achievable sensitivity varied somewhat. All antiserums exhibited precipitin lines against native DNA on Ouchterlony analysis, as well as by counterimmunoelectrophoresis (CIE).[5] None reacted with RNA by either technique. All exhibited >90% binding in a modified Farr assay using synthetic dsDNA[13] and also reacted with this antigen by CIE. Two of the four reacted with heat-denatured DNA by CIE. Preabsorption of all antisera with calf thymus DNA eliminated reactivity with both human DNA and dAT (the alternating copolymer of deoxyadenylate and deoxythymidylate) by CIE. Antiserums were heated at 56° for 30 min and clarified by centrifugation before use.

Methods

The method to be described is the semimicro technique we initially reported[12] and which requires only 3 μl of sample per assay. Subsequently a number of modifications were made which will also be commented on.

Agarose (0.7%) was dissolved in barbital buffer (pH 8.0, $\mu = 0.05$) in a boiling water bath. To 3 ml of this solution was added, after cooling to 50°, an experimentally determined volume of antiserum (in the range of 2

[12] C. R. Steinman, J. Immunol. Methods 31, 373 (1979).
[13] C. R. Steinman, U. Deesomchok, and H. Spiera, J. Clin. Invest. 57, 1330 (1976).

to 100 μl/ml of melted agarose). After thorough mixing, the solution was poured into a mold made from glass slides sealed with silicone grease and then allowed to solidify. Its dimensions were 75 $\times$ 30 $\times$ 1.2 mm. The covering glass plate was carefully removed and, using a hollow trochar, a series of wells, 2.5 mm in diameter, was punched in a row 7 mm from a long edge and 4 mm apart.

When serum or plasma specimens were to be assayed, they were first heated at 56° and clarified by centrifugation. They were then mixed with an equal volume of 5 M guanidine hydrochloride. A 6-μl portion of each specimen was then pipetted into a well. Apparently similar results were obtained if the 3-μl test sample was introduced directly into the well followed by 3 μl of 5 M guanidine, a somewhat simpler procedure. Each slide carried, in addition to test specimens, at least three standards made from DNA at known concentrations in phosphate buffered saline (PBS, 0.10 M NaCl, 0.05 M potassium phosphate, pH 7.0). These standards had been prepared by adding native DNA to normal, unheated human plasma, then heating at 56° for 30 min and clarifying by centrifugation. Filter paper wicks were then applied and electrophoresis carried out with the samples at the cathodal end for 16 hr at 4° and 1–2 V/cm. The gel was then washed overnight in PBS and stained for 30 min at room temperature in ethidium bromide, 40 μg/ml of PBS. After two 30 min washes with gentle agitation in PBS, the gel was transferred to a thin glass petri dish, covered with PBS, and photographed in a darkroom using a Polaroid MP-3 camera with a 5-in. lens, Polaroid type 47 film and a Wratten 25A filter at 1:1 magnification. The dish containing the gel was placed directly on a UVL-22 lamp (Ultraviolet Products, Inc.) resting face-up under the camera. Particular care was taken to remove or mask all sources of extraneous fluorescent material that might add to background fluorescence either directly or by reflection from the gel. Exposures were determined empirically and ranged from 2 to 15 min. The film was developed for twice the recommended time and the lengths of the precipitate arcs were measured on the film and quantitated by reference to the coelectrophoresed standards in the usual manner.

A standard curve for DNA in plasma is shown in Fig. 1. For this assay antiserum RR was incorporated into the agarose at a concentration of 33 μl/ml. The inset shows an extension of this assay to a DNA concentration of 0.125 μg/ml using antiserum RR at 8 μl/ml. At this level the assay quantitates 375 pg of DNA per 3 μl of original plasma sample present in each well.

To determine assay reproducibility DNA was added to 11 different normal plasma specimens at a concentration of 7 μg/ml. Each specimen was then assayed on a separately run gel, but under identical conditions. The mean length of the rockets thereby obtained was 8.3 mm with a stan-

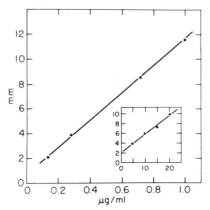

FIG. 1. Standard assay curves for DNA in plasma using antiserum RR as described in the text. From C. R. Steinman, *J. Immunol. Methods* **31**, 373 (1979).

dard error of the mean of 1.97 mm. Treatment of DNA-containing plasma with DNase I, after addition of $MgCl_2$, 0.01 M, prevented formation of a detectable precipitate. Pretreatment with heat-denatured DNase I did not affect the assay. RNA was not detectable with two different antiserums. The presence of rheumatoid factor in the test plasma had no apparent effect on the assay. Heat-denatured DNA gave a poorly outlined rocket with one antiserum.

The use of a second antiserum to assay DNA in plasma standards and in two normal serums is shown in Fig. 2. In this regard it should be noted that although normal plasma generally contains no DNA detectable by CIE (with a sensitivity of ca. 0.05 μg/ml), normal serum contains relatively high but variable concentrations of DNA.[5,15]

Comments

1. The assay is most reliable at concentrations of DNA of about 1 μg/ml and higher. At lower concentrations only exceptional antisera give reproducible results. Possibly a hybridoma-derived antibody[14,16] with well-defined properties will prove to be a more appropriate reagent should one become available.

2. The ability to photograph faintly fluorescent rockets depends largely on minimizing extraneous background fluorescence so as to maxi-

[14] B. Hahn, F. Ebling, S. Freeman, B. Clevinger, and J. Davie, *Arthritis Rheum.* **23**, 942 (1980).
[15] G. L. Davis and J. S. Davis, IV, *Arthritis Rheum.* **16**, 52 (1973).
[16] C. Andrjewski, Jr., B. D. Stollar, T. M. Lalor, and R. S. Schwartz, *J. Immunol.* **124**, 1499 (1980).

FIG. 2. Photograph of rockets obtained using antiserum Yu at 7 μl/ml of agarose. The wells contained, from left to right, plasma standards containing 0, 1, 2, and 3 μg/ml, respectively, of human DNA and normal serum from two individuals. The latter contained 3.1 and 2.3 μg/ml of DNA as determined by reference to the standards shown, a result similar to that obtained previously by semiquantitative CIE.[5] No rockets were visible on the gel, either under visible or ultraviolet light. From C. R. Steinman, *J. Immunol. Methods* **31**, 373 (1979).

mize contrast. In addition to the obvious need for meticulous cleanliness in this regard, it is necessary to keep the agarose moist since, when dry, the gels themselves are fluorescent. Other than this, the major source of background fluorescence is the plasma sample. This problem is easily solved by continuing electrophoresis long enough to cause non-DNA components to migrate off the far end of the gel. In this regard, it is necessary that the agarose exhibit a minimum of nonspecific protein binding as well as low electroendosmosis. Agarose obtained from Marine Colloids, FMC Corporation [grade HGT (P)] appears, in initial studies, to have such properties.

3. Distorted rockets occasionally result from erratic drying of sample wells during electrophoresis. A convenient means of preventing this is, after loading the wells with test samples, to fill them completely by carefully layering barbituate buffer over them and then placing a small strip of Saran wrap over the gel and the sample wells so as to exclude air from the

wells and prevent evaporation. Another cause of distorted rockets is poor adhesion of the gel to the glass backing, particularly around the sample wells, allowing partial leakage of test samples. Use of a hydrophilic plastic backing (Gel Bond from the Marine Colloids Division, FMC Corp.) over the glass has been found to eliminate this problem. However, because the plastic backing is faintly fluorescent, it must be removed before photographing. This can be accomplished by repeatedly flexing the plastic-backed gel and then gently sliding the gel off the backing manually. No significant distortion of the rockets has resulted from this manipulation.

4. The effect of molecular weight of the DNA on the rockets has not yet been systematically explored. It is clear, however, that high-molecular-weight DNA gives falsely low assay results apparently because then the sieving effect of the agarose acquires significant influence on mobility of the DNA and hence on the kinetics of its interaction with antibody. Thus intact lambda DNA with a chain length of about 49 kilobase pairs fails to produce a measurable rocket under routine conditions. Treatment of such samples by mild ultrasonication appears to eliminate this problem. Very low-molecular-weight DNA such as can be obtained by very intensive sonication of human native DNA, for example, also gives anomalous results. Although the lower size limit for reproducible assays has not yet been clearly established, preliminary size measurement of sonicated DNA specimens by gel electrophoresis suggests that significant anomalies do not occur until the DNA reaches a size of several hundred base pairs or less. As the DNA gets shorter, routine electrophoresis results in slightly longer, less well-defined rockets that yield progressively less EB fluorescence at fixed total DNA concentrations. Presumably, this is due to the requirement of a higher degree of antibody binding in order to immobilize short chain DNAs so that the final, immobilized complex has a higher antibody: antigen ratio than would longer chain DNA. This hypothetical explanation would account for both the slightly larger rockets formed as well as the weaker EB fluorescence (since antibody and EB might be expected to compete for binding to some extent).

Because the effect of DNA chain length on assay results has not yet been systematically explored, DNA used for calibration standards should have approximately the same chain length as does the unknown DNA if such information is available. If the size is unknown, it seems prudent for careful work to use a calibration standard of defined length and to consider the assay results to be of operational validity relative to the standards used rather than to be an absolute measure of DNA concentration, at least until the importance of this variable is further defined.

5. Since the initial description of this procedure, a number of simplifications have been introduced. Molten gels may be poured onto glass (or

plastic) slides rather than molded, if done on a level surface and with sufficient speed to prevent premature gel formation. When using such gels we
avoid punching wells within 6 mm of each end where the gel thickness
decreases. Addition of a low concentration of ethidium bromide (1 μg/ml)
directly to the molten gel (and also to the buffer in the electrophoresis
chamber) seems not to interfere with assays at $\geq$1 μg/ml of dsDNA and
greatly simplifies the procedure as no further staining or washing is necessary provided the plasma proteins are completely removed electrophoretically. Although one might expect antibody and dye to compete for similar
binding sites in some cases, this has not been a practical difficulty. Further, we have found that electrophoresis can be carried out at room temperature if tap water is circulated through a cooling platon under the gel.

If adequate amounts of test samples are available, use of larger wells
and larger sample volumes simplifies the procedure greatly. With some
antiserums it has been possible to complete electrophoresis in 4 hr at 3
V/cm. Exclusion of air and prevention of drying of sample wells as described above may be particularly important when using larger wells at
higher voltage gradients. Finally, we have found that some antiserums
yield rockets that do not begin at the well but rather at some distance ano-
dally. In such cases, adequate quantitation has been achieved by measuring the length of the rockets actually formed, ignoring the distance between the well and the base of the rocket.

Acknowledgments

This work was supported in part by NIH grant number AM15544 and by grants from the
SLE Foundation of America, the Hearst Foundation, and The Gina Finzi Lupus Research
Fund. The assistance of Nana Chan, Jae Kang, and Anita Sardo is gratefully acknowledged.

[14] Detection and Semiquantitation of DNA by Counterimmunoelectrophoresis (CIE)

By CHARLES R. STEINMAN

Introduction

Counterimmunoelectrophoresis (CIE) can, with minor modification
and suitable precautions, be used for detecting and semiquantitating DNA
in plasma and presumably other biological fluids. As little as 20 ng/ml of
dsDNA in plasma can be unambiguously detected by this means if an adequate antiserum is used. The major modification of the usual CIE method-

ology is addition of 1 M sodium chloride to the antigen well immediately prior to electrophoresis. This procedure enhances the sensitivity of the assay by more than an order of magnitude, presumably by disrupting noncovalently bound DNA–protein aggregates. If salt is not added in this way, a marked discrepancy is found between detectability of DNA in plasma and DNA in phosphate-buffered saline at the same concentration.[1,2] If electrophoresis is performed promptly the added salt does not appear to significantly increase the conductivity of the gel, even when multiple specimens are simultaneously assayed. Further, because of electroendosmosis (EEO) this added salt migrates away from the area of antigen–antibody interaction and therefore does not interfere with precipitate formation.

Since our experience with this assay has been confined largely to detecting DNA in human plasma, this procedure will be described in detail.[2] However, the method would be expected to be generally applicable. The procedure to be described is itself a modification of that described by Davis and Davis.[1]

Materials

Agarose purchased from Sigma with "medium EEO" has been employed for most of these studies although recently we have noted some variability in assay sensitivity that appears attributable to use of different batches of agarose. DNA is from human placenta (Calbiochem) and is quantitated by optical density at 260 nm assuming $E_{1\,cm}^{1\,\%} = 200$.

In order to avoid artifactual detection of DNA, precautions must be closely adhered to in collecting plasma specimens. Probably the major cause of false-positive results in plasma is inapparent partial clotting due to inadequate or delayed mixing of blood and anticoagulant (since DNA appears to be released into serum erratically during clotting[1,2]) although it is difficult to clearly demonstrate this to be the cause. The assay sensitivity is such that the DNA content of approximately 0.1% of the leukocytes in normal blood would be detectable if released into the fluid phase so that precautions for avoiding unnecessary trauma to the cells are routinely taken (although we have found in a limited series of experiments that intentional graded shearing of normal anticoagulated whole blood by passage through a 20-gauge needle generally results in visible hemolysis well before DNA release is detected). These routine precautions include collection of blood through a 19-gauge needle into a plastic syringe, followed

[1] G. L. Davis, Jr., and J. S. Davis, *Arthritis Rheum.* **16**, 52 (1973).
[2] C. R. Steinman, *J. Clin. Invest.* **56**, 512 (1975).
[3] C. R. Steinman, U. Deesomchok, and H. Spiera, *J. Clin. Invest.* **57**, 1330 (1976).

by prompt, gentle mixing (by repeated inversion) with an EDTA anticoagulant solution [0.07 ml containing 10.5 mg of K_3EDTA per 7 ml of whole blood as is contained in a standard "lavendar top" Vacutainer tube (Becton Dickinson) is adequate], centrifugation at 2000 g for 30 min, careful removal of the supernatant with a Pasteur pipet leaving a 1 cm layer of undisturbed plasma above the buffy coat, and finally recentrifugation of the supernatant as before, again leaving undisturbed plasma over the sediment at the bottom of the centrifuge tube. With this method, fewer than 3% of normal plasma specimens exhibit DNA by CIE.

The antisera found to yield the indicated level of sensitivity have all produced precipitin lines against dsDNA by routine double diffusion and, in addition, bind essentially 100% of double stranded DNA in direct binding assays.[3] However, not all sera with these characteristics are adequate for use in CIE so that empirical selection of appropriate antisera is also necessary. Further, potential reactivity with other antigens such as single stranded DNA, RNA, and other potential plasma components must be explored in defining the specificity of the serum chosen.

Methods

To conserve antiserum, we have generally used a semimicro technique. Rectangular glass plates 110 × 90 mm are covered, while on a level surface, with 18 ml of 0.7% agarose in barbital buffer (pH 8.0, μ = 0.05). Paired antigen and antibody wells are cut with a trochar, 2 mm in diameter, and 5 mm apart (center to center). Adjacent pairs are 10 mm apart so that as many as eight pairs can be cut in a row on a standard plate while still allowing for a 1 to 2 cm margin at each end. Three such rows may be cut on each plate with the top and bottom rows 20 mm from their respective edges to allow room for the wicks. Plasma samples to be assayed are first heated at 56° for 30 min and then clarified by centrifugation. For each of the paired rows, the wells on the cathodal side contain antigen (i.e., the test specimen) with antibody in those on the anodal side. Five microliters of each test sample is placed in an antigen well. Then the same volume of antiserum is placed in each antibody well. Finally, using a Hamilton syringe with a repeating dispenser, 1 μl of 5 M sodium chloride is added to each antigen well immediately before electrophoresis. No further attempt is made to mix this solution with the antigen already present in the well. Electrophoresis is then carried out at room temperature at 10 V/cm for 30 min. We routinely employ as positive controls standard solutions of human placental DNA (Calbiochem) in normal plasma (heat inactivated after the addition of DNA) in each corner of the plate, at 0.1 and 0.05 μg/ml. Plates are read blindly within an hour of electrophoresis.

Because the wells are small and the precipitin lines faint, at least at low DNA concentrations, the plates are read by indirect illumination from below by a focused bright pocket flashlight and with the aid of a hand lens. Precipitin lines generally are located close to the antibody well. Only sharply defined straight lines are considered positive. Artifactual precipitates are generally easy to identify either by their curved configuration or their indistinct outline. By this means even very faint lines, if sufficiently sharp, can with confidence be identified as immune precipitates, as can be confirmed by appropriately controlled studies. For critical work we have employed the following sets of controls for each positive plasma specimen. First, the specimen is digested for 30 min at 37° with electrophoretically purified pancreatic DNase I (Worthington) in the presence of 5 mM Mg^{2+} (in excess of the amount chelated by EDTA). A similar incubation may be carried out using DNase I that had been inactivated by heating at 100° for 10 min. A second set of controls employs antiserum that had been preabsorbed with an excess of either dsDNA or RNA in similar amounts. In this way specificity for DNA is established both by absorption with dsDNA (but not by RNA) and by susceptibility to digestion by a highly specific enzyme. The appearance of a CIE plate carrying a set of such controls (modified as indicated to facilitate reproduction) is shown in Fig. 1.

For semiquantitation, serial twofold dilutions of plasma are made in normal, DNA-free (by CIE) plasma and examined as before. The final dilution giving a visible precipitin line is taken as the end point. A very crude approximation of absolute DNA concentration can be made by determining the limit of DNA detectability using a series of standards containing decreasing concentrations of DNA in plasma and converting endpoint dilution as determined above to absolute concentration of DNA. An example of use of such an assay is illustrated in Fig. 2.

Comments

1. If enough material is available, larger wells and volumes can be used, greatly simplifying the reading of the plates.
2. Some antiserums can be diluted with buffered saline without apparent loss of sensitivity, thereby preserving these reagents.
3. Preelectrophoresis of plates with only the antibody wells filled (at 10 V/cm for 30 min), followed by addition of antigen and salt and then by routine electrophoresis, results in formation of immune precipitates further from the edges of the antibody wells so that they can be more easily seen. A similar result is obtained by using agarose with "high" electroendosmosis. However, at least in the latter case (and on ocassion even

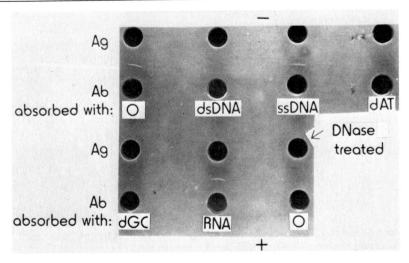

FIG. 1. Specificity controls for antiserum Yu. Antibody (Ab) wells all contained this anti-serum, absorbed as indicated and diluted twofold with buffered saline before use. Each absorbant was employed at 100 μg/ml of undiluted antiserum and was incubated at 37° for 1 hr before use. Antigen (Ag) wells all contained human DNA at 0.4 μg/ml of normal human plasma. In one Ag well, as indicated, this material had been digested with DNase I as described in the text. Abbreviations are as follows: 0, buffer control; dsDNA, native calf thymus DNA; ssDNA, heat denatured, quick-cooled calf thymus DNA: dAT, the alternating copolymer of deoxythymidylate and deoxyadenylate, a synthetic dsDNA: dGC, the corresponding copolymer of deoxyguanidylate and deoxycytidylate; RNA, yeast RNA. It should be noted that for ease of reproduction the illustrated CIE plate differs from that described in the text in that larger dimensions and volumes were employed. In addition the antibody was subjected to preelectrophoresis before proceeding with the assay in order to allow precipitin lines to form further from the antibody wells as is also described in the text.

under routine conditions) heavy diffuse artifactual precipitates are formed erratically on the cathodal side of the antibody wells with consequent interference with visualization of precipitin lines. In most such cases the precipitates appear to be euglobulins as they are removed by washing overnight at room temperature in buffered saline. Immune precipitates are not removed by this procedure so that such plates generally can be adequately interpreted in this way.

   4. Attempts to increase the sensitivity of the CIE procedure by staining the precipitin lines either with ethidium bromide followed by photography of the UV-illuminated plates as described elsewhere[4] or with bromophenol blue have not been successful. However, the former procedure can be used for additional confirmation of the DNA content at least of the

---

[4] C. R. Steinman, *J. Immunol. Methods* **31**, 373 (1979).

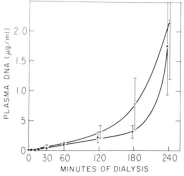

FIG. 2. Use of CIE as a semiquantitative method, as described in the text, to follow the course of release of DNA into the circulation during the course of hemodialysis. Each point represents the mean ± SEM of assay results on timed plasma specimens collected from six patients during the course of hemodialysis. Open circles indicate the result on plasma derived from blood traveling from dialysis coil to patient and closed circles from patient to coil. The examination of sonicated specimens, as described in the text, is also necessary to eliminate the remote possibility that such data might simply reflect a change in size of circulating DNA present at a fixed concentration throughout dialysis. From C. R. Steinman, *Am. J. Med.* **62**, 693 (1977).

stronger precipitin lines, if desired, since its relatively specific red fluorescence is clearly visible following such staining.

5. The effect of DNA chain length on assay sensitivity has only been partially explored. Some very large DNAs of uniform size such as intact λ phage (49 kilobase pairs long) are difficult to detect in this assay presumably because the gels sieving effect, even at 0.7% agarose, slows its migration sufficiently to prevent detection by this method. Therefore, if such DNA is suspected, test samples should first be subjected to ultrasonication. For example, using a Sonifier (Bronson) and the "standard" microtip with the sample contained within a 0.5 ml closed plastic centrifuge tube (Sarstedt) the sample is subjected, while submerged under water, to 60 sec of sonation at a power setting of "6."

In this procedure the probe is positioned (under water) so that it is *outside* the sealed sample-containing tube but within 5 mm of its contents. Enough ultrasonic energy is transmitted in this way to fragment the relatively brittle long dsDNA molecules while allowing multiple specimens to be treated without cross-contamination by the probe. By following this procedure, large dsDNA molecules are reduced to a size range of about 300 to 2000 base pairs (as determined by agarose gel electrophoresis) which is readily detectable at concentrations as low as 0.02 μg/ml with several of our antiserums. Further, these conditions for sonication leave a wide margin for error since an apparently similar size distribution is achieved by sonication times ranging from 30 to 120 sec. In this regard, it should be noted that the commercially obtained human DNA used as described above for positive controls is heterogeneous in size and ranges from approximately 400 to 15,000 base pairs long. Other commercially available eukaryotic DNA preparations may be of much higher molecular weight (frequently >20 kilobase pairs) and may therefore give anomalous CIE results unless first sonicated as described. The possibility of using preelectrophoresis of the antigens as an alternative to sonication in examining DNA of high molecular weight has not been explored.

At the lower end of DNA chain length we have found, using one antiserum, that the 79-base pair fragment of the HindII restriction endonuclease digest of ϕX 174 DNA forms a visible precipitate in a two-dimensional electrophoresis system where the first dimension was used to separate the fragments by size and the second to detect them by a "CIE-like" electrophoretic system with antiserum contained in a trough rather than a well. However, no reliable estimate of the sensitivity of the routine CIE assay for DNA of this size could be obtained in this way. Recent data on binding inhibition by dsDNA suggest that most such antisera bind dsDNA that is about 100 base pairs long but not 20 base pairs long,[5] a result consistent with our own findings.

6. High concentrations of DNA may result in falsely negative results by resulting in precipitation within the antibody well. Concentrations of DNA greater than about 3 to 5 μg/ml (depending on the antiserum) have generally given such a result. If these concentrations are suspected, appropriately diluted samples should be examined.

7. Since most SLE antiserums containing high concentrations of anti-dsDNA also react with ssDNA, these two partially cross-reacting antigens cannot be distinguished with certainty by such reagents. Although we have not attempted it, this distinction might be achieved by using appropriately absorbed antiserums and/or ssDNA-specific nuclease (e.g., S_1-nuclease) treatment of these specimens. A possible future alternative, should it become available, might be the use of a monoclonal antibody specific for the helical secondary structure of dsDNA and that might therefore not react with ssDNA although even then one would expect the short helical regions present in most mammalian DNA after heat denaturation to be recognized.

Acknowledgments

This work has supported in part by NIH grant number AM15544 and by grants from the SLE Foundation of America, the Hearst Foundation, and The Gina Finzi Lupus Research Fund. The assistance of Nana Chan, Jae Kang, and Anita Sardo is gratefully acknowledged.

[5] B. D. Stollar and M. Papalian, *J. Clin. Invest.* **66**, 210 (1980).

[15] Enzyme-Linked Immunosorbent Assay for Antibodies to Native and Denatured DNA

By Joan L. Klotz

Introduction

A variety of assays for measurement of antibodies to DNA have been described, the most sensitive being those utilizing radioisotopically labeled DNA, or a fluorescently labeled second antibody. Assays utilizing enzyme conjugated second antibodies have proved to be a valuable alternative to fluorescent and radioimmunoassays in a variety of serological tests, and this approach is easily adapted to detect antibodies to DNA. In addition to sensitivity and high specificity, enzyme-linked immunosorbent assays (ELISA) offer the advantages of low cost and simplicity. The enzyme-linked second antibody is a generally useful reagent for detecting a variety of different antigens and antibodies,[1-3] has a long shelf life and is not hazardous to use.

An enzyme linked immunoassay for antibodies to DNA was first described by Pesce *et al.* in 1974.[4] However, due to poor adsorbtion of native DNA (nDNA) to the polystyrene solid phase,[5] early assays detected primarily antibodies to single-stranded denatured DNA (dDNA). This difficulty is easily overcome by precoating the solid phase with a positively charged compound such as protamine sulfate.

Preparation of Alkaline Phosphatase Conjugated Antiimmunoglobulin Antibodies

I describe here the method for preparing alkaline phosphatase linked anti-human immunoglobulin. The conjugate can be made specific for human IgG, IgM, or other immunoglobulin classes by substitution of the appropriate antiserum for the anti-human immunoglobulin serum described here. Similar methods may be used to prepare conjugated antibody to immunoglobulins of other species such as the mouse.

---

[1] E. Engvall and P. Perlmann, *Immunochemistry* **8**, 871 (1971).

[2] E. Engvall and P. Perlmann, *J. Immunol.* **109**, 129 (1972).

[3] A. Voller, *Diagn. Horiz.* **2**, 1 (1978).

[4] J. Pesce, N. Mendoza, I. Boreisha, M. A. Gaizutis, and V. E. Pollak, *Clin. Chem.* **20**, 353 (1974).

[5] E. Engvall, *Lancet* **2**, 1410 (1976).

Preparation of a Glutaraldehyde Cross-Linked Immunoadsorbent

Immunoglobulins are precipitated from 50 ml of pooled normal human sera by addition of $(NH_4)_2SO_4$ to a final concentration of 45% saturation. The precipitated proteins are collected by centrifugation, redissolved in phosphate-buffered saline, 0.01 M phosphate with 0.15 M NaCl pH 7.4 (PBS), and dialyzed overnight against PBS. The precipitation and dialysis is repeated to minimize trapped serum protein contaminants, and the final solution adjusted to contain 50 mg protein/ml. Ten milliliters of this solution is used to form a glutaraldehyde cross-linked immunoadsorbent according to the procedure of Avrameas and Ternynck.[6]

Preparation of Specific Anti-Human Immunoglobulin Antibodies

Ten milliliters of antiserum to human immunoglobulins is added to the cross-linked immunoadsorbent and the mixture is stirred at room temperature for 60 min. The mixture is centrifuged at 2000 g for 10 min at 4°, and the gel washed several times with cold PBS until, after passing through a 0.8 μm Millipore filter, the absorbance of the supernatant, at 280 nm, is less than 0.05. The gel is then dispersed in 0.2 M HCl–glycine buffer pH 2.8 and stirred slowly for 15 min at 4° to elute the bound antibodies. After centrifugation, the supernatant is collected and neutralized with 1 M K_2HPO_4, the gel eluted a second time with the low pH buffer and the supernatant fractions combined. If it is to be used again, the gel is eluted a third time using 0.2 M HCl–glycine pH 2.2, neutralized with 10 ml K_2HPO_4 plus 40 ml distilled water, and finally resuspended in PBS containing 0.01% sodium azide. Antibody contained in the supernatant pool from the elutions at pH 2.8 is concentrated to 5 mg/ml. Concentration of protein without concomitant concentration of salts is conveniently accomplished using an Amicon pressure cell.

Conjugation of Specific Antibody to Alkaline Phosphatase[2]

Conjugation is carried out using 1.5 mg alkaline phosphatase and 0.5 mg immunoadsorbent-purified antibody. Using Sigma Type VII calf intestine alkaline phosphatase, 5 mg/ml in 3.2 M $(NH_4)_2SO_4$ (Sigma Chemical Co., St. Louis, MO) the procedure is as follows. Three-tenths milliliter of the alkaline phosphatase suspension is centrifuged, 1000 g for 10 min at 4°. The supernatant is removed and 0.1 ml of the antibody preparation added to the pellet. After dialysis overnight, the volume is determined (there will be some increase) and 10 μl of glutaraldehyde added

---

[6] S. Avrameas and T. Ternynck, *Immunochemistry* **6**, 53 (1969).

such that the final glutaraldehyde concentration is 0.2%. The mixture is incubated at room temperature for 2 hr and then diluted to 1 ml with PBS and dialyzed overnight against 1 liter of PBS. Finally, the conjugate is diluted to 10 ml with 50 mM Tris–HCl buffer pH 8.0 containing 5% bovine serum albumin (BSA), 0.02% NaN$_3$, and 1 mM MgCl$_2$. The dilute conjugate should be stored at 4°, it is not stable to freezing and defrosting.

Alkaline phosphatase conjugated antibodies, specific for the immunoglobulins of several different species, are now available commercially. The data presented in Fig. 2 and the table were obtained with an anti-human immunoglobulin conjugate purchased from Microbial Assoc. (Walkersville, MA).

Preparation of DNA

Since assays for anti-DNA activity are frequently performed using sera from systemic lupus erythematosus patients which may contain antibodies to a variety of nuclear proteins in addition to DNA, we customarily subject DNA obtained commercially to additional purification. DNA is dissolved at 4–5 mg/ml in 0.2 M Tris–HCl pH 7.4 containing 50 mM Na$_3$EDTA. Sodium dodecyl sulfate is added to a final concentration of 1%. Proteinase K (E. M. Laboratories, Inc., Darmstadt, G.F.R.) is added to a final concentration of 200 μg/ml and the solution incubated at 37° for 16–20 hr. This solution is extracted twice with chloroform–isoamyl alcohol (24:1) and the DNA recovered from the aqueous phase by spooling, after the addition of 2 volumes of absolute ethanol. The spooled DNA is washed twice in 2 liters of 70% ethanol, air dried, and dissolved in 5 mM Tris–HCl pH 7.5 containing 0.5 mM Na$_3$EDTA, to a final concentration of about 3 mg DNA/ml. The DNA solution is stored at 4°.

For use in the ELISA procedure DNA is diluted to the desired concentration, usually 25 μg/ml, with 50 mM Tris–HCl pH 7.5 containing 10 mM Na$_3$EDTA and 10 mM EGTA. Denatured DNA is prepared by boiling the dilute solution for 15 min followed by a rapid cooling on ice.

Preparation of Assay Tubes

One milliliter of 1% aqueous protamine sulfate (Sigma) is pipetted into 12 × 75-mm polystyrene tubes. The protamine sulfate solution is discarded after 90 min, the tubes washed 3 times with distilled water, inverted, and allowed to dry. These tubes have been stored for as long as 3 months before use, with no change in activity.

Assay for Anti-DNA Activity

For each serum dilution to be tested 4 tubes are used. Three of these contain dilute denatured or native DNA, the fourth containing the Tris–HCl buffer with no DNA added, serves as a zero DNA control. Solutions are added to, and removed from, the tubes according to the following sequence:

One milliliter dilute DNA or 50 mM Tris–HCl, 10 mM EDTA, 10 mM EGTA buffer. Incubate for 60 min.

One milliliter serum, diluted with 50 mM phosphate buffer containing 10 mM Na$_3$EDTA, 10 mM EGTA, 0.85% NaCl, 0.05% Tween 20, 1% BSA and 0.02% NaN$_3$, pH 7.8 (phosphate diluent). Incubate for 5 hr.

One milliliter conjugate diluted wih phosphate diluent. Incubate 16–18 hr.

One milliliter carbonate buffer, pH 9.8 with 1 mM MgCl$_2$, containing the alkaline phosphatase substrate, p-nitrophenyl phosphate (5 mg/ml). Incubate 30 min.

All incubations are at room temperature on a reciprocating shaker. Between each addition the tubes are emptied by aspiration and washed 3 times with PBS containing 0.05% Tween 20 (PBS-Tween). The enzymatic reaction is stopped at 30 min by addition of 0.1 ml 2 M NaOH. Hydrolysis of the substrate is monitored by measuring the absorbance of the final solution at 410 nm.

EDTA and EGTA are included in the Tris buffer and in the phosphate diluent to chelate magnesium and calcium ions and thus inhibit DNase activity. Phosphate-buffered physiological saline is used to dilute sera in order to avoid artifactual binding of immunoglobulins to DNA.[7] The Tween-20 and BSA are included to reduce nonspecific protein binding to the tubes.

For each batch of conjugate the appropriate end dilution must be determined empirically since there is variation from batch to batch. The avidity of the antisera from which the specific antibodies are purified is in large part responsible for this variation, but some variation is obtained between batches of conjugate prepared from the same immunoadsorbent purified antibody. In our laboratory, plasma obtained by plasmapheresis of a patient in SLE crisis is always used as a positive control when assaying human sera. Using this plasma diluted $\frac{1}{100}$, our standard conjugate dilution is chosen such that at the end of the assay, after 30 min incubation with substrate, the absorbance at 410 nm is between 1.0 and 1.5. A pool of sera from 6 month and older (New Zealand Black × New Zealand White)

---

[7] T. Pincus, *Arthritis Rheum.* **14**, 623 (1971).

F_1 mice is used as a positive control when assaying mouse sera. These positive controls are stored in aliquots at $-70°$.

S_1 Nuclease Treatment

When assaying sera for antibodies to nDNA, single-stranded contaminants are removed by S_1 nuclease treatment of the tubes after adsorption of DNA. Following incubation with dilute nDNA, the tubes are washed 3 times with PBS-Tween, and 1 ml 50 mM sodium acetate buffer pH 4.5 containing 50 mM NaCl, 5% glycerol, 1 mM ZnSO$_4$, and 100 units S_1 nuclease (Sigma Type III) is added to each tube. The tubes are incubated at 37° on a reciprocating shaker for 1 hr. Following this, the tubes are washed with PBS-Tween, dilute sera are added, and the assay for anti-DNA activity carried out as usual. Figure 1 illustrates the effect of incubating tubes coated with nDNA and dDNA with varying concentrations of S_1 nuclease. Different biochemical supply companies define units of S_1 nuclease differently, and measure the activity under different conditions. These differences must be considered when using S_1 nuclease from sources other than Sigma.

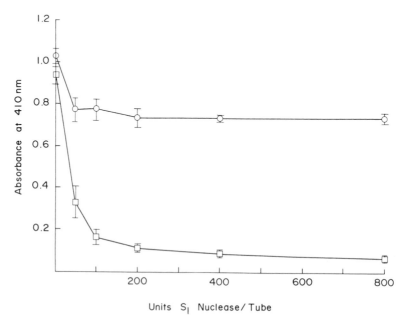

Units S_1 Nuclease/Tube

FIG. 1. The effect of incubating DNA-coated tubes with S_1 nuclease prior to the addition of SLE plasma, on the assay for anti-DNA activity. All tubes were incubated with 25 μg DNA prior to the addition of S_1 nuclease. O——O, nDNA adsorbed. □——□, dDNA adsorbed. Conjugate A $\frac{1}{200}$. The data represent the mean ± SEM of triplicate assays.[10]

Discussion

The effects of increasing concentrations of DNA during the antigen coating step, and of precoating the polystrene tubes with protamine sulfate are shown in Fig. 2. In tubes not coated with protamine sulfate, very little anti-nDNA activity is detectable, and then only at the highest DNA concentrations used. Although anti-dDNA activity is detectable in the

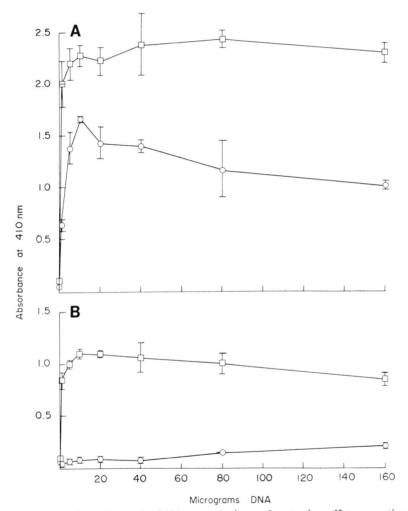

FIG. 2. The effect of increasing DNA concentration, and protamine sulfate precoating of the polystyrene tubes, on the assay for anti-DNA activity. (A) dDNA (B) nDNA. □——□, tubes precoated with protamine sulfate; O——O, tubes not precoated with protamine sulfate. Conjugate B $\frac{1}{500}$. The data represent the mean ± SEM of triplicate assays.

tubes not treated with protamine sulfate, the sensitivity of the assay is increased when this treatment is included (see the table). High concentrations of nDNA appear to have an inhibitory effect. This effect is probably due to release of antigen from the solid phase during the course of the assay, and a resulting competition for antibody between the released and bound DNA.

One advantage of the solid phase method over soluble phase assays when assaying for anti-nDNA activity is the ease with which one can remove single-stranded DNA contaminants by S_1 nuclease treatment. A significant amount of single-stranded DNA contamination is present in DNA purified by conventional techniques.[8] This problem can be avoided by using closed circular DNA, but commercial supplies of this are expensive and as yet not available in radiolabeled form. An alternative immunofluorescent assay using DNA in the kinetoplast of the hemoflagellate *Cirthidia luciliae* as antigen has proved useful in a number of laboratories.[9]

We have used the ELISA method for assaying anti-DNA activity in a large number of mouse sera, and a smaller number of human sera.[10] Most often the final readings obtained with the zero DNA controls are below 0.1, but occasionally we find sera, usually with high anti-DNA activity, which give higher values than this. Although the cause of these high control values has not been determined, we offer two possible explanations: the presence of DNA/anti-DNA complexes in the sera which bind to the protamine sulfate, or antibodies to basic nuclear proteins in the sera which are cross-reactive with the protamine sulfate.

The assay may also be carried out in polystyrene microtiter plates

TITRATION OF SLE PLASMA FOR ANTI-dDNA ACTIVITY,
COMPARING DATA OBTAINED WITH PROTAMINE SULFATE
PRECOATED AND NONPRECOATED TUBES[a]

	Absorbance at 410 nm	
Plasma dilution	+ Protamine sulfate	− Protamine sulfate
1/100	2.246 ± 0.051[b]	1.434 ± 0.150
1/1000	0.444 ± 0.029	0.187 ± 0.014
1/2000	0.200 ± 0.010	0.032 ± 0

[a] Using conjugate B 1/500.
[b] ± SEM of triplicate assays.

[8] R. I. Samaha and W. S. Irvin, *J. Clin. Invest.* **56**, 466 (1975).
[9] L. A. Aarden, E. R. deGroot, and T. E. Feltkamp, *Ann. N.Y. Acad. Sci.* **254**, 505 (1975).
[10] J. L. Klotz, R. M. Minami, and R. L. Teplitz, *J. Immunol. Methods* **29**, 155 (1979).

(Cooke Lab Products, Alexandria, VA), which allows one to screen a large number of sera simultaneously using small quantities of reagents. Although many laboratories lack the equipment for making spectrophotometric readings on such small volumes, positive and negative wells are easily distinguished by visual examination.

Acknowledgments

Figure 1 is reproduced from Klotz et al. (1975).[10] The careful technical assistance of R. M. Minami is greatfully acknowledged. This work was supported by grant GA-77-009 State of California, Department of Health, Preventive Medical Services Branch, Chronic Disease Control Section.

[16] Passive Hemagglutination and Hemolysis Tests for Anti-DNA Antibody

By TAKESHI SASAKI

Introduction

Several techniques are currently employed to detect the antibodies against DNA, including immunoprecipitation,[1] counterimmunoelectrophoresis (CIE),[2] DNA-coated bentonite particle flocculation,[3] immunofluorescence,[4] complement fixation,[5] and radioimmunoassay.[6-8] The usefulness of these assays is often limited by the lack of sensitivity, the reactions, and the complexity of the difficulty in separating single-stranded (ss) DNA from double-stranded (ds) DNA. Passive hemagglutination (PHA) is a sensitive and practical method for detection of antibodies and was applied to DNA by Jokinnen and Julkunen.[9] Clark et al.[10] reported a method for detecting the cells responding to ssDNA by rosette and plaque formation. We have modified their method using $CrCl_3 \cdot 6H_2O$ for the detection of antibodies against ssDNA and dsDNA and found it

[1] M. Seligma, C. R. Acad. Sci. Paris 245, 243 (1957).

[2] A. Klajuman, R. Farkarsh, and B. D. Myers, J. Immunol. 111, 1136 (1973).

[3] J. Bozicevich, J. P. Nason, and D. E. Kayhoe, Proc. Soc. Exp. Med. 103, 636 (1960).

[4] W. Crowe and I. Kushner, Arthitis Rheum. 20, 811 (1977).

[5] R. Ceppellini, E. Polli, and F. Celada, Proc. Soc. Exp. Biol. Med. 96, 572 (1957).

[6] R. T. Wold, F. E. Young, E. M. Tan, and R. S. Farr, Science 161, 806 (1968).

[7] R. I. Carr, D. Koffler, V. Agnello, and H. G. Kunkel, Clin. Exp. Immunol. 4, 527 (1969).

[8] T. Pincus, P. H. Schur, J. A. Rose, J. L. Recker, and N. Talal, New Engl. J. Med. 281, 701 (1969).

[9] E. J. Jokinen and H. Julkunen, Ann. Rheum. Dis. 24, 477 (1965).

[10] C. Clark, D. A. Bell, and J. H. Vaugman, J. Immunol. 109, 1143 (1972).

had many clinical and experimental advantages.[11,12] This article describes the current methods of the passive hemagglutination and passive immune hemolysis (PHL) tests for the detection of antibodies against ssDNA and dsDNA.

Methods

Materials

Buffers. Phosphate-buffered saline: 0.15 M sodium chloride containing 0.01 M phosphate buffer (PBS), pH 7.2: NaCl, 8.2 g; $Na_2HPO_4 \cdot 12H_2O$, 2.5 g; KH_2PO_4, 0.4 g; aq. dist. ad., 1000 ml.

Stock solution of subsequent veronal buffer (5 × VB): NaCl, 85.0 g; sodium diethylbarbiturate, 5.75 g; diethylbarbituric acid, 3.05 g; aq. dist. ad., 2000 ml.

Gelatin veronal buffer containing 0.0005 M Ca^{2+} and 0.00015 M Mg^{2+}(GVB^{2+}), pH 7.5: 5 × VB, 200 ml; 0.03 M $MgCl_2 \cdot 2H_2O$, 5 ml; 0.1 M $CaCl_2 \cdot 6H_2O$, 5 ml; 2% gelatin, 50 ml; aq. dist. ad., 1000 ml.

$MgCl_2$, $CaCl_2$, and gelatin solution are autoclaved and stored at 4°.

Alsever's solution: $C_6H_5O_7Na_3 \cdot 2H_2O$, 8.0 g; NaCl, 4.2 g; $C_6H_{12}O_6$, 20.5 g; $C_9H_5O_7 \cdot H_2O$, 0.55 g; aq. dist. ad., 1000 ml.

Sheep Red Blood Cells (SRBC). SRBC are aseptically collected in sterile modified Alsever's solution and kept at 4°. The cells are used after a period of 2–3 weeks following collection and are washed 3 times by physiological saline before use. The buffy coats on the surface of the packed cells are carefully removed.

Chromic chloride. Chromic chloride ($CrCl_3 \cdot 6H_2O$) powder is stored at 4° in a stoppered, light-proof container. A 20 mg/ml solution in physiological saline is freshly prepared each time in a clean tube.

Nuclease S_1. Nuclease S_1 from *Aspergillus oryzae* is purchased from Seikagaku Co. Ltd., Tokyo. The enzyme activity is stable at least for 1 year.

Complement. Guinea pig serum is used as a complement source. Lyophilized powders are obtained commercially and stored at 4°. Before assay these are dissolved in 0.1% GVB^{2+} and absorbed with an equal volume of 20% SRBC for 30 min on ice to remove natural hemolytic activity against SRBC.

Test Sera. All sera are inactivated by heating at 56° for 30 min before use. Then natural anti-SRBC antibodies are absorbed by incubating the

[11] T. Sasaki, Y. Ono, and N. Ishida, *J. Immunol.* **119**, 26 (1977).
[12] T. Sasaki, S. Ishida, S. Onodera, T. Saito, T. Furuyama, and K. Yoshinaga, *J. Immunol. Methods* **22**, 327 (1978).

serum with an equal volume of 20% washed SRBC at 37° for 1 hr, followed by an incubation at 4° overnight. The serum is recovered after centrifugation at 400 g for 5 min and diluted one-half in GVB^{2+}.

Antigen 1. A highly polymerized DNA from calf thymus (Sigma) is dissolved at a concentration of 500 $\mu g/ml$ in PBS.

2. Sonication of DNA (soDNA): DNA solution is sonically disrupted using a sonicator (Tomy Model UR-150P) in an ice bath. Sonication is carried out at 20 kc for 1 min and stopped to avoid heat degradation of DNA during sonication. This procedure is repeated 10 times. This treatment makes the DNA homogenous in size (Fig. 1) and is necessary to eliminate the false positive reaction in PHA test.

3. Preparation of ssDNA: Sonicated DNA is heated in a boiling water bath for 15 min and cooled rapidly in an ice bath. This treatment is repeated at least twice.

4. Preparation of dsDNA: Commercially available native DNA is not homogeneous with respect to double strandness. Chromatography on methylated bovine albumin Kieselguhr (MAK) or hydroxyapatite (HAT) has been widely employed to separate ssDNA from dsDNA.[13-15] How-

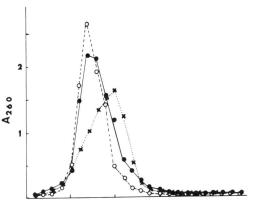

Fraction Number

FIG. 1. Sucrose density-gradient centrifugation of DNA. One milliliter of each DNA solution was layered on a 50 ml linear 5 to 20% (w/v) sucrose gradient containing PBS and centrifuged at 24,000 rpm for 16 hr at 4° in a RPS 25-2 swingbucket rotor in a Hitachi 65-P centrifuge. ●———●, DNA (500 $\mu g/ml$) sonicated for 10 min; ×········×, DNA (200 $\mu g/ml$) sonicated for 5 min; ○--------○, dsDNA (500 $\mu g/ml$) digested with nuclease S_1.

[13] M. Tan and M. V. Epstein, *J. Lab. Clin. Med.* **81**, 122 (1973).
[14] D. Alarcon-Segovia, E. Fishbein, and S. Estrada-Parra, *J. Rheumatol.* **2**, 172 (1975).
[15] R. J. Harbeck, E. J. Bardana, P. F. Kohler, and R. I. Carr, *J. Clin. Invest.* **52**, 789 (1973).

ever, Locker *et al.*[16] have pointed out that dsDNA had to be purified using a column of benzoylated-naphthoylated-DEAE cellulose if the antibodies specific for dsDNA are to be detected. Figure 2 shows that soDNA also contains ssDNA or ss regions in the dsDNA molecules. dsDNA is prepared by pretreatment of soDNA with nuclease S_1, an endonuclease which selectively splits the phosphodiester bond in ssDNA and RNA. The procedure is performed according to the method of Shishido and Ando.[17] A mixture consisting of 1 ml of soDNA (500 $\mu g/ml$), 0.2 ml of 0.2 M acetate buffer (pH 5.0) supplemented with 0.0001 M $ZnCl_2$, 0.1 ml of nuclease S_1 solution (10000 units/ml), and 0.7 ml of distilled water is incubated at 37° for 30 min. The mixture is then cooled to stop the reaction and dialyzed against several changes of PBS at 4° for 24 hr. With the procedure described above, 80–85% of the DNA is recovered, as measured

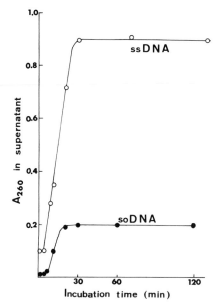

FIG. 2. Digestion of sonicated DNA by nuclease S_1. Each DNA was treated with 100 μg of nuclease S_1 as described under Methods. To the incubation mixture 0.1 ml of bovine albumin (1 mg/ml) and 2.2 ml of 10% perchloric acid were added and chilled for 20 min in an ice bath. After centrifugation, absorption of the supernatant of soDNA (●--------●) or heat-denatured ssDNA (○---------○) was measured at $A_{260\ nm}$. Reproduced from T. Sasaki, *J. Immunol. Methods* **22**, 327 (1978).

---

[16] J. D. Locker, M. E. Medof, B. M. Bennet, and S. Sukuhupunyaraksa, *J. Immunol.* **118**, 694 (1977).

[17] K. Shishido and T. Ando. *Biochim. Biophys. Acta* **287**, 477 (1972).

by OD_{260} in the restored solution. dsDNA is stored frozen at $-20°$ until use.

Coupling

One milliliter of ssDNA (500 μg/ml) is mixed with 0.8 ml of 0.9% saline and 0.1 ml of $CrCl_3 \cdot 6H_2O$ (20 mg/ml). dsDNA (200 μg/ml) is similarly mixed with 0.8 ml of 1.2% saline and 0.1 ml of $CrCl_3 \cdot 6H_2O$. To this mixture is added 0.2 ml of a 50% SRBC. The suspensions are mixed rapidly and kept at room temperature for 5 min. Several milliliters of cold physiological saline are added to avoid spontaneous nonspecific clumps. The cells are then centrifuged for 10 min at 12,000 rpm. They are washed three times as completely as possible to eliminate free DNA antigen and finally made up to 10 ml with GVB^{2+} to give a 1% cell suspension. The cells are stored at $4°$ in veronal buffer and can be used for at least 5 days. The suitable concentration of ssDNA in the sensitizing procedure is 500 μg/ml. The use of dsDNA at concentrations higher than 200 μg/ml often leads to nonspecific agglutination of SRBC while concentration lower than 100 μg/ml leads to less effective antigen specificity.

Hemagglutination Test (PHA)

The test is carried out in a microtiter system using V-type microtitration plates (Cooks Engineering Co., Alexandria, VA). Twofold serial dilutions of the tested serum are prepared with GVB^{2+} in a final volume of 25 μl per well (starting at a dilution of 1:8). Then 25 μl of 1% DNA–SRBC suspension is added to each well and mixed by gentle rotating. The plates are covered and placed in a $37°$ incubator. The settling patterns frequently are observed after 2 hr, but they are optimal following incubation overnight at $4°$. A complete carpet at the bottom of the well is recorded as 3^+, an even mat with a compact bottom is scored as 2^+, and a clearly defined small clump is scored 0. The last well giving a definite hemagglutination reaction (2^+, 3^+) is taken as the highest positive titer. Controls are included in each assay as follows: (a) negative control, serial dilutions of a negative serum and DNA–SRBC suspension; (b) positive control, serial dilutions of a positive serum of known titer and DNA–SRBC suspension; (c) serum control, 25 μl of tested serum and 25 μl of $CrCl_3 \cdot 6H_2O$ treated SRBC; (d) antigen control, 25 μl of buffer and 25 μl of DNA–SRBC suspension.

Hemolysis Test (PHL)

Twenty-five microliter of serum is diluted serially with a microdiluter in a U plate. With agitation 50 μl of guinea pig serum (complement source) diluted fivefold with GVB^{2+} followed by 25 μl of 1% DNA coated cell suspension are added to the diluted sera in each well. After the plates are well shaken, they are incubated at 37° for 1 hr and then the hemolytic patterns are read (Fig. 3). The degree of hemolysis of the indicated cells are recorded as a 0–4 scale (4^+, complete hemolysis; 2^+, 50% hemolysis; 0, no hemolysis). The titers of sera are expressed as the highest dilutions of serum resulting in 2^+, or higher hemolysis. Controls are as follows: (a) negative control, serial dilutions of a healthy serum, complement (C'), and DNA–SRBC suspension; (b) positive control, serial dilutions of a positive serum, C' and DNA–SRBC suspension; (c) serum control, serial dilutions of tested serum, C' and 25μl of $CrCl_3 \cdot 6H_2O$ treated cells; (d) antigen control, 25 μl of GVB^{2+}, C' and DNA–SRBC suspension; (e) complement control; 25 μl of positive serum, heat-inactivated C' or C' with the addition of 0.01 M disodium ethylenediamintetraacetic acid (EDTA) and DNA–SRBC suspension. Anticomplementary activity seems to be negligible in PHL tests on the basis of comparison with results of micro-complement fixation test (Fig. 4).

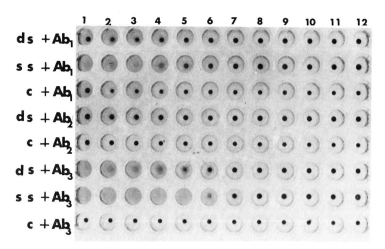

Fig. 3. A hemolysis titration plate incorporating red cells coated with DNA. Negative hemolysis in the first and fourth rows indicates absence of antibodies against dsDNA in Ab_1 and Ab_2, which were obtained from two rabbits hyperimmunized with methylated bovine serum albumin and ssDNA complexes. Second row (ss + Ab_1) shows positive hemolysis up to a dilution of 1:64 (twofold serial dilutions of the tested serum starting at 1:8). Row 6 or 7 illustrates, respectively, the presence of antibodies against dsDNA and ssDNA in Ab_3, serum from an active SLE patient. ss, ssDNA–SRBC suspension; ds, dsDNA–SRBC suspension; c, $CrCl_3 \cdot 6H_2O$ treated SRBC suspension; c + Ab, serum control; 12, antigen control.

Characteristics of the Tests

Antigenic Specificity of DNA–SRBC Suspension

Heat-denatured DNA contains nuclease S_1 resistant portions in its structure,[17] and free ssDNA diminishes plaque formation by anti-dsDNA producing cells in some NZB/W F_1 mice,[12] suggesting that there are immunologically available ds regions in heat-denatured DNA. Similarly soDNA–SRBC suspensions are capable of reacting with the antibodies against dsDNA and ssDNA.[18] Commercially available ^{14}C-labeled KB cell DNA and the DNA prepared by HAP or MAK column chromatography restore dsDNA and dsDNA with ss regions.[16] Since the solution containing dsDNA with ss regions is capable of binding to the antibodies against ssDNA, its use induces false positive results for the test of anti-dsDNA antibodies.[16] On the other hand, dsDNA digested with nuclease S_1 is specifically reactive with the antibodies with dsDNA specificity, based on inhibition in the PHA test, plaque assay,[12] and solid phase radioimmunoassay.[19]

Sensitivity

These tests described above are sensitive enough to estimate the antibodies against DNA clinically and experimentally. Comparison with other quantitative immunochemical techniques (CIE, conventional microcomplement fixation test) is given in Fig. 4. Two mercaptoethanol (2-ME) sensitive and resistant antibodies against DNA are detected using these assays (Fig. 5).

Incidence of Nonspecific Agglutination

It is well known that serum substances other than antibodies against DNA may bind to DNA and contribute to false positive results.[20] Procedures given in this chapter are necessary to minimize the incidence of nonspecific reactions. Use of tanned human type O, Rh-negative red cells is reported to avoid nonspecific agglutination without preabsorption of the sera with the cells.[18] However, human, chicken, bovine, and horse red cells treated with $CrCl_3 \cdot 6H_2O$ often induces nonspecific agglutination, which can be eliminated by preabsorption. Human RBC seems to be difficult to coat and standardize, compared to other RBCs. Buffers containing Ca^{2+} are recommended because these inhibit the nonspecific binding of

---

[18] Y. H. Inami, R. M. Nakamura, and E. M. Tan, *J. Immunol.* **109**, 1143 (1972).
[19] T. Sasaki and S. Ishida, *Scand. J. Immunol.*, in press.
[20] D. Koffler, V. A. Agnello, R. Thoburn, and H. G. Kunkel, *J. Exp. Med.* **134**, 169s (1971).

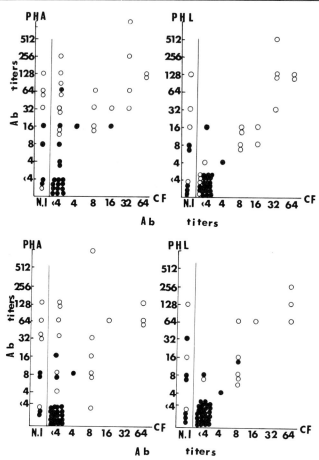

FIG. 4. Comparison of the titers of anti-ssDNA (upper) and anti-dsDNA (lower) antibodies judged by CIE, PHA, PHL and microcomplement fixation (CF) tests. ○, CIE positive; ●, CIE negative; N.I, The titers of the antibody were not identified by the anticomplementary activity in CF test. Reproduced from T. Sasaki, *J. Immunol. Methods* **22,** 327 (1978).

immunoglobulins to DNA.[21] Moreover, false positive reactions are reduced by using sonicated DNA.[12,18]

Applications

Anti-DNA Plaque Assay

DNA–SRBC suspensions can be also used for *in vivo* and *in vitro* quantitation of antibody-secreting cells or plaque-forming cells (PFC)

[21] A. Wakizaka and E. Okuhara, *J. Immunol. Methods* **25,** 119 (1979).

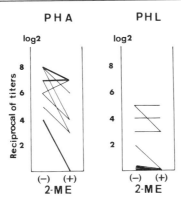

Fig. 5. 2-Mercaptoethanol sensitivity of agglutinating and hemolytic antibodies against DNA. Twofold diluted sera are incubated with an equal volume of 0.2 *M* 2-mercaptoethanol (2-ME) at 37° for 2 hr and then dialyzed against PBS for 48 hr.

against ssDNA and dsDNA. PFC to DNA are detected by the plaque assay of Jerne and Nordin.[22] The procedure is as follows: Eagles minimum essential medium is used in this assay. To the mixtures of 0.1 ml of a suitable dilution of the cell suspension being tested plus 0.1 ml of a 20% DNA–SRBC suspension or a 20% SRBC suspension are added 0.8 ml of 0.7% purified agar (Difco) containing 0.001% DEAE-dextran (Sigma). They are immediately transferred and dispersed in a 60 × 15-mm petri dish containing 10 ml of 1.4% purified agar. The dishes are left undisturbed at 37° for 1 hr. Then 1 ml of 20-fold diluted guinea pig serum previously absorbed with SRBC is added to the monolayer as a complement source. If indirect plaques are to be assayed, appropriate antiserum is added as well. After incubation at 37° for 45 min, the number of PFC is scored using a low-power dissecting microscope with dark-field illumination. Hemolytic plaques against DNA are small but fine. Incubation for longer than 2 hr or use of SRBC from older stock may lead to microplaques, resulting from cell breakdown. A high-power phase contrast microscope is used for the examination of doubtful plaques. The number of background PFC to SRBC must be subtracted from that of the corresponding DNA–PFC because DNA–SRBC suspensions react not only with the antibodies against DNA but also with natural antibodies against SRBC. Figure 6 shows the incidence of ssDNA and dsDNA PFC in the spleen from NZB/W F_1 mice which are known to develop autoimmune disease and lupus nephritis.

[22] N. K. Jerne and A. A. Nordin, *Science* **140**, 405 (1963).

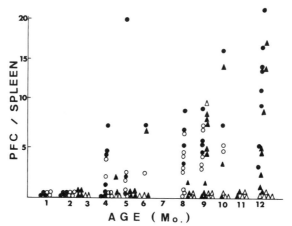

FIG. 6. Incidence of anti-ssDNA (○) and anti-dsDNA (△) producing cells in the spleen from NZB/W F_1 mice. Close symbols, female mouse; open symbols, male mouse.

Clinical Application

These tests are simple, convenient, and sensitive. PHL is a complement-mediated hemolysis and therefore capable of detecting complement-fixing antibodies to DNA. Measurement of complement-fixing antibody against DNA may be important in the management of systemic lupus erythematosus, especially with lupus nephritis.[12,23,24] The usefulness of complement-fixing test is, however, limited because of the presence of anticomplementary activity. Since PHL can be used without such interference, these tests are useful in diagnosing systemic lupus erythematosus, following its activity and evaluating the effect of treatment.

[23] T. Tojo and G. J. Frio, *Science* **161**, 904 (1972).
[24] P. H. Schur, E. Monroe, and N. Rothfield, *Arthritis Rheum.* **15**, 174 (1972).

[17] Radioimmunoassay of Antibody to DNA·RNA Hybrids of Naturally Occurring Sequence

By HIROSHI NAKAZATO

The RIA procedure described here was developed for the analysis of antibody specific for DNA·RNA hybrids of naturally occurring sequence.[1] The procedure is based on the use of (a) DNA·[³H]RNA hybrid

[1] H. Nakazato, *Anal. Biochem.* **98**, 74 (1979).

synthesized *in vitro* using single-stranded circular DNA of bacteriophages and *Escherichia coli* RNA polymerase as a radiolabeled antigen, (b) protein A-bearing *Staphylococcus aureus* cells as an adsorbent to separate antibody bound antigen from free antigen, and (c) a glass microfiber filter to collect the staphylococci for the measurement of the radioactivity of the antigen bound to the antibody. Because DNA·RNA hybrids of naturally occurring sequence and the hybrids of synthetic homopolymers are usually serologically distinguishable, it was important to use a DNA·RNA hybrid of naturally occurring sequence as a radiolabeled antigen.

It must be kept in mind that free IgGs of certain animals[2–4] and of some subclasses[5,6] of a given animal have very weak affinity for protein A. Although the apparent higher affinity of antigen–antibody complexes for protein A-bearing *S. aureus* cells[7] might allow the use of the cells in RIA even when the antibody is from an animal whose IgGs have low overall affinity to protein A, care must be taken to make sure that this is the case.

Preparation of DNA·[³H]RNA Hybrid[1]

Reagents
Tris·HCl, 1 M, pH 8.0
KCl, 1 M
$MnCl_2$, 1 M
Dithiothreitol (DTT), 0.1 M
Single-stranded circular DNA of ϕX174 or fd virus (Miles), 20 A_{260}
units/ml
RNA polymerase (nucleosidetriphosphate:RNA nucleotidyltransferase, EC 2.7.7.6) of *E. coli* (Miles), about 2500 units/ml
CTP, GTP and UTP, 20 mM solutions of each
[³H]ATP, 2 mM (2.5 mCi/μmol)
10% (w/v) sucrose in NET (0.1 M NaCl, 0.01 M EDTA, and 0.01 M
Tris–HCl, pH 7.2)
15% (w/v) sucrose in NETS (NET containing 1% SDS)
30% (w/v) sucrose in NETS
Care must be taken to avoid RNase contamination by autoclaving solutions, dissolving heat-labile reagents in autoclaved H_2O and heating glassware extensively (200°, overnight).

[2] G. Kronvall, U.S. Seal, J. Finstad, and R. C. Williams, Jr., *J. Immunol.* **104**, 140 (1970).
[3] I. Lind, I. Live, and B. Mansa, *Acta Pathol. Microbiol. Scand. Section B* **78**, 673 (1970).
[4] J. J. Langone, *J. Immunol. Methods* **24**, 269 (1978).
[5] M. R. Mackenzie, N. L. Warner, and G. F. Mitchell, *J. Immunol.* **120**, 1493 (1978).
[6] G. A. Madgyesi, G. Füst, J. Gergely, and H. Bazin, *Immunochemistry* **15**, 125 (1978).
[7] S. W. Kessler, *J. Immunol.* **115**, 1617 (1975).

Purification of ³H-labeled Precursor

[³H]ATP used as a precursor requires purification for maximal synthesis of the hybrid. [³H]ATP (5 mCi) in 10 ml of 50% ethanol is diluted with two volumes of H_2O and loaded on a Sephadex A-25 column (0.7 × 0.5 cm) which has been washed with 2 ml each of 0.1 N NaOH, 0.1 N HCl, and 5 ml of H_2O. After washing with 5 ml of H_2O and 5 ml of 0.05 M NaCl, [³H]ATP is eluted with 0.3 M NaCl and collected in 1-ml fractions. After measuring radioactivity in 1-μl aliquots, the peak fractions are pooled and 3 vol of ethanol are added. After standing overnight at $-20°$, [³H]ATP is pelleted by centrifugation (11,000 g, 1 hr) and dissolved in H_2O for RNA synthesis. Recovery is usually more than 90%. [³H]GTP can be purified similarly.

Procedure

The reaction mixture[8] (200 μl) contains in order of addition, 82 μl of H_2O, 17 μl of Tris–HCl, 10 μl of KCl, 20 μl of DTT, 15 μl of DNA, 4 μl of CTP, 4 μl of GTP, 4 μl of UTP, 40 μl of [³H]ATP, 2 μl of $MnCl_2$, and 2 μl of RNA polymerase. After 2 hr at 37° the reaction is stopped by addition of 20 μl of 0.2 M EDTA. The mixture is loaded on 0.8 ml of 10% sucrose in NET which has been layered on top of 10.7 ml of a 15% to 30% linear sucrose density gradient in NETS. The 10% sucrose layer is necessary to prevent the formation of insoluble potassium dodecyl sulfate. It is centrifuged for 17 hr at 26,000 rpm at 21° in a Spinco SW41 rotor. After fractionation and measurement of trichloroacetic acid-insoluble radioactivity, peak fractions sedimenting at about 21 S in the case of ϕX174 hybrid (Fig. 1A) and 23 S in the case of fd hybrid (not shown) are pooled and precipitated with 2 vol of ethanol overnight at $-20°$. After centrifugation for 1 hr at 11,000 g, the precipitate is dissolved in 360 μl of ice cold H_2O. Immediately after it is dissolved, 40 μl of 0.1 M Tris–HCl buffer (pH 7.4) containing 1 M NaCl, 50 mM EDTA, 0.2% NaN_3, and 0.67 μM of pentachlorophenol is added. The solution is recentrifuged to remove any insoluble aggregate and the supernatant is stored at 4°. Freezing and thawing should be avoided to prevent aggregation of the hybrid. The purity of the hybrid can be assessed by equilibrium density gradient centrifugation (Fig. 1B).

Quantity of the Hybrid

The quantity of the hybrid is expressed in terms of picomoles of nucleotide residues. The relationship of the radioactivity to the amount of

[8] J. G. Stavrianopoulos, J. D. Karkas, and E. Chargaff, *Proc. Natl. Acad. Sci. U.S.A.* **69**, 2609 (1972), with slight modifications.

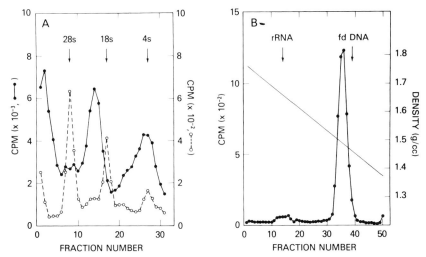

FIG. 1. (A) Sedimentation of *in vitro* transcription products of φX174 DNA through sucrose density gradient. See text for details. (●——●) trichloroacetic acid-insoluble radioactivity of ^{3}H-labeled transcription products. (○---○) ^{3}H-labeled HeLa cell cytoplasmic RNA markers sedimented in a parallel run. (↓) marker peaks. (B) Equilibrium density gradient centrifugation of φX174 DNA·RNA hybrid. Cs_2SO_4 (2.64 g) was dissolved in 3.5 ml of 0.02 M Tris–HCl buffer, pH 7.0 containing 0.15 M NaCl, 0.001 M EDTA, a small aliquot of pooled hybrid [fraction 11 to 17 in (A)] and ^{14}C-labeled HeLa cell cytoplasmic RNA. Centrifugation was for 2 days at 34,000 rpm at 18° in a Spinco SW 50L rotor. (●——●) ^{3}H radioactivity, (——) density. (↓) fd DNA and *E. coli* rRNA peaks in a parallel run. From Nakazato.[1] (Reproduced from H. Nakazato, *Anal. Biochem.* **98**, 74 (1979), with the permission of Academic Press, New York.)

hybrid is as follows: 5550 dpm corresponds to 1 pmol of [^{3}H]AMP incorporated; 1 pmol of [^{3}H]AMP incorporated corresponds to 2/0.328 pmol of φX174 DNA·RNA synthesized or 2/0.341 pmole of fd DNA·RNA synthesized (0.328 and 0.341 represent the TMP content of φX174[9] and fd[10] DNA, respectively; 2 corrects for the double-strandedness of the hybrid). The single-stranded DNA regions left untranscribed[11] are assumed to have similar base composition as the other parts of the DNA and/or are small enough to be neglected.

Preparation of Protein A-Bearing S. aureus Cells

Reagents and Materials

Heat killed, formalin-fixed, protein A-bearing *S. aureus* cells, see below

[9] R. L. Sinsheimer, *J. Mol. Biol.* **1**, 37 (1959).
[10] H. Hoffman-Berling, D. A. Marvin, and H. Dürwald, *Z. Naturforsch.* **18**, 876 (1963).
[11] A. Bassel, M. Hayashi, and S. Spiegelman, *Proc. Natl. Acad. Sci. U.S.A.* **52**, 796 (1964).

Buffer A: 0.02 M sodium phosphate buffer, pH 7.4 containing 0.15 M NaCl, 5 mM EDTA, 0.02% NaN$_3$, 0.067 μM of pentachlorophenol and 0.1% Tween 20 (Sigma)

Buffer A (0.5%): same as above except that it contains 0.5% Tween 20 instead of 0.1% Tween 20

Glass microfiber filter: grade GF/F (Whatman) (effective retention 0.7 μm) or its equivalent, 2.4 or 2.5 cm in diameter

Filter holder

Bovine serum albumin (BSA)

Ovalbumin

^{125}I-labeled IgG (0.5 ~ 1 × 10^6 cpm/μg)[12]

IgG (or serum) from the same animals

5% TCA (w/v)

95% ethanol

Sources and Pretreatment of S. aureus Cells

Heat-killed, formalin-fixed, protein A-bearing *S. aureus* cells are now commercially available from a few sources. Although the author has experience only with Pansorbin (Calbiochem), there is no reason to believe that the others are not as good. As is shown below, the IgG binding capacity of the Pansorbin was usually 2- to 3-fold less than that of the cells prepared by individual investigators.[7,13]

A 10% suspension of Pansorbin is centrifuged for 10 min at 11,000 g to pellet cells.[7] The cells are suspended in the equal volume of buffer A (0.5%) and left at room temperature for 15 min. They are pelleted again and washed twice with buffer A. Finally, they are suspended to 10% in buffer A containing 1 mg/ml of ovalbumin and 0.2 mM phenylmethylsulfonylfluoride (PMSF) and kept at 4° for a maximum of 2 weeks. Older preparations tend to give a higher background. A stock suspension (not washed) was kept at 4° for more than 6 months without trouble. Freezing and thawing of either suspension should be avoided.

Measuring the IgG Binding Capacity of the S. aureus Cells

The reaction mixture (in 150 μl of buffer A containing 200 μg of BSA) contains either (a) ^{125}I-labeled IgG (0.1 μg), varying amounts of homologous unlabeled IgG, and 50 μl of a 10% suspension of *S. aureus* cells; or (b) ^{125}I-labeled IgG (0.1 μg), 25 μg of unlabeled IgG, and varying amounts of *S. aureus* cell suspension.

---

[12] W. M. Hunter, *in* "Handbook of Experimental Immunology" (D. M. Weir, ed.), 3rd ed., p. 14. Blackwell, Oxford, 1978.

[13] S. Jonsson and G. Kronvall, *Eur. J. Immunol.* **4**, 29 (1974).

After standing for more than 10 min at room temperature, about 1 ml of buffer A is added and the mixture is vortexed vigorously. The mixture is poured onto a sampling manifold holding a glass microfiber filter which has been presoaked for about 30 min in buffer A containing 0.1 mg/ml each of BSA and polyvinyl sulfate. The emptied tube is rinsed with 3 ml of buffer A and the wash is poured onto the filter. This is repeated three times. The filter is further washed twice with 3 ml of buffer A and once with cold 5% TCA and 95% ethanol. The radioactivity on the dried filter is counted in a gamma counter.

In the case of the preparation used in the experiment shown (Fig. 2), 50 μl of the 10% suspension of *S. aureus* cells bound 25 to 30 μg of rabbit IgG. In both experiments, at most, 82–86% of ^{125}I-labeled rabbit IgG was bound to *S. aureus* cells and retained on a GF/F filter. The relatively low binding compared to Kessler's report[7] seems to be due to the degradative effects of iodination on the IgG of our preparation.

Note. A glass microfiber filter of larger pore size (e.g., GF/C) is not suitable for the experiment because of the poor retention of the cells.[1] A nitrocellulose filter should not be used because it binds proteins, single-stranded DNA, and DNA·RNA hybrid nonspecifically and also is clogged easily with small amounts of the cells.

The maximal workable volume of the 10% suspension of the cells is about 200 μl.

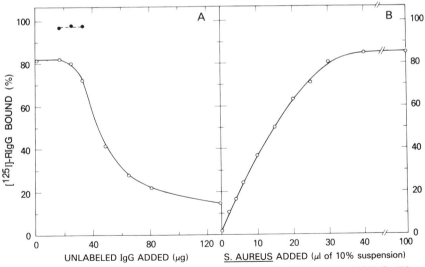

FIG. 2. Binding of ^{125}I-labeled rabbit IgG to *S. aureus* cells. ^{125}I-labeled rabbit IgG, 100 ng (75,500 cpm), was used. See text for details. From Nakazato.[1]

The nonspecific retention of the ^{125}I on the filter should be less than 2%.

By substituting the unlabeled IgG with serum, the amount of cells necessary to handle a volume of the serum can be obtained.

Characterization of the Antibody by Competition RIA

Reagents
Buffer A (see above)
Antibody (serum, immunoglobulin or IgG fraction)
Nonimmune serum, immunoglobulin or IgG fraction
Ovalbumin, 20 mg/ml in buffer A
DNA·[³H]RNA hybrid
Unlabeled nucleic acids: DNA·RNA hybrid is prepared as above using only unlabeled precursors. Single-stranded and double-stranded DNA of the viruses, single-stranded or double-stranded synthetic homopolymers, rRNA, and tRNA are commercially available.

Titration of Antibody

A reaction mixture (0.55 ml of buffer A) contains 40 μg of ovalbumin, 5 pmol (ca.2000 cpm) of DNA·[³H]RNA hybrid, and a varying amount of antibody. Nonimmune serum, the immunoglobulin fraction, or the IgG fraction is added to make the total protein content the same in each reaction.

After 30 min at 37°, an appropriate amount of a 10% suspension of *S. aureus* cells is added. After more than 10 min at room temperature, the cells are collected on a GF/F filter and washed as described above. A dried filter is heated for 30 min at 80° with 0.5 ml of Protosol (New England Nuclear) in a glass scintillation vial closed tightly with a cap with a cone-shaped plastic liner. After cooling, 10 ml of Econofluor (NEN) or toluene-based scintillation cocktail is added and the radioactivity is measured in a scintillation spectrometer.

Note. PMSF which had been included in the reaction mixture of the original report[1] is omitted because it is not necessary under normal circumstances.

It should be kept in mind that EDTA, a potent inhibitor of DNase, is also somewhat inhibitory in some antigen–antibody reactions.[12,14]

The nonspecific retention of DNA·[³H]RNA is about 1.5%.

The maximal retention of the antigen in the presence of saturating amounts of antibody depends on the extent of the purity of the antigen.

[14] H. Nakazato, *Biochem. Biophys. Res. Commun.* **96**, 400 (1980).

More than 90% of the ϕX174 DNA·[³H]RNA or fd DNA·[³H]RNA prepared as above should be retained (Fig. 3). The better binding of fd DNA·[³H]RNA hybrid than that of ϕX174 DNA·[³H]RNA hybrid at low antibody concentration seems to reflect the larger size of the former than that of the latter.

Because the retention pattern was sigmoidal (Fig. 3) and unlike the binding pattern where the binding of one antibody molecule results in the precipitation of one nucleic acid molecule,[15] the unequivocal quantitation of the antibody was not possible by this method.[16] In order to estimate the antibody activity of a sample,[16] one relative unit of antibody is defined as the amount of the antibody that binds 50% of the input hybrid.

By the method described above, three kinds of antibody could be detected.[16] One was single-stranded DNA-specific, another was DNA·RNA hybrid and double-stranded RNA-specific, and a third was specific for DNA·RNA hybrid. The first antibody could be detected because the viral

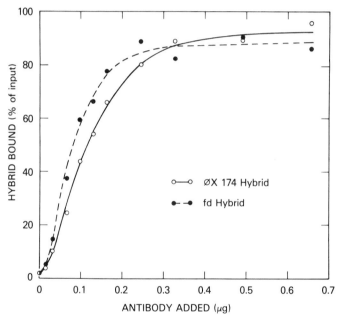

FIG. 3. Binding of ϕX174 DNA·RNA hybrid and fd DNA·RNA hybrid by anti-ϕX174 DNA·RNA hybrid antibody. ϕX174 DNA·[³H]RNA hybrid (5.16 pmol) (1880 cpm) or 5.38 pmol of fd DNA·[³H]RNA hybrid (2040 cpm) and the IgG fraction of anti-ϕX174 DNA·RNA hybrid antibody were reacted as described in the text. Nonimmune rabbit IgG was added to make the total rabbit IgG amount to 1 μg in each reaction. From Nakazato.[1]

[15] L. A. Aarden, F. Lakmaker, and E. R. DeGroot, *J. Immunol. Methods* **11**, 153 (1976).
[16] H. Nakazato, *Biochemistry,* **19**, 2835 (1980).

DNA·[³H]RNA prepared as mentioned above had a single-stranded DNA region.[11] The serological relatedness of the DNA·RNA hybrid and the double-stranded RNA was demonstrated by the production of antibodies cross reactive with DNA·RNA hybrid such as poly(dT)·poly(A) by poly(A)·poly(U) or poly(C)·poly(I).[17] φX174 DNA·RNA elicited an antibody which reacted with both DNA·RNA hybrid and double-stranded RNA.[16]

Thus, it is essential to assess the specificity of the antibody by competition RIA.

Competition RIA

Competition RIA is performed essentially the same as a titration of the antibody except: (1) A competitor is added and the reaction mixture is mixed well before adding the antibody. (2) The reaction is started by add-

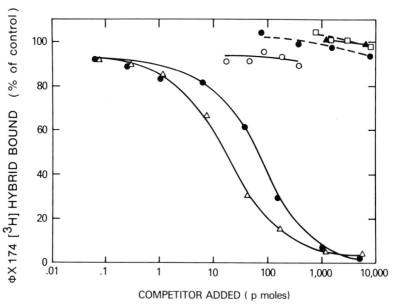

FIG. 4. Competition RIA of single-stranded DNA-specific antibody. Immunoglobulin fraction of anti-φX174 DNA·RNA hybrid was absorbed with poly(dT)·poly(A). After pelleting the immunoprecipitate, the supernatant was submitted to competition RIA. The binding of 5.5 pmol of φX174 DNA·[³H]RNA hybrid to 1.44 μg of the supernatant immunoglobulin was competed with: ●——●, φX174 DNA·RNA; △——△, φX174 single-stranded DNA; ○——○, poly(dC)·poly(I); ▲——▲, poly(dT)·poly(A); ●---●, poly(A)·poly(U); and □---□, rRNA. Without competition, 57% of the antigen was bound.

[17] B. D. Stoller, Crit. Rev. Biochem. 3, 45 (1975).

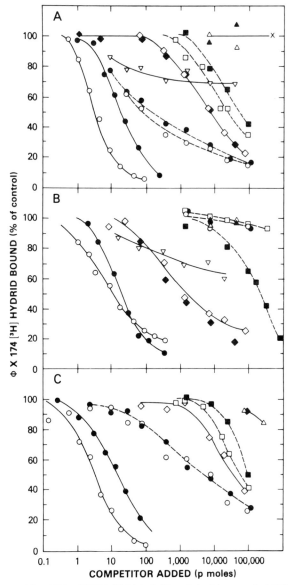

FIG. 5. Competition RIA of DNA·RNA hybrid and double-stranded RNA-specific antibody and DNA·RNA hybrid-specific antibody. (A) 4.6 ng; (B) 36 ng; (C) 5.4 ng of fractionated antibody was used as an antibody. The majority of (A) or (C) was DNA·RNA hybrid and double-stranded RNA-specific. (B) was a minor component of (A); isolated by absorbing the major activity to poly(A)·poly(U)-agarose. Binding of ϕX174 DNA·[³H]RNA (5.4 pmol) without a competitor was 49.8, 42.2, and 58.5%, respectively. Competitors were ●——●, ϕX174 DNA·RNA; △——△, ϕX174 single-stranded DNA; ▲——▲, ϕX174 RF double-stranded DNA; X, calf thymus single or double-stranded DNA; ○——○, poly(dC)·poly(I); ◇——◇, poly(dT)·poly(A); ◆——◆, poly(dI)·poly(C); ▽——▽, poly(dG)·poly(C); ●---●, poly(A)·poly(U); ○---○, poly(C)·poly(I); □---□, E. coli rRNA; and ■---■, E. coli tRNA. From Nakazato.[16]

ing an amount of the antibody which gives 40 to 50% binding of the input DNA·[³H]RNA hybrid.

As examples, the competition patterns of single-stranded DNA-specific antibody (Fig. 4), DNA·RNA hybrid, and double-stranded RNA-specific antibody (Fig. 5A and 7C) and DNA·RNA hybrid-specific antibody (Fig. 5B) are shown. For the numerical expression, the index of relative potency of a competitor is defined[16] as (amount of a competitor needed to inhibit the binding of DNA·[³H]RNA hybrid by 50%)/(Amount of unlabeled homologous DNA·RNA hybrid needed to inhibit the binding of DNA·[³H]RNA by 50%).

Note. Convenient amounts of a competitor for a preliminary test are 1, 10, 100, 1000, 10,000, and 100,000 pmol.

Care must be taken in pipetting to avoid carrying over a trace amount of a competitor from the tube containing the largest amount of one competitor to the next tube containing the least amount of another competitor.

If an antiserum has a mixed population of antibodies of the specificity mentioned above antibodies can be fractionated as described if necessary.[16] However, because the recovery of DNA·RNA hybrid-specific antibody may be low after fractionation,[16] the following method is recommended. First, absorb single-stranded DNA-specific antibody to single-stranded DNA-cellulose (P. L. Biochemicals or prepare[18]), then absorb double-stranded RNA reactive antibody to poly(A)·poly(U)-agarose or poly(C)·poly(I)-agarose. The fraction unabsorbed to either resin contains DNA·RNA hybrid-specific antibody and nonspecific immunoglobulin.

Poly(A)·poly(U)-agarose was prepared[16] as follows. A small column was packed with 1 ml of poly(U)-agarose (P. L. Biochemicals) and washed extensively with buffer A until the A_{260} of the wash fell to less than 0.01. About 80 A_{260} units of poly(A) in 0.4 ml of buffer A was added to the column and allowed to bind to the column for 10 min. After extensive washing, poly(A)·poly(U)-agarose was transferred to a graduated centrifuge tube and stored in the presence of 0.02% NaN₃ at 4°. About 50 A_{260} units of poly(A) were bound to 1 ml of agarose which originally contained about 70 A_{260} units of poly(U).

¹⁸ G. Herrick and B. Alberts, this series, Vol. 21D, p. 198.

Section V

Toxins

[18] The Enzyme-Linked Immunosorbent Assay (ELISA) for the Detection and Determination of *Clostridium botulinum* Toxins A, B, and E

By S. NOTERMANS, A.M. HAGENAARS, and S. KOZAKI

Introduction

The enzyme-linked immunosorbent assay (ELISA), developed recently, represents a significant addition to the existing detection techniques for botulinum toxins.

ELISA is similar in design to the radioimmunoassay (RIA), however, instead of a radioactive label an enzyme label is used. The enzyme label may be conjugated either to the antigen (e.g., to the toxin in a competitive ELISA) or to the antibody (e.g., to the IgG in the sandwich ELISA). When determining botulinum toxins labeling the toxin with an enzyme is less advisable, as handling highly toxic substances, like botulinum toxins in relatively high concentrations, requires extreme care and active immunization of laboratory personnel. Furthermore, sufficient quantities of antibodies can easily be obtained by immunizing rabbits with relatively small quantities of toxin. Therefore, labeling the antibodies is preferable.

In the following two ELISA sandwich techniques for the determination of botulinum toxins are described in detail. Details of the production of the different reagents to be used in the ELISA are also given. Finally some results with special reference to sensitivity, cross-reactivity, and applicability of ELISA for the determination of botulinum toxins are presented and discussed.

Sandwich Techniques for the Determination of Botulinum Toxins

In Fig. 1 both the sandwich ELISA and the "double sandwich" ELISA are represented schematically. The sandwich ELISA, originally described by Engvall and Perlmann,[1] is performed using polystyrene tubes which are coated with anti-botulinum IgG, e.g., from rabbits. After incubation with the toxin, the amount of adsorbed toxin is measured using anti-botulinum IgG conjugated to an enzyme. The amount of enzyme is determined spectrophotometrically after the addition of a suitable substrate.

In the so-called "double-sandwich" ELISA a second anti-botulinum

---

[1] E. Engvall and P. Perlmann, *Immunochemistry* **8**, 871 (1971).

METHODS IN ENZYMOLOGY, VOL. 84

Sandwich technique:

Rabbit IgG-Toxin-Rabbit IgG ∞ Enzyme

Double-sandwich technique:

Rabbit IgG-Toxin-Horse antitoxin-rabbit antihorse IgG ∞ Enzyme

FIG. 1. Schematic presentation of ELISA techniques for determination of botulinum toxins

Ig is bound to the adsorbed toxin. The antibodies in this case have to be obtained from an animal species other than was used for the production of the coating antibodies (e.g., horse). The amount of adsorbed antibodies is measured with anti-horse Ig–enzyme conjugate. For the first sandwich technique each toxin-specific antiserum has to be conjugated to the enzyme. An advantage of the "double sandwich" technique is that the same anti-Ig–enzyme conjugate can be used for the detection of all botulinum toxins.

Production of Reagents to Be Used in the ELISA

Production of Antiserum

Seven types (A, B, C, D, E, F, and G) of *Clostridium botulinum* strains are currently recognized on the basis of antigenically distinct toxins. Types A, B, and E strains are the principal causes of botulism in man. *C. botulinum* type A produces at least three different progenitor toxins of molecular size 19 S, 16 S, and 12 S, respectively. *C. botulinum* type B produces two progenitor toxins (16 S and 12 S). For type E only a 12 S progenitor toxin can be demonstrated.[2-6] The different progenitor toxins can be shown in specific culture media and they can be obtained in pure form. Dissociation of the molecules occurs when the toxins are exposed to pH higher than 7.5.[3,4,6] The 16 S toxin, known as Large-toxin (L-toxin), is dissociated into two components: a nontoxic component with

[2] C. Lamanna and G. Sakaguchi, *Bacteriol Rev.* **32**, 242 (1971).
[3] S. Kozaki, S. Sakaguchi, and G. Sakaguchi, *Infect. Immun.* **10**, 750 (1974).
[4] S. Sugii and G. Sakaguchi, *Infect. Immun.* **12**, 126 (1975).
[5] S. Kozaki and G. Sakaguchi, *Infect. Immun.* **11**, 932 (1975).
[6] M. Kitamura, S. Sakaguchi, and G. Sakaguchi, *Biochim. Biophys. Acta* **168**, 207 (1968).

hemagglutinin activity and the toxic component (7 S). The 12 S toxin, known as Medium-toxin (M-toxin), is dissociated into two components: a nontoxic component without hemagglutinin activity and a toxic one (7 S). This toxic component (7 S) is known as derivative toxin. Sakaguchi *et al.*[7] have shown that the nontoxic components of the various antigenically distinct toxins have some common immunological properties.

For the specific detection of the various types of botulinum toxins type-specific antisera are needed. They can be prepared by immunization of rabbits with the purified derivative toxin. They are produced from L- or M-toxins by DEAE-Sephadex chromatography as described by Sugii and Sakaguchi[4] for toxin type A, by Kozaki *et al.*[3] for toxin type B and Kitamura *et al.*[6] for toxin type E, resulting in purified derivative toxins.

Antisera against the derivative toxins are prepared by the immunization of rabbits with formalinized derivative toxin. The toxoid is emulsified in an equal volume of Freunds's complete adjuvant. Two doses of 1 ml (0.1 mg of toxoid) are injected subcutaneously into rabbits at 2-day intervals. Four weeks later, a booster injection without adjuvant is given subcutaneously (1 ml of 0.1 mg/ml toxoid). The rabbits are bled 1 week after the booster injection. From the serum the IgG fraction is isolated by the method of Steinbuch and Audran.[8]

Anti-horse IgG and IgG(T) sera are prepared by the immunization of rabbits with a washed immune precipitate of 100 Lf tetanus toxoid and 100 IU of horse tetanus antitoxins, emulsified in Freund's complete adjuvant. Two additional injections with the same mixture are given at an interval of 2 months. The rabbits are bled 10 days after the last injection.

Preparation of Antibody–Enzyme Conjugate

The enzyme used in this study was horse radish peroxidase as it was easily available, inexpensive, and stable. Other enzymes frequently used are alkaline phosphatase, β-galactosidase, and glucosidase. Coupling of the enzyme peroxidase to the antibody can be done by several different methods: i.e., use of bifunctional reagents like benzochinon[9] or glutaraldehyde,[10] or chemical modification of one of the components.[11] The method of Nakane and Kawaoi[11] was found to be very suitable if performed as follows: Antibodies are isolated from antisera using caprylic acid as described by Steinbuch and Audran[8]: all proteins are precipitated,

---

[7] G. Sakaguchi, S. Sakaguchi, S. Kozaki, S. Sugii, and I. Ohishi, *Jpn. J. Med. Sci. Biol.* **27**, 161 (1974).
[8] M. Steinbuch and R. Audran, *Arch. Biochem. Biophys.* **134**, 279 (1969).
[9] T. Ternynck and S. Avrameas, *Immunochemistry* **14**, 767 (1977).
[10] S. Avrameas and T. Ternynck, *Immunochemistry* **8**, 1175 (1971).
[11] P. K. Nakane and A. Kawaoi, *J. Histochem. Cytochem.* **22**, 1084 (1974).

except the immunoglobulins. This caprylic acid method is simple and gives satisfactory results. The immunoglobulin G fraction is dialyzed and freeze-dried.

The peroxidase is activated by 0.08 M sodium periodate after blocking the amino groups with dinitrofluorobenzene. The excess periodate is removed by using ethylene glycol, after which the solution is dialyzed against 0.01 M sodium carbonate pH 9.5. This solution is added to the freeze-dried IgG and gently shaken for 2.5 hr. Coupling is the result of a reaction between the free amino groups on the immunoglobulin and the aldehyde groups of the activated peroxidase. In this way conjugates of high molecular weight are formed. Such conjugates were compared with conjugates prepared according to the two-step method of Avrameas.[10] A higher activity was found for Nakane conjugates than for Avrameas conjugates[12] and for this reason, only Nakane conjugates were used. Boorsma and Streefkerk[13] obtained the same results.

Conjugates are freeze-dried in the presence of 1% bovine serum albumin and 2% lactose and stored at 4°; they remain stable for at least 1 year.

Substrate

The substrate used for the determination of peroxidase in the ELISA was 80 mg 5-aminosalicylic acid dissolved in 100 ml H_2O. After adjusting the pH to 6.0 one volume of 0.05% H_2O_2 is added to 9 volumes of the solution.

Other substrates are available, e.g., o-phenylenediamine, o-toluidine etc., and some of them are very chromogenic, thus increasing the sensitivity of the ELISA, but all have the disadvantage of being potentially carcinogenic, mutagenic, or to have sensitizing activity or irritating properties. Even 5-aminosalicylic acid has some of these properties although to a lesser extent.

ELISA Procedure

In Table I[14-16] a detailed description of the procedure of the double-sandwich ELISA is presented.

[12] A. M. Hagenaars, A. J. Kuipers, and J. Nagel, *Proc. workshop Enzym. Tech.* Pisa (1979).
[13] D. M. Boorsma and J. G. Streefkerk, *J. Immunol. Methods* **30,** 245 (1979).
[14] S. Notermans, J. Dufrenne, and M. van Schothorst, *Jpn. J. Med. Sci. Biol.* **31,** 81 (1978).
[15] S. Notermans, J. Dufrenne, and S. Kozaki, *Appl. Environ. Microbiol.* **37,** 1173 (1979).
[16] S. Kozaki, J. Dufrenne, A. M. Hagenaars, and S. Notermans, *Jpn. J. Med. Sci. Biol.* **32,** 199 (1979).

TABLE I

PROCEDURE FOR THE "DOUBLE-SANDWICH" ELISA FOR BOTULINUM TOXINS[a,b]

Absorption to solid surface	One milliliter of rabbit anti-botulinum IgG, diluted in 0.07 M phosphate buffer pH 7.2 containing 0.15 M NaCl (PBS) is added to polystyrene tubes (50 × 11 mm, LKB-produktor AB., Stockholm, Sweden). The tubes are incubated overnight in a rotary shaker at room temperature. The tubes are washed twice with PBS containing 0.05% Tween 20 (Merck)
Incubation of samples	One milliliter of the test sample diluted with 0.05 M phosphate buffer pH 6.5 containing 0.2% gelatin and 5 mM EDTA is incubated at 37° for 90 min in a rotary shaker. After 90 min the tubes are washed as described above
Serum incubation	One milliliter of horse anti-botulinum serum diluted in PBS is added. The tubes are incubated at 37° for 90 min in a rotary shaker and washed as described above
Conjugate incubation	To each tube 1 ml of in PBS diluted rabbit anti-horse IgG, conjugated with the enzyme horseradish peroxidase, is added. The tubes are incubated at 37° for 90 min in a rotary shaker and washed as described above
Substrate reaction	An amount of 80 mg of 5-aminosalicylic acid (Merck) is dissolved in 100 ml of distilled water at 70°. Prior to use the pH of this solution (20°) is adjusted to 6.0 with N NaOH. A 10-ml aliquot of 0.05% H_2O_2 (BDH, Chemicals) is added to 90 ml of the solution. After an incubation time of 30 min the peroxidase activity is measured spectrophotometrically at 449 nm

[a] Adapted from Notermans et al.[14,15] and Kozaki et al.[16]
[b] In the sandwich ELISA the serum incubation is omitted. Anti-botulinum IgG-peroxidase is used as the conjugate.

Adsorption to Solid Phase (Coating)

Immunoglobulins adhere to polystyrene surfaces spontaneously. Adsorption, however, does not occur in the presence of detergents like Tween 20. Therefore, the polystyrene tubes must not be cleaned with a surface active agent before coating. The adsorption of proteins to a solid phase is dependent on time, temperature, and pH. Prolonged adsorption (overnight) appears to give a more uniform coating.

For each application, the optimal IgG concentration needs to be established by checker-board titration. For the determination of botulinum toxins a rule of thumb is that for coating, the quantity of IgG used is that which will neutralize ca. 1000–10,000 mouse ip LD_{50} of toxin. After coating and after each incubation step the tubes have to be washed. The addition of Tween 20 to the PBS prevents nonspecific reactions and adherence of other proteins to the polystyrene surface during the next steps. Washing of the tubes consists of emptying the tubes and completely filling them with washing fluid. This procedure has to be repeated at least once. It may also be done by flushing them for 1 min with an excess of washing fluid under pressure (1–2 atm.).[17]

Incubation of Samples

Samples to be tested vary from culture fluids to sera and pure botulinum toxins. Samples have to be diluted in solutions which do not degrade the toxin. Botulinum toxins are most stable at pH 4.5–6.5,[18] and since immunological reactions are optimal at neutral pH the best for the dilution fluid is 6.5. The addition of 0.2% gelatin is also recommended to stabilize the toxin.

Proteolytic enzymes present in samples may interfere with the ELISA. This interference, by one or more of these enzymes, can be inhibited by the addition of 5 mM EDTA to the diluent[16] as it was found that Ca^{2+} was required by a proteolytic enzyme produced by *Clostridium botulinum* type B.[19]

When using the *in vivo* methods for the determination of botulinum toxins, the toxicity is increased by the addition of trypsin.[20] However, tryptic activation of the samples to be tested with ELISA should be avoided because it may also affect the coating IgG, and although trypsin increases the toxicity it does not increase the antigenicity.

Serum Incubation

In the "double-sandwich" ELISA the bound toxin has to be reacted with botulinum antibodies. These antibodies must be obtained from an animal species other than that which provided the antibodies for the coating of the tubes. If for example the tubes are coated with rabbit anti-botulinum IgG, then horse anti-botulinum can be used for the other step. Commercial horse anti-botulinum serum may, however, contain antibacterial substances such as cresoles which may influence the results of the ELISA. Notermans *et al.*[15] and Kozaki *et al.*[16] showed that horse serum from Behringwerke AG (Marburg, Lahn, Federal Republic of Germany) and from Wellcome Reagent Limited (Beckenham, United Kingdom) can be used satisfactorily, as can polyvalent horse antibotulinum serum (Behringwerke).

Before incubation the sera are diluted so that 1 ml of the dilution neutralizes 1000–10,000 mouse ip LD_{50} of botulinum toxin. The toxin neutra-

[17] Anonymous, *Bull. WHO* **54**, 129 (1976).
[18] L. Spero, *Arch Biochem. Biophys.* **73**, 484 (1958).
[19] B. R. DasGupta and H. Sugiyama, *Biochim. Biophys. Acta* **268**, 719 (1972).
[20] J. Gerwig, C. E. Dolman, and D. A. Arnott, *J. Bacteriol.* **81**, 819 (1961).

lizing antibodies in horse serum belong to the IgG and the T-globulins. It is not necessary to isolate them before use.

Conjugate Incubation

The conjugate is composed of rabbit IgG antibodies against horse serum Ig and peroxidase. The IgG may be produced using various species of animal. It is important that the IgG is prepared by immunizing animals with both the IgG and IgG (T) fraction of the serum, since both immunoglobulins show antibody activity. The optimal conjugate dilution to be used in the assay should be determined by checker-board titration just as in the foregoing steps. In these experiments the conjugate was diluted 150 to 300 times. It was found that low extinction values of the blanks were obtained when a low conjugate concentration is used.

Substrate Reaction

To quantify the peroxidase bound to the tubes 5-aminosalicylic acid was used. However, as the reagent may be carcinogenic it is advisable to continue the search for alternative reagents.

The enzyme-substrate reaction should be stopped within the linear phase of the reaction curve. For quantitative purposes a standard incubation time of at least 30 min should be allowed. When several dilutions are tested, endpoint titers are determined which permit the results to be read visually. When results are based on one dilution only, readings should be made with a spectrophotometer. The results are expressed as extinction values. The height of a certain extinction value should be assessed against a standard curve.

Sensitivity, Cross-reactivity, and Practical Application of the ELISA

Sensitivity

The lowest detectable quantity of toxin depends on the type of ELISA technique applied. Kozaki et al.[16] found that for toxin type B the sandwich ELISA is about 10 times less sensitive than the "double sandwich" ELISA. This was also found to be the case for toxin types A and E. With the double sandwich ELISA ca. 100 mouse ip LD_{50}/ml of both toxin type A and type E are detectable.[14,15] For toxin type B the lowest detectable level was ca. 400 mouse ip LD_{50} of toxin per ml. Yolken and Stopa[21] could improve the sensitivity of ELISA by using a substrate (4-methylumbelliferyl phosphate) that yielded a fluorescent product upon enzyme action.

[21] R. H. Yolken and P. J. Stopa, *J. Clin. Microbiol.* **10**, 317 (1979).

Performed in this way, the ELISA was approximately 100 times more sensitive than the normal ELISA for the detection of human rotavirus in a stool suspension. In principle, the fluorigenic substrate is also suitable for the detection of botulinum toxins, when alkaline phosphatase is used instead of peroxidase, with 4-methylumbelliferyl phosphate as substrate.

Cross-reactivity

Slight cross-reactions were found in the ELISA for botulinum toxins A, B, and E. Notermans *et al.*[14,15] and Kozaki *et al.*[16] have shown that those cross-reactions were limited to culture filtrates of some *C. botulinum* strains. In the ELISA for toxin type B, cross-reactions were observed with culture filtrates of *C. botulinum* type A. In the ELISA for toxin type E, cross-reactions were found with culture filtrates of both *C. botulinum* type A and type B. No cross-reactions were observed with either *C. botulinum* type C, *C. sporogenes*, or *C. perfringens* strains. Cross-reactions with the *C. botulinum* strains mentioned were eliminated by dilution of the samples, but this resulted in a loss of sensitivity.

All *in vitro* systems to detect botulinum toxins require at least one specific component. As far as the "double sandwich" ELISA is concerned specific antisera have to be used. In the reports of Notermans *et al.*[14,15] and Kozaki *et al.*[16] such antisera were obtained by immunizing rabbits with highly purified toxoids (see Production of Reagents to Be Used in the ELISA).

Nonspecific reactions which occur in the ELISA can be caused by cross-reacting antibodies and/or by nonimmunological binding of proteins. Cross-reacting antibodies may be present if the toxic component used for immunization is not completely free from the nontoxic components. These can be removed by absorption of the antiserum with nonneutralizing botulinum toxins containing the nontoxic components.[6] An example of nonimmunological binding is the interaction between IgG and protein A produced by some bacterial species such as *Staphylococcus aureus*. The binding of protein A to IgG is prevented by using pepsin digested antibodies, F(ab')$_2$ fragments, instead of native IgG antibodies[22] as protein A binds exclusively to the Fc-fragment of IgG.

In Fig. 2 the results of the assay of different purified progenitor botulinum toxins in the double sandwich ELISA for toxin type E are presented. ELISA experiments were performed in duplicate and the tubes were coated with rabbit anti-botulinum type E IgG. The results show that about 100 mouse ip LD_{50}/ml of pure toxin type E is detectable. No differences in extinction values were found with trypsin-treated and nontrypsin-

---

[22] G. A. Stewart, R. Varro, and D. R. Stanworth, *Immunology* **35,** 785 (1978).

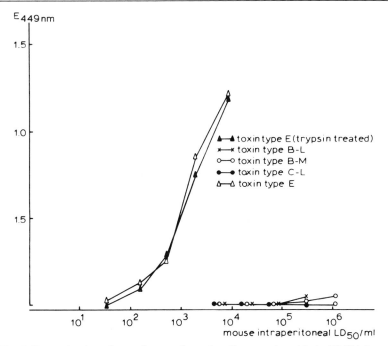

FIG. 2. Determination of several types of pure botulinum toxins with the ELISA for toxin type E

treated toxin type E. The trypsin-treated toxin was used after the addition of soybean trypsin inhibitor. This indicates that trypsin has no effect on the immunological activity of the toxin. No significant cross-reactions occurred with other pure botulinum toxins, even at concentrations up to 10^6 mouse ip LD_{50}/ml.

In Table II results are presented obtained with culture supernatants of different clostridial strains and a strain of *Staphylococcus aureus*, producing protein A, in the ELISA testing for toxin type E. Culture supernatants were obtained by incubation in liver broth as described by Haagsma[23] with spores of the different bacterial types. The cultures were incubated at 30° for 4 days. *S. aureus* was incubated aerobically in brain heart infusion broth at 30° for 2 days.[24] The "double sandwich" ELISA with a coating of either rabbit anti-botulinum type E IgG, or specific IgG indicated as S(IgG), or specific pepsin digested IgG indicated as S[F(ab′)₂] was used. F(ab′)₂ was prepared by digesting the IgG with pepsin.[24] Cross-reacting

[23] J. Haagsma, Thesis, Rijksuniversiteit Utrecht, The Netherlands (1973).
[24] J. W. Koper, A. M. Hagenaars, and S. Notermans, *J. Food Safety* **2**, 35 (1980).

TABLE II

SCREENING OF CULTURE SUPERNATANTS OF VARIOUS CLOSTRIDIAL STRAINS AND
Staphylococcus aureus IN THE DOUBLE-SANDWICH ELISA[a] WITH DIFFERENT
COATINGS TESTING FOR TOXIN TYPE E

Culture type	Coating material[b] and dilution								
	IgG			S(IgG)			S[F(ab')₂]		
	1/4	1/16	1/64	1/4	1/16	1/64	1/4	1/16	1/64
C. botulinum type A strain 62A[c]	++[d]	+	–	++	+	–	++	+	–
strain 141A[c]	++	+	–	++	+	–	++	+	–
strain Av4[c]	++	+	–	++	+	–	++	+	–
C. botulinum type B strain SNB77[c]	++	+	–	++	+	–	+	+	–
strain CDI 1	+	–	–	+	–	–	+	–	–
strain RIV 1	+	–	–	+	–	–	+	–	–
C. botulinum type C strain Cα	–	–	–	–	–	–	–	–	–
C. botulinum type E strain RIV 1	++	++	++	++	++	++	++	++	++
C. perfringens strain L 6539	–	–	–	–	–	–	–	–	–
C. sporogenes	–	–	–	–	–	–	–	–	–
S. Aureus	++	++	++	++	++	++	–	–	–

[a] Test samples were diluted in 0.05 M phosphate buffer, pH 6.5, containing 0.2% gelatin and 5 mM EDTA.
[b] For abbreviations see text.
[c] Proteolytic strains of C. botulinum.
[d] ++, OD 449 > 0.8; +, OD 449 = 0.1–0.8; –, OD 449 < 0.1.

antibodies were removed from IgG and from F(ab')$_2$ by absorption on to the culture filtrates of C. *botulinum* type A strain 141A and C. *botulinum* type B strain SNB77 immobilized on CNBr-activated Sepharose 4B (Pharmacia Fine Chemicals AB, Uppsala, Sweden). The coupling reacting was performed according to the recommendations of the manufacturers using 5 ml culture filtrate of both strains and 15 ml of swollen gel. Immunoabsorption was performed by incubating 1 mg pepsin-digested rabbit anti-botulinum IgG, or 1 mg rabbit antibotulinum IgG, diluted in 10 ml PBS for 2 hr at 20° and shaken end over end.

From the results it is clear that cross-reactions do still occur when either IgG, S(IgG), or S[F(ab')$_2$] are used. However, with a culture filtrate of S. *aureus* no cross-reaction was observed when S[F(ab')$_2$] was used for coating, indicating that no protein A-like interference was present. The cross-reactions with proteolytic C. *botulinum* strains were stronger than those obtained with the culture supernatants of nonproteolytic strains. No cross-reactions were observed with the other *Clostridium* strains tested.

The degree of cross-reaction proved not to be related to the amount of the toxin present in the samples. They increased however, with increasing incubation time of the culture supernatants of the proteolytic strains used. Cross-reactions with dilutions of the culture supernatants of the proteolytic strains also occurred when the tubes were coated with normal rabbit IgG, in which case the extinction values were comparable to those obtained with rabbit anti-botulinum IgG. This suggests that the proteolytic enzymes present in the culture supernatant were bound nonimmunologically to the coat of rabbit IgG.

The cross-reactions could not be prevented by dilution of the test samples in 0.05 M phosphate buffer containing 5 mM EDTA to which 0.1 or 1% normal IgG was added. Addition of 0.05% Tween 20 or 0.1% bovine serum albumin to the dilution fluid of the samples also failed to prevent cross-reactions. At present no solution for this problem has been found.

Practical Application of ELISA

Simultaneous Determination of Botulinum Toxins. For some investigations it is not strictly necessary to determine botulinum toxins as single immunological types; it may be sufficient to assess the presence or absence of toxins dangerous to man. An assay in which several types of toxins can be detected simultaneously would be preferable. Kozaki *et al.*[16] and Notermans *et al.*[15] have already shown that for the serum incubation trivalent antitoxin serum can be used. If in the ELISA the tubes are coated with trivalent IgG one single toxin or a combination of toxins can be determined. Results of a "simultaneous" ELISA for toxins type A, B, and E are presented in Fig. 3. In these experiments tubes were coated

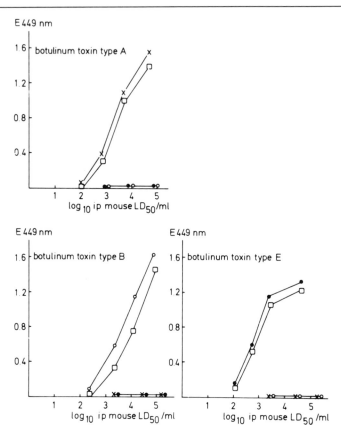

FIG. 3. Simultaneous determination of botulinum toxins. Tubes are coated with rabbit IgG directed against toxin type A (×——×), toxin type B (○——○), toxin type E (●——●), and toxin types A, B, and E (□——□).

with a mixture of rabbit IgG against toxin types A, type B, and type E. In the test pure botulinum toxins were used.

From the results it is clear that the single botulinum toxins are detectable both with a coat of trivalent IgG and monovalent IgG. No cross-reactions among the different types of toxins were observed.

Determination of Botulinum Toxins in Enrichment Cultures. The presence of *C. botulinum* in samples is determined by culturing these samples in an enrichment medium. Presence of toxin in the culture fluid indicates the presence of *C. botulinum* in the original sample. The detection of toxins as well as the determination of the immunological type are normally performed using the mouse bioassay. To detect the toxins and to determine their type specificities 10–20 mice are used. In this chapter experiments are described for testing enrichment cultures for the presence of botulinum toxins with ELISA to reduce the use of mice.

Tubes containing 30 ml of freshly prepared liver broth were inoculated with different samples containing spores of *C. botulinum*. The tubes were heated at 70° for 20 min and incubated anaerobically at 30° for 5 days. Following incubation the cultures were examined for the presence of the various toxins, using the mouse bioassay and the ELISA.

With the mouse bioassay 2 ml samples of the supernatants were drawn and diluted with 6 ml of sterile 0.05 *M* phosphate buffer, pH 6.5, which contained 0.2% gelatin. Two-milliliter aliquots were mixed with 0.5 ml of anti-botulinum serum (Institute Pasteur, Paris). The type(s) of the added antiserum corresponded to the toxin expected. Two mice (18–20 g) were injected intraperitoneally with 0.5 ml of each mixture and observed for 4 days. Titers of botulinum toxins were determined by injecting mice intraperitoneally with 0.5 ml each of serial 3-fold dilutions, 3 mice were used per dilution. The intraperitoneal LD_{50} was calculated as described by Reed and Muench.[25]

Culture fluids tested in ELISA were diluted 1:4 with 0.05 *M* phosphate buffer pH 6.5 containing 0.2% gelatin and 5 m*M* EDTA.

In Table III for instance results are presented of the detection of botulinum toxins in a culture supernatant of a mud sample to which spores of *C. botulinum* were added. The toxicity was determined using the mouse bioassay and the ELISA. For the ELISA the polystyrene tubes were

TABLE III

DETECTION OF *C. botulinum* TOXINS IN CULTURE SUPERNATANTS OF A MUD SAMPLE FROM A LAKE ARTIFICIALLY CONTAMINATED WITH SPORES OF *C. botulinum*

Quantity of mud cultured (g)	Type and number of *C. botulinum* spores added	Log_{10}ip mouse LD_{50} of toxin/ml	Extinction values in ELISA[a]	
			Coat anti-B	Coat anti-E
0.0	E 5000	4.9	0.1	2.0
0.1	E 5000	5.1	0.1	2.1
1.0	E 5000	4.4	0.0	1.8
0.0	B 3000	5.3	1.8	0.2
0.1	B 3000	3.1	0.3	0.0
1.0	B 3000	1.2	0.0	0.0

[a] Culture supernatants were diluted 1:4 in 0.05 *M* phosphate buffer pH 6.5 containing 0.2% gelatin and 5 m*M* EDTA.

[25] L. J. Reed and H. Muench, *Am. J. Hyg.* **27**, 493 (1938).

coated either with rabbit anti-botulinum type B IgG or with rabbit anti-botulinum type E IgG. The results show that *C. botulinum* type E (strain RIV 2) produced more than 10,000 mouse ip LD_{50} of toxin per ml culture fluid. Cross-reactions with the ELISA for toxin type B was negligible. Toxin production of *C. botulinum* type B (strain SNB 77) depended on the quantity of mud cultured. When 1 g of mud, inoculated with ca. 3000 spores of *C. botulinum* type B, was incubated in liver broth about 16 mouse ip LD_{50} of toxin were detected per ml culture supernatant, whereas the ELISA failed to detect toxin.

In Table IV results are presented of the detection of botulinum toxins in culture supernatants of two mud samples taken from a fishery pond naturally contaminated with *C. botulinum* type E. From each sample different quantities of mud were inoculated into liver broth. After incubation the presence as well as the types of toxin were determined using the mouse bioassay and the ELISA. The ELISA was performed either with a coat of rabbit anti-botulinum type B, type E, or with a mixture of rabbit anti-botulinum type A, B, and E. It is evident again that the results of the ELISA correspond well with the results of the mouse bioassay, although the ELISA failed to detect toxin in the one positive culture supernatants of sample 2. Cross-reactions were not observed.

Finally in Table V results of the ELISA carried out to assess the presence of botulinum toxins in culture supernatants obtained from several

TABLE IV

DETECTION OF *C. botulinum* TOXINS IN CULTURE SUPERNATANTS OF MUD
SAMPLES FROM A FISHERY POND

	Amount of mud cultures (g)	Mouse bioassay	Extinction values in ELISA[a]		
			Coat anti-B	Coat anti E	Coat anti-A,B,E
Sample 1	1.0	+ (type E)	0.0	0.2	0.2
	0.1	+ (type E)	0.0	0.3	0.2
	0.01	+ (type E)	0.0	0.4	0.3
	0.001	−	0.0	0.0	0.0
	0.0001	+ (type E)	0.0	0.5	0.3
Sample 2	1.0	+ (type E)	0.0	0.0	0.0
	0.1	−	0.0	0.0	0.0
	0.01	−	0.0	0.0	0.0
	0.001	−	0.0	0.0	0.0
	0.0001	−	0.0	0.0	0.0

[a] See Table III.

TABLE V
DETECTION OF *C. botulinum* TOXINS IN CULTURE SUPERNATANTS OF SEVERAL
SAMPLES NATURALLY CONTAMINATED WITH *C. botulinum*

Samples investigated	Mouse bioassay	Log_{10}ip mouse LD_{50} of toxin/ml	Extinction values of ELISA[a]		
			Coat anti-B	Coat anti-E	Coat anti-A,B,E
1 g coagulation mud from a water production plant	+ (type B and E)	1.70	0.0	0.0	0.0
1 g mud from a fishery pond	+ (type E)	4.16	0.0	1.0	0.9
1 g feces from a cow	+ (type B)	2.15	0.0	0.0	0.0
0.1 g feces from a cow	+ (type B)	2.20	0.0	0.0	0.0
0.01 g feces from a cow	+ (type B)	3.20	0.4	0.0	0.4
0.001 g feces from a cow	+ (type B)	3.80	0.8	0.2	0.7
1 g mud from a lake	+ (type C)	3.20	0.0	0.0	0.0
1 g grass silage	+ (type B)	1.35	0.0	0.0	0.0

[a] See Table IV.

naturally contaminated samples are presented. The samples were tested
in the mouse bioassay to determine the quantity of toxin as well as the
type of toxin. Again it is clear that the amount of toxin present in the enrichment cultures may be below the detection level of the ELISA. Cross-
reactions were not observed in any of the ELISA systems used.

General Remarks

The ELISA is a very promising technique for the detection of botulinum toxins. It is a more quantitative reaction than the hemagglutination
reaction[26] and is readily usable in laboratories where the facilities for
working with radioisotopes (radioimmunoassay) are limited. In comparison with the mouse bioassay, however, the sensitivity of the ELISA is
still too low. It might be improved by using a substrate that yields a fluorescent product.[21] If the ELISA can be improved in this way supernatants
of enrichment cultures could be tested for the presence of toxin.

In vitro techniques for detection of botulinum toxins can only be successful if highly specific antisera are available. However, false positive
reactions do sometimes occur due to nonimmunological binding of proteins present in the test samples. There was no indication as to the nature
of the interfering substances present in for example the culture superna-

[26] G. M. Evancho, D. H. Ashton, E. J. Briskey, and E. J. Schantz, *J. Food Sci.* **38**, 764 (1972).

tants of proteolytic strains of *C. botulinum*. It may be speculated that proteolytic enzymes are bound to the IgG molecules, and more research is required to avoid such interference.

As long as there is no certainty about the specificity of the reactions it is recommended that duplicate samples be tested in tubes coated with IgG of normal rabbit serum, instead of rabbit antibotulinum IgG.

[19] Determination of *Escherichia coli* Enterotoxin and Cholera Toxin by Radioimmunoassay

By MIROSLAV CESKA

Several radioimmunoassay procedures for assessment of cholera toxin and *E. coli* enterotoxin are described in the literature. This chapter describes mainly the radioimmunoassay procedures that have been developed and used in the author's laboratory.

Preparation of *E. coli* Enterotoxin and Cholera Toxin Antiserum (a-ECT and a-VCT)

Various animals have been reported to give good quality antiserum against *E. coli* enterotoxin (ECT) and cholera toxin (VCT).[1–4] Depending on animal species, these toxins (in complete as well as in incomplete Freund's adjuvants) were used for immunization in quantities anywhere from 25 μg to 40 mg per immunization dose. The most commonly used immunization routes are subcutaneous, intradermal, and intramuscular, with the toxins usually being introduced on multiple sites at 2 to 3 week intervals. Titers of antisera are most commonly evaluated by a single radial immunodiffusion technique.[5] Antisera in suitable aliquots can be stored at low temperatures for at least a year. Cholera toxin antiserum is available for distribution from the National Institute of Allergy and Infectious Diseases, Bethesda, MD and from the Swiss Serum Institute, Bern.

[1] D. J. Evans, Jr. and D. G. Evans, *J. Clin. Microbiol.* **5**, 100 (1977).
[2] J. W. Petterson, K. E. Hejtmancik, D. E. Markel, J. P. Craig, and A. Kurosky, *Infect. Immun.* **24**, 774 (1979).
[3] J. Holmgren and A. M. Svennerholm, *Curr. Microbiol.* **2**, 55 (1979).
[4] R. A. Finkelstein, *Infect. and Immun.* **2**, 691 (1970).
[5] W. Becker, *Immunochemistry* **6**, 539 (1969).

Purification of Toxin Antibodies from Immune Sera

Various protein purification methods such as salt precipitation, ion exchange chromatography, and immunosorbent techniques are recommended for purification of toxin antibodies intended for use in radioimmunoassay procedures. For the double ligand technique it is essential that the antibody be purified to a high degree by an immunosorbent technique prior to its radiolabeling. It is also preferable to use highly purified toxin antibody for the preparation of solid-phase conjugates. This will increase the total binding yield between antigen and antibody due to the increase in the concentration of specific antibody per given area of solid phase (Fig. 1).

As an example the immunosorbent purification method for the prepa-

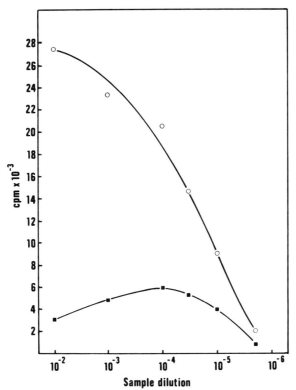

FIG. 1. Uptake of radioiodinated cholera toxin by (a) solidified cholera toxin antiserum, ■——■; (b) solidified, immunosorbent purified cholera toxin antibody, O——O. Polystyrene tubes were coated with antiserum, or antibody at the indicated dilutions (10^0 = undiluted sample).

ration of specific cholera toxin antibody is described here. Cholera toxin (Schwarz Mann, lot Nr. EZ-3336) is coupled to activated CH-Sepharose 4B at pH 8.0 to the extent of 0.3 mg VCT/ml suspension. Twenty milliliters of this immunosorbent is filled into a column 5 cm in diameter (this gives a gel bed height of 1 cm). After column equilibration with phosphate-buffered saline pH 7.4 (containing 0.05% NaN$_3$), 30 ml of sheep or rabbit cholera toxin antibody (purified from antiserum by ammonium sulfate precipitation and by ion-exchange chromatography on DEAE-cellulose) is placed on the column. After thorough washing with the above mentioned buffer, the specific cholera toxin antibody is eluted with either 0.05 or 0.1 M acetic acid solution. The eluted antibody solution is immediately neutralized with 1.0 M NaOH and dialyzed against 0.1 M NaCl solution containing 0.1% NaN$_3$, concentrated by vacuum dialysis, aliquoted, and stored at $-70°$.

On the basis of immunological cross-reactivity between cholera toxin and *E. coli* enterotoxin, the above described method can also be used to prepare the specific *E. coli* enterotoxin antibody from *E. coli* enterotoxin antiserum.

Radioiodination of Cholera Toxin[6]

Iodination of cholera toxin can conveniently be done by the chloramine-T method[7] as described below. Na^{125}I purchased from the Radiochemical Centre, Ltd., Amersham, England is used for radiolabeling. Cholera toxin (15 μl) (Schwarz Mann, lot BZ2484) (1 mg/ml) in phosphate-buffered saline, pH 7.2 is used. A 25-μl amount of iodination buffer (0.5 M sodium phosphate buffer, pH 7.5) is added, then 50 μl of Na^{125}I and 50 μl of chloramine-T (100 μg). The iodination is performed at room temperature ($\sim 22°$) for a period of 30 sec and terminated by the addition of sodium metabisulfite. After the addition of carrier potassium iodide and human serum albumin (HSA) the mixture is chromatographed on a Sephadex G-15 column (plastic syringe 10–12 ml). The eluted ^{125}I-labeled cholera toxin is diluted to desired concentration with 0.05 M sodium phosphate buffer pH 7.4 containing 0.9% NaCl, 0.3% HSA, and 0.05% NaN$_3$. After aliquoting, the radiolabeled cholera toxin (VCT-^{125}I) is stored at $-20°$. Specific radioactivity of radiolabelled cholera toxin is on average around 40 mCi/mg.

[6] M. Ceska, F. Effenberger, and F. Grossmüller, *J. Clin. Microbiol.* **7**, 209 (1978).
[7] F. C. Greenwood, W. M. Hunter, and J. S. Glover, *Biochem. J.* **89**, 114 (1963).
[8] H. B. Greenberg, D. A. Sack, W. Rodriguez, R. B. Sack, R. G. Wyatt, A. R. Kalica, R. L. Horswood, R. M. Chanock, and A. Z. Kapikian. *Infect. Immun.* **17**, 541 (1977).

Radioiodination of *E. coli* Enterotoxin[9]

Radioiodination of *E. coli* enterotoxin can be done according to the iodination method given above for cholera toxin.[9]

Radioiodination of Immunosorbent Purified Cholera Toxin Antibody

Immunosorbent purified sheep-cholera toxin antibody is labeled with radioiodine by the chloramine T-method.[7] For iodination 25 μg of a-VCT in 5.0 μl of 0.5 M sodium phosphate buffer, pH 7.5 is mixed with 25.0 μl of iodination buffer (0.5 M sodium phosphate buffer, pH 7.5). Then 20 μl of Na[125]I (specific radioactivity $\leq$ 14 mCi/μg) and 250 μg chloramine-T in 50 μl is added. Total volume of the iodination mixture is 100 μl. The iodination is performed at room temperature ($\sim$ 22°) for a period of 60 sec and is terminated by the addition of 335 μg sodium metabisulfite in a volume of 50 μl. After the addition of 20 μl of carrier KI (2 mg) and 200 μl of HSA (10 mg), the mixture (370 μl) is chromatographed on a Sephadex G-25 column (made from a 10–12 ml plastic syringe). The radiolabeled immunoglobulin is diluted to 15 ml with 0.05 M sodium phosphate buffer, pH 7.4, containing 0.9% NaCl, 0.3% HSA, and 0.05% NaN$_3$, aliquoted in 300-μl amounts and stored at $-$ 20°. Specific radioactivity of radiolabeled cholera toxin antibody is on the average around 90 mCi/mg.

Repurification of Radiolabeled Toxin and Radiolabeled Antitoxins

Radioiodinated toxins and toxin antibodies can be used over an extended time period provided that they are repurified from their split products prior to being used. The repurification can easily be done on Sephadex G-15 or G-25 columns.

Preparation of Solid Phase Bound Antibody Conjugates

The procedure for the preparation of solid-phase bound antibodies is the same for competitive as well as for double-ligand techniques. However the antibody concentration used for the preparation of conjugates for a competitive assay is usually much smaller than that used for a double ligand technique. The optimal concentration of antibody needed for a given assay has to be established experimentally by a "binding experiment" (Fig. 1). A few examples of preparations of toxin antibody and toxin–receptor conjugates are given.

[9] M. Ceska, F. Grossmüller, and F. Effenberger, *Infect. Immun.* **19**, (1978).

Binding of Cholera Toxin Antibody to Plastic Materials by Adsorbtion

Polystyrene tubes (11 by 70 mm) are used for coating with antiserum or with antibodies by a modification of a procedure published previously.[6] Polystyrene tubes are left to coat overnight (~ 15 hr) at 4° with 0.5 ml of the antiserum or purified antibody diluted in 0.05 M sodium carbonate buffer, pH 9.6. After the termination of the coating procedure, the tubes are washed thoroughly two times with 3-ml portions of 0.90% NaCl containing 0.50% Tween 20 and then three times with incubation buffer that consists of 0.05 M sodium phosphate buffer, pH 7.4, containing 0.90% NaCl, 0.30% HSA, 0.05% NaN_3, and 0.50% Tween 20.

Also, other plastic material such as poly(vinyl chloride) (microtiter plates) can be used as a suitable solid-phase for coupling cholera toxin antibody.[10]

Covalent Binding of Cholera Toxin Antibody to BrCN-Activated Sepharose 4B

Lyophilized cholera toxin antiserum (75 mg) (Swiss Serum Institute) is diluted in 40 ml of coupling buffer pH 9.0 (0.1 M $NaHCO_3/Na_2CO_3$, 0.5 M NaCl, 0.05% NaN_3). This solution is added to 8 g washed BrCN-activated Sepharose 4B and incubated by end-over-end rotation at room temperature for about 3 hr. The Sepharose–antibody conjugate is washed on a sintered glass filter four times with 100-ml portions of the above mentioned coupling buffer. The remaining active groups on BrCN-Sepharose 4B are blocked by incubation with 50 ml of 0.5 M glycine in coupling buffer pH 9.0 overnight at 4° by end-over-end rotation. Sepharose–antibody conjugate is then washed four times with 100-ml portions of 0.1 M NaAc/AcH, pH 4.0 containing 0.5 M NaCl and four times with 100-ml portions of coupling buffer. Nine final washings are made with storage buffer (0.05 M sodium phosphate buffer, pH 7.4 containing 0.9% NaCl and 0.05% NaN_3). Coupling efficiency: 78.5% or 1.76 mg cholera toxin antiserum is bound to 1 ml Sepharose 4B.

Binding of "Cholera Toxin Receptor" to BrCN-Sepharose

Ammonium sulfate- or acid-precipitated material of supernatant obtained from homogenized rabbit intestinal mucosal tissue was shown to bind reversibly radioiodinated cholera toxin in a fashion similar to cholera toxin antibody. Other animal species as well as tissues other than intestinal mucosa were also shown to bind radioiodinated cholera toxin revers-

---

[10] K. E. Hejtmancik, J. W. Peterson, D. E. Markel, and A. Kurosky, *Infect. Immun.* **17**, 621 (1977).

ibly and can therefore be employed in the radioreceptor assay of this toxin as well.

Five milliliter of the "cholera toxin receptor" material (~ 50 mg protein/ml) was diluted with coupling buffer and added to 3 g of washed BrCN-activated Sepharose 4B. The reaction mixture was incubated by end-over-end rotation at room temperature for around 3 hr. The other preparation steps are identical to those described for binding of cholera toxin antibody to BrCN-activated Sepharose. Coupling efficiency: 93.6% or 4.52 mg "cholera toxin receptor" is bound to 1 ml Sepharose 4B.

Radioimmunoassay and Radioreceptor Procedures Used for the Determination of Cholera Toxin and *E. coli* Enterotoxin

For the sake of simplification, the schematic representation of the various assay systems shown below as well as the methods of description are shown only for cholera toxin. It should be realized that the same assay systems with the identical procedures can also be employed for the detection of *E. coli* enterotoxin. Furthermore, on the basis of immunological cross-reactivity between these two toxins, it is possible for example to use cholera toxin antibody for the determination of *E. coli* enterotoxin and vice versa. Some examples of this will be shown in a few figures.

Competitive Radioimmunoassay, Using Polystyrol–Antibody Conjugates[8,9]

This assay is similar to one previously reported for hormones.[11,12] A 500-μl amount of VCT-containing samples as well as 500 μl of VCT-^{125}I are added to cholera toxin antibody coated tubes. The tubes are left to incubate at 4° for overnight (~ 15 hr). After the termination of the incubation, the liquid is removed by aspiration, and the tubes are washed three times with 3-ml portions of 0.90% NaCl solution containing 0.5% Tween 20. After the last wash, the tubes are dried by suction and placed in a gamma counter. The schematic representation of this system is given in Fig. 2.

Figure 3 shows a standard curve, that is the inhibition of VCT-^{125}I absorption to its antibody by an increasing amount of unlabelled cholera toxin. The effect of different concentrations of cholera toxin antibody (used for coupling) on the cholera toxin radioimmunoassay is seen in Fig. 4.

The effect of consecutive and simultaneous addition of ^{125}-I labeled *E*.

[11] K. Catt and G. W. Tregear, *Science* **158**, 1570 (1967).
[12] M. Ceska, F. Grossmüller, and U. Lundkvist, *Acta Endocrinol.* **64**, 111 (1970).

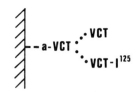

FIG. 2. Schematic representation of a solid-phase competitive radioimmunoassay of cholera toxin using cholera toxin antibody bound to inner surface of polystyrol tubes.

coli enterotoxin (ECT-^{125}I) to its solidified antibody is seen in Fig. 5 (consecutive addition: 24 hr incubation of solidified antibody with unlabeled *E. coli* enterotoxin, before addition of radiolabeled *E. coli* enterotoxin). The consecutive addition gives an assay of a slightly higher sensitivity.

The effect of incubation temperature and time on the reaction between *E. coli* enterotoxin and anti-*E. coli* enterotoxin is seen in Fig. 6. It can be seen that raising the temperature allows the incubation time to be shortened considerably. Figure 7 shows a cholera toxin standard curve and the

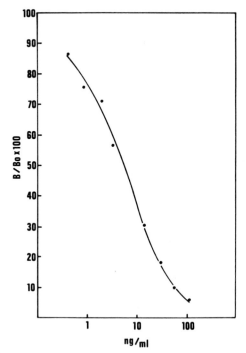

FIG. 3. Solid-phase competitive radioimmunoassay of cholera toxin, a typical standard curve. Cholera toxin antibody at dilution of 1:30,000 was used for coating polystyrene tubes. From Ceska *et al.*[6]

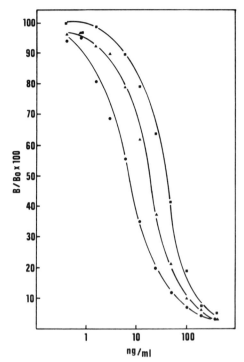

FIG. 4. Solid-phase competitive radioimmunoassay of cholera toxin. Polystyrene tubes were coated with the following concentrations of anti-cholera toxin antibody: ■——■, 1:30,000; ▲——▲, 1:100,000; and ●——●, 1:500,000.

result obtained by testing for immuno-cross-reactive *E. coli* enterotoxin contained in 4 different experimental *E. coli* batches (H10407).

Competitive Radioassay, Using Sepharose-4B-"Receptor" Conjugate

A 500 μl amount of VCT-containing samples (diluted in incubation buffer) is added to 10 μl of Sepharose 4B–"receptor" conjugate. After 30 min incubation at room temperature, 250 μl of VCT-^{125}I ($\sim$ 100,000 cpm $\cong$ 1 ng VCT-^{125}I) is added and the tubes are left to incubate at 4° for overnight ($\sim$ 15 hr). After terminating the incubation, the liquid is removed by aspiration and the tubes are washed six times with 3-ml portions of 0.90% NaCl solution containing 0.5% Tween 20. After the last wash the tubes are dried by suction and placed in a gamma counter. The schematic representation of this assay system is given in Fig. 8. Figure 9 shows a cholera toxin standard curve and the results obtained by testing for immuno-cross-reactive *E. coli* enterotoxin contained in an experimental *E. coli* batch No. 146 (H10407).

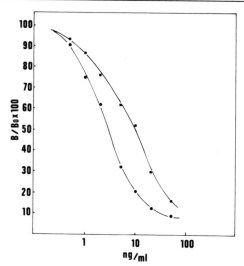

FIG. 5. Solid-phase competitive radioimmunoassay of *E. coli* enterotoxin. Solidified *E. coli* exterotoxin incubated with ECT-^{125}I simultaneously (■——■) and consecutively (●——●). From Ceska *et al.*[9]

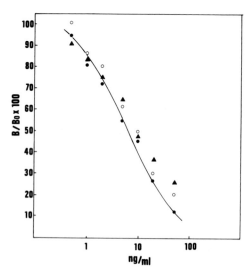

FIG. 6. Solid-phase competitive radioimmunoassay of *E. coli* enterotoxin. The following incubation conditions were used: ▲——▲, 2 hr at 45°; ○——○, 4 hr at 45°; and ●——●, 24 hr at 4°. From Ceska *et al.*[9]

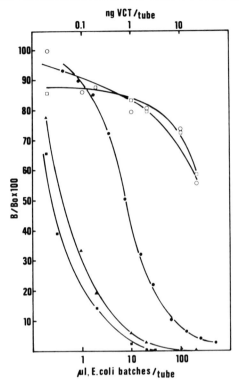

FIG. 7. Solid-phase competitive radioimmunoassay of cholera toxin and *E. coli* entero-toxin contained in four different experimental batches (all H10407). ●——●, cholera toxin standard curve; □——□, *E. coli* batch No. 178; ○——○, *E. coli* batch No. 181; ▲——▲, *E. coli* batch No. 146; and ■——■, *E. coli* batch No. 169.

Double-Ligand Radioimmunoassay Using Polystyrol–Antibody Conjugate

To polystyrol–antibody conjugate are added 500 μl toxin standards (or unknown samples containing either cholera toxin or *E. coli* entero-toxin) diluted in PBS–BSA–NaN$_3$ buffer pH 7.4. The tubes are left to in-cubate at 4° overnight (~ 15 hr) and then they are washed three times with 3-ml portions of PBS–BSA–NaN$_3$ buffer pH 7.4. After the last wash the

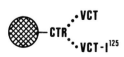

FIG. 8. Schematic representation of a solid-phase competitive radioreceptor assay of cholera toxin, using cholera toxin and *E. coli* enterotoxin receptor from rabbit intestinal mu-cosal tissue bound to BrCN-activated Sepharose 4B.

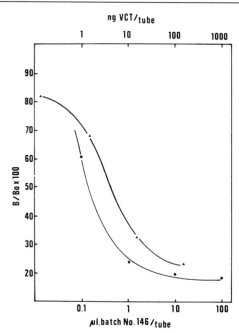

FIG. 9. Solid-phase competitive radioreceptor assay of cholera toxin (●——●) and *E. coli* enterotoxin (contained in *E. coli* experimental batch No. 146) (▲——▲).

tubes are dried by suction. Five hundred microliters of immunosorbent purified a-VCT-^{125}I (or a-ECT-^{125}I) are added (~200,000 cpm) and the tubes are left to incubate at 4° overnight (~15 hr). After terminating the incubation the tubes are washed as previously described. The schematic representation of this assay system is given in Fig. 10. Figure 11 shows a cholera toxin standard curve and the results obtained by testing for an immuno-cross-reactive *E. coli* enterotoxin contained in an experimental *E. coli* batch No. 146 (H 10407).

Double-Ligand Radioimmunoassay Using Sepharose-Antibody Conjugate

To 25 μl of Sepharose–antibody conjugate are added 250 μl of toxin containing sample (VCT or ECT), diluted in PBS–BSA–NaN$_3$ buffer pH 7.4. The reaction mixture is left to incubate overnight (~15 hr) at +4°. The conjugate is then washed six times with 0.9% NaCl containing 0.5% Tween 20. After the last washing and centrifugation, 250 μl of immunosorbent purified a-VCT-^{125}I (or a-ECT-^{125}I) diluted in PBS–BSA–NaN$_3$ buffer, pH 7.4 are added. The tubes are then left to incubate at +4° overnight (~15 hr). After washing the conjugate according to the procedure

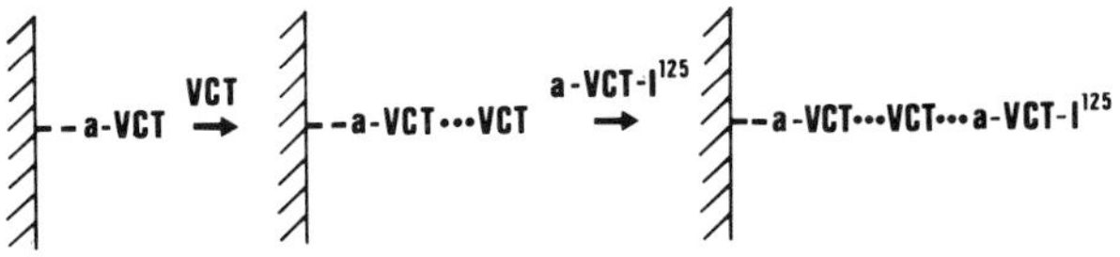

$$--a\text{-}VCT \xrightarrow{\text{VCT}} --a\text{-}VCT\cdots VCT \xrightarrow{a\text{-}VCT\text{-}I^{125}} --a\text{-}VCT\cdots VCT\cdots a\text{-}VCT\text{-}I^{125}$$

FIG. 10. Schematic representation of a solid-phase double-ligand radioimmunoassay of cholera toxin using cholera toxin antibody bound to inner surface of polystyrol tubes.

previously described, the samples are placed in a gamma counter. The schematic representation of this assay system is given in Fig. 12. Figure 13 shows cholera toxin standard curve and the results obtained by testing for an immuno-cross-reactive *E. coli* enterotoxin contained in an experimental *E. coli* batch No. 146 (H10407).

Double-Ligand Radioimmunoassay Using Polyvinyl–Antibody Conjugate[8]

Schematically this system is identical to the system shown in Fig. 10. Polyvinyl microtiter plates are coated with anti-cholera serum (diluted in phosphate-buffered saline pH 7.4). (The optimal serum dilution is found by titration.) After incubating the plates at room temperature for 12 hr, the wells are washed five times with PBS and then filled with PBS containing 1% BSA and are again incubated for 12 hr.

After the removal of PBS–BSA, the plates are washed twice with PBS. Fifty microliters of the appropriate antigen is added to the wells and

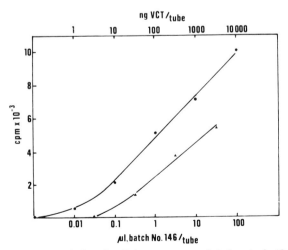

FIG. 11. Solid-phase double-ligand radioimmunoassay of cholera toxin (●——●) and *E. coli* enterotoxin (contained in *E. coli* experimental batch No. 146) (▲——▲).

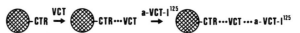

FIG. 12. Schematic representation of a solid-phase, double-ligand radioimmunoassay of cholera toxin using cholera toxin antibody bound to BrCN-activated Sepharose-4B.

allowed to incubate overnight at room temperature. The wells are then washed five times with PBS and 50 μl of a-VCT-^{125}I is added to each well. After 4 hr incubation at 37°, the plates are again washed five times with PBS and cut up with scissors and the separate wells are placed in gamma-counting tubes for analysis. A modified procedure using poly(vinyl chloride) microtiter plates was published by Hejtmancik et al.[10]

Double-Ligand Radioassay, Using Sepharose 4B-Toxin–Receptor Conjugate

To 25 μl of Sephadex–receptor conjugate are added 250 μl of toxin containing sample diluted in PBS–BSA–NaN$_3$ buffer, pH 7.4. The reac-

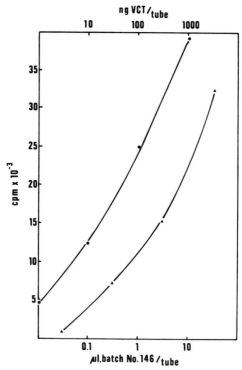

FIG. 13. Solid-phase double-ligand radioimmunoassay of cholera toxin (●———●) and *E. coli* entertoxin contained in *E. coli* experimental batch No. 146 (▲———▲).

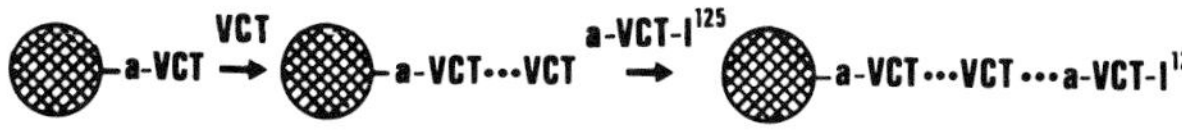

VCT   a-VCT-I¹²⁵
⬡-a-VCT → ⬡-a-VCT···VCT → ⬡-a-VCT···VCT···a-VCT-I¹²⁵

FIG. 14. Schematic representation of a solid-phase double-ligand radioreceptor assay of cholera toxin using receptor from rabbit intestinal mucosal tissue bound to BrCN-activated Sepharose-4B.

tion mixture is left to incubate overnight (~15 hr) at +4°. The following washing procedure and the addition of a-VCT-^{125}I is done according to the procedure previously described. The schematic representation of this assay system is given in Fig. 14. Figure 15 shows a cholera toxin standard curve and the results obtained by testing for immuno-cross-reactive *E. coli* enterotoxin contained in an experimental *E. coli* batch No. 146 (H10407).

Determination of Antitoxin Antibodies by Radioimmunoassay

The radioimmunoassay systems described above for the determination of cholera toxin and *E. coli* enterotoxin can also be used for the determination of toxin antibodies. As an example two solid-phase competitive procedures are given below.

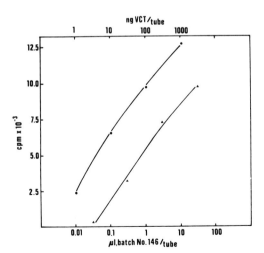

FIG. 15. Solid-phase double-ligand radio-receptor assay of cholera toxin (●——●) and *E. coli* enterotoxin contained in *E. coli* experimental batch No. 146 (▲——▲).

FIG. 16. Schematic representation of a solid-phase competitive radioimmunoassay of cholera toxin antibody using anticholera toxin antibody bound to inner surface of polystyrol tubes.

Competitive Radioimmunoassay (Using polystyrol–Antibody Conjugates) for the Determination of Cholera Toxin and E. coli Enterotoxin Antibodies[8]

A 500-μl amount of anti-cholera toxin in serial 10-fold dilutions, as well as 500 μl of VCT-^{125}I ($\sim$2 ng/ml) are added to antibody-coated tubes. The tubes are left to incubate at 4° overnight ($\sim$15 hr). After the termination of the incubation, the liquid is removed by aspiration and the tubes are washed three times with 3-ml portions of 0.90% NaCl solution containing 0.5% Tween 20. After the last wash the tubes are dried by suction and placed in a gamma counter. The schematic representation of this sys-

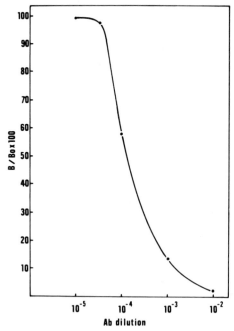

FIG. 17. Solid-phase competitive radioimmunoassay of cholera toxin antibody. From Ceska et al.[9]

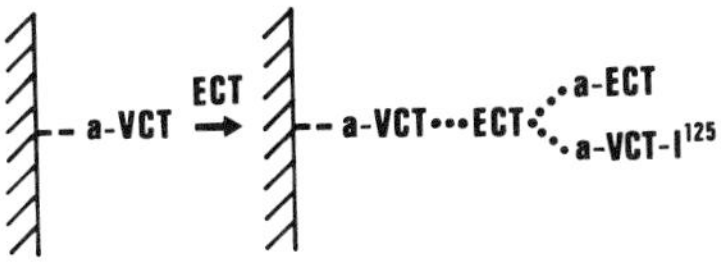

ECT
-- a-VCT → -- a-VCT···ECT···$^{..}$a-ECT
$^{..}$a-VCT-I^{125}

FIG. 18. Schematic representation of a solid-phase double-ligand radioimmunoassay of *E. coli* enterotoxin antibody bound to inner surface of polypropylene microtiter plates.

tem is given in Fig. 16. Figure 17 shows competition between solidified and soluble cholera toxin antibody for cholera toxin. Cholera toxin antiserum from the Swiss Serum Institute was used as a standard.

Competitive Radioimmunoassay (Using Polyvinyl–Antibody Conjugates) for the Determination of Cholera Toxin and E. coli Enterotoxin-Antibodies[13]

Polyvinyl microtiter plates are coated with anti-cholera toxin serum (50 μl per well) diluted 1/2000 in phosphate-buffered saline. After incubation of the plates at room temperature for 12 hr, the wells are washed and filled with phosphate-buffered saline containing 1% BSA. For testing each well is inoculated with 25 μl of *E. coli* crude antigen and the plates are incubated for 12 hr at room temperature. The plates are then washed, 40 μl of serial twofold dilutions of sera to be tested for antibody or buffer are added to the wells, and the plates are incubated at room temperature overnight. A 10-μl sample of radiolabeled anti-cholera toxin IgG (~200.000 cpm) is then added to each well and the incubation is continued for 3 hr at 37°. The plates are then washed and cut up with scissors, and individual wells are placed in gamma-counting tubes for assay. A rabbit anti-crude *E. coli* enterotoxin serum is used as a standard (standardized against Swiss Serum Institute cholera toxin antiserum[13]). The schematic representation of this system is given in Fig. 18.

Acknowledgments

I am grateful to the American Society for Microbiology for permission to reproduce illustrations and quotations from previously published work.

[13] H. B. Greenberg, M. M. Levine, M. H. Merson, R. B. Sack, D. A. Sack, J. R. Valdesuso, D. Nalin, D. Hoover, R. M. Chanock, and A. Z. Kapikian, *J. Clin. Microbiol.* **9**, 60 (1979).

[20] Radioimmunoassay of Staphylococcal Enterotoxin C

By N. Dickie and M. Akhtar

Introduction

The enterotoxins elaborated by *Staphylococcus aureus* are a class of simple proteins that are responsible for the clinical manifestations of staphylococcal food poisoning. Five antigenic variants of the staphylococcal enterotoxins (types A, B, C_1, C_2, D, and E) have been identified to date.[1] Enterotoxin C_1 and C_2 are cross-reacting variants which differ in their isoelectric point. Enterotoxin C_1 is the more alkaline variant.[2] Partially purified staphylococcal enterotoxins have been available sufficient for use in the microslide test[3] which is the method most frequently used to detect enterotoxin in food. The availability of highly purified enterotoxins has made it possible to develop radioimmunoassay systems requiring relatively simple sample preparation and being able to detect as little as 1 ng of enterotoxin per gram of food in food extracts.[4]

This chapter contains a detailed description of the solid-phase radioimmunoassay used in our laboratory and its application to the detection of staphylococcal enterotoxin C_2 (SEC_2) in food. It is a modification of our previously reported assay[5] and is about 10 times more sensitive. It makes use of a different antiserum, different incubation conditions, and a better defined tracer. Techniques of radioimmunoassay employing solid-phase coupled antibodies offer several potential advantages. These have been reviewed by Wide[6] and include: simple separation procedures which are usually suitable for automation, nondisruptive separation of bound and free antigen, high precision, a small misclassification error, and a low dissociation rate constant which is an advantage when late addition of labeled ligand is used to increase assay sensitivity.

[1] M. S. Bergdoll, *in* "Food-Borne Infections and Intoxications" (H. Riemann and F. L. Bryan, eds.), 2nd ed., p. 443. Academic Press, New York, 1979.
[2] C. R. Borja and M. S. Bergdoll, *Biochemistry* **6**, 1467 (1967).
[3] E. P. Casman and R. W. Bennett, *Appl. Microbiol.* **13**, 181 (1965).
[4] M. S. Bergdoll and R. Reiser, *J. Food Protect.* **43**, 68 (1980).
[5] N. Dickie, C. E. Park, and H. Robern, *J. Clin. Pathol.* **29**, 833 (1976).
[6] L. Wide, *in* "Radioimmunoassay and Related Procedures in Medicine," Vol. 2, p. 143. International Atomic Energy Agency, Vienna, 1978.

Method of Radioimmunoassay

Preparation and Storage of Reagent

Enterotoxin C_2 from *Staphylococcus aureus* strain 361 and rabbit anti-enterotoxin C_2 serum were prepared in our laboratory as described[7] and were stored in freeze-dried form. *Buffer A*, used for diluting the antiserum is 0.1 M sodium bicarbonate (pH 9.6) containing 0.05% (w/v) sodium azide. *Buffer B*, used for diluting the enterotoxin, is 0.05 M sodium phosphate buffer (pH 7.4), containing 0.15 M sodium chloride, sodium azide, and 0.5% bovine serum albumin (RIA grade, Sigma Chemical Co.). The diluted reagents are stored in sealed containers at 4°.

Preparation of ^{125}I-SEC_2

The iodination procedure is similar to the technique used for angiotensins I and II.[8] The reaction was carried out with continuous magnetic stirring in a 20-ml glass scintillation vial cooled in an ice-bath. Two and one-half milliliters of 0.05 M sodium phosphate buffer (pH 7.5) was added to the vial, followed by 1.0 mCi carrier-free $Na^{125}I$ in 0.1 N NaOH (usually in 10 μl) and 15 μl SEC_2 (1.0 mg/ml dissolved in H_2O). One-half milliliter of a freshly prepared solution of chloramine-T (30 μg/ml in the phosphate buffer) was then added drop-wise over a period of 1 min, and after 10 min, 0.7 ml of a freshly prepared solution of sodium metabisulfite (30 μg/ml in the phosphate buffer) was added. After another 10 min of stirring, the reaction mixture was filtered through a 3-ml column of Dowex-1 resin, equilibrated with the phosphate buffer, to remove unreacted iodide[9] and fractions of 0.5 ml were collected. The seven or eight tubes containing greatest radioactivity were pooled.

Gel filtration was used for assessing the degree of iodination and for determining the extent of damage or degration of ^{125}I-SEC_2. Two milliliters of the freshly labeled SEC_2 was applied to a 2.5 × 33-cm column of Sephadex G-100 which was equilibrated and run in buffer B at room temperature. Fractions of approximately 4.0 ml were collected at the flow rate of 10 to 15 ml/hr and 50- to 100-μl aliquots were counted in a well type gamma counter for 1 min. For most SEC_2 iodinations a small amount (1% or less) of high-molecular-weight radioactive material appears shortly

---

[7] H. Robern, S. Stavric, and N. Dickie, *Biochim. Biophys. Acta* **393**, 148 (1975).

[8] A. E. Freedlender, F. Fyhrquist, and H. J. G. Hollemans, *in* "Methods of Hormone Radioimmunoassay" (B. M. Jaffee and H. R. Behrman, eds.), p. 455. Academic Press, New York, 1974.

[9] B. A. Miller, R. F. Reiser, and M. S. Bergdoll, *Appl. Environ. Microbiol.* **36**, 421 (1978).

after the void volume on the Sephadex column (peak I of Fig. 1). This material most likely represents aggregated or damaged enterotoxin which only poorly binds to the antibody. Fractions from the second peak, corresponding to the purified enterotoxin, were selected to minimize contamination with material from peak I and peak III, presumably free [125]I. The selected fractions were pooled and stored at 4°. The iodination efficiency was calculated from the elution pattern. In the iodination procedure about 30% of the radioactivity is incorporated into the protein. The labeled enterotoxin is usable for up to 2 months, after which the assay tends to become unstable resulting in erratic replicates and reduced dose–response.

Assay Procedure

The salient features of the radioimmunoassay are as follows: antibody-coated plastic tubes are incubated with known amounts of SEC_2 or the

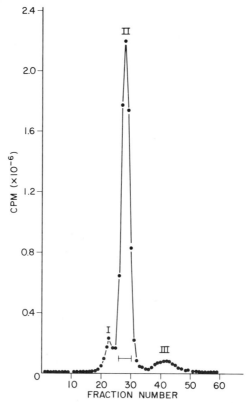

FIG. 1. The pattern of elution of radioactivity after gel filtration (Sephadex G-100) of ^{125}I-SEC_2. The fractions (4.0 ml) were pooled as indicated.

test sample at 4°. The tube contents are removed, labeled SEC_2 is added to each tube, and the incubation is continued briefly at room temperature. The antibody bound and free label are separated by aspiration of the tubes.

Preparation of Antibody-Coated Tubes. The coating procedure follows the principles of Catt and Tregear,[10] who elaborated in detail the requirements for coating plastic tubes with rabbit γ-globulin. Tubes were coated at room temperature. A range of antiserum dilutions prepared in buffer A was used and the coating time was 10 min. At the antiserum dilution where 25 to 30% of the tracer (15,000 cpm) was bound in the absence of unlabeled SEC_2 (corresponding to our assay conditions), 0.078 ng of unlabeled SEC_2 displaced about 15% of the bound tracer from solid-phase antibody. This dilution of antiserum was used for preparation of standard curves.

The interior of polypropylene tubes (12 × 75 mm, No. 9950, Canada Wide Scientific Ltd., Ottawa) was coated with antibody by addition of 0.20 ml of anti-SEC_2 antiserum (diluted 1:2,000, in this case) in buffer A. The tubes containing the diluted antibody were then stored at room temperature for 10 min. The antibody solution was removed by decantation, and the tubes were washed twice with buffer B by vortex mixing for a few seconds. To reduce nonspecific adsorption of SEC_2 to the polyethylene surface, 1 ml of 0.5% (w/v) bovine serum albumin (buffer B) was added to each tube and the tubes were stored at room temperature for 30 min. The buffer was then removed from the tubes as completely as possible by decanting and draining of the tubes. The tubes were used for the assay on the same day, usually within 1 hr of preparation.

The Standard Curve. Serial dilutions of standard SEC_2 from a starting concentration of 12.5 ng/ml were prepared in 100 µl of buffer B and dispensed to the coated tubes to give five concentrations in the range 0.078 to 1.25 ng/tube. Then buffer B (100 µl) was added to make a final volume of 200 µl. The mixture was shaken gently and the tubes were allowed to remain at 4° for 20 hr. Each standard was assayed in duplicate, the zero dose tubes in triplicate. After this preincubation, the contents were discarded, the tubes were washed twice with ice-cold buffer B (1 ml) on a vortex mixer and inverted for complete drainage. In the next step, 200 µl (15,000 cpm) of ^{125}I-SEC_2 in buffer B was added to each tube and the incubation was continued for 2 hr at room temperature. At the end of the assay incubation, the contents were removed by vacuum suction and each tube was washed twice with ice-cold 0.85% (w/v) sodium chloride. The incubation tubes (antibody-bound label) were counted in an automatic

[10] K. Catt and G. W. Tregear, *Science* **158**, 1570 (1967).

well scintillation gamma counter at about 70% efficiency. Each tube was counted for 5 min. The binding was expressed as the ratio of bound counts to total counts (B/T) after a correction for background radioactivity (150 cpm) had been made.

Calculation of Results. The standard curve was drawn manually by plotting the ratio of bound counts to total counts (B/T) on the vertical axis against the arithmetic dose of enterotoxin on the horizontal axis. Figure 2 shows a typical standard curve of the system described. The advantage of this approach is that it presents a reasonably complete picture of the characteristics of any one assay run. For instance, the amount of labeled enterotoxin bound in the zero dose tubes can be read directly from the curve; if this is low, it suggests deterioration in either the labeled enterotoxin or the antibody. Unknown values can be obtained by interpolation on the standard curve.

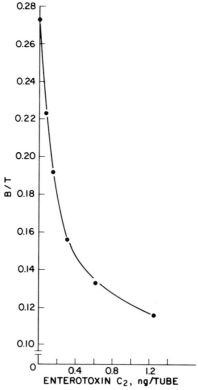

Fig. 2. Enterotoxin C_2 standard curve.

Some Characteristics and Results of the Assay

Reproducibility of Standard Curves

Standard curves were established on twelve different days with two different lots of labeled SEC_2. Mean values and standard deviations of the percentage bound counts ($B/B_0 \times 100$) for each standard were calculated and are shown in Table I.

Precision

Intraassay precision was examined by calculating the variance of replicate responses for several doses on the standard curve and describing the relationship between the variance and mean of the dose–response (B/T) by regression analysis, as suggested by Rodbard et al.[11] The power function (their Eq. 5) was fit to the data without weighting and by the use of a hand-held calculator. The standard deviation of the SEC_2 estimate at any point on the standard curve was estimated from the standard deviation of the response variable and the slope of the dose–response curve calculated at this point.[12] Precision, expressed as the predicted coefficient of variation of the dose, varies with the part of the dose-response curve covered (Fig. 3). It is less at the extreme of low SEC_2 concentration s. In the region of the standard curve corresponding to 30 to 65% inhibition of binding, precision of duplicate samples is about 10%.

TABLE I
REPRODUCIBILITY OF STANDARD CURVES

Added SEC_2/tube (ng)	$B/B_0 \times 100$
0.078	83.5 ± 3.3[a]
0.156	70.0 ± 2.4
0.312	59.6 ± 4.7
0.625	45.1 ± 2.0
1.25	35.1 ± 2.0

[a] Each value represents the mean and standard deviation of the percentage bound ($B/B_0 \times 100$) for 12 standard curves.

[11] D. Rodbard, R. H. Lennox, H. L. Wray, and D. Ramseth, Clin. Chem. 22, 350 (1976).
[12] D. Rodbard, in "Radioimmunoassay and Related Procedures in Medicine," Vol. 2, p. 21. International Atomic Energy Agency, Vienna, 1978.

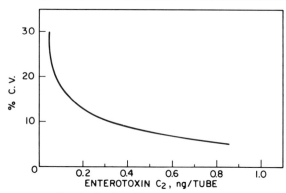

Fig. 3. Predicted intraassay precision.

Sensitivity

In this assay, the counts bound in the absence of unlabeled SEC_2 could usually be distinguished statistically ($p < 0.05$, one-tailed t test) from the counts bound in the presence of an unknown, assayed in duplicate, if 6% of the maximum binding is displaced. In an incubation volume of 0.20 ml, this displacement corresponds to about 30 pg of SEC_2, as estimated by dose interpolation on the standard curve. This gives a working sensitivity for the assay of 0.3 ng/ml with 100 μl as the largest sample size.

Specificity

Specificity of the antiserum was assessed by comparing the ability of other related proteins such as staphylococcal enterotoxins A, B, and E at various concentrations to compete with ^{125}I-SEC for binding sites on the antiserum. When staphylococcal enterotoxins A,[13] B (Makor Chemicals, Jerusalem), and E,[14] were added at 50 ng/tube, about 15% of labeled SEC_2 was displaced from the antibody with enterotoxins A and E, 30% with enterotoxin B. Increasing the concentration of each of these toxins to 100 ng/tube did not increase the percentage of labeled SEC_2 displaced from the antibody. This effect is a reflection of the specificity of the antiserum and may be different for other antisera.

An important criterion for specificity is a linear decrease in measured immunoreactive SEC_2 levels with dilution. This was demonstrated with the supernatant fluid of cultures of *S. aureus* strain 361. Serial dilutions were prepared to encompass the picogram dose range of the assay. Sam-

[13] H. Robern, S. Stavric, and N. Dickie, *Biochim. Biophys. Acta* **393**, 159 (1975).
[14] C. R. Borja, E. Fanning, I-Y. Huang, and M. S. Bergdoll, *J. Biol. Chem.* **247**, 2456 (1972).

ples were analyzed in duplicate and the apparent amount of material detected per tube was calculated. Analysis of an experiment, following Rodbard et al.,[15] used the $\log_{10}$ transformation of all data to correct for heterogeneity of variance. These data are presented in Fig. 4. Calculation of a straight line by means of simple least-squares regression analysis gave a slope of 0.962 and a correlation of 0.985, which indicates that good parallelism was achieved.

Assay of Food Extracts

For preparation of sample, solid food material was extracted in two volumes of liquid per weight of food by the acid–base extraction method of Robern et al.[16] In brief, the extract is adjusted to pH 4.5 and to clear it centrifuged. The pH of the mixture is then adjusted to pH 7.4 and again centrifuged. When oily specimens were examined the final supernatant

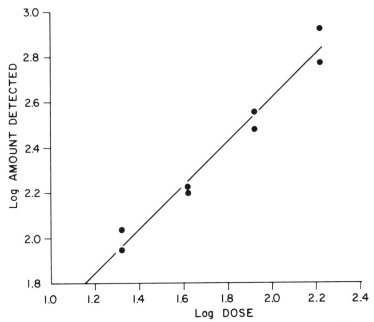

FIG. 4. Testing parallelism of standard and unknown (strain 361 culture fluid) dose–response curves using simple linear regression. Log (pg SEC_2 detected) versus log sample volume (10^{-5} μl).

[15] D. Rodbard, P. J. Munson, and A. De Lean, in, "Radioimmunoassay and Related Procedures in Medicine," Vol. 1, p. 469. International Atomic Energy Agency, Vienna, 1978.
[16] H. Robern, T. M. Gleeson, and R. A. Szabo, Can. J. Microbiol. **24**, 436 (1978).

fluid was filtered through a Millex disposable filter unit (0.45 μm, Millipore) and the filtrate used for analysis.

Accuracy was determined by recovery studies of SEC_2 added to food extract containing less than 0.3 ng/ml of endogenous SEC_2 (determined by radioimmunoassay). Enterotoxin C_2 was added to the extract at 12.5 ng/ml and at halving dilutions of 6.2 and 3.1 ng/ml. Buffer B was the diluent and 100 μl was assayed in duplicate at each dilution. Miller et al.[9] reported that recovery of enterotoxin varies from food to food. We therefore tested recoveries in a number of foods of high protein content in which interference could be anticipated. Table II shows that this was not the case and that the radioimmunoassay was satisfactory.

Unknown samples are regularly tested at multiple dilutions and a mean SEC_2 concentration is calculated from these dilutions. The radioimmunoassay results are verified by comparing the results of serial dilutions of unknowns with serial dilutions of standards. The curves should be parallel. Identification of low levels of SEC_2-like immunoreactivity could present a problem in that it may be due to SEC_2 or due to nonspecific inhibition of the antigen–antibody reaction by protein or other substances in the extract. To overcome this problem, further purification and concentration are required. For this purpose, we used hydrophobic interaction chromatography. This technique has recently been adapted to the concentration of dilute protein solutions for analytical purposes[17] but, to our knowledge, it has not been used to concentrate staphylococcal enterotoxin.

TABLE II
RECOVERY OF ADDED STAPHYLOCOCCAL
ENTERTOXIN C_2 FROM SOME FOODS

Food	r^a	Recovery (%)
Salmon	0.993	88
Sardine	0.990	70
Salami	0.999	67
Ham	0.999	86
Cheese	0.995	83

[a] The SEC_2 (pg/tube) versus SEC_2 added (pg/tube) was analyzed by regression which permitted r values to be calculated. Salmon = $1.124x - 129$; sardine = $0.868x - 90$; salami = $0.725x - 27$; ham = $0.909x - 29$; cheese = $1.099x - 144$.

[17] O. Vesterberg and L. Hansen, *Biochim. Biophys. Acta* **534**, 369 (1978).

TABLE III
RECOVERY OF SEC_2 FROM ITALIAN SALAMI EXTRACT TO WHICH KNOWN AMOUNTS
OF SEC_2 HAD BEEN ADDED

SEC_2 in 35 ml of extract applied to phenyl-Sepharose (ng)	Volume of final eluate (ml)	Reduction in ^{125}I uptake by 100 μl of eluate		Estimated SEC_2 (ng)	Recovery (%)
		%	Equivalent SEC_2 (ng)		
0	2.5	0	0	0	—
5	2.6	27.3	0.16	4.2	83.2
10	2.2	45.4	0.37	8.1	81.0
20	3.5	46.3	0.39	13.6	68.0
50	3.0	61.7	1.05	31.5	63.0

Mean ± SD = 73.8 ± 9.8

Use of Hydrophobic Interaction Chromatography to Concentrate SEC_2

Small columns of phenyl-Sepharose CL-4B (Pharmacia Fine Chemicals) were prepared in 10-cm^3 glass syringe barrels (bed volume of 4.0 ml). The columns were washed with 100 ml of H_2O and then equilibrated with 40 ml of 2.0 M $(NH_4)_2SO_4$. All chromatography was performed at room temperature at a flow rate of 0.2 ml/min. The purification method was tested by dissolving labeled SEC_2 (125,000 cpm) in 35 ml of Italian salami extract. Before application to the column, solid ammonium sulfate (10.56 g) was added to the extract so that the final concentration was 2.0 M. After remaining at room temperature for 1 hr, the extract was clarified by centrifugation at 4° for 90 min at 23,500 g. The supernatant fluid containing approximately 120,000 cpm was applied to the column, treated with two bed volumes of 2.0 M $(NH_4)_2SO_4$ and the appearance of radioactivity, if any, was monitored in a gamma counter. Then a solution of ethylene glycol (1 : 1 in H_2O) was applied to the column and fractions of 0.5 ml were collected. Of the 65% SEC_2 initially bound, 94% was eluted with ethylene glycol in fractions 8 through 11. Similarly, 5 to 50 ng of unlabeled SEC_2 in 35 ml of salami extract was applied to columns, and radioimmunoassay was used to monitor the recovery of SEC_2 after elution with ethylene glycol. This eluate (fractions 8 through 11) was desalted on a small disposable column of Sephadex G-25 prior to radioimmunoassay. Initial experiments with labeled SEC_2 and phenol red were used to determine how much eluant (buffer B) was required to elute the protein fraction. When the column (PD-10, Pharmacia) had been calibrated in this way, a series of similar columns were prepared to desalt the fractions as they left the phenyl-Sepharose column. The high-molecular-weight fractions were straw-colored due to the presence of food components so that elution from Sephadex G-25 could also be followed visually. The fractions were pooled and 100 μl duplicates were used for the assay of SEC_2.

Recoveries, calculated from the standard curve, were about 70% (Table III), which was sufficient to enable the detection of as little as 0.14 ng SEC_2/ml in 35 ml of the extract. This level of SEC_2 could not be detected directly by the radioimmunoassay as it is currently performed. The added sensitivity accrues from an approximate 10-fold concentration of the toxin. The technique is applicable to other staphylococcal enterotoxins such as types A, B, and E. For example, with enterotoxin A we detected as little as 0.2 ng/g of naturally contaminated food (unpublished results).

[21] Solid-Phase Radioimmunoassay for Bacterial Lipopolysaccharide

By Diane M. Jacobs and Jan A. Gutowski

Lipopolysaccharide (LPS) is the endotoxic constituent of gram-negative bacteria. Present in up to 20–30% it, together with various lipids and proteins, makes up the outer membrane of the cell. Gram-negative bacteria when allowed to invade and multiply in an animal host release LPS into the circulatory system which in turn can give rise to a multitude of pathophysiological manifestations such as fever, shock and even death.[1]

A number of biological,[2-4] chemical,[5,6] and enzymatic[7] assays are at present available for estimating endotoxin levels. However, these tend to be semiquantitative at best and interpretable only when applied under specific and restrictive test conditions.

The development of a radioimmunoassay for lipopolysaccharide has in the past been hindered by the unique chemical and physical characteris-

---

[1] S. Kadis, G. Weinbaum, and S. J. Ajl (eds.), "Microbial Toxins," Vol. V. Academic Press, New York, 1971.

[2] K. C. Milner and R. A. Finkelstein, J. Infect. Dis. 116, 529 (1966).

[3] W. R. Keene, H. R. Silberman, and M. Landy, J. Clin. Invest. 40, 295 (1961).

[4] R. E. Pierone, E. J. Broderick, A. Bundeally, and L. Levine, Proc. Soc. Exp. Biol. Med. 133, 790 (1970).

[5] V. S. Waravdekar and L. D. Saslow, J. Biol. Chem. 234, 1945 (1959).

[6] Y. D. Karkhanis, J. Y. Zeltner, J. J. Jackson, and D. J. Carlo, Anal. Biochem. 85, 595 (1978).

[7] R. Nandan and D. R. Brown, J. Lab. Clin. Med. 89, 910 (1977).

METHODS IN ENZYMOLOGY, VOL. 84

tics of the molecule. Thus the absence of residues which lend themselves to iodination by conventional labeling techniques has necessitated the attachment of various tyrosine-like ligands; the difficulty in raising an antiserum of sufficiently high titer has led to the establishment of an immunization schedule specific for this molecule; and its high nonspecific adherence to precipitated protein has prompted the development of a solid-phase assay system.

The technique of radioimmunoassay provides a sensitive and quantitative method of measuring lipopolysaccharide unaffected by various inhibitors or cross-reactive molecules afflicting other presently available assays. Here we present a radioimmunoassay for *E. coli* 055:B5 lipopolysaccharide.[8] No data will be given as to the cross-reactivity of 055:B5 antiserum in detecting other strains of LPS.

Principle

The radioimmunoassay for lipopolysaccharide is based on the competition of binding to anti-LPS IgG coated polystyrene tubes of ^{125}I-labeled and native LPS. After incubation, bound and free LPS are separated by aspiration and the tubes are counted after being washed with saline.

Thiobarbituric Acid (TBA) Determination of LPS

The thiobarbituric acid method[5] is used for quantitating LPS and derivatized LPS in column fractions and pooled solutions using purified *E. coli* 055:B5 LPS as standard. It has been modified by Morrison and Leive,[9] and is used in our laboratory as described below.

Reagents
2-Thiobarbituric acid, pH 2.0. Mix 0.71 g crystalline thiobarbituric acid in 90 ml H_2O and add 0.7 ml 1.0 N NaOH. Shake in warm water until solution is clear. Adjust to 100 ml and filter. Solution is said to be stable for 30 days, but is usually made fresh weekly. Dilute 1:2 in water just before use
Periodic acid reagent: 0.025 N sodium periodate in 0.125 N H_2SO_4
Arsenite solution: 2% reagent grade sodium arsenite in 0.5 N HCl
0.2 N H_2SO_4
10 N NaOH
Procedure. Add 0.2 ml sample (1–100 μg LPS) to 0.1 ml 0.2 N H_2SO_4. Heat in boiling water bath 15 min using condensers (marbles) on

[8] J. A. Gutowski and D. M. Jacobs, *Immunol. Commun.* **8**, 347 (1979).
[9] D. C. Morrison and L. Leive, *J. Biol. Chem.* **250**, 2911 (1975).

tubes. Add 0.2 ml periodic acid reagent and heat at 55° for 22 min. Add 0.4 ml arsenite solution and mix immediately. Add 1.6 ml thiobarbituric acid. Heat 12 min in boiling water bath. Allow to cool. Read at 532 nm. If a precipitate has formed in any tubes, add 2 drops NaOH to all tubes before reading.

Isolation of LPS

This phenol-water extraction procedure was first described by Westphal and Jann[10] and subsequently modified by Morrison and Leive.[9] The latter modifications have been adopted and used in our laboratory for extracting *E. coli* 055:B5 as follows.

Reagents
Bacteria: 50 g wet weight in 50 ml saline or 12.5 g dry weight suspended in 50 ml saline (lyophilized bacteria obtained from the New England Enzyme Center, Boston, MA.)
90% Phenol (Mallinckrodt). Place 45 g phenol in 100 ml graduated cylinder. Carefully add 4 ml boiling distilled water. Warm to 56°. Adjust to 50 ml. Heat to 68° in water bath
Sepharose 4B (Pharmacia) column, 0.1 M Tris buffer, pH 8.0 with 0.05% azide
1 M $MgCl_2$
Ribonuclease A (5× crystallized, Type 1-A, Sigma)
Pronase (protease from *Streptomyces griseus*, repurified, Type VI, Sigma)

Procedure. Rapidly add equal volume 90% phenol to bacterial suspension and mix well at 68° for 30 min. Transfer suspension to 25-ml screw-capped Corex tubes and centrifuge at 10,000 rpm for 30 min at 4°. Remove and reserve top aqueous phase. Wash pellet and phenol phase with 50 ml saline, and centrifuge as before. Remove aqueous phase and combine with aqueous phase from first extraction. Discard phenol phase which is very dark and has large, sticky pellet. Dialyze aqueous phase against several changes of several liters of saline until no odor of phenol is discernible. Add 0.5 ml 1 M $MgCl_2$ for each 200 ml. Add RNase to 20 μg/ml, and incubate 1 hr at 37°. Add Pronase to 20 μg/ml, incubating overnight at 37°. Concentrate to 25 ml by Amicon ultrafiltration using PM30 filter.

Load preparation onto 3-liter Sepharose 4B column. Diameter/height ratio is not crucial—laboratory column is 10 × 40 cm. Elute with 0.1 M

[10] O. Westphal and K. Jann, *in* "Methods in Carbohydrate Chemistry" (R. L. Whistler, ed.), p. 83. Academic Press, New York, 1965.

Tris buffer, pH 8.0 containing 0.05% azide, collecting 15–20 ml fractions per tube. Read fractions at 260 nm to detect nucleic acids. Assay 20 μl alternate fractions with TBA for LPS. Peaks should overlap only minimally. Pool and concentrate LPS peak and dialyze against pyrogen-free water.

Preparation of IgG Fraction of Antiserum

Immunization and Preparation of Serum. Bacteria are grown in minimal medium for 18 hr and heated at 100° for 2.5 hr. For immunization, the suspension is adjusted with saline to $OD_{600\ nm} = 0.7$, approximately 10^9 organisms per milliliter. New Zealand white rabbits are injected intravenously with 1 ml suspension three times per week for at least 4 months. If dried bacteria are available, 0.2 ml of a suspension of 1 mg/ml in saline can be used. Rabbits are allowed to rest for 10 days before bleeding. (A test bleed is recommended.) Blood is collected, allowed to clot at room temperature, and held at 4° overnight. Serum is separated from the clot after centrifugation. Chronic immunization is necessary to induce high titers of IgG anti-LPS antibody. High titered antiserum will give good precipitation lines when tested by immunodiffusion. Ouchterlony plates made up of 1.5% Ionagar No. 2 in barbital acetate buffer, pH 8.6 (5.4 g sodium barbital, 4.3 g sodium acetate trihydrate, and 58.2 ml 0.1 N HCl per liter) containing 0.5 M glycine are used. Wells are filled with undiluted antiserum and LPS samples at 2 mg/ml. Line can be seen after incubation overnight at room temperature.

Fractionation of Serum. The IgG fraction of serum is prepared by chromatography on DEAE Affi-Gel Blue (Bio-Rad Laboratories) using the directions supplied with the resin. In our laboratory we separate 3 ml serum on a 21 ml column equilibrated with a buffer composed of 0.02 M Tris–HCl, 0.011 M NaCl, pH 7.2. The serum is dialyzed against the buffer before application to the column and eluted with three bed volumes of the starting buffer. Fractions are monitored at 280 nm and peak fractions pooled for use. Protein concentration of the pool is determined by any routine laboratory procedure. Use of this resin produced IgG which gave more reproducible sigmoid curves in the final RIA than IgG prepared on regular DEAE.

Preparation of Derivatized LPS

Reagents

p-Hydroxyphenylacetic acid (pHPAA) (Aldrich Chemical Co.)
4-Hydroxyphenethylamine (Tyramine, T) (Sigma Chemical Co.)

1-Ethyl-3-(3-dimethylaminopropyl)carbodiimide-HCl (EDC) (Pierce Chemical Co.)
Glutaraldehyde, 50% (Fisher Scientific Co.)
Sephadex G-75, Sepharose 4B (Pharmacia Fine Chemicals)
0.01 N NaOH
Sodium phosphate buffer: 0.1 M, pH 6.5
Sodium acetate buffer: 0.5 M, pH 5.0
Glycine: 2 M, pH 8.0
Ethanol

Procedure

p-Hydroxyphenylacetic Acid-LPS (pHPAA-LPS).[11,12] To a solution of pHPAA (11.4 mg) in 1.25 ml of water is added 6 mg EDC. The pH is adjusted to 5–6 with 0.01 N NaOH, and the mixture is incubated at 37° for 8 hr in the dark with stirring. LPS (3 mg) in 40 μl of water is now added and the incubation is continued in the dark at 37° for a further 24 hr with stirring, maintaining pH 5–6 with 0.01 N NaOH. The pHPAA-LPS is purified on a Sephadex G-75 column (1 × 20 cm) equilibrated in water, collecting 1-ml fractions. The desired material is eluted in the first of two peaks (fractions 7–14) and pooled. Yield is ca. 2.5 mg.

Tyramine-LPS (T-LPS).[13] To a pale yellow solution of tyramine (16.4 mg), dissolved with warming in 3 ml of ethanol, is added 2 ml of sodium phosphate buffer (0.1 M, pH 6.5), 3 mg of LPS, and 27 mg of glutaraldehyde. The mixture is agitated gently for 3 hr at room temperature at which time the reaction is terminated by addition of 200 μl of sodium acetate buffer (0.5 M, pH 5.0) over 30 sec. Unreacted aldehyde groups are blocked by addition of 2.0 ml of 2 M glycine, pH 8.0, and mixing is continued for 2 hr at room temperature. A slight precipitate which occasionally forms during the reaction is removed by centrifugation and the derivatized LPS is purified on a column (1 × 50 cm) of Sepharose 4B equilibrated in water, collecting 1.3-ml fractions. The derivatized LPS is eluted in the first of two peaks (fractions 5–11) and pooled. Yield is about 2.0 mg.

Both pHPAA-LPS and T-LPS are indistinguishable from the underivatized endotoxin by various biophysical, immunological, and biological criteria.[8]

[11] J. C. Sheehan and G. P. Hess, *J. Am. Chem. Soc.* **77,** 1067 (1955).
[12] H. G. Khorana, *Chem. Ind. (London)* 1087 (1955).
[13] D. J. Ford, R. Radin, and A. J. Pesce, *Immunochemistry* **15,** 237 (1978).

Iodination of Derivatized LPS[14]

Reagents
Na[125]I: ≥ 17 mCi/μg, carrier-free IMS-30 (Amersham/Searle)
Phosphate buffers: 0.1 and 0.5 M sodium phosphate, pH 7.2
Chloramine-T: 1 mg/ml in 0.1 M phosphate buffer (Eastman-Kodak)
Sodium metabisulphite: 2 mg/ml in 0.1 M phosphate buffer (Fisher Scientific Co.)
Borate buffer: prepared by adjusting pyrogen free saline (0.9% NaCl, Travenol Laboratories) to 50 mM in sodium borate, 50 mM in $CaCl_2$, and 0.02% NaN_3, pH 8.0
Derivatized LPS (T-LPS or phPAA-LPS), 500 μg/ml in 0.1 M phosphate buffer
Sephadex G-75 (Pharmacia Fine Chemicals)
For transfer of low concentrations of LPS glass equipment must be used throughout.

Procedure. Forty microliters 0.5 M phosphate buffer, and 10 μg derivatized LPS are added to the vial containing 1 mCi Na[125]I and mixed by vortexing. The mixture is then treated with 20 μl of chloramine-T over 30 sec with vigorous vortexing. The reaction is stopped with 50 μl sodium metabisulfite followed by 100 μl of borate buffer and vortexed. The ^{125}I-LPS is purified on a column (10 ml glass pipette) of Sephadex G-75 equilibrated in borate buffer and 1-ml fractions collected. The purified material is eluted in fractions 6–7 (Peak I) and free iodide in fractions 10–11 (Peak II). It is stored at 4° until use.

Normally, about 10% of recovered ^{125}I is incorporated into the derivatized endotoxin which has a final minimum specific activity of ca. 2–5 mCi/mg based on a total recovery of column applied material.

As it is not possible to accurately measure this small amount of LPS in the eluate by the TBA assay, the calculated specific activity is only a minimum estimate. We have, however, used a more direct approach to determine the specific activity by increasing the amount of LPS used in the labeling procedure. When a column is loaded with 170 μg phPAA-LPS and the eluted fractions assayed by the TBA method, 8–10% of the starting material is recovered in one peak corresponding to Peak I. When the same amount of derivatized LPS is iodinated and fractionated, Peak I contains 8–10% of the starting material and Peak II, 26–30% of the starting material. Peak I therefore contains 21–33% of recovered radioactivity. (Data based on routine iodinations of this type carried out over a 2-year period.)

---

[14] F. C. Greenwood, W. M. Hunter, and J. S. Glover, *Biochem. J.* **89**, 114 (1963).

These data suggest that (1) more iodide is incorporated into LPS when a larger amount is labeled and (2) 90% of the applied LPS sticks to the column nonspecifically, whether or not it is iodinated. On the assumption that unrecovered radioactivity is all in column-adherent LPS, we can estimate the amount of LPS recovered in Peak I and calculate the specific activity which is 2–5 mCi/mg. If the same fraction of LPS is recovered when 10 μg is iodinated, the actual specific activity could well be an order of magnitude higher.

Radioimmunoassay

Reagents
12 × 75 mm polystyrene tubes (Falcon No. 2052)
Borate-buffered saline (BBS) with azide: prepared by adjusting pyrogen-free saline (0.9% NaCl, Travenol Laboratories) to 50 mM in borate and 0.02% NaN$_3$ at pH 8.0 ("pH 8.0 BBS") and pH 9.5 ("pH 9.5 BBS")
Bovine serum albumin (BSA): 0.5% in pH 8.0 BBS (Sigma, RIA grade)
IgG fraction of rabbit anti-LPS diluted in pH 9.5 BBS
37° shaker/water bath or a shaker (R-2 reciprocator, New Brunswick Scientific Co., set at 200 strokes/min) in a 37° warm room
Procedure

Coating of Tubes with Antibody. One-half milliter of an appropriate dilution (see next section) of the IgG fraction of rabbit anti-LPS made up in pH 9.5 BBS is carefully pipetted into the bottom of each polystyrene tube. After 1 hr at room temperature the antibody solution is removed by aspiration and pooled for reuse. The tubes are incubated with 1 ml of 0.5% BSA for a further 30 min at room temperature. After the BSA is discarded, the tubes are washed twice with 3 ml of pH 8.0 BBS and drained inverted over absorbent tissue at 4°. Normally the tubes are used within a few hours, but they will maintain their binding capacity for 24–36 hr when stored at 4°. The pooled antiserum has a shelf life of at least 1 month when kept at 4°, but should be regularly monitored for 100% binding efficiency if used repeatedly. Contamination with extraneous endotoxin is usually undetectable.

Coating with 0.5% BSA does not normally improve the assay. Nonetheless, this procedure is recommended to preclude any unexpected nonspecific binding of ^{125}I-LPS to tube surfaces. Likewise, increasing the tube-coating time from 1 hr up to 4 hr does not appreciably increase the amount of IgG bound.

Determination of Antibody Titer. Anti-LPS IgG fraction is serially diluted in pH 9.5 BBS in 1 ml volumes in glass tubes. Polystyrene tubes are coated with 0.5 ml each dilution as described above. Coated tubes are subsequently incubated with ca. 15,000 cpm ^{125}I-derivatized LPS in 600 μl pH 8.0 BBS and incubated with shaking for 4 hr at 37°. The solutions are removed by aspiration, the tubes are washed once with 3 ml of saline, and bound radioactivity determined in a gamma scintillation spectrometer.

From a plot of percentage of added radioactivity bound against $\log_{10}$ concentration of anti-LPS IgG the antibody titer is estimated as the concentration of IgG which results in a 50% binding of tracer relative to the maximum observed binding. This IgG concentration is used in coating RIA tubes for detection of LPS.

Detection of LPS. LPS samples are serially diluted in pH 8.0 BBS in glass tubes. Four hundred microliters of each sample and 100 μl pH 8.0 BBS are incubated overnight at 4° in antibody-coated polystyrene tubes on a reciprocating shaker. One hundred microliters of ^{125}I-derivatized LPS (ca. 15,000 cpm) are added, and the tubes are incubated for a further 2 hr at 37°. The reaction mixture is then removed by aspiration, the tubes are washed once with 3 ml of saline, and counted.

Two controls are run concurrently: B_0 (100% binding) contains 500 μl of pH 8.0 BBS and 100 μl of ^{125}I-LPS; NSB (nonspecific binding) contains the same addition as B_0 but uses tubes coated with 0.5% BSA only and no antibody. Samples and controls are run at least in duplicate, preferably triplicate.

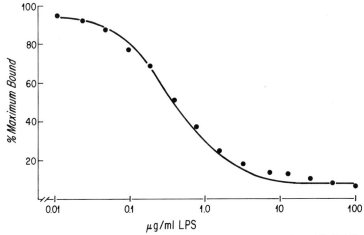

FIG. 1. Detection of LPS in radioimmunoassay using IgG prepared on Affi-Gel Blue dextran.

Percentage bound is calculated from the relationship:

$$\frac{\text{Sample cpm} - \text{NSB cpm}}{B_0 \text{ cpm}} \times 100\%$$

Standard curves are plotted as percentage tracer bound against $\log_{10}$ LPS concentration in μg/ml (Fig. 1).

LPS content of unknown is estimated from the dilution factor at 50% binding relative to that of the standard sample.

In the case of *E. coli* 055:B5 LPS, a sensitivity range of approximately 10–500 ng/ml was obtained.[8] Nonspecific binding and 100% binding were about 5 and 60%, respectively, of the total counts added.

[22] Radioimmunoassays of Thyroxine (T₄), 3,5,3'-Triiodothyronine (T₃), 3,3',5'-Triiodothyronine (Reverse T₃, rT₃), and 3,3'-Diiodothyronine (T₂)

By WILMAR M. WIERSINGA and INDER J. CHOPRA

Introduction

Although there has been a decided improvement in our understanding of the field of endocrinology during the last decade because of the many rapid developments, the resulting increase in complexity of certain issues has led to confusion in some instances. For example, we now know that substances synthesized within an endocrine cell are not necessarily those conventionally viewed as the hormone, but merely are precursors (e.g., preproparathormone and proparathormone in case of parathormone). Along the same line, the substance secreted by an endocrine cell into the bloodstream is not necessarily the one that ultimately provokes the hormonal effect in the target tissues but only a prehormone (e.g., testosterone for dihydrotestosterone). The hormonal effects in the target cells may be modulated not just by circulating hormone levels but also by the receptor state (e.g., number and affinity of receptors for the hormone), and in some cases also by postreceptor alterations.

Recent discoveries involving thyroxine (T₄) provide an excellent illustration of these complexities, First, T₄ synthesized in the thyroid gland is secreted unchanged but is also converted intrathyroidally to a more active triiodothyronine (T₃).[1] Second, T₄ released into circulation is monodeiodinated in the extrathyroidal tissues into either the more potent hormone

[1] H. Ishii, K. Tanaka, K. Naito, M. Nishikawa, and M. Inada, *Annu. Meet., 62nd, Endocr. Soc., Washington, D.C.* p. 202, Abstr. 512 (1980).

METHODS IN ENZYMOLOGY, VOL. 84

T_3 or to the calorigenically inactive reverse T_3 (rT_3).[2,3] Subsequently, T_3s (T_3 and rT_3) are sequentially deiodinated in extrathyroidal tissues into three diiodothyronines (T_2s), two monoiodothyronines (T_1s), and finally into thyronine (T_0), presumably the predominant end product of iodothyronine metabolism (Fig. 1). Third, the nuclear receptors for thyroid hormones can change in number or affinity.[4,5] Last, modulation of the thyroid hormone effect on the target tissues may involve a postreceptor modulation.[6]

The availability of sensitive and specific radioimmunoassays (RIAs) for the study of various iodothyronines has been a key factor responsible for this new information. The present report reviews the RIAs of T_4, T_3, rT_3, and $3,3'$-T_2 and their application to the diagnosis of thyroid dysfunction in thyroidal and nonthyroidal diseases.

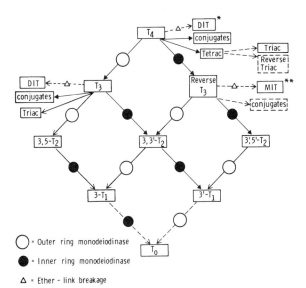

FIG. 1. Sequential deiodination of T_4. Dashed lines indicate pathways which have not yet been demonstrated conclusively. *, DIT could also be produced by breakage of ether linkage of $3,5$-T_2. **, MIT could also be produced by breakage of ether linkage of $3,3'$-T_2 and 3-T_1. Modified from I. J. Chopra, D. H. Solomon, U. Chopra, S. Y. Wu, Y. Nakamura, and D. A. Fisher, *Recent Prog. Horm. Res.* **34**, 521 (1978).

[2] L. E. Braverman, S. H. Ingbar, and K. Sterling, *J. Clin. Invest.* **49**, 855 (1979).
[3] I. J. Chopra, *J. Clin. Invest.*, **58**, 32 (1976).
[4] L. J. DeGroot, A. H. Colyoni, P. A. Rue, H. Seo, E. Martino, and S. Refetoff, *Biochem. Biophys, Res. Commun.* **79**, 173 (1977).
[5] J. Bernal, S. Refetoff, and L. J. DeGroot, *J. Clin. Endocrinol. Metab.* **47**, 1266 (1978).
[6] H. L. Schwartz, M. A. Forciea, C. N. Mariash, and J. K. Oppenheimer, *Endocrinology* **105**, 41 (1979).

Radioimmunoassays (RIAs) of Iodothyronines: General Principles

RIAs are based on the principle of saturation analysis, also described as the competitive-inhibition principle which can be applied equally as well to competitive protein binding assays (CPBAs). The principle pertains to a competition between a nonradioactive compound (in the standards or unknown specimens) and the same (or a very similar) compound in a labeled (e.g., radioactive) form for a limited number of high-affinity binding sites on an antibody (in RIA) or on another (nonantibody) binding protein (in CPBA) capable of specifically binding the compound.

It therefore seems evident that an RIA can be developed of any compound provided three basic reagents are available: (1) the compound itself in a pure nonradioactive form; (2) an antibody capable of specifically binding the compound; and (3) a tracer (radioactive) form of the compound itself or a similar agent that is recognized by the specific compound-binding sites on the antibody. Given these reagents, a concentration (titer) of antibody that binds about 30–50% of a tracer amount of the radioactive compound is first determined. The binding of the tracer to this concentration (or dilution) of antibody is then determined in the presence of various concentrations of standards (nonradioactive pure compound) in order to prepare a standard curve. The data on binding of the tracer to antibody in the presence of an unknown can then be employed to determine, from the standard curve, the concentration of the compound in the unknown specimen.

The structural formulas of various iodothyronines are presented in Fig. 2. These compounds differ in the number and the location of the iodine atoms attached to the thyronine skeleton. Although these variations in iodothyronines are seemingly extremely minimal, they are associated with enormous differences in their biological and immunological activities. T_3, which is the most potent iodothyronine, has two iodine atoms in the inner (or α) ring and one in the outer (or β) ring. Reverse T_3 has two iodine atoms in the outer ring and one in the inner ring of thyronine. In a sharp contrast with T_3, rT_3 is virtually devoid of any calorigenic or TSH-suppressive activity. T_2s, T_1s, and T_0 also possess little or no metabolism-stimulating (calorigenic) or TSH suppressive properties.

There are marked differences in the concentrations of the various iodothyronines in human serum (Table I). Therefore, it is necessary that the iodothyronine binding antibody used in RIA be highly specific. The methods of preparation and purification of tracer (radioactive) iodothyronine, production of iodothyronine binding antibodies, and preparation of standards and of iodothyronine-free serum (for use in the standards to render their protein concentration similar to the unknowns) are very similar for the assays of T_4, T_3, rT_3, and $3,3'-T_2$. After a discussion of these, a detailed description of each RIA follows.

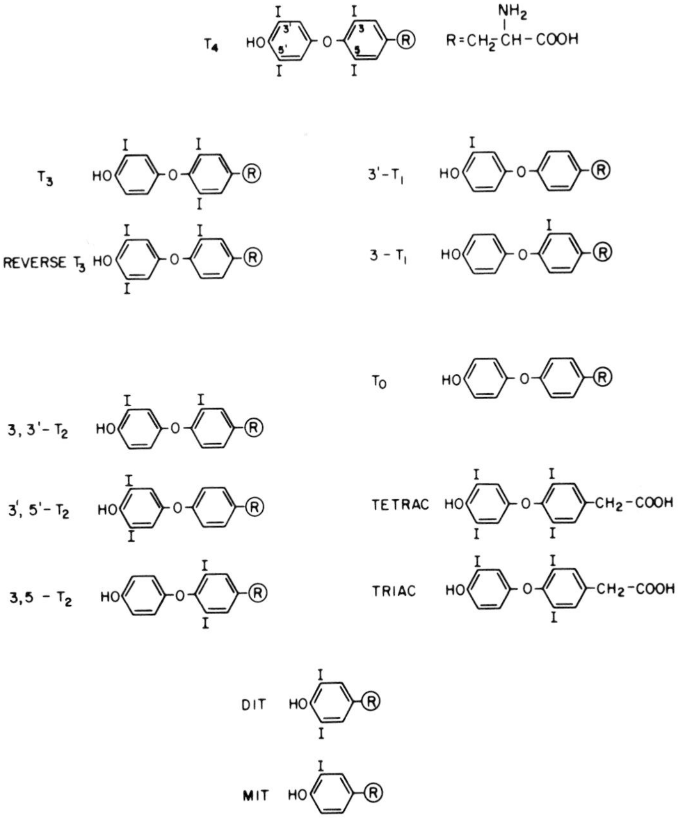

FIG. 2. Molecular constitution of iodothyronines and their analogs detectable in human circulation.

Preparation of Radiolabeled Iodothyronines

Until recently radiolabeled thyroid hormones have mainly been prepared by the iodide exchange reaction described by Gleason in 1955.[7] The radioiodothyronines produced by the exchange of iodine atoms are generally of low specific activity, much below the theoretical maximal level (Table II). A low specific activity of T_4 tracer is not a major practical disadvantage in T_4 RIA. T_4 is present in human serum in rather high concentrations and a highly sensitive T_4 RIA for measurement of T_4 in serum of healthy people is not required, even in most, if not all, hypothyroid patients. The serum concentrations of other iodothyronines, e.g., T_3, rT_3, $3,3'-T_2$, however, are quite low (Table I), and therefore high specific activity tracers are required to facilitate RIAs of appropriate sensitivity.

[7] G. I. Gleason, J. Biol. Chem. **213**, 837 (1955).

TABLE I
MEAN VALUES OF VARIOUS IODOTHYRONINES IN NORMAL
HUMAN SERUM

Iodothyronine	n	ng/dl	nmol/liter
T_4	26	8600	111
T_3	16	118	1.82
rT_3	27	40.5	0.62
$3,3'-T_2$	44	4.1	0.08
$3,5-T_2$	30	7.3	0.14
$3',5'-T_2$	53	6.4	0.12
$3'-T_1$	25	1.4	0.04
TETRAC[a]		120	1.60
TRIAC	II	5.5	0.089
TRIPROP	II	<5.0	<0.080

[a] These data are from A. Burger, P. Nicod, P. Sutter, M. B. Valloton, A. Vagenakis and L. E. Braverman, *Lancet* 1, 653 (1976); other data are from authors' laboratories.

Burger and Ingbar[8] applied the chloramine-T method for radioiodination of T_3 and T_4 and their thyroacetic acid analogs. It has subsequently been applied to radioiodination of T_2s and T_1s[9] (Table III). Interestingly, iodothyronines labeled in the inner ring of the molecule have not been encountered in this procedure and all radioactive iodine atoms incorporated into the iodothyronine are found in the outer ring of the molecule. Furthermore, the procedure results in the addition of a radiodine atom in the outer ring much more frequently than exchange labeling of already existing outer ring iodine atoms; this phenomenon, if applied appropriately, can be used to produce radioactive iodothyronines of a much greater spe-

TABLE II
THEORETICAL MAXIMAL SPECIFIC ACTIVITY
(μCi/μg) FOR RADIOIODOTHYRONINES

Iodothyronine	Number of ^{125}I atoms/molecule	
	1	2
T_4	3000	6000
rT_3	3500	7000
T_3	3500	—
$3,3'-T_2$	4400	—

[8] A. Burger and S. H. Ingbar, *Endocrinology* 94, 1189 (1974).
[9] Y. Nakamura, I. J. Chopra, and D. H. Solomon, *J. Nucl. Med.* 18, 1112 (1977).

TABLE III
Nature of Products of Iodination and Their Specific Activity When
Iodothyronines Are Radioiodinated with the Chloramine-T Method[a]

Substrate	Product	Percentage of ^{125}I used[b]	Specific activity[c] $\mu Ci/\mu g$	Specific activity[c] Ci/mmol
3'-T$_1$	3'-T$_1$	15.7	400	158
	3',5'-T$_2$	64.5	3000	1569
	NaI	19.8	—	—
3-T$_1$	3,3'-T$_2$	59.0	2700	1412
	rT$_3$	21.6	3000	1950
	NaI	19.4	—	—
3,3'-T$_2$	3,3'-T$_2$	18.0	200	105
	rT$_3$	65.0	2700	1755
	NaI	17.0	—	—
3,5-T$_2$	T$_3$	52.4	1500	975
	T$_4$	21.6	1800	1399
	NaI	26.0	—	—
T$_3$	T$_3$	21.3	200	130
	T$_4$	60.7	1300	1010
	NaI	18.0	—	—

[a] Data from Y. Nakamura, I. J. Chopra, and D. H. Solomon, *J. Nucl. Med.* **18**, 1112 (1977).
[b] The relative proportions of radioactivity incorporated into iodothyronines and that remaining unreacted were first determined by column chromatography on Sephadex LH-20, and relative distribution of radioactivity in various products of radioiodination was then determined by paper chromatography.
[c] Based on the specific activity of ^{125}I at the time of use, the maximum expected specific activity varied between 1700 and 2100 Ci/mmol in all cases except T$_3$, where it was 1400 Ci/mmol.

cific activity than the labeled compounds produced by exchange labeling of iodine atoms already present in the iodothyronine. Thus, production of radiolabeled iodothyronines of high specific activity is to be expected when the substrate for iodination is an iodothyronine that has one iodine atom less in the outer ring than the desired radioactive iodothyronine. Using this strategy, 6 out of the 9 thyronines in Fig. 2 have been radioiodinated and used as tracers in their RIAs;[9,10] the exceptions are 3,5-T$_2$, 3-T$_1$, and T$_0$ which do not possess an iodine atom in the outer ring of the molecule.

Many radiolabeled iodothyronines (e.g., T$_4$, T$_3$, and rT$_3$) are now available commercially (e.g., from Abbott Laboratories in Illinois, New England Nuclear and Amersham Companies, Boston, and Industrial Nu-

[10] N. Kochupillai and R. S. Yalow, *Endocrinology* **102**, 128 (1978).

clear Company, St. Louis, Missouri). Purification of the commercial product may be accomplished by Sephadex chromatography (e.g., on a Sephadex G-25 fine column, elution buffer 0.01 M NaOH in distilled water). It is of practical importance to realize that iodothyronines with a high molecular weight may elute later than those with lower molecular weight, and different iodothyronines with the same molecular weight (e.g., T_3 and rT_3) may elute at different volumes. This is so because Sephadex chromatography separates iodothyronines not just by gel filtration (a process in which larger molecules elute before the smaller ones) but also by differential absorption of iodothyronines into the gel. In several experiments we calculated the relative retardation of iodothyronines by a Sephadex G-25 fine column. Mean ratios V_x/V_t (V_x = elution volume of substance x; V_t = total volume of the column) were: ^{125}I-labeled albumin, 0.39; ^{125}I, 0.66; ^{125}I-3,3'-T_2, 1.33; ^{125}I-T_3, 1.70; ^{125}I-rT_3, 2.97; and ^{125}I-T_4, 3.88. The absorption of iodothyronines onto the Sephadex gel is illustrated by the finding that the ratio V_x/V_t is greater than 1. It has been reported that column adsorption is influenced by the phenolic hydroxyl group and by the number and location of iodine atoms in the iodothyronine molecule; in general, retardation by the gel increases with an increase in the number of iodine atoms in the outer ring of iodothyronine.[11-13]

To prepare ^{125}I (or ^{131}I)-labeled iodothyronine, the following convenient and practical procedure is suggested. In this method radioiodination of iodothyronine is accomplished using chloramine-T, and unreacted radioiodide is removed by application of the iodination mixture onto a small Sephadex LH-20 column. The various radioiodothyronines in the eluate are then separated by either Sephadex or descending paper chromatography.

Reagents

a. Carrier-free Na^{125}I (specific activity ~ 17 Ci/mg) is adjusted to 40 mCi/ml with 0.4 M phosphate buffer (pH 6.2) upon arrival.

b. Chloramine-T and sodium metabisulfite which are commercially available, are dissolved in 0.05 M phosphate buffer (pH 6.2) just before use.

c. Thyronines to be iodinated are dissolved in either absolute methanol or 0.1 M NaOH and then diluted to 5×10^{-4} M with absolute methanol prior to use.

Radioiodination

a. Ten microliters of methanol containing 10^{-9} mol of the substrate

[11] S. Lissitsky, J. Bismuth, and M. Rolland, *Clin. Chim. Acta* **7**, 183 (1962).
[12] E. H. Mougey and J. W. Mason, *Anal. Biochem.* **6**, 223 (1963).
[13] F. Blasi and R. V. De Masi, *J. Chromatogr.* **28**, 33 (1967).

and mixed with 25μl of 0.4 M phosphate buffer (pH 6.2) containing approximately 1 mCi of ^{125}I in 10 × 75 mm disposable glass culture tubes.

b. The iodination reaction is started by adding 10 μl of 0.05 M phosphate buffer (pH 6.2) containing 4 μg of chloramine-T.

c. The iodination reaction is stopped after 2 min by adding 20 μl of a solution prepared by 1:10 dilution of saturated sodium metabisulfite in 0.05 M phosphate buffer (pH 6.2).

d. The reaction mixture is transferred to a Sephadex LH-20 column (1 × 2.5 cm), made in a 3- or 5-ml disposable plastic syringe.

e. The reaction vessel is washed once with 50 μl of 0.05 M phosphate buffer (pH 6.2) and washings are transferred on to the column.

f. Unreacted radioioide is removed by washing the column with 1 ml of 0.05 M phosphate buffer (pH 6.2) followed by 4 ml of water and 0.6 ml of methanol–2 N NH$_4$OH (99:1, v/v).

g. The radioiodinated compounds are eluted with 4 ml of methanol–2 N NH$_4$OH (99:1, v/v).

h. The eluate is dried under a thin stream of nitrogen or air (preferably former) and the residue is applied after being dissolved in 100 μl methanol –2 N NH$_4$OH (99:1) to a paper strip for descending paper chromatography.

Descending Paper Chromatography

a. Descending paper chromatography may be performed using hexane:tertiary amyl alcohol:2 N NH$_4$OH (1:5:6) system[14] and 3 MM Whatman chromatographic paper.

b. Chromatography is allowed to proceed for 15–20 hr.

c. Chromatograms are run in parallel of 20–30 μg of each of the nonradioactive substrates and expected products of radioiodination.

d. The radioactive spots on the paper chromatogram are cut corresponding to various nonradioactive compounds (identified in the parallel chromatogram by either ultraviolet light or by staining with diazotized sulfanilic acid) and eluted with 30–40 ml of methanol:2 N NH$_4$OH (99:1).

e. The eluate is dried under nitrogen or air, and the residue is dissolved in 1–3 ml of 50% propylene glycol; store at 4°.

As stated above for Sephadex chromatography, chromatographic separation of iodothyronines depends on the number of iodine atoms as well as their location in the iodothyronine molecule (v.s.).[15] Thus, less iodine in the outer ring is associated with greater migration (e.g., T$_3$ migrates farther than T$_4$ and 3,5-T$_2$ farther than T$_3$). Similarly, alterations in the side chain (e.g., deamination, decarboxylation) of iodothyronine is also associated with a greater migration of the compound in chromatography (Table IV).

[14] D. Bellabarba, R. E. Petersen, and K. Sterling, *J. Clin. Endocrinol. Metab.* **28**, 305 (1960).

[15] L. G. Plaskett, *Chromatogr. Rev.* **6**, 91 (1964).

TABLE IV

R_f Values of Iodide, Iodotyrosines, Iodothyronines, and Their Derivatives in Descending Paper Chromatography Using 1:5:6 Hexane:Tertiary Amyl Alcohol:2 N NH$_4$OH[a]

Compounds	R_f value
3,5-Diiodo-L-tyrosine (DIT)	0.02
3-Iodo-L-tyrosine (MIT)	0.04
L-Tyrosine	0.06
3′,5′ Diiodo-DL-thyronine	0.07
Iodide	0.10
3,3′,5′-Triiodo-L-thyronine (reverse T$_3$)	0.12
3′-Iodo-L-thyronine (3′-T$_1$)	0.13
3,5,3′,5′-Tetraiodo-L-thyronine (thyroxine, T$_4$)	0.17
DL-Thyronine (T$_0$)	0.24
3,3′-Diiodo-L-thyronine (3,3′-T$_2$)	0.26
3,5,3′,5′-Ettraiodothyroacetic acid (TETRAC)	0.31
3,5,3′,5′-Tetraiodopropionic acid (TETRAPROP)	0.31
3-Iodo-L-thyronine (3-T$_1$)	0.40
3,5,3′-Triiodo-L-thyronine (T$_3$)	0.42
3,5-Diiodo-L-thyronine (3,5-T$_2$)	0.48
3,5,3′-Triiodothyroacetic acid (TRIAC)	0.65
3,5,3′-Triiodothyropropionic acid (TRIPROP)	0.65
3,5-Diiodothyroacetic acid (DIAC)	0.71
3,5-Diiodothyropropionic acid (DIPROP)	0.71

[a] Data from Y. Nakamura, I. J. Chopra, and D. H. Solomon, J. Nucl. Med. 18, 1112 (1977).

The specific activity of compounds prepared as just described is close to the theoretical maximum (Tables II and III). Failure to reach the maximal specific activity can be due to contamination of the substrate with other iodothyronines or iodide and to inadequate separation of the radioiodinated products. It may be recalled that stable radioiodothyronines of high specific activity are produced consistently successfully when the substrate for iodination has one less iodine atom in the outer ring than the one desired.

Preparation of Antibodies against Iodothyronines

In the 1960s, the iodothyronines were generally considered nonantigenic, apparently because of their low molecular weights. However, early

[16] B. L. Brown, R. P. Ekins, S. M. Ellis, and W. S. Reith, Nature (London) 226, 359 (1970).
[17] I. J. Chopra, J. C. Nelson, D. H. Solomon, and G. N. Beall, J. Clin. Endocrinol. Metab. 32, 299 (1971).
[18] H. Gharib, W. E. Mayberry, and R. J. Ryan, J. Clin. Endocrinol. Metab. 33, 509 (1971).

in the 1970s several groups obtained antibodies to iodothyronines by immunization of experimental animals (e.g., rabbits) with conjugates of iodothyronines with either albumin or another large molecular weight carrier protein.[16-18] Of particular interest was the finding of both T_3 and T_4 antibodies in sera of rabbits immunized with thyroglobulin;[17] this finding may be explained by the consideration that T_3 and T_4 exist in a natural covalent linkage within the thyroglobulin molecule. Curiously, some rabbits immunized with thyroglobulin produced antisera directed mainly against T_3 with only minimal reactivity with T_4, while other animals produced antisera with a much greater specificity for T_4 than T_3. Although the procedure for producing T_3 and T_4 antibodies simultaneously was convenient and consistently produced adequate antibody reagent for T_4 RIA, the avidity of the antisera thus raised for T_3 was frequently suboptimal. Therefore, antisera directed against T_3 or other iodothyronines besides T_4 are preferably produced by immunization of rabbits with conjugates of iodothyronines with a carrier protein, e.g., human or bovine serum albumin, hemocyanin, or thyroglobulin. T_4-binding antibodies are produced when rabbits are immunized with either bovine, porcine, or human thyroglobulin. Bovine and porcine thyroglobulins are available commercially (Sigma Chemical Co., St. Louis, Missouri). The advantage of immunizing animals with human thyroglobulin is that the antisera thus produced may also be used in a RIA of human thyroglobulin.[19] Preparation of human thyroglobulin and the procedure for conjugation of iodothyronines to a carrier protein are described below in detail.

Preparation of Human Thyroglobulin. The following procedure is based on the original method of Derrien *et al.*[20]

1. Normal thyroid glands obtained at autopsy or surgery are cut into 1-cm pieces, and washed with 0.15 M NaCl to remove adherent blood; they are extracted with 3 volumes (w/v) of 0.15 M NaCl overnight at 4° while stirring. Up to 50 g of thyroid tissue can be processed conveniently.
2. The suspension is filtered through muslin, and the filtrate is centrifuged at 7500 rpm for 20 min.
3. Saturated ammonium sulfate (72.5 ml) per 100 ml supernatant is added at room temperature to bring the salt concentration in the mixture to 42%; it is stirred slowly to avoid localized high salt concentrations. Crude thyroglobulin precipitates at this stage.
4. It is centrifuged at 13,000 rpm for 20 min and the supernatant is discarded.

[19] A. J. Van Herle and R. P. Uller, *J. Clin. Invest.* **56,** 272 (1975).
[20] Y. R. Derrien, R. Michel, and J. Roche, *Biochim. Biophys. Acta* **2,** 454 (1948).

5. The precipitate is resuspended in 45% ammonium sulfate (i.e., 45 ml of saturated solution plus 55 ml of water) using 50 ml or more for each 50 g of thyroidal tissue.
6. Water (10.8 ml) per 50 ml suspension is added under continuous stirring to reduce the concentration of ammonium sulfate to 37%. It is left stirring for 2 to 3 hr at room temperature and then centrifuged to eliminate insoluble impurities.
7. Sufficient saturated ammonium sulfate is added slowly with stirring to the supernatant to get a final salt concentration of 42% (5.3 ml saturated ammonium sulfate/50 ml suspension, obtained in step 5).
8. It is centrifuged at 13,000 rpm for 20 min, the supernatant is discarded, and the precipitate is washed with 42% ammonium sulfate and centrifuged again.
9. The precipitate is suspended in 10 ml of 0.15 M phosphate buffer, (pH 7.2) and dialyzed against running tap water overnight, then against four changes of distilled water at 4°, each time using 4 liters for 12 hr.
10. Finally, the solution of thyroglobulin obtained in step 9 is dialyzed against 4 liters of 0.15 M NaCl and then stored in aliquots at $-20°$.
11. The protein concentration may be determined by diluting an aliquot 10 times with 0.15 M NaCl and measuring the optical density at 280 nm in a 1-cm silica cell. A solution of 100 μg thyroglobulin has, under these conditions, an optical density of 0.10. The yield of thyroglobulin by this method usually varies between 30 and 50 mg/g of thyroidal tissue. The thyroglobulin solution so obtained has been found adequate for use as an antigen for production of T_4 antiserum. However, it should be purified further by Sephadex G-200 chromatography for analytical use.

Conjugation of Iodothyronines to a Carrier Protein. A convenient protocol for conjugation, derived from the method of Oliver *et al.*[21] is as follows:

1. Protein (50 mg) (thyroglobulin or human or bovine serum albumin) is dissolved in 25 ml of phosphate-buffered saline (PBS) (0.01 M sodium phosphate, 0.14 M NaCl, pH 7.4).
2. While stirring, 20 mg of iodothyronine is added dissolved in 4 ml of dimethylformamide; 30 mg of 1-cyclohexyl-3-(2-morpholino-ethyl)carbodiimide metho-*p*-toluenesulfonate (Morpho-CDI, Aldrich Chemical Co., Inc., Milwaukee, Wis.) is then added.

[21] G. C. Oliver, Jr., B. M. Parker, D. L. Brasfield, and C. W. Parker, *J. Clin. Invest.* **47**, 1035 (1968).

3. An additional 10 mg of Morpho-CDI is added after 10 min. The solution is continually stirred at room temperature in the dark for 18 hr.

4. The reaction mixture is dialyzed against three changes of PBS each time. using 4 liters for 24 hr at $4°$.

5. The solution containing the conjugate may be stored frozen at -10 to $-20°$.

Immunization of Rabbits. For production of T_4-antisera, an emulsion of 10 mg of thyroglobulin in 0.5 or 1 ml phosphate-buffered saline is injected (0.14 M sodium chloride, 0.01 M sodium phosphate, pH 7.4), and an equal volume of complete Freund's adjuvant (Perrin's modification, Calbiochem Laboratories, La Jolla, Ca.) is injected subcutaneously (sc) at five sites in the back of female New Zealand rabbits. For production of T_3-, rT_3-, and $3,3'$-T_2-binding antisera, and emulsion of equal volumes of complete Freund's adjuvant (~ 1 ml) and of an appropriate solution of conjugate containing about 1 mg of conjugated protein is similarly injected sc. Rabbits are reinjected approximately 7 times at 10- to 20-day intervals. Rabbit sera are obtained at 10 days after the last injection and stored at $-20°$. Antisera so stored retain their ability to bind iodothyronines essentially unchanged for up to 10 years.

Selection of Antisera for RIAs. Immunization of rabbits by the method described above yields useful antisera in the majority of cases. Over 90% of trace amount ($\sim 10-50$ pg) of radiolabeled iodothyronines may bind to an excess of antiserum. The optimal antiserum dilution for use in RIA is determined by studying the binding of tracer amounts of radioactive antigen to various dilutions (1/100 to 1/1,000,000) of rabbit antiserum. An RIA is usually most sensitive when antiserum is used in a dilution that binds $\sim 30-40\%$ of a tracer amount of radiolabeled iodothyronine.[22] Although an antiserum which can be used at a very high dilution (/100,000–1/1,000,000) is of much practical advantage, a lower titer of the antiserum does not necessarily imply a low-affinity antiserum or an otherwise poor antiserum for RIA.

The most crucial step in selecting antisera is evaluation of the specificity of the antiserum for the substance under study. In case of an ideal antiserum, no compound (especially the iodothyronines closely related to the iodothyronine under study) should inhibit the binding of tracer quantity of radioactive iodothyronine to the antiserum while even small (picogram) quantities of the test substance should show an appreciable inhibition. Be-

[22] R. S. Yalow and S. A. Berson, *in* "Radioisotopes in Medicine: In Vitro Studies" (R. L. Hayes, F. A. Goswitz, and B. E. P. Murphy, eds.), p. 7. U.S. Atomic Energy Commission, Oak Ridge, 1968.

cause of the marked similarity in chemical structure of various iodothyronines, it has not been possible so far to obtain antisera with absolute specificity. Hence, the need to choose, among various antisera, the one which will be least influenced in its binding of test iodothyronine by other compounds. The degree to which any cross-reactivity is permissible (or acceptable) is dictated mainly by the relative concentration of test iodothyronine and cross-reacting compounds in the assay. In studies of human serum that has a high concentration of T_4 relative to other iodothyronines (Table I), cross-reactivity of T_4 is of prime importance in the RIAs of each of the lower iodothyronines; on the other hand, even considerable cross-reactivity of tri- or diiodothyronines in T_4 RIA will not affect the results seriously, and can therefore be neglected.

TABLE V

CROSS-REACTIVITY OF IODOTHYRONINES AND RELATED COMPOUNDS IN
RADIOIMMUNOASSAYS OF T_4, T_3, rT_3, AND $3,3'-T_2$ [a]

Compound	T_4 RIA[b]	T_3 RIA[c]	rT_3 RIA[d]	$3,3'-T_2$ RIA[e]
L-T_4	100*	0.05	0.06	0.006
D-T_4	133	0.20	0.045	<0.0001
L-T_3	10	100*	0.02	0.2
D-T_3	19	140	0.006	0.07
(D)L-rT_3	—	<0.1	100*	0.04
$3,5-T_2$	<0.3	0.60	<0.002	0.01
$3,3'-T_2$	—	3.0	10.0	100*
$3',5'-T_2$	—	—	—	<0.001
$3-T_1$	<0.1	—	0.24	12.0
$3'-T_1$	—	—	—	0.4
T_0	<0.3	—	<0.002	0.008
TETRAC	8	0.80	0.034	<0.0001
TETRAPROP	14	—	—	<0.0001
TRIAC	—	40	0.003	0.01
TRIPROP	—	19	0.005	0.04
3,5-DIPROP	—	—	0.005	0.007
DIT	<0.06	<0.0002	0.03	<0.0001
MIT	<0.06	<0.001	<0.02	0.003
KI	<0.001	<0.0001	<0.0003	<0.0001
Tg	—	<0.0001	<0.0003	—

[a] The data are expressed on a weight basis relative to a value of 100 (*) for the iodothyronine under study.
[b] I. J. Chopra, D. H. Solomon, and R. S. Ho, J. Clin. Endocrinol. Metab. 33, 865 (1971).
[c] I. J. Chopra, R. S. Ho, and R. Lam, J. Lab. Clin. Med. 80, 729 (1972).
[d] I. J. Chopra, J. Clin. Invest. 54, 583 (1974).
[e] S. Y. Wu, I. J. Chopra, Y. Nakamura, D. H. Solomon, and L. R. Bennett, J. Clin. Endocrinol. Metab. 43, 682 (1976).

The data on specificity of some representative assays of T_4, T_3, rT_3, and $3,3'-T_2$ are shown in Table V. From the product of the percentage of cross-reactivity and the serum concentration of various test iodothyronines and other related compounds (in so far as data are available), one can derive that cross-reactivity of related substances in various available RIAs of iodothyronines (Table V) is unlikely to seriously affect the results of measurements of T_4, T_3, rT_3, or $3,3'-T_2$ under most physiological and pathological conditions. One exception may be the influence of cross-reactivity of T_4 in rT_3 RIA; it appears that approximately 10% of the apparent rT_3 immunoreactivity in normal human serum may derive from circulating T_4. Cross-reactivity of $3,3'-T_2$ in rT_3 RIA, while much greater than of T_4, is of little practical importance since $3,3'-T_2$ concentration in normal human serum is very low, only about 4.1 ng/dl. Similarly, the cross-reactivity of other agents in RIAs in Table V is of little consequence in practical terms.

One may attempt to lower cross-reactivity of an antiserum by preabsorption with the cross-reacting agent; this usually results in losses in sensitivity of RIA. Some have approached the problem of T_4 cross-reaction in the rT_3 RIA by correction of the measured rT_3 concentration for T_4 cross-reactivity. Correction for cross-reaction, however, is a complicated matter, which is permissible only after the characteristics of antibody binding of test agent and the cross-reacting agent have been tested meticulously.[23] One of the requirements for applying such a correction is that the degree of cross-reactivity of the compound in question in an RIA must remain the same over a wide range of concentrations of the compound.

Minimization of the Effect in RIA of Iodothyronine-Binding Proteins in Serum

Iodothyronines circulate in human serum tightly bound to serum proteins. The proportion of unbound iodothyronines in serum is very small, only about 0.03% for T_4, 0.3% for T_3, 0.2% for rT_3, and 0.7% for $3,3'-T_2$. To measure total concentration of iodothyronines in serum by RIA, measures must be undertaken to "free" the iodothyronines bound to serum proteins to make them available for reaction with iodothyronine-binding sites on the antibody used in RIA. These measures will also serve to minimize the binding by serum proteins of tracer (radioactive) iodothyronine used in the RIA.

The three proteins that bind iodothyronines in human serum are thyroxine-binding globulin (TBG), thyroxine-binding prealbumin (TBPA),

[23] H. Mathur, R. P. Ekins, B. L. Brown, P. G. Malan, and A. B. Kurtz, *Clin. Chim. Acta* **91**, 317 (1979).

and albumin. TBG and TBPA serve as low-capacity, high-affinity binding proteins for T_4, T_3, and rT_3, whereas albumin is the high-capacity, low-affinity binding protein for all iodothyronines; $3,3'$-T_2 is bound predominantly by albumin. The binding of iodothyronines to prealbumin can be inhibited by barbital ions; therefore, the RIAs are generally set up in barbital buffer. Albumin interferes modestly in RIAs and its effect can be reduced substantially simply by diluting serum in RIA tubes. Several agents are available to minimize binding of iodothyronines to TBG. These include 8-anilino-l-naphthalenesulfonic acid (ANS),[24,25] sodium salicylate,[26] thiomerosal,[27] and dilantin,[28] although ANS is probably the most potent[29] and convenient to use because of its easy solubility in aqueous media. Other procedures that have been employed to reduce the influence in RIA of binding of iodothyronines to the serum proteins include alkalinization of serum to a pH of ≥ 11.0[30,31] or heating of serum.[32] Extraction procedures using ethanol or Sephadex column chromatography[31] have been employed for RIAs but are not yet popular probably because they are more time consuming than working with unextracted serum.

Preparation of Standards for RIA of Iodothyronines: Use of Iodothyronine-Free Serum

Reagent grade iodothyronines can be obtained commercially (e.g., Sigma Chemical Co., St. Louis, Missouri; Henning Berlin GmbH, Berlin, West Germany). The iodothyronines are dissolved and diluted to a concentration of 10 μg/ml in 0.01 M NaOH. Further dilutions for use in the standard curve (standard solutions or "standards") are made in barbital buffer (0.075 M barbital buffer, pH 8.6, with 1% normal rabbit serum and 0.1% sodium azide) or another buffer deemed appropriate for RIA. It is important that the standards are either not contaminated or only minimally contaminated with other iodothyronines. Further chromatographic procedures may be required in some cases to purify the standard. Questions about the degree of purity of the iodothyronines have been raised at times when evaluating cross-reactivity in RIAs of various iodothyro-

[24] I. J. Chopra, *J. Clin. Endocrinol. Metab.* **34**, 938 (1972).
[25] I. J. Chopra, R. S. Ho, and R. Lam, *J. Lab. Clin. Med.* **80**, 729 (1972).
[26] P. R. Larsen, *J. Clin. Invest.* **51**, 1939 (1972).
[27] H. Gharib, W. E. Mayberry, and R. J. Ryan, *J. Clin. Endocrinol. Metab.* **33**, 509 (1971).
[28] J. Lieblich and R. D. Utiger, *J. Clin. Invest.* **51**, 157 (1972).
[29] I. J. Chopra, in "Handbook of Radioimmunoassay" (G. E. Abraham, ed.), p. 679. Dekker, New York, 1977.
[30] M. Hüfner and R. D. Hesch, *Clin. Chim. Acta* **44**, 101 (1973).
[31] J. Faber, T. Friis, C. Kirkegaard, and K. Siersbaek-Nielsen, *Acta Endocrinol.* **87**, 313 (1978).
[32] K. Sterling and P. O. Milch, *J. Clin. Endocrinol. Metab.* **38**, 866 (1974).

nines.[33] Purity of iodothyronines is clearly an issue in studies of specificity of a RIA when iodothyronines from different sources behave differently in the RIA.

To ensure that the milieu in the tubes containing the standards is as similar as possible to that of the test samples, it is necessary in RIA of unextracted serum that iodothyronine-free serum be added to the tubes for standards in a volume equal to the unknowns. Iodothyronine-free serum can be obtained from severely hypothyroid humans or animals, or it can be prepared by "stripping" the iodothyronines from the serum by charcoal or an ion-exchange resin.

We have previously used serum of a sheep, obtained at 6 weeks after surgical total thyroidectomy (serum T_4 was <1 $\mu g/dl$ and serum T_3 <25 ng/dl). Pooled human serum treated with Norit-A decolorizing pharmaceutical charcoal works equally as well and can be obtained more conveniently. The procedure involves the addition of 7.5 g of charcoal/100 ml of serum, stirring overnight, followed by centrifugation as 25,000 g for three successive 1-hr periods[34] or passage through a Seitz filter to remove the charcoal.

Iodothyronine-free serum may also be prepared by treating pooled human serum with a resin, e.g., AG 1-X_2 (Analytical Grade Anion Exchange Resin, 200–400 mesh chloride form, Bio-Rad). A procedure is as follows: 200 g of AG 1-X_2 is washed consecutively 10 times with 600 ml of distilled water, three times with 400 ml 0.225 M sodium barbital pH 8.6, and 10 times with 0.075 sodium barbital, pH 8.6; 500 ml of pooled serum is mixed with 50 g of washed AG 1-X_2 resin under rotating conditions for 1 hr. After centrifugation the same procedure is applied to the supernatant for an additional three times. The efficiency of the procedure for removing iodothyronines may be checked by experiments in which known amounts of radiolabeled iodothyronines are added to the pooled serum. We found that the procedure allowed removal of 95–98% of 3,3'-T_2, 99.2% of rT_3, 99.5% of T_3, and 99.3% of T_4 added to serum. Standard curves made in iodothyronine-free serum prepared by the charcoal or resin technique or in serum from a severely hypothyroid sheep were essentially superimposable.

Radioimmunoassays (RIAs) of T_4, T_3, rT_3, and 3,3'-T_2

This section describes the RIAs of each of the various iodothyronines as performed in the authors' laboratories. The choice of the many variables in each assay (e.g., concentration of reagents, incubation times)

[33] B. N. Premachandra, *J. Clin. Endocrinol. Metab.* **47,** 746 (1978).
[34] P. R. Larsen, *Metab. Clin. Exp.* **20,** 609 (1971).

may be facilitated by mathematical approximation of the optimal conditions for reaching equilibrium between antigen and its binding by sites on the antibody;[35] however, we have determined the optimal conditions by trial and error.

Since an appropriate RIA method must be sensitive, precise, and specific, let us examine some of these terms from a practical standpoint. The precision of measurement of a hormone concentration (X) in an RIA is dependent on the slope of the RIA standard curve and the error of measurement of the response metameter [i.e., the percentage of tracer bound to antibody (B) when the dose corresponds to hormone concentration X]. When X is zero, the variations in B_0 (percentage of tracer bound to antibody in the absence of added nonradioactive standard) can help define the detection limit or "sensitivity" of the RIA system. Sensitivity may be defined as the minimum amount of iodothyronine that can be distinguished from no or zero dose with a 95% confidence; this is taken as the iodothyronine concentration indicated by 2 SD away from the initial binding i.e., $B_0 - 2$ SD. The intraassay coefficient of variation (C. V.) when measured as $(SD/X \times 100\%)$ is an indication of the precision of measurements of each dose on the standard curve. The C.V. for all doses in an RIA may be averaged to obtain a value of mean C.V. within an assay.

Due to a combined immunological and chemical nature of reactions in an RIA, nonspecific effects may occur as a result of cross-reactivity of related hormonal compounds and of nonhormonal effects, e.g., pH, ionic environment, temperature, variations in the damage of labeled hormone and possibly anticoagulants, if plasma samples are tested in the RIA. Nonhormonal, nonspecific effects may be estimated by setting up control incubation mixtures or "blanks," which contain no antiserum but are otherwise identical to an incubation mixture containing test samples. Further care is needed to render serum used in the standard curve similar to unknown test sera in all respects except that the serum in the standards is free of the iodothyronine to be measured. Further tests for delineating the validity of an RIA include demonstration of parallelism (or an absence of clear nonlinearity) between the standard curve and dilution curves of unknown samples, and of quantitative recovery of varying doses of iodothyronines added to incubation mixture.

Quality control of RIA determination of concentration of iodothyronines may be estimated by determining the reproducibility of the results of several test samples in several assays followed by calculation of interassay coefficient of variation of the assay. The formula to determine interassay C.V. is the same as mentioned previously for calculation of intraassay C.V. A given assay is subject to rejection if hormone concentra-

[35] R. P. Ekins, *Br. Med. Bull.* **30**, 3 (1974).

tions of several samples in it are outside the range of the mean ± 2 SD of those in the same samples in previous experiments.

Radioimmunoassay of T_4[24,29]

Reagents

1. *Incubation Buffer.* The incubation buffer is made up of 0.075 barbital buffer, pH 8.6, containing 1% normal rabbit serum and 0.1% sodium azide. The barbital buffer is commercially available (B-2 buffer, Beckman Inc., Fullerton, Ca.). As mentioned earlier, barbital buffer serves to inhibit TBPA in test serum; normal rabbit serum is a source of carrier γ-globulin that aids in precipitation of radioactive T_4 bound to anti-T_4 antibody in the final step of RIA. Sodium azide is used as a bacteriostatic agent.

2. T_4-*Standard.* Reagent grade sodium L-thyroxine may be obtained from Mann Research Laboratories. It is dissolved and diluted to 10 $\mu g/dl$ in 0.1 M NaOH. Dilutions are made for use in RIA (0.25–40 $\mu g/dl$ T_4) in iodothyronine-free serum for a 10- to 14-point standard curve.

3. *Anti-T_4 Antiserum.* The anti-T_4 antiserum may be obtained from a rabbit immunized with human thyroglobulin (v.s.). The final dilution of antiserum in a previously published RIA[24] was 1:300; at this dilution, anti-T_4 bound approximately 50% of a tracer amount ($\sim$ 100–200 pg) of radioactive T_4. Other antisera have shown similar binding at a dilution of 1/1000.

4. T_4-*Tracer.* Radioactive ^{125}I-T_4 (specific activity 75–100 $\mu Ci/\mu g$) in 50% propylene-glycol may be obtained from Abbott Laboratories, North Chicago, Ill. or Industrial Nuclear, St. Louis, Mo. This preparation is diluted in barbital buffer to get 10,000 cpm/50 μl.

5. *ANS.* ANS may be obtained from K & K Laboratories, Hollywood, Ca. A fresh solution is prepared in barbital buffer (150 μg/50 μl or 3 mg/ml) for each assay.

Assay Procedure. Various reagents are pipetted in 10 × 75-mm disposable glass tubes in the following order.

1. Assay buffer up to a final volume of 0.5 ml.
2. 50 μl ANS (150 μg/tube).
3. 50 μl ^{125}I-T_4 (10,000 cpm, $\sim$ 100–200 pg T_4).
4. 25 μl unknown serum, or T_4 standards in iodothyronine free serum.
5. 100 μl antiserum (working dilution 1:60, final dilution 1:300). Tubes are swirled on a Vortex mixer and incubated for 1 hr at room temperature and 5 min in a water bath at 4°.
6. 30–50 μl of a previously titered goat anti-rabbit γ-globulin (Antibodies Inc., Davis, Ca.).

Tubes are swirled again, incubated for 20 hr at 4°, centrifuged at 2000 rpm for 30 min, and aspirated to remove the supernatant. The radioactivity is determined in the precipitate (containing the fraction of ^{125}I-T$_4$ that is bound to anti-T$_4$- in a gamma scintillation counter).

Each unknown serum sample and standard point is assayed at least in duplicate. Steps 1 through 3 can be performed conveniently by pipetting 375 μl of a premixed solution containing barbital buffer, ANS and ^{125}I-T$_4$ using an automatic pipet. Nonspecific binding is determined in simultaneously incubated tubes that contain all reagents except anti-T$_4$ antibody which is replaced by an equal volume of assay buffer; the counts in precipitates of these tubes indicate nonspecifically bound or trapped radioactivity that should be subtracted from counts in all other tubes. Each assay should also include two tubes that contain all reagents except nonradioactive T$_4$. The counts in precipitates of these tubes are assigned an arbitrary value of 100% and the counts are expressed in the other tubes as a percentage of counts in zero-T$_4$ or 100% tubes.

A standard curve can be constructed by plotting the percentage of bound radioactivity on the ordinate and the T$_4$ concentration (in μg/dl) of the standards on the abscissa. Knowing the percentage of bound radioactivity in the unknowns, the T$_4$ concentrations can be read directly in μg/dl from the standard curve.

Some Notes of Interest on T$_4$-RIA. The detection limit of T$_4$ RIA approximates 0.5 μg/dl, and the intraassay coefficient of variation is about 4.3%. The data on specificity of T$_4$ RIA are shown in Table V; some other features are presented in Table VI.

Serum concentrations in hypothyroid, euthyroid, and hyperthyroid subjects measured by a T$_4$ RIA correlate well with those measured by a competitive protein binding assay (CPBA).[24,36] However, T$_4$ RIA values in sera of hyperthyroid patients may exceed T$_4$ CPBA values by more than what may be expected from losses during extraction of samples for T$_4$ CPBA. The reason for this discrepancy is not clear; however, the presence in serum of T$_4$ covalently linked to serum proteins,[24] an incomplete extraction of T$_4$ in CPBA at high T$_4$ concentrations, and problems with techniques of separation of bound from free radioactivity in T$_4$ RIA have been suggested.[24,37,38] However, some investigators do not find this discrepancy in T$_4$ values of hyperthyroid patients tested by RIA and CPBA.[39] In any case, this variable observation does not limit the diagnostic accuracy of T$_4$ RIA.[24]

[36] B. E. P. Murphy and C. J. Pattee, *J. Clin. Endocrinol Metab.* **24,** 187 (1964).
[37] N. P. Kubasik, H. E. Sine, and M. H. Murray, *Clin. Chem.* **19,** 1307 (1973).
[38] B. N. Premachandra and S. I. Ibrahim, *Clin. Chim. Acta,* **70,** 43 (1976).
[39] P. R. Larsen, J. Dockalova, D. Sipula, and F. M. Wu, *J. Clin. Endocrinol. Metab.* **37,** 177 (1973).

TABLE VI

Some Features of Radioimmunoassays for Various Iodothyronines, as Performed in the Authors' Laboratories

Iodothyronine	Sensitivity (ng/dl)	Intraassay variation (%)	Nonspecific binding (%)	Parallelism	Recovery (%)	Interassay variation (%)
T₄ RIA	500	4.3	4–7	Present	105	7.1
T₃ RIA	15–25	5.8	4–7	Present	99.5	9.8
rT₃ RIA	10	5.6	1.5–3	Present	93	10
3,3'-T₂ RIA	1.2	6.0	4–7	Present	91	11

We have used the double-antibody system for separation of bound and unbound hormone fractions in T_4 RIA. However, dextran-coated charcoal,[39] ammonium sulfate,[40] or polyethylene glycol (Carbowax 6000[41]) may also be employed. We find the charcoal method to be less precise than the "second" antibody method. Ammonium sulfate and polyethylene glycol methods yield higher nonspecific binding than the "second" antibody method. The main disadvantages of the latter system are its cost and the need for overnight incubation before the results can be obtained. If polyethylene glycol is combined with the "second" antibody, as described recently,[42] the results can be obtained the same day. Also, nonspecific binding is low and comparable to the method in which the "second" antibody described above is used alone to separate bound and free fractions in the RIA.

T_4 CPBA essentially displaced the protein-bound iodine (PBI) method for assessing circulating T_4 in the 1960s because it provided the advantage of not being affected by iodine contamination of the serum. T_4 RIA now provides substantial advantages over the T_4 CPBA method. It is more sensitive, more convenient, and more specific. Also RIA permits testing of multiple samples at one time because it is more suitable for automation than CPBA. The T_4 RIA obviates the extraction step necessary in CPBA. The linear range of the standard curve is from 1.2 to 40 μg/dl for RIA and from 3.0 to 20 μg/dl for CPBA. Unsaturated fatty acids,[43,44] halofenate,[45] and orphenadrine[46] all interfere *in vitro* in the T_4 CPBA but not in T_4 RIA. Ethanol extracts of sera of patients taking these agents contain appreciable quantities of these compounds or their metabolites and this may lead to spuriously elevated 'thyroxine values when the method of testing is CPBA using TBG[36] and not when the RIA method is used.

Radioimmunoassay of T_3[25]

Reagents

1. *Assay Buffer.* The assay buffer is barbital buffer, as described for T_4 RIA.
2. *T_3-Standard.* Reagent grade sodium L-triiodothyronine may be obtained from Mann Research Laboratories or Aldrich Chemical

[40] D. Mayes, S. Furuyama, D. C. Kem, and C. A. Nugent, *J. Clin. Endocrinol. Metab.* **30**, 682 (1970).
[41] B. Desbuquois and G. D. Aurbach, *J. Clin. Endocrinol. Metab.* **33**, 732 (1971).
[42] I. J. Chopra, *J. Clin. Endocrinol. Metab.* **51**, 117 (1980).
[43] K. Rootwelt, *Scand. J. Clin. Lab. Invest.* **35**, 649 (1975).
[44] W. Shaw, J. L. Hubert, and F. W. Spiertoo, *Clin. Chem.* **22**, 673 (1976).
[45] F. E. Karch, J. P. Morgan, and N. P. Kubasik, *J. Clin. Endocrinol. Metab.*, **43**, 26 (1976).
[46] W. M. Wiersinga, A. J. Fabius, and J. L. Touber, *Acta Endocrinol.* **86**, 522 (1977).

Company, Inc., Milwaukee, Wisconsin. Stock solution (10 μg/ml in 0.01 M NaOH) is diluted in barbital buffer to concentrations of 0.1–100 ng/ml.

3. T_3-*Antiserum.* Antiserum obtained by immunization of a rabbit with a conjugate of T_3 and thyroglobulin (or serum albumin) (v.s.) is diluted e.g., 1:4,000, so that it binds about 30–40% of a tracer amount ($\sim$100 pg) of ^{125}I-T_3.

4. T_3-*Tracer.* Radioactive ^{125}I-T_3 (specific activity $\geq$80 μCi/μg) may be obtained from Abbott Laboratories, North Chicago, Illinois, and diluted in barbital buffer to obtain 10,000 cpm/100 μl buffer.

5. *ANS.* ANS is dissolved in barbital buffer to a concentration of 200 μg/100 μl (2 mg/ml) immediately prior to an assay.

Assay Procedure. The various reagents are added in 10 × 75-mm disposable glass tubes in the following order:

1. Assay buffer to adjust final volume to 1 ml.
2. 100 μl ANS (200 μg/tube).
3. 200 μl unknown serum, or 100μl of standard solutions.
4. 200 μl iodothyronine-free serum in tubes for standard curve.
5. 100 μl antiserum (working dilution 1:400;[25] final dilution 1:4,000).
6. 100 μl ^{125}I-T_3 ($\sim$10,000 cpm, $\sim$100 pg T_3).

Tubes are swirled after steps 4, 5, and 6 and then incubated for 24 hr at 4°.

7. 60–100 μl of a potent goat anti-rabbit γ-globulin is added. After swirling, tubes are reincubated for 20–24 hr at 4° to allow precipitation of the antibody-bound radioactive T_3. Tubes are centrifuged and the radioactivity in pellet using a gamma scintillation counter is determined.

Tubes are prepared without antiserum ("blanks") and some are prepared without nonradioactive T_3 ("100% tubes") and the standard curve is plotted as described above for T_4 RIA.

Some Notes of Interest on T_3 RIA. The detection limit of T_3 RIA is $\sim$15 to 25 ng/dl, and the intraassay coefficient of variation approximates 5.8%. The data on specificity of RIA are shown in Table V. Of the various cross-reacting substances, only T_4 may contribute slightly to the T_3 measured by RIA using an antiserum described in Table V; however, this will account for only about 5 ng/dl of T_3 in normal serum (T_4 concentration $\sim$10 μg/dl) out of a total of about 113 ng/dl. Other characteristics of RIA using this antiserum are shown in Table VI.

Since cross-reactivity of T_4 in T_3 RIA is a potentially serious difficulty, it seems important to recall reports suggesting that T_4 reactivity in T_3 RIA may vary with different T_4 batches of T_3s;[28] this indicates contamination of T_4 preparations with T_3 rather than a real cross-reaction of T_4 with T_3 binding sites on the anti-T_3 antibody. It may be necessary, therefore, to purify commercial T_4 prior to study of its cross-reaction in T_3 RIA.

T_3 RIA can be performed in a shorter time than indicated previously by incubating assay tubes at 37° for a few hours[47] and by using polyethylene glycol in separation of bound from free radioactivity in the final step of RIA.[41,42] Similarly, if necessary, sensitivity of the RIA can also be improved substantially by using a high specific activity ($\sim$ 3,000 μCi/μg) radioactive T_3 as the tracer;[48] such a tracer may permit use of a more dilute antiserum in RIA than may be otherwise feasible with lower specific activity tracers.

T_3 RIA has been a major tool in unraveling of the role of T_3 in health and disease in the 1970s. It is more convenient and accurate and less cumbersome and timeconsuming than the various competitive protein-binding techniques that were available for measuring serum T_3 in the 1960s;[49,50] the previous techniques were also subject to methodological artifacts not encountered in direct RIA.[51,52]

Radioimmunoassay of rT$_3$[53]

Reagents

1. *Assay Buffer*. The assay buffer is barbital buffer as described for T_4 RIA.

2. *rT$_3$ Standard*. Reagent grade L-rT$_3$ can be obtained from Henning Berlin GmbH, West Germany. The only available form of rT$_3$ at the time of our initial study,[53] was DL-rT$_3$; it was obtained through the courtesy of Dr. Robert L. Meltzer of Warner-Lambert Research Institute, Morris Plains, New Jersey. The rT$_3$ powder is dissolved and diluted to 10 μg/ml in 0.01 M sodium hydroxide; this stock solution is diluted in barbital buffer to concentrations of 50 pg to 30 ng/ml for use in RIA.

3. *rT$_3$ Antiserum*. The rT$_3$ antiserum produced by immunization of rabbits is used with a rT$_3$-human serum albumin conjugate. In the appropriate assay dilution, it should bind $\sim$ 30–40% of a tracer

[47] J. C. Standefer, P. H. Lenz, G. H. Wien, and J. M. Toth, *Clin. Chem.* **22,** 396 (1976).
[48] J. Weeke and H. Örskov, *Scand. J. Clin. Lab. Invest.* **35,** 237 (1975).
[49] J. A. Nauman, A. Nauman, and S. C. Werner, *J. Clin. Invest.* **46,** 1346 (1967).
[50] K. Sterling, B. Bellabarba, S. I. Newman, and M. R. Brenner, *J. Clin. Invest.* **48,** 1150 (1969).
[51] D. A. Fisher and J. H. Dussault, *J. Clin. Endocrinol. Metab.* **32,** 675 (1971).
[52] P. R. Larsen, *Metabolism* **20,** 609 (1971).
[53] I. J. Chopra, *J. Clin. Invest.* **54,** 583 (1974).

amount of ^{125}I-rT$_3$ (10–50 pg); in one study the final dilution of anti-serum was only 1:250.[53]

4. *rT$_3$ Tracer.* Tracers of high specific activity (~3000 μCi/μg) may be prepared by radioiodination of 3,3'-T$_2$ as described above. Dilute tracer in barbital buffer to obtain approximately 10,000 cpm (and ~10–30 pg rT$_3$) per 100 μl.

5. *ANS.* This was not employed in our rT$_3$ RIA[53] because the nature of the binding of rT$_3$ to serum proteins was unknown at the time. However, we have continued to find the original system very convenient and practical, and it is this system, therefore, that is described below. In this system rT$_3$ is measured in ethanol extracts of sera which are prepared by mixing of 1 ml of 95% ethanol with 0.5 ml of serum, followed by centrifugation at 1000 g for 10 min; the supernatant is used directly in the assay.

Assay Procedure. The following reagents are pipetted consecutively in 10 × 75-mm disposable glass tubes:

1. Assay buffer to make up the final volume to 1 ml.
2. 300 μl ethanol extract of unknown test sample or of iodothyronine-free serum.
3. 100 μl of standard solutions (in tubes for standard curve only).
4. 100μl of anti-rT$_3$ antiserum (working dilution, 1:25;[53] final dilution, 1:250).
5. 100 μl ^{125}I-rT$_3$ solution containing 10,000 cpm (~10–30 pg rT$_3$). The tubes are mixed and incubated at 4° for 24 hr.
6. Add 60–100 μl of a previously titered goat anti-rabbit γ-globulin.

Subsequent procedures are performed including centrifugation of incubation mixtures, counting of precipitates, and correction for nonspecific binding etc., as described previously for T$_4$ RIA.

Some Notes on rT$_3$ RIA. The reverse T$_3$ assay[53] detected 10 pg rT$_3$/tube; it permitted measurement of serum rT$_3$ in a concentration as low as 10 ng/dl when 300 μl of ethanol extract representing 100 μl of serum was assayed. The 3,3'-T$_2$ cross-reacted maximally in rT$_3$ RIA (Table V) but its serum concentration is usually too low, only about 10% that of rT$_3$, to significantly influence rT$_3$ values measured in RIA. On the other hand, even though T$_4$ showed little (0.06%) cross-reaction in rT$_3$ RIA, its serum concentration relative to rT$_3$ is very high and therefore even a modest cross-reaction of T$_4$ can influence rT$_3$ values considerably. Studies showed that the smallest dose of T$_4$ that appreciably reduced the binding of rT$_3$ tracer to antibody was 15 ng/tube; thus, T$_4$ in serum would influence rT$_3$ values measured by our RIA when its concentration in serum is 15 μg/dl or higher. The nonspecific binding in rT$_3$ RIA varied between 1.5 and 3%. Dilution curves of unknown sera paralleled the standard curve. The mean recovery of nonradioactive rT_3 added to serum prior to the ethanol extraction was 93%; the interassay variation was 10.3% (Table VI).

Several other RIAs of rT_3 have now been described that employ unextracted serum and ANS to reduce binding of rT_3 to rT_3 binding proteins in serum[54–57]; readers may refer to the original publications if they wish to set up such an RIA system. Separation methods other than second antibody method described here, i.e., charcoal, ammonium sulfate, polyethylene glycol, etc., may also be applied to rT_3 RIA in a manner similar to that for T_4 (or T_3) RIA described above.

Radioimmunoassay of 3,3'-T_2[58,59]

Reagents

1. *Assay Buffer.* The assay buffer is 0.15 M Tris buffer (pH 8.2) containing 0.25% normal rabbit serum and 0.1% sodium azide.
2. *3,3'-T_2 Standard.* Reagent grade 3,3'-L-T_2 may be purchased from Henning Berlin GmbH, West Germany. T_2 used in our studies was either from this source or was a gift from Dr. Paul Block Jr., of River Research, Toledo, Ohio. Standard solutions, containing 0.1– 10 ng/ml T_2, are prepared by diluting stock solution (10 μg/ml in 0.01 M NaOH) in assay buffer.
3. *3,3'-T_2 Antiserum.* Antiserum may be produced by immunization of rabbits with a conjugate of 3,3'-T_2 with human serum albumin. The titer of an anti-T_2 antiserum is then determined. The antiserum selected for our studies bound 46–56% of a tracer amount of ^{125}I-T_2 (25–50 pg) in a final dilution of 1:15,000.
4. *3,3'-T_2 Tracer.* High specific activity (2700 μCi/μg) ^{125}I-3,3'-T_2 may be prepared by radioiodination of 3-T_1 by the chloramine-T method described previously. Tracer is diluted in assay buffer to obtain approximately 15,000 cpm/100 μl buffer.

---

[54] P. Nicod, A. Burger, V. Staeheli, M. B. Vallotton, *J. Clin. Endocrinol. Metab.* **42**, 823 (1976).

[55] K. D. Burman, R. C. Dimond, F. D. Wright, J. M. Earll, J. Bruton, and L. Wartofsky, *J. Clin. Endocrinol. Metab.* **44**, 660 (1977).

[56] W. A. Ratcliffe, J. A. Marshall, and J. G. Ratcliffe, *Clin. Endocrinol. (Oxford)* **5**, 631 (1976).

[57] R. S. Griffiths, E. G. Black, and R. Hoffenberg, *Clin. Endocrinol. (Oxford)* **5**, 679 (1976).

[58] S. Y. Wu, I. J. Chopra, Y. Nakamura, D. H. Solomon, and L. R. Bennett, *J. Clin. Endocrinol. Metab.* **43**, 682 (1976).

[59] F. Geola, I. J. Chopra, D. H. Solomon, and R. M. B. Maciel, *J. Clin. Endocrinol. Metab.* **48**, 297 (1979).

5. *ANS*. A solution containing 10 mg/ml ANS in assay buffer is prepared.

Assay Procedure. The various reagents are added in 10 × 75-mm disposable glass tubes in the following order:

1. Assay buffer to adjust final volume to 1 ml.
2. 100 μl ANS.
3. 400 μl of unknown serum or of iodothyronine-free serum in tubes for standard curve.
4. 100 μl of standard solutions in tubes for standard curve.
5. 100 μl antiserum (working dilution 1 : 1500, final solution 1 : 15,000).
6. 100 μl ^{125}I-3,3'-T$_2$ (10-15,000 cpm, ~25–50 pg T$_2$). Tubes are mixed by swirling after steps 3, 4, 5, and 6 and then incubated for 24 hr at 4°.
7. 40–60 μl or previously titered goat anti-rabbit γ-globulin are added. Subsequent steps of separation of bound from free radioactivity and counting of assay are identical to those described previously for T$_4$-RIA.

The detection limit of the T$_2$ RIA was ~1.2 ng/dl. Other features of the RIA are noted in Table VI, and data on specificity of the 3,3'-T$_2$ RIA are listed in Table V. The cross-reaction of T$_4$ and T$_3$ would not be expected to account for any more than 10% of the observed mean normal serum 3,3'-T$_2$ concentration of 4.1 ng/dl. Other iodothyronines did not cross-react to the extent that their usual concentrations in the serum of man would influence 3,3'-T$_2$ levels measured by RIA. Readers may review other published RIAs of 3,3'-T$_2$ in unextracted serum that employ ANS to minimize the influence of T$_2$ binding proteins in RIA.[60,61] Small Sephadex (G-25 fine) columns have also been used in T$_2$ RIA,[62] but experience with this system remains limited to date.

Sample Handling and Storage

It is recommended that only a moderate venous compression be applied while collecting blood for determination of iodothyronines; venous stasis associated with prolonged compression may result in some artifactual increase in serum concentration of iodothyronines.[63] After application of a tourniquet for 5 min, we found a mean increase of 14% for serum

[60] K. D. Burman, D. Strum, R. C. Diamond, Y. Y. Djuh, F. D. Wright, J. M. Earll, and L. Wartofsky, *J. Clin. Endocrinol. Metab.* **45,** 339 (1977).
[61] A. Burger and C. Sakoloff, *J. Clin. Endocrinol. Metab.* **45,** 384 (1977).
[62] J. Faber, C. Kirkegaard, I. B. Lumholtz, K. Siersbalk-Nielsen, and T. Friis, *J. Clin. Endocrinol. Metab.* **48,** 611 (1979).
[63] S. J. Judd, J. N. Carter, J. M. Corcoran, and L. Lazarus, *Br. Med. J.* **4,** 735 (1975).

rT_3, of 6% for serum T_3, and of 9% for serum T_4; free T_4 index did not change in these studies. There is only a modest ($\sim 10\%$) diurnal variation in serum concentration of T_4 and T_3; this is probably a result of variations in hemoconcentration associated with postural changes.

We have not observed differences in the concentration of T_4, T_3 or rT_3 when measured in serum or plasma of heparinized blood. However, it is possible that other anticoagulants in plasma may influence the RIAs. Serum specimens can be stored frozen for prolonged periods; we have not encountered any appreciable changes in concentrations of iodothyronines in serum specimens stored at $-20°$ for up to 5 years.

Serum Iodothyronine Concentrations in Health and Disease

Physiological Studies

Normal values of serum T_4, T_3, rT_3, and $3,3'$-T_2 in adult subjects, as determined by RIAs, are listed in Table VII. No significant differences in serum concentration of iodothyronines are found between males and females except during pregnancy, at which time serum total concentrations of T_4, T_3, and rT_3 are increased. This is due to an increase in the serum concentration of TBG that avidly binds these iodothyronines; serum concentrations of free T_4, T_3 or rT_3 remain normal in pregnancy, and serum T_2 concentration does not change appreciably. The latter finding is probably a reflection of the fact that $3,3'$-T_2 does not bind to TBG as well as T_4, T_3, and rT_3.

Age has little effect on serum iodothyronine concentration in adult life; low serum T_3 and high serum rT_3 concentrations often reported in the elderly population may be a result of an associated (often unsuspected) systemic illness rather than of old age.[64] It is difficult to interpret recent reports of moderately reduced serum $3,3'$-T_2 levels in old age;[65,66] no data are yet available on the effect of age on kinetics of metabolism of $3,3'$-T_2.

In contrast with adult life, remarkable alterations in the serum iodothyronine concentrations are evident at the time of birth and soon thereafter (Table VII). Thus, serum T_4, rT_3, and $3,3'$-T_2 concentrations are increased, while serum T_3 is decreased in the newborn. Kinetic studies in sheep[67] suggest that these changes are due to an increased production rate of T_4, a reduced production of T_3 from T_4, and a decreased meta-

---

[64] T. H. Olsen, P. Laurberg, and J. Weeke, *J. Clin. Endocrinol. Metab.* **47**, 1111 (1978).

[65] A. Burger and C. Sakoloff, *J. Clin. Endocrinol. Metab.* **45**, 384 (1977).

[66] J. Faber, C. Kirkegaard, J. B. Lumholtz, K. Siersbaek- Nielsen, and T. Friis, *J. Clin. Endocrinol. Metab.* **48**, 611 (1979).

[67] I. J. Chopra, J. Sack, and D. A. Fisher, *Endocrinology*, **97**, 1080 (1975).

TABLE VII

MEAN SERUM CONCENTRATIONS OF T_4, T_3, rT_3, AND $3,5'-T_2$ IN HEALTH AND DISEASE

	Serum concentration of							
	$T_4{}^{a,c}$		$T_3{}^{a,c}$		$rT_3{}^{a,c}$		$3,3'-T_2{}^{b-d}$	
Group	n	Mean ± SD (μg/dl)	n	Mean ± SD (ng/dl)	n	Mean ± SD (ng/dl)	n	Mean ± SD (ng/dl)
Normal subjects	26	8.6 ± 1.6	16	118 ± 34	27	40.5 ± 10.4	30	4.1 ± 2.4[b,d]
Pregnancy	5	15.0 ± 6.4	5	195 ± 126	5	54.2 ± 6.5	51	2.6 ± 0.36[b,d]
Newborn (cord serum)	7	11.9 ± 2.4	7	24 ± 9.7	7	136 ± 29	18	9.3 ± 1.92[b]
Hyperthyroidism	22	23.3 ± 5.7	22	744 ± 245	22	103 ± 48.6	13	13.2 ± 8.6
Hypothyroidism	12	1.7 ± 1.0	11	33 ± 11	12	18.6 ± 9.2	5	<1.25 ± 0[b]
Hypothyroidism treated with T_4	12	14.4 ± 6.3	12	174 ± 74	12	54.8 ± 18.6	—	—
Nonthyroidal Illness	12	5.8 ± 1.9[c]	12	53 ± 28[c]	12	76 ± 11.2[c]	12	2.6 ± 1.7[d]

[a] Data from I. J. Chopra, J. Clin. Invest. **54**, 583 (1974);
[b] F. Geola, I. J. Chopra, D. H. Solomon, and R. M. B. Maciel, J. Clin. Endocrinol. Metab. **48**, 297 (1979);
[c] F. Geola, I. J. Chopra, and D. L. Geffner, J. Clin. Endocrinol. Metab. **50**, 336 (1980);
[d] Unpublished data.

bolic clearance rate of rT_3 in the fetus and the newborn. Altered serum levels of iodothyronines in the newborn become comparable to the adult within a few weeks after birth.[68] Some studies suggest that serum T_3 and T_4 levels are lower in summer than in winter.[69]

Serum Concentration of Iodothyronines in Diseases of the Thyroid

Serum concentrations of all iodothyronines are typically increased in hyperthyroid patients and decreased in hypothyroid patients (Table VII). Serum levels of T_3 are relatively increased to a greater degree than T_4 in hyperthyroidism. On the other hand, serum T_3 is relatively decreased to a lesser extent than T_4 in hypothyroidism. These findings are consistent with a relative overproduction of T_3 compared with T_4, both in hyperthyroidism and in primary hypothyroidism in which the thyroid gland (or its remnant in hypothyroidism) is stimulated excessively by circulating thyroid stimulators. Interestingly, there are some ($\sim 5–10\%$) hyperthyrid patients in whom serum T_4 is normal but serum T_3 is clearly high; this has been referred to as T_3 toxicosis. In the same vein, there may occur in hypothyroid patients, a normal serum T_3 in association with clearly decreased serum T_4.

Overall, the measurement of serum T_3 is a more sensitive test in the diagnosis of hyperthyroidism than is T_4, whereas serum T_4 measurement is a much more sensitive test than serum T_3 in the diagnosis of hypothyroidism. Measurements of serum rT_3 and $3,3'$-T_2 are usually not needed in the diagnosis of hyperthyroidism and hypothyroidism except in special circumstances, e.g., in the setting of a nonthyroid illness (vide infra).

It is noteworthy that considerable confusion may arise in the evaluation of serum levels of iodothyronines in the presence of anti-T_3 or anti-T_4 antibodies in a serum sample. Such anti-iodothyronine antibodies have been described in a few patients with chronic thyroiditis or Graves' disease. The results of serum T_4 or T_3 in sera with antibodies are influenced markedly by the separation method used in RIAs. One may observe spuriously increased values when the second antibody method is used and spuriously reduced values when polyethylene glycol or activated charcoal is used in the final step of RIA to separate bound from free radioactivity.

Serum Concentration of Iodothyronines in Nonthyroidal Diseases

Recent studies have documented several abnormalities in the serum concentration of iodothyronines in adult patients with a wide variety of

---

[68] I. J. Chopra, J. Sack, and D. A. Fisher, *J. Clin. Invest.* **55**, 1137 (1975).

[69] A. G. H. Smals, H. A. Ross, and P. W. C. Kloppenborg, *J. Clin. Endocrinol. Metab.* **44**, 998 (1977).

nonthyroidal diseases (Table VII). Thus, a combination of subnormal serum T_3 and normal serum T_4 is found in approximately two-thirds of patients with a nonthyroidal illness; serum rT_3 concentration is high normal or clearly high in these patients.[70,71] This so called "low T_3 syndrome" may be found in almost any disease and is related in general to the severity of the disease.[3,70-76]

Kinetic studies in patients with hepatic cirrhosis,[3,74] renal failure,[75] or diabetes mellitus[76] indicate impaired production rates of T_3 and normal or slightly increased production rates of rT_3; production rates of T_4 were either clearly normal or minimally subnormal in these studies. Thus, high serum rT_3 in nonthyroidal illness is not due so much to an increase in the production of rT_3 as it is to a decrease in its metabolic clearance rate. The low T_3, on the other hand, is mainly a result of decreased production of T_3 from the metabolism of T_4. Serum $3,3'-T_2$ levels may also be low in systemic illnesses,[77,78] presumably as a result of reduced activity of iodothyronine 5'-deiodinase that catalyzes conversion of rT_3 to $3,3'-T_2$ as well as T_4 to T_3.

Serum T_4 may be subnormal in approximately one-third of patients with nonthyroidal disease.[71] The major determinant of this fall in serum T_4 is a decrease in T_4-binding potency of serum proteins. There may be a reduction in the serum concentration of all T_4 binding proteins in serum, albumin, TBPA as well as TBG, although the extent to which each protein is affected is variable. In addition, there is in the circulation of some "sick" patients a nondialyzable, heat-sensitive factor that inhibits the binding of T_4 to serum proteins.[79]

Despite substantial reduction in serum T_3, patients with nonthyroidal

[70] I. J. Chopra, U. Chopra, S. R. Smith, M. Reza, and D. H. Solomon, *J. Clin. Endocrinol. Metab.* **41**, 1043 (1975).
[71] I. J. Chopra, D. H. Solomon, G. W. Hepner, and A. A. Morgenstein, *Ann. Intern. Med.* **90**, 905 (1979).
[72] W. A. Burr, R. S. Griffiths, E. G. Black, R. Hoffenberg, H. Meinhold, and K. W. Wenzel, *Lancet* **2**, 1277 (1975).
[73] A. G. Vagenakis, A. Burger, and G. I. Portnay, *J. Clin. Endocrinol. Metab.* **41**, 191 (1975).
[74] S. Nomura, C. S. Pittman, J. B. Chambers, Jr., M. W. Buck, and T. Shimizu, *J. Clin. Invest.* **56**, 643 (1975).
[75] V. S. Lim, V. S. Fang, A. S. Katz, and S. Refetoff, *J. Clin. Invest.* **60**, 522 (1977).
[76] C. S. Pittman, A. K. Suda, J. B. Chambers, Jr., H. G. McDaniel, G. Y. Ray, and B. K. Preston, *J. Clin. Endocrinol. Metab.* **48**, 854 (1979).
[77] I. B. Lumholtz, J. Faber, M. Buck Sorensen, C. Kirkegaard, K. Siersbaek-Nielsen, and T. H. Friis, *Horm. Metab. Res.* **10**, 566 (1978).
[78] F. L. Geola, I. J. Chopra, and D. L. Geffner, *J. Clin. Endocrinol. Metab.* **50**, 336 (1980).
[79] I. J. Chopra, G. N. Chua Teco, A. H. Nguyen, and D. H. Solomon, *J. Clin. Endocrinol. Metab.* **49**, 63 (1979).

illnesses appear euthyroid clinically; an essentially normal serum TSH response to intravenous administration of thyrotropin-releasing hormone (TRH) also favors euthyroidism. The significance of alterations in thyroid hormone metabolism and circulating thyroid hormone levels in nonthyroidal disease, if any, is unclear at present. Studies to date, however, do not favor use of exogenous thyroid hormones in the treatment of patients with low serum T_3 (or even T_4) (and unlike true hypothyroidism, high or high normal serum rT_3) in the setting of a nonthyroidal illness.

Effect of Some Pharmacological Agents on Serum Concentration of Iodothyronines

Stimulation of the thyroid gland by exogenous TSH (e.g., 10 U bovine TSH im) or by endogenous TSH released following administration of TRH (200 μg iv) results in an increased thyroidal secretion of T_3 and T_4 and an acute increase in serum T_3 and T_4 levels, while serum rT_3 and $3,3'$-T_2 do not change appreciably.[3,29] Absence of an increase in serum rT_3 and T_2 may reflect that thyroidal secretion contributes very modestly to rT_3 or $3,3'$-T_2 in circulation of normal man. Treatment of euthyroid or hypothyroid subjects with replacement doses (50–75 μg/day) of T_3 is associated with low serum T_4 and rT_3 levels because of a reduction in thyroidal secretion of T_4; serum T_3 levels vary between normal and high in subjects treated with T_3. Treatment of hypothyroid or euthyroid subjects with replacement doses (~ 0.2 mg/day) of T_4, on the other hand, results in an essentially normal serum T_4, T_3, rT_3, and $3,3'$-T_2.

Administration of several drugs to man including dexamethasone, sodium ipodate, iopanoic acid, amiodarone, propranolol, and propylthiouracil is associated with a reduction in serum T_3 concentration with little or no change in serum T_4 concentration and whenever studied, an increase in serum concentration of rT_3.[80-87] These drugs inhibit activity of iodothyronine 5'-monodeiodinase in extrathyroidal tissues resulting in a dimi-

---

[80] I. J. Chopra, D. E. Williams, J. Orgiazzi, and D. H. Solomon, *J. Clin. Endocrinol. Metab.* **41,** 911 (1975).

[81] J. Abuid and P. R. Larsen, *J. Clin. Invest.* **54,** 201 (1974).

[82] M. Saberi, F. H. Sterling, and R. D. Utiger, *J. Clin. Invest.* **55,** 218 (1975).

[83] D. L. Geffner, M. Azukizawa, and H. M. Hershman, *J. Clin. Invest.* **55,** 224 (1975).

[84] A. Burger, D. Dinichert, P. Nicod, M. Jenny, Th. Lemarchand-Beraud, and M. B. Vallotton, *J. Clin. Invest.* **58,** 255 (1976).

[85] H. Bürgi, C. Wimpfheimer, A. Burger, W. Zaunbauer, H. Rösler, and Th. Lemarchand-Beraud, *J. Clin. Endocrinol. Metab.* **43,** 1203 (1976).

[86] S. Y. Wu, I. J. Chopra, D. H. Solomon, and L. R. Bennett, *J. Clin. Endocrinol. Metab.* **46,** 691 (1978).

[87] W. M. Wiersinga and J. L. Touber, *J. Clin. Endocrinol. Metab.* **45,** 293 (1977).

nution in the production of T_3 from T_4 and in the metabolic clearance rate of rT_3.[88,89]

Acknowledgments

Supported by USPHS Grants AM-16155 and AM-17251 from NIH and a grant from The Netherlands Organization for Advancement of Pure Research (Z.W.O. to WMW).

[88] I. B. Lumholtz, K. Siersbaek-Nielsen, J. Faber, C. Kirkegaard, and Th. Friis, *J. Clin. Endocrinol. Metab.* **47**, 587 (1978).
[89] I. B. Lumholtz, M. Busch-Sorensen, J. Faber, Th. Friis, C. Kirkegaard, and K. Siersbaek-Nielsen, *Acta Med. Scand. Suppl.* **624**, 31 (1979).

[23] Chemiluminescence Immunoassay for Serum Thyroxine

By Hartmut R. Schroeder

Introduction

Numerous hormones, drugs, and other substances present at low levels in complex body fluids are commonly measured by protein-binding radioassays. These methods employ a binding protein with high affinity and specificity (such as an antibody) and a radiolabeled ligand. Usually, the ligand of interest and a radiolabeled analog of the ligand compete for a limited number of antibody-binding sites, and then the amount of free or antibody-bound radiolabeled ligand is determined and related to the level of unlabeled ligand via standards.

Most commercial tests for serum thyroxine employ competitive protein-binding reactions monitored with radiolabeled thyroxine.[1-3] Such assays pose some inconvenience (licensing and safety regulations) and stability problems (2-4 month shelflife) which can be avoided through the use of nonradioisotopic labels. Some aminophthalhydrazide compounds are useful alternatives to radiolabels since they can produce light with high quantum efficiency through simple oxidation reactions, which allows

[1] L. E. Braverman, A. G. Vagenakis, A. E. Foster, and S. H. Ingbar, *J. Clin. Endocrinol.* **32**, 497 (1971).
[2] I. J. Chopra, *J. Clin. Endocrinol. Metab.* **34**, 938 (1972).
[3] T. Mitsuma, J. Collucci, L. Shenkinan, and C. S. Hollander, *Biochem. Biophys. Res. Commun.* **46**, 2107 (1972).

their detection at picomolar levels.[4] The ligand of interest is covalently coupled through an aminoalkyl arm to the amino group of isoluminol (6-amino-2,3-dihydrophthalazine-1,4-dione) to give a stable nonradioisotopic labeled ligand that can be used to monitor an immunoassay. Isoluminol is preferred because alkylation of the amino group increases quantum efficiency whereas similar substitution on luminol (5-amino-2,3-dihydrophthalazine-1,4-dione) causes a substantial loss.[5-7]

An immunoassay for thyroxine in serum has been developed in which a thyroxine–isoluminol conjugate is monitored by chemiluminescence. The sensitivity of competitive-binding assays is determined partially by the lower limit of detection of the label. Consequently, derivatives of 7-aminonaphthalene-1,2-dicarboxylic acid hydrazide have also been employed as label because they have very high quantum efficiency[8] and can be detected at lower concentrations than those of isoluminol.

Preparation of Reagents

Water used for preparation of reagents is glass distilled from charcoal-filtered deionized water. Except for disposable glassware, all glass containers that are used for reagent storage are cleaned with concentrated nitric acid and rinsed well with distilled water. Reagents are prepared and diluted in glass containers. These solutions are dispensed from disposable glass pipets or pneumatic pipets with disposable plastic tips.

Synthesis of Chemilumigenic Compounds

The synthesis of both an aminophthalhydrazide and an aminonaphthylhydrazide and their coupling to thyroxine have been described previously.[4] These reactions are shown schematically in Figs. 1 and 2.

Chemilumigenic Thyroxine Conjugates

A stock solution (1 mM) of conjugate (VII) (Fig. 1) is prepared in 0.1 M Na$_2$CO$_3$ buffer at pH 10.5. The conjugate is stable in this buffer at 4° for several months. A 0.3 mM solution of conjugate (XIII) (Fig. 2) is made daily in 0.1 M Na$_2$CO$_3$–0.15 M NaOH, pH 12.6. Both conjugates are di-

[4] H. R. Schroeder, R. C. Boguslaski, R. J. Carrico, and R. T. Buckler, this series Vol. 57, p. 425.
[5] R. B. Brundrett, D. F. Roswell, and E. H. White, *J. Am. Chem. Soc.* **94**, 7536 (1972).
[6] K.-D. Gundermann and M. Drawert, *Chem. Ber.* **95**, 2018 (1962).
[7] R. B. Brundrett and E. H. White, *J. Am. Chem. Soc.* **96**, 7497 (1974).
[8] K.-D. Gundermann, W. Horstmann, and G. Bergman, *Justus Liebigs Ann. Chem.* **684**, 127 (1965).

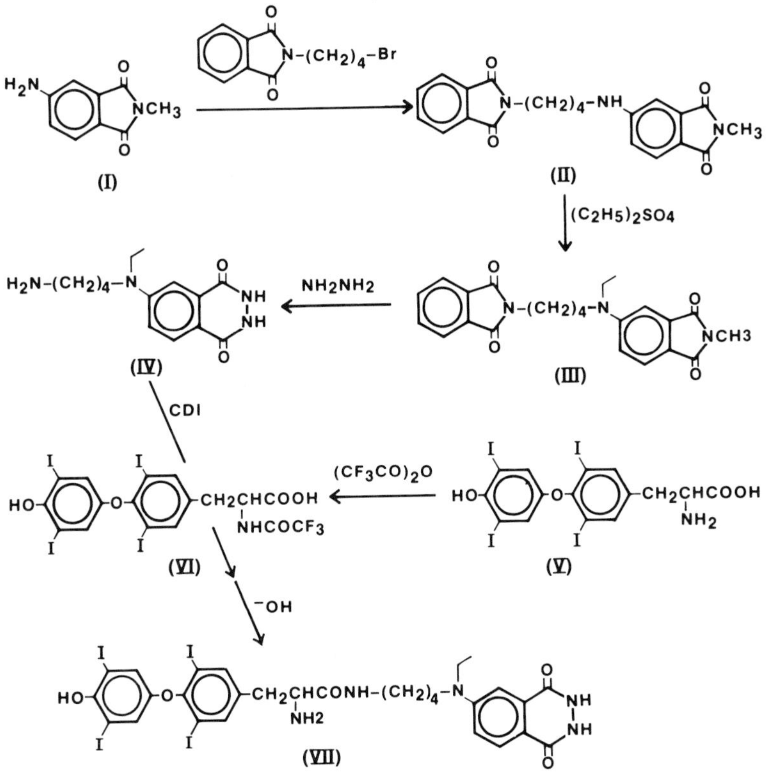

FIG. 1. Synthesis of 6-{N-ethyl-N-[6-(thyroxinylamido)butyl]amino}-2,3-dihydrophthalazine-1,4-dione.

luted in 0.1 M NaOH or 75 mM sodium barbital (5,5-diethyl barbiturate) buffer, pH 8.6, as needed.

Microperoxidase

One milligram of microperoxidase (Sigma Chemical Corp., St. Louis, MO) is dissolved in 2.5 ml of 10 mM Tris–HCl, pH 7.4. This 200 μM stock solution is stable at 4° for more than a month. It is diluted daily to 2 μM in 75 mM barbital buffer, pH 8.6.

Thyroxine Standard

A solution of 0.5 mg/ml thyroxine (Sigma Chemical Corp., St. Louis, MO) is prepared in 6.4 mM NaOH, aliquoted, and stored at −20°. The stock solution is diluted 10-fold in 0.1 M NaOH and the thyroxine concentration is determined spectrophotometrically using the molar extinction of

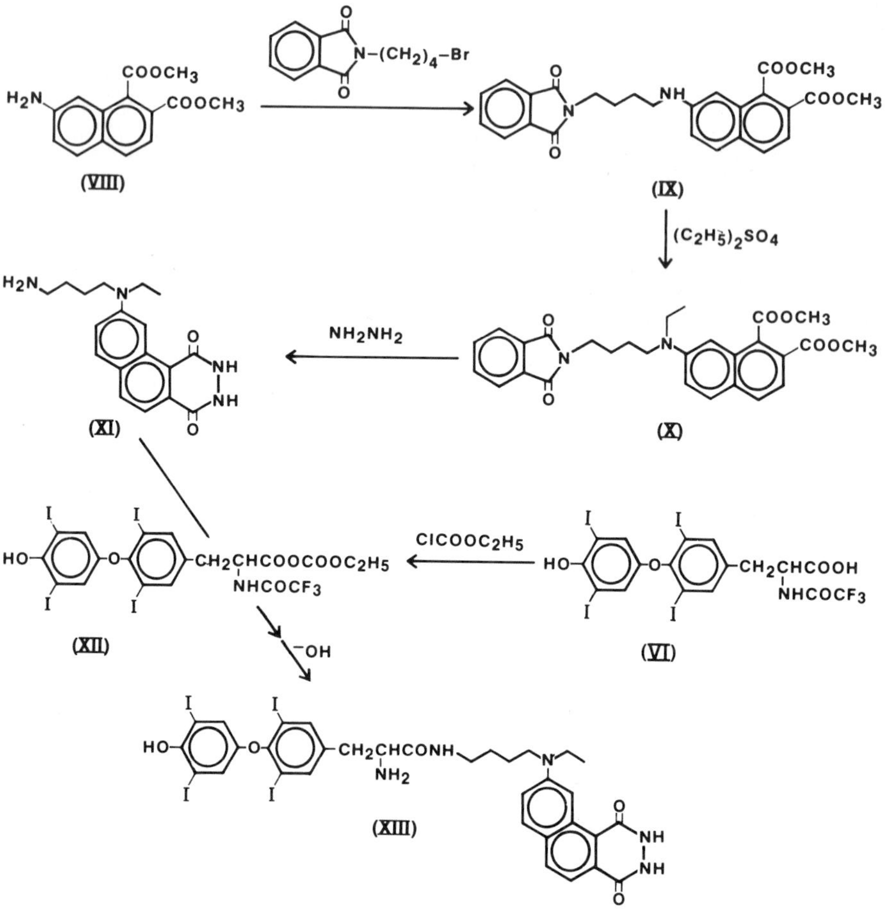

FIG. 2. Synthesis of 7-[N-ethyl-N-(4-thyroxinylamido)butyl]aminonaphthalene-1,2-dicarboxylic acid hydrazide.

6210 at 325 nm.[9] Further dilutions are made as needed in 75 mM barbital, pH 8.6. For serum assays on Sephadex columns, the standards are prepared by diluting various volumes of the thyroxine stock solution in 10% serum (stripped of thyroxine according to Mitsuma *et al.*[3]).

Antibody to Thyroxine

Rabbits are immunized with a thyroxine-bovine serum albumin conjugate to produce antibody[10] which is then purified by a published procedure.[11]

[9] C. L. Gemill, *Arch. Biochem. Biophys.* **54,** 359 (1955).
[10] N. N. Alexander and V. F. Jennings, *Clin. Chem.* **20,** 1353 (1974).
[11] D. Livingston, this series Vol. 34, p. 725.

Preparation of Sephadex Columns

Seralute thyroxine RIA columns (1 × 6.5 cm) (Ames Co., Miles Laboratories, Inc., Elkhart, IN) containing either 1.0 or 1.5 ml bed volume of Sephadex G-25 are washed successively as follows: six times with 4 ml of 7% acetic acid, twice with 4 ml of water and six times with 4 ml of 0.1 M NaOH.

In addition, the columns filled with 1.5 mL of Sephadex are finally washed once with 75 mM barbital, pH 8.6, for separating free and antibody-bound conjugate (**XIII**) in solutions.

Serum Samples

Clinical serum samples were obtained through the South Bend Medical Foundation, South Bend, IN and kept at $-20°$ until assayed. Reference thyroxine values were ascertained by the Tetralute assay procedure (Ames Co., Div. Miles Laboratories, Inc., Elkhart, IN).

Light Measurements

Peak light intensities from chemiluminescent reactions in this study were automatically recorded with a Model 760 Luminescence Biometer (E. I. DuPont DeNemours & Co., Inc., Wilmington, DE) at a sensitivity setting of 820. For some measurements, a custom-built microprocessor connected to the Biometer monitored the signal at 10-msec intervals and printed out the peak light intensity and total light production in arbitrary units. The microprocessor is activated when the signal from the photomultiplier exceeds a threshold level that is five times greater than the electronic noise.

The light emissions can also be quantitated with instruments available from Aminco-Bowman, Inc., SAI Technology, Packard Instruments, Inc., and Lumac or by scintillation spectrometer.

Chemiluminescent Detection Reaction

Light producing reactions are carried out in 6 × 50-mm disposable glass tubes (Kimble, Div., Owens-Illinois, Toledo, OH; Order No. 73500) mounted in the Biometer.

Microperoxidase (2 μM in 75 mM barbital buffer, pH 8.6) and 0.2 M NaOH are combined in a ratio of 2:3.5 (v/v). Reaction mixtures are assembled in the glass tubes by the addition of a 55-μl aliquot of this catalyst solution to 95 μl of chemilumigenic compound in 75 mM barbital buffer, pH 8.6 (final pH 12.6). After 10 min (mixtures are generally stable for at least 1 hr) tubes are placed in the Biometer and 10 μl of 90 mM H_2O_2 in

10 mM Tris–HCl, pH 7.4, is added to initiate chemiluminescence. The peak light intensity is attained within about 1 sec and measured. Reported peak light intensity values represent the average from three replicate reactions.

Alternatively, reactions are carried out directly in 75 mM barbital, pH 8.6 and using 3 mM H_2O_2 instead, otherwise as above.

Generally, the H_2O_2 is injected manually with a 25-μl Hamilton syringe. However, determinations involving clinical serums were performed with a custom-built automatic injection device. The pump consisting of a Teflon cylinder and sapphire piston is equipped with Teflon valves (Series 2, Model 1 Teflon solenoid valve, General Valve Corp., East Hanover, NJ) and operated by solenoids. The H_2O_2 solutions passes through 0.95-mm i.d. Teflon tubing to a 0.78-mm i.d. light-tight glass orifice mounted 4 cm directly above the surface of the reaction mixture. The device delivers a 10 ± 0.06 μl aliquot of H_2O_2 in 30 msec.

Detection Limits for Chemilumigenic Labels and Their Thyroxine Conjugates

The sensitivity of chemiluminescence measurements is based on the efficiency of light production. The quantum yield of the aminophthalhydrazides and aminonaphthylhydrazides depends on the oxidation conditions used and various greatly with the chemical structure. Among numerous oxidation systems, the most sensitive employ H_2O_2 and a heme catalyst.[4,12] Microperoxidase (heme attached to an 11 amino acid chain) is the best catalyst since it is stable and equally effective from pH 8.6 to 13, while other heme containing catalyst are efficient only at high pH. In order to determine sensitivity, luminol at various concentrations was oxidized and the light produced was measured. A plot of either peak light intensity or total light production vs luminol concentration indicated a signal 1.5 times the control value in the absence of luminol (lower detection limit) at 1 pM luminol.

Structural changes in the aminophthalhydrazides can produce steric and electronic effects that greatly alter their chemiluminescence quantum yield.[5–7] To provide for covalent linking of ligands, an aminobutyl bridging arm is attached to the amino group of isoluminol since alkylation at this position (Fig. 1) increases quantum efficiency.[4] The peak light intensity of the fully alkylated label (IV) in a H_2O_2–hematin system was nearly equal to that of luminol. Attachment of thyroxine to the label gave a conjugate (VII) that exhibited 40-fold reduction in peak light intensity. Nonetheless, in the H_2O_2–microperoxidase system at pH 12.6, this conjugate

[12] H. R. Schroeder and F. M. Yeager, *Anal. Chem.* **50**, 1114 (1978).

(VII) is detectable at 0.1 nM (Fig. 3), which is adequate for use in immunoassays. Lower detection limits are similar as above with either total light measurements and with conjugates employing different bridging arms.[4] Since addition of thyroxine at a concentration 1000 times greater than the label concentration does not affect the light emission, the losses observed with conjugate (VII) may be due to quenching by iodine via an intramolecular route.[13,14] Coupling of the ligand biotin to a similar label did not change the peak light intensity produced in the H_2O_2–microperoxidase system. Therefore, most other ligands should have more moderate effects on quantum yield when they are coupled to isoluminol derivatives.

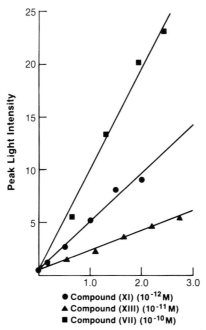

● Compound (XI) (10^{-12} M)
▲ Compound (XIII) (10^{-11} M)
■ Compound (VII) (10^{-10} M)

FIG. 3. Quantitation of chemilumigenic compound by light produced during oxidation. Reaction mixtures (150 μl) contained the (■) thyroxine–isoluminol conjugate (VII), (●) aminonaphthylhydrazide (XI), or its (▲) thyroxine conjugate (XIII) at indicated concentrations, 0.27 μM microperoxidase, 57.5 mM sodium barbital, and 50 mM NaOH (final pH 12.6). Manual addition of 10 μl of 90 mM H_2O_2 in 10 mM Tris–HCl, pH 7.4, initiated light production. Peak light intensities of reactions performed in triplicate were measured and averaged.

[13] G. K. Radda, in "Methods in Membrane Biology" (E. D. Moru, ed.), Vol. 4, p. 140. Plenum, New York, 1975.
[14] G. R. Fleming, A. W. E. Knight, J. M. Morris, R. J. S. Morrison, and G. W. Robinson, *J. Am. Chem. Soc.* **99**, 4306 (1977).

Aminonaphthylhydrazides possess greater quantum efficiency[8] and can be measured quantitatively at lower levels than the aminophthalhydrazides. For example, (**XI**) (Fig. 2) can be determined at concentrations as low as 0.1 pM (0.015 fmol) with the H_2O_2–microperoxidase system at pH 12.6 (Fig. 3). However, at pH 8.6, the detection limit is only about 0.5 pM, which may be related to the poorer solubility of (**XI**) in aqueous solvents which are not strongly basic. Thyroxine in the conjugate (**XIII**) quenches the light also, but only about half as much as in (**VII**). As a result the thyroxine aminonaphthylhydrazide conjugate (**XIII**) can be measured at substantially lower levels than the corresponding aminophthalhydrazide conjugate (**VII**).

Kinetics of Light Emission

Light emission produced by N,N-diethylisoluminol and thyroxine conjugate (**VII**) in the microperoxidase system at pH 8.6 reached peak intensity 220 msec after H_2O_2 was added (determined with the microprocessor). Reaction was slower at pH 12.6 and peak light intensity was attained at 680 msec with N,N-diethylisoluminol and 1140 msec with the less efficient (**VII**). Then the light intensities decreased with half-lives of 0.5 and 4.5 sec at pH 8.6 and 12.6, respectively.[4,12]

Reproducibility of Light Measurements

The reproducibility of chemiluminescence measurements was evaluated in the H_2O_2–microperoxidase system.[4,12] Peak light intensities were measured for 15 identical reactions containing 1.67 nM (**VII**) at pH 12.6. With manual addition of H_2O_2, the average peak light intensity was 29.0 ± 5.2 (SD), i.e., ±18%. The coefficient of variation (CV) decreased to ±5% when triplicate results were averaged.

Reactions conducted with 11.3 pM N,N-diethylisoluminol gave a mean peak light intensity of 29.3 ± 3.7 (SD), i.e., ±12.7%, for triplicates. Since reactions with N,N-diethylisoluminol attained peak light intensity nearly twice as rapidly as those with (**VII**), the greater variability may be related to the speed of the reaction. Use of the automatic injector, which eliminates contact of the H_2O_2 with metal, lowered the CV to ±5%. Recently, when a 50-μl Hamilton syringe with a Teflon-tipped plunger was used for addition of H_2O_2 by hand, the CV also decreased to ±5%. With the automatic injector reproducibility of peak light intensities with (**VII**) was similar at pH 8.6 to that at 12.6 and when total chemiluminescence rather than peak light intensity was measured.

Finally, a rapid mixing flow system was designed to measure low levels of light with improved reproducibility.[15] A reagent containing chemilumigenic compound, microperoxidase and base in one syringe was rapidly mixed in an 8-μl custom-made flow cell with H_2O_2 from a second syringe to produce light which was measured with the Biometer. The lower detection limit for N,N-diethylisoluminol was 1 pM. The reaction period for each measurement was less than 10 sec and the results were unaffected above a minimum flow rate. The mean light intensity from six identical reactions containing 30.8 pM N,N-diethylisoluminol was 22.6 $\pm$ 0.45 (SD), i.e., $\pm 2\%$, with a background signal of 0.5. Thus, highly sensitive chemiluminescent measurements can be made reproducibly.

Chemiluminescence Immunoassay for Thyroxine in Serum Using a Thyroxine–Aminophthalhydrazide Conjugate

A good immunoassay for thyroxine requires sensitive detection of the thyroxine–isoluminol conjugate (**VII**) and high specificity of the antibody for the thyroxine moiety. These prerequisites are met in the excellent sensitivity of the chemiluminescence reaction (Fig. 3) and degree of cross-reactivity of antibody for (**VII**). The partially purified antibody cross-reacted 35% with (**VII**) as determined from the levels of thyroxine and (**VII**) required to inhibit binding of ^{125}I-labeled thyroxine by 50% in the Tetralute procedure.

With serum samples several considerations are relevant in choosing an assay format utilizing (**VII**). Binding of thyroxine by prealbumin, albumin, and thyroxine-binding globulin in serum[16] can interfere in the competitive protein-binding reaction. Furthermore, serum in the detection reaction quenches chemiluminescence. When 3 nM (**VII**) was oxidized with the H_2O_2–microperoxidase system at pH 12.6 in the presence of 0.1 and 1.0% human serum, the peak light intensities decreased by 20 and 75%, respectively, from the control value measured without serum.[17] In addition, the alkaline pH of the detection reaction dissociated the antibody (**VII**) complex. These observations make separation of bound and free conjugate (heterogeneous assay) imperative. Therefore, a Sephadex column format was used for the chemiluminescence immunoassay to avoid these problems. The merits of this procedure in radioassays are convenience, quantitative extraction of thyroxine from serum, and elimination of serum interference problems.[1,10] In adapting this method to the present assay,[17] it

[15] H. R. Schroeder and P. O. Vogelhut, *Anal. Chem.* **51**, 1583 (1979).
[16] S. M. Snyder, R. R. Cavalieri, I. D. Goldfine, S. H. Ingbar, and E. C. Jorgensen, *J. Biol. Chem.* **251**, 6489 (1976).
[17] H. R. Schroeder, F. M. Yeager, R. C. Boguslaski, and P. O. Vogelhut, *J. Immun. Methods* **25**, 275 (1979).

was discovered that the thyroxine–isoluminol conjugate (VII) migrates through the gel slower than ^{125}I-labeled thyroxine, used as a control (table). To compete effectively in binding reactions, thyroxine and (VII) should occupy the same column segment. Therefore, the conjugate is washed into the gel bed with base prior to addition of the thyroxine specimen. During optimization, the volume of base was varied to maximize competition in identical protein-binding reactions in the immunoassay described later.

Titration of (VII) with Antibody

Binding reactions are carried out directly on columns containing 1 ml of Sephadex G-25. The thyroxine–isoluminol conjugate (VII) (5 pmol) is applied to a set of columns and washed into the gel with 1 ml of 0.1 N NaOH, followed by 4 ml of 75 mM sodium barbital buffer, pH 8.6. Rabbit antibody to thyroxine at various dilutions in 300 μl of the same buffer is added and columns are incubated for 1 hr at room temperature. Then, the antibody-bound fraction of (VII) is eluted with 0.8 ml of barbital buffer, while unbound (VII) is retained by the Sephadex. Ninety-five-microliter aliquots of each column effluent are oxidized with the H_2O_2–microperoxidase system, pH 12.6, and monitored by chemiluminescence. Results in Fig. 4 show that 1 μl of antibody binds 50% of the thyroxine conjugate (VII) under these assay conditions.[17]

Chemiluminescence Immunoassay for Thyroxine with (VII)

The immunoassay procedure is similar to the titering of the antibody. Aliquots (200 μl) of base containing 5 pmol of (VII) are applied to a number of Sephadex G-25 columns and washed into the gel with 1 ml of 0.1 N NaOH. A 200-μl sample consisting of thyroxine standard or serum speci-

MIGRATION OF ^{125}I-LABELED THYROXINE AND THYROXINE–ISOLUMINOL CONJUGATE
(VII) ON SEPHADEX G-25[a]

	Elution volume	
Eluant	^{125}I-labeled thyroxine	(VII)
0.1 N NaOH	3 ml peak	6 ml peak
75 mM barbital buffer,	Start 7 ml,	Start 8 ml,
	13 ml broad peak	broad streak

[a] Two hundred microliters of 50 nM (VII) or 15 nM ^{125}I-labeled thyroxine in 37.5 mM barbital–50 mM NaOH was applied to a column containing 1 ml of Sephadex at alkaline pH and eluted with either base or buffer as indicated above. Eluted fractions were analyzed for chemiluminescence as in Fig. 3 or for radioactivity.

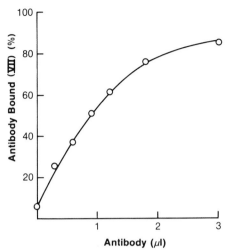

FIG. 4. Titering of antibody with (**VII**). The conjugate (**VII**) (5 pmol) in 200 μl of 0.1 M NaOH was applied to a set of columns containing 1 ml of Sephadex G-25 in 0.1 M NaOH. One milliliter of 0.1 N NaOH was added to each column and followed with 4 ml of 75 mM sodium barbital buffer, pH 8.6. Various amounts of antibody to thyroxine were diluted in 300 μl of the same buffer and added to individual columns. After 1-hr incubation at room temperature, the antibody-bound fraction was eluted with 0.8 ml of barbital buffer. Aliquots (95 μl) of the effluent were combined with 55 μl of 0.73 μM microperoxidase in 27 mM barbital–127 mM NaOH. Then 10 μl of 90 mM H_2O_2 in 10 mM Tris–HCl was delivered automatically and chemiluminescence was measured. A standard curve as in Fig. 3 was used to related peak light intensity values to concentrations of (**VII**). Results are expressed as a percentage of the total (**VII**) that was antibody bound. Reported values are averages from triplicate binding reactions which were each analyzed in triplicate.

mens diluted 1 : 10 in 0.1 N NaOH is added and the column is washed with 4 ml of 75 mM sodium barbital buffer, pH 8.6. The competitive protein-binding reaction is initiated with addition of 300 μl of an antibody solution which is capable of binding 60% of (**VII**) under conditions of the assay. After 1-hr incubation at room temperature, the antibody-bound thyroxine species are eluted with 0.8 ml of barbital buffer. The fraction of (**VII**) bound to antibody is determined from the peak light intensity produced by 95-μl aliquots of the eluate in the H_2O_2–microperoxidase system at pH 12.6.

The results shown in Fig. 5 demonstrate that the assay determines thyroxide quantitatively from 25 to 200 nM, which covers the clinically significant range.[17] The intraassay variability in peak light intensity values from binding reactions performed in triplicate is ±5% (CV). A 1-hr incubation is convenient, although 0.5–2 hr gives equally good results.

The specificity of this immunoassay can be confirmed by substituting

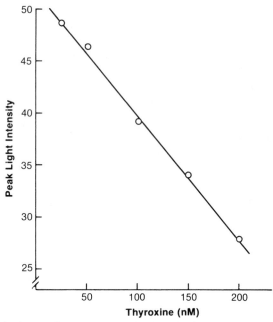

Fɪɢ. 5. Standard curve for serum thyroxine by chemiluminescence immunoassay. The procedure was as in Fig. 4 except that after washing 5 pmol of (**VII**) into each column bed with 1 ml of base, a 200-μl solution containing the thyroxine serum standards in 90 m*M* NaOH (Preparation of Reagents) was added. Also, 1.2 μl of antibody in 300 μl of buffer was used to initiate binding reactions. Concentrations of competing thyroxine were related directly to the peak light intensities of (**VII**) in the antibody-bound fraction by means of a curve. Twenty-microliter aliquots of clinical specimens were diluted to 200 μl in 0.1 *M* NaOH and analyzed as the standards above. The thyroxine values were determined from a curve generated with thyroxine standards prepared in serum.

isoluminol for the thyroxine–isoluminol conjugate (**VII**). Isoluminol (0.7 pmol) binds tightly to Sephadex and is not eluted by the buffer or by antibody to thyroxine. In addition, exogenously added thyroxine has no effect on light production even at a concentration 1000 times greater than that of (**VII**). Also, antibody at levels employed here has no effect on chemiluminescence.

Radioimmunoassays generally measure ligands at concentrations greater than the labeled ligand. In contrast, this assay with 5 pmol of (**VII**) determined thyroxine at levels below 5 pmol per column (Fig. 5). Thus, the lower cross-reactivity of (**VII**) resulted in a more sensitive assay.

Thyroxine levels of 28 clinical serums have been determined with the chemiluminescence immunoassay and compared to those obtained with the Tetralute assay.[17] As shown in Fig. 6, results give reasonably good statistical correlation ($r = 0.98$) for the two methods. The assay covers

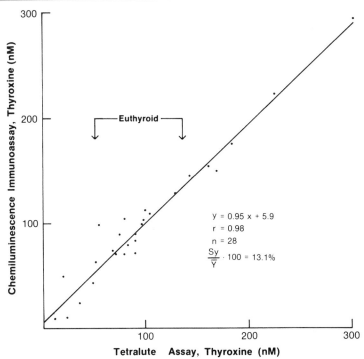

FIG. 6. Comparison of thyroxine levels in serum determined by the chemiluminescence immunoassay and a reference procedure. Twenty-eight clinical serums were assayed in triplicate as in Fig. 5 and in duplicate by the Tetralute method. In both assays, determinations with hypo- and hyperthyroid serums were repeated at twice and one-half the recommended volume of serum, respectively. One set of thyroxine standards was used with both tests.

the clinically significant range for thyroxine (25 to 200 n*M*) and is a valid analytical procedure for determining thyroxine in serum. Reproducibility is adequate, but limited by column characteristics, and the accuracy of making chemiluminescence measurements. The overall assay time is similar to that of conventional radioassays. However, the assay involves more steps and requires greater care in execution, although automation may moderate these problems.

Chemiluminescence Immunoassay for Thyroxine Using a
Thyroxine–Aminonaphthylhydrazide Conjugate

The thyroxine–aminonaphthylhydrazide conjugate (**XIII**) is attractive for improving the sensitivity of the chemiluminescence immunoassay because it can be quantitated at lower levels than the corresponding aminophthalhydrazide (**VII**) (Fig. 3). A partially purified antibody cross-reacted 20% with (**XIII**), which appears adequate.

Chemiluminescence Immunoassay for Thyroxine with (XIII)

Competitive binding reactions (200 μl) containing 2.5 nM of (**XIII**), various concentrations of thyroxine as shown in Fig. 4 and 8 μl of antibody are assembled in 75 mM sodium barbital buffer, pH 8.6. The reaction mixtures are incubated at room temperature for 10 min and then a 150-μl aliquot of each is applied to individual columns containing 1.5 ml Sephadex G-25 that is equilibrated with the barbital buffer. The columns are washed with 1.5 ml of barbital buffer to separate the free and antibody-bound portions of (**XIII**). Aliquots (95 μl) of the effluent which contain the antibody-bound (**XIII**) are assayed with the H_2O_2–microperoxidase system, at pH 12.6.

Results in Fig. 7 show a rapid highly sensitive assay that will quantitatively measure thyroxine from about 1 to 16 nM. Previously, a similar assay employing 19.3 nM of (**VII**) and a 1-hr incubation gave a usable range from 10 to 60 nM thyroxine.[4] However, in the present assay the peak light intensities produced with (**XIII**) and antibody-bound (**XIII**) decrease nearly linearly by about 30% during the first hour after dilution. This apparent lack of stability of (**XIII**) appears to be related to lower solubility in solutions not highly alkaline. Nonetheless, by using a short incubation period for the competitive binding assay, a usable assay is possible. Perhaps the aminonaphthylhydrazide labels can be used to advantage in immunoassays employing ligands more water soluble than thyroxine.

Comments

The ability of chemilumigenic labels to monitor competitive protein-binding reactions has been demonstrated.[4,17,18] Ligands are coupled to aminophthalhydrazides or aminonaphthylhydrazides to produce conjugates that can be measured at picomolar levels by chemiluminescence. The immunoassay for thyroxine in serum described here employs a Sephadex column procedure. This format not only accomplishes separation of antibody-bound and free (**VII**), but also eliminates the interference by serum in the binding reaction and the subsequent chemiluminescent detection reaction. Alternatively, an antibody-coated tube format may provide greater convenience.

Current chemiluminescence measurements are variable and inconvenient. An automatic injector provides some convenience and improves

---

[18] H. R. Schroeder, P. O. Vogelhut, R. J. Carrico, R. C. Boguslaski, and R. T. Buckler, *Anal. Chem.* **48**, 1933 (1976).

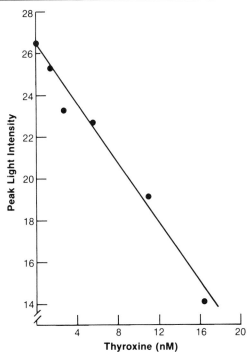

F_{IG}. 7. Standard curve for a competitive protein-binding assay for thyroxine. Triplicate reaction mixtures (200 μl) containing 2.5 nM of (**XIII**), the indicated levels of thyroxine and 8 μl of antibody to thyroxine (added last) in 75 mM sodium barbital buffer, pH 8.6, were incubated at room temperature for 10 min. This level of antibody was sufficient to bind 63% of (**XIII**) in the absence of competing thyroxine under conditions of the assay. A 150-μl aliquot of each reaction was applied to a set of columns packed with 1.5 ml of Sephadex G-25 equilibrated with the same buffer. The fraction of (**XIII**) bound to antibody was eluted with 1.5 ml of the barbital buffer. Three 95-μl aliquot of each effluent were oxidized as in Fig. 4 and the peak light intensities were measured and results were averaged.

the reproducibility of both the volume of H_2O_2 delivered and mixing. It also avoids contact of metal with the H_2O_2 which can affect the efficiency of the chemiluminescent reaction. Perhaps the highly reproducible rapid flow system could be automated. It is noteworthy that reaction mixtures containing chemilumigenic compound and microperoxidase are quite sensitive to vapors of H_2O_2 which cause premature light emission. When separated, all reagents are stable.

Acknowledgments

The author is sincerely grateful to numerous people who contributed to this work and are cited in the references. Also, review of the manuscript by R. Moore and D. Morris is greatly appreciated.

Section VI

Endogenous Compounds of Low Molecular Weight

[24] Radioimmunoassay of Serum Bile Acids

By P. E. Ross

Introduction

Bile acids, synthesized in the liver from cholesterol, are characterized by a unique enterohepatic circulation with efficient conservation. Reviews of the major reactions forming these C_{24} acids have appeared elsewhere,[1] as have detailed descriptions of the enterohepatic circulation[2] and consequently only a brief summary will be considered here.

In man two bile acids are synthesized from cholesterol, the primary bile acids cholic acid ($3\alpha,7\alpha,12\alpha$-trihydroxy-5β-cholanoic acid) and chenodeoxycholic acid ($3\alpha,7\alpha$-dihydroxy-5β-cholanoic acid). These bile acids are conjugated to either glycine or taurine, which increases their water solubility and decreases the pK_a from 6–6.5 for unconjugated bile acids to 4.0–5.5 for glycine conjugates and 1.5–2.0 for taurine conjugates. Consequently the four species of primary bile acid conjugate are present as salts in biological fluids. The hepatocyte secretes these bile salts into bile canaliculi by a process as yet incompletely understood, and they pass into the gall bladder. Bile salts secreted to aid digestion are themselves reabsorbed from the intestinal lumen by active and passive mechanisms, the combined processes accounting for about 90% of the bile salts secreted.

Passive absorption occurs throughout the intestine but is important in the proximal small intestine where less polar bile acids with higher pK_a values are preferentially absorbed. Consequently dihydroxy glycine-conjugated bile acids will be better absorbed in the proximal ileum while the remaining bile salts undergo active transport in the distal ileum. Bile salts which are not absorbed in the ileum enter the colon and are extensively deconjugated and may be metabolized to secondary bile salts by gut flora before passive absorption. The most common metabolic step is 7α-dehydroxylation leading to deoxycholic acid ($3\alpha,12\alpha$-dihydroxy-5β-cholanoic acid) and lithocholic acid (3α-hydroxy-5β-cholanoic acid). Ursodeoxycholic acid ($3\alpha,7\beta$-dihydroxy-5β-cholanoic acid) is found frequently in biological fluids but at present the origin is not clear.

This mixture of conjugated and unconjugated primary and secondary bile salts returns to the liver protein-bound in the portal blood and is then

---

[1] H. Danielsson and J. Sjövall, *Annu. Rev. Biochem.* **44**, 233 (1975).

[2] A. F. Hofmann, *in* "Advances in Internal Medicine" (G. H. Stollerman, ed.), Vol. 21, p. 501. Year Book Medical Publ., Chicago, Illinois, 1976.

METHODS IN ENZYMOLOGY, VOL. 84

absorbed by an active transport mechanism by the hepatocyte. Little is known of the details of liver uptake in man, although 90% trihydroxy and about 70% of dihydroxy bile salts are extracted. In man secondary bile salts are not rehydroxylated but reconjugation occurs before secretion into the bile canaliculi. The small proportion of bile salts not extracted from portal blood enters the peripheral circulation where the concentration will reflect intestinal absorption and hepatic uptake of these molecules. The bile salts commonly found in serum are listed in Table I.

Serum concentrations of bile salts are low in normal subjects but show marked increases in hepatobiliary disorders and have potential use as a sensitive liver function test. However, early methods were unreliable and although GLC proved sufficiently sensitive the procedures are not suited to the large studies necessary to evaluate a potential liver function test. The problem was resolved when the first bile salt radioimmunoassay was reported by Simmonds et al.,[3] allowing sensitive assay of large numbers of samples. This chapter describes the procedures necessary to establish

TABLE I
SERUM BILE SALTS COMMONLY FOUND IN MAN

Trivial name	IUPAC name
Cholic acid[a]	$3\alpha,7\alpha,12\alpha$-Trihydroxy-5β-cholan-24-oate
Cholyltaurine	Tauro-$3\alpha,7\alpha,12\alpha$-Trihydroxy-5β-cholan-24-oate
Cholylglycine	Glyco-$3\alpha,7\alpha,12\alpha$-Trihydroxy-5β-cholan-24-oate
Chenodeoxycholic acid[a]	$3\alpha,7\alpha$-Dihydroxy-5β-cholan-24-oate
Chenodeoxycholyltaurine	Tauro-$3\alpha,7\alpha$-Dihydroxy-5β-cholan-24-oate
Chenodeoxycholylglycine	Glyco-$3\alpha,7\alpha$-Dihydroxy-5β-cholan-24-oate
Deoxycholic acid[a]	$3\alpha,12\alpha$-Dihydroxy-5β-cholan-24-oate
Deoxycholyltaurine	Tauro-$3\alpha,12\alpha$-Dihydroxy-5β-cholan-24-oate
Deoxycholylglycine	Glyco-$3\alpha,12\alpha$-Dihydroxy-5β-cholan-24-oate
Ursodeoxycholic acid[a]	$3\alpha,7\beta$-Dihydroxy-5β-cholan-24-oate
Ursodeoxycholyltaurine	Tauro-$3\alpha,7\beta$-Dihydroxy-5β-cholan-24-oate
Ursodeoxycholylglycine	Glyco-$3\alpha,7\beta$-Dihydroxy-5β-cholan-24-oate
Lithocholic acid[a]	3α-Hydroxy-5β-cholan-24-oate
Lithocholyltaurine	Tauro-3α-Hydroxy-5β-cholan-24-oate
Lithocholylglycine	Glyco-3α-Hydroxy-5β-cholan-24-oate
Sulfolithocholyltaurine[a]	Tauro-3α-Sulfo-5β-cholan-24-oate
Sulfolithocholylglycine[a]	Glyco-3α-Sulfo-5β-cholan-24-oate

[a] Sulfated and unconjugated bile salts are present in minimal concentrations in serum from subjects without hepatobiliary disease.

[3] W. J. Simmonds, M. G. Korman, V. L. W. Go, and A. F. Hofmann, Gastroenterology 65, 705 (1973).

and validate radioimmunoassays of bile salts and discusses potential future developments.

Production of Antisera

Bile salts have molecular weights below 1000 and are not immunogenic per se but, like steroids, can be covalently bonded to protein. This imparts immunogenicity and the antibodies produced may be specific for the bile salt haptens.

Preparation of Bile Salt–Protein Conjugates

Peptide bonds have generally been used for conjugation of steroids to proteins which requires the introduction of carboxyl groups to the steroid nucleus. However, bile salts, both conjugated and unconjugated, already contain a carboxyl group at the end of the side chain and so far all workers have used this group to form a covalent bond with the protein. Two methods have been used, based on the carbodiimide method of Goodfriend et al.[4] and the mixed anhydride method of Erlanger et al.[5]

Carbodiimide Method. This method uses a two-step reaction to form a peptide bond between the carboxyl group of the bile salt and the terminal amino group of lysine residues of bovine serum albumin (BSA).

Bile salt (14 μmol) and 30 μmol of 1-ethyl-3-(3-dimethylaminopropyl) carbodiimide hydrochloride (ECDI) are dissolved in 0.5 ml of 0.1 M sodium bicarbonate. BSA (40 μmol) dissolved in 1 ml of distilled water is added and the pH adjusted to 7 with 0.1 M potassium dihydrogen phosphate. ^{3}H-labeled bile salt (0.1 μCi, 50 pmol) is added to determine the conjugation efficiency and the solution is then stirred overnight at 20°. The carbodiimide–BSA conjugates are purified by dialysis at constant volume against 0.9% w/v sodium chloride, using an Amicon 8MC microultrafiltration cell. The membrane has a molecular weight limit of approximately 30,000. Dialysis is continued until no further radioactivity appears in the filtrate, after which an aliquot of the conjugate (5%) is used to determine the molar ratio of bile salt:BSA.

Mixed Anhydride Method. This more complex procedure also forms a peptide bond between the carboxyl group of bile salts and terminal amino groups of lysine residues of BSA. Bile salt (0.22 mmol) is dissolved in the minimum volume of 1,4-dioxane, using gentle reflux. After complete dissolution the solution is transferred to a cold-room and carefully cooled to 8°, ensuring that no crystallization occurs.

---

[4] T. C. Goodfriend, L. Devine, and G. D. Fasman, *Science* **144**, 1344 (1964).
[5] B. F. Erlanger, F. Borek, S. M. Beiser, and S. Lieberman, *J. Biol. Chem.* **228**, 713 (1957).

[3]H-labeled bile salt is again added, together with 0.22 mmol tri-n-butyl-amine and 0.22 mmol of isobutyl chloroformate and after stirring, the mixture is left for 20 min at 8°. BSA (4 μmol) is dissolved in 8 ml distilled water, mixed with 0.5 ml of 0.5 M sodium hydroxide and then 8 ml of 1,4-dioxane. After equilibration at 4° the bile salt solution is added dropwise to this albumin solution which is stirred continuously. During this addition the pH should be monitored and adjusted as necessary within the range 8.7–9.2. When all the bile salt solution is added the reaction is allowed to proceed for 1 hr at 4° and the pH again checked. After adjustment of the pH, if necessary, within the same range the solution is left stirring at 4° for a further 2 hr.

Purification of the conjugate is complicated in this case because the reaction mixture contains appreciable amounts of 1,4-dioxane. The effect of this solvent is minimized by dilution of the reaction mixture with four volumes of 0.9% sodium chloride, after which the solution can be concentrated to a convenient volume in the ultrafiltration cell before extensive dialysis as described for the carbodiimide method above. Alternatively, column chromatography has been used successfully to resolve protein-bound and unbound bile acid. After dilution with an equal volume of 0.01 M phosphate buffer (pH 7.4) the solution is applied to a column (20 × 1.5 cm) of Sephadex G-25 equilibrated with 0.01 M phosphate buffer (pH 7.4). The protein peak, determined by UV absorption of the column eluate, is found in the void volume. Other workers have used Bio-Gel P2 (Bio-Rad) and again the albumin–bile salt compound is recovered in the void volume.[3] As the small molecules used in the reactions but not conjugated to protein are nonimmunogenic it can be argued that the purification of bile salt–protein conjugates is unnecessary. Cowen and co-workers[6] successfully raised antisera to lithocholylglycine 3-sulfate without such purification steps.

Discussion. The carbodiimide method has been widely used for the conjugation of steroids and proteins and subsequently for bile acids and proteins. The reaction is carried out in aqueous medium and can give bile salt–protein molar ratios ranging from 10:1 at pH 7.4 to 25:1 at pH 5.4. However, our own studies have produced molar ratios of 5:1–8:1, below the range 8:1–30:1 used as immunogens for steroid immunoassays. This may be predicted on theoretical grounds, for the optimal pH of this reaction is pH 4–5 but bile acids show poor solubility at this pH and protein amino groups are not reactive below pH 8.[7]

The carbodiimide reaction has also been criticized as leading to cross-

---

[6] A. E. Cowen, M. G. Korman, A. F. Hofmann, J. Turcotte, and J. A. Carter, *J. Lipid Res.* **18,** 698 (1977).

[7] K. L. Carraway and D. E. Koshland, this series, Vol. XXV, p. 616.

linking within the albumin molecule and between albumin molecules,[8] which could reduce the molar ratio of bile acid: protein. However, while these strictures may explain our failure with this reaction it is difficult to reconcile with other workers' success.

Niswender and Midgley[9] found that steroid–protein conjugates containing less than 10 steroid molecules per mole of BSA produced antisera of low titer, which may explain the low titers reported by several groups where the molar ratio of bile salt: protein immunogen was about 12 : 1.[10,11] These reports indicated that the carbodiimide reaction in our hands produced immunogens which were unlikely to form high titer antisera.

The mixed anhydride reaction gave hapten protein ratios of 16 : 1 for cholylglycine and 25 : 1 for chenodeoxycholylglycine, and by repeating the reaction after dialysis of the cholylglycine–albumin compound the ratio increased to 20 : 1. Lithocholylglycine conjugation gives ratios consistently over 30 : 1. These hapten–albumin ratios, similar to those reported for steroids, indicate that about half of the 59 terminal amino groups of lysine residues can be conjugated to bile acids by this method.

The advantages of using an aqueous medium and smaller quantities of materials make the carbodiimide reaction attractive. To overcome the problems of using inappropriate pH conditions we have reversed the normal approach by synthesizing a bile acid which terminates with a primary amine (see Radioactive Tracer, ^{125}I Tracers) and conjugating this with carboxyl groups in the albumin molecule.[12] These "bile amines" are readily soluble at pH 4–5 and give hapten–albumin ratios in excess of 55 : 1 with the carbodiimide reaction described. Glycine-conjugated bile acids may be converted in the same way to terminate in a primary amine group and such derivatives also yield hapten–albumin ratios in excess of 55 : 1. However, these high molar ratio complexes failed to produce antibodies to the hapten when used as immunogens. Hapten–albumin ratios of 25 : 1 to 30 : 1 appear to be optimal for production of antibodies while ratios below 10 : 1 and exceeding 36 : 1 fail to produce antibodies. This question is discussed in detail in a recent publication from this laboratory.[12]

The Site of Conjugation. The entire literature of bile acid radioimmunoassay has utilized the natural group, the carboxyl at C_{24} or the carboxyl of the glycine conjugated to the C_{24} carboxyl. Consequently no informa-

[8] V. H. T. James and S. L. Jeffcoate, *Br. Med. Bull.* **30**, 50 (1974).
[9] G. D. Niswender and A. R. Midgley, *in* "Immunologic Methods in Steroid Determination" (F. G. Peron and B. V. Caldwell, eds.), p. 149. Appleton, New York, 1970.
[10] S. Matern, R. Krieger, and W. Gerok, *Clin. Chim. Acta* **72**, 39 (1976).
[11] A. A. Mihas, J. G. Spenney, B. I. Hirschowitz, and R. G. Gibson, *Clin. Chim. Acta* **76**, 389 (1977).
[12] A. Hill, P. E. Ross, and I. A. D. Bouchier, *Steroids* **37**, 393 (1981).

tion is available to define the importance of the site of conjugation although some general comments can be made by extrapolation from data provided by steroid radioimmunoassays. Steroid–protein conjugates were frequently prepared by reaction of the groups at C_3 or C_{17}, the A or D ring of the steroid. It was soon noticed that the specificity of antisera produced was greatest for the part of the molecule furthest from the protein, so that steroids differing only in D ring groups showed high cross-reactivity for antisera prepared with C_{17} conjugates.[13] In the light of this experience bile acid molecules covalently bound to protein through the glycine molecule at C_{24} of conjugated bile acids would be expected to produce antisera which cannot discriminate readily between glycine and taurine conjugates. This prediction is largely supported by results, for most antisera determine glycine and taurine conjugates equally.[3,14–16]

This characteristic is advantageous for the methods using such antisera report total conjugated bile salts which in most cases is equivalent to the total bile salt present. This point will be discussed further (see Accuracy and Precision, Accuracy). However, Demers and Hepner[17] have produced antisera which are relatively specific for glycine conjugates (20% cross-reactivity with taurine conjugates) while Mäentausta and Jänne[18] found marked differences between antisera and Mihas et al.[11] reported different cross-reactivity at different antiserum dilutions. This suggests that some antisera may be produced which detect differences in the molecule at the site of conjugation to protein. Reference to the steroid radioimmunoassay literature shows an excellent example of sensitivity to the site of conjugation when Gilby and Jeffcoate[19] reported affinity constants of 0.3, 1.6, and 5.9×10^9 liter/mol for testosterone, testosterone oxime, and testosterone oxime-ϵ-lysine, respectively.

Bile acids are characterized by hydroxyl groups on the A, B, and C rings and of course the D ring side chain carboxyl group, and sites for conjugation apart from the carboxyl group are restricted. As all bile salts have a 3α-hydroxyl conjugation through this group might impart specificity for glycine or taurine conjugates without loss of specificity for the 7α- or 12α-hydroxyls. At present this approach has not been reported.

[13] S. L. Jeffcoate, in "Radioimmunoassay Methods" (K. E. Kirkham and W. M. Hunter, eds.), p. 151. Churchill, Edinburgh, 1971.
[14] G. M. Murphy, S. M. Edkins, J. W. Williams, and D. Catty, Clin. Chim. Acta 54, 81 (1974).
[15] S. W. Schalm, G. P. van Berge Henegouwen, A. F. Hofmann, A. E. Cowen, and J. Turcotte, Gastroenterology 73, 285 (1977).
[16] Y. A. Baqir, J. Murison, P. E. Ross, and I. A. D. Bouchier, J. Clin. Pathol. 32, 560 (1979).
[17] L. M. Demers and G. Hepner, Clin. Chem. 22, 602 (1976).
[18] O. Mäentausta and O. Jänne, Clin. Chem. 25, 264 (1979).
[19] E. D. Gilby and S. L. Jeffcoate, J. Endocrinol. 57, xlvii (Abstr.), (1973).

Immunization

Although experience with bile acid protein immunogens is limited most reports indicate the rabbit as the preferred animal. The reason for this choice is probably a balance between the volume of serum which can be obtained and the cost per animal, because there are few instances where the species is important.[20] Each group of workers appears to use individual immunization schedules; the only criterion by which these schedules are judged is by production, or otherwise, of useful antisera. No comparisons have been made between different immunization schedules, different proteins conjugated to the bile salt hapten or preparation of emulsion. The main reason for this is the variability of response in different animals which indicates that many animals would be required to resolve this problem. Consequently the methods of immunization seem likely to remain something of an art although a few indicators can be derived from the limited studies of this nature reported by steroid biochemists. A study by Walker *et al.*[21] suggested that the optimal immunogen is a conjugate with bovine serum albumin while the best adjuvant studied was Freund's complete adjuvant. Freund's complete adjuvant, a mixture of mineral oil and emulsifier "complete" with suspension of mycobacterium butyricium, is almost universally used as the vehicle for immunization. This is true for all reports of radioimmunoassay of bile salts. There are several procedures available for preparation of the emulsion, well reviewed by Herbert.[22] By trial and error we have found the double syringe method[23] the simplest and most reliable where comparatively small amounts of emulsion are to be prepared. A Silverson rotary emulsifier was efficient but wasteful of immunogen, in our experience.

The immunogen is diluted to 1 mg/ml of protein with 0.9% sodium chloride and placed in one Luer-lock syringe. Adjuvant (1–1.5 vol) is placed in the other syringe and the two syringes are joined by a double-hubbed needle, taking care to exclude all air. (The combined volume of adjuvant and immunogen must not exceed the capacity of either syringe.) The aqueous immunogen is then passed into the adjuvant and the mixture passed to and fro to form a water-in-oil emulsion. The emulsion should be checked by allowing a few drops to fall onto a beaker of cold water. Satisfactory emulsions remain as an intact drop; if the drop breaks up showing an oil-in-water emulsion it is ineffective for immunization.[24] Where infor-

[20] B. A. L. Hurn, *Br. Med. Bull.* **30**, 26 (1974).
[21] C. S. Walker, S. J. Clark, and H. H. Wotiz, *Steroids* **21**, 259 (1973).
[22] W. J. Herbert, *in* "Handbook of Experimental Immunology" (D. M. Weir, ed.), p. 1207. Blackwell, Oxford, 1967.
[23] B. S. Berlin and R. W. McKinney, *J. Clin. Invest.* **52**, 657 (1958).
[24] J. Freund, *Annu. Rev. Microbiol.* **1**, 291 (1947).

mation is given, bile salt immunogens appear to have been mixed 1 : 1 with adjuvant. This is more difficult to form into a satisfactory emulsion and consequently we use a ratio of approximately 1.5 adjuvant : 1 immunogen. This is less than the ratio of 2–3 : 1 suggested by Hurn[20] but we have not tried ratios as high as this. We have adopted the procedure of Jeffcoate *et al.*[25] based on an original proposal of Vaitukaitis *et al.*[26] for administration to rabbits. The guiding principle behind this decision was the requirement for small amounts of immunogen together with rapid production of antibodies. The primary immunization is given as 10 intradermal immunizations, each of 0.2 ml at 5 sites on each flank of the rabbit. Although this procedure may not respond to booster immunizations[20] 1 ml immunizations in the magnus/adductor muscle of each thigh raised the titer (Fig. 1). Possibly this difference relates to the comparatively low titer obtained for

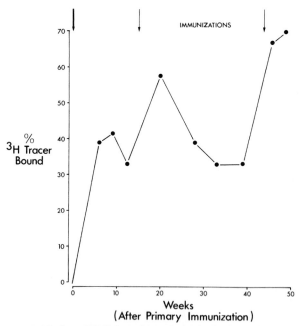

FIG. 1. Changes in binding of [^{3}H]chenodeoxycholylglycine by antiserum during immunization. Antiserum dilution is 1 : 7000. The heavy arrow indicates primary immunization and lighter arrows booster immunizations. From J. Murison, M.Sc. Thesis, University of Dundee, 1978.

[25] S. L. Jeffcoate, D. T. Holland, H. M. Fraser, and A. Gunn, *Immunochemistry* **11**, 75 (1974).

[26] J. Vaitukaitis, J. B. Robbins, E. Nieschlag, and G. T. Ross, *J. Clin. Endocrinol. Metab.* **33**, 988 (1971).

bile salt antisera compared with steroid antisera. An improvement of 10% in binding is important at a titer of 1 : 800 but is less valuable at 1 : 100,000. Certainly other reports confirm the value of booster immunization for bile salts.[3,11]

Collection

Blood removed from a marginal ear vein 6 weeks after the primary immunization usually shows the presence of antibodies. Antisera to glycochenodeoxycholate, glycocholate, glycoursodeoxycholate, and glycolithocholate are always produced, although the titer varies. Antisera to glycodeoxycholate have proved less reliable in our experience and at present, where rabbits have produced antisera to this particular bile salt, the cross-reactivity to cholylglycine has been too high for use in radioimmunoassay. Others have been more successful.[10,17] Antibody production shows a rapid rise after 6 weeks but may plateau or fall after 9–12 weeks. A booster immunization will usually increase the titer. Blood samples are best removed 10–14 days after booster immunization. The changes in binding of [³H]chenodeoxycholylglycine by 1 : 7,000 dilution of antisera are shown in Fig. 1. These changes are peculiar to each rabbit and therefore administration of booster immunizations are best determined by monitoring titer every week. After a fall in titer booster immunization shows an increase of 20% binding in the example shown. This increase seems most marked in young rabbits and reduces about 1 year after primary immunization.

Purification of Antisera

Antisera to low-molecular-weight compounds probably contain heterogeneous antibodies and may also contain antibodies to the protein used for conjugation. Murphy et al.[14] used polymerized BSA to remove antibodies to BSA formed together with antibodies to cholylglycine. No other report indicates an attempt to purify the antibodies and, while this approach is satisfactory when titer, affinity and specificity are adequate, it is possible that purification might provide one fraction of the antibody population with improved characteristics for the assay.

Purification of antisera is considered important where solid-phase radioimmunoassay is used, as the purification will maximize the binding of relevant antibodies to the solid-phase. However, Van den Berg et al.[27] conjugated antibodies to Sepharose (Pharmacia) without purification and developed a successful assay. The assay may have been improved if anti-

[27] J. W. O. van den Berg, M. van Blankenstein, E. P. Bosman-Jacobs, M. Frenkel, P. Horchner, O. I. Oost-Harwig, and J. H. P. Wilson, *Clin. Chim. Acta* **73**, 277 (1976).

bodies were purified before conjugation but no data are available for comparison.

Storage of Antisera

Antisera to bile salts have not shown any change in affinity or titer when stored in aliquots at $-20°$ for periods of up to 2 years. We have no data for periods exceeding this length of time. Published reports of bile salt radioimmunoassays include one report where antisera is freeze-dried for storage[17] but no information was given to indicate stability of antisera stored in this way. However, care may need to be taken as there is some evidence that antisera deteriorate with time when freeze-dried.[28] This factor may be particularly relevant as RIA kits for bile salts become more widely used.

Development of Assay

Characterization of antisera cannot be properly carried out until the optimal conditions for incubation and separation of antibody-bound antigen and free antigen have been determined. Separation of antibody-bound and free steroids includes adsorption of free antigen by dextran-coated charcoal. This was unsuccessful when first applied to bile salt radioimmunoassay[3] and this procedure has been used only once for bile salt assay.[29] The alternative, simple, approach involves fractional precipitation of the antibody-bound fraction, in which case the concentration of γ-globulin is important. Simmonds et al.[3] added γ-globulin by using charcoal-extracted serum (bile salt free serum) in the incubation while Murphy et al.[14] simply added porcine γ-globulin to the incubation buffer. In the absence of a clear choice between the two methods we assessed both using ammonium sulfate and polyethylene glycol 6000 (PEG) for precipitation.

Incubation Buffer

Charcoal-Extracted Serum. Although methods for the preparation of charcoal-extracted serum may vary in minor detail all are essentially the same.[3] Bile acids are removed from 50 ml of pooled serum with approximately 2.5 g of activated charcoal (Sigma Chemical Co. untreated powder). After 30 min thorough mixing and centrifugation (10,000 g for 20 min) the process is repeated twice and the serum filtered through a Millipore membrane (0.22 μm). In common with other reports, this procedure

[28] Y. Y. Al-Tamer, A. Taylor, and V. Marks, *Clin. Chem.* **23**, 2175 (1977).
[29] J. G. Spenney, B. J. Johnson, B. I. Hirschowitz, A. A. Mihas, and R. Gibson, *Gastroenterology* **72**, 305 (1977).

removed 95% of radioactive tracer bile salt. Charcoal could be replaced by a resin such as XAD 7 (Rohm and Hass U.K. Ltd) which has been used to extract bile salts from serum[30] and may prove easier to use than charcoal.

The incubation used with this source of protein comprises 0.1 ml charcoal-extracted serum (diluted 1:2 with buffer), 0.1 ml of radioactive tracer, 0.1 ml of antiserum (suitably diluted), and 0.7 ml of 0.01 M phosphate buffer, pH 7.4. A change in the amount of protein is effected by addition of different dilutions of charcoal-extracted serum. Total radioactivity added is determined by substituting buffer for antiserum so that no binding occurs. This is the same as counting 0.1 ml tracer providing nonspecific binding of tracer bile salt is negligible.

Albumin-γ-Globulin Buffer. The incubation differs from that described above in that the charcoal-extracted serum is omitted and the buffer (0.01 M phosphate buffer pH 7.4) contains 0.1% w/v bovine serum albumin, 0.9% w/v sodium chloride, and 0.03% w/v sodium azide. Various amounts of γ-globulin are dissolved in the buffer so that 0.5 ml of buffer contain from 50 to 500 μg of γ-globulin to determine the optimal concentration. The incubation in this case comprises 0.3 ml antisera (diluted appropriately with buffer) and 0.2 ml of radioactive tracer also diluted with buffer (0.05 μCi/ml). In the absence of sample 0.1 ml of phosphate buffer is added to give an incubation volume of 0.6 ml. Total radioactivity is again determined by substitution of buffer for antiserum provided that the porcine γ-globulin does not bind significant amounts of tracer bile salt.

Precipitation Procedure

The two well-established precipitation methods for separation for free and bound bile salts are saturated ammonium sulfate and PEG. The latter procedure[31] with two exceptions[14,16] has been the method of choice for bile salt radioimmunoassays.

After incubation 0.5 ml of a 37.5% w/v aqueous solution of PEG is added to the incubation (1.0 ml) to give a final concentration of 12.5% w/v PEG in a total volume of 1.5 ml. This addition must be carried out at 4° and consequently if incubation is carried out at a different temperature, time must be allowed for equilibration at 4° before precipitation.

Addition of saturated ammonium sulfate (0.6 ml) to the incubation volume (0.6 ml in this case) gives a 50% saturated solution of ammonium sulfate which precipitates the bound fraction. On rare occasions the pH of saturated ammonium sulfate is below 4.5 in which case precipitation is

[30] S. Barnes and A. Chitranukroh, *Ann. Clin. Biochem.* **14**, 235 (1977).
[31] B. Desbuquois and G. D. Aurbach, *J. Clin. Endocrinol. Metab.* **33**, 732 (1971).

unreliable. We routinely check the pH of saturated ammonium sulfate solutions before use.

Incubation

The conditions required to establish equilibrium between antibody-bound and free bile salt vary dramatically between different laboratories[14,17] and must therefore be determined for each antiserum.

A preliminary experiment must be carried out to determine an antiserum dilution which gives about 50% binding under conditions selected from the literature. Incubation conditions can then be studied systematically by varying the γ-globulin concentration, temperature, and time.

The use of different concentrations of γ-globulin in incubation showed an unexpected divergence between the two precipitants. Use of PEG required 0.2 mg γ-globulin/tube for optimal precipitation without nonspecific binding, while ammonium sulfate precipitation required only 50 µg/tube. Another observation from this study was that binding to antisera was apparently lower at the same dilution when precipitation was by PEG rather than ammonium sulfate (Table II). These observations were also noted when an anti-chenodeoxycholylglycine serum was assessed.

Investigation of the length of time required for incubation showed that after 1 hr there is no increase in binding with either precipitant and a similar study at 4°, room temperature and 37° showed that temperature was unimportant with our antisera (Table III).

The combination of reduced binding and increased requirement for γ-globulin led us to adopt ammonium sulfate precipitation. Bile salt free

TABLE II
EFFECT OF γ-GLOBULIN CONCENTRATION ON ANTIBODY BINDING USING PEG AND SATURATED AMMONIUM SULFATE PRECIPITATION[a,b]

Concentration of γ-globulin (µg/assay tube)	PEG		Saturated (NH₄)₂SO₄	
	% Binding	Nonspecific binding (%)	% Binding	Nonspecific binding (%)
25	—		68.2	1.5
50	57.4	0	70.6	0.9
100	63.0	2.2	74.4	4.7
200	64.3	2.8	75.0	6.0
400	66.1	4.0	76.0	7.0
500	65.8	3.6	—	—

[a] J. Murison, M.Sc. Thesis, University of Dundee, 1978.
[b] The cholylglycine antiserum dilution is 1:400 and [³H]cholylglycine added per tube is 3 pmol (7000 dpm). Nonspecific binding is the percentage ³H label precipitated in the absence of antiserum.

TABLE III
EFFECT OF TEMPERATURE AND TIME ON BINDING BY
CHOLYLGLYCINE ANTISERUM, USING AMMONIUM SULFATE
PRECIPITATION FOR SEPARATION OF ANTIBODY-BOUND AND
FREE BILE SALT[a]

Time	[³H]Cholylglycine bound (%)		
	4°	Room temperature	37°
30 min	52.9	54.9	54.0
1 hr	55.1	55.7	54.3
2 hr	54.7	54.6	55.7

[a] J. Murison, M.Sc. Thesis, University of Dundee, 1978.

serum has not been tried with PEG under our conditions but was compared with albumin buffer containing 0.05 mg γ-globulin/tube using ammonium sulfate precipitation. There was no difference and since albumin γ-globulin buffer is simple to prepare, this system is used routinely.[16]

The differences in conditions for incubation and separation reported for radioimmunoassay of bile salts show clearly the importance of optimizing the conditions for each antiserum. Failure to do this may mean use of antisera at a low titer which can be improved fairly simply. For example a preliminary study indicated 50% binding with 1:300 dilution of anticholylglycine serum but this improved to 50% binding at 1:750 dilution with subsequent refinement of conditions for equilibration and separation of antibody bound and free bile salt.[32]

Characterization of Antisera

After definition of the assay conditions and method of separation of free and bound bile salt, antisera must be assessed for titer, affinity, and specificity.

Titer

This is the dilution of antisera required to give 50% binding of radioactive tracer and is dependent on the conditions of assay, mass of tracer, etc. Therefore comparison of antisera titers can be made only where conditions are identical, which usually means within the same laboratory. However, the significance of titer is limited because titer indicates the total number of assays possible from a sample of antiserum. Antisera performance can only be assessed by the affinity and specificity.

[32] J. Murison and P. E. Ross, Unpublished (1977).

Affinity

Reports of bile salt radioimmunoassays, including our own, have failed to define the affinity constants for the antisera used. The most common procedure for determination of antibody affinity is by the Scatchard plot based on the equation $R = K (q - B)$ where R is the ratio of free:bound labeled antigen, K is the affinity constant, q is the concentration of antigen binding sites, and B is the concentration of bound antigen. When R is plotted against B the graph is linear with slope $-K$ and x intercept $= q$ provided the antibody contains a single binding site and only one ligand is present.[33] A mathematical model for two binding sites with one ligand produces a curve from which asymptotes give affinity constants for the two binding sites.[34]

Scatchard plots of our standard curves were linear over the range of acceptable precision (see Response Curves) and gave the affinity constants and molar concentrations of antibody binding sites shown in Table IV.

Although the affinity constant allows mathematical prediction of assay conditions we approached the problem practically. Ekins[35] has shown mathematically that the greatest sensitivity is found using antibody concentration of $3/K$ (which corresponds to 50% binding) and isotope concentration of $4/K$ provided various assumptions are valid. We calculated the amount of isotope required to provide adequate precision of counting and then determined the antisera dilution required to give 50% binding. This approach does not require knowledge of the affinity constant.

TABLE IV

AFFINITY CONSTANTS AND MOLAR CONCENTRATIONS OF ANTIBODY BINDING SITES DERIVED BY SCATCHARD PLOT[a]

Antisera	Affinity constant (liters/mol)	Concentration of antibody binding sites/assay (nmol/liter)
Cholylglycine	1×10^8	14.8
Chenodeoxycholylglycine	6.6×10^8	5.7
Lithocholylglycine	0.8×10^8	16.0
Ursodeoxycholylglycine	7.9×10^8	3.5

[a] G. Scatchard, *Ann. N.Y. Acad. Sci.* **51,** 660 (1949).

[33] H. A. Feldman and D. Rodbard, in "Principles of Competitive Binding Assays" (W. D. Odell and W. H. Daughaday, eds), p. 158. Lippincott, Philadelphia, Pennsylvania, 1971.
[34] H. A. Feldman, *Anal. Biochem.* **48,** 317 (1972).
[35] R. P. Ekins, in "Hormone Assays and Their Clinical Application" (J. A. Loraine and E. T. Bell, eds.), p. 1. Churchill, Edinburgh, 1976.

When the affinity constant is low separation procedures which adsorb free antigen (e.g., dextran-coated charcoal) alter the equilibrium and cannot be used,[36] which probably explains why Simmonds et al.[3] found charcoal adsorption to be unreliable. Knowledge of the affinity constant may help predict such interference and avoid the problem but careful evaluation of the method should demonstrate inaccuracies caused by stripping bile salt from the antibody binding sites.

Synthesis of ^{125}I-labeled bile salts produces high specific activity compounds which allow addition of small masses of tracer while retaining good counting precision. This does not alter the binding affinity per se but the molecular changes made to introduce ^{125}I may effect a change. Chemical modifications (see next section) to the carboxylic acid group of the glycine conjugate may alter the affinity constant if the antibody recognizes the bridge joining bile salt to protein in the immunogen. Spenney et al.[29] found that displacement of [^{125}I]cholylglycylhistamine was most sensitive to cholylglycylhistamine and the concentrations of cholylglycine were approximately 10 times greater to produce comparable displacement. Cross-reactivities were different for glycine and taurine conjugates, supporting the view that their antibody recognizes the bile salt–protein bridge.

The antiserum used for our ^{125}I assay for conjugated ursodeoxycholic acid does not discriminate between glycine and taurine conjugates and the affinity constant is the same for both ursodeoxycholylglycine and N,1-ursodeoxycholylglycyl-N,2(4-hydroxybenzoyl)-1,2-diaminoethane.

When the assay was established at 50% binding the standard curve was linear over the range 1–5.0 pmol. This narrow range is not helpful when sample concentrations vary by as much as 200×, for dilution is critical if the sample is to produce data within this narrow linear range. By increasing the antibody concentration to give a zero antigen binding of 80% the useful range of the standard curve was extended to cover 1–40 pmol/assay tube. With the combination of antibody concentration and labeled antigen concentration used in this assay the standard curve of percentage bound against displacing antigen concentration showed a reduced slope at low concentrations, causing reduced precision in this region. This is not important under the conditions used as this reduced precision occurs at concentrations below those found in serum samples. The consequences of manipulation of assay conditions are developed theoretically in the excellent review by Feldman and Rodbard[33] and should allow precise development of the assay conditions to suit particular analytical problems.

---

[36] F. J. Auletta, B. V. Caldwell, and G. L. Hamilton, in "Methods of Hormone Radioimmunoassay" (B. M. Jaffe and H. R. Behrman, eds.), p. 359. Academic Press, New York, 1974.

Specificity

Bile salt radioimmunoassays do not use chromatographic purification of samples prior to assay and specificity is therefore solely dependent on the specificity of the antigen–antibody reaction. If structurally similar compounds compete for antibody binding sites they are said to cross-react and will lead to falsely high measurements. This is assessed by addition of increasing amounts of such compounds under the defined assay conditions and determination of the percentage tracer bound to antibody. This allows presentation of cross-reactivity as what is effectively a standard curve for each cross-reacting compound. An alternative presentation requires determination of the mass of cross-reacting compound required to displace 50% of the tracer bound to the antibody, again under assay conditions. Cross-reactivity (%) is then expressed as $x/y \times 100$ where x is the mass of specific bile acid and y the mass of interfering compound required to displace 50% of bound tracer. Although this latter method gives a precise figure for cross-reactivity it must be remembered that cross-reactivity assessed in this way relates to 50% displacement of tracer and will be different at any other displacement.

In the absence of chromatographic purification prior to radioimmunoassay cross-reactivities of bile acids must be carefully assessed. Other molecules with a steroid nucleus do not seem to cross-react but it is important that the concentration in serum is considered. For example, unesterified cholesterol may be present at concentrations exceeding that of cholylglycine by a factor of 2000 and cross-reactivity must be assessed at concentrations appropriate to this excess.

It is also important that cross-reactivities are studied under the conditions to be used for the assay. Mihas *et al.*[11] found that cross-reactivities changed markedly as the antiserum dilution was increased from 1 : 500 to 1 : 700, an observation difficult to explain on theoretical grounds. Other reports have not commented on this phenomenon, which may indicate that other workers have only studied cross-reactivity at the dilution used for assay. Our own antisera do not change cross-reactivity appreciably at different dilutions. The nature of precipitant may also influence cross-reactivity, for Spenney *et al.*[29] found cross-reactivity of 26% for glycochenodeoxycholate using dextran-coated charcoal to adsorb the free antigen. With PEG precipitation of antibody-bound antigen the cross-reactivity increased to 92% . Other bile salts also showed marked increases in cross-reactivity which the authors attribute to the high affinity binding by charcoal which "strips" bile salt from the low-affinity binding sites of the antibody (see Antisera Affinity). Although we assessed both saturated ammonium sulfate and PEG for separation of free and bound antigen, cross-reactivities were only determined under normal assay conditions with ammonium sulfate precipitation. Values are reported in Table V.

TABLE V
CROSS-REACTIVITIES[a]

Cross-reacting compound	Antiserum			
	Cholylglycine	Chenodeoxycholylglycine	Lithocholylglycine	Ursodeoxycholylglycine
Cholylglycine	100	<1	<1	<1
Cholic acid	19	<1	<1	<1
Chenodeoxycholylglycine	7	100	5	<1
Chenodeoxycholic acid	<1	10	<1	<1
Deoxycholylglycine	5	<1	1	<1
Deoxycholic acid	<1	<1	<1	<1
Ursodeoxycholylglycine	<1	1	2	100
Ursodeoxycholic acid	<1	<1	<1	10
Lithocholylglycine	1	1	100	<1
Lithocholic acid	<1	<1	5	<1
Cholesterol	ND[b]	ND	ND	ND
Progesterone	ND	ND	ND	ND
Estradiol	ND	ND	ND	ND

[a] Our antisera do not discriminate between glycine and taurine conjugates.
[b] ND, Not detected.

Radioimmunoassay reports agree that antisera to bile salts are specific for the bile salt structure but show variable specificities toward other bile salts, both conjugated and unconjugated. Our antisera do not discriminate between glycine and taurine conjugates and the major cross-reactivity is found for the unconjugated bile salt. Cholylglycine antisera shows the poorest specificity with cross-reactivities of the conjugated bile salts ranging from 10% for glycochenodeoxycholate to 1% for glycolithocholate. Specificities for the dihydroxy bile salts and monohydroxy bile salts are generally better than for cholate although so far we have not produced an antisera for deoxycholylglycine with acceptable cross-reactivities. In general, cross-reactivities reported by other workers show considerable variations and it is therefore important that cross-reactivity is carefully examined. A high cross-reactivity may not preclude use of the antisera but the implications to results of the assay must be considered. This is considered further in relation to the accuracy of radioimmunoassay (see Accuracy and Precision, Accuracy).

Addition of serum to the radioimmunoassay medium may lead to interference from molecules in the serum which alter the equilibrium without competing for antibody binding sites. This is usually investigated by assaying serial dilutions of a serum sample. If concentrations in serum are not constant after correction of the dilution factor then purification of the sample will be necessary. Two methods[14,18] have reported increasing "concentrations" of glycocholate as the serum samples were progressively diluted. These authors refluxed 1 ml of serum gently with 19 ml of ethanol and, after cooling, the precipitated proteins were sedimented by centrifugation at 3000 g for 15 min. The supernatant was dried *in vacuo*, dissolved in 1 ml of ethanol and 0.1-ml aliquots taken to dryness and then dissolved in buffer for assay. This procedure resolved the interference which Murphy *et al.*[14] attribute to competitive binding of bile salts to both albumin and antibody. This suggests that a low-affinity antibody is being used by these authors. Precipitation of protein in serum did not effect the concentrations of glycocholate or glycochenodeoxycholate measured by radioimmunoassay while dilution of serum samples gives a constant concentration of bile salt after correction for dilution.[32] In the glycoursodeoxycholic acid assay samples must be diluted 1 : 10 or more if this linearity is to be maintained, but samples require this dilution for assay on the precise region of the standard curve (see Response Curves).

Apart from ourselves[16] and Murphy *et al.*[14] other bile salt radioimmunoassays have used charcoal-extracted serum for preparation of samples. With our antisera there was no difference between use of charcoal-extracted serum and the use of buffer containing albumin and γ-globulin. It must, however, be borne in mind that study of a few samples which show no dilution effect does not indicate that all samples are free of nonspecific interference. Ideally, all samples should be analyzed at two dilutions to preclude this effect. Specificity should ideally be studied by comparison of radioimmunoassay with another assay such as GLC. Unfortunately normal fasting levels of individual bile salts are close to the detection limits of GLC methods and few laboratories have compared the two methods. During development of the radioimmunoassays all have been carefully compared with the GLC method also developed in this department[37] and good agreement shown between both methods. (see Accuracy and Precision, Accuracy).

Radioactive Tracer

[3]H Tracers

The primary ^{3}H-labeled bile salts conjugated to glycine are readily available commercially, although the specific activities are low (2–5 Ci/mmol). This does not necessarily restrict precision if the radioactive dose is calculated, allowing for binding and dilutions etc., to give acceptable counting precision. We add approximately 7000 dpm which gives a coefficient of variation of 4% with a counting time of 10 min, when the binding is 55% with zero nonradioactive antigen. Any increase in the dpm added initially would allow reduction of the counting time but would not significantly alter the assay performance because, as counting precision improves, assay precision is limited by experimental error. However, the low specific activity of bile salt tracers means that the mass of tracer added is 1–3 pmol, figures approaching the lowest points on the standard curves, and optimal antiserum dilution is low with such large masses of tracer. This can be improved by iodination of the bile salt molecule by a procedure similar to that for steroids. This procedure can, of course, be applied to bile salts which are not commercially available as ^{3}H tracers, extending the range of bile salts which can be assayed.

[125]I Tracers

We have prepared ^{125}I-labeled bile salts by the following procedure.[12] Other methods have been reported.[18,29,38]

Ursodeoxycholylglycine (2 mmol) and tributylamine (2 mmol) are dissolved in 10 ml of 1,4-dioxane, cooled to 10° and mixed with isobutyl-chloroformate for 15–20 min. Diaminoethane (2 ml; 36.8 mmol) dissolved in water is added quickly and the mixture stirred vigorously for 1 hr. Evaporation at 40° *in vacuo* leaves an oily residue which is mixed with

[37] P. E. Ross, C. R. Pennington, and I. A. D. Bouchier, *Anal. Biochem.* **80**, 458 (1977).
[38] G. J. Beckett, J. E. T. Corrie, and I. W. Percy-Robb, *Clin. Chim. Acta* **93**, 145 (1979).

10 ml 0.1 M NaOH and extracted three times with diethyl ether to remove the tributylamine. The aqueous layer is then extracted with 2 × 5 ml n-butanol and the combined extracts reduced in volume to 0.5 ml approximately. The crude product is precipitated by addition to diethyl ether and, after filtration and drying, is redissolved in M HCl (10 ml). This acidic solution is extracted several times with ethyl acetate, made alkaline with 5 M NaOH, and the diaminoethane derivative extracted into n-butanol (2 × 5 ml). The combined extracts are again reduced in volume to 0.5 ml and after precipitation with diethyl ether the product is collected by filtration. The yield of N,1-ursodeoxycholylglycyl-1,2-diaminoethane is 75%.

The mixed anyhydride of 4-hydroxybenzoic acid is then prepared by dissolving 250 μmol of the acid and 500 μmol of tributylamine in 5 ml of 1,4-dioxane. After cooling to 10°, 250 μmol of isobutylchloroformate are added and the mixture stirred for 15 min before addition of 250 μmol of N,1-ursodeoxycholylglycyl-1,2-diaminoethane in 5 ml of ethanol. This addition is carried out with vigorous stirring. After 2 hr continuous stirring 5 ml of M NaOH are added and the tributylamine extracted with 2 × 5 ml of diethyl ether. The aqueous layer is made acid with 5 ml of 2 M HCl and the product extracted with 2 × 5 ml of ethyl acetate. After drying *in vacuo* the product is purified by TLC, using the systems ethyl acetate : methanol : toluene : triethylamine, 50 : 25 : 25 : 10 (a) and ethyl acetate : methanol : toluene : glacial acetic acid 50 : 25 : 25 : 10 (b).

System (a) resolves 2 bands, shown by ultraviolet (UV) absorption (4-hydroxybenzoic acid) and phosphomolybdate (bile acid). The band with higher R_f (comprising 4-hydroxybenzoic acid and bile salt-4-hydroxybenzoic acid derivative) chromatographed in system (b) shows 2 bands by UV absorption but only one band with phosphomolybdate reagent. After elution of this band with ethanol and concentration to 0.1 ml (approximately) the product, N-1-ursodeoxycholylglycyl-1,2-diaminoethane-N-2,-4-hydroxybenzoic acid is obtained by precipitation with diethyl ether. The yield is 16%.

Iodination

N,1 - Ursodeoxycholylglycyl - N,2(4 - hydroxybenzoyl) - 1,2 - diaminoethane (20 μg in 10 μl of ethanol) is added to 10 μl of 0.5 M phosphate buffer (pH 7.4) and 5 μl (500 μCi) of Na^{125}I solution (as supplied by the Radiochemical Centre, Amersham, U.K.). Chloramine-T (5 μg) in 5 μl of water is added and, after stirring for 30 sec, the reaction is stopped by addition of 100 μg of sodium metabisulfite in 100 μl of 0.5 M phosphate buffer (pH 7.4). Sodium iodide (100 μg in 1.0 ml 0.5 M phosphate buffer) is added and the solution transferred to a stoppered glass tube. After extraction with 5 ml ethyl acetate the organic layer is dried with 1 g of anhydrous sodium sulphate and then evaporated to dryness *in vacuo*. The residue is dissolved in 100 μl ethanol and the iodinated derivative separated by TLC using a system comprising ethyl acetate:methanol:toluene:triethylamine, 50:25:25:10. A radiochromatogram scanner shows a single peak (R_f 0.25) well separated from the non-iodinated compound (R_f 0.5). Elution of this radioactive zone with 10 ml ethanol yields the iodinated compound showing 75% incorporation of [125]I. This method is a modification of that used by Hunter.[39]

The iodinated bile acid derivative is stable when stored at room temperature in this ethanolic solution and the dilute solution (30,000 dpm/0.1 ml) is stable when prepared in the albumin γ-globulin buffer. Fresh tracer is prepared after two half-lives of [125]I.

Assay Procedure

[3]H Tracers

Antiserum is diluted with 0.01 M phosphate buffer (pH 7.4) containing 0.3 g/liter sodium azide, 1 g/liter bovine serum albumin, and 0.25 g/liter porcine γ-globulin.

0.1 ml sample + 0.3 ml ^{3}H-labeled bile salt + 0.2 ml antiserum
Incubate 1 hr at room temperature
Precipitation with 0.6 ml saturated ammonium sulfate (45 min at 4°)
Centrifugation of precipitate (1500 g for 30 min at 4°)
Liquid scintillation counting: 0.5 ml supernatant + 4.5 ml NEN BIOFLUOR

[125]I Tracers

Antiserum, tracer and samples are diluted with 0.01 M phosphate buffer containing 0.3 g/liter sodium azide, 3.3 g/liter albumin, and 1.6 g/liter porcine γ-globulin.

0.1 ml sample + 0.1 ml ^{125}I-labeled bile salt + 0.2 ml antiserum
Incubate 1 hr at room temperature
Precipitation by addition of 0.4 ml saturated ammonium sulfate (45 min at 4°)
Centrifugation of precipitate (1500 g for 30 min at 4°)
Count 0.5 ml of supernatant

---

[39] W. M. Hunter, *in* "Handbook of Experimental Immunology" (D. M. Weir, ed.), p. 608. Blackwell, Oxford, 1967.

Response Curves

Under the conditions defined for each radioimmunoassay concentrations in the range 0–100 pmol/assay tube are sufficient to assess the practical limits of concentration which can be measured reliably. Tracer labeled with ^{125}I allows reduction of this range to 0–50 pmol/assay tube. Mean and standard deviations are shown for 10 successive standard curves of conjugated chenodeoxycholic acid and conjugated ursodeoxycholic acid (Fig. 2a and b). Standard deviations are greater at low concentrations of added standard when using ^{3}H tracer bile salt but decrease as the concentration of nonradioactive bile salt increases. This is a reflection of the low specific activity of ^{3}H-labeled bile salts which limits the amount of radioactivity which can be used for each assay and thus limits counting precision. When free bile salt is counted the zero standard will show the greatest coefficient of variation and this will reduce as the concentration of nonradioactive bile salt increases. Conversely, if the bound fraction is counted the zero standard will show the lowest coefficient of variation and this will increase as the standard concentration increases. With low specific activity ^{3}H-labeled bile salts the best counting precision will therefore be obtained by counting the free fraction when titer is maximized by using the lowest amount of radioactivity commensurate with acceptable counting precision. Ekins'[35] theoretical predictions for optimal assay sensitivity require counting of both free and bound bile salt but this does not apply with such low specific activity tracers, for the increased coefficient of variation of measuring bound bile acid will reduce assay performance more than measurement of the free fraction alone.

The high specific activity of ^{125}I-labeled bile acid allows addition of more radioactivity and, as shown in Fig. 2b, the standard deviations are constant over the standard curve. Either fraction can be determined reliably and although only free counts are used for calculation of results, the sum of free and bound counts provide a useful check that adsorption of ^{125}I-labeled bile salt has not occurred.

The shape of the standard curves shows that methodological errors will exert different effects on assay precision at different points on the standard curve. Mathematical treatment of standard curves can yield linear relationships which are attractive for calculation of results but it must be noted that the extremities of these lines will be imprecise relative to the central region.[40] Probably the most common method of converting the standard curve to a linear function is the logit plot[41] of $\log Z/1 - Z$ against

[40] W. H. C. Walker, *Clin. Chem.* **23**, 384 (1977).
[41] D. Rodbard and J. E. Lewald, *Acta Endocrinol. Suppl.* **64, 147**, 79 (1970).

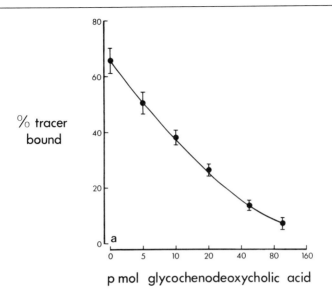

p mol glycochenodeoxycholic acid

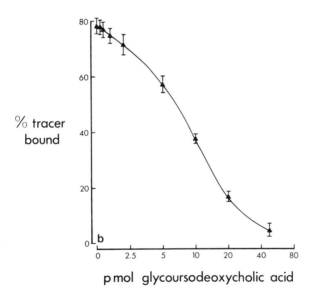

p mol glycoursodeoxycholic acid

FIG. 2. (a) Standard curve for chenodeoxycholylglycine. Standard deviations shown as a bar are determined from 10 consecutive assays using an antiserum dilution of 1:7000. The tracer is [³H]chenodeoxycholylglycine (2 pmol). (b) Standard curve for ursodeoxycholylglycine. Standard deviations shown as a bar are determined from 10 consecutive assays using an antiserum dilution of 1:2000. The tracer is [¹²⁵I]-N,1-ursodeoxycholylglycyl-N,2(4-hydroxybenzoyl)-1,2-diaminoethane (15 fmol).

log antigen concentration, where Z = bound:free ratio/bound:free ratio at zero concentration.

The authors[41] recommend use of a computer weighting technique to improve precision at the ends of the line but such an approach is rarely adopted. We have used an alternative approach which defines practically the acceptable limits of the standard curve and samples are diluted to fall within this range. Duplicate samples falling within four distinct regions of the standard curve for ursodeoxycholic acid conjugates have been used to determine the coefficient of variations applicable over this standard curve, using the formula $CV = (\Sigma d^2/2N)^{1/2}$ where d = [(highest value of duplicate/lowest value of duplicate) − 1]× 100. The results of this study (Table VI) show good precision between 1 and 20 pmol/assay tube but concentrations above or below this range are not acceptable and samples must be diluted appropriately and the assay repeated. Similar studies with ^{3}H-labeled bile acid tracers define the limits for these assays as 5–40 pmol/assay tube, although sample dilutions usually fall within the range 5–30 pmol/assay tube in our assays.

Sensitivity of the assays can be defined as a function of the slope of the standard curve or as the minimum amount of bile salt significantly different from zero. The slope of the standard curve alone may not measure sensitivity[40] and the latter definition has been applied to our data. In each case the sensitivity by this definition is less than the lower limit defined by acceptable precision and is not important practically. The limit of 5 pmol/assay tube allows detection of 0.1 μmol/liter with ^{3}H tracer where serum dilution is not critical. With ^{125}I tracer 0.1 μmol/liter can be determined with serum diluted 1:10 with buffer.

TABLE VI

COEFFICIENTS OF VARIATION[a]
DETERMINED AT FOUR DIFFERENT
REGIONS OF THE STANDARD CURVE
FOR URSODEOXYCHOLYLGLYCINE

Concentration (pmol/assay tube)	Coefficient of variation (%)
0.25–1	17.9
1.0–5.0	6.6
15–20	8.2
>20	19.8

[a] Calculated from 10 duplicate analyses for each concentration range.

Stability of Standards and Samples

Standards

Although stock solutions (0.5 μmol/liter) of bile acids are stable when stored at 4° the dilute solutions suitable for the preparation of standard curves are not stable in phosphate buffer (0.01 M, pH 7.4). This problem is resolved by preparation of these standards in charcoal extracted serum or albumin buffer (see Development of Assay, Incubation Buffer). When stored at 4° these standards (0–1 μmol/liter) are stable for at least 1 month.

Samples

Serum samples are stable when stored at $-20°$ but repeated thawing and freezing leads to decreasing concentrations of bile salts measured by radioimmunoassay.[32] It is therefore important that samples are not thawed and refrozen between assays for the different bile salts.

Accuracy and Precision

Precision

Radioimmunoassay of bile salts is intended to provide clinically useful information and must therefore be shown to be reliable. Both precision and accuracy must be defined for a particular assay. Intraassay precision we have defined in two ways. First, a pool of normal serum is used to provide 10 replicate samples which are assayed together and from these concentrations the mean and coefficient of variation calculated. This value is then compared with coefficients of variation calculated from duplicate samples assayed routinely. As the precision of an assay is not fixed throughout the range of the standard curve this latter approach allows coefficients of variation to be determined throughout the standard curve without preparing numerous pools of serum. A pool of serum is also used to provide a quality control sample which is assayed with every routine assay. The coefficient of variation of this sample is a measure of interassay precision. Precision data for our assays are shown in Table VII.

Accuracy

Accuracy must also be assessed, although this is difficult at low concentrations. Our approach has been to add bile salt standards to charcoal-extracted serum, although this is not entirely satisfactory because char-

TABLE VII
PRECISION OF RADIOIMMUNOASSAYS

			Coefficient of variation (%)	
			Duplicates (10 pairs)	
Assay	Intraassay	Interassay replicates (N = 10)	Low[a]	High[b]
1. Cholylglycine	13.6	11.8	7.3	6.7
2. Chenodeoxycholylglycine	10.1	8.6	4.6	5.0
3. Lithocholylglycine	18.3	12.0	6.7	6.9
4. Ursodeoxycholylglycine	13.5	7.0	6.6	8.2

[a] Low represents samples falling between 5 and 10 pmol/assay tube for assays 1–3 and between 1 and 5 pmol/assay tube for assay 4.

[b] High represents samples with concentrations between 25 and 30 pmol/assay tube for assays 1–3 and 15–20 pmol/assay tube for assay 4.

coal removes many other molecules together with bile salts and the resultant fluid is not truly representative of serum. Similarly, serial dilution of serum is not satisfactory because the higher dilutions will contain less protein than would normally be present at low serum concentrations of bile salts. The protein concentration has been shown to affect results of bile salt radioimmunoassays.[14,18]

We have used serum samples fortified with known amounts of standard bile salts but the main determinant for accuracy has been the correlation with values obtained by GLC. The GLC method developed in this department[37] is sufficiently sensitive to measure the primary bile salts in normal fasting sera and to monitor postprandial changes[42] which has allowed extensive validation of the methods both at normal fasting and elevated concentrations. Our data (Table VIII) show good linearity with gradients close to unity and intercepts less than 0.5 μmol/liter with the exception of our cholylglycine assay at higher serum concentrations. All three factors are necessary to define fully the agreement between two methods.

The importance of such comparisons is illustrated by the report showing that a highly specific antiserum may not necessarily produce an accurate radioimmunoassay for steroids.[43] This may reflect differences between cross-reactivity studies which use only tracer and a single cross-reacting compound, and assay of serum which involves simultaneous competition between all cross-reacting compounds present.

[42] C. R. Pennington, P. E. Ross, and I. A. D. Bouchier, *Digestion* **17,** 56 (1978).
[43] H. Adlercreutz, T. Leikinen, and M. Tikkanen, *J. Steroid Biochem.* **7,** 105 (1976).

TABLE VIII
CORRELATION OF GLC ASSAY AND RADIOIMMUNOASSAY FOR SERUM BILE SALTS

Assay	Concentration range (μmol/liter)	Equation of line of best fit[a]	Correlation coefficient
Cholylglycine	0–12	$y = 1.04x + 0.2$	0.968 ($N = 20$)
	6.5–150	$y = 0.92x + 4.3$	0.953 ($N = 24$)
Chenodeoxycholylglycine	0–12	$y = 0.94x + 0.05$	0.976 ($N = 20$)
	9.0–110	$y = 1.06x - 0.2$	0.980 ($N = 20$)
Lithocholylglycine	0–4	$y = 1.08x + 0.12$	0.943 ($N = 12$)

[a] Line of best fit determined by least-squares method. y = radioimmunoassay results; x = GLC results.

Methodological differences between GLC and radioimmunoassay can lead to difficulties during the study of assay accuracy. The secondary bile salts are present in serum at lower concentrations than the primary bile salts and in the case of lithocholyl conjugates the concentration is frequently below the detection limits of GLC. Consequently the range of samples which can be compared by these methods is limited although radioimmunoassay can detect lithocholyl conjugates in most samples. Matern et al.[10] found a similar problem with cholyl conjugates, for only 5 of 8 samples contained detectable cholic acid by GLC in their correlation study.

Gas–liquid chromatography cannot differentiate between conjugated and unconjugated bile salts unless the hydrolysis step is omitted. Usually the concentration of unconjugated bile salt is too low to be detected by GLC and this procedure is rarely carried out. However, as radioimmunoassay usually measures only conjugated bile salts the two methods cannot be expected to correlate where serum does contain significant amounts of unconjugated bile salt. This proved to be important during validation of a radioimmunoassay for ursodeoxycholyl conjugates. Correlation carried out for a selection of sera showed good agreement for most samples but some showed underestimation of up to 30% by radioimmunoassay. The samples showing this variation all came from patients taking ursodeoxycholic acid for gallstone dissolution and, when GLC was carried out for unconjugated ursodeoxycholate, significant levels were found. Insufficient samples have been assayed by GLC for unconjugated ursodeoxycholic acid and consequently correlation of radioimmunoassay and GLC for conjugated ursodeoxycholic acid is not yet available.

Specificity must also be considered, for cross-reactivity of other bile salts must affect accuracy adversely. The major cross-reaction usually occurs with the relevant unconjugated bile salt but where the cross-reactivity exceeds 10% for a conjugated bile salt this can significantly alter the measured concentration, particularly in cases of hepatobiliary disease where ratios of bile salt concentrations are altered.[44]

Antisera for secondary bile salts must show low cross-reactivity toward the primary bile salts because the secondary bile salt concentrations tend to be lower than primary bile salt concentrations, particularly in hepatobiliary disease.[44] Our antisera for lithocholyl conjugates and ursodeoxycholyl conjugates meet this criteria but we have not yet produced an antisera to deoxycholyl conjugates with acceptable specificity. One report of radioimmunoassay has not recorded cross-reactivities for either cholic acid conjugates or chenodeoxycholic acid conjugates[17] although the low concentrations found in the presence of high concentrations of primary bile salts suggests little, if any, cross-reactivity. Mäentausta and Jänne[18] report no significant cross-reactivity with primary bile salts for their deoxycholylglycine assay and accuracy should not be affected by cross-reaction.

Future Developments

The future role of serum bile salt as a liver function test will modulate further development of immunoassays but already some improvements have been reported.

Antisera

All antisera have been prepared by conjugation of the side chain carboxyl group which fails to produce antisera capable of clear discrimination between glycine and taurine conjugates and which suffers from cross-reaction with unconjugated bile salts. Choice of an alternative site for conjugation may improve specificity and allow, for example, precise study of glycine–taurine ratios for each bile salt.

The antisera currently available could possibly be combined to improve specificity. For example the addition of anti-chenodeoxycholylglycine serum may improve specificity of our assay for conjugates of cholic acid. A theoretical model for this approach has been discussed.[45]

Separation Procedures

Solid-phase radioimmunoassay has been reported for bile salt assay, using antiserum covalently bound to Sepharose.[27] At present, large volumes of antiserum are necessary for such an assay because these authors

---

[44] C. R. Pennington, P. E. Ross, and I. A. D. Bouchier, *Gut* **18,** 903 (1977).
[45] D. Rodbard, *Steroids* **29,** 149 (1977).

found that 10 ml of antiserum was sufficient for 600 assays of cholylgly-cine compared with 50,000 assays from 10 ml of antiserum used at 1 : 1000 dilution with γ-globulin precipitation. Isolation of antibodies may reduce the currently excessive requirement for antiserum while use of larger animals would lead to increased supply of antisera, although at greater cost.

Enzyme Immunoassay

This procedure requires conjugation of bile salt to enzyme so that enzyme activity is retained and the bile salt–enzyme equilibrates with antibody and bile salt in the same way as radioactive bile salt. One such method requires separation of bound and free antigen,[46] a method recently improved by combination of enzyme-immunoassay and solid-phase separation.[47] However, we have developed a method where antibody inhibits the enzyme–bile salt and thus no separation procedure is necessary.[48] At present this homogeneous enzyme immunoassay shows sensitivity similar to radioimmunoassay but the concentration range which can be assayed without dilution is wider. The assay presently requires larger volumes of antiserum than radioimmunoassay but this disadvantage is offset by the simplicity of measurement (spectrophotometer) and seems likely to be developed further.

Acknowledgments

I would like to thank three postgraduate students, J. Murison, Y. A. Baqir, and A. Hill for important contributions and also Dr. G. Wilson and Dr. A. F. Hofmann for helpful discussions during development of this work.

[46] S. Ozaki, A. Tashiro, I. Makino, S. Nakagawa, and I. Yoshizawa, *J. Lipid Res.* **20**, 240 (1979).
[47] Y. Maeda, T. Setoguchi, T. Katsuki, and E. Ishikawa, *J. Lipid Res.* **20**, 960 (1979).
[48] Y. A. Baqir, P. E. Ross, and I. A. D. Bouchier, *Anal. Biochem.* **93**, 361 (1979).

[25] Radioimmunoassay of Ecdysteroids

By M. A. DELAAGE, M. H. HIRN, and M. L. DE REGGI

Introduction

Ecdysone or 20-OH-ecdysone (Fig. 1) can be rendered immunogenic by coupling to proteins via a large variety of positions. Up to now the 2,3 and 22 secondary hydroxyl groups,[1-3] the 6 keto group,[4,5] and the 26 carbon[6] have been successfully used. The antibodies raised against each immunogen show comparable affinity for their hapten, but the cross-reactivity for ecdysone analogs may differ considerably, depending on the position of coupling to the carrier. Several radioimmunoassays have been set up with the antibodies thus obtained, using as labeled analog, either tritiated ecdysone or better an iodinated derivative. The radioimmunoassay described here belongs to the latter category and is now widely used. It was developed according to the general strategy of Oliver *et al.*[7] for digoxin in which the keystone is the succinylation of the hapten.

Succinylation of 20-OH-Ecdysone

Thirty milligrams of succinic anhydride is dissolved in 1 ml of dioxane, with optional addition of 10^7 cpm of [^{14}C]succinic anhydride (New England Nuclear). Next, 6 to 8 mg of 20-OH-ecdysone (Sigma) is added and the reaction is started with 100 μl of triethylamine. The mixture is stirred at 30° for 2 hr. The reaction is stopped by addition of 2 ml of water and the mixture is loaded on a QAE-Sephadex column (1 × 30 cm, chloride form). Elution is done by a ionic strength gradient: 300 ml of 0.02 M NaCl, 300 ml of 0.4 M NaCl without any added buffer. A typical elution profile is shown in Fig. 2. Monosuccinyl-20-OH-ecdysone is identified by

---

[1] R. C. Lauer, P. Solomon, K. Nakanishi, and B. F. Erlanger, *Experientia* **30**, 560 (1974).

[2] M. L. de Reggi, M. Hirn, and M. A. Delaage, *Biochem. Biophys. Res. Commun.* **66**, 1307 (1975).

[3] D. S. Horn, J. S. Wilkie, B. S. Sage, and J. D. O'Connor, *J. Insect. Physiol.* **22**, 901 (1976).

[4] D. W. Borst and J. D. O'Connor, *Steroids* **24**, 637 (1974).

[5] P. Porcheron, J. Foucrier, C. Gros, P. Pradelles, P. Cassier, and F. Dray, *FEBS Lett.* **61**, 159 (1976).

[6] K. D. Spindler, C. Beckers, U. Gröschel-Stewart, and H. Emmerich, *Hoppe Seylers Z. Physiol. Chem.* **359**, 1269 (1978).

[7] G. C. Oliver, Jr., B. M. Parker, D. L. Brasfield, and C. W. Parker, *J. Clin. Invest.* **47**, 1035 (1968).

METHODS IN ENZYMOLOGY, VOL. 84

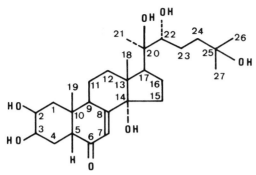

FIG. 1. Numbering of positions on ecdysone. Standard and IUPAC names of the main derivatives quoted in this paper: ecdysone = $2\beta,3\beta,14\alpha,22R$, 25-pentahydroxy-5$\beta$-cholest-7-en-6-one; 20-OH-ecdysone = $2\beta,3\beta,14\alpha,20R,22R,25$-hexahydroxy-5$\beta$-cholest-7-en-6-one; inokosterone = $2\beta,3\beta,14\alpha,20R,22R,26$-hexahydroxy-5$\beta$-cholest-7-en-6-one; makisterone A = $2\beta,3\beta,14\alpha,20R,22R,25$-hexahydroxy-24-methyl-5$\beta$-cholest-7-en-6-one.

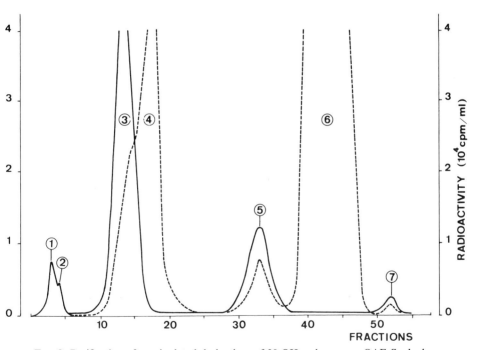

FIG. 2. Purification of succinylated derivatives of 20-OH-ecdysone on QAE-Sephadex. (1) Dioxane; (2) 20-OH-ecdysone; (3) succinyl-20-OH-ecdysone; (4) unidentified by-product; (5) dissuccinyl-20-OH-ecdysone; (6) succinate; (7) trisuccinyl-20-OH-ecdysone. Optical density at 248 nm, ——; radioactivity,------.

its position and by the molar ratio of succinate to 20-OH-ecdysone as determined from ^{14}C radioactivity and the UV spectrum (molar extinction coefficient of ecdysone at 248 nm: 11,600). The compound consists of a mixture of the 2- and 3-hemisuccinates, the former being predominant due to the higher reactivity of the equatorial 2-OH function.[8] The 22-hemisuccinate is present in negligible amounts. Other peaks were likewise identified as di- and trisuccinyl-20-OH-ecydsone and by-products. Succinyl-20-OH-ecdysone is desalted without further purification according to the protocol established for succinyl-cyclic AMP.[9] The same QAE-Sephadex column used for the purification step is washed with 0.02 M NaCl before applying the whole peak of succinyl-20-OH-ecdysone. Then 0.01 M HCl is applied and succinyl-20-OH-ecdysone is eluted as a sharp peak at a pH of around 2.8. The product is then neutralized with NaOH and lyophilized. A small calculable amount of NaCl remains with succinyl-20-OH-ecdysone.

Synthesis of Succinyl-20-OH-Ecdysone Tyrosine Methyl Ester (SETME)

In the cold room, 8 μmol of succinyl-20-OH-ecdysone is dissolved in 400 μl of freshly distilled dimethylforamide. The medium is alkalinized by 12 μl of a 1/11 dilution of triethylamine in in dimethylformamide. Freshly distilled ethylcholoroformate is diluted 1/16 in dimethylformamide immediately before use and 12 μl of this dilution is added to the incubation mixture which is vortexed and left for 10 min. Five milligrams of tryosine methyl ester is dissolved in 200 μl of dimethylformamide containing 5 μl of triethylamine. The coupling is achieved by mixing 50 μl of this solution and 100 μl of the solution of activated succinyl-20-OH-ecdysone. After half an hour, the mixture is loaded on a Sephadex LH-20 column and eluted with methanol-chloroform (1 : 1, v/v). The first peak contains only SETME; by-products are eluted later. The fractions are pooled, dried under vacuum, and redissolved in water. After checking of the UV spectrum, which must be identical to that of an equimolecular mixture of 20-OH-ecdysone and tyrosine methyl ester, part of the solution is diluted to 0.1 mM in 0.1 M phosphate buffer, pH 7.5. This solution is distributed into 20-μl aliquots ready for iodination and stored at $-30°$. In these conditions SETME can be stored several years without any noticeable damage.

---

[8] M. N. Galbraith and D. H. S. Horn, *Aust. J. Chem.* **22**, 1045 (1969).
[9] H. L. Cailla and M. A. Delaage, *Anal. Biochem.* **48**, 62 (1972).

Iodination

The Hunter and Greenwood method[10] is used with minor modifications.[11] Iodination is performed by adding 20 μl of 0.1 mM SETME, 10 μl (1 mCi) of [^{125}I]iodide (New England Nuclear NEZ 033A or Amersham IMS-30), and 10 μl of a 1 mg/ml solution of chloramine-T in 0.05 M phosphate buffer pH 7.5. The mixture is vortexed and left for 2 min. The reaction is stopped by 20 μl of a solution of 1 mg/ml sodium metabisulfite in water. Then 500 μl of citrate buffer (0.1 M citrate pH 6.2, 0.5 mg/ml sodium azide) is added and the mixture is loaded on a Sephadex G-25 column (0.9 × 58 cm) and eluted with the same citrate buffer. A typical elution profile is presented in Fig. 3. Usually over 90% of iodide has been incorporated. Free iodide is just a shoulder before the main peak which represents monoiodo-SETME (ISETME). The smaller peak eluted afterward is the diiodo derivative. The fractions containing monoiodo derivative are pooled and diluted in 0.1 M citrate buffer (pH 6.2) plus 0.5 mg/ml sodium azide and 2 mg/ml serum albumin (bovine or human). The final dilution is adjusted to approximately 3 × 10^5 cmp/ml, the solution is distributed into 3-ml aliquots, then frozen and stored at −30°. The iodinated derivative is thus perfectly stable and retains full immunoreactivity up to 6 months.

Immunogen

The same batch of succinyl-20-OH-ecdysone activated by ethylchloroformate and used for synthesizing SETME is also used for coupling 20-OH-ecdysone to albumin. Six milligrams of lyophilized human albumin is added to 300 μl of the activation mixture described above and 10 μl of triethylamine diluted 1/11 in dimethylformamide. After half an hour of stirring the incubation medium is diluted with 1 ml water and chromatographed on Sephadex G-25. The fractions containing the 20-OH-ecdysone –albumin conjugate are pooled and stored at −30°. The yield of coupling is determined from the analysis of the UV spectrum using the following molar coefficients:

$$\text{ecdysone: } E_{248} = 11600 \qquad E_{280} = 654$$
$$\text{albumin:}$$
$$E_{248} = 22105 \qquad E_{280} = 35425$$

[10] W. M. Hunter and F. C. Greenwood, *Nature (London)* **194**, 495 (1962).
[11] M. A. Delaage, D. Roux, and H. L. Cailla, in "Advances in Pharmacology and Therapeuthics" (J. C. Stoclet, ed.), Vol. 3, p. 193. Pergamon, Oxford, 1979.

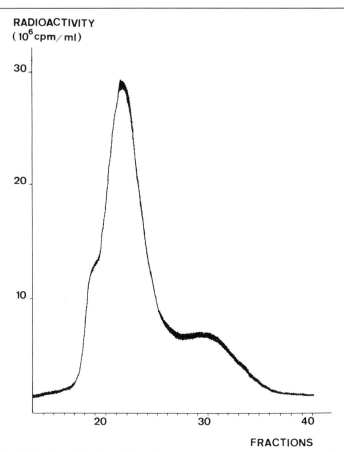

FIG. 3. Purification of iodinated SETME on G.25 Sephadex (58 × 0.9 cm), fraction volume 2.8 ml. Elution is done by 0.1 M citrate pH 6.2, with 0.5 mg/ml sodium azide. Free iodide is seen as a shoulder around fraction 19. The main peak centered on fraction 22 is [^{125}I]SETME; the second peak around fraction 30 is [^{125}I]diiodo-SETME. Both are fully immunoreactive.

The protocol gives a molar ratio of about 11 mole ecdysteroid/mole albumin. Large variations around this value are acceptable.

Immunization

The immunization procedure was that of Ross et al,[12] but other classical schemes can be used. Briefly, 1 mg of immunogen is dissolved in 4 ml

[12] G. T. Ross, J. L. Vatukaitis, and J. B. Robbins, in "Structure Activity Relationships of Protein and Polypeptide Hormones" (M. Marcoulies and F. L. Greenwood, eds.), Vol. 1, p. 153. Excerpta Medica, Amsterdam, 1971.

of water with 8 mg of lyophilized BCG (Institut Pasteur). The mixture is emulsified with 4 ml of complete Freund's adjuvant and injected into the back of four rabbits (2 ml each) in about 50 intradermal injections. Booster injections are administered monthly in the same way. Normally high titers of antiserum (a workable dilution of over 1/50,000, for example) are obtained after the first booster injection. Ready-for-use dilutions of antiserum in 0.1 M citrate pH 6.2, containing 1 g/liter albumin and 0.5 mg/ml sodium azide, are divided into 3-ml aliquots and stored at $-30°$.

Specificity of Antibodies

The average equilibrium constants have been determined for a large set of ligands (Fig. 4).[13] Some representative results are listed in the table. The following conclusions can be drawn:

1. The best ligands are those most clearly ressembling the immunogen: monosuccinyl-20-OH ecdysone and its tyrosylated derivative, with a K_d value around 4.6×10^{-10} M. Surprisingly, ecdysone is better recognized than 20-OH-ecdysone, their K_d values being 6.3×10^{-10} and 1.2×10^{-9} M, respectively.
2. Major modifications on the side chain result in loss of immunoreactivity. For example antibodies fail to recognize 22–25 dideoxyecdysone or poststerone which differs from ecdysone by the absence of the side chain.
3. Minor modifications on the distal carbons of the side chain preserve immunoreactivity. For example 26-OH-ecdysone has a K_d value of 6.1×10^{-10} M.
4. Modifications in the nuclei can also alter immunoreactivity: 3-dehydroecdysone is 30 times less well recognized than ecdysone probably because of the deformation of the A cycle.

EQUILIBRIUM DISSOCIATION CONSTANTS (K_d) OF SOME ECDYSONE ANALOGS WITH ANTIBODIES

Compound	K_d (M)
Ecdysone	6.1×10^{-10}
20-OH-Ecdysone	1.3×10^{-9}
Monosuccinyl-20-OH-ecdysone	4.6×10^{-10}
26-OH-Ecdysone	6.1×10^{-10}
3-Dehydroecdysone	1.8×10^{-8}
2,3,22-Trisuccinyl-20-OH-ecdysone	1.5×10^{-7}

[13] M. H. Hirn, Thesis of Doctorat, Marseille, 1978.

5. When both the nuclei and the side chain are modified, a complete loss of immunoreactivity usually results. This is the case for 2, 3, 22-trisuccinyl-20-OH-ecdysone. Cholesterol and immediate derivatives are not recognized by antibodies.

Incubation and Separation of Free and Bound Fractions

The most trustworthy method either for testing antibodies or for radioimmunological measurements is equilibrium dialysis.[14] Incubations are performed in dialysis cells of 2×200 μl or of 2×20 μl (Biologie Appliquée, Marseille), fitted with large pore cellulose membranes, Sartorius SM 11533 or Schleicher and Schüll RC53. On one side of the membrane the antibody solution is placed; on the other, a mixture of equal volumes of the iodinated derivative and sample to be assayed (standard solution or biological extract). Microdialysis is carried out in a special apparatus, the compartments of which are open. They are filled with ordinary disposable pipets (20 μl). After 18 hr of gentle rotation in a humidified chamber each sample is recovered with 15 μl calibrated microcapillaries from Drummond or Corning which are then directly counted in the gamma counter.

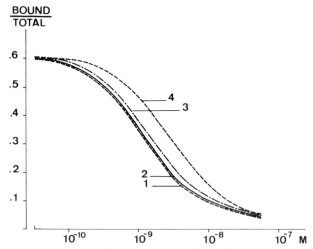

FIG. 4. Displacement curves of [125I]SETME by different ligands. Abscissa indicates logarithm of molar concentration in dialysis. Ordinate indicates bound radioactivity divided by total radioactivity in the antibody compartment. (1) SETME; (2) succinyl-20-OH-ecdysone; (3) ecdysone; (4) 20-OH-ecdysone.

[14] H. L. Cailla, G. S. Gros, E. J. P. Jolu, M. A. Delaage, and R. C. Depieds, *Anal. Biochem.* **56**, 383 (1973).

Extraction of Ecdysteroids from Biological Samples

Ecdysteroids can be extracted in perchloric acid and directly assayed. When a chromatographic separation of the different ecdysteroids is needed for individual quantification another extraction system based on solubilisation in methanol is preferred.

Perchloric Acid Extraction. About 20 mg of tissue or biological fluid is homogenized in 200 μl of 1 N perchloric acid. If possible this homogenization is directly performed in a small plastic tube usable in the centrifuge. Proteins and cell debris are pelleted (2 min at 15,000 g). One hundred and fifty microliters of the supernatant is neutralized by 15 μl of 6 M $K_2 CO_3$. A 1-min centrifugation eliminates the $KClO_4$ formed and the supernatant can be directly assayed by equilibrium dialysis as described, after suitable dilution in citrate buffer.

Alcoholic Extraction. The biological sample, up to 20 mg, is homogenized in 200 μl of methanol by brief sonication. Insoluble material is discarded by a 2-min centrifugation at 15,000 g. The supernatant is evaporated under reduced pressure in order to decrease the volume. The sample is then applied to a silica gel plate (Merck 5554) which is developed for 1 hr with a mixture of chloroform–methanol 80 : 20 (v/v). In this system, ecdysone and 20-OH-ecdysone have R_f values of 0.20 and 0.15, respectively. Other immunoreactive products can be found on the plate. The so-called "low polarity products" are found at high R_f; some of them have been identified as 2,22-dideoxyecdysone and 2,22,25-trideoxyecdysone.[15] At low R_f sulfo and gluco conjugates of ecdysone[16] are found. In some cases (crustaceans) one can find products comigrating with 20-OH-ecdysone: inokosterone (25-deoxy-20, 26-dihydroxyecdysone) and makisterone (24-Me-20-OH-ecdysone). In this case more refined TLC methods must be used. Each immunoreactive spot can be eluted by scraping the corresponding area and extracting it with ethanol. After drying the sample is redissolved in citrate buffer for radioimmunoassay. Ecdysteroids can also be separated by HPLC.[17,18]

Other solvents can be used for extraction: for example is dioxane which extracts both ecdysone and juvenile hormone in a good yield. Then the two hormones can be easily separated from each other by solvent partition and assayed in their respective radioimmunoassay.[19]

[15] C. Hetru, M. Lagueux, Luu Bang, and J. A. Hoffmann, *Life Sci.* **22**, 2141 (1978).
[16] J. Koolman, J. A. Hoffmann, and P. Karlson, *Hoppe Seylers Z. Physiol. Chem.* **353**, 1043 (1973).
[17] M. Hori, *Steroids* **14**, 33 (1969).
[18] R. Lafont, G. Martin-Sommer, and J. C. Chambet, *J. Chromatogt.* **170**, 185 (1979).
[19] C. Strambi, A. Strambi, M. L. de Reggi, M. H. Hirn, and M. A. Delaage, *Eur. J. Biochem.* **118**, 401 (1981).

Comments

The radioimmunoassay of ecdysone is one of the easiest to undertake because of the specific qualities of the molecule, i.e., its optimal size as a hapten, its solubility in water as well as in less polar solvents, its high UV absorption, and its resistance to oxydizing reagents and iodine. The sensitivity threshold for microincubations is around 5×10^{-15} mol, which represents less than 1 μl of hemolymph in any insect at basal activity. Ecdysone can be assayed in only a few eggs of as small an animal as *Drosophila melanogaster*. Thus the radioimmunoassay of ecdysone is generally not the limiting factor of a program. If necessary, sensitivity can be sacrificed for accuracy gained by using a relative excess of antibody. Moreover, even the substitution of tritiated ecdysone for the iodinated derivative, which results in a 4-fold loss in sensitivity, is perfectly tolerable in most cases.

Regarding the antibody specificity it should be emphasized that the search for absolute specificity is a pipedream. At the maximum affinity level attainable with antibodies (K_d between 10^{-11} and 10^{-10} M) the noncovalent interactions are always compatible with small variations in the positions or even the number of functional groups. Nevertheless, antibody recognition requires the preservation of the main characteristics of ecdysone: compounds lacking either the en-one structure or the side chain are not recognized. Thus all immunoreactive material found in a biological sample can be considered as belonging to the ecdysone pathway. An assay performed on a crude extract gives a result which, if expressed in "ecdysone-equivalent," represents a fair evaluation of the activity of the whole ecdysone metabolism. Each radioimmunoassay opens a "window" onto the original hapten as well as its close precursors and metabolites. Fortunately, immunoreactivity and chromatographic behavior generally obey different criteria, it is thus easy to combine a TLC (or HPLC) step with a single radioimmunoassay and to follow in this way three or four consecutive metabolites in the same biological sample.

[26] Radioimmunoassay of Motilin

By JOHN C. BROWN and JILL R. DRYBURGH

Introduction

Gastric motor activity is regulated by several control mechanisms initiated when the gastric contents pass into the duodenum. It has been suggested[1] that one of these controlling mechanisms is a "pH-dependent dual humoral mechanism." A linear oligopeptide with 22 amino acid residues subsequently named motilin has been isolated from the small intestine of hogs.[2,3] The amino acid sequence Phe-Val-Pro-Ile-Phe-Thr-Tyr-Gly-Glu-Leu-Gln-Arg-Met-Glu-Glu-Lys-Glu-Arg-Asn-Lys-Gly-Gln has been reported[3] and a 14-glutamine motilin has also been described.[4] The amino acid sequence has been confirmed by synthesis[5] and identical biological potencies have been ascribed to the synthetic analogue (13-norleucine-motilin) and the natural porcine polypeptide.

In vivo studies with motilin have shown it to be a powerful stimulator of motor activity in both fundic and antral pouches of the stomach of dogs, in doses as small as 50 ng/kg. Continuous infusion of motilin resulted in the demonstration of tachyphylaxis. A reduction in the response to motilin in the presence of atropine has led to the suggestion that motilin may act via an intermediate, possibly acetylcholine and that tachyphylaxis might be due to depletion of this intermediate.[6]

Itoh *et al.*[7,8] have suggested a role for motilin in the control of the interdigestive migrating myoelectric complex. They demonstrated that motilin was without effect on this myoelectric complex when administered during the digestive state, but during the interdigestive state a pattern of activity resembling the naturally occurring contractions could be induced.

---

[1] J. C. Brown, *Gastroenterology* **52**, 225 (1967).

[2] J. C. Brown, M. A. Cook, and J. R. Dryburgh, *Gastroenterology* **62**, 401 (1972).

[3] J. C. Brown, M. A. Cook, and J. R. Dryburgh, *Can. J. Biochem.* **51**, 533 (1973).

[4] H. Schubert and J. C. Brown, *Can. J. Biochem.* **52**, 7 (1974).

[5] E. Wünsch, J. C. Brown, K-H. Deimes, F. Drees, E. Jaeger, J. Musiol, R. Scharf, H. Stocker, P. Thamm, and G. Wendlberger, *Z. Naturforschung* **28**, 235 (1973).

[6] M. A. Cook, Ph.D. Thesis, University of British Columbia, Vancouver (1972).

[7] Z. Itoh, R. Honda, K. Hiwatashi, S. Takeuchi, I. Aizawa, R. Takayanagi, and E. F. Couch, *Scand. J. Gastroenterol.* **11**, Suppl. 39, 93 (1976).

[8] Z. Itoh, S. Takeuchi, I. Aizawa, R. Takayanagi, H. Mori, T. Taminato, Y. Seino, H. Imura, and N. Yanaihara, *in* "Gastrointestinal Hormones and Pathology of the Digestive System" (M. Grossman, V. Speranza, N. Basso, and F. Lezouche, eds.). Plenum, New York, 1978.

METHODS IN ENZYMOLOGY, VOL. 84

Wingate *et al.*[9] also observed that motilin injection can stimulate premature migrating complexes in the spontaneous interdigestive sequence. A role for motilin in the regulation of gastric emptying is disputed. In dogs, gastric emptying of liquid, but not solid meals was accelerated by motilin infusion,[10] whereas in man, the synthetic analog (13-nleu-motilin) delayed gastric emptying of a liquid meal[11] while porcine motilin stimulated the emptying of solids.[12]

The contractile response of strips of rabbit duodenal and colonic circular muscle and fundic muscle from human stomach was potently increased by motilin[13,14] and a potentiation of acetylcholine induced contractions by 13-nleu-motilin has been observed.[15] Strunz *et al.*[16] concluded that motilin acted directly on the muscle cell when they were unable to demonstrate an inhibition of motilin-induced contraction by ganglionic blockade with hexamethonium or tetrodotoxin. In isolated strips of duodenal muscle 13-nleu-motilin produced membrane depolarization.[17]

The Radioimmunoassay

Preparation of Purified Motilin

Motilin was purified in collaboration with Prof. V. Mutt at the Karolinska Institute Stockholm, by techniques already described.[2,3] A final purification stage using Sephadex G-25 fine was introduced and the final product was referred to as M5. Homogeneity of the final product was established by polyacrylamide gel electrophoresis, high voltage electrophoresis of M5 and M5 following tryptic digestion, and amino acid analysis of the fractions obtained in the final purification stage. M5 was used throughout the assay for standards, iodination, and for control samples.

[9] D. L. Wingate, H. Ruppin, W. E. R. Green, H. H. Thompson, W. Domschke, E. Wünsch, L. Demling, and H. D. Ritchie, *Scand. J. Gastroenterol.* **11**, Suppl. 39, 111 (1976).

[10] H. T. Debas, T. Yamagishi, and J. R. Dryburgh, *Gastroenterology* **73**, 777 (1977).

[11] H. Ruppin, S. Domschke, W. Domschke, E. Wünsch, E. Jaeger, and L. Demling, *Scand. J. Gastroenterol.* **10**, 199 (1975).

[12] S. R. Bloom, N. D. Christofides, I. Modlin, and M. L. Fitzpatrick, *Gastroenterol.* **74**, 1010 (Abst.) (1978).

[13] T. Segewa, M. Nakano, Y. Kai, H. Kawatani, and H. Yajima, *J. Pharm. Pharmacol.* **28**, 650 (1976).

[14] U. Strunz, W. Domschke, D. Domschke, P. Mitznegg, E. Wünsch, E. Jaeger, and L. Demling, *Scand. J. Gastroenterol.* **11**, 199 (1976).

[15] U. Strunz, W. Domschke, S. Domschke, P. Mitznegg, E. Wünsch, E. Jaeger, and L. Demling, *Scand. J. Gastroenterol.* **11**, Suppl. 39, 29 (1976).

[16] U. Strunz, W. Domschke, P. Mitznegg, S. Domschke, E. Schubert, E. Wünsch, E. Jaeger, and L. Demling, *Gastroenterology* **68**, 1485 (1975).

[17] J. Riemer, K. Kölling, and C. J. Mayer, *Pflügers Arch.* **372**, 243 (1977).

Preparation and Storage of Standards and Controls

An amount of natural porcine motilin (M5) between 1.0 and 2.0 mg was accurately weighed on a Cahn microbalance, dissolved in deionized water to give a final concentration of 1.0 μg/100 μl. Aliquots of 100 μl were pipetted into siliconized glass tubes (10 × 75 mm), lyophilized, covered, and stored at −20° until required. At monthly intervals an aliquot (1.0 μg) was reconstituted in 0.04 M phosphate buffer, pH 6.5, containing 0.25% Trasylol and 5% bovine serum albumin to a concentration of 80 ng/ml. One milliliter amounts of this solution was stored in polyethylene tubes at −20°. Standards were prepared from this aliquot, usually concentrations of 6.25 to 400 pg/100 μl were set up by serial dilution. The 80 ng/ml aliquots were used only once.

A control was established for determination of inter- and intraassay variability. Natural porcine motilin, M5, was dissolved in 0.04 M phosphate buffer, pH 6.5, containing 0.5% bovine serum albumin with 2% Trasylol to a concentration of 100 pg/100 μl. The control solution was aliquotted into polyethylene tubes and stored at −20°. Controls were used only once then discarded.

Production of Antisera to Motilin

Preparation of Immunogen. Both guinea pigs and rabbits have been successfully immunized with purest porcine motilin, conjugated to bovine serum albumin.[18,19] Conjugation has been achieved using the carbodiimide condensation reaction.[20] In a conjugation procedure, 1.0 mg porcine motilin was dissolved in 200 μl deionized water and to this was added approximately 10 × 10⁴ cpm ¹²⁵I-labeled motilin, to follow the degree of conjugation. One milligram of bovine serum albumin in 100 μl of water and 10 mg l-ethyl-3-(-3-diethylaminopropyl)carbodiimide in 100 μl were added, and the reaction mixture was stirred for 60 min. The reaction was terminated either by dialysis of the solution against distilled water for 16 hr at 4° or the reaction mixture was applied to a Sephadex G-25 fine column and eluted with 0.2 M acetic acid. An approximation of the degree of conjugation could be assessed from the chromatogram obtained following gel filtration (Fig. 1).

Immunization Schedule. The animals were initially immunized with 50 μg of conjugated M5 (25 μg unconjugated peptide) dissolved in deionized water (250 μl) and emulsified with an equal volume of Freund's com-

[18] J. R. Dryburgh and J. C. Brown, *Gastroenterology* **68**, 1169 (1975).
[19] J. R. Dryburgh and J. C. Brown, in "Methods of Hormone Radioimmunoassay" (B. M. Jaffe and H. R. Behrman, eds.), 2nd ed. Academic Press, New York, 1979.
[20] T. L. Goodfriend, L. Levine, and G. D. Fasman, *Science* **144**, 1344 (1964).

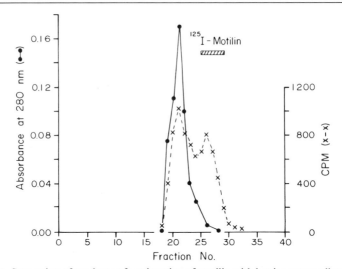

FIG. 1. Separation of products of conjugation of motilin with bovine serum albumin, on a 0.9 × 100 cm column, Sephadex G-25, eluting buffer 0.2 M acetic acid. Fraction size 1.5 ml. Column was calibrated with ^{125}I-labeled motilin prior to use.

plete adjuvant. Each rabbit received multiple intradermal injections in the suprascapular region and the guinea pigs received multiple subcutaneous injections in the abdominal and inner thigh regions.

Subsequent immunizations, usually at 1-month intervals, were performed with conjugated motilin emulsified with Freund's incomplete adjuvant. In rabbits, blood was obtained by bleeding from the marginal ear veins and from the guinea pigs by cardiac puncture. Bleedings were performed between 10 and 14 days following the second and subsequent immunizations, the samples were allowed to clot at 4° for 20 min then serum obtained by centrifugation. After assay for titer, antisera were stored lyophilized in aliquots at −20°. When required, lyophilized aliquots were reconstituted in the appropriate volume of deionized water, diluted with assay diluent buffer to a 1:10 dilution and stored in 100-μl aliquots at −20°.

Iodination of Motilin

Determination of the amino acid composition and confirmation of the structure by synthesis have revealed that motilin contains a single tyrosine residue in position 7, with no tryptophan, no methionine, and no histidine residues. Incorporation of ^{125}I into motilin has been accomplished employing the chloramine-T method of Hunter and Greenwood.[21] The

[21] W. M. Hunter and F. C. Greenwood, *Biochem. J.* **89,** 114 (1963).

end product is a relatively stable ^{125}I-labeled motilin of specific activity in the range of 250–450 mCi/mg. Successful iodination can also be achieved using the lactoperoxidase method of Miyachi et al.[22] The polypeptide is not easily fragmented by oxidizing agents as evidenced by homogeneity on polyacrylamide gel electrophoresis, following exposure to chloramine-T for up to 20 min.

   Chloramine-T Method. The following reagents were prepared freshly for each iodination: (1) natural porcine motilin (M5) 2.0 or 4.0 μg dissolved in 50 μl 0.2 M phosphate buffer pH 7.5; (2) Na^{125}I—1.0 mCi, carrier-free; (3) chloramine-T—40 μg in 10 μl deionized water; (4) sodium metabisulfite—100 μg in 20 μl of distilled water.

   The Na^{125}I and motilin in phosphate buffer are added together in a 10 × 75-mm siliconized glass tube and gently mixed. Chloramine-T is added and mixed with gentle bubbling, the reaction being allowed to proceed for 15 sec before addition of sodium metabisulfite to terminate the reaction. The reaction mixture was immediately transferred to a column of Sephadex G-25 fine (0.6 × 30 cm) which had been equilibrated with the eluting buffer for 2 hr prior to use. The eluting buffer was 0.2 M acetic acid containing 0.5% BSA and 100 KIU Trasylol per ml. Fractions of 400 μl were collected and a 10-μl aliquot was removed and counted for 0.1 min in an automatic gamma counter. Specific and nonspecific binding of the column fractions, diluted to 5000 cpm/100 μl, were determined following incubation for 24 hr at 4°, with and without antiserum (Fig. 2). Fractions showing the highest binding, and lowest nonspecific binding were pooled, diluted with the eluant buffer, and aliquotted to 2 × 10^6 cpm/2 ml for storage at at −20°. Iodinated motilin, prepared and stored in this manner, has remained stable for up to 3 months. Alternative methods of storage which have been proven to be satisfactory include lyophilization, or dilution in acid ethanol (50 ml 99% ethanol with 0.75 ml concentrated HCl) to 2 × 10^6 cpm/100 μl then storage at −20°.

   Lactoperoxidase Method. Alternatively, a gentler and more easily controlled method oxidation involving the use of the enzyme lactoperoxidase can be employed.[22] The following reagents were freshly prepared: (1) natural porcine motilin—4 μg in 50μl 0.05 M acetate buffer pH 5.0; (2) Na^{125}I—1.0 mCi, carrier free; (3) lactoperoxidase—500 ng in 10μl acetate buffer; (4) hydrogen peroxide—0.86 nM in deionized water (3 × 10 μl at 5-min intervals).

   The above reagents were mixed in a 10 × 75-mm siliconized glass tube in the order presented. After 15 min, 5 min after the last addition of hydrogen peroxide, the mixture was transferred to a Sephadex G-25 fine col-

[22] Y. Miyachi, J. L. Vaituikaitis, E. Nieschlag, and M. B. Lipsett, *J. Clin. Endocrinol. Metab.* **34**, 23 (1972).

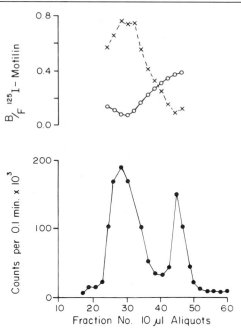

FIG. 2. Column profile of separation of ^{125}I-labeled motilin from free iodide (chloramine-T) on 0.9 × 30 cm column of Sephadex G-25 fine with 0.2 M acetic acid and bovine serum albumin and trasylol in eluting buffer. The peptide:iodine ratio was 2.0 μg:1 mCi. Counts per 0.1 min (●——●), maximum binding (×——×), nonspecific binding (○——○).

umn and eluted, monitored, assayed, and stored as described for the chloramine-T technique (Fig. 3). The specific activity of ^{125}I-labeled motilin prepared by either of the techniques described is usually in the range 300–400 mCi/mg. The binding kinetics of the antiserum to natural porcine motilin have been shown to be almost identical to those obtained for ^{125}I-labeled motilin.[18]

Assay Procedure

The diluent buffer which was used in all dilutions and for correction of the final volume to 1.0 ml was 0.04 M phosphate containing 5.0% charcoal-extracted plasma and 0.25% Trasylol. The composition of the incubation mixture was:

100 μl ^{125}I-labeled motilin (approximately 5 × 10^3 cpm)
100 μl motilin standard (6.25–400 pg)
or 100 μl inter- and intraassay control
or 50–200 μl sample to be assayed
100 μl antiserum—appropriately diluted
Diluent buffer to 1.0 ml

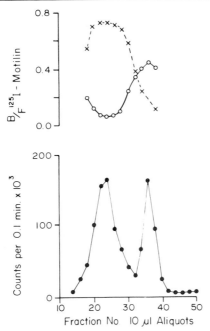

FIG. 3. Column profile of separation of ^{125}I-labeled motilin from free iodide (lactoperoxidase) on 0.9 × 30 cm column of Sephadex G-25 fine with 0.2 M acetic acid and bovine serum albumin and trasylol in eluting buffer. The motilin:iodine ratio was 2 μg:1 mCi. Counts per 0.1 min (●——●), maximum binding (×——×), nonspecific binding (○——○).

All assays were set up with standards in triplicate and unknowns in at least duplicate at 4° in 10 × 75-mm glass culture tubes. Incubation was of the equilibrium type, carried out at 4° for 48 or 72 hr. Nonspecific binding was measured for the standards, inter- and intraassay controls, each group of serum or plasma samples, and all other unknowns. Adsorption of peptides and iodinated substances to glass can occur. The plasma content of the diluent buffer and presiliconization of assay tubes with 1% (v/v) dichlorosilane in benzene have been studied and both have been found to be unnecessary in the radioimmunoassay for motilin. The addition of extra plasma or serum to the standard curve tubes to compensate for possible excess protein effects when plasma or serum samples were measured was also found to be unnecessary.

Separation Procedure

Both specific and nonspecific methods have been employed for the separation of the antigen–antibody complex from the free antigen. The nonspecific method of adsorption of the free antigen to dextran-coated charcoal has been successfully employed in the radioimmunoassay for motilin.

Sodium phosphate buffer, 0.04 M, pH 6.5, containing 2% aged plasma was cooled to 4°. Dextran T-70 (Pharmacia Ltd.) is added to a concentration of 2.5 mg/ml and thoroughly mixed to produce a uniform suspension. Charcoal (carbon decolorizing C-170, Fisher Scientific) is then suspended in the mixture, to a final concentration of 12.5 mg/ml. The dextran-charcoal suspension is mixed gently at 4° for at least 30 min prior to use, 200 μl is added to each tube, vortexed, and immediately centrifuged for 20 min at 2800 rpm. The supernatants are discarded and the charcoal pellet counted in an automatic gamma counter. All results are expressed as percentage bound, after correction for nonspecific binding. Large volumes of dextran-coated charcoal can be prepared in phosphate buffer, mixed 3–4 hr, and stored at 4°. An appropriate volume can be removed as required, plasma added, and the suspension mixed for 15 min before use. This stock suspension keeps well for up to 2 weeks.

Affinity Chromatography

Preparation of Antibody–Sepharose Complex. Activation of Sepharose 4B with cyanogen bromide was performed by the method of Cuatrecasas[23] and could be stored as a moist slurry at 4° for up to 14 days. The activated Sepharose 4B was suspended in an equal volume of 0.1 M $NaHCO_3$. Antiserum to motilin was diluted in an equivalent volume of 0.1 M $NaHCO_3$ to a final concentration of 30 μl antiserum per gram of Sepharose 4B. The reaction mixture was stirred gently for 24 hr at 4° after which time the antiserum-Sepharose complex was washed well with 20 volumes cold deionized water. Antibody titers were determined on aliquots of antiserum prior to coupling and the initial wash after the reaction was complete to determine the efficiency of the coupling reaction. Figure 4 illustrates the antiserum dilution curves obtained from these 2 samples. Almost complete coupling of antibody was achieved. Unreacted active groups on the gel matrix were blocked by treatment with excess ethanolamine. Ethanolamine was adjusted to pH 9.0 by the addition of 5.0 M HCl, and added to the antibody–Sepharose suspension, to a final concentration of 1.0 M. After 4 hr at 4° the excess ethanolamine was removed by washing with 10 alternating cycles of 0.1 M sodium acetate, pH 4.0 and 0.1 M sodium phosphate pH 8.0. The final product was stored in an equal volume of 0.1 M $NaHCO_3$ at 4°, with 0.01% sodium azide added as a preservative.

Application to Radioimmunoassay. Serial dilutions of the antiserum

[23] P. Cuatrecasas, *J. Biol. Chem.* **245**, 3059 (1970).

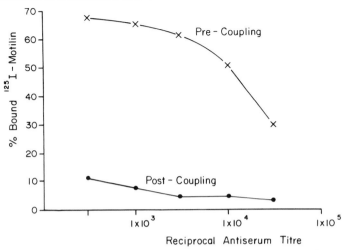

Reciprocal Antiserum Titre

Fig. 4. Motilin antiserum dilution curves comparing the activity of the antiserum prior to coupling with Sepharose 4B with activity remaining in supernatant following completion of the coupling procedure.

coupled to Sepharose 4B were incubated with ^{125}I-labeled motilin in a total volume of 1.0 ml, for 48 hr at 4°. All other conditions were identical to those employed in the RIA. The tubes were then centrifuged at 2800 rpm for 30 min, the supernatants discarded, and the pellets counted. Figure 5 shows a comparison of the antiserum–Sepharose coupled dilution curve with an antiserum dilution curve obtained with the same antiserum before coupling, in a routine RIA. No significant loss in activity was produced by coupling the antibody to a solid matrix.

Almost identical standard curves could be produced with the Sepharose–antibody and the normal conditions. Separation of bound and free in the former was achieved simply by centrifugation. Reproducibility was poorer with the Sepharose–antibody complex probably because of lack of homogeneity of the suspension.

Secretion of IR Motilin (Immunoreactive Motilin)

In dogs the introduction of alkaline Tris buffer into the duodenum resulted in a rapid increase in serum IR-motilin levels from 294 ± 44 to 498 ± 100 pg/ml at a time when duodenal pH increased by 0.7 pH units. This rise was accompanied by increased motor activity in denervated pouches of the body of the stomach. Maximum circulating motilin levels, 916 ± 96 pg/ml, were reached within 5 min when motor activity was also maximal, but duodenal pH (at the site of the recording electrode) had returned to preinfusion levels.[18,24] The effect of alkalinization of the duodenum on circulating IR-motilin levels in man is contradictory. Hellemans *et*

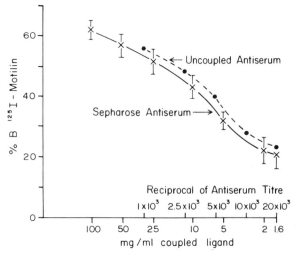

FIG. 5. Antibody dilution curves on aliquots of antiserum taken before and after coupling of the antiserum to CNBr-activated Sepharose 4B.

al.[25] have described elevated serum IR-motilin in response to antral perfusion with sodium bicarbonate and delayed increases to sodium hydroxide. Mitznegg *et al.*[26] and Track *et al.*[27] have described decreased IR-motilin levels following duodenal alkalinization in man. Although species differences may in part account for the conflicting observations with duodenal pH, the site of alkalinization and antiserum employed may also contribute. In response to feeding a small decrease in IR-motilin is usually observed occurring 30–45 min after either glucose or mixed meal,[28] whereas increased motilin levels have been described following ingestion or intravenous infusion of fat.[29]

Acknowledgment

Supported by research grants from the Medical Research Council of Canada.

[24] J. C. Brown and J. R. Dryburgh, *in* "Gut Hormones" (S. R. Bloom, ed.), Churchill Livingstone, Edinburgh, 1978.

[25] J. Hellemans, G. Van-trappen, and S. R. Bloom, *Scand. J. Gastroenterol.* **II**, Suppl. 39, 67 (1976).

[26] P. Mitznegg, S. R. Bloom, N. Christofides, H. Besterman, W. Domschke, S. Domschke, E. Wünsch, and L. Demling. *Scand. J. Gastroenterol.* **II**, Suppl. 39, 53 (1976).

[27] N. S. Track, S. Collins, and T. Lewis, *in* "Gut Hormones" (S. R. Bloom, ed.), Churchill Livingstone, Edinburgh, 1978.

[28] J. R. Dryburgh, Ph.D. Thesis, University of British Columbia, Vancouver (1977).

[29] S. R. Bloom, N. D. Christofides, H. S. Besterman, T. E. Adrian, and M. A. Ghatei, *Gastroenterology* **74**, 1010 (Abst.) (1978).

[27] Radioassay of Vitamin B_{12} and Other Corrinoids

By J. J. PRATT and M. G. WOLDRING

I. Introduction

The term "corrinoid" in the title of this review refers to a specific set of chemical compounds related to vitamin B_{12} with structures shown in Fig. 1. As we shall see later, corrinoids possessing vitamin B_{12} activity are heterogeneous: a compound with vitamin activity in one organism need not have that activity in another (vitamin B_{12} antagonists are also known). The use of one organism (for instance a bacterium) to assay vitamin B_{12} activity in another (e.g., a mammal) can give rise to subtle analytical problems. Similar problems arise when a "chemical" assay method, such as competitive binding radioassay, is used to estimate vitamin activity.

Because they are biologically active in such small amounts, biological, biochemical, and clinical laboratories need to assay picogram amounts of nonradioactive corrinoids. Competitive binding assay is one of the more convenient methods of achieving this. This review is meant to explore the application of corrinoid radioassay generally and not just in the clinical laboratory. Each type of laboratory has its own requirements for analyzing these compounds so that no straight forward "recipe" can be given. The assays are split into different steps in the review which follows and the analyst must decide which combination of procedures is most suitable for his requirements and which criteria of accuracy he should use.

II. Chemistry of the Corrinoids

Figure 1 shows the general structure of the cobalamins, the most important form of vitamin B_{12} in mammals. Figure 3 depicted later shows a stereo pair of the carbon skeleton of 5'-deoxy-5'-adenosylcobalamin. The three-dimensional structure is important when considering the specificity of corrinoid binding proteins and the design of suitable immunogens for the production of antibodies (Section VII).

A. Nomenclature

i. Corrin. The basic ring structure in Fig. 1. without side chains and without the cobalt atom.

ii. Corrinoid. Any compound containing the corrin ring structure.

iii. Cobyrinic Acid. The corrin ring plus the cobalt atom plus the 8-methyl groups plus the carboxyethyl- and carboxymethyl groups.

METHODS IN ENZYMOLOGY, VOL. 84

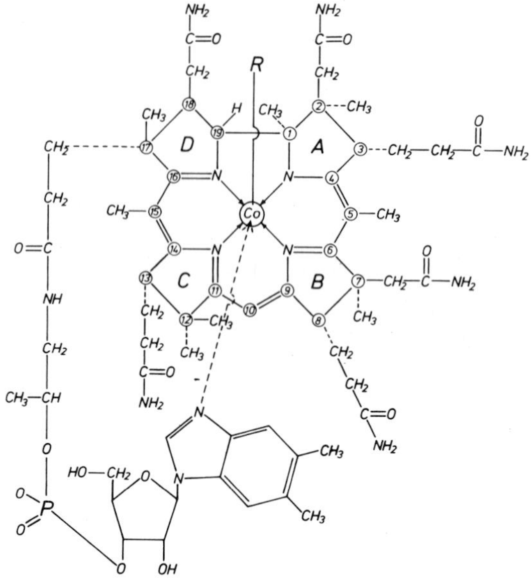

FIG. 1. Covalent structure of cobalamin.

iv. Cobyric Acid. Cobyrinic acid with carboxyethylamide (carboxy-methylamide) substituted at carbons 2, 3, 7, 8, 13, and 18.

v. Cobinic Acid. Cobyrinic acid with D-1-amino-2-hydroxypropane in amide linkage with the carboxyethyl substituent at C-17.

vi. Cobinamide. The hexaamide of cobinic acid.

vii. Cobamic Acid. Cobinic acid with D-ribofuranose 3-phosphate in phosphoric ester linkage with the aminopropanol residue.

viii. Cobamide. The hexaamide of cobamic acid.

viv. Cobalamin. Cobamide with 5,6-dimethylbenzimidazole in α-glycosidic linkage with the ribose moiety, i.e., the complete structure shown in Fig. 1.

x. Vitamin B_{12}. Like all vitamins, vitamin B_{12} is defined as an organic factor in food or medium required in small amounts to prevent the occurrence of or to alleviate the symptoms of its deficiency. The biologically important corrinoids and their relationships with vitamin B_{12} are described in Section III. In mammals, the major, perhaps only, compounds with vitamin B_{12} activity found in the body are substituted cobalamins. These are: Coβ-methylcobalamin (methylcobalamin, R = methyl in Fig. 1), Coβ-hydroxy cobalamin (hydroxycoabalamin, R = OH$^-$ in Fig. 1), and Coβ-5'-deoxy-5'adenosylcobalamin ("vitamin B_{12} coenzyme," R = 5'-dexoyadenosine attached via C'-5 in Fig. 1). *In vitro*, these ligands can be replaced by others such as sulfite, nitrite, and, in particular, cyanide to form more stable complexes. These artificial compounds have vitamin B_{12} activity in animals and microorganisms since they are metabolically converted back to the natural forms. The nomenclature of the corrinoids is explained in greater detail in the recent IUPAC-IUB rules.[1]

B. Properties

Cobalt containing corrinoids as isolated are all in the III oxidation state. The resulting differences in charge can form the basis for chromatographic separation (Section VI). Corrinoids are freely soluble in water but most are sufficiently hydrophobic to be extracted into polar organic solvents (Sections V and VI). These compounds have very characteristic absorption spectra with very high extinction coefficients at wavelengths ranging from 260 to 500 nm.[2] These properties, together with the unique presence of very tightly bound cobalt which can be labeled with a variety of radioactive isotopes, mean that a very wide variety of biochemical methods can be used for extraction, purification, and analysis.

The general chemistry of these compounds is discussed in a number of reviews.[2-7]

TABLE I

EQUILIBRIUM BINDING CONSTANTS AND
DISSOCIATION RATE CONSTANTS OF SOME
$CO\beta$-LIGANDS OF COBALAMIN

$Co\beta$-ligand	Log K for displacement of H_2O (M^{-1})[5]	k_{-1} (sec^{-1})[7]
NH_3	7	
Imidazole	4.6	
OH^-	6.3	
N_3^-	4.9	3×10^{-2}
CN^-	12	
Cl^-	0.1	
I^-	1.5	35
SO_3^{2-}	7.3	10^{-5}

[1] I.U.P.A.C.-I.U.B. rules, *Pure Appl. Chem.* **48,** 495 (1976).
[2] W. Friedrich, "Vitamin B-12 und verwandte Corrinoide" (Fermente, Hormone, Vitamine und die Beziehungen dieser Wirkstoffe zueinander, 3^e Auflage, Band III/2, herausgegeben von R. Ammon, W. Dirscherl). Thieme, Stuttgart, 1975.
[3] R. Bonnett, *Chem. Rev.* **63,** 573 (1963).
[4] D. G. Brown, *in* "Current Research Topics in Bioinorganic Chemistry (Progress in Inorganic Chemistry, Vol. 18)" (S. J. Lippard, ed.), p. 177. Wiley, New York, 1973.
[5] J. M. Pratt, Inorganic Chemistry of Vitamin B-12. "Academic Press, New York, 1972.
[6] H. A. Hill, *in* "Inorganic Biochemistry" (G. I. Eichhorn, ed.), p. 1067. Elsevier, Amsterdam, 1973.
[7] H. P. C. Hogenkamp, *in* "Cobalamin: Biochemistry and Pathophysiology" (B. M. Babior, ed.), p. 21. Wiley, New York, 1975.

i. Exchange of the Coβ-Ligand. The Coβ-ligand is relatively labile. Table I shows that cyanide forms the most stable complex. It is usual practice to convert corrinoids to the cyanoform before the assay to eliminate uncertainties concerning the influence of these ligands on the protein binding affinity.

ii. Dilute acid will hydrolyze the carboxyamide groups of the molecule.[7-9] Monocarboxylic acids may be used to prepare suitable immunogens (Sections VII,C,iv). More concentrated acid will hydrolyze the phosphate ester to yield various cobinic acid amides.[9,11] Epimerization at C-8 and C-13 occurs in concentrated acid.[7,10] Corrinoids are generally not stable in alkali.

iii. Exchange of the Coα-Ligand. The various cobalamin analogs differ widely in their capacity to form the "base-off" form to leave the lower (α) coordination position free for other ligands. These differences have important consequences for the specificity of binding to protein because at least one important binding protein binds to the α-position. If the α-coordination position is blocked by, e.g., cyanide, this is no longer possible (Section VII). A high pH, 5,6-dimethyl benzimidazole dissociates from the cobalt of cobalamin which may then exist in the dicyano form. The lower cyanide is lost on lowering the pH below 8.0. Analogs containing bases other than benzimidazole derivatives (Section III,A), such as adenine and 2-methyladenine displace the cyanide much less readily so that the dicyano form will persist in the presence of excess cyanide[7] (see Section VII,C,i). In the absence of cyanide, thiols such as glutathione and cysteine in quite low concentrations will bind to the Co α-position of aquocobalamin.[12] Such complex formation may have important consequences when tissue samples are to be analyzed.

iv. Adenosylcobalamin and methylcobalamin are highly sensitive to light. A few seconds exposure to daylight is sufficient to destroy these forms. When it is necessary to quantitate them separately, it is usual to work either in the dark or under dim red light. Cyanocobalamin, the most stable form of cobalamin, is only moderately light sensitive (10% loss per 0.5 hr in "direct sunlight"). Analysis of cyano forms in subdued daylight is a sufficient precaution.

---

[8] D. Thusius, *J. Am. Chem. Soc.* **93**, 2629 (1971).

[9] K. Bernhauer, F. Wagner, H. Beisbarth, P. Reitz, and H. Vogelmann, *Biochem. Z.* **344**, 289 (1966).

[10] K. Bernhauer, H. Vogelmann, and F. Wagner, *Hoppe Seylers Z. Physiol. Chem.* **349**, 1281 (1968).

[11] J. G. Buchanan, A. W. Johnson, J. A. Mills, and A. R. Todd, *J. Chem. Soc.* p. 2845 (1950).

[12] P. Y. Law and J. M. Wood, *J. Am. Chem. Soc.* **95**, 914 (1973).

III. Biochemistry and Ecology of the Corrinoids

Corrinoids are neither synthesized nor required by higher plants and algae. They are however synthesized in large amounts by many bacteria, fungi, protozoa, and lower algae. Many other microorganisms and all metazoa require certain corrinoids for growth but are unable to synthesize them themselves and must obtain their vitamin directly or indirectly from microorganisms. There exists, therefore, a variety of food chains along which the corrinoids pass. Most microorganisms produce a variety of corrinoids, only a few of which are active as vitamins in other organisms. The biochemistry of corrinoid biosynthesis and utilization is intimately connected with the ecology of these compounds and both have consequences for corrinoid assay whether by bioassay or radioassay.

A. Biosynthesis

The biosynthesis of the corrinoids has been reviewed in detail by Friedrich[2] and Friedmann.[13] Cobyrinic acid is the simplest corrinoid that can be isolated. It is found in small amounts in the sources mentioned below. Carboxy-groups are amidated in a specific sequence to yield cobyric acid[8,14] which can be found in large amounts in sewage sludge, feces, and presumably all primary sources of corrinoids.[15] The aminopropyl group is usually incorporated after amidation of cobyrinic acid. Cobinamide occurs in large amounts in rumen, colon, and cecal contents and in bacterial cultures.[16] Synthesis of cobalamin and its analogs proceeds via cobinamide phosphate, guanosine diphosphate cobinamide, and cobalamin 5'-phosphate.

B. Naturally Occurring Corrinoids

i. Coβ-Ligands. The chemistry of the labile Coβ-ligands is discussed in Section II. The biochemistry of ligand exchange is discussed later (Section III,E).
ii. "Incomplete Corrinoids." The occurrence of intermediates of cobalamin biosynthesis has already been outlined. These intermediates may be secreted by the living cell or released on the death of the cell. The possibility that carboxy derivatives of cobalamin (which also occur naturally) may arise as degradation products has been raised[17] (Section IX).

[13] H. C. Friedmann, *in* "Cobalamin: Biochemistry and Pathophysiology" (B. M. Babior, ed.), p. 75. Wiley, New York, 1975.
[14] K. Bernhauer, F. Wagner, H. Michna, P. Rapp, and H. Vogelmann, *Hoppe Seylers Z. Physiol. Chem.* **349**, 1297 (1968).
[15] B. Bartosinski, B. Zagalak, and J. Pawelkiewicz, *Biochim. Biophys. Acta* **136**, 581 (1967).
[16] J. E. Ford, S. K. Kon, and J. W. G. Porter, *Biochem. J.* **50**, 9P (1952).
[17] W. Friedrich, *in* "Antivitamins" (J. C. Somogyi, ed.), p. 178. (Karger, Basel, 1966.

iii. Coα-Ligand Analogs. As mentioned previously, many but not all microorganisms have little discriminatory power in choosing the "lower" nucleotide. The lack of specificity reflects both the synthesis of the α-ribosylnucleoside-5'-phosphate from β-ribofuranosylnicotinic acid 5'-phosphate and the enzymatic displacement of GMP from the corrin. 5,6-Dimethylbenzimidazole may be replaced by adenine (attached via N-7 or N-9), 2-methyladenine, 2-methylmercaptoadenine, 5-hydroxybenzimidazole, 5-methylbenzimidazole, 6-methylbenzimidazole, hypoxanthine, phenol, and cresol to name only the most important naturally occurring aglycons. More than 100 such analogs are known. Many of these are produced by "directed synthesis" by growing the organism in the presence of large amounts of the relevant aglycon and must be regarded as artificial analogs (for a review see Friedrich[2]).

iv. Cobalt-Free Analogs. Corrinoids without cobalt are produced by some bacteria.[18,19]

v. Degradation Products. Little is known about the mechanism of corrinoid degradation but we may assume that products of partial degradation are widespread. Besides the deamidocobalamins mentioned previously, yellow corrinoids, presumable degradation products, have been isolated from several bacterial species.[20] Interestingly, they appear to be unique in that they have vitamin B_{12} activity in man (on injection) but not in any of the microorganisms used in bioassays, including *E. coli* mutants which are otherwise the least exacting organisms in their corrinoid requirements.

vi. Peptide Complexes. As we shall see later, biologically active corrinoids are almost always found bound to specific proteins of high molecular weight. This binding is largely dependent on London and van der Waals forces and "hydrophobic binding." Other, less specific complexes are known: hydroxycobalamin forms coordination complexes with serum albumin via a histidine residue.[21,22] Sporadic reports continue to appear on complexes of cobalamin with lower molecular weight peptides from bacteria,[23,24] liver,[25-27] and serum.[28] The biological significance of these

[18] J. I. Toohey, this series, Vol. 18, p. 71.
[19] K. Sato, S. Shimizu, and S. Fukui, *Biochem. Biophys. Res. Commun.* **39**, 170 (1970).
[20] K. Helgaland, J. Jonsen, S. Laland, T. Lygren, and O. Rømcke, *Nature (London)* **199**, 604 (1963).
[21] R. T. Taylor and M. L. Hanna, *Arch. Biochem. Biophys.* **141**, 247 (1970).
[22] E. L. Lien and J. M. Wood, *Biochim. Biophys. Acta* **264**, 530 (1972).
[23] K. Hausmann, L. Ludwig, and K. Mulli, *Acta Haematol.* **10**, 282 (1953).
[24] J. G. Heathcote and F. S. Mooney, *in* "Vitamin B-12 und Intrinsic Factor" (H. C. Heinrich, ed.), p. 540. Enke Verlag, Stuttgart, 1962.
[25] K. Mulli and J. O. Schmid, *Z. Vitam. Horm. Fermentforsch.* **8**, 225 (1957).
[26] A. Hedbom, *Biochem. J.* **74**, 307 (1960).
[27] J. Skupin, K. Nowakowska, and E. Pedziwilk, *Bull Acad. Pol. Sci. Ser. C* **23**, 435 (1975).
[28] E. Gizis and L. M. Meyer, *Nature (London)* **217**, 272 (1968).

complexes is not clear. Cyanocobalamin can be released from these complexes prior to bio- or radioassay by simple treatment with cyanide and all the available evidence argues for the presence of cobalt coordination complexes with histidine and cysteine residues.

C. Natural Occurrence and the Food Chain

The richest primary sources of corrinoids in nature are soil, sewage, sludge, feces, and rumen, cecum, and colon contents, i.e., those sources where microbial growth is greatest. The corrinoids from these primary sources are heterogeneous and only a small fraction is biologically active in man (Section III,D). Most herbivores obtain their vitamin B_{12} from contaminated vegetable matter or by coprophagy. The effectiveness of these sources is attested by the difficulty in obtaining deficient laboratory animals.[29] Fresh, untreated water contains between 0.1 and 5.0 ng ml^{-1} of biologically active corrinoid[30] and this may be a major source of supply for some animals. Ruminants obtain abundant supplies from the rumen microflora. Seawater contains between 2 pg ml^{-1} (winter) and 0.2 pg ml^{-1} (summer) of dissolved, corrinoid largely in the form of cobalamin analogs.[31] Direct adsorption must be the source of the vitamin for many lower algae. For instance, the marine alga *Stichococcus* requires cobalamin and responds with half maximum growth to 2 pg ml^{-1} of cobalamin.[32] The effectiveness of the cobalamin concentrating mechanism must therefore be extremely great.

Absorbed cobalamin may be transfered via many steps along the food chain. This has the important consequence that the spectrum of heterogeneity of the corrinoids becomes progressively simpler the further along the food chain we go. Corrinoids are absorbed and metabolized by active processes requiring binding proteins (Section VII) and enzymes (Section III,E). The corrin is repeatedly subjected to forms of "proofreading" as it passes from one organism to the next and from one tissue to the next, different parts of the molecule are "examined" by binding proteins of different specificity. This has the consequence that cobalamin is by far the most important corrinoid found in higher organisms, even though several analogs are biologically active when injected (Section III,D). The analysis of animal tissues is fairly straightforward but the analysis of primary sources of corrinoid by radioassay or bioassay presents formidable problems.

[29] R. Green, S. V. van Tonder, G. J. Oettle, G. Cole, and J. Metz, *Nature (London)* **254**, 148 (1975).
[30] W. J. Robbins, A. Hervey, and M. E. Stebbins, *Science* **112**, 455 (1950).
[31] C. B. Cowey, *J. Mar. Biol. Assoc. U.K.* **35**, 609 (1956).
[32] R. A. Lewin, *J. Gen. Microbiol.* **10**, 93 (1954).

D. Corrinoids as Vitamins B_{12}

Corrinoids other than cobalamin dominate the corrinoid fraction in the primary sources mentioned above. For instance, the production of "vitamin B_{12} by the rumen microflora in the sheep assayed with *Ochromonas malhamensis* (which is more or less specific for cobalamin) is ca. 1 mg per day whereas a less specific radioassay gave estimates of ca. 4 mg per day.[33,34] Extensive data on the specificity of requirements are available only for man, chicks, and those microorganisms frequently used for the bioassay of vitamin B_{12}. Table II[35–43] shows the activities of some naturally occurring analogs for these species. It should be noted that the specificities for man and chicks are for injected material. The specificity for oral vitamin is much greater because many analogs are rejected by the gastrointestinal absorption mechanism.

We see that we can only define the vitamin B_{12} activity with respect to the organism in question, a particularly difficult task in some instances because the specificity of the requirements of the majority of (micro)organisms is quite unknown. The picture is greatly complicated by the occurrence of antagonists. For instance, naturally occurring cobalt-free corrinoids and monocarboxycobalamins act as antagonists for *E. coli* 113-3, *Euglena gracilis*, and *Lactobacillus leichmannii*.[44–46] The biologist should be aware of the possibility that many of the naturally occurring analogs may have an antibiotic function. To add to the confusion, the possibility exists that a corrinoid may act as an antagonist in one species but be a vitamin in another. Some monodeamidocobalamin isomers act as antag-

[33] A. L. Sutton and J. M. Elliot, *J. Nutr.* **102**, 1341 (1972).
[34] M. F. Hedrich, J. M. Elliot, and J. E. Lowe, *J. Nutr.* **103**, 1646 (1973).
[35] M. E. Coates, J. E. Ford, and G. F. Harrison, *in* "Vitamin B-12 and Intrinsic Factor" (H. C. Heinrich, ed.), p. 276. Enke, Stuttgart, 1962.
[36] Kj. Blumberger, P. Petrides, and K. Bernhauer, *in* "Vitamin B-12 and Intrinsic Factor" (H. C. Heinrich, ed.), p. 82. Enke, Stuttgart, 1957.
[37] M. E. Coates and S. K. Kon, *in* "Vitamin B-12 and Intrinsic Factor" (H. C. Heinrich, ed.), p. 72. Enke, Stuttgart, 1957.
[38] H. Dellweg, E. Becher, and K. Bernhauer, *Biochem. Z.* **327**, 422 (1956).
[39] K. Bernhauer and W. Friedrich, *Angew. Chem.* **66**, 776 (1954).
[40] F. B. Brown and J. C. Cain, E. L. Smith, *Biochem. J.* **59**, 82 (1955).
[41] J. W. G. Porter, *in* "Vitamin B-12 and Intrinsic Factor" (H. C. Heinrich, ed.), p. 43. Enke, Stuttgart, 1957.
[42] J. M. McLauchlan, C. G. Rogers, E. J. Middleton, and J. A. Campbell, *Can. J. Biochem. Physiol.* **36**, 195 (1958).
[43] J. F. Adams and F. McEwan, *J. Clin. Pathol.* **24**, 15 (1971).
[44] A. Simon, *Zentralbl. Bacteriol. Abt. II* **123**, 586 (1969).
[45] D. Perlman and J. I. Toohey, *Nature (London)* **212**, 300 (1966).
[46] D. Perlman and J. I. Toohey, *Arch. Biochem. Biophys.* **124**, 462 (1968).

TABLE II
SPECIFICITY OF VITAMIN B_{12} REQUIREMENTS FOR SEVERAL ORGANISMS[a]

Corrinoid	E. coli 113-3	Lactobacillus leichmannii	Euglena gracilis	Ochromonas malhamensis	Chick (injection)	Man (injection)
Cobalamin	+	+	+	+	+	+
5-Hydroxybenzimidazoylcobamide	+ (35)	+ (35)		+ (35)	+ (35)	+ (36)
5-Methylbenzimidazoylcobamide	+ (37)	+ (35)		+ (35)	+ (35)	+ (36)
6-Methylbezimidazoylcobamide		+ (35)		+ (35)	+ (35)	
Benzimidazoylcobamide	+ (37)	+ (38)		+ (37)		+ (36)
Adenyl[7]cobamide	+ (37)	+ (37)	+ (37)	− (37)		− (39)
Hypoxanthine[7]cobamide		+ (40)		− (40)		
2-Methyladenyl[7]cobamide	+ (37)	+ (37)	+ (37)	− (37)		± (37)
Cobinamide	+ (37)	− (42)	− (43)			− (41)

[a] Numbers in parentheses are reference numbers.

onists in *E. coli*[44] and baboons (on injection)[47] but as vitamins in chick embryos.[48]

Antagonists may exert their effect at various levels. The initial binding and concentration step or the active absorption step may be the target for the antagonist. Several analogs were found to be antagonistic for chicks when given orally, but not when given by injection.[35] Such findings are significant for those analysts dealing with foodstuffs. Alternatively, one of several enzyme reactions may be the target.[47]

Depending on the species, several substances may bypass the vitamin B_{12} requirement. Thus, methionine can bypass the B_{12} requirement in *E. coli* 113-3 and 2'-deoxyribonucleotides can completely replace the vitamin in *L. leichmannii*. Methionine alone cannot support the growth of *O. malhamensis* but it can spare the vitamin requirement when the vitamin is limiting as it is in bioassays[49] and give rise to erroneous results. There is a report of a noncorrinoid "activity" which can bypass the vitamin requirement of both man and *O. malhamensis*.[50] These considerations are important for the interpretation of bioassays with which radioassays are usually compared (Section IX).

This subsection shows that we must have a clear idea of not only what we measure but also of why we measure it.

E. Corrinoids within Cells and Tissues

In healthy higher animals, the major corrinoid in the tissues (including blood) is cobalamin complexed with various $Co\beta$-ligands. The $Co\beta$-liganded forms of cobalamin are metabolically interconvertible. In blood, quantitatively the most important cobalamins are 5'-deoxy-5'-adenosylcobalamin and methylcobalamin which arise from the tissues. The third major cobalamin in blood is hydroxycobalamin (aquocobalamin). Hydroxycobalamin itself has no known biological role: it may be a product of some transmethylation reactions and may arise from the adenosyl and methyl forms by photolysis *in vivo*. Cyanocobalamin is found in only very small amounts.[51] The known functions of the adenosyl- and methylcobalamins in higher animals are as coenzymes for methylmalonyl-CoA mutase and methyltransferases, respectively. Neither of these functions can account for the symptoms caused by B_{12} malabsorption in man (which include pernicious anemia, neuropathy, and ultimately death) because ge-

[47] R. C. Siddons, *Nature (London)* **247**, 308 (1974).
[48] E. C. Naber and J. F. Elliot, *J. Nutr.* **104**, 28 (1974).
[49] Y. Sugimoto and S. Fukui, *FEBS Letters* **43**, 261 (1974).
[50] H. Baker, O. Frank, and C. M. Leevy, *Experientia* **18**, 458 (1962).
[51] D. M. Mathews, J. C. Linnel, *in* "The Cobalamins" (H. R. V. Arnstein and R. J. Wright, eds.), p. 23. Churchill, Edinburgh, 1971.

netically determined defects in both of these enzyme systems are known in man and these do not lead to the symptoms described (for reviews see Barker,[52] Mahoney and Rosenberg,[53] Beck,[54] and Chanarin[55]). Chanarin has postulated the existence of an as yet undiscovered cobalamin responsible for the prevention of these symptoms. There are reports of other corrinoids in mammalian liver (identified as growth factors for *E. coli* 113-3[56,57] but these have yet to be distinguished from the peptide complexes mentioned earlier (Section III,B,vi).

Many of the Coα-ligand analogs are found in their coenzyme forms in bacterial cells where each analog may have a separate function in the wide range of enzyme reactions involving corrinoids in these cells.[52]

IV. Design of Corrinoid Binding Assays

A. Assay of Coenzyme Forms

This is the strategy likely to be adopted in the biochemical laboratory. Corrinoids are extracted from bacterial cells, tissues, or enzyme preparation and separated chromatographically before conversion into a single Coβ-ligand form (usually the cyanide) and radioassay. The special feature of this type of analysis is that all operations up to the conversion to the cyanide form have to be carried out in the dark or under very dim red light because the coenzyme forms are extremely photolabile. A product of photolysis is hydroxycobalamin but it is doubtful if this is the only product.[43] Suitable extraction and chromatography procedures are given in Sections V and VI. A highly specific binding protein is not required for the radioassay because the samples will have been purified chromatographically, indeed a nonspecific binding protein may be desirable if noncobalamin analogs are expected.

B. Assay of Coα-Ligand and Other Analogs

Assay of these compounds finds application in the study of corrinoids in microorganisms, soil, feces, sewage sludge, rumen and cecal contents,

[52] H. A. Barker, *Annu. Rev. Biochem.* **41**, 55 (1972).

[53] M. J. Mahoney and L. E. Rosenberg, *in* "Cobalamin: Biochemistry and Pathophysiology" (B. M. Babior, ed.), p. 369. Wiley, New York, 1975.

[54] W. S. Beck, *in* "Cobalamin: Biochemistry and Pathophysiology" (B. M. Babior, ed.), p. 403. Wiley, New York, 1975.

[55] I. Chanarin, *in* "Essays in Medical Biochemistry" (V. Marks and C. N. Hales, eds.), p. 1. Biochemical Society, London, 1977.

[56] G. Tortolani and V. Mantovani, *J. Chromatogr.* **92**, 201 (1974).

[57] E. V. Quadros, D. M. Mathews, I. J. Wise, and J. C. Linnell, *Biochim. Biophys. Acta* **421**, 141 (1976).

and in sea and fresh water. Analysis of foodstuffs also falls into this category.

In this strategy, corrinoids are extracted and concurrently converted into a stable form, usually mono- or dicyanides before chromatographic purification and radioassay. The use of a nonspecific binding protein is essential. It must be remembered that each corrinoid will bind to the protein with a different affinity so that each substance must be measured against a chemically identical standard and that some corrinoids may not bind at all and so be missed.

C. Direct Assays

These methods involve concurrent extraction and conversion to a single form (usually the cyanide) before radioassay. This strategy is used only when the binder is specific for just one corrinoid in the sample. The specificity requirements for the binding protein are very exacting. These methods are usually used in the clinical laboratory where speed, precision, and cost effectiveness are important considerations. Here it is assumed that the binding protein used "recognizes" only cyanocobalamin. Unfortunately, this has not always turned out to be the case (Section IX).

V. Extraction Procedures

A. Extraction of Coenzyme forms

In the assay of coenzyme forms, extraction with a denaturing organic solvent is usually followed by evaporation to dryness and repeated solvent extractions to give a small volume of clean product suitable for chromatography. Bacteria may be extracted with boiling 80% ethanol in water,[58] 70% acetone in water,[59] or 20% pyridine.[60] Boiling ethanol can also be used for extracting liver[56,61,62] and a wide variety of other animal tissues.[57] Four to ten milliliters of 80% ethanol is added to 1 ml of 1:2 homogenate and boiled for 30 to 60 min. The suspension is passed through a cotton wool filter and then through paper filters of decreasing pore size until a clear solution is obtained. The solvent is evaporated to dryness under reduced pressure.

[58] H. A. Barker, R. D. Smyth, H. Weissbach, A. Munch-Petersen, J. I. Toohey, J. N. Ladd, B. E. Volcani, and R. M. Wilson, *J. Biol. Chem.* **235**, 181 (1960).
[59] E. Irion and L. Ljungdahl, *Biochemistry* **7**, 2350 (1968).
[60] R. H. Abeles and H. A. Lee, *J. Biol. Chem.* **236**, 2347 (1961).
[61] J. I. Toohey and H. A. Barker, *J. Biol. Chem.* **236**, 560 (1961).
[62] K. Lindstrand, *Acta Chem. Scand.* **19**, 1785 (1965).

The following solvent partitioning steps serve both to clean up the extract and to concentrate the corrinoid before chromatography.

i. The residue is dissolved or suspended in a little water and the pH adjusted to 3.0 and any precipitate removed by centrifugation. Liquid phenol or m-cresol is added in very small aliquots with vigorous mixing until two phases appear. The corrinoids partition into the phenol (m-cresol) phase.

ii. A more effective and versatile modification of the above method is to extract the aqueous solution, pH 3.0, with the 0.1 volumes of m-cresol–carbon tetrachloride (0.05 volumes of m-cresol, 0.05 volumes of CCl_4).[63] Corrinoids partition into the organic phase. The organic phase is washed with water; 0.15 volumes of water, 0.125 volumes of 2-butanol, and 0.05 volumes of CCl_4 are added to the organic phase. The lower organic phase containing lipids is discarded.

iii. An alternative is to dissolve sodium sulfate in the aqueous solution to a concentration of 20% (w/v) and extract 3 times with 0.1 volume of benzyl alcohol[64] One-half volume of chloroform is added to the organic phase which is then extracted three times with 0.1 volume of water. Corrinoids pass back into the aqueous phase and lipids remain in the chloroform phase.

The extraction procedures (ii) and (iii) may be repeated to obtain the corrinoid in a smaller volume but with greater procedural losses.

B. Extraction as the Mono- or Dicyano-Derivatives

The above procedures are modified simply by adding NaCN to the original tissue extract to give a concentration of 10 μg ml^{-1}. Sufficient time should be allowed for the displacement of the Coβ-ligand by cyanide because the rate constant of the reaction may be rather low (see Table I). NaCN at a concentration of 1 μg ml^{-1} should be used for all aqueous solutions to maintain the corrinoid in the cyano-form.

C. Extraction with Aqueous Solutions

Extraction with aqueous buffers is suitable for analyses only where reconcentration and chromatography are unnecessary. This requirement restricts its use to assays on blood serum and animal tissue extracts (e.g., 1 : 100 homogenates).

Blood serum is extracted by boiling for 20 to 30 min with a high ionic strength buffer ($\tau/2 = 0.25-0.5$) containing cyanide and with a final pH of 4 to 6. The high ionic strength ensures maximum precipitation of protein

[63] P. Reisenstein, *Blood,* **29,** 494 (1967).
[64] A. L. Beck and J. L. Brink, *Environ. Sci. Tech.* **10,** 173 (1966).

ENDOGENOUS COMPOUNDS OF LOW MW

and a slightly acid pH stabilizes the cyanocobalamin released. The cyanide gives more complete extraction, presumably by displacing the corrinoid from peptides and coordination complexes with albumin (Section III).

i. Acetate. Human serum (0.2 ml) (or tissue homogenate) is added to 1 ml of 1.0 *M* sodium acetate buffer, pH 4.4 containing 25 μmol of KCN per liter. The mixture is heated to 100° for 30 min, cooled, and centrifuged. The supernatant can be used directly for radioassay.[65]

ii. Glycine. Serum (0.2 ml) is added to 1.4 ml of 0.35 *M* glycine hydrochloride buffer, pH 3.4, containing 75 mmol of NaCl and 100 μmol of KCN per liter. The mixture is heated to 100° for 15 min, neutralized, and used without separation of the denatured proteins.[66]

iii. Glutamic Acid. Serum (0.2 ml) is added to 0.8 ml of a solution containing the following constitutents per liter: L-glutamic acid (6 g), DL-malic acid (2 g), ammonium succinate (1.2 g), KH_2PO_4 (0.6 g), and KCN (20 mg). The mixture is heated to 100° for 15 min and cooled. Denatured proteins do not precipitate so that the solution is used directly for the radioassay.[67] Even for such a consistent sample as blood serum, a host of factors related to the extraction procedure can affect the subsequent radioassay. These are discussed in detail in Section IX.

The more complex extraction procedures lead to considerable procedural losses. The recovery should be estimated using a very small amount of radioactive corrinoid. It is important that the recovery tracer be the same as the corrinoid to be assayed because there is evidence that the recovery of adenoyslcobalamin after a very simple extraction is considerably less than that of cyanocobalamin from the same sample.[43] The production of radioactive corrinoids is outlined in Section VIII.

Extraction methods should be tested for completeness. Completeness of extraction is not the same as 100% recovery of added exogenous corrinoid. Cobalamins bound to protein or complexed with peptides (Section III) may or may not be extracted together with added labeled cobalamin so that a result corrected for procedural losses may still be inaccurate. Completeness of extraction may be tested for by reextracting the residue from the first extraction after additional treatment with papain and high concentrations of cyanide.[68] An ideal method to test for completeness of extraction is to inject the animal with [^{57}Co]cyanocobalamin and allow the radioactive material to equilibrate with endogenous cobalamin, a process

[65] E. Mortensen, *Clin. Chem.* **18,** 895 (1972).
[66] D. S. Ithakissios, D. O. Kubiatowicz, D. S. Windorski, and J. H. Wicks, *Clin. Chem.* **23,** 2043 (1977).
[67] L. Wide and A. Killander, *J. Clin. Lab. Invest.* **27,** 151 (1971).
[68] J. L. Raven and M. B. Robson, *J. Clin. Pathol.* **28,** 531 (1975).

which takes several weeks.[63] After equilibration, cobalamin is extracted and the radioactivity measured.[69]

VI. Separation of the Corrinoids

A very wide range of chromatographic and electrophoretic methods is available for corrinoid separation. Separation is necessary for two reasons:

i. Co-enzyme forms of cobalamin from animal tissues and analogs from primary corrinoid sources and foodstuffs must be separated before their separate assay.

ii. During the development of assays for cobalamin without chromatography, comparison must, at some stage, be made with an assay which does include chromatography in order to test for specificity (Section VIII).

A. Separation of Coenzyme Forms

5'-Deoxy-5'-adenosyl- and methylcobalamin bear a formal positive charge at the cobalt atoms as well as a negative charge at the phosphate. The delocalization of the formal positive charge is dependent on the nature of the Coβ-ligand. Cyanide and hydroxide ions show differing degrees of polarisation of the Co-ligand bond. Moreover, the hydroxide can accept a proton (pK_a = 6.8) so that further distinction can be made between hydroxide and cyanide forms. These properties are responsible for the very effective separations possible using ion-exchange chromatography.

Amberlite CG 50/11 ($-COO^-$), eluted with water (to remove impurities) and with 0.5 M sodium acetate to elute methylcobalamin has been used with liver extracts.[62] Complete separation and recovery of the most important cobalamins can be obtained with SP (sulfopropyl)-Sephadex. Elution is first with water to remove impurities and then with 0.05 M sodium acetate buffer pH 5.0.[56,70] Peptide-containing cobalamins have been separated with QAE (quaternary aminoethyl)-Sephadex, eluted with increasing concentrations of NaCl in water.[56] The separation of the coenzyme forms of a wide range of Coα-ligand analogs has been described.[71] The bacterial extract was cleaned and concentrated by extraction into phenol-trichloroethylene and separated on a Dowex 50-W ($-SO_3^-$) col-

[69] S. V. van Tonder, J. Metz, and R. Green, *Clin. Chim. Acta* **63**, 285 (1975).
[70] G. Tortolani, P. Bianchini, and V. Mantovani, *J. Chromatogr.* **53**, 577 (1970).
[71] J. I. Toohey, D. Perlman, and H. A. Barker, *J. Biol. Chem.* **236**, 2199 (1961).

umn eluted with sodium and ammonium acetate mixtures of increasing pH. Very good separation of the coenzyme forms and cyanocobalamin can be obtained by electrophoresis on cellulose acetate with 0.5 M acetic acid as solvent.[72]

Differences in the polarity of the coenzyme forms can be exploited in reverse-phase high performance liquid chromatography[73] and on unsubstituted polystyrene resin (Amberlite XAD-2).[74] In the latter case, the separation of a very large number of analogs as well as the coenzyme forms on elution with a concentration gradient of isobutanol in water was described.

Thin-layer chromatography methods are not well adapted to subsequent radioassay of the separated components. Thin-layer chromatography of coenzyme forms of cobalamin in serum and tissue extracts has been reviewed.[51,75]

B. Separation of Analogs in the Cyanide Form

The nature of the Coα-ligand exerts a strong influence on the polarization of the other cobalt bonds and on the delocation of any charge. Ion-exchange chromatography and electrophoresis are also suitable for separation of cobalamin and its analogs. All of the methods outlined above for the coenzyme forms are probably adaptable for the separation of analogs. Early work with ion-exchange resins has been reviewed by Pawelkiewicz.[76] A very wide range of Coα-analogs and "incomplete" corrinoids from feces can be separated in one run on ion-exchange resin (Dowex-50), eluted with aqueous buffers of increasing pH.[77] Good separation of deamidocobalamins is obtained on DEAE-cellulose columns eluted with a concentration gradient of acetic acid in water, or on Ecteola paper.[78] Phosphocellulose has been used to separate cobalt-free analogs. Elution was with 0.1 M acetic acid.[19] Very great care must be taken when using strongly acidic ion-exchangers such as Dowex-50 or phosphocellulose because they will hydrolyze the carboxyamide groups of the corrin and produce artifacts if the separation is not done quickly at 0°.

---

[72] G. Tortolani and P. G. Ferri, J. Chromatogr. **88**, 430 (1974).

[73] F. Pellerin, J.-F. Letavernier, and N. Chanon, Ann. Pharm. Fr. **35**, 413 (1977).

[74] H. Vogelman and F. Wagner, J. Chromatogr. **76**, 359 (1973).

[75] D. I. Bilkus and L. Mervyn, in "The Cobalamins" (H. V. R. Arnstein and R. J. Wrighton, eds.), p. 17. Churchill, Edinburgh, 1971.

[76] J. Pawelkiewicz, in "Vitamin B-12 and Intrinsic Factor" (H. C. Heinrich, ed.), p. 280. Enke, Stuttgart, 1962.

[77] K. H. Menke, in "Vitamin B-12 and Intrinsic Factor" (H. C. Heinrich, ed.), p. 74. Enke, Stuttgart, 1962.

[78] K. Bernhauer, H. Vogelmann, and F. Wagner, Hoppe Seylers Z. Physiol. Chem. **349**, 1271 (1968).

Paper electrophoresis is a very effective way to separate analogs. Acetic acid (1 M),[79] 0.05 M phosphate buffer, pH 6.5,[80] and 0.05 M Na_2CO_3[81] have been used for various separations. Electrophoresis in one direction and chromatography in the second direction is also possible for difficult separations.[82]

Thin-layer chromatography (silica gel, eluted with 5% water in methanol)[83] and paper chromatography (Whatman No. 1), eluted with 2-butanol saturated with 3% acetic acid[63]; Whatman No. 3MM, eluted with 2-butanol:acetic acid:water, 8800:82:4250, containing HCN (6.2 mM),[84] can both be used.

Reverse-phase high performance liquid chromatography is eminently suitable for the separation of analogs. n-Alkane (C-18)-coated silica as stationary phase may be used with either aqueous buffer–methanol[64] or 12% methylcyanide in water[85] as eluant. A new chromatography method uses pig "R-binding proteins" (Section VII), attached to agarose beads as the stationary phase.[86] Endogenous corrinoids are washed from the column with 80% phenol in water which does not permanently denature the binding protein. Phenol is removed with water. A very large volume of corrin solution is slowly passed through the column. Impurities are removed with aqueous buffers and corrinoids are eluted with 85% phenol in water. To date the method has been used only for the isolation of corrinoids but obviously it has potential for concentrating corrinoids from a very large volume and then separating them by fractional elution. We envisage the use of other binding proteins from bacteria or protozoa (Section VII) for similar purposes.

VII. Corrinoid Binding Proteins

The widespread occurrence of binding proteins gives the assayist a very wide choice of binding proteins, each with differing affinity and specificity. The use of naturally occurring binding proteins for radioassay is very important because radioimmunoassays for cobalamins are a quite recent development and have not been tested adequately for specificity.

[79] E. S. Holdsworth, Nature (London) 171, 148 (1953).
[80] J. B. Armitage, J. R. Cannon, A. W. Johnson, L. F. J. Parker, E. L. Smith, W. H. Stafford, and A. R. Todd, J. Chem. Soc. p. 3849. (1953).
[81] W. Friedrich and K. Bernhauer, Chem. Ber. 90, 154 (1957).
[82] S. K. Kon, Biochem. Soc. Symp. 13, 17 (1955).
[83] D. B. Endres, K. Painter, and G. D. Niswender, Clin. Chem. 24, 460 (1978).
[84] J. F. Kolhouse and R. H. Allen, J. Clin. Invest. 60, 1381 (1977).
[85] C.-Y. Wu and J. J. Wittick, Analyt. Chim. Acta 79, 308 (1975).
[86] J. F. Kolhouse and R. H. Allen, Analyt. Biochem. 84, 486 (1978).

A. Sensitivity

Radioassay sensitivity is determined largely by the affinity of the binding protein for its ligand and only indirectly by the specific activity of the labeled ligand.[87-92] Corrinoids are usually present in very low concentrations in nature so that high affinity is a desirable property in the binding proteins.

Table III[93-99] gives affinity constants of the most widely used binding proteins for cyanocobalamin.

B. Specificity

Interpretation of published data on specificity is complicated by the fact that investigators have either preincubated the analog with binding protein before adding radioactive cobalamin or have incubated a mixture of analog, radioactive cobalamin, and binding protein for a period too short to allow equilibration. Both of these circumstances will grossly overestimate the cross-reaction of the analog to an extent depending on the exact assay conditions. In the first case, binding protein will be almost saturated with analog during the initial reaction:

$$10 \text{ B} + 20 \text{ A} \xrightarrow{\text{fast}} 10 \text{ B} : \text{A} + 10 \text{ A}$$

$$10 \text{ B} : \text{A} + 10 \text{ A} + 20 \text{ Cobl*} \xrightarrow{\text{slow}} 2 \text{ B} : \text{A} + 8 \text{ B} : \text{Cobl*} + 18 \text{ A} + 12 \text{ Cobl*}$$
Equilibrium state

[87] R. Yalow and S. A. Berson, in "Radioisotopes in Medicine: In Vitro Studies" (R. L. Hayes, F. A. Goswitz, and B. E. P. Murphy, eds.), p. 7. U.S. Atomic Energy Commission, Oak Ridge, Tennessee, 1968.

[88] R. P. Ekins, G. B. Newman, and J. L. H. O'Riordan, in "Radioisotopes in Medicine: In Vitro Studies" (R. L. Hayes, F. A. Goswitz, and B. E. P. Murphy, eds.), p. 59. U.S. Atomic Energy Commission, Oak Ridge, Tennessee, 1968.

[89] H. Feldman and D. Rodbard, in "Principles of Competitive Protein Binding Assays" (W. D. Odell and W. H. Daughaday, eds.), p. 158. Lippincott, Philadelphia, 1971.

[90] H. Feldman, D. Rodbard, and D. Levine, Anal. Biochem. 45, 530 (1972).

[91] R. P. Ekins, G. B. Newman, R. Piyasena, P. Banks, and J. D. H. Slater, J. Steroid Biochem. 3, 289 (1972).

[92] W. H. C. Walker and P. M. Keane, in "Handbook of Radioimmunoassay" (G. E. Abraham, ed.), p. 87. Dekker, New York, 1977.

[93] L. Mervyn, in "Comprehensive Biochemistry" (M. Florkin and E. H. Stotz, eds.), p. 153. Elsevier, Amsterdam, 1970.

[94] E. Hippe and H. Olesen, Biochim. Biophys. Acta 243, 83 (1971).

[95] R. H. Allen and C. S. Mehlman, J. Biol. Chem. 248, 3660 (1972).

[96] R. H. Allen and C. S. Mehlman, J. Biol. Chem. 248, 3670 (1972).

[97] E. Hippe, Scand. J. Clin. Lab. Invest. 29, 59 (1972).

[98] E. Hippe and H. Olesen, Scand. J. Clin. Lab. Invest. 35, 577 (1975).

[99] R. A. Giannella, S. A. Broitman, and N. Zamcheck, J. Clin. Invest. 50, 1100 (1971).

TABLE III
AFFINITY CONSTANTS OF SOME
CYANOCOBALAMIN-BINDING PROTEINS

Binding protein	K_{eq} (M)	Reference
Intrinsic factor		
Human	0.67×10^{-10}	93
	0.17×10^{-10} (26°)	94
	0.67×10^{-10} (4°)	95
Pig	0.67×10^{-10} (4°)	96
Tranocobalamin II		
Human	0.33×10^{-11} (26°)	94
R-binding proteins		
Human blood	0.33×10^{-11} (26°)	94
Human saliva	0.67×10^{-10}	97
Dog stomach	0.53×10^{-11}	98
Antibodies	Sufficient for radioassay	
Bacteria		
B. subtilis	0.5×10^{-9}	99
E. coli ("wild type")	0.6×10^{-9}	99
S. paratyphi	2.0×10^{-9}	99

where B = binding protein, A = analog, and Cobl* = radioactive cyano-cobalamin. If the second incubation is relatively short the influence of the analog on the binding of radioactive cyanocobalamin will be exaggerated. In the second case, binding protein binds very rapidly with the first corrinoid that it meets:

$$10 \text{ B} + 20 \text{ A} + 20 \text{ Cobl*} \xrightarrow{\text{fast}} 5 \text{ B} \vdots \text{ A} + 5 \text{ B} \vdots \text{ Cobl*} + 15 \text{ A} + 15 \text{ Cobl*}$$

$$5 \text{ B} \vdots \text{ A} + 5 \text{ B} \vdots \text{ Cobl*} \xrightarrow{\text{slow}} 2 \text{ B} \vdots \text{ A} + 8 \text{ B} \vdots \text{ Cobl*} + 18 \text{ A} + 12 \text{ Cobl*}$$
$$+ 15 \text{ A} + 15 \text{ Cobl*} \qquad\qquad\qquad \text{Equilibrium state}$$

Again the influence of the analog is exaggerated. Only the equilibrium situation reflects the true cross-reaction.[100] The rate of attaining equilibrium is determined by the dissociation rate constant of the binder–analog complex. This rate constant is very small for corrinoid binding proteins.[95,96] This increase in cross-reaction has been described for steroids.[101] These considerations are important because we have reason to believe that the presence of analogs seriously affects the results of some cobalamin radioassays.[102] If these assays were performed under equilibrium conditions, systematic errors would be reduced.

[100] J. J. Pratt, W. Koops, M. G. Woldring, and T. Wiegman, *Eur. J. Nucl. Med.* **1**, 37 (1976).
[101] J. J. Pratt and M. G. Woldring, *Clin. Chim. Acta* **68**, 87 (1976).
[102] J. F. Kolhouse, H. Kondo, N. C. Allen, E. Podell, and R. H. Allen, *New Engl. J. Med.* **299**, 785 (1978).

Reliable data on the specificity of binding are hard to come by because, until recently, crude preparations of binding protein have been used. The cross-reaction curves obtained using a mixture of binding proteins can be complex. Figure 2 is a schematic representation of the cross-reaction curves using (a) a protein showing no cross-reaction with an analog, (b) a single binding protein showing high cross-reaction with the analog, and (c) a mixture of the two binding proteins. In each instance, it is assumed that small amounts of binding protein with a slight excess of radiolabeled cyanocobalamin are incubated to equilibrium in the presence of increasing amounts of nonlabeled cyanocobalamin or analog. Protein-bound radioactivity is separated and the amount given by the ordinate, the amount of added cobalamin or analog is given by the abscissa. We see that the error produced by the analog when a mixture of binding proteins is used is not equal to the mean of the errors found when the pure proteins are used but rather that the error produced by small amounts of analog is determined by the least specific binding protein in the mixture. The significance of these considerations for the specificity of radioassays has been discussed in detail in connection with radioimmunoassays using heterogeneous antibodies.[103,104] Data on the specificity of some binding proteins are shown in Table IV.

C. Properties of Binding Proteins

i. Intrinsic Factor. Intrinsic factor is a glycoprotein produced by the stomach of mammals. The general properties of the protein are given in an

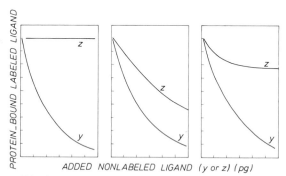

FIG. 2. Effect of binding protein heterogeneity on the specificity of corrinoid radioassay. For explanation, see text.

[103] H. Feldman, D. Rodbard, and D. Levine, *Anal. Biochem.* **45**, 530 (1972).
[104] J. J. Pratt, M. G. Woldring, R. Boonman, and J. Kittikool, *Eur. J. Nucl. Med.* **4**, 159 (1979).

TABLE IV

CROSS-REACTION OF SEVERAL CORRINOIDS WITH CYANOCOBALAMIN (100%) USING SEVERAL PROTEINS[a]

Analog, cq. change in cyanocobalamin	Human intrinsic factor	Human TC-II	Human granulo- cyte R-protein	Dog stomach binder	Human saliva
None (cyanocobalamin)	100	100	100	100	100
3-Carboxyethyl-	0.06	0.6	10		
8-Carboxyethyl-	30	30	60		
13-Carboxyethyl-	0.4	30	70		
13-Epibenzimidazoyl-	80	70	50		
cobamide	50	80	70		
2-Methyladenyl-5- cobamide	0.05	5	70		
Adenyl-5-cobamide (pseudo-vitamin B$_{12}$)	0.007	10	10	+(98)	100 (95)
Cobinamide	< 0.0001	0.07	10	+(98)	
B-ring lactam	+(105)			+(98)	

[a] Data are from reference 84 unless otherwise as given in parentheses.

excellent review by Gräsback.[106] Intrinsic factor binds cobalamins re-leased by digestive processes in the duodenum and jejunum. The glyco-protein is not attacked by digestive enzymes, but is adsorbed onto the sur-face of the ileal mucosa. Within the mucosal cells the complex of intrinsic factor is dissociated by an active process which is possibly mediated by a distinct cobalamin binding protein.[107] Cobalamin is finally released into the portal blood. Intrinsic factor is the first and probably the most impor-tant mechanism the mammal uses for discriminating against cobalamin analogs. Only the intrinsic factors of man and the pig have found use in radioassays. The intrinsic factor of ruminants should also be usable since indirect evidence indicates that it is very efficient at selecting cobalamin from the mass of analogs produced by the rumen microflora.[33,34] The "stomach cobalamin binder" of the dog is homogeneous but shows the specificity typical of "R-binding proteins."[98] The "intrinsic factor" of chicken proventriculi may be less specific than its mammalian counterpart since pseudo-vitamin B$_{12}$ interferes with the absorption of cobalamin in this species.[37]

There is some evidence that a histidyl residue of intrinsic factor dis-

[105] E. L. Lien, L. Ellenbogen, P. Y. Law, and J. M. Wood, J. Biol. Chem. 249, 890 (1974).
[106] P. Gräsbeck, in "Progress in Hematology" (E. B. Brown and C. V. Moore, eds.), Vol. 6, p. 233. Grune & Stratton, New York, 1969.
[107] G. Marcoulis, Biochim. Biophys. Acta 499, 373 (1977).

places the 5,6-dimethylbenzimidazoyl group from the Coα-position of co-balamin to form a coordination complex.[108] It is therefore possible that intrinsic factor discriminates between cobalamin and purine analogs, not by "binding" the imidazole derivative but by a chemical displacement reaction (see Section II,B,ii). Such a mechanism would explain the rather unusual specificity requirements of intrinsic factor with respect to the Coα-base: hindered benzimidazole derivatives such as linear naphthimi-dazoylcobamide bind well to intrinsic factor but unhindered purines such as 7-purinoylcobamide do not. Table IV shows that human intrinsic factor is specific for the substituents on the A- and C-rings and possibly for the "lower" base. Changes in the B-ring have little effect on binding. We have no information on the effect of changes in the D-ring. We see from Figs. 1 and 3[109] that the binding site of intrinsic factor is comparatively large. We may ask ourselves whether it is possible that the hapten-binding site of antibodies can be large enough to mimic intrinsic factor.

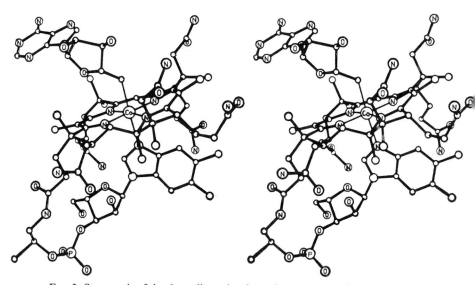

FIG. 3. Stereo pair of the three-dimensional covalent structure of 5'-deoxy-5'-adenosyl-cobalamin. The 3-D effect may be obtained with a viewer or by using the "crossed-eyes" technique. The molecule is viewed from slightly under the plane of the corrin ring and from between the C- and B-rings; the structure can therefore be directly compared with that in Fig. 1. Reproduced with modification and with permission from Hamilton,[109] copyright 1970 by the American Association for the Advancement of Science.

[108] E. L. Lien, L. Ellenbogen, P. Y. Law, and J. M. Wood, Biochem. Biophys. Res. Commun. 55, 730 (1973).
[109] W. C. Hamilton, Science 169, 133 (1970).

Because of its high specificity (Table IV), intrinsic factor is the binding factor of choice for the assay of cobalamin in mammalian tissues. Its relatively low affinity (Table III) accounts for the relatively low maximum attainable sensitivity reported by several workers. Intrinsic factor is a very stable protein so that it is likely that reports of its unsuitability in this respect probably reflect the fact that it is readily adsorbed onto glass surfaces when used in the absence of a substantial concentration of "protecting" protein.[110]

Gastric juice or crude stomach extracts (depepsinized by raising the pH to 10.5 for 30 min before readjusting to pH 7.5) invariably contain large amounts of "R-binding protein" activity. R-binding proteins are nonspecific (see below) and must be removed before use in radioassay. Intrinsic factor may be purified by a complex series of chromatographic steps,[106,111] however, the yield is low. A single stage, high yield, purification of intrinsic factor by affinity chromatography is possible.[95,96] The product of such a purification has been used for a highly specific cobalamin assay.[102] However, for use in radioassay it is only necessary to remove the R-proteins. This may simply be done by treating the mixture with trypsin which destroys the R-protein and leaves intrinsic factor intact[112] or by treating with antileukocyte or antihuman saliva sera which contain blocking antibodies which cross-react with other "R-binding proteins."[113,114] A much simpler method would be to add a gross excess of an analogue such as pseudo-vitamin B$_{12}$ or cobinamide to the mixture to block the binding by R-proteins but leave the specific binding activity of intrinsic factor intact. Such an approach has proved successful in steroid immunoassay where addition of a related steroid has been used to block nonspecific binding by plasma transport proteins[115-117] or the least specific antibodies in an antiserum.[104]

ii. Plasma Binding Proteins. Human blood plasma contains three cobalamin binding proteins: transcobalamin I (TC-I), transcobalamin II (TC-II), and transcobalamin III (TC-III). TC-II has the function of transporting cobalamins from the liver and possibly the ileum to the periferal tissues. The function of TC-I is less well defined: it may be involved in

---

[110] P. Newmark and N. Patel, *Blood* **38**, 524 (1972).

[111] L. Ellenbogen and D. R. Highly, *J. Biol. Chem.* **242**, 1004 (1967).

[112] R. H. Allen, B. Seetharam, E. Podell, and D. H. Alpers, *J. Clin. Invest.* **61**, 47 (1978).

[113] R. Wolff and J. P. Nicolas, *Lancet* **1**, 1008 (1967).

[114] R. Gräsbeck and H. Aro, *Biochim. Biophys. Acta* **252**, 217 (1971).

[115] H. Jurjens, J. J. Pratt, and M. G. Woldring, *J. Clin. Endocrinol. Metab.* **40**, 19 (1975).

[116] J. J. Pratt, T. Wiegman, R. E. Lappöhn, and M. G. Woldring, *Clin. Chim. Acta* **59**, 337 (1975).

[117] J. J. Pratt, R. Boonman, M. G. Woldring, and A. J. M. Donker, *Clin. Chim. Acta* **84**, 329 (1978).

transport of the cobalamins back to the liver, in sequestering cobalamin and its analogs in a form unavailable to invading microorganisms and parasites,[118] or simply as a source of nonspecific binding activity capable of intercepting cobalamin analogs during their transport through the body. TC-I has the properties of an "R-protein" (see below). TC-III has the properties of an R-protein and is present in small amounts in serum. There is evidence that its presence in serum and plasma is an artifact due to release by granulocytes *in vitro*. The chemical and physiological properties of plasma binding proteins are given in an excellent review by Allen.[119]

The usefulness of TC-II for radioassay appears to be limited because its specificity for cobalamin is intermediate between intrinsic factor and "R-binding proteins" (Table IV). The only advantage possessed by TC-II is that its affinity for cobalamin is substantially greater than that of intrinsic factor (Table III). Transcobalamin II may be purified from human plasma by affinity chromatography.[120] Alternatively, resort may be made to the sera of other animals which have been reported to contain only TC-II-like binding activity.[121]

iii. R-Proteins. A class of binding proteins found in all human tissues and body fluids have been given the name "R-binding-proteins." These proteins cross-react immunologically with each other and the assumption is that they all react identically with corrinoids, for which there is some evidence (Table IV). Considerable differences in physical properties are found between R-proteins from different tissues but these differences can be ascribed to differences in carbohydrate content and to different states of aggregation. The amino acid compositions of the R-proteins are similar if not identical. For our purposes, we may treat the R-proteins as a single entity.

The very high affinity for corrinoids (Table III), low specificity (Table IV), and exceptional stability of these proteins make them eminently suitable for corrinoid radioassay after chromatographic separation. Exceptionally sensitive radioassays have been developed after careful opimisation of assay conditions.[88]

There is little need to purify these proteins before use in radioassay because their most important property, lack of specificity, will be realised even if specific binding proteins are also present (see above). R-binding proteins in human serum and, presumably, other sources, are largely saturated with endogenous cobalamin.[119] Useful increase in the maximum at-

---

[118] H. S. Gilbert, *Blood* **44**, 38 (1974).
[119] R. H. Allen, *in* "Progress in Hematology" (E. B. Brown, ed.), Vol. 9, p. 57. Grune & Stratton, New York, 1975.
[120] R. H. Allen and P. W. Majerus, *J. Biol. Chem.* **247**, 7709 (1972).
[121] A. E. Finkler and P. D. Green, *Biochim. Biophys. Acta* **329**, 359 (1973).

tainable sensitivity could be achieved by removing this endogenous material either by dialyzing the soluble protein against 7.5 N guanidinium hydrochloride[119,122] or by treating solid-phase-bound R-protein with 85% phenol in water.[86] Human saliva is a cheap, reliable source of R-protein readily available to every analist.

iv. Antibodies. Two laboratories have reported the production of cobalamin-binding antibodies and their use in radioimmunoassays.[83,123,124, 125] Van de Wiel elicited the production of cyanocobalamin-binding proteins on injection of a noncovalent complex of cyanocobalamin with a commercial preparation of pig intrinsic factor which was probably heavily contaminated with pig R-proteins. A multi-emulsion in oil with properties similar to Freund's incomplete adjuvant was used. The cobalamin-binding activity was relatively low and showed the curious property of rapidly diminishing with a half-life less than that of IgG in the rabbit. A second immunogen was prepared by coupling deamidocyanocobalamin (prepared by brief treatment with nitrous acid) to bovine serum albumin using a carbodiimide. Only 0.23 molecule of cyanocobalamin per molecule albumin was incorporated. The adjuvant was again a "multiple emulsion." The binding was judged to be due to antibodies on the basis of electrophoretic mobility. The same rapid disappearance of binding activity was found. The authors suggested that this might be due to saturation with endogenous cobalamin: rabbits eat half of their daily fecal output so that their intake of cobalamin arising from the fecal microflora is very great indeed. A second explanation put foreward was that cells producing massive amounts of cobalamin-binding antibodies may be depriving themselves of the vitamin—a kind of suicide. In man, hypogammaglobulinemia is a symptom of vitamin B_{12} deficiency.[126]

Endres *et al.*[83] prepared monocarboxylic acids by acid hydrolysis of cobalamin and coupled this mixture to bovine serum albumin using isobutyl chloroformate. The extent of incorporation of hapten was not stated. Freund's complete adjuvant was used. The binding activity was judged to be due to antibody because it was coprecipitated with rabbit γ-globulin and sheep anti-rabbit γ-globulin. Unfortunately, no data on antibody con-

---

[122] J. A. Begley and C. A. Hall, *Blood* **45**, 281 (1975).
[123] D. M. F. van de Wiel, W. Th. Goedemans, and M. G. Woldring, *Clin. Chim. Acta* **56**, 143 (1974).
[124] D. M. F. van de Wiel, W. Th. Goedemans, J. A. de Vries, and M. G. Woldring, "Radioimmunoassay and Related Procedures in Medicine," Vol. II, p. 185. International Atomic Energy Agency, Vienna, 1974.
[125] G. D. Niswender and J. M. Hudson, *U.S. Patent* No. 3 981 863, Sept. 1976.
[126] C. K. V. van Dommelen, G. Slagboom, G. T. Meesters, and S. K. Wadman, *Acta Med. Scand.* **174**, 193 (1963).

centration in the course of immunization were given. Neither group reported data on cross-reactions with analogs.

Antibodies raised against hapten-protein conjugates are specific for that part of the free hapten which was the furthest from the protein in the immunogen[127-129] (see also other reviews in this volume). We can predetermine the specificity of the antisera by careful choice of position of attachment of hapten to the protein immunogen. When we come to consider the ideal specificity of antisera against cobalamin for use in assays in mammalian tissue we realize that we know nothing about the nature of biologically inactive analogs which are known to interfere with some existing radioassays (Section IX). If 2-carboxymethyl or 7-carboxymethyl groups which project above the plane of the corrin-ring (Figs. 1 and 3) are important analogs, we should attach cobalamin to the protein immunogen via a group which projects below the plane of the ring. If, however, Coα-liganded base analogs are important we should choose carboxygroups which project above the plane as positions of attachment. In this case we could also consider attachment of the cobalamin to the protein via the cobalt itself (see below).

There is one possibility that would raise serious immunochemical difficulties. We have seen that Coα-analogs, such as pseudo-vitamin B$_{12}$ (Coα base = adenine), are ubiquitous in the biosphere (Section III). It is possible that such analogs are responsible for the errors dealt with in Section IX. We shall see later that several benzimidazole derivatives are also found as the Coα-base and that these analogs are absorbed by man and that they are biologically active (Section VIII,E,iii, see also Section VII,C,i) and may be an important source of vitamin under some circumstances. We would then face the problem that an antiserum designed to discriminate against purine analogs would also discriminate against these benzimidazole derivatives.

We will make a number of suggestions for the production of suitable immunogens. Some of the derivatives listed have been used for affinity chromatography columns. None has yet been used for the successful production of antisera.

 a. Specific carboxy-groups. The best antisera were raised to conjugates of bovine serum albumin with mixed monocarboxy acids. We have no information on the sequence of deamidation of cobalamin by nitrous acid. We know however, that 13-carboxyethyl and 18-carboxymethyl de-

[127] K. Landsteiner, "Specificity of Serological Reactions." Harvard Univ. Press, Cambridge, Massachusetts, 1945.

[128] S. J. Gross, D. H. Campbell, and H. H. Weetall, *Immunochemistry* **5,** 55 (1968).

[129] G. D. Niswender and A. R. Midgley, *in* "Immunologic Methods in Steroid Determination" (F. G. Peron and B. V. Caldwell, eds.), p. 149. Appleton, New York, 1979.

rivative predominate in the monocarboxy fraction after acid hydrolysis.[10] We see from Fig. 3 that the 13-carboxyethyl substituent projects below the plane and rests next to the Coα-base. The isolated 18-carboxymethyl derivative (axial to the D-ring) should be more suitable for the preparation of an immunogen. Simple methods are available to separate these derivatives[10] which could be used to produce a wide range of well-defined immunogens.

b. *The phosphate group.* Direct attachment of hydroxycobalamin to polylysine and albumin using carbodiimide has been reported.[130] The conclusion of the authors that attachment took place via the phosphate group is based on very slender evidence. Any antisera produced against such conjugates would be specific for the β-face of the corrin-ring.

c. *Coβ-derivatives.* Coβ-carboxypropylcobalamin has been synthesised and an attempt made to couple it to polylysine and to raise antibodies.[131] No information was given as to the incorporation of cobalamin to form the immunogen. It is not clear whether the small amount binding activity measured in the serum was due to antibodies or to serum R-proteins. A similar derivative has been prepared for use in affinity chromatography of intrinsic factor.[105] A bromoalkane derivative of agarose was mixed with hydroxycobalamin in aqueous solution, sodium borohydride was added to reduce the cobalamin to the Co(I) state. Co(I) cobalamin reacted with the bromalkyl agarose and the product was allowed to autoxidise to give the desired Coβ-derivative. The reduction step must be performed under strictly anaerobic conditions in the dark or under subdued red light. If such a method were adapted for the production of a hapten–protein immunogen, all subsequent operations including immunization should be performed under subdued red light to prevent photolysis. Immunization with an emulsion in Freund's complete adjuvant should be via the intramuscular route to avoid *in vivo* photolysis at the more usual intracutaneous sites.

d. *5'-Deoxyadenosyl group.* Direct attachment of polyglutamic acid to 5'-deoxy-5'-adenosylcobalamin via the adenine amino group using carbodiimide has been claimed.[130] The product was not well characterized but appeared to be impure with respect to the position of attachment.

e. *The ribose moiety of cyanocobalamin.* A mixture of hemisuccinates was prepared for physiological experiments in bacteria.[132] Such a product might be used to prepare a protein complex via the hemisuccinate carboxyl groups.

v. *Binding Proteins from Bacteria and Other Microorganisms.* Many, but not all, bacteria and microorganisms contain corrinoid binding pro-

[130] H. Olsen, E. Hippe, and E. Haber, *Biochim. Biophys. Acta* **243**, 66 (1971).
[131] H. Gershman, N. Nathanson, R. H. Abeles, and L. Levine, *Arch. Biochem. Biophys.* **153**, 407 (1972).

teins. Table III shows the cyanocobalamin-binding properties of binding proteins from several species. Little is known of the specificity of these binding proteins. We may expect that the specificity of the corrinoid concentrating mechanism would be no more specific than the growth requirements of that organism. This is borne out by the observation that *E. coli* mutants will bind cyanocobinamide just as well as cyanocobalamin.[133] The binding by *O. malhamensis* is specific for "complete" corrinoids but Coα-analogs of cobalamin, such as pseudo-vitamin B_{12}, are bound as strongly as cyanocobalamin.[134] In *O. malhamensis* the specificity for benzimidazole-containing corrinoids for support of growth is presumably imparted by a second "proofreading" process.

Binding proteins from microorganisms are suitable for corrinoid radioassay after chromatographic separation but cannot be considered for direct cobalamin radioassay in mammalian tissue. This conclusion is supported by the observations of Van de Wiel *et al.*[135] who found large *individual* discrepancies between a bioassay and a radioassay, both using *L. leichmannii* when human serum was analyzed.

The original description of bacterial binding proteins[136] makes the potentially very useful suggestion that bacteria could be used to concentrate and purify cobalamin from dilute sources. The authors of this article came very close to being the first to describe a competitive protein binding radioassay.

vi. Other Binding Proteins. As stressed in Section III, corrinoids exist in nature largely in protein-bound form. A range of binding proteins may be expected in all species which require vitamin B_{12}. This huge field constitutes an interesting area of investigation for the biologist and those seeking a binding protein suitable for radioassay. Binding proteins in the blood of some amphibians have been investigated.[137] Binding proteins from fish sera[138-140] and chicken serum[141,142] are present in very high con-

[132] T. Toraya, K. Ohashi, H. Ueno, and S. Fukui, *Vitamins (Japan)* **48,** 557 (1974).
[133] J. C. White, P. M. di Girolamo, M. L. Fu, Y. A. Preston, and C. Bradbeer, *J. Biol. Chem.* **248,** 3978 (1973).
[134] J. E. Ford, *J. Gen. Microbiol.* **19,** 161 (1958).
[135] D. M. F. van de Wiel, J. A. de Vries, M. G. Woldring, and H. O. Nieweg, *Clin. Chim. Acta* **55,** 155 (1974).
[136] R. L. Davis and B. F. Chow, *Science* **115,** 351 (1952).
[137] D. W. Sonneborn and H. J. Hansen, *Proc. Soc. Exp. Biol. Med.* **136,** 903 (1961).
[138] H.-R. Kim, *Korean J. Biochem.* **7,** 65 (1975).
[139] D. O. Kubiatowicz, D. S. Ithakissios, and D. C. Windorski, *Clin. Chem.* **23,** 1037 (1977).
[140] J. W. Buchanan, P. A. McIntyre, U. Scheffel, and N. Wagner, *J. Nucl. Med.* **18,** 394 (1977).
[141] P. A. Newmark, R. Green, A. M. Musso, and D. L. Mollin, *Br. J. Haematol.* **25,** 359 (1973).
[142] R. Green, P. A. Newmark, A. M. Musso, and D. L. Mollin, *Br. J. Haematol.* **27,** 507 (1974).

centrations and have been used for radioassays. Unfortunately, the results obtained on analyzing human sera indicate that fish- and chicken-binding proteins have properties similar to human R-proteins.

Parasites face particularly acute problems in obtaining their supplies of the vitamin. They must compete with mammalian binding proteins such as intrinsic factor and the transcobalamins for limited supplies of the vitamin and at the same time avoid being overwhelmed by analogs rejected by the host. On teleological grounds, we may expect parasites to be a source of specific binding proteins of high affinity.

VIII. Radioassay Technique

A. Principle

Dilutions of binding protein (binding capacity, about 70 pg per tube) are incubated with a slight excess of radioactive corrinoid, usually cyano-cobalamin (about 100 pg per tube) and with standard solution or sample. Nonlabeled corrinoid competes with the labeled material as described in other reviews in this volume. Protein-bound radioactivity is physically separated from the "free" fraction. A graph of the amount of standard is prepared. Unknowns can be read off this curve. Using higher concentrations of binding protein and radioactive cobalamin, binding sites on the protein are saturated at all times and some authors have used a simple isotope dilution formula to calculate the results without the use of a standard curve.[143] Reducing the concentrations of reactants to near the equilibrium constant of binding of cobalamin with its binder increases the sensitivity greatly, but does mean that a standard curve has to be used for each assay batch. The theory and general principles of competitive protein binding assay together with statistical aspects of the calculation of results are dealt with in a number of reviews.[87−92,144−148]

B. Radioactive Corrinoids

³H-, ¹⁴C-, and ³²P-labeled corrinoids have been prepared by growing bacteria in the presence of radioactive precursors. The usefulness of these

[143] K. S. Lau, C. Gottlieb, L. W. Wasserman, and V. Herbert, *Blood* **26,** 202 (1965).

[144] A. Zettner, *Clin. Chem.* **19,** 699 (1973).

[145] G. E. Abraham, ed., "Handbook of Radioimmunoassay." Dekker, New York, 1977.

[146] T. Chard, "An Introduction to Radioimmunoassay and Related Techniques." North-Holland Publ., Amsterdam, 1978.

[147] R. P. Ekins, G. B. Newman, and J. L. H. O'Riordan, *in* "Statistics in Endocrinology" (J. S. McArthur and T. Colton, eds.), p. 345. M.I.T. Press, Cambridge, Massachusetts, 1970.

[148] D. Rodbard and G. R. Frazier, this series, Vol. 37, p. 3.

isotopes is limited by the reduction in specific activity which occurs by dilution of the precursors. By taking the proper precautions to ensure cobalt deficiency, radioactive cobalt can be incorporated biosynthetically with little loss of specific activity. The isotope of choice is ^{57}Co because it can be produced cheaply at a high specific activity and has convenient radiation characteristics. ^{57}-Co-labeled corrinoids are usually measured in an automatic well-type counter but liquid scintillation counting can also be used.[149]

[^{57}Co]Cyanocobalamin is the usual commercially available form of cobalamin. Other forms of cobalamin are required for recovery estimates (Section V). Radioactive hydroxycobalamin may be prepared by isolating cyanocobalamin chromatographically, to remove excess cyanide from commercial preparations, and subjecting a neutral aqueous solution to direct sunlight for several hours. Radioactive hydroxycobalamin is then isolated chromatographically from unreacted cyanocobalamin and other products of photolysis. Methyl- and 5'-deoxy-5'-adenosylcobalamin may be prepared by reducing [^{57}Co]cyanocobalamin (purified of excess cyanide) with sodium borohydride, adding methyl chloride or 5'-deoxy-5'-chloroadenosine in excess, and allowing the mixture to autoxidise as described for the preparation of similar nonradioactive derivatives (Section VII,C,iv,c). Radioactive analogs may be prepared biosynthetically on a small scale from ^{57}CoCl$_2$ using *Propionibacterium arabinosum*.[84] A novel radiolabeled derivative is ^{125}I-labeled tyrosine methyl ester in amide linkage with monocarboxy derivatives of cyanocobalamin.[83] Such a label is useable only for radioimmunoassays where the tyrosine methyl ester is attached to the same position on the cobalamin as was the protein used as immunogen for the production of the antisera. This is because the antibodies are unable to "recognize" chemical changes in the hapten if these changes are made at the same position as in the immunogen (Section VII,C,iv). The ability to attach a radioiodinated group to a hapten for use in radioimmunoassay automatically means that nonradioactive "labels" can be used similarly. It should be possible to develop enzyme-, chemiluminescence-, coenzyme-, or bacteriophage-linked immunoassays for cyanocobalamin.

C. Incubation

As outlined in Section VII incubation to equilibrium is desirable where interfering analogs may be present, by which we mean complete equilibration of all components, including standards and analogs, and not merely incubation until maximum binding of labeled cobalamin has been attained.

[149] S. Gutcho, J. Johnson, and H. McCarter, *Clin. Chem.* **19,** 998 (1973).

D. Separation of "Bound" and "Free"

The earliest competitive protein binding assay for cobalamin used dialysis for the removal of non-protein-bound radioactivity.[150] This method is not suitable for routine use. Active charcoal suspensions can be used in conjunction with intrinsic factor,[102,143] serum "R-proteins,"[151] and saliva "R-proteins"[152] to adsorb "free" radioactive cobalamin leaving the "bound" fraction in the supernatant. The supernatant is decanted and the radioactivity is measured. Separation by adsorption does not disturb the equilibrium appreciably because the half-life of dissociation of most cobalamin binders is very long. Active charcoal separation in the presence of high protein concentrations can be used for "robust" methods and large assay series. Before about 1972, it was customary to use charcoal suspensions in buffers containing little or no protein. Residual proteins present after simple extraction with acetate buffers influenced the separations of "bound" and "free" producing erratic results. The effect of residual protein can be reduced by performing the assay in buffers containing 1 mg ml^{-1} of bovine serum albumin.[65] This concentration of albumin also prevents absorption of intrinsic factor onto glass. Residual protein after serum extraction has a second influence on the results: the protein binds radioactive cyanocobalamin giving rise to falsely low results, whatever method is used to separate "bound" from "free." This source of error is reduced by using high ionic strength buffer at a pH optimum for protein precipitation (Section V,C,i). Extraction methods which do not involve precipitation of the denatured protein (Section V,C) are especially prone to this sort of error and are best avoided. The effect of residual binding protein in the extract can be further reduced by performing the assay in the presence of additional bovine serum albumin as above.

Adsorption of transcobalamin-bound radioactivity onto DEAE-cellulose has been used.[153,154] This method cannot be used in conjunction with buffers containing the high protein concentrations which are necessary to avoid errors (see above). In order to avoid this limitation, Van de Wiel et al.[155] used standards prepared in "vitamin-B_{12}-free serum."

Many "solid-phase" methods have been used to separate "bound" from "free" radioactivity, largely in an attempt to avoid difficulties with

[150] R. M. Barakat and R. P. Ekins, Lancet 2, 25 (1961).
[151] N. Grossowicz, D. Sulitzeanu, and D. Merzbach, Proc. Soc. Exp. Biol. Med. 109, 604 (1962).
[152] R. Carmel and C. A. Coltman, J. Lab. Clin. Med. 74, 967 (1969).
[153] E. P. Frenkel, S. Keller, and M. S. McCall, J. Lab. Clin. Med. 68, 510 (1966).
[154] M. Wagstaff and A. Broughton, Br. J. Haematol. 21, 581 (1971).
[155] D. F. M. van de Wiel, W. Th. Goedemans, and M. G. Woldring, Clin. Chim. Acta 56, 143 (1974).

earlier charcoal separation methods. Adsorption onto the inner surface of plastic test tubes has been used for intrinsic factor[156] and antibodies.[83] Unfortunately this method cannot be combined with the use of the high albumin concentrations, necessary to avoid nonspecific interference, because exchange of bovine serum albumin for cobalamin binder at the plastic surface occurs.

Covalent complexes of binding protein with solid particles offer the best alternative to charcoal. Radioassays using intrinsic factor covalently attached to cellulose,[67] Sephadex,[67] Sepharose,[157] and many others have been used. There is no reason why any one of the many available solid phase systems should not be used in combination with either intrinsic factor or R-proteins. Binding proteins in bacteria provide the experimenter with a ready source of "nonspecific" binding protein. In this case, the intact bacterium serves as the "solid phase."[135]

E. Criteria of Accuracy

No matter how specific the binding protein or how thorough the chromatographic purification, unexpected cross-reactions (Section VII) or nonspecific interferences (see above) may still be encountered. It is always necessary to test an analytical method according to several criteria. As we shall see in the next section, cobalamin assays in human blood serum have not been rigorously tested.

In the steroid field, specific cross-reactions and nonspecific interferences due to positive and negative blanks, sample destruction, incomplete recovery, or interference in the competitive binding step by proteins or lipids were recognised at the outset. The problems involved in corrinoid analysis are no less complex so that the same criteria of accuracy must be applied.[158,159]

i. Recovery of pure, unlabeled corrinoid added to representative samples should be 100% throughout the analytical range. [This recovery estimate is distinct from the use of a small amount of radioactive corrinoid to correct for procedural losses in some purifications (Section V).] In the case of animal tissues or human serum, the recovery of added unlabled methyl- and 5'-deoxy-5'-adenosylcobalamin and not cyanocobalamin should be tested because these compounds may not behave in the same way as cyanocobalamin.[43]

ii. The graph of amount of analyte, estimated after all corrections for

[156] J. R. Rubini, in "In Vitro Procedures with Radioisotopes in Medicine," p. 355. International Atomic Energy Agency, 1970.
[157] P. J. Brombacher, A. H. J. Gijzen, and M. P. J. Soons, Clin. Chim. Acta **52**, 311 (1974).
[158] Anonymous, J. Endocrinol. **72**, 1 (1977).
[159] J. J. Pratt, Clin. Chem. **23**, 1190 (1977).

procedural losses, set out against the volume or weight of sample must be linear. This is equivalent to saying that the sample dilution curve must run parallel to the standard dilution curve.

iii. The assay in its proposed form must give the same results as a more complex assay which involves an additional purification. The more complex "reference method" usually involves extraction and at least one chromatographic purification of each sample using individual recovery estimates using a very small amount of radioactive analyte. We should be aware of the possibility that biologically *active* analogs may be present in human tissue. Benzimidazoyl-, 5-methylbenzimidazoyl-, and 6-methylbenzimidzoylcobamides occur naturally, are absorbed via the intrinsic factor mechanism, and are active in man[2,160,161] (Table II).

iv. The proposed method should give results similar to those obtained using a totally independent method. In the case of cobalamin radioassay, a suitable bioassay is the obvious choice for such a comparison. In cases of discrepancies, it should not be forgotten that the "reference method" might be in error (Section IX,B,iii).

IX. Vitamin B_{12} Radioassay in the Clinical Laboratory

By far the most important applications of corrinoid radioassay are in the clinical laboratory. Assays on serum are very important aids in diagnosis and therapy. Assays on tissue biopsies[69,162] are restricted to clinical research. The requirements of the clinical laboratory are dominated by the need to provide clinically useful results, i.e., to provide a clear distinction between vitamin B_{12} deficiency and repletion. This requirement is more important than even the provision of an analytically valid assay. The choice of assay method for the clinical laboratory is made on the basis of cost effectiveness so that only methods which avoid the use of chromatography and individual recovery estimates can be considered.

A. Clinical Applications

Increased serum cobalamin concentrations are found in acute degenerative liver disease and myeloproliferate disorders, but vitamin B_{12} assays play a relatively minor role in such cases. Vitamin B_{12} deficiency in man results in intestinal malabsorption, megaloblastic anemia (which is reversible), and degeneration of the spinal cord and cerebrum (which may

[160] A. L. Latner and L. Raine, *Nature* (*London*) **180,** 1197 (1957).
[161] E. K. Blackburn, H. T. Swan, G. R. Trudhope, and G. M. Wilson, *Br. J. Haematol.* **3,** 429 (1957).
[162] E. P. Frenkel, J. D. White, C. Galey, and C. Croy, *Am. J. Clin. Pathol.* **66,** 863 (1976).

or may not be reversible). Untreated deficiency invariably leads to incapacity and often in death. Deficiency has a variety of causes. The most important cause for which the name "pernicious anemia" is generally reserved is an autoimmune reaction against intrinsic factor or the parietal cells of the stomach. Intestinal malabsorption due to excess acid production by the stomach, pancreatic insufficiency, intestinal infestation by bacteria or helminths, or to alcohol abuse is an important cause of deficiency. Clinical aspects of deficiency are discussed in a number of reviews.[55,163-166] The correlation of serum vitamin B_{12} assays with total body reserves is generally valid.[167-169] Discrepancies due to a number of causes are however found, so that a number of other clinical tests must be performed to confirm any diagnosis. One of the causes of these discrepancies is inaccuracy of the serum vitamin determination, so that a reliable assay method would reduce the necessity for the more complex confirmatory tests.

B. Errors in Clinical Assay Methods

The history of clinical assay methods for vitamin B_{12} is a sad one caused by lack of appreciation of sources of error.

i. Errors Caused by Residual Protein. The amount of residual protein present after extraction, especially at low ionic strength, varies from one individual serum to the next. Falsely low results caused by this protein have been discussed in the previous section. In early work published before about 1972, these lowered results tended to mask the falsely raised values caused by the presence of analogs (see below) so that reasonable "correlation" with bioassays was found. Correction of the protein error gave higher results, so that greatly increased reference values has to be used.

ii. Biologically Inactive Analogs. The presence of cobalamin analogs in human serum which interfere with radioassays was suggested some years ago.[170] Kolhouse *et al.*[102] have recently provided evidence for the presence of analogs which react with "R-proteins" but not with pig intrinsic factor. The possibility that these analogs are biologically inactive in

[163] I. Chanarin, "The Megaloblastic Anaemias." Blackwell, Oxford, 1969.

[164] S. Waxman, *Med. Clin. N. Am.* **57**, 315 (1973).

[165] V. Herbert, *in* "The Pharmacological Basis of Therapeutics" (L. S. Goodman and A. Gilman, eds.), 5th ed., p. 1324. Macmillan, New York, 1975.

[166] E. Neuman, *Med. Lab.* **29**, 184 (1976).

[167] K. G. Ståhlberg, S. Radner, and Å. Nordén, *Scand. J. Haematol.* **4**, 312 (1967).

[168] J. F. Adams, H. I. Tankel, and F. MacEwen, *Clin. Sci.* **39**, 107 (1970).

[169] E. Magnus, *Scand. J. Gastroenterol.* **29**, 47 (1974).

[170] S. P. Rothenberg, *Metabolism* **22**, 1075 (1973).

man or may even be cobalamin antagonists was discussed. These analogs are particularly important in cases of vitamin B$_{12}$ deficiency. Their presence appears to be responsible for the high values found by radioassays and, worse, for the overlap between clinical deficiency and repletion found using these methods.

These conclusions are not as black and white as these authors suggest. Many investigators have reported satisfactory separation of deficiency and repletion using less specific methods.[152,171] Discrimination may also depend on factors such as quality of reagents, robustness of the assay technique, and the use of a favorable part of the standard curve (see below).

The occurrence of analogs in the biosphere (Section III) and the lack of specificity of "R-proteins" (Section VII) have long been known and should have warned against assuming that assays on unfractionated material were accurate. Several early publications indicated the presence of analogs in mammalian tissue.[33,34,172] We may only speculate as to the nature of the analogs. Coα-purine analogs produced by bacteria may possibly be absorbed to a small extent from the colon. We know that cyanocobalamin can be absorbed in small amounts independently of intrinsic factor, even when given rectally.[173-175] Alternatively, we may postulate that cobalamin is degraded very slowly in the body, possibly by phosphodiesterases or amidases known to be present in the liver.[176,177] Whatever their source, analogs, once in the body, are only very slowly removed.[178]

iii. Comparison of Methods. Results obtained with radioassays have often been compared with those obtained by bioassay, using *O. malhamensis, Euglena gracilis,* or *L. leichmannii.* Accurate assays should give a 1:1 correspondence as a necessary, but not sufficient, criterion of specificity. In the field of cobalamin radioassays on serum such a correspondence has yet to be demonstrated convincingly. This is because there is reason to doubt the analytical validity of both radioassays and bioassays (see below).

Even correlation about the line of identity, which has seldom been achieved, can mask large *individual* discrepancies. By individual discrepancies we mean that sera from several patients may give values of 200 pg

[171] D. M. Mathews, R. Gunasegaram, and J. C. Linnell, *J. Clin. Pathol.* **20,** 683 (1967).
[172] J. E. Ford, *Br. J. Nutr.* **7,** 299 (1953).
[173] A. Doscherholmen and P. S. Hagen, *J. Clin. Invest.* **36,** 1551 (1957).
[174] H. Berlin, R. Berlin, and G. Brante, *Acta Med. Scand.,* **184,** 247 (1968).
[175] I. L. Mackenzie and R. M. Donaldson, *Fed. Proc. Fed. Am. Soc. Exp. Biol.* **28,** 41 (1969).
[176] H. G. Bray, S. P. James, W. V. Thorpe, and M. R. Wasdell, *Biochem. J.* **47,** 294 (1950).
[177] A. R. Williamson, M. R. Salaman, and H. W. Kreth, *Ann. N.Y. Acad. Sci.* **209,** 210 (1973).
[178] H. C. Heinrich and E. E. Gabbe, *in* "Vitamin B-12 und Intrinsic Factor" (H. C. Heinrich, ed.), p. 252. Enke, Stuttgart, 1962.

ml^{-1} with a radioassay but give values between 150 and 300 pg ml^{-1} with bioassay; other sera may give values of 400 pg ml^{-1} with bioassay but a range of 300 to 600 pg ml^{-1} with radioassay. Such discrepancies are well outside even the most pessimistic estimates of assay reproducibility. Several publications record individual discrepancies of this order of magnitude.[179-181] Any discrepancies should be investigated not only because they affect the diagnostic usefulness of the method but also because they are interesting in themselves. Raven et al.[179] give detailed evidence that discrepancies are large and are consistent in any one patient over a period of several months. Clinical chemistry deals with highly variable material and should not allow individual patients to be buried in statistics.

Owing to the apparent ease of most radioassays, such determinations are requested by the physician early in the diagnostic process. This has the consequence that nearly all sera received by the laboratory are normal. Many investigators have been content to use this material retrospectively for comparison of methods. Experience with hormone radioassays teaches us the pattern of interfering materials in abnormals is different from that in normal people and has, moreover, a greater influence on the results. The situation with vitamin B_{12} radioassay now appears to be similar.[102] Clinical chemistry is the science of abnormality and reliance should not be placed in comparisons using material from normal people or unselected patients.

Bioassays are themselves subject to systematic errors due to (a) compounds which can bypass the cobalamin requirement (Section III), (b) the presence of analogs which may replace cobalamin or may act as antagonists (Sections III and VII), (c) incomplete extraction, or (d) vitamin destruction during extraction. Incomplete extraction[43,68] and vitamin destruction during extraction[182,183] will affect bio- and radioassays equally so that even a 1:1 correspondence of results is not a sufficient criterion of accuracy. Interpretation of bioassays is made even more difficult by the occurrence of very large *individual* differences between results in assays using *Euglena gracilis* and *L. leichmannii*.[179,184]

[179] J. L. Raven, M. B. Robson, J. O. Morgan, and A. V. Hoffbrand, *Br. J. Haematol.* **22,** 21 (1972).

[180] C. E. Voogd and M. J. Mantel, *Clin. Chim. Acta* **54,** 369 (1974).

[181] A. Killander and L. Weiner, *Scand. J. Gastroenterol. Suppl.* **29,** 43 (1974).

[182] D. H. Orrell and A. D. Caswell, *J. Clin. Pathol.* **25,** 181 (1972).

[183] V. Herbert, E. Jacob, K.-T. J. Wong, J. Scott, and R. D. Pfeffer, *Am. J. Clin. Nutr.* **31,** 253 (1978).

[184] R. E. Davis, J. Moulton, and A. Kelly, *J. Clin. Pathol.* **26,** 494 (1973).

C. The Future of Clinical Radioassays for Vitamin B_{12}

There is now an urgent need for a specific, well-documented radioassay method for serum vitamin B_{12}. Such an assay would probably use purified intrinsic factor (Section VII,C,i). Only very preliminary data on the use of antisera have yet been published (Section VII,C,iv). Any radioimmunoassays must be thoroughly tested because the sources of nonspecificity are many and the consequences of error are great. We are now in a position to give the minimal chemical and clinical criteria by which any proposed assay should be judged (Sections VII,E and IX,B).

One difficulty which may encountered with intrinsic factor as binding agent is its moderate affinity for cyanocobalamin which restricts the maximum achievable sensitivity (Section V,A). High sensitivity is desirable for a number of reasons. We have seen that residual protein in the sample is liable to affect the results obtained with intrinsic factor (Section IX,B,i). A high ratio of extractant to serum (e.g., 1.8 ml acetate-cyanide to 0.2 ml of serum) reduces the amount of residual protein and simultaneously allows us to perform the assay in duplicate to detect the very occasional erratic results to which radioassays are prone. In addition the sensitivity of the assay should be adjusted to give the maximum discriminatory power, defined in the proper statistical terms,[88,89] in the region of 100 to 200 pg ml^{-1} of serum. It is in this region that the most important clinical decisions concerning deficiency or repletion are made. These considerations mean that 10 to 20 pg of cobalamin per tube should fall in the middle of the radioassay curve.

X. Conclusion

Assay of corrinoids from primary sources requires preliminary extraction and chromatographic purification. If the concentrations are large, methods other than radioassay might be considered. High performance liquid chromatography (HPLC) of corrinoids has a detection limit of a few nanograms which is sufficient for many purposes. An interesting possibility is to use atomic absoprtion spectrometry in conjunction with HPLC. The detection limit of such a method should overlap the analytical range of radioassays. Mammalian ''R-proteins'' and bacteria are suitable binders for radioassay after chromatographic separation.

The advantage of simple radioassays over bioassays in the clinical laboratory is that they are less exacting for the laboratory and personnel. Modifcation of the L. leichmannii bioassay which uses chloramphenicol containing media and chloramphenicol-resistant bacteria to avoid the need for sterile technique[184] should not be encouraged because of the risk of introducing multiple-resistance plasmids into the hospital environment. The belated recognition of the importance of analogs in human tissue sets us the task of providing a standardized, well-documented radioassay method. Experience with clinical radioassay for cobalamin highlights many sources of error and falacy in other binding assays, including the various forms of immunoassay.

Acknowledgments

We wish to thank Miss Willy Börger for expert secretarial assistance and Mr. A. Rijskamp for drawing the figures.

Section VII

Drugs

A. Antineoplastic Agents
Articles 28 through 34

B. Drugs Active on the Nervous System and Neurotransmitters
Articles 35 through 41

C. Other Drugs
Articles 42 through 45

[28] Radioimmunoassay of Methotrexate, Leucovorin, and 5-Methyltetrahydrofolate

By JOHN J. LANGONE

Methotrexate (MTX; Fig. 1) and related folic acid antagonists have been used for over 30 years as antineoplastic agents.[1-4] Recently, improved treatment of certain cancers has been achieved by using high-dose MTX followed by "rescue" with leucovorin[2-4] (5-formyltetrahydrofolate or citrovorum factor; Fig. 1). 5-Methyltetrahydrofolate (5-MTHFA; Fig. 1) is a metabolite of leucovorin and may play a role in its therapeutic effectiveness.[5-7] In addition, 5-MTHFA is the major circulating and storage form of folate.[8]

General Considerations

Several reports of radioimmunoassays (RIA) for MTX have appeared since 1974[9-19] including a fully automated system.[20] The immunizing conjugates, species of animals immunized, radiolabeled tracers used, and methods of separating antibody-bound from free labeled tracer are summarized in Table I. In each case, a carbodiimide reaction was used to couple MTX through its carboxyl group(s) to the free amino groups of the

---

[1] S. Farber, L. K. Diamond, R. D. Mercer, R. F. Sylvester, and J. A. Wolff, *New Engl. J. Med.* **238**, 787 (1948).

[2] B. A. Chabner and M. Slavik, *Cancer Chemother. Rep.* **6** (Pt. 3), 1 (1975).

[3] I. Djerassi, *Cancer Chemother. Rep.* **6** (Pt. 3), 3 (1975).

[4] J. S. Penta, *Cancer Chemother. Rep.* **6** (Pt. 3), 7 (1975).

[5] J. Perry and I. Chanarin, *Br. J. Haematol.* **18**, 329 (1970).

[6] V. M. Whitehead and H. A. Stein, *Biochem. Soc. Trans.* **4**, 918 (1976).

[7] B. M. Mehta, A. L. Gisolfi, D. J. Hutchinson, A. Nirenberg, M. G. Kellick, and G. Rosen, *Cancer Treat. Rep.* **62**, 345 (1978).

[8] V. Herbert, A. R. Larrabee, and J. M. Buchanan, *J. Clin. Invest.* **41**, 1134 (1962).

[9] L. Levine and E. Powers, *Res. Commun. Chem. Pathol. Pharmacol.* **9**, 543 (1974).

[10] C. Bohuon, F. Duprey, and C. Boudene, *Clin. Chim. Acta* **57**, 263 (1974).

[11] V. Raso and R. Schreiber, *Cancer Res.* **35**, 1407 (1975).

[12] V. Raso, *Cancer Treat. Rep.* **61**, 585 (1977).

[13] L. J. Leoffler, M. R. Blum, and M. A. Nelsen, *Cancer Res.* **36**, 3306 (1976).

[14] J. Hendel, L. J. Sarek, and E. F. Hvidberg, *Clin. Chem.* **22**, 813 (1976).

[15] G. W. Aherne, E. M. Piall, and V. Marks, *Br. J. Cancer* **36**, 608 (1977).

[16] J. W. Paxton and F. J. Rowell, *Clin. Chim. Acta* **80**, 563 (1977).

[17] J. W. Paxton, and F. J. Rowell, and C. M. Cree, *Clin. Chem.* **24**, 1534 (1978).

[18] R. S. Kamel and J. Gardner, *Clin. Chim. Acta* **89**, 363 (1978).

[19] J. J. Langone, *J. Immunol. Methods* **24**, 269 (1978).

[20] R. Kamel, J. Landon, and G. C. Forrest, *Clin. Chem.* **26**, 97 (1980).

METHODS IN ENZYMOLOGY, VOL. 84   ISBN 0-12-181984-1

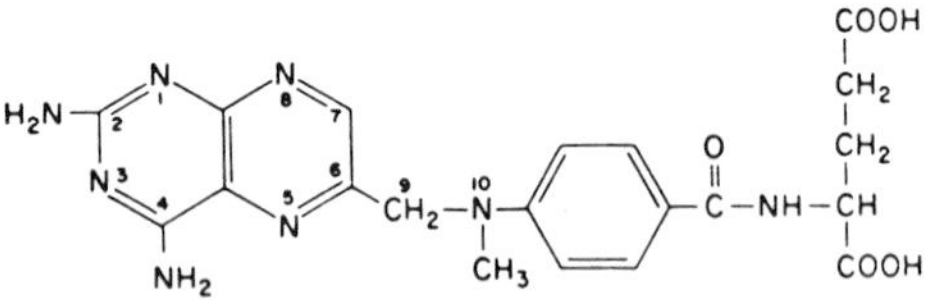

MTX

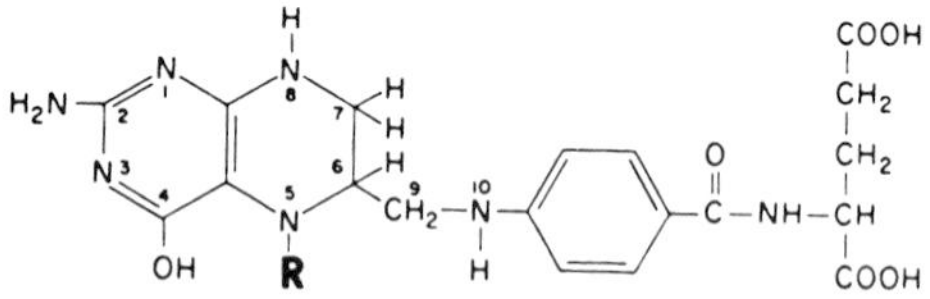

Leucovorin : R = CHO

5 − MTHFA : R = CH$_3$

FIG. 1. Chemical structures of methotrexate (MTX), leucovorin (i.e., 5-formyltetrahy-drofolate or citrovorum factor), and 5-methyltetrahydrofolate (5-MTHFA).

protein or polypeptide carrier. Tracer molecules include [³H]MTX or MTX derivatives labeled with ^{125}I or^{75}Se. ^{125}I-labeled protein A (^{125}I-PA) is the only tracer that does not rely on radiolabeled MTX or a labeled drug derivative. Since the use of ^{125}I-PA obviates the need for labeled hapten, it also has been used in sensitive and specific solid phase assays for leuco-vorin[21] and 5-MTHFA.[22] These drugs are relatively labile, and sufficiently stable derivatives of high specific activity have not been available for routine development of classical RIAs.

Development of the Assays

Preparation of Immunogens

We used keyhole limpet hemocyanin (KLH) as the carrier and 1-ethyl-3-(3-dimethylaminopropyl) carbodiimide (EDAC) as the coupling agent to prepare immunogens from MTX, leucovorin, and 5-MTHFA.[1,21,22] The following procedure is representative.

[21] J. J. Langone and L. Levine, *Anal. Biochem.* **95,** 472 (1979).
[22] J. J. Langone, *Anal Biochem.* **104,** 347 (1980).

Reagents
   25 mg KLH (Schwarz/Mann, Orangeburg, NY)
   12 mg MTX, leucovorin, or 5-MTHFA (purchased from Sigma
   Chemical Co., St. Louis, MO)
   25 mg EDAC (Bio-Rad Laboratories, Rockville Center, NY)
   Procedure. Separate solutions of hapten and EDAC in 0.5 ml water
and KLH in 2.5 ml water, pH 8.0, are prepared just before use. The hapten solution, then the EDAC is added to the protein and the reaction mixture allowed to stir at room temperature for 4 hr. The protein peak containing the conjugate is isolated by chromatography on a column of
Sephadex G-50 M (1.5 × 30 cm) wetpacked and eluted with 0.05 M phosphate-buffered isotonic saline, pH 7.2.
   Other carriers and proportions of reagents have been used to prepare
MTX conjugates (Table I). In several cases, BSA was used after the carboxyl groups were methylated to prevent direct reaction with EDAC.
MeBSA is prepared by the following procedure.[14]
   MeBSA. Dissolve 2 g BSA in 500 ml dehydrated methanol. Allow dry
hydrogen chloride to bubble through the solution at 4° with stirring for
3 hr. After an additional 24 hr at 4°, the precipitate is collected and
washed with dry diethyl ether. If centrifugation is used, be extremely cau-
tious. Methanol and ether are highly flammable and may ignite if exposed
to heat or spark from any electrical device.
   When MeBSA is used, the product from the coupling reaction probably will be a precipitate. Low-molecular-weight components may be removed by dialysis using sodium bicarbonate[11] or other suitable buffer.
   Incorporation of Hapten. The molar ratio of MTX to carrier has been
estimated in two ways: by determining the incorporation of tracer
amounts of [³H]MTX added to the reaction mixture[13,16] and by relating the
absorption by the conjugate at 308 and/or 380 nm to absorption by a standard solution containing MTX.[9,10,13,15,18] Depending on the reaction procedure, the incorporation ranged from 4 to 85 mol MTX/mol carrier (Table
I). The incorporation of leucovorin or 5-MTHFA was not determined.

Immunization

   Rabbits, goats, and sheep have been immunized to produce antibodies
suitable for radioimmunoassay of MTX (Table I). Conjugates emulsified
in complete Freund's adjuvant have been administered sc, id, ip, or im.
Booster injections of insoluble immunogen suspended in saline have been
given iv.[11] Although there are several successful schedules, no comparison has been made to determine optimum properties of the immunogen or
the best immunization procedure.

TABLE I

DEVELOPMENT OF RADIOIMMUNOASSAYS FOR METHOTREXATE, LEUCOVORIN, AND 5-METHYLTETRAHYDROFOLATE

Immunogen carrier[a]	Moles hapten/mole carrier[b]	Species immunized (route)[c]	Radiolabeled tracer	Method of separating bound from free hapten	Reference
Methotrexate					
KLH	N.G.	Rabbit, goat (sc)	[^{3}H]MTX	Double-antibody	d
MeBSA	1–23	Rabbit (sc)	[^{3}H]MTX	Double-antibody	e
MeBSA	N.G.	Rabbit (id, sc, then iv)	[^{3}H]MTX	Nitrocellulose membrane filtration	f,g
HSA, poly(L-lysine) (MW 85,000)	45–49 (HSA) 81–85 [poly(L-lysine)]	Rabbit (id, then sc)	[^{3}H]MTX	Dextran-coated charcoal	h
MeBSA	N.G.	Rabbit (id)	[^{3}H]MTX	Polyethylene glycol (Carbowax 6000)	i
Ovalbumin	4	Rabbit (not specified) Sheep (im)	[^{3}H]MTX	Dextran-coated charcoal	j
BSA	34	Rabbit (ip, im, and sc)	[^{3}H]MTX, ^{75}Se-labeled MTX derivative, ^{125}I-labeled tyrosine methyl ester-MTX derivative	Double-antibody	k,l
poly(Ala)-poly(Lys)	20	Sheep (id)	^{125}I-labeled derivatives prepared with Bolton–Hunter reagent or histamine	Antibody-coated magnetisable particles (cellulose/iron dioxide)	m,n
KLH	N.G.	Rabbit, goat (sc)	^{125}I-labeled protein A from *Staphylococcus aureus*	Methotrexate-coated polyacrylamide beads	o

Leucovorin				
KLH	N.G.	Rabbit (sc)	^{125}I-labeled protein A from *S. aureus*	Leucovorin-coated poly-acrylamide beads	p
5-Methyltetrahydrofolate				
KLH	N.G.	Rabbit (sc)	^{125}I-labeled protein A from *S. aureus*	5-MTHFA-coated poly-acrylamide beads	q

[a] KLH, hemocyanin from giant keyhole limpet; MeBSA, methylated bovine serum albumin; HSA, human serum albumin; BSA, bovine serum albumin MTX, methotrexate; 5-MTHFA, 5-methyltetrahydrofolate.

[b] Hapten–carrier conjugates were prepared using 1-ethyl-3-(3-dimethylaminopropyl)carbodiimide except in Ref. (e) where 1-cyclohexyl-3-2-morpholinoethyl)carbodiimide was used.

[c] sc, subcutaneous; id, intradermal; iv, intravenous; im, intramuscular.

[d] L. Levine and E. Powers, *Res. Commun. Chem. Pathol. Pharmacol.* **9**, 543 (1974).

[e] C. Bohuon, F. Duprey, and C. Boudene, *Clin. Chim. Acta* **57**, 263 (1974).

[f] V. Raso and R. Schreiber, *Cancer Res.* **35**, 1407 (1975).

[g] V. Raso, *Cancer Treat. Rep.* **61**, 585 (1977).

[h] L. J. Loeffler, M. R. Blum, and M. A. Nelsen, *Cancer Res.* **36**, 3306 (1976).

[i] J. Hendel, L. J. Sarek, and E. F. Hvidberg, *Clin. Chem.* **22**, 813 (1976).

[j] G. W. Aherne, E. M. Piall, and V. Marks, *Br. J. Cancer* **36**, 608 (1977).

[k] J. W. Paxton and F. J. Rowell, *Clin. Chim. Acta* **80**, 563 (1977).

[l] J. W. Paxton, F. J. Rowell, and G. M. Cree, *Clin. Chem.* **24**, 1534 (1978).

[m] R. S. Kamel and J. Gardner, *Clin. Chim. Acta* **89**, 363 (1978).

[n] R. S. Kamel, J. Landon, and G. C. Forrest, *Clin. Chem.* **26**, 97 (1980).

[o] J. J. Langone, *J. Immunol. Methods* **24**, 269 (1978).

[p] J. J. Langone, *Anal. Biochem.* **95**, 472 (1979).

[q] J. J. Langone, *Anal. Biochem.* **104**, 347 (1980).

The following schedule for production of rabbit anti-5-MTHFA sera also can be used to prepare antibodies to MTX and leucovorin.[22] Bleedings collected after one or two booster injections may be satisfactory. Of course, all bleedings should be tested to determine which is most suitable in terms of titer, which usually reaches an optimum after several weeks or months leading to a more sensitive assay, and specificity which may broaden during the course of immunization making the assay less specific (see below, assay for 5-MTHFA).

Procedure. New Zealand albino rabbits were injected subcutaneously at multiple sites on the neck and back with 0.6 ml immunogen (2.0 mg protein) emulsified in 0.6 ml complete Freund's adjuvant. After 2 weeks the animals were given similar booster injections and bled 6 (bleeding 1) and 20 days (bleeding 2) later. One month later they were immunized again and bled after 7, 9, and 11 days; these bleedings were pooled (bleeding 3). After resting for 5 months, the rabbits were immunized again and bled 14 days later (bleeding 4). Two weeks later they were boosted and bled after 7, 9, 11 days (pooled bleeding 5).

Radiolabeled Tracers

In classical RIAs for MTX, [³H]MTX which is available commercially from Amersham/Searle and New England Nuclear has been used most often. However, derivatives of ^{125}I-labeled MTX made with tyrosine methyl ester (sp. act. 264 Ci/mmol),[17] histamine,[18] or N-succinimidyl-3-(4-hydroxyphenyl)-propionate (Bolton–Hunter reagent)[18] have been prepared and used successfully. A MTX derivative labeled with ^{75}Se also has been tested.[17]

We used [³H]MTX in a standard double antibody radioimmunoassay. However, we also have assayed MTX, leucovorin, and 5-MTHFA by a solid phase method[19] in which ^{125}I-labeled protein A serves as the tracer for specific antibody bound to immobilized hapten. The basic steps including the preparation of ^{125}I-PA were described in this series, Vol. 70.

Double Antibody Radioimmunoassay

The buffer used to make all dilutions was 0.092 M phosphate containing 4 mM magnesium chloride and 0.1 mM dithiothreitol, pH 7.2. One-tenth milliliter goat anti-MTX, diluted 1/1000 [³H]MTX (9.5 Ci/mmol; 20,000 cpm), and either buffer (to determine maximum binding) or samples (either standard MTX or text sample) are incubated at 37° for 1 hr in a total volume made up to 1.3 ml with buffer.

Normal goat serum (1/100, 0.1 ml) is added as carrier followed by an optimal dilution of rabbit anti-goat γ-globulin (0.1 ml). The solution is mixed and allowed to incubate at 4° overnight. The immune precipitate is collected by centrifugation at 1000 g at 4° for 30 min, dissolved in 0.2 ml 0.1 N sodium hydroxide, then transferred to vials containing scintillation fluid for counting.

Solid Phase Radioimmunoassay

Briefly, the test is performed in two steps: (1) fluid phase antibody (either whole serum or the IgG fraction) is incubated with immobilized hapten, then the beads are washed; (2) excess ^{125}I-PA is added and a similar incubation and wash procedure is carried out. The number of cpm bound is a measure of IgG antibody fixed specifically to the immunadsorbent. Inhibition of antibody binding by known amounts of fluid phase hapten gives a dose-related decrease in ^{125}I-PA binding. This standard curve is used to estimate levels of drug in test samples based on the observed degree of inhibition.

Immobilized Hapten

Reagents
Buffer, 0.003 M phosphate, pH 6.35
300 μg MTX, leucovorin, or 5-MTHFA
500 mg polyacrylamide beads derivatized with ethylamino groups
(Affi-gel 701, Bio-Rad Labs, Rockville Center, NY)
6 mg EDAC (0.3 ml of 20 mg/ml solution in buffer)
Procedure. The reaction and subsequent washings are carried out at 4°. Hapten dissolved in 0.5 ml buffer is mixed with the Affi-gel 701 suspended in 9.0 ml buffer. EDAC solution is added and the mixture allowed to rock overnight protected from light. The beads are centrifuged (1500 g; 10–15 min), the supernatant liquid is decanted, and the pellet washed by resuspension in 10 ml coupling buffer. This procedure is repeated two more times. The beads are then washed three times with 20 ml cold 5 M guanidine hydrochloride, pH 7.2, followed by 8–10 washes with phosphate-buffered isotonic saline, pH 7.2. The pellet is mixed thoroughly to give a uniform suspension in 25 ml phosphate or veronal-buffered saline containing 0.001 M Mg^{2+}, 0.00015 M Ca^{2+}, 0.1% gelatin, and 0.04% sodium azide. The beads are stored at 4°.
Absorption of Anti-5-MTHFA. Later bleedings of anti-5-MTHFA were absorbed to remove antibodies reacting with leucovorin or the linkage group common to the immunogen and the hapten-bead conjugate used in immunoassay. Early bleedings were relatively specific for 5-MTHFA but also contained antibridging group antibodies that could be removed by absorption.

Preparation of Absorbent. The procedure given above for the hapten-bead conjugate was used except 100 mg leucovorin and 300 mg EDAC were reacted with 1 g Affi-gel 701 (capacity 240 μmol ligand). These beads were stored in 10 ml phosphate-buffered saline, pH 7.2, containing 0.04% sodium azide.

Absorption Procedure. One milliliter of absorbent suspension is washed twice with phosphate-buffered saline, pH 7.2, by centrifugation at 3000 g for 10 min. One milliliter of antiserum containing 0.02% azide is added and the pellet resuspended by vigorous vortex mixing. The mixture is allowed to rock 1 hr at 20°, and then centrifuged at 3000 g for 10 min. This procedure is repeated two more times. The serum is filtered through a Millipore filter (0.22 μm pore size) and stored at $-20°$. Other buffers can be used to store the beads, including buffers containing EDTA to exclude divalent cations. However, a protein source free of IgG must be added as stabilizer.

Immunoassay. The buffer is 0.01 M veronal-buffered 0.15 M saline containing 0.01 M EDTA and 0.1% gelatin, pH 7.2. Glass tubes (13 × 100 mm) are recommended since the beads may not form a tight pellet when centrifuged in plastic tubes. In a typical procedure aliquots (0.1 ml) containing 330 μg MTX beads, rabbit anti-MTX (1/1200), and either buffer (to determine maximum binding) or sample containing either standard MTX or test solution are mixed and incubated at 30° for 1 hr. The beads are washed with two 3-ml portions of buffer by centrifugation at 1500 g at 4° for 5 min, then incubated with 0.1 ml ^{125}I-PA; 40,000 cpm should supply sufficient reagent to saturate the IgG binding sites. After 1 hr at 30°, the beads are washed as above and radioactivity on the bead pellet is counted in a gamma counter.

This assay is illustrated using rabbit antibodies since rabbit IgG reacts efficiently with PA. Although free goat IgG normally reacts poorly[19] goat anti-MTX also has been used[23] since the ability of these antibodies to bind PA is enhanced significantly once they are bound to immobilized MTX.

Sensitivity and Specificity

A set of typical standard curves is shown in Fig. 2 for anti-MTX, anti-leucovorin, and anti-5-MTHFA using the homologous ligands as inhibitors. These results were obtained with the respective solid phase (^{125}I-PA) assays and rabbit antisera. Similar curves were constructed with known amounts of structurally related compounds to determine specificity. The results are summarized in Table II as the quantity of ligand required to inhibit immune binding by 50%. Only 0.8 ng MTX, 0.28 ng leucovorin, or

[23] J. J. Langone, *J. Immunol. Methods* **34**, 93 (1980).

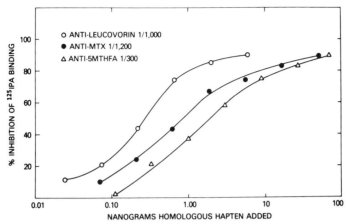

FIG. 2. Typical inhibition curves obtained with rabbit antisera to methotrexate (●), leucovorin (○), or 5-methyltetrahydrofolate (△) in the presence of differing amounts of homologous hapten. Each assay used the appropriate immunadsorbent and ^{125}I-labeled protein A as the tracer.

TABLE II

SPECIFICITY OF ANTIBODIES AGAINST METHOTREXATE,
LEUCOVORIN, AND 5-METHYLTETRAHYDROFOLATE

	ng required for 50% inhibition of ^{125}I-labeled protein A binding to		
Inhibitor	Anti-methotrexate[a]	Anti-leucovorin[b]	Anti-5-methyl-tetrahydrofolate[c]
Methotrexate	0.80	>10,000[d]	>6,700[e]
Leucovorin	>10,000[e]	0.28	1,850
5-Methyltetrahydrofolic acid	>10,000[d]	160	1.6
Tetrahydrofolic acid	>10,000[d]	38	500
Dihydrofolic acid	>10,000[d]	600	6,800
Folic acid	>10,000[d]	1,900	>6,700[d]
Aminopterin	100	>10,000[d]	>6,700[e]

[a] Antiserum diluted 1/1200 and 330 μg MTX beads; 12,500 cpm bound.
[b] Antiserum diluted 1/1000 and 100 μg leucovorin beads; 12,200 cpm bound.
[c] Antiserum (absorbed with immobilized leucovorin; see text) diluted 1/300 and 500 μg 5-MTHFA beads; 7100 cpm bound.
[d] 20–40% inhibition at this level.
[e] <10% inhibition at this level.

TABLE III

CLINICAL APPLICATIONS OF METHOTREXATE RADIOIMMUNOASSAY[a]

Dose of MTX and route of administration[b]	Leucovorin rescue	MTX determined in		Reference
100 mg iv	–	Plasma	Levels followed up to 72 hr	c
150 mg/kg iv	+			
10–100 mg im	–			
5 mg intrathecally	–	Serum or spinal fluid	Levels after iv dose followed up to 8 hr	d
50 mg iv	–			
1.4 g bolus iv, then 0.66 g infusion over 12 hr	+	Plasma	Levels followed up to about 180 hr; similar results obtained with enzyme assay using dihydrofolate reductase	e
1 g infused iv over 6 hr	+	Plasma	Levels followed to about 30 hr after treatment. RIA more sensitive than spectro-fluorometric procedure. Fluorescence method may be subject to error when leucovorin is also given	f
60 mg weekly im	–	Plasma and erythrocytes	Levels measured just before next dose. In one patient, the concentration/time curve was followed for 24 hr	g

Dose/route[b]		Fluid	Comments
Alternating weekly between 60 mg im and 20 mg intrathecally plus 40 mg im	—		Levels followed over 25 hr
200 or 300 mg bolus iv	—	Serum and urine	Levels followed over 5 hr
20 mg orally	—		Levels followed up to 35 hr after infusion.[h]
150 mg infused iv over 24 hr; or 1 g infused over 6 hr	+	Serum	Leucovorin given im (4 × 12 mg given over 6 hr) did not deflect MTX levels.[i]
2 g infused iv	—	Serum	Levels followed for > 3 days[j]
25 mg orally or im; or 50 mg im	—	Serum or plasma	—[k]

[a] Of course measured levels differ widely depending on several factors including dose, route, and time course of administration, and the fluid tested. The sensitivity of these assays generally is in the 1–10 ng MTX range. Serum concentrations of MTX in the μg/ml range are common.

[b] iv, intravenous; im, intramuscular.

[c] L. Levine and E. Powers, Res. Commun. Chem. Pathol. Pharmacol. 9, 543 (1974).

[d] C. Bohuon, F. Duprey, and C. Boudene, Clin. Chim. Acta 57, 263 (1974).

[e] V. Raso and R. Schreiber, Cancer Res. 35, 1407 (1975).

[f] L. J. Loeffler, M. R. Blum, and M. A. Nelsen, Cancer Res. 36, 3306 (1976).

[g] J. Hendel, L. J. Sarek, and E. F. Hvidberg, Clin. Chem. 22, 813 (1976).

[h] G. W. Aherne, E. M. Piall, and V. Marks, Br. J. Cancer 36, 608 (1977).

[i] J. W. Paxton and F. J. Rowell, Clin. Chim. Acta 80, 563 (1977).

[j] J. W. Paxton, F. J. Rowell, and G. M. Cree, Clin. Chem. 24, 1534 (1978).

[k] R. S. Kamel, J. Landon, and G. C. Forrest, Clin. Chem. 26, 97 (1980).

1.6 ng 5-MTHFA gave 50% inhibition and the assays were specific enough for use in routine analysis of physiological fluids. RIAs for MTX using [³H]MTX or labeled MTX derivatives (see Table I) show similar sensitivity and specificity for this drug.

Results

Clinical Applications of MTX Assays

The classical radioimmunoassays for MTX have been used to follow drug levels in physiological fluids and erythrocytes of drug-treated cancer patients. Information regarding these experiments is summarized in Table III. Since large amounts (i.e., up to gram quantities during infusion followed by leucovorin "rescue") of MTX are administered, there has been little trouble in following MTX levels in blood, urine, or spinal fluid, sometimes up to several days after treatment. In those cases where leucovorin was given, there was no apparent effect on the disappearance of MTX.

Applications of Solid Phase (¹²⁵I-PA) Assays

The solid phase assays also have found practical application. The curves in Fig. 3 show concentrations of MTX and leucovorin in plasma of

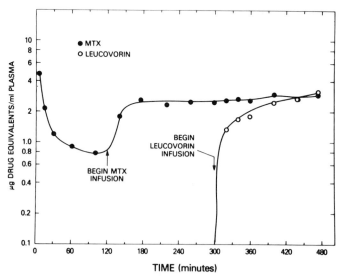

Fig. 3. Levels of MTX (●) or leucovorin (○) in the plasma of a dog (8.8 kg) given a bolus iv injection of 1 mg MTX/kg followed by an infusion of MTX (10 μg/kg/min for 360 min) and an infusion of leucovorin (20 μg/kg/min during the last 180 min of the MTX infusion). This experiment was performed in collaboration with Dr. James Straw (1978).

a dog given an initial bolus iv injection of MTX (1 mg/kg), then an infusion of MTX (10 μg/kg/min over 360 min) followed by an infusion of leucovorin (20 μg/kg/min for 180 min). Leucovorin was started 3 hr into the MTX infusion. The initial MTX level fell rapidly, then increased to a plateau value of 2.5–3.0 μg/ml by 40 min after the MTX infusion began. Since the antibodies are specific leucovorin levels could be determined in unprocessed samples without interference even at this high concentration of MTX. The plateau concentration of leucovorin was close to 3.0 μg/ml plasma.

Similarly, the appropriate assays were used to follow the disappearance of leucovorin (Fig. 4) or 5-MTHFA (Fig. 5) from the serum of rabbits given either drug iv at a dose of 1 mg/kg.[21,22]

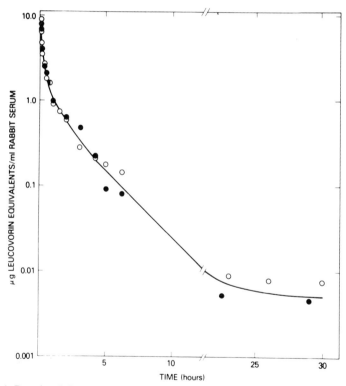

FIG. 4. Drug levels in sera of 2 rabbits given 1 mg leucovorin/kg as a single injection in the right marginal ear vein. Before injection and at different times after 2 ml blood was collected from the left ear, allowed to clot, and the serum analyzed using the appropriate solid phase (^{125}I-PA) assay. A standard curve was obtained in the presence of a corresponding amount of prebleed serum and was identical to the curve obtained in buffer alone. Results are from J. J. Langone and L. Levine, *Anal. Biochem.* **95,** 472 (1979).

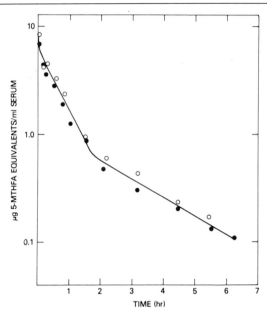

FIG. 5. 5-Methyltetrahydrofolate levels in sera of 2 rabbits given 1 mg/kg as a single injection in the right marginal ear vein. Other details are given in legend to Fig. 4. Results from J. J. Langone, *Anal. Biochem.* **104**, 347 (1980).

Conclusion

Radioimmunoassays for MTX have proved useful for monitoring drug levels in physiological fluids of cancer patients. The ability to use ^{125}I-PA as a tracer for antibody has allowed the development of practical solid phase assays for MTX as well as leucovorin and 5-MTHFA. These assays may be useful in metabolism studies as well as clinical applications.

[29] Preparation of ^{125}I-Labeled Methotrexate and Its Use in Magnetizable Particle Solid-Phase Radioimmunoassays

By R. S. KAMEL AND J. LANDON

Two different immunogens were used for raising antisera to methotrexate (MTX) in sheep. One was prepared by coupling MTX to a branched-chain synthetic polypeptide, poly(Ala)-poly(Lys) (PAPL) and the other by coupling MTX to keyhole limpet hemocyanin (KLH). The antisera obtained were subsequently used to develop radioimmunoassays for MTX employing ^{125}I-labeled MTX derivatives as tracers. Two techniques for tracer preparation are described, involving covalent linkage of MTX to either ^{125}I-labeled N-succinimidyl-3-(4-hydroxyphenyl) propionate (Bolton and Hunter reagent) or ^{125}I-labeled histamine. The latter proved the more immunoreactive.

A rapid, sensitive, and specific RIA was developed, employing antibodies covalently linked to magnetizable particles. Results obtained for serum samples correlated closely with those using an enzymatic (dihydrofolate reductase) competitive protein binding assay ($r = 0.97$), and with an independent RIA employing ^{3}H-labeled MTX ($r = 0.98$).

A fully automated continuous-flow RIA for MTX has been developed using a Technicon automated radioimmunoassay developmental system. Separation of the fraction bound to magnetizable particles was achieved with an electromagnetic field. The assay is rapid and specific, and has a small incubation volume (approx 160 μl). The assay of each sample takes only 15 min, with a sample throughput of at least 30/hr. The working range is wide (1–100 ng/ml) and the within-assay coefficient of variation less than 2.5%. There is no significant carryover between samples of high and low concentration. Results by the automated method correlate closely with both a manual assay using the same reagents and separation technique ($r = 0.99$) and a competitive protein binding assay ($r = 0.96$).

Introduction

The main purpose of monitoring drug therapy is to achieve optimum therapeutic effectiveness and minimize side effects. Its value varies according to the relationship between blood levels and therapeutic or toxic effects for the particular drug.

Methotrexate (MTX) (Fig. 1) has been used widely, either alone or in

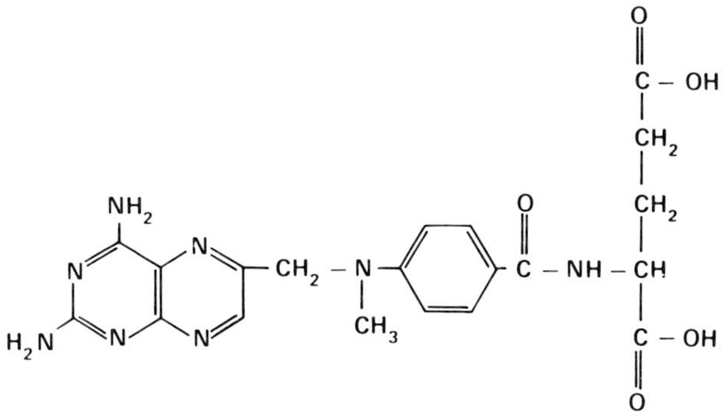

FIG. 1. The chemical structure of MTX.

combination therapy, for a wide variety of malignancies.[1] In recent years high dose MTX regime with folinic acid rescue has been used successfully.[2] However, in a survey of 498 patients receiving such therapy, Van Hoff and co-workers[3] reported a 6% drug related mortality. MTX has also been used for the treatment of some nonneoplastic diseases and is considered the drug of choice for severe psoriasis.[4] Toxicity can be predicted and patients at high risk identified from elevated MTX levels 48 hr after administration of the drug.[5-8] Chan and co-workers[9] suggested that pretherapy serum MTX levels should be determined if high dose MTX is administered at weekly intervals. Indeed, other workers[10-11] suggest that high doses of MTX should not be administered in the absence of facilities for monitoring serum levels. Therefore, a specific, precise, rapid, and simple assay for routine monitoring is required.

Existing methods for MTX assay include microbiological,[12] fluorimetric,[13] high performance liquid chromatography,[14] competitive protein binding radioassay,[15] and spectrophotometric enzyme inhibition procedures.[16] Recently, radioimmunoassays,[17-19] enzymoimmunoassays,[20-21]

[1] W. A. Bleyer, Cancer Treat. Rev. **4,** 87 (1977).

[2] N. Jaffe and D. Paed, Cancer **30,** 1627 (1972).

[3] D. D. Von Hoff, J. S. Penta, L. J. Helman, and M. Slavik, Cancer Treat. Rep. **61,** 745 (1977).

[4] G. D. Weinstein, Ann. Intern. Med. **86,** 199 (1977).

[5] L. C. Falk, T. F. Long, and S. M. Kalman, Pharmacologist **17,** 201 (1975).

[6] Y. Wang, E. Lantin, and W. W. Satow, Clin. Chem. **22,** 1053 (1976).

[7] R. G. Stoller, K. R. Hande, S. A. Jacobs, S. A. Rosenberg, and B. A. Chabner, New Engl. J. Med. **297,** 630 (1977).

[8] A. Nirenberg, C. Mosende, B. M. Mehta, A. L. Gisolfi, and G. Rosen, Cancer Treat. Rep. **61,** 779 (1977).

[9] H. Chan, W. E. Evans, and C. B. Pratt, Cancer Treat. Rep. **61,** 797 (1977).

[10] S. Salasoo, M. G. Irving, R. Lam-Po-Tang, D. O'Gorman-Hughes, and A. Freedman, Med. J. Aust. **1,** 777 (1976).

[11] W. H. Isacoff, F. Eilber, H. Tabbarah, P. Klein, M. Dollinger, S. Lemkin, P. Sheehy, L. Cone, B. Rosenbloom, L. Sieger, and J. B. Bloch, Cancer Treat. Rep. **62,** 1295 (1978).

[12] W. C. Noble, P. M. White, and H. Baker, J. Invest. Dermatol. **64,** 69 (1975).

[13] M. V. Freeman, J. Pharmacol. Exp. Ther. **12,** 1 (1957).

[14] J. A. Nelson, B. A. Harris, W. J.. Decker, and D. Farquhar, Cancer Res. **37,** 3970 (1977).

[15] C. E. Myers, M. E. Lippman, H. M. Eliot, and B. A. Chabner, Proc. Natl. Acad. Sci. U.S.A. **72,** 3683 (1975).

[16] L. C. Falk, D. R. Clark, S. M. Kalman, and T. F. Long, Clin. Chem. **22,** 785 (1976).

[17] L. Levine and E. Powers, Res. Commun. Chem. Pathol. Pharmacol. **9,** 543 (1974).

[18] G. W. Aherne, E. M. Piall, and V. Marks, Br. J. Cancer **36,** 608 (1977).

[19] J. W. Paxton, F. J. Rowell, and G. M. Cree, Clin. Chem. **24,** 1534 (1978).

[20] V. Marks, M. J. O'Sullivan, M. N. Al-Bassam, and J. W. Bridges, Proc. Int. Symp. Enzyme Labelled Immunoassay Horm. Drugs, Ulm, July (1978).

[21] J. B. Gushaw, and J. S. Miller, Clin. Chem. **24,** 1032 (1978).

an immunoassay employing a general ^{125}I-labeled protein A,[22] and an immunoassay based on inhibition of passive immune hemolysis[23] have been published. All appear suitable for routine purposes because of their specificity, sensitivity, practicality, and wide applicability.

For establishing an immunoassay for MTX, the three major requirements are antibody, tracer, and a separation technique.

Small molecules such as drugs are not immunogenic and will only elicit antibody formation when linked as haptens to a carrier protein,[24] such as bovine serum albumin (BSA), thyroglobulin, fibrinogen, keyhole limpet hemocyanin (KLH), and synthetic polypeptides. In this work KLH-MTX and poly(Ala)-poly(Lys) MTX conjugates were used as immunogens for the preparation of antisera in sheep.

The second important requirement is production of a labeled drug or analog. Assays employing a radioactive label are termed radioimmunoassays (RIA) and those using no isotopic labels are named according to the label employed, such as enzymoimmunoassay and fluoroimmunoassay. Radioimmunoassay for MTX has been developed employing tritium-labeled MTX tracer, but requires use of an expensive beta scintillation counter, costly liquid scintillant, long counting times (due to their low specific activity) and the need to correct blank quenching. It is much more convenient to employ the gamma emitting isotope, ^{125}I, and another gamma emitting isotope, ^{75}Se, has also been used,[19,25]

Direct iodination of small molecules is seldom possible, either because they lack suitable residues or because the direct introduction of the large iodine atom impairs their immunoreactivity. Thus conjugation labeling is usually employed and two approaches are possible. First, conjugation of the drug to a substance which can be readily iodinated and subsequent iodination of the conjugate by a conventional method. Second, the drug may be conjugated to a reactive, preiodinated substance. Among suitable carriers for conjugation labeling are tyrosine methyl ester,[19] tyramine,[26–27] and proteins such as BSA.[28] Recently the fluorescein isothiocyanate or N-acetyl-L-histidine derivatives of nortriptyline and demethylnortriptyline have been used.[29] In this work, ^{125}I-labeled histamine and N-succinimidyl-

[22] J. J. Langone, *J. Immunol. Methods* **24**, 269 (1978).
[23] T. Borsos, V. C. Dunkel, and J. J. Langone, *J. Immunol. Methods* **32**, 105 (1980).
[24] S. M. Beiser, V. P. Butler, and B. F. Erlanger, in "Textbook of Immuno-Pathology" (P. A. Miescher, ed.), 2nd ed., p. 15. 1976.
[25] G. W. Aherne, E. M. Piall, and V. Marks, *Ann. Clin. Biochem.* **15**, 331 (1978).
[26] H. van Vunakis, D. S. Freeman, and H. B. Djika, *Res. Commun. Chem. Pathol. Pharmacol.* **2**, 379 (1975).
[27] E. M. Weiler and M. H. Zenk, *Clin. Chem.* **25**, 44 (1979).
[28] J. M. Wal, G. Bories, S. Mamas, and F. Dray, *FEBS Lett.* **57**, 9 (1975).
[29] R. S. Kamel, J. Landon, and D. S. Smith, *Clin. Chem.* **26**, 1997 (1979).

3-(4-hydroxyphenol) propionate (Bolton and Hunter reagent) were employed for the preparation of ^{125}I-labeled MTX tracers.

All radioimmunoassays require separation of the bound and free labeled antigen.[30] The only two universal techniques are second antibody and solid phase procedures. Solid phase separation was introduced by Catt and his colleagues[31] and, in this work, we employed first antibody covalently linked to magnetizable particles for the development of manual and fully automated radioimmunoassays for MTX. The other separation method employed was magnetizable charcoal. The use of magnetizable particles allows the elimination of centrifugation and enables the development of specific, rapid, and complete separation.

Materials and Methods

MTX solid (96.4% purity) and MTX injections (25 g/liter MTX) were from Lederle Laboratories Division, Cyanamid of Great Britain Limited (Gosport, Hampshire, UK), 7-OH-MTX was the gift of Dr. David G. Johns of the Division of Cancer Treatment NCI (Bethesda, Md), and Freund's complete and incomplete adjuvants were from Difco Laboratories (Detroit, Mich).

Aminopterin, folic acid, dihydrofolic acid, tetrahydrofolic acid, folinic acid, N-5-methyltetrahydrofolic acid, N-(p-aminobenzoyl)-L-glutamic acid, histamine (free base), bovine serum albumin (BSA) (type V), isobutyl chloroformate, ethylenediamine dihydrochloride, acrylamide, and 1-ethyl-3-(3-dimethylaminopropyl)carbodiimide (CDI) were from Sigma (Poole, Dorset, UK); Na[125]I from the Radiochemical Centre (Amersham, Bucks, UK); tri-n-butylamine from Eastman Kodak (Rochester, NY); 1,4-dioxane, N,N-methylene bisacrylamide, N,N,N',N'-tetramethylethylene-diamine, ethanolamine, and dimethylformamide (DMF) from BDH (Poole, Dorset, UK); multichain poly(Ala)-poly(Lys) (PAPL) (molecular weight 90,000) from Miles Laboratories (Stoke Poges, Berks, UK); N-succinimidyl-3-(4-hydroxyphenol) propionate and keyhole limpet hemocyanin (KLH) from Calbiochem Co. (San Diego, Calif.), Sephadex G-15, Dextran T70, and Sephadex LH20 from Pharmacia Fine Chemicals (Uppsala, Sweden). Cyanogen bromide was from Aldrich Chemical Co. (Gillingham, Dorset, UK), cellulose power was from Whatman (Maidstone, Kent, UK), and ferric oxide (Fe_3O_4) was from Fisons Distol Reagents (Sydney, Australia).

---

[30] J. G. Ratcliffe, Br. Med. Bull. **30**, 32 (1974).
[31] K. J. Catt, H. D. Niall, G. W. Tregear, and H. S. Barger, J. Endoctinol. Metab. **28**, 121 (1968).

Vinblastine sulfate and vincristine sulfate were from Eli Lilly and Co. (Baskingstoke, Hampshire, UK), cytarabine from Upjohn Ltd (Sussex, UK), fluorouracil from Roche Products (Welwyn Garden City, Herts UK), and cyclophosphamide from W. B. Pharmaceuticals (Bracknell, Berks, UK).

A 6 × 8-in. multipolar magnet was obtained from Magnet Applications, City Road, London.

Assay diluent was 100 mM sodium phosphate buffer, pH 7.5, containing 1 g/liter BSA and 1 g/liter sodium azide.

MTX standards were prepared by diluting MTX (25 g/liter injectable MTX, Lederle Laboratories Division, Cyanamid of Great Britain, Gosport, Hampshire, UK) in pooled normal serum to give concentrations of 1, 2.5, 5, 10, 25, 50, and 100 ng/ml. These were aliquoted and stored at −20° until required. The drugs and analogs used for cross-reactivity studies were also prepared in normal human serum to give concentrations of 1, 5, 10, 25, 50, 100, 250, 500, and 1000 ng/ml, calculated as free base or free acid.

Serum or plasma samples were collected from patients receiving MTX. Some had been assayed by a competitive protein binding assay (CPBA) by Dr. Calvert (using the method of Myers *et al.*[15]) from the Royal Marsden Hospital, Sutton, Surrey, and others by an RIA by Dr. Aherne (using the method of Aherne *et al.*[18]) from the Department of Biochemistry, University of Surrey, Guildford, Surrey. Lipemic samples and various normal human serum samples were obtained from Technical Laboratory Services, London.

Preparation and Assessment of Immunogens

MTX-Poly(Ala)-Poly(Lys) Immunogen (MTX-PABL)

The immunogen was prepared by coupling MTX to a branched synthetic polypeptide, PAPL, following the method of Jaton and Ungar-Waron.[32] MTX (12.5 mg) was dissolved in 1 ml of 50 mmol/liter sodium bicarbonate buffer, pH 8.3, and mixed with 125 mg of PAPL dissolved in 50 ml of the same buffer. Then 25 mg of CDI was added and stirred for 18 hr at room temperature in the dark. The reaction mixture was dialyzed against two changes of 1 liter distilled water and then lyophilized.

MTX-Keyhole Limpet Hemocyanin Immunogen (MTX-KLH)

The immunogen was prepared by coupling MTX to KLH. MTX (25 mg) was dissolved in 1 ml of DMF and mixed with 25 mg of CDI in

---

[32] J. C. Jaton and H. Ungar-Waron, *Arch. Biochem. Biophys.* **122**, 157 (1967).

1 ml of saline. After 30 min, 100 mg KLH in 4 ml saline was added, and mixing continued for 1 hr. The reaction mixture was dialyzed against two changes of 1 liter of distilled water for 2 hr and then lyophilized.

The PAPL immunogen was assessed qualitatively by electrophoresis and chromatography and quantitatively by spectrophotometry.

Electrophoresis. Electrophoresis was performed in 200 mM sodium bicarbonate buffer, pH 9.0. MTX solution (10 mg/ml) and MTX-conjugate solutions were applied in 2 μl amounts to chromatography paper strips, and electrophoresis performed at about 10 V/cm for 1 hr. The yellow color of the two solutions was visible on the strip with the MTX spot having moved further toward the anode.

Chromatography. Paper chromatography was also performed to characterize the conjugates. As before, 2 μl amounts were spotted onto chromatography paper, which was then developed with 200 mmol/liter sodium bicarbonate, pH 9.0. Yellow spots were seen corresponding to both the MTX and conjugate solutions, with the MTX spot moving with the solvent front (R_f 1.0), while the conjugate had R_f 0.8. On the basis of these two methods, the conjugate product contained no free MTX.

Spectrophotometry. The absorption spectrum of MTX extends to 450 nm, and the UV spectra of MTX, carbodiimide and poly(Ala)-poly(Lys) were measured.

The MTX-conjugate was freeze-dried, yielding a very pale yellow solid. A known weight was taken and dissolved in bicarbonate buffer, and the absorption spectrum determined. An approximate ratio of 20 MTX : 1 polymer was calculated. For the KLH-MTX immunogen, a result of about 730 MTX per KLH was calculated, assuming no loss of protein during preparation.

Preparation of Antisera

Seven sheep were immunized intramuscularly with 1 mg of the MTX-PAPL conjugate (containing approximately 100 μg MTX) emulsified with Freund's complete adjuvant. Booster injections were given 6 weeks later, and 4 ml of blood taken from each sheep 14 days after the booster injection. Antiserum dilution curves were performed on each sample and the sheep with the best initial titer received booster injections (as above) at fortnightly intervals over a period of 3 months, during which time it was bled repeatedly, and the serum pooled. This antiserum was used throughout this study, unless otherwise stated.

Three sheep were immunized intramuscularly and subcutaneously (several sites) with 1.65 mg of the KLH conjugate (containing approximately 175 μg MTX) suspended in 2 ml distilled water and emulsified with 6 ml Freund's complete adjuvant were given 6 weeks later, and 10 days later a test bleed (about 20 ml) was taken from each animal. All the sheep received booster injections at fortnightly intervals and bled (about 1000 ml blood) 10 days following the second boost. This procedure was continued for 10 months.

Preparation of ¹²⁵I-Labeled Methotrexate

MTX Linked to Bolton and Hunter Reagent (BH-MTX)

Ethylenediamine dihydrochloride (45 mg in 0.5 ml distilled water) and CDI (60 mg in 0.5 ml distilled water) were added simultaneously to MTX (30 mg) in 1 ml DMF, while mixing. The reaction mixture was incubated in the dark at 4° for 24 hr and the MTX derivative purified by preparative paper chromatography (Whatman No. 17) in 50 mmol/liter ammonium bicarbonate, pH 9.0. The product, which had a lower mobility than pure MTX and was identifiable by its yellow color, was eluted from the paper twice with 50 mmol/liter ammonium bicarbonate, pH 9.0, lyophilized, and stored at −20°.

N-Succinimidyl-3-(4-hydroxyphenyl) propionate (0.25 μg) was iodinated with 2 mCi Na¹²⁵I, as described by Bolton and Hunter.[33] The iodinated reagent was then reacted with the MTX derivative (5 μg) in 50 μl ice cold sodium borate buffer (50 mmol/liter, pH 8.5) for 15 min. A further 100 μl of borate buffer was added before purification on a Sephadex G-15 column (10 × 0.8 cm) in 50 mM sodium bicarbonate buffer, pH 9.0. After collection of 20 × 0.5 ml fractions containing unreacted Na¹²⁵I, distilled water was applied to the column and further 0.5 ml fractions collected. A peak containing the desired ¹²⁵I-labeled MTX eluted between fractions 6 and 10. Incorporation of ¹²⁵I into the MTX derivative was approximately 20%.

MTX Linked to ¹²⁵I-Labeled Histamine (¹²⁵I-hist-MTX)

The preparation of ¹²⁵I-labeled histamine linked to MTX was a modification of the technique described by Hunter and his colleagues.[34] MTX (1.5 mg) in 100 μl DMF was activated by incubation with 10 μl tri-n-butylamine (diluted 1:5 v/v in dioxane) and 10 μl isobutyl chloroformate (also diluted 1:5 in dioxane) in an ice/water bath at 8–10° for

---

³³ A. E. Bolton and W. M. Hunter, *Biochem. J.* **133**, 529 (1973).
³⁴ W. H. Hunter, P. W. Nars, and F. J. Rutherford, *in* "Steroid Immunoassay" (E. H. D. Cameron, S. G. Hillier, and K. Griffiths, eds.), p. 141. Alpha Omega Publ., Cardiff, U.K., 1975.

30 min. For preparation of ^{125}I-labeled histamine, histamine (2.2 μg) in 500 mM sodium phosphate buffer, pH 8.9 (10 μl) was added to 1 mCi Na^{125}I (10 μl) followed by chloramine-T (50 μg) in 10 μl of distilled water. After agitation for 45 sec at room temperature, 10 μl distilled water containing 300 μg sodium metabisulfite was added. The activated MTX solution was diluted in 1.5 ml DMF, and 50 μl added to the ^{125}I-labeled histamine preparation. Sodium hydroxide (10 μl, 0.1 M) was added, and the solution mixed and incubated at 0° for 2 hr.

For purification, 450 μl 100 mM HCl and 1.0 ml ethyl acetate were added to the reaction mixture, mixed gently, and the ethyl acetate (upper) layer discarded. The aqueous layer was neutralized with 450 μl 100 mM NaOH, and reextracted with a further 1 ml ethyl acetate. The aqueous layer was applied to a Sephadex LH20 column (10 × 0.8 cm), eluted with 100 mM sodium phosphate buffer, pH 7.5, and 500 μl fractions collected. A peak of radioactivity containing the desired product eluted between fractions 62 and 90 and represented incorporation of about 50% of ^{125}I activity into ^{125}I-labeled histamine-linked MTX.

Preparation of Magnetizable Solid-Phase Antibodies

Anti-MTX serum was covalently coupled by the cyanogen bromide method of Axen et al.[35] to cellulose/iron oxide particles, prepared by a modification[36] of the method of Robinson and co-workers[37] as follows.

ZnCl$_2$ (500 g) was dissolved in distilled water (500 ml) by gentle stirring and cellulose (50 g) was added. The temperature was raised to approximately 75°, iron (50 g) was added, and the mixture stirred, first at high speed to ensure complete dispersion of the iron oxide and then gently for 1 hr, during which the viscosity increased noticeably. After standing overnight it was dispensed by syringe into 200 mM HCl. The resulting hard gel was washed under running tap water for 2 hr, dried by placing in a hot air oven for 48 hr, and then ground in a coffee grinder to a size of approximately 800 μm. It was again washed with water, stirred in EDTA solution overnight to remove Zn, washed repeatedly, and dried again for 48 hr in the oven. Finally, the dried product (cell/Fe$_3$O$_4$) was placed in a ball mill (Pascal No. 2) containing a stainless-steal ball charge (10 kg) and milled for 24 hr to form a uniform particle size distribution of 1–10 μm.

After activation of the magnetizable particles with cyanogen bromide,[35] whole anti-MTX serum was coupled, as described by Wide,[38] at a coupling ratio of 2 ml/g solid-phase, unless otherwise stated. The product

[35] R. Axen, J. Porath, and S. Ernback, *Nature* (*London*) **214**, 1302 (1967).
[36] M. T. Pourfarzaneh, Ph.D. Thesis, London University, 1980.
[37] P. J. Robinson, P. Dunnill, and M. D. Lilly, *Biotechnol. Bioeng.* **15**, 603 (1973).
[38] L. Wide, *Acta Endocrinol.* **142**, 207 (1969).

was finally washed three times with assay diluent buffer, resuspended in 50 g/liter, and stored at 4°.

Preparation of Magnetizable Charcoal

This was prepared by a modification of the method described previously.[39] Acrylamide (38 g), N,N-methylenebisacrylamide (2 g), iron oxide (38 g), and $N,N,N'N'$-tetramethylethylenediamine (10 ml) were added to a suspension of charcoal Norit OL (38 g in 250 ml distilled water). Ammonium persulfate (12 g in 20 ml distilled water) was then added with vigorous stirring (stirring should be stopped as soon as the gel is formed). The homogeneous gel was left to cool at room temperature, passed through a meat grinder, then left to dry in a hot air oven (80°) overnight. The dried product was ground in a Braun (coffee) grinder and the fine granules were suspended in distilled water (about 1 g/15 ml) and milled in a ball mill (Pascall Engineering Company, Sussex, UK) for 3 days. The particles were washed four times with distilled water, the supernatant being discarded each time after sedimenting the magnetizable particles on a magnet. Finally, the particles were suspended in phosphate buffer (1 g/40 ml) containing 1 g/liter sodium azide and stored at 4°.

Antiserum Dilution Curves

Antisera were screened by constructing antiserum dilution curves using ^{125}I-hist-MTX (100 μl, diluted to give 1000 cps) and NHS (50 μl), added to each of 10 doubling dilutions of antisera in 100 μl assay diluent from an initial dilution of 1/100. After 1 hr incubation at room temperature, separation was achieved using magnetizable charcoal.

Standard Curves

Manual Magnetizable Particle Solid-Phase RIA

To 50 μl patient's serum or standard was added ^{125}I-labeled MTX (100 μl diluted to give 1000 cps) and 50 μl magnetizable solid-phase coupled anti-MTX (0.8 mg). Control tubes contained either no unlabeled MTX, or magnetizable solid-phase coupled to normal sheep serum. The tubes were incubated at room temperature in a shaker (Luckham Ltd) for 30 to 60 min, 1 ml of assay diluent or distilled water added, and antibody-bound separated from free antigen by rapid sedimentation on a magnet. The supernatants were aspirated and the precipitates counted for 10 sec in a gamma counter. A similar standard curve was obtained for the MTX standards in saliva.

[39] C. Dawes and J. Gardner, *Clin. Chim. Acta* **86,** 353 (1978).

A more sensitive assay was developed using 100 μl of standard or patient's sample, 10 μl of ^{125}I-hist-MTX and 10 μl of the magnetizable solid-phase suspension. The assay tubes were incubated for 30 min and separated as above. When antiserum against MTX-KLH immunogen was used, only 0.2 mg of magnetizable solid-phase antibody was required to obtain a similar standard curve.

Magnetizable Charcoal Separation RIA

[125]I-hist-MTX (100 μl) (diluted to give 1000 cps) was added to 50 μl patient's serum or standard, followed by 100 μl of anti-MTX serum diluted 1:100. After 1 hr incubation at room temperature, separation was achieved by addition of 100 μl magnetizable charcoal suspension (2.5 mg) or 500 μl dextran-coated charcoal (2.5 mg). Then 1 ml of the assay diluent was added and sedimentation of the charcoal achieved by means of a magnet or by centrifugation (2500 rpm, 5 min, 4°), respectively. Supernatants were aspirated and the tubes counted for 10 sec.

A Fully Automated, Continuous Flow Radioimmunoassay

Automated equipment previously described[40] was used with a standard glass mixing coil giving 10 min incubation and a plastic counter coil providing 80 sec counting time. Tracer and solid-phase volumes were each approximately 50 μl and the experimental system ran at 30 samples/hr with a 1:2 sample to wash ratio.

The same tracer (^{125}I-hist-MTX) and solid-phase antibody reagents proved applicable to both the automated and the manual assay (described above) and a standard curve covering the concentration range of 1–100 ng/ml was selected. A sample volume of about 60μl was adopted, giving a total incubation volume of approximately 160 μl. ^{125}I-labeled MTX concentration was selected to give a recorded count of 800–1000 cps for the highest standard, thereby minimizing statistical counting errors.

The concentration of solid-phase antibody was chosen to give a standard curve spanning the selected range. Since the weight of solid phase used (0.62 mg in 50 μl) was insufficient to activate the instrument's detector system, normal sheep IgG-linked particles (1.5 mg in 50 μl) were added to increase particle density.

A series of standards was aspirated at the beginning and end of each assay and quality control sera were inserted in duplicate after every 10 samples. At the beginning of each run a normal human serum sample was run, at least in duplicate, to determine the binding of the ^{125}I-labeled MTX to antibody in the absence of unlabeled drug (Co). Counts obtained for the

[40] G. C. Forrest, *Ann. Clin. Biochem.* **14**, 1 (1977).

standards were expressed as a percentage of Co and plotted vs MTX concentration in the standards.

Results

Assessment of ^{125}I-Labeled Methotrexate Products

Determination of the Immunoreactivity of ^{125}I-Labeled MTX Tracers. The immunoreactivities of the ^{125}I-BH-MTX and ^{125}I-hist-MTX tracers were assessed by using excess amount of antibody and separation of the bound and free fractions by magnetizable charcoal. Results showed 65 and 90% immunoreactivities for ^{125}I-BH-MTX and ^{125}I-hist-MTX, respectively. Similar results were obtained when magnetizable solid-phase was used. Nonspecific binding was less than 4%.

The immunoreactivity of ^{125}I-hist-MTX tracer depends on the purification procedure followed. When a Sephadex LH20 column was used alone, without a preextraction step, maximum immunoreactivity was only 70%. With the ethyl acetate preextraction steps higher immunoreactivity (85–90%) was achieved. Assessment of the ethylacetate extracts showed a 5 and 23% immunoreactivity for acidic and neutral extracts, respectively, which represented about 20% of the total radioactivity employed.

Assessment of the Stability of ^{125}I-Labeled MTX Preparations

The stability of ^{125}I-hist-MTX has been studied under various conditions. When pooled column fractions were stored at $-20°$, the tracer was found to be stable over a period of 5 months, with no significant loss in maximum binding and no detectable increase in nonspecific binding. When diluted, the freeze-dried tracer was stored at 4° and was stable over a period of 40 weeks without any significant loss in its immunoreactivity, as shown in Fig. 2. When the freeze-dried tracer was stored at room temperature for 2 months there was also no loss in its immunoreactivity or increase in its nonspecific binding.

When a freeze-dried product was kept for 3 days at 45° after reconstitution, the tracer showed no detectable loss in immunoreactivity and nonspecific binding was constant.

Determination of the Specific Activity of ^{125}I-hist-MTX Tracer

This was assessed by the self-displacement technique.[41] Four sets of standard curves containing various amounts of ^{125}I-hist-MTX (254, 1000,

[41] D. J. Gocke, J. Gerten, L. Sherwood, and J. H. Laragh, *Circ. Res.* **24-25** (Suppl 1), 131 (1969).

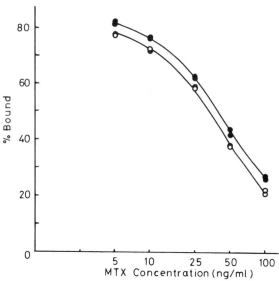

FIG. 2. Typical standard curves for the magnetizable solid-phase RIA of MTX using a freshly prepared ^{125}I-hist-MTX tracer (●) and ^{125}I-hist-MTX tracer freeze-dried and stored at 4° for 40 weeks (○).

3705, and 7380 cps/tube) were run as described previously, using isopropanol separation.

For the calculation of the specific activity these steps were followed:

1 Ci $\equiv$ 3.7 $\times$ 10^{10} disintegration per second (dps)

Let x = measured cps

Let y = mass of "sample" per tube giving x counts

x counts = $(x/75) \times 100$ disintegration, since the efficiency of the 1600 NE counter is 75%, as given by the company's manual. Thus, 1.33 is used as a conversion factor from count/s to disintegration/s.

$$\text{Specific activity} = \frac{\text{dps}}{\text{mass of sample} \times 3.7 \times 10^{10}} \text{ Ci/mass unit used}$$

$$= \frac{\text{cps} \times 1.33}{\text{mass} \times 3.7 \times 10^{4}} \mu\text{Ci/mass unit used}$$

$$= \frac{\text{cps}}{\text{mass}} \times 3.6 \times 10^{-5}$$

$$= \frac{x}{y} \times 3.6 \times 10^{-5}$$

The mean values of difference in mass displaced by certain differences in cps between two standards are:

1.075 ng ≡ 5955 cps
0.46 ng ≡ 2975 cps
0.115 ng ≡ 646 cps

Specific activity for $= \dfrac{5955}{1.075} \times 3.6 \times 10^{-5} = 0.199 \ \mu\text{Ci/ng}$
first set of standards

Specific activity for $= \dfrac{2975}{0.48} \times 3.6 \times 10^{-5} = 0.232 \ \mu\text{Ci/ng}$
second set of standards

Specific activity for $= \dfrac{646}{0.115} \times 3.6 \times 10^{-5} = 0.202 \ \mu\text{Ci/ng}$
third set of standards

The mean value of specific activity = 211 μCi/μg. Thus 1000 cps used in the assay represents 150 pg of the tracer used in each assay tube.

To ensure the equal immunoreactivity of both the tracer and the native MTX, three antibody dilution curves were constructed, one with 0.05 ng tracer (expressed as MTX content), another with 0.3 ng tracer, and the third having both tracer (0.05 ng) and unlabeled MTX (0.25 ng) in each tube. Figure 3 shows the equal immunoreactivity of the labeled and unlabeled MTX.

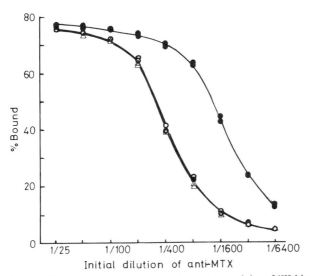

FIG. 3. Antiserum dilution curves to assess the immunoreactivity of ^{125}I-hist-MTX. The three curves represent antiserum dilutions in the presence of 0.05 ng ^{125}I-labeled MTX (●), 0.3 ng ^{125}I-labeled MTX (○), and 0.05 ng ^{125}I-labeled MTX + 0.25 ng unlabeled MTX (△). This indicates equal immunoreactivity of the labeled MTX with unlabeled MTX.

Assessment of Antisera

Figure 4 shows a typical antibody dilution curve using magnetizable charcoal and bleed 6 of the best sheep immunized with PAPL-MTX immunogen, and bleed 3 of the sheep immunized with KLH-MTX immunogen. The latter operated at a much higher titer.

Antibody dilution curves for sheep anti-MTX linked to magnetizable solid phase (300 μl/g for the sheep immunized with KLH-MTX immunogen and 2 ml/g for the sheep immunized with PAPL-MTX immunogen) were constructed. ^{125}I-hist-MTX tracer (100 μl) and NHS (50 μl) were added to each of seven doubling dilutions of magnetizable anti-MTX solid-phase in 50 μl, with (2.5 mg) in the first tube. Nonspecific binding was assessed using the same solid-phase coupled to normal sheep serum (2 ml/g).

Figure 5 shows that the titer of the magnetizable solid-phase antiserum prepared with KLH-MTX immunogen is higher (despite the smaller volume of antiserum coupled) than that of antiserum prepared with PAPL-MTX immunogen. Nonspecific binding was not significant when 1.25 mg/tube or less solid-phase was used.

Magnetizable Solid-Phase Antibody Radioimmunoassay

Figure 6 shows typical manual RIA standard curves using magnetizable solid-phase antibody.

Effect of the Amount of Anti-MTX Serum Coupled to the Solid-Phase

To determine the maximum capacity of solid-phase for anti-MTX serum, raised against the MTX-PAPL immunogen, different amounts (0.5, 1, and 2 ml) were coupled to 1 g of magnetizable cellulose solid-phase, as described above. Figure 7 shows antiserum dilution curves using these solid-phase preparations. The highest coupling ratio (2 ml antiserum/1 g solid-phase) provides the best binding capacity and allows reasonable amounts of solid-phase to be used in the assay, which helps keep nonspecific binding low.

With antiserum prepared using the MTX-KLH immunogen, a 300 μl/g solid-phase coupling ratio showed binding even higher than that using PAPL-MTX immunogen and a 2 ml/g coupling ratio. A similar standard curve was obtained using only 0.2 mg of this magnetizable solid-phase anti-MTX.

Effect of Mixing and Nonmixing on the Magnetizable Solid-Phase RIA

Standard curves were constructed with ^{125}I-hist-MTX tracer and a magnetizable solid-phase (0.8 mg). Tubes were incubated for various pe-

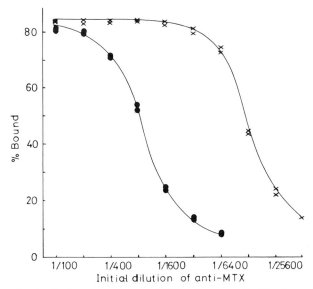

FIG. 4. Antibody dilution curves for anti-MTX using magnetizable charcoal separation for antisera obtained from sheep immunized with KLH-MTX immunogen after three boosts (×) and sheep immunized with PAPL-MTX immunogen after six boosts (●).

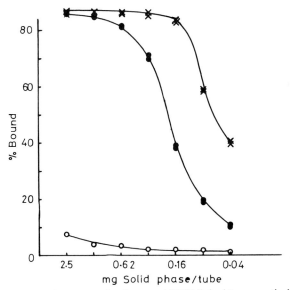

FIG. 5. Antibody dilution curves for sheep anti-MTX linked to magnetizable solid-phase [300 μl/g for sheep immunized with KLH-MTX immunogen (×), and 2 ml/g for sheep immunized with PAPL-MTX immunogen (●)] using ^{125}I-hist-MTX tracer. Normal sheep serum linked to magnetizable solid-phase (2 ml/g) was used to determine the solid-phase nonspecific binding (○).

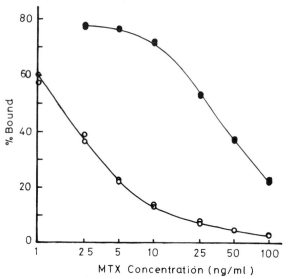

Fig. 6. Typical normal (●) and sensitive (○) RIA standard curves for MTX using ^{125}I-hist-MTX tracer and magnetizable solid-phase anti-MTX.

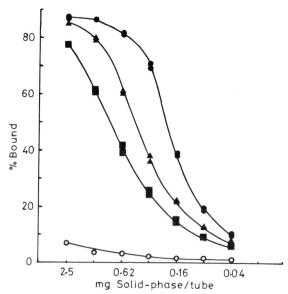

Fig. 7. Antibody dilution curves for sheep anti-MTX against MTX-PAPL linked to magnetizable solid phase when using 2 ml (●), 1 ml (▲), and 0.5 ml (■) of the anti-MTX, or 2 ml of normal sheep serum (○) per gram solid-phase and ^{125}I-hist-MTX tracer.

riods; one set was continuously mixed on a shaker, the other identical set was not mixed. Diluent buffer (1 ml) was added and separation performed by a magnet after 10, 30, 60, and 120 min incubation time. From a comparison of the mixing and nonmixing sets of standard curves, it is apparent that incubation for 30 min is required to achieve equilibrium without mixing, while this was achieved within the first 10 min of incubation with continuous mixing.

The kinetics of binding of the tracer at room temperature in the presence of zero and top concentration standard are illustrated in Fig. 8. A steady state was reached after a 5-min period.

For a more sensitive assay, 100 μl of standard or patient's sample were mixed with 10 μl of tracer and 10 μl of the solid-phase (0.165 mg). Experiments similar to those described above were performed and showed that 30-min incubation at room temperature was sufficient to achieve equilibrium without mixing. No detectable change in nonspecific binding over various incubation periods was noted.

Manual Magnetizable Solid-Phase RIA Precision

Intraassay precision with ^{125}I-hist-MTX was measured by assaying eight patients' samples 10 or 20 times using the magnetizable solid-phase separation technique. The coefficients of variation ranged from 2.1 to 7.6% (Table I).

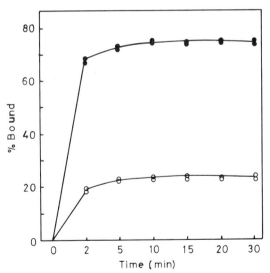

FIG. 8. Kinetics of binding of ^{125}I-hist-MTX with anti-MTX linked to magnetizable solid-phase (0.6 mg) in the presence of zero concentration (●) and 100 ng/ml MTX (○) standards for various timed intervals with continuous mixing.

TABLE I
WITHIN-ASSAY PRECISION STUDIES

Mean amount assayed per tube (ng/ml)	Number of analyses (in duplicate)	Standard deviation (ng/ml)	Coefficient of variation (%)
With ^{125}I-hist-MTX tracer			
18.2	10	0.84	4.6
31.6	20	0.66	2.1
44.8	10	2.4	5.3
50.5	20	1.6	3.2
70.4	10	4.3	6.2
Sensitive assay			
2.5	10	0.19	7.6
5.0	10	0.26	5.1
7.2	10	0.5	7.2

Recovery

The accuracy of the assay was assessed by the following methods: (a) Known amounts of methotrexate were added to aliquots of patients' sera and NHS. The sera were assayed before and after the addition of the methotrexate and recoveries ranged from 94 to 95.7%. (b) Eight sera were assayed at two different dilutions (1:20 and 1:40) and the results, which were in good agreement, are given in Table II.

Cross-reactivity

The cross-reactivity of the anti-MTX serum coupled to magnetizable solid-phase was assessed with structurally related folate derivatives and

TABLE II
MEASUREMENT OF PATIENT SAMPLES AT
VARIOUS DILUTIONS

Serum	Dilution (ng/ml) 1:20	1:40
A	34.2	17.2
B	21.0	9.5
C	100.0	51.0
D	47.5	26.0
E	24.1	12.8
F	41.0	21.7
G	61.5	34.0
H	82.0	39.8

drugs which might be used in combination with MTX. The percentage cross-reactivity was calculated from the curves by applying the formula suggested by Abraham.[42]

Table III shows that all but aminopterin showed negligible cross-reactivity with the antiserum when tested at concentrations up to 25 μg/ml for the analogs and up to 1 mg/ml for the other drugs.

Correlation Studies

With Competitive Protein Binding Assay. For the ^{125}I-hist-MTX and magnetizable solid-phase separation assay, results obtained for 16 patients' samples correlated closely with those obtained by the use of a competitive protein binding assay[15] in another laboratory (Royal Marsden Hospital, Dr. Calvert) (Fig. 9, correlation coefficient, $r = 0.97$).

With ^{3}H-RIA. Results for 25 patients' samples measured by a [^{3}H]MTX RIA[18] correlated closely with those of the present assay (correlation coefficient 0.98) (Fig. 9).

Magnetizable Charcoal Radioimmunoassay

Figure 10 shows a typical RIA standard curve using magnetizable charcoal separation.

TABLE III

CROSS-REACTION MTX-RIA USING MAGNETIZABLE SOLID-PHASE SEPARATION AND ^{125}I-hist-MTX TRACER

Substance	Cross-reactivity
MTX	100%
7-OH-MTX	0.5%
Aminopterin	18%
Folic acid	
Dihydrofolic acid	No detectable displacement of
Tetrahydrofolic acid	labeled MTX at 25 μg/ml serum
Folinic acid	concentration
N-5-Methyltetrahydrofolic acid	
N-(p-Aminobenzoyl)-L-glutamic acid	
Vincristine	
Vinblastine	No detectable displacement of
Cytarabine	labeled MTX at 1 mg/ml serum
Fluorourocid	concentration
Cyclophosphamide	

[42] G. E. Abraham, *in Proc. Tenovus Workshop, 5th, Cardiff* April 1974.

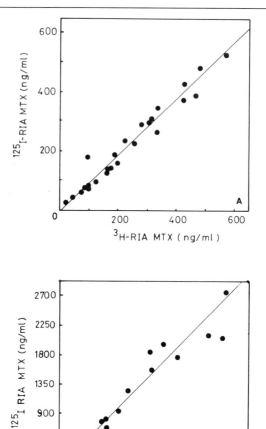

FIG. 9. Correlations between ^{125}I-magnetizable solid-phase RIA and (A) ^{3}H-RIA and (B) CPBA methods for MTX determination.

Correlation of Magnetizable Charcoal with Magnetizable Solid-Phase RIA Methods

[125]I-hist-MTX tracer was used in the assay of 20 serum samples from patients receiving MTX treatment by two procedures. For the magnetizable charcoal the antiserum was used at 1:100 initial dilution. For the magnetizable solid-phase separation, the normal assay procedure was employed. Results correlated closely ($r = 0.99$).

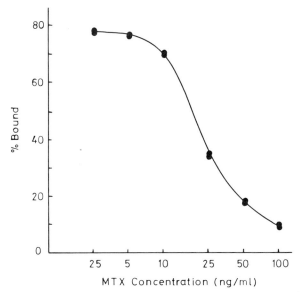

FIG. 10. Typical standard curves for MTX using ^{125}I-hist-MTX tracer and magnetizable charcoal.

Automated Assay

Figure 11 shows a typical automated RIA standard curve.

Carryover. This was assessed by repeatedly assaying the zero and highest (ng/ml) standards alternatively, in duplicate. Thus, data for the zero standard preceded by both a 100 ng/ml and a zero standard were obtained, and likewise data for the 100 ng/ml standard preceded by both a zero and a 100 ng/ml standard. The results, shown in Table IV, demonstrate that carryover is insignificant.

Precision. Two serum specimens from patients receiving MTX were each assayed 24 times to assess within-assay precision. Mean results were 32.6 and 14.7 ng/ml, with CV of 2.4 and 2.1%, respectively.

Accuracy. MTX was added at 50 ng/ml to 15 individual samples of normal human serum, and to 5 lipemic sera (triglyceride levels between 1.9 and 4.0 mmol/liter). Average analytical recovery of MTX from the normal sera was 101.4 ± 5.4%, and from the lipemic specimens 102.2 ± 3.8% (mean ± SD in each case). Patients' samples were assayed either undiluted or diluted in normal human serum. Since MTX levels may vary greatly, dilution of the samples is often necessary.

The correlation coefficients (r) for the fully automated RIA vs manual RIA and fully automated RIA vs competitive protein binding assay were 0.99 ($y = 1.16x + 9.5$) and 0.96 ($y = 1.2x - 122$), respectively.

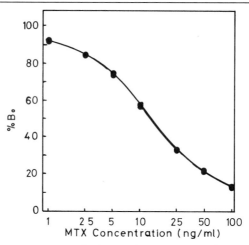

Fig. 11. A typical standard curve for automated MTX assay.

Specificity. Cross-reactivity of a series of MTX analogs (aminopterin, folic acid, folinic acid, dihydrofolic acid, tetrahydrofolic acid), and of drugs which may be given in combination therapy (vincristine, vinblastine, cytarabine, fluorouracil, and cyclophosphamide) was determined using cross-reactant concentrations up to 1 mg/liter. Only in the case of aminopterin was any significant cross-reaction found (18%).

Discussion

One aim of the present study was to investigate two methods of preparing ^{125}I-labeled MTX. The methods chosen, covalent coupling of MTX to Bolton and Hunter reagent and covalent coupling to ^{125}I-labeled histamine, are simple and widely applicable. The ^{125}I-labeled histamine-linked

TABLE IV
Carryover Data for Successive Sampling of High and Low
Standards in Duplicate

	High standard (100 μg/liter)		Zero standard (0 μg/liter)	
	After low standard	After high standard	After high standard	After low standard
Mean cps				
($n = 10$)	31.0	30.9	130.6	130.6
SD	0.79	0.73	1.55	1.5
CV	2.6	2.3	1.2	1.2

MTX gave a higher yield at iodination (50%), was of greater immunoreactivity (90% bound in the presence of excess antibody), and allowed development of a more rapid and precise assay. Unlike most steroid ^{125}I-tracer preparations, in which the product is extracted into the organic layer of ethyl acetate, ^{125}I-hist-MTX remained in the aqueous layer, which may account for the production of a tracer with high immunoreactivity.

Using the Bolton and Hunter method, yield at iodination was only 20% and attempts to improve this were unsuccessful. Immunoreactivity of the ^{125}I-labeled product (65% bound in the presence of antibody excess) was inferior and the assay was less rapid. However, this method of labeling small molecules may be preferred where convenient primary amine groups exist.

A recently described ^{125}I-RIA[19] used ^{125}I-labeled tyrosine methyl ester-MTX tracer and dextran-coated charcoal as a separation step. The iodo-histamine tracers in general are believed to be more stable than iodo-phenols, notably iodotyrosine derivatives.[43-44] The nonspecific binding interference which may occur with ^{125}I-labeled tyrosine derivative tracer can be almost eliminated with the use of labeled histamine derivatives.[45] This interference might occur due to thyroid binding proteins present in serum samples. The ^{125}I-labeled histamine MTX preparation described here is, therefore, considered to be a superior tracer for the RIA of MTX.[46]

The LH20 column technique, used for the purification of radioiodination products for MTX, is safer and simpler than commonly used TLC procedures for haptens, and should be preferred in cases in which the greater resolving power of TLC separations is not necessary.

Antisera were raised against MTX covalently linked to a polymer and to KLH. Negligible cross-reactivity occurred with folic acid, which differs from MTX by only one methyl and one amine group and only 18% cross-reactivity was observed with aminopterin, which differs from MTX by only one methyl group. The major MTX metabolite, 7-OH-MTX, does not cross-react significantly. A further measure of specificity was a comparison of results obtained for serum samples assayed by this assay and by a competitive protein binding assay based on dihydrofolate reductase[15] (DHFR). MTX produces its antineoplastic activity by binding and inhibiting DHFR. If MTX metabolites existed in patients' sera which had high affinity for the antibody and low affinity for DHFR, consistently higher

[43] R. Edwards, E. D. Gilby, and S. L. Jeffcoate, in "RIA and Related Procedures in Medicine," Vol II. International Atomic Energy Agency, Vienna, 1974.
[44] J. Maclouf, H. Sors, P. Pradelles, and F. Dray, Anal. Biochem. 87, 169 (1978).
[45] K. Painter and C. R. Vader, Clin. Chem. 25, 797 (1979).
[46] D. Goldman, Cancer Treat. Rep. 61, 549 (1977).

results should have been obtained by the ^{125}I-RIA. The close correlation observed implies that the assay is highly specific.

The antisera raised using the MTX-KLH immunogen had a higher titer than those raised against the poly(Ala)-poly(Lys)-MTX immunogen. Indeed, the three sheep immunized with the former all showed higher titer than any of the seven sheep immunised with poly(Ala)-poly(Lys)-MTX immunogen. This may be due to the KLH carrier (KLH-MTX) having higher immunogenicity than the poly(Ala)-poly(Lys).

The first antibody magnetizable particles are the most reliable. They can replace any existing separation technique, and are adaptable to manual, semi-automated, and fully automated immunoassay systems. Precision is often better than that achieved using conventional separation techniques, possibly due to the greater ease and reliability of the aspiration step when the solid-phase is firmly held in the bottom of the assay tubes by the magnetic field, and the clear visibility of the black particles. The likelihood of misclassification error should, thus, be reduced. Magnetizable charcoal represents a simple and reliable separation technique and results obtained correlated closely with those obtained using a magnetizable solid-phase antibody system. Thus, the assays could be used for routine monitoring of MTX.

For maximum antineoplastic activity, intracellular MTX concentrations must be maintained at higher levels than those necessary to saturate the enzyme DHFR.[46] Calvert and co-workers[47] suggest that serum levels in excess of 1 μM should be maintained for treatment of human breast tumours. Salasoo and associates[10] estimated that serum MTX levels of between 100 μM and 1 M should be maintained for 12 hr for treatment of osteogenic sarcoma.

For monitoring toxicity, Stoller and co-workers[7] claim that patients at high risk are identifiable from serum levels. In their study, serum levels above 900 nM (about 450 ng/ml), 48 hr after MTX infusion, identified patients in the high risk group. Highly sensitive standard curves are not necessary and, in the present work, standard curves were constructed to cover a less sensitive range to avoid inaccuracies incurred by large serum dilutions. Nonetheless, sensitivity could be a requirement for pharmocokinetic studies (e.g., monitoring renal clearance), and a more sensitive assay was developed. It was possible, by virtue of the high specific activity of the ^{125}I label and the avidity of the antisera, to extend assay sensitivity below levels ever previously attained by RIA.

It is concluded that an RIA incorporating ^{125}I-labeled histamine-linked MTX as tracer and a magnetic separation step (magnetizable solid-phase

[47] A. H. Calvert, P. K. Bondy, and K. R. Harrap, *Cancer Treat. Rep.* **61,** 1647 (1977).

particles or magnetizable charcoal) offers a simple, rapid, accurate, and specific manual method for routine monitoring of serum MTX levels.

The automated assay described here provides a rapid means of measuring large numbers of such sera with a high degree of accuracy and precision, and represents the only fully automated assay for MTX described to date. Any automated assay for MTX, using either direct or continuous-flow principles, must be capable of handling the wide range of concentrations commonly encountered in patients' sera. In the present system, known high samples are prediluted prior to assay so that concentrations greater than 100 ng/ml are not encountered. Fortunately samples can be readily diluted into the appropriate range if data on time and dosage are available. Carryover, which was a major problem with earlier continuous-flow automated systems employing long incubation periods, is not a problem and the sampling rate of 30 samples/hr could be doubled in view of the absence of carryover and the use of high specific activity tracer.

The relatively high concentration of BSA in the assay buffer was used to minimize the possible effects of intersample variation in protein content. A key feature of the automated system is the discrete, synchronized addition of both tracer and solid-phase to sample containing segments only. During the wash cycle, these reagents are recirculated and not pumped into the assay stream, which ensures that they are utilized with maximal economy.

Acknowledgments

We wish to thank Dr. J. Gardner and Dr. G. Forrest for their help and useful discussion and Dr. T. Merrett for help with antiserum preparation.

We would like to thank Dr. G. W. Aherne and Dr. H. Calvert for supplying patients' samples and results for correlation with our assays.

R. S. Kamel is grateful for a grant from the Ministry of Higher Education and Scientific Research, Iraq.

[30] Competitive Protein Binding Assay of Methotrexate

By CHARLES ERLICHMAN, ROSS C. DONEHOWER, and CHARLES E. MYERS

Introduction

Folic acid analogs with antimetabolic activity have been used in cancer chemotherapy for several decades. The principle antifolate agent in current clinical use is methotrexate (2,4-diamino-N-10-pteroylglutamic acid), which exerts its major effect by potent inhibition of the enzyme dihydrofolate reductase (DHFR) (EC1.5.1.3 tetrahydrofolate dehydrogenase). Inhibition of this enzyme prevents the conversion of dihydrofolate (FH_2) to tetrahydrofolate (FH_4) (Fig. 1). The cellular pool of reduced folates is necessary for the synthesis of thymidylate from deoxyuridylate, *de novo* synthesis of purine nucleotides, and protein synthesis. Although methotrexate affects the synthesis of nucleic acids and proteins, cells which are most sensitive are those actively synthesizing DNA.[1] Despite a detailed understanding of the cellular mechanism of drug action and the inclusion of methotrexate in many therapeutic regimens, only a rudimentry appreciation of pharmacokinetic correlates with the drug's toxicity was available. The introduction of therapy with high-dose infusions of methotrexate indicated a clear need for rapid, sensitive, and specific methods for the measurement of methotrexate to aid in the identification of patients at high risk for toxicity. In using such high dosages of methotrexate, severe and potentially lethal drug toxicity can be averted only by the subsequent administration of leucovorin (*N*-5-formyltetrahydrofolate) as a rescue agent. The dose and duration of this rescue should ideally be based on the measurement of circulating levels of methotrexate. Several methods which satisfy the needs of a clinical assay for methotrexate have been described. The assays in common clinical use currently are the radioimmunoassay[2-4] and the competitive protein binding assay[5,6] which will be the subject of this discussion. Other methods for the measurement of methotrexate have been described, including a DHFR inhibition

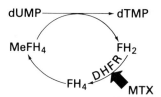

Fɪɢ. 1. Site of biochemical action of MTX. dUMP, deoxyuridine monophosphate; dTMP, deoxythymidine monophosphate; FH_2, dihydrofolate; FH_4, tetrahydrofolate; $MeFH_4$, methylene tetrahydrofolate; DHFR, dihydrofolate reductase; MTX, methotrexate.

[1] D. G. Johns and J. R. Bertino, *in* "Cancer Medicine" (J. F. Holland and E. Frei, eds.), p. 739. Lea & Febiger, Philadelphia, Pennsylvania, 1973.
[2] G. W. Aherne, E. M. Piall, and V. Marks, *Br. J. Cancer* **36**, 608 (1977).
[3] L. J. Loeffler, M. R. Blum, and M. A. Nelsen, *Cancer Res.* **36**, 3306 (1976).
[4] V. Raso and R. Schreiber, *Cancer Res.* **35**, 1407 (1975).
[5] C. E. Myers, M. E. Lippman, H. M. Eliot, and B. A. Chabner, *Proc. Natl. Acad. Sci. U.S.A.* **72**, 3683 (1975).
[6] E. Arons, S. P. Rothenberg, M. DeCosta, C. Fischer, and M. P. Iqbal, *Cancer Res.* **35**, 2033 (1975).

assay,[7] an enzyme-linked immunoassay, and high-pressure liquid chromatography techniques.

This paper will describe the competitive protein binding assay (CPBA) for methotrexate, discuss its clinical utility, and point out its usefulness in studying the interaction of tight binding competitive inhibitors of DHFR with this enzyme.

Materials and Methods

Competitive protein binding assays using enzymes rather than antibodies as receptor proteins have been developed for several chemotherapeutic agents. The measurement of unknown quantities of an inhibitor (MTX) by making use of its specific binding to its target enzyme (DHFR) is similar in principle to radioimmunoassays. In the case of methotrexate, the advantages that the use of an enzyme receptor protein confers on this method is that the laborious preparation of an appropriate antibody is avoided, the variability of antibody–drug affinity from one lot of antibody to another is eliminated, and, depending on the source of DHFR, meaningful information regarding the action of methotrexate on this enzyme can be derived from *in vitro* studies. The requisite specificity and sensitivity for successful use of this method and inherent assumptions in the theory of ligand binding assays are similar to those of the radioimmunoassay and have been described previously.[8]

The receptor protein in the CPBA of methotrexate described by Myers *et al.*[5] is DHFR partially purified from a strain of dichloromethotrexate-resistant *Lactobacillus casei* with a specific activity of 4.6 μmol of tetrahydrofolate/min/mg protein at pH 7.5 and 37°. It binds a maximum of 1.9 nmol of methotrexate per mg protein at pH 6.2 and 23° as determined by incubating the enzyme with excess radiolabeled methotrexate. Enzyme from this source is commercially available from the New England Enzyme Center, Boston, MA. Prior to assay, the stock DHFR is diluted 3500-fold in 0.5 M potassium phosphate buffer pH 6.2 containing freshly prepared NADPH (Sigma Chemical Co.) resulting in a solution with 5 pmol of methotrexate binding capacity and 2.4 μmol NADPH per ml. Other sources of DHFR have been utilized such as that derived from L1210 leukemia cells.[9]

Unlabeled MTX used in the determination of a standard curve and [³H]MTX are both purified by an ammonium bicarbonate gradient elution from a DEAE-cellulose column similar to that previously described.[10] The

[7] W. C. Werkheiser, S. F. Zakrewski, and C. A. Nichol, *J. Pharmacol. Exp. Ther.* **137**, 162 (1962).

[8] R. P. Ekins, *Br. Med. Bull.* **30**, 3 (1974).

[9] S. P. Rothenberg, M. DaCosta, and M. P. Iqbal, *Cancer Treat. Rep.* **61**, 575 (1974).

concentration of the unlabeled drug is determined spectrophotometrically utilizing an extinction coefficient of $7.0 \times 10^{-3} \, M^{-1} \, cm^{-1}$ at 370 nm in 0.05 M Tris–HCl pH 7.5.[1] Solutions of known MTX concentration are then prepared in 0.15 M potassium phosphate pH 6.2. [^{3}H]Methotrexate (Amersham-Searle) with specific activity of 9.5 Ci/mmol is diluted in 0.1 M potassium phosphate buffer pH 6.2 with 330 units of heparin per ml in a final concentration of 0.033 μCI/ml prior to assay.

To separate unbound methotrexate from that bound to DHFR, a charcoal slurry is prepared by combining 10 g of activated charcoal, 2.5 g of bovine serum albumin fraction V, and 0.1 g of high-molecular-weight dextran in 100 ml of distilled water. The pH is then adjusted to 6.2 with 1 N HCl. These reagents are purchased from Sigma Chemical Co. This slurry is highly effective in binding the MTX which is not bound to DHFR in the assay tube. Less than 3% of the unbound [^{3}H]methotrexate remains after addition of 50 μl of the charcoal slurry.

Plasma samples containing 100 units of heparin per ml are acidified by the addition of 20 μl of 1 N HCl per ml of plasma to achieve a pH of 6.2. Spinal fluid is similarly adjusted to pH 6.2 by the addition of 10 μl of 1 N HCl to each ml of fluid. Optimal binding of MTX to *L. casei* DHFR occurs at pH 6.2.

Assay Method. The assay is performed by the rapid sequential addition of (1) 150 μl of the dilute [^{3}H]methotrexate solution; (2) 200 μl of serially diluted unknown sample or standard solution of unlabeled methotrexate; and (3) 100 μl of the diluted solution of DHFR and NADPH. The assay is immediately agitated and 50 μl of charcoal slurry is added. Assay tubes are mixed again and centrifuged at 700 g for 30 min. A 200-μl aliquot of the supernatant fluid is removed, being careful not to disturb the charcoal pellet, then added to 10 ml of Aquasol (New England Nuclear Corp.) and counted in a liquid scintillation counter. The supernatant fluid contains the enzyme-bound methotrexate—both labeled and unlabeled.

The amount of bound [^{3}H]methotrexate will vary inversely with respect to the amount of unlabeled methotrexate present. Therefore, using known concentrations of methotrexate in the range of 1×10^{-9} to $1 \times 10^{-7} M$, a sigmoid standard curve may be constructed by plotting percentage of maximum bound, i.e., when no unlabeled methotrexate is present, versus concentration of methotrexate on semilogarithmic paper (Fig. 2). It is possible to increase the linear portion of the curve by plotting the logit transformation on the ordinate (inset, Fig. 2). The logit transformation is calculated as follows:

$$\text{logit} = \ln \left[\frac{B/B_0}{1 - B/B_0} \right] \tag{1}$$

[10] V. T. Oliverio, *Anal. Chem.* **33**, 263 (1961).

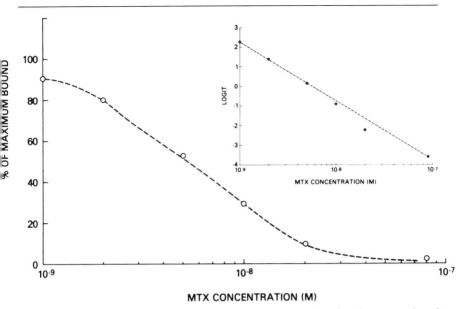

Fig. 2. Standard curve generated by varying the concentration of methotrexate plotted on x axis and [³H]methotrexate bound as a percentage of maximum bound in the absence of methotrexate on the y axis. Inset is the plot of logit versus methotrexate concentration of the same data points ($r^2 = 0.990$).

where B_0 is the amount of [³H]methotrexate bound to DHFR in the absence of the competing unlabeled methotrexate and B is the amount of the [³H]methotrexate bound to DHFR in the presence of the known or unknown amounts of unlabeled methotrexate in each assay tube.

Assay Results

The sensitivity of this assay for methotrexate is 0.3 pmol per assay tube as originally described. This translates into measurable levels of methotrexate in biological fluids as low as 1.5 nM. This level of sensitivity is possible because of the high binding affinity of methotrexate for DHFR. The association constant which gives a measure of this affinity is 2.1 × 10^8 M^{-1} for the methotrexate–DHFR reaction. Furthermore, the availability of [³H]methotrexate with high specific activity allows small quantities of [³H]methotrexate to be employed in the assay and still have significant radioactivity to permit statistically acceptable counting of samples. The reproducibility of the assay is good with a coefficient of variation of less than 10% between concentrations of methotrexate of 1.5 to 50 nM for duplicate samples.

The specificity of [³H]methotrexate for DHFR is high. This is demonstrated by the abolition of binding of [³H]methotrexate in the presence of 2000-fold excess unlabeled methotrexate. Under these conditions, the competition for receptor sites by the ligands is greatly in favor of the unlabeled methotrexate.

Assessment of interference which may occur in this assay can be carried out in the presence of increasing concentrations of folate analogs and known metabolites of methotrexate. The degree of interference that any of these substances may cause in the assay can be conveniently expressed as the concentration of the substance which inhibits binding of [³H]methotrexate to 50% of its maximum under the conditions of the assay (I_{50}). The compounds of greatest concern are 7-hydroxymethotrexate, 2,4-diamino-N-10-methylpteroic acid (DAMPA), and N-5-formyltetrahydrofolic acid (leucovorin) (Fig. 3). 7-Hydroxymethotrexate, which is a hydroxylation product of the liver, has an I_{50} of 40 nM, 8-fold greater than labeled methotrexate. DAMPA, the carboxypeptidase G_1 cleavage product of methotrexate, has an I_{50} of 440 nM, 88-fold greater than the unlabeled methotrexate. Both have been identified in the plasma of patients following high-dose infusion. Leucovorin, used in the "rescue" of patients receiving high-dose methotrexate infusions has an I_{50} of 0.1 mM or 24,000 times higher than for methotrexate itself. Other folate analogs which have been tested in this assay are 5-methyltetrahydropteroylglutamate, the major circulating folate and N-10-methylfolic acid, a contaminant of the commercially available methotrexate. A summary of the I_{50} results for all

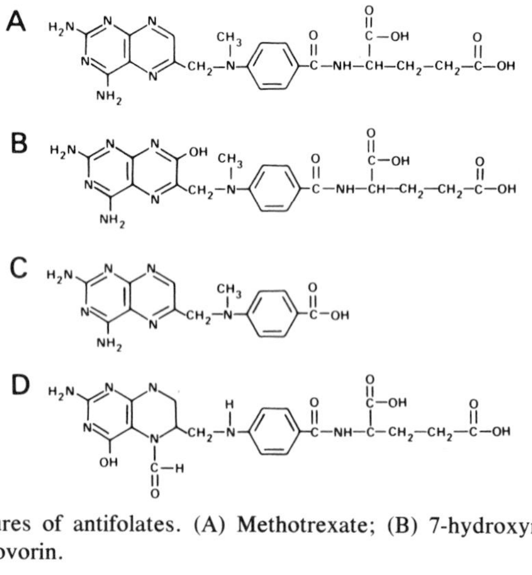

Fig. 3. Structures of antifolates. (A) Methotrexate; (B) 7-hydroxymethotrexate; (C) DAMPA; (D) leucovorin.

these agents is listed in the table. These findings suggest that folate analogs are potential sources of error in the interpretation of the methotrexate levels obtained by this assay. However, the available data suggest that the clinical significance of this interference is likely to be minimal since only in a small number of cases do levels of metabolites accumulate which may interfere with the assay.

Results obtained by the CPBA have been compared to those of an enzyme inhibition assay for methotrexate[5] and at least two radioimmunoassays.[11,12] The values determined in plasma and CSF by the CPBA and the enzyme inhibition method show a good correlation between the two methods. Comparison of the CPBA to a commercially available radioimmunoassay for methotrexate suggested that plasma samples may give divergent results in the two assays. The divergence was found to be maximal in those samples obtained 48 to 72 hr after the start of the drug infusion and indicated a consistently higher level of methotrexate as measured by the RIA. Further studies revealed that 2,4-diamino-N-10-methylpteroic acid may be a significant metabolite in some patiens which would interfere with the tested radioimmunoassay. The I_{50} for this metabolite in the radioimmunoassay was only 2.3 times greater than that of unlabeled methotrexate.[11] Therefore, small amounts of 2,4-diamino-N-10-methylpteroic acid could contribute significantly to the measured level of methotrexate by this radioimmunoassay. Since antibody specificity may vary from one radioimmunoassay method to another depending on the technique and species in which the antibody is raised, cross-reaction of this nature must be determined in each case.

INTERFERENCE OF FOLATE ANALOGS IN
THE CPBA AS MEASURED BY $I_{50}{}^{a}$

Compound	I_{50} (nM)
Methotrexate	5.0
7-Hydroxy-MTX	40
DAMPA	400
N-10-Methylfolic acid	100
Leucovorin	100,000
5-methyltetrahydropteryolglutamate	1,300,000

^{a} I_{50} is the concentration of a compound required to decrease the binding of [^{3}H]methotrexate to 50% of its maximum.

[11] R. C. Donehower, K. R. Hande, J. C. Drake, and B. A. Chabner, *Clin. Pharmacol. Ther.* **26,** 63 (1979).
[12] R. Virtanen, E. Iisalo, M. Pavinen, and E. Nordman, *Acta Pharmacol. Toxicol.* **44,** 296 (1979).

Applications of the Competitive Protein Binding Assay for Methotrexate

The competitive protein binding assay has a major role in the clinical monitoring of therapy with high-dose methotrexate. Monitoring multiple infusions given in a standard fashion has allowed the identification of a plasma parameter which will predict with a high degree of probability whether a patient will develop clinical toxicity. After a 6-hr infusion of 50–250 mg/kg, patients having levels of methotrexate greater than 0.9 μM in plasma at 48 hr were identified as having a significant chance of developing methotrexate systemic toxicity.[13] The ability to rapidly determine whether a plasma level exceeding 0.9 μM exists in any given patient on such a regimen allows the physician to increase the dose of leucovorin and thus decrease the likelihood of toxicity. The use of this assay to determine the pharmacokinetic behavior of a small test dose of methotrexate in a given patient in short order has allowed one group of investigators to individualize the patient's high dose of methotrexate using a computer modeling system.[14] Such routine monitoring and potential for tailoring a course of therapy has been greatly simplified by the availability of such a rapid, simple, and inexpensive assay as the competitive protein binding assay.

The competitive protein binding technique can also be used to study the interaction of methotrexate and dihydrofolate reductase. As mentioned previously, methotrexate is considered a tight binding inhibitor of DHFR. Difficulties arise in applying steady-state rate equations in determining the dissociation constants of tight binding inhibitors. Certain assumptions and approximations often requiring complex mathematical analyses must be made in order to obtain an estimate of the K_i value of methotrexate for DHFR.[15,16] Even after acknowledging these difficulties, application of different techniques may result in large variations in the K_i obtained.[15] However, using equilibrium binding analysis of methotrexate–DHFR interaction, an estimate for methotrexate binding affinity to DHFR can be arrived at simply. Such experiments have been carried out under similar conditions as those for the CPBA. Two series of assay tubes were prepared containing DHFR from *L. casei* with 7 pmol of methotrexate binding capacity, 0.24 μmol of NADPH, 0.075 μmol of potassium phosphate buffer, and varying amounts of [^{3}H]methotrexate from 1 to 20

[13] R. G. Stoller, K. R. Hande, S. A. Jacobs, S. A. Rosenberg, and B. A. Chabner, *New Engl. J. Med.* **297**, 630 (1977).
[14] S. Monjanel, J. P. Rigault, J. P. Cano, Y. Carcassonne, and R. Favre, *Cancer Chemother. Pharmacol.* **3**, 189 (1979).
[15] W. P. Greco and M. T. Hakala, *J. Biol. Chem.* **254**, 12104 (1979).
[16] S. Cha, *Biochem. Pharmacol.* **24**, 2177 (1975).

pmol. In the second set of tubes, 5 nmol of unlabeled methotrexate was added to all the reaction mixtures. The final volume was 450 μl. The excess unlabeled methotrexate which was added to the second set of tubes is sufficient to prevent specific binding of [³H]methotrexate to DHFR, allowing assessment of nonspecific binding of the labeled methotrexate to protein. Separation of bound and free [³H]methotrexate was carried out as previously described using the charcoal slurry, and the supernatant containing the bound counts was counted. Plotting the results initially as bound versus total [³H]methotrexate and then according to the method of Scatchard[17] lead to results shown in Fig. 4. This Scatchard plot suggests that there is a single homogeneous site on DHFR for methotrexate binding and the slope of this line is equivalent to the -K_A or association constant. The inverse of the K_A gives a K_D (dissociation constant) of 4.76 × 10^{-9} M. This result is obtained by carrying out the binding studies during a time period when equilibrium exists for formation of the enzyme inhibitor complex and in the absence of dihydrofolate. Thus, the need for steady-

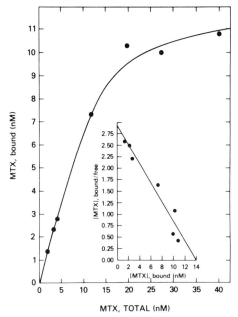

FIG. 4. Plot of bound [³H]methotrexate versus total [³H]methotrexate. Experimental conditions are described in text. Inset is the same data replotted according to the method of Scatchard[17] using an unweighted least-squares fit (r = 0.97) (from Ref. 5).

[17] G. Scatchard, *Ann. N.Y. Acad. Sci.* **51,** 660 (1949).

state assumptions of kinetic studies is not applicable and solution of a "stiff" differential equation-defined model is not required.

Extension of the equilibrium binding studies allows the determination of the off rate or k_{off} for methotrexate from DHFR. Such kinetic binding studies have been performed for the DHFR derived from L1210 leukemia cells and *Escherichia coli* MB 1428.[18,19] Assuming that the dissociation of methotrexate from the dihydrofolate reductase is a unimolecular event described by

$$EI \xrightarrow{k_{off}} E + I$$

where E is DHFR and I is methotrexate, then a plot of ln (bound methotrexate) versus time is linear with a slope of k_{off}. The experimental design involves modification of the CPBA for methotrexate and is outlined for the determination of the k_{off} for DHFR purified from L1210 leukemia. DHFR purified from L1210 leukemia cells with specific activity 33 units/mg protein was used in these studies. The reaction volume of 450 μl consisted of 0.1 units of DHFR, 1 μM [^{3}H]methotrexate (specific activity 6.4 Ci/mmol), 0.1 mM NADPH in 0.05 M potassium phosphate pH 7.4, and 0.15 M KCl. After incubation for 20 min at 37°, 50 μl of the charcoal slurry prepared as previously described was added to the assay tube. Each tube was agitated and then centrifuged at 4° and 700 g for 45 min. The supernatant which contained the enzyme–methotrexate complex is removed and equally divided into a series of tubes containing 100 μM unlabeled methotrexate and 0.1 mM NADPH. At specific time points, 50 μl of the charcoal slurry is added and centrifuged again. The radioactivity in the supernatant is determined by scintillation counting. The counts represent the remaining DHFR–[^{3}H]methotrexate complex at each time point. Plotting these results yields an off rate of 0.0014 min^{-1}. Similar methodology was applied to determine the k_{off} of methotrexate from DHFR from *E. coli*.[19] However, separation of bound and free [^{3}H]methotrexate was carried out using a minicolumn technique. The k_{off} determined in this fashion was 0.014 min^{-1}.

The versatility of the CPBA for methotrexate is demonstrated by the ability to adapt this assay to measure levels of other antifolate agents. Since this assay depends on the competition by [^{3}H]methotrexate and unlabeled methotrexate, other antifolate agents which bind to DHFR can be substituted for the unlabeled methotrexate if these agents bind to the same site as methotrexate. The sensitivity of the assay for any given anti-

[18] M. Cohen, R. A. Bender, R. Donehower, C. E. Myers, and B. A. Chabner, *Cancer Res.* **38**, (1978).
[19] J. C. White, *J. Biol. Chem.* **254**, 10889 (1979).

folate will depend on the binding constant of the antifolate to be measured. Determination of the I_{50} for the antifolate to be measured will give an estimate of the range over which the assay will be useful because a linear portion of the sigmoidal curve generated by adding varying amounts of the antifolate to be assayed is centered about the I_{50} concentration. The assay range will usually extend above and below the I_{50} concentration by one log concentration. The application of this general approach has been demonstrated for several antifolate agents.[20] Aminopterin has an I_{50} concentration of $5 \times 10^{-8} M$ which is 10-fold higher than methotrexate in the CPBA. Therefore, this assay can be easily adapted to measure levels of aminopterin as low as 5 nM. Similarly, methasquin, another antifolate, with an I_{50} of $5.4 \times 10^{-8} M$ can be assayed to levels of approximately 5 nM also. An attempt to utilize the L. casei DHFR to assay triazinate (another antifolate) revealed that the I_{50} was $1 \times 10^{-5} M$, thus making the sensitivity of this assay only 1 μM. However, this agent has a higher affinity for the DHFR derived from methotrexate-resistant L1210 leukemia cells[21] with an I_{50} concentration of $1 \times 10^{-7} M$. Therefore, by using another source of DHFR, the assay can be adapted to improve the sensitivity of measuring triazinate to levels of $1 \times 10^{-8} M$. This principle of changing the source of the DHFR depending on its affinity for the ligand of interest in order to maximize assay sensitivity is possible because of the comparative studies which have been carried out with DHFR from a variety of sources.[22] These studies have shown a significant difference in concentration of a given antifol to inhibit DHFR activity from varying sources by 50%. Trimethoprim, an antifolate used in microbial infections, causes 50% inhibition of DHFR activity derived from E. coli, S. aureus, and P. vulgaris in nanomolar concentrations, whereas concentrations of approximately 0.3 mM is required in order to cause similar inhibition of DHFR from human liver or rat liver. Rothenberg et al.[9] have demonstrated that assay of pyrimethamine, an antimalarial antifolate, can be carried out using DHFR from L1210 leukemia cells with a sensitivity of approximately 0.1 nmol per assay tube.

Conclusion

We have demonstrated that the competitive protein binding assay for methotrexate utilizing principles of equilibrium binding analysis is a remarkably versatile technique for study of DHFR and antifolates. The

[20] C. E. Myers, H. M. Eliot, and B. A. Chabner, Cancer Treat. Rep. 60, 615 (1976).
[21] A. R. Cashmore, R. T. Skeel, D. R. Makulu, E. J. Gralla, and J. R. Bertino, Cancer Res. 35, 17 (1975).
[22] J. J. Burchall, Ann. N.Y. Acad. Sci. 186, 143 (1971).

assay, described herein, is simple and rapidly performed with sensitivity comparable to published radioimmunoassay techniques for methotrexate and excellent specificity in the clinical setting. Its applications in the routine monitoring of patients treated with high-dose methotrexate and in pharmacokinetic modeling are presented. In addition, applications of this technique to the study of binding affinity and kinetics of binding of methotrexate to DHFR are outlined. Finally, data describing the ability of this assay to be modified for measurement of other antifolate agents have been discussed. The broad spectrum of uses to which this assay may be put makes the CPBA for methotrexate extremely useful as a laboratory method.

[31] Ligand-Binding Radioassay of N^5-Methyltetrahydrofolate and Its Application to N^5-Formyltetrahydrofolate

By MARIA DA COSTA and SHELDON P. ROTHENBERG

Introduction

N^5-Methyltetrahydrofolate (mTHF), the major intracellular and circulating folate cofactor, can now be measured by ligand-binding radioassay using any one of the natural binding proteins isolated from human or animal tissues or biologic fluids. For most of the reported methods, the folate-binding protein present in cow's milk[1] has been used as the binder and radioisotopically labeled folic acid (FA), or some derivative of FA, as the tracer. Most of these naturally occurring folate-binding proteins, however, have greater affinity for FA than for mTHF or other naturally occurring folates. Consequently, if the classical competitive inhibition radioassay is used with the standard mTHF and tracer competing simultaneously for binding sites on the protein, the sensitivity of the dose–response curve will be diminished. There will also be an inherent error in the total folate concentration when a mixture of folates, each with varying affinity for the binding protein, is measured by a competitive inhibition radioassay. This error can be minimized by carrying out the radioassay at pH 9.3 at which the affinity of FA and mTHF for the binder appears to be equal,[2]

---

[1] J. Ghitis, *Am. J. Clin. Nutr.* **20**, 1(1967).
[2] J. K. Givas and S. Gutcho, *Clin. Chem.* **21**, 427(1975).

or by using a sequential noncompetitive system[3] in which the tracer is added to the reaction to titrate only the unoccupied binding sites on the protein after all the other folates (standard mTHF or folates in the sample being assayed) have reacted with the binding protein. We prefer the non-competitive sequential radioassay because of its extremely high sensitivity, and this method, using the folate binder partially purified from cow's milk and [³H]FA as the tracer, is described in detail.

Reagents

0.01 M borate buffer, pH 8.0

0.05 borate–Ringer's solution containing 2 g of L-ascorbic acid per liter. The ascorbate is added just before use and the pH adjusted to 8.0 with 2 N HCl

0.067 M sodium phosphate, pH 7.0 (0.027 M monobasic sodium phosphate and 0.04 M dibasic sodium phosphate), containing 0.08 M NaCl

Sodium borohydride

[³H]FA is diluted in borate-Ringer's buffer, from which ascorbic acid is omitted, to a concentration of 5,000 pg/ml. Commercial sources of [³H]FA have a specific activity ranging from 10 to 50 Ci/mmol, so that 250 pg in the reaction mixture provide adequate radioactivity counts (cpm) if the liquid scintillation counter has an efficiency greater than 40%. The purity of the [³H]FA is determined by coprecipitation with unlabeled FA and zinc sulfate.[4] If the purity is less than 85%, the preparation is not suitable and can be purified by DEAE chromatography[5]

mTHF: This is commercially available, and it is prepared by dissolving the lyophilized powder in the borate–Ringer's–ascorbate buffer. The spectrum should have a maximum absorption at 289 nm and minimum at 245 nm. The concentration is established spectrophotometrically using $E_{1\ cm}^{1\%}$ of 655 at 290 nm

2.5 g of neutral Norit A charcoal suspended in an aqueous solution of 0.5% dextran (molecular weight ≈40,000) or 0.25% bovine hemoglobin

6% ovalbumin treated with 15 mg of Norit A charcoal per ml. After shaking for a few minutes, the charcoal is pelleted by centrifugation at 30,000 g. This solution is diluted before use to 1% in the borate–Ringer's buffer

Folate-binding protein partially purified from cow's milk

---

[3] S. P. Rothenberg, M. da Costa, and Z. Rosenberg, *New Engl. J. Med.* **286**, 1335 (1972).

[4] S. P. Rothenberg, *Anal. Biochem.* **16**, 176 (1966).

[5] P. F. Nixon and J. R. Bertino, this series, Vol. 18, p. 661.

Preparation of Binding Protein from Milk

Pasteurized or unpasteurized whole cow's milk is warmed to 37° and 5% acetic acid is slowly added until the milk is completely curdled (pH 4.5–5.0). The curd is separated by filtration through a Büchner funnel and, in order to increase the folate-binding capacity of the final preparation, the pH of the filtrate is lowered to 2.0–2.5 with concentrated HCl to dissociate the bound endogenous folates from the binding protein. The dissociated folates are then removed by adding 15 mg of neutral Norit A charcoal per ml of solution and the suspension stirred continuously for 2 hr at 4°. The charcoal is pelleted by centrifugation for 30 min at 30,000 g at 4°, and the pH of the supernate raised to 7.4 with concentrated NaOH. Solid $(NH_4)_2SO_4$ is added to this solution to 45% of saturation, and after standing for 1–2 hr at 4°, the surface scum is removed and the precipitated proteins are pelleted by centrifugation and then discarded. Additional $(NH_4)_2SO_4$ is added to this supernatant solution to raise its concentration to 60% of saturation and the mixture allowed to incubate at 4° overnight. The precipitate, containing the binding protein, is collected by centrifugation, dissolved in 0.01 M borate buffer, pH 8.0, and dialyzed for 18 hr against the same buffer. After centrifugation, to remove any insoluble debris, it is stored in aliquots at −20° as the stock solution.

The Determination of the Binding Titer of the Stock Solution

A dilution of the stock solution of binder protein which binds 50–60% of the [³H]FA (usually 200–250 pg) is determined as follows. The dilutions are prepared in the borate–Ringer's buffer containing 1% ovalbumin, and 100 μl of each dilution is incubated with 100 μl of [³H]FA (200–250 pg) in 300 μl of borate–Ringer's–ascorbate buffer, pH 8.0, for 30 min. Four hundred microliters of the charcoal suspension is then added to adsorb the free folate, the mixture vortexed briefly, and the charcoal pelleted by centrifugation. An aliquot of the supernatant solution containing the bound [³H]FA is added to 15 ml of any scintillation cocktail used for counting aqueous solutions and the radioactivity determined with a liquid scintillation counter.

To correct for radioactivity in the [³H]FA preparation which does not adsorb to charcoal, a "charcoal blank" is prepared by treating 100 μl of the [³H]FA in 400 μl of the buffer with 400 μl of the charcoal suspension. The radioactivity of this charcoal blank is subtracted from the radioactivity of the assay reactions to determine the net radioactivity bound to the binding protein. The percentage of [³H]FA bound is then calculated as follows:

$$\frac{\text{cpm bound in assay reaction } - \text{ cpm in charcoal blank}}{\text{total cpm in the assay reaction } - \text{ cpm in charcoal blank}} \times 100$$

The Standard mTHF Dose–Response Curve

For reasons presented earlier, a two-phase sequential incubation reaction is used to enhance the sensitivity of the radioassay, and to minimize the difference in the affinity of the binding protein for [³H]FA, the mTHF standards, and the folates which are being assayed. mTHF standards (25–300 pg) contained in 100 μl of borate–Ringer's–ascorbate solution is incubated with 100 μl of the previously determined dilution of binding protein in 200 μl of borate–Ringer's–ascorbate buffer at room temperature for 30 min and then at 4° for 30–120 min. This phase of the radioassay may also incubate overnight if this is more convenient. The purpose of lowering the temperature is to minimize dissociation of the bound folate. In the second phase of the radioassay [³H]FA contained in 100 μl is added to the reaction to titrate the binding sites unoccupied by the mTHF. After 30 min of additional incubation at 4°, 400 μl of the charcoal suspension is added as described above, and the bound radioactivity in an aliquot of the supernate is determined after centrifugation of the charcoal.

A dose–response curve is graphed by plotting the ratio of bound [³H]FA to free [³H]FA (B/F) against the mTHF standards.

Preparation and Assay of Samples

The biologic material to be assayed may require preliminary extraction if the endogenous folates are protein bound or if the sample contains an endogenous unsaturated folate-binding protein.

1. Serum containing 2.5 mg of ascorbate per ml is assayed untreated if it contains little or no unsaturated folate-binding protein. It is stored at −20°.

2. Serum extraction: If the serum contains unsaturated binding protein (vide infra), it should be diluted in three volumes of Ringer's solution containing 250 mg of ascorbate per 100 ml, pH 4.5, and boiled for 5 min. The precipitated proteins are pelleted by centrifugation and the supernatant solution assayed.

3. Whole blood: One volume of whole blood containing 2.5–5.0 mg of ascorbic acid per ml is diluted with 24 volumes of distilled water, and this hemolysate is then diluted again with an equal volume of 0.05 M borate–Ringer's–ascorbate buffer, pH 8.0. An aliquot of this hemolysate is then assayed without further extraction. Whole blood containing ascorbic acid can be stored at 4° for several days.

4. Spinal fluid: It is stored at $-20°$ containing 2.5 mg of ascorbate per ml and is assayed unextracted.

5. Urine: It is stored containing 5 mg of ascorbic acid per ml at $-20°$ and is assayed unextracted.

6. Tissues: These are homogenized in ten volumes of 1–2% ascorbic acid, pH 6.0, and boiled for 5–10 min. The precipitated proteins are pelleted by centrifugation and the supernatant solution assayed.

An aliquot of each of the samples to be assayed is substituted for the standard mTHF and the total volume of the reaction is always adjusted to 400 μl with the borate–Ringer's–ascorbate buffer. For serums, as little as 10–50 μl can be used, and for red cells, spinal fluid, and tissue extracts, which contain a much greater concentration of folate, smaller volumes or dilutions of the preparation are made in the borate–Ringer's–ascorbate buffer and assayed.

A correction blank will be necessary when assaying unextracted serum to correct for any binding of [³H]FA by endogenous unsaturated binder. This is frequently found in some serums from patients with liver disease, folate deficiency, leukemia, cancer, etc. Accordingly, the same aliquot of test sample is incubated with the [³H]FA, but without the folate-binding protein, in the borate–Ringer's–ascorbate buffer to a total volume of 500 μl. The radioactivity remaining in the supernate after the addition and separation of the charcoal suspension is subtracted from the assay reactions, as described earlier for the correction with the charcoal buffer blank. If this correction is not made for samples containing a substantial amount of unsaturated binder (i.e., sufficient to bind 20–30% of the tracer), then a falsely lower folate concentration will be assayed. Serums which contain a concentration of such an unsaturated binder which can bind more than 20% of the tracer should be extracted as described above and reassayed.

All samples are assayed in two or more dilutions to ensure the accuracy of the measurement. If the concentration per unit volume for each dilution differs by more than $\pm 15\%$ from the mean concentration per unit volume for the dilutions assayed, then the sample should be reassayed.

Radioassay for N^5-Formyltetrahydrofolate (f⁵THF)

The folate binder from cow's milk does not bind f⁵THF and, therefore, it is measured only after it is quantitatively converted to mTHF. On acidification, f⁵THF loses water to become N^5,N^{10}-methenyltetrahydrofolate which is reduced to mTHF by an excess of borohydride, the intermediate analog probably being N^5,N^{10}-methylenetetrahydrofolate.[6]

[6] I. Chanarin and J. Perry, *Biochem. J.* **105,** 633 (1967).

The test sample, to which 5 mg of ascorbic acid per ml is added, is mixed with an equal volume of 1 N HCl, boiled for 10 min to accelerate the formation of N^5,N^{10}-methenyltetrahydrofolate, and then mixed with 6.5 ml of 0.67 M sodium phosphate solution, pH 7.0, containing 0.08 M NaCl and 20 mg sodium borohydride. The reaction mixture is incubated for 60 min at 25° and any precipitate formed is removed by centrifugation. The mTHF formed is protected by the ascorbate and is then assayed as described above. If the test sample also contains mTHF, it is measured before and after the acid-borohydride conversion, the difference in values being the equivalent of the f⁵THF. An appropriate correction must be made for the total dilution used in acidification and reduction when computing the concentration of f⁵THF in the original sample.

[32] Radioimmunoassay of Bleomycin

By ALAN BROUGHTON

Introduction

Classical methods used to determine drug concentration in serum and other biological fluids are usually very time consuming, often requiring expensive equipment and highly skilled personnel, and in many instances these methodologies are insensitive and imprecise. The introductions of radioimmunoassays initially for insulin by Berson and Yalow[1] and later for digoxin by Smith and others[2] and more recently for bleomycin by Broughton and Strong[3] has allowed the laboratory worker to measure accurately and precisely these compounds. Also, this has produced data for the clinician to more precisely diagnose and treat his patients. This chapter will describe the development of the radioimmunoassay for the drug bleomycin and briefly discuss some of the clinical applications of this assay.

Antibody Production

The antiserum was produced using the clinical preparation of bleomycin sulfate (BLM) blexane R. This preparation contains mainly bleomycin A2 (55–70%) and B2 (25–32%). Bleomycin has a molecular weight of

---

[1] S. A. Berson and R. S. Yalow, *J. Clin. Invest.* **39**, 1157 (1960).
[2] T. W. Smith, V. P. Butler, Jr., and E. Haber, *New Engl. J. Med.* **281**, 1212 (1969).
[3] A. Broughton and J. E. Strong, *Cancer Res.* **36**, 1418 (1976).

1500 and would, therefore, not be a satisfactory immunogen when injected alone but when covalently linked to a macromolecule should elicit a satisfactory immune response. Bleomycin was conjugated to bovine serum albumin (BSA) using 1-ethyl 3-(dimethylaminopropyl)carbodiimide to form an amide bond as described by Goodfriend and others.[4] The conjugate was prepared as follows: 20 mg of BSA was dissolved in 1 ml of phosphate-buffered saline (PBS) (0.15 M NaCl in 0.011 M phosphate buffer with 1 g/liter sodium azide added as a preservative), then 45 mg of BLM and 0.1 ml of carbodiimide (900 ng/ml) was added slowly to the mixture with constant stirring.

This solution was mixed at room temperature for 1 hr and then at 4° for 3 days. Dialysis against PBS for 18 hr at 4° was performed and the resulting mixture purified using column chromatography (Sephadex G-25). Spectrophotometry was used to determine the molar incorporation of the bleomycin to the albumin and was found to be 28:1. The animal chosen for immunization was the New Zealand white rabbit and three of these were immunized with the conjugate using the following procedure: The rabbits each received 0.5 mg of conjugate emulsified in complete Freund's adjuvant by intramuscular injection into all limbs. This was followed by monthly injections of 0.5 mg of the conjugate to either fore or hind limbs. The animals were bled after three booster injections and the resulting antisera evaluated as discussed below.

Iodination Procedures

A modified Hunter and Greenwood[5] chloramine-T technique was used to iodinate the bleomycin. This was performed as follows: 10 μl Na^{125}I (100 mCi/ml, 0.1 N NaOH), 10 μl of chloramine-T 5 mg/ml solution in 0.1 M borate buffer (pH 9.0), and 10 ml/BLM (1 mg/ml solution in 0.1 M borate buffer (pH 9.0) were mixed and then incubated at room temperature for 1 min. After 1 min 10 μl of sodium metabisulfate (12 mg/ml solution in 0.1 M borate buffer pH 9.0) was added to stop the reaction followed by 10 μl of potassium iodide (20 mg/ml in 0.1 M borate buffer, pH 9.0). The use of more alkaline media for the iodination procedure probably results in the iodination of the bleomycin occurring in the imidazole ring, this can be compared to histamine iodination in certain polypeptide immunoassay systems. Purification of the iodinated material was achieved using column-chromatography (Sephadex C1-25, 30 × 0.9 cm), a linear gradient of ammonium formate (pH 6.4) from 0.1 to 1.0 M was used to elute the iodinated product, and the specific activity of the iodin-

[4] G. L. Goodfriend, L. Levine, and D. Fasma, *Science* **144**, 1344 (1974).
[5] W. M. Hunter and F. C. Greenwood, *Nature* (*London*) **194**, 495 (1962).

ated bleomycin was approximately 5 $\mu Ci/\mu g$ with an incorporation of 3.5×10^{-5} atoms of iodide per molecule of bleomycin.

Radioimmunoassay Method

A typical competitive protein binding assay using ^{125}I-labeled bleomycin, unlabeled bleomycin, and antibody was used in all subsequent experiments described below. The antibody bound drug was separated from the free drug by the use of a dextran-coated charcoal separation procedure. A typical system for the assay is as follows: 200 μl of 1% gelatin in phosphate-buffered saline (PBS), 100 μl sample of appropriate solution, 100 μl ^{125}I-labeled bleomycin diluted to approximately 30,000 cpm in PBS, and 100 μl of a suitable dilution of antisera were mixed and incubated for 10 min at room temperature at 37° and 10 min at 4°. Following this incubation, 200 μl of dextran-coated charcoal plus 400 μl of phosphate-buffered saline at 4° were added and the mixture was then centrifuged at 2000 g for 10 min at 4°. The amount of radioactivity in the supernatant was determined using an automatic gamma counter.

Assay Validation

The assay system was validated using the following protocol. The antisera was evaluated for avidity and specificity. The iodinated bleomycin was evaluated for radiation damage and specific activity; finally the assay itself was validated using the dose–response curve and comparing it to a microbiological assay. The titer of the antisera defined as that dilution of antisera which would bind 50% of the labeled bleomycin (2 ng) was 1:1000. The avidity as determined by the equilibrium technique of Odell and others[6] was 1.3×10^9 liters mol^{-1}. The antisera was evaluated for specificity in cross-reactivity experiments using the following drugs: adramycin, 1-(2-chloroethyl)-3-cyclohexyl-1-nitrosourea, 5-fluorouracil, arabinolsyl, cytosine, vincristine, and prednisolone. The cross-reactivity was determined as the concentration at which 50% binding to antibody sites occurs, expressed as a percentage related to the concentration of bleomycin. All the above drugs cross-reacted less than 1%.

The cross-reactivity of the bleomycin was of great interest. The bleomycins differ from each other in their terminal amine moieties and the cross-reactivity studies suggest that the presence and nature of this part of the molecule determines much of the immunoreactivity. Illustrated in Table I is the percentage cross-reactivity of some bleomycin analogs with

---

[6] W. D. Odell, G. Abraham, and H. R. Rand, *Acta Endocrinol. (Copenhagen)* **142** (Suppl.), 54 (1969).

TABLE I
IMMUNOREACTIVITY OF BLEOMYCIN AND ITS ANALOGS WITH RABBIT ANTISERA[a]

Bleomycin	Mean concentration (pmole)	Standard deviation (pmole)	Percentage cross-reactivity
Sulfate	3.45	±0.11	100
A_2	2.99	±0.22	115.4
B_2	4.63	±0.17	74.5
A_5	29.50	±5.14	11.7
Iso-A_2	13.70	±0.29	25.2
Acid	>1000.00	—	<1.0
BU-2231B	>1000.00	—	<1.0
Desamido A_2	8.26	0.51	44.6
B_1	214.2	15.9	1.6

[a] The indicated quantity of bleomycin analog was required to produce a 50% inhibition of ^{125}I-labeled bleomycin sulfate bound to antibody. Results are expressed as the mean concentration, in picomoles, of the bleomycin analog determined in five analyses.

the bleomycin antisera. It should be noted that A2 and B2 cross-react more effectively than any other analogs. Although B2 has a terminal moiety differing somewhat from A2, there is a structural resemblance between the two side chains as shown with molecular models. Furthermore, bleomycin A5 which is a spermidine moiety in the terminal amine competes only 11.7% effectively against the antibody, whereas, isobleomycin differing from A2 only in translocation of the carbamoyl group in the mannose sugar, competed with 25% of the efficiency of bleomycin sulfate.

Another compound studied was bleomycinic acid which is devoid of terminal amine groups; this failed to react with the antibody. Bleomycin B1 has only an amine group in lieu of a terminal amino acid and reacts only 1.6% as effectively. A recent metabolite described by Umezawa,[7] desamidobleomycin, has intact terminal amines but the amine group on the β-amine alanamide has been removed by enzyme action. This compound was also studied and found to exhibit a cross-reactivity of approximately 44% with the bleomycin antibody. Conversely an antibody raised against a desamidobleomycin conjugate cross-reacted 100% against bleomycin sulfate.

My co-workers and I were very fortunate during these experiments to have available a new generation of bleomycin, the tallysomycins. Experiments with these drugs produced further evidence to support the immuno dominance of the terminal amines and the tertiary structure of the *nucleus*. These compounds in addition to the existing nucleus contained a new amino sugar tallyosamine and specific terminal amino moieties (Fig.

[7] H. Umezawa, S. Hori, T. Sawa, T. Yoshiova, and T. Takench, *J. Antibiot.* **27**, 419 (1974).

1). The compounds tallysomycin A and B were tested against the bleomycin antibody and exhibited no cross-reactivity. The compound tallysomycin A was conjugated to albumin by the carbodiamide technique, then an antibody raised to this conjugate by the technique described previously. Using the criteria listed above, cross-reactivity of this antibody with bleomycin was less than 1% as with desamidobleomycin and the parent compound pheomycin. Tallysomycin A with or without chelated copper, and a fragment of the molecule consisting of the terminal amine and half the nucleus cross-reacted 100% with the new antisera. Tallysomycin B without copper competed only 10% effectively. When copper was added to this molecule there was 40% cross-reactivity. The chelation of copper by bleomycin has been studied by Umezawa (1976)[8] who showed strong evidence that copper(II), a square planar atom, ligates from the nitrogen atom situated in the β-amine alanamide, imidazole ring, 4-aminopyrimidine residue, and the sugar carbamyl function. This ligation probably produces a confirmational change of the molecule enabling antibody to bind more effectively.

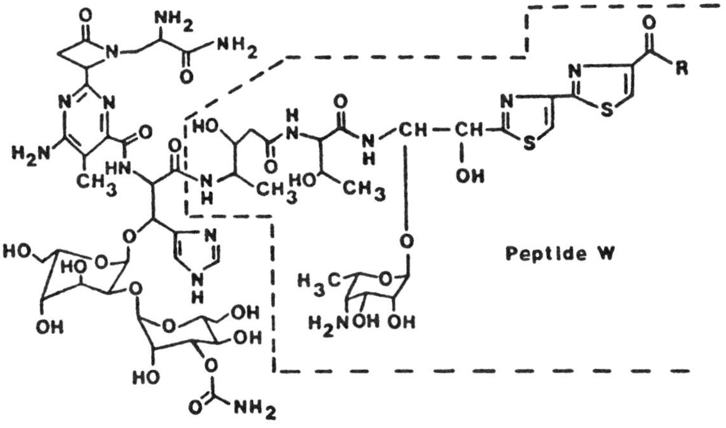

Tallysomycin A $R = -NH-(CH_2)_3-CH-CH_2-CO-NH-(CH_2)_3NH(CH_2)_4NH_2$
                                                     |
                                                    NH_2

Tallysomycin B $R = -NH(CH_2)_3NH(CH_2)_4NH_2$

Tallysomycin S_2B $R = -NH-(CH_2)_3-\overset{+}{S}\overset{CH_3}{\underset{CH_3}{<}}$

Tallysomycin E_1a $R = -NH-(CH_2)_2CH_2NH_2$

Fig. 1. Structure of tallysomycins.

[8] H. Umezawa, *Gann Monogr. Cancer Rev.* **19,** 3 (1976).

A similar situation may account for the loss of immunoreactivity by antibleomycin against isobleomycin A-2 which is also copper free. These studies using specific analogs of known chemical structure may serve as a model for the future development of immunoassays to small molecules and provide investigators with a way of producing specific analytes and antibodies. The use of the hybridoma technique for monoclonal antisera offers an ideal opportunity for this type of antisera production.

The iodinated bleomycin was studied for its immunoreactive stability and the radiation damage incurred during the iodination procedure. The four radioactive peaks obtained from the column chromatographic purification procedure were studied. The first two peaks were not immunoreactive and probably represent inorganic iodide. The other two peaks did exhibit immunoreactivity and represented some form of labeled bleomycin. The first of these peaks was unstable with nonspecific radioactivity, i.e., that not bound to bleomycin increasing to 20% of the total radioactivity in 4 days. However, the second of these peaks was stable with low nonspecific activity for at least one half-life of the iodine.

Because the process of iodination often results in radiation damage to the parent molecule, it is important to assess the damage prior to using in an assay system. This can be done by two competitive binding experiments: First by maintaining a constant amount of iodinated label when increasing amounts of noniodinated drug are mixed with antibody; and second increasing amounts of radioactive-labeled material are mixed with the antibody. The results of this experiment should produce two superimposable curves, if there is no radiation damage; diversion of the curves suggests radiation damage of the tracer which should not be used in a radioimmunoassay.

Finally, the assay system itself was validated using the following protocol. The data were reduced using a logit–log transformation as described by Robard et al.[9] The resulting dose–response curve was evaluated for sensitivity and precision—sensitivity being defined as that amount of material that can be statistically differentiated from zero. The precision of the assay was determined using the dose–response curve and the method of Midgeley et al.[10] They used the index lambda, which is defined as the ratio between the standard deviation of each y value and the slope of regression line of the mean values of logit y and the corresponding log X. This is found to be 0.024 at 5 ng, a value which compares favorably with those reported for many polypeptide hormone assays.

Recovery experiments were performed by adding bleomycin in known concentrations to serum specimens and measuring these by radioimmuno-

[9] D. Robard, W. Boidson, and P. Raford, *J. Lab. Clin. Med.* **74,** 770 (1969).
[10] A. R., Midgeley, G. D. Niswender, and R. W. Regar, *Acta Endocrinol.* **63,** 163 (1977).

assay. The mean recovery was 102.6% ± 3.3% standard error. The assay was then studied for comparison with the standard microbiological assay system using *Bacillus subtilus* ATC6633 as a test organism. The coefficient of correlation obtained between the two assays was 0.987. The assay was also compared to another radioimmunoassay described by Elson and others,[11] and here the coefficient of correlation with plasma samples was 0.93. Elson *et al.* obtained a coefficient of correlation of 0.96 when they compared their own assay to the microbiological assay.

Clinical Applications of the Radioimmunoassay of Bleomycin

The assay described here has been used extensively in clinical pharmacological studies of patients receiving the drug by a variety of routes and regimes.[12,13]

This drug is administered by intravenous infusion and it is of interest to note that the elimination pharmacokinetics of bleomycin as determined by radioimmunoassay differ depending on the type of infusion given.

Protocols for treatment of patients with testicular cancer requires either the continuous infusion of bleomycin over a long period of time or a bolus injection. My colleagues and I[12] were fortunate to study many patients receiving bleomycin on these two protocols. Table II shows the results of the elimination pharmacokinetics. If all the blood levels obtained experimentally are used in the computations, marked differences in parameters are apparent, but if the plasma levels below 10 μg/ml bleomycin are excluded the kinetics are similar for both protocols. Interpretation of

TABLE II
PHARMACOKINETIC DATA FOR BLEOMYCIN INFUSIONS

	Group A[a]	Group B[b]	Group C[c]
Elimination half-life ($t_{1/2}$)	14.5	2.98	2.03
Volume of distribution (1/kg)	2.25	0.458	0.354
Volume of central compartment (1/kg)	0.251	0.145	0.142
Renal clearance (ml/min/1.73^2)		61.4	76.9
Total body clearance (ml/min/1.73^2)		119.0	128.0

[a] Group A, five-day continuous infusion, all data points.
[b] Group B, five-day continuous infusion data points 10 U/ml excluded.
[c] Group C, short-term (15 min) infusion.

[11] M. K. Elson, M. M. Oken, and R. B. Shafer, *J. Nucl. Med.* **18**, 296 (1977).
[12] A. Broughton, J. E. Strong, P. Holoyoe, and C. W. M. Bedrossian, *Cancer* **40**, 2772 (1977).
[13] S. T. Crooke, R. L. Comis, L. H. Einhorn, J. E. Strong, A. Broughton, and A. W. Prestayko, *Cancer Treatment Rep.* **67**, 1631 (1977).

these data suggests three possible mechanisms. Although most evidence points to a two-compartment system, there may be an additional deep compartment which binds bleomycin releasing it slowly during the long infusion. The release continues after completion of the therapy thus prolonging the elimination half-life of the drug.

Another interpretation of these results is the appearance of a metabolite of bleomycin. Conjecture suggests that desamidobleomycin, which cross-reacts 44% with the antibody, may be such a metabolite. Obviously this metabolite would not be detectable following a bolus injection because of the rapid excretion of bleomycin into the urine, but after a continuous infusion lasting 5 days, the level of the metabolite may be of great significance. The development of a specific radioimmunoassay for desamidobleomycin and subsequent studies might answer that question.

The most clinically significant information to arise from the study of the pharmacology of bleomycin radioimmunoassay is related to the excretion by the kidney and the creatinine clearance rate. Data from patients receiving combination therapy of bleomycin infusions and *cis*-platinum, a nephrotoxic drug, show that when the creatinine clearance falls to 25 ml/min or below the $t_{1/2B}$ (the half-life of excretion) of the drug increases exponentially, but at creatinine clearance levels above 25, the $t_{1/2}$ is normal.[14] This suggests that dose regimes do not need to be modified until the creatinine clearance level falls below this value. The absolute blood levels obtained during infusion of the drug are of very little value for the interpretation of pharmacological data, and any use of this assay in a routine sense must be related to the study of its elimination kinetics and not to its peak plasma values.

[14] S. T. Crooke, F. Luft, A. Broughton, J. E. Strong, K. Casson, and L. Einhorn, *Cancer* **39**, 1430 (1977).

[33] Radioimmunoassay of 1-β-D-Arabinofuranosylcytosine[1]

By TADASHI OKABAYASHI and JOHN G. MOFFATT

A variety of assay methods for 1-β-D-arabinofuranosylcytosine (Ara-C) have been developed in the course of pharmacokinetic and dose schedule studies on this clinically useful antileukemic drug. Some of these

[1] Contribution No. 551 from the Institute of Organic Chemistry, Syntex Research, 3401 Hillview Avenue, Palo Alto, California 94304.

methods have proved to be useful for routine plasma assays,[2-4] but have limitations which are inherent to their assay principles.

As will be described in other chapters, radioimmunoassay has the advantage of adaptability to a wide variety of requirements, which may vary depending upon the experimental purposes. This is achieved by selecting antigens for animal immunization, antibodies having adequate specificity, proper radioligands, and assay conditions. The method described here is an example of an Ara-C radioimmunoassay which seems to meet the requirements for many kinds of animal experiments, and probably also for determination of plasma Ara-C levels in human leukemic patients undergoing therapy.

Preparation of Materials for Ara-C Radioimmunoassay

The key intermediate 1-(5-O-succinyl-β-D-arabinofuranosyl)cytosine (III, S-Ara-C, see Fig. 1) was prepared via selective 5'-O-acylation of Ara-C hydrochloride with 2,2,2-trichloroethylsuccinyl chloride (Ib) in dimethylacetamide according to the general method of Gish et al.[5] The resulting 1-[5-O-(2,2,2-trichloroethylsuccinyl)-β-D-arabinofuranosyl]cytosine (II) was then deblocked by treatment with zinc and acetic acid in dimethylformamide to give pure S-Ara-C (III) in 59% yield.

2,2,2-Trichloroethyl Hemisuccinate (Ia)

A solution of 2,2,2-trichloroethanol (15.0 g; 0.10 mol), succinic anhydride (12.0 g; 0.12 mol), and triethylamine (8.7 ml, 0.12 mol) in ethyl acetate (100 ml) was heated under reflux for 1 hr and then evaporated. A solution of the residue in 5% aqueous sodium bicarbonate was washed twice with ether and then brought to pH 2 with sulfuric acid. The resulting solid was washed with water, dried *in vacuo,* and crystallized from chloroform–hexane giving 19.2 g (77%) of Ia with mp 88–89° (reported[6] mp 88–90°): NMR'CDCl$_3$, 2.77 ppm (s, 4, COCH_2), 4.86 (s, 2, OCH$_2$); ir (KBr) 3200, 1700 cm^{-1}.

[2] R. L. Momparler, A. Labitan, and M. Rossi, *Cancer Res.* **32**, 408 (1972).

[3] (i) B. M. Mehta, M. B. Meyers, and D. J. Hutchison, *Cancer Chemother. Rep. Pt. 1* **59**, 515 (1975). (ii) L. J. Hanka, S. L. Kuentzel, and G. L. Neil, *Cancer Chemother. Rep.* **54**, 393 (1970).

[4] R. L. Furner, R. W. Gatson, J. D. Strobel, S. El Dareer, and L. B. Mellet, *J. Natl. Cancer Inst.* **52**, 1521 (1974).

[5] T. D. Gish, R. C. Kelly, G. W. Camiener, and W. J. Wechter, *J. Med. Chem.* **14**, 159 (1971).

[6] D. C. Bishop, S. C. R. Meacock, and W. R. N. Williamson, *J. Chem. Soc.* **C**, 670 (1966).

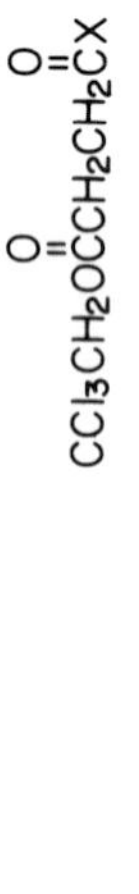

Fig. 1. Structures of materials used in Ara-C radio-immunoassay.

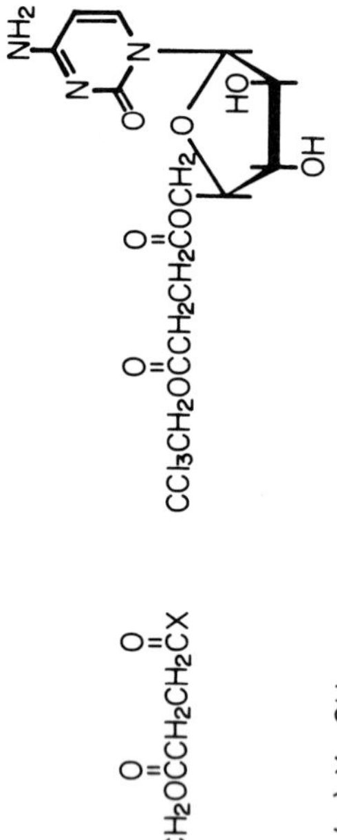

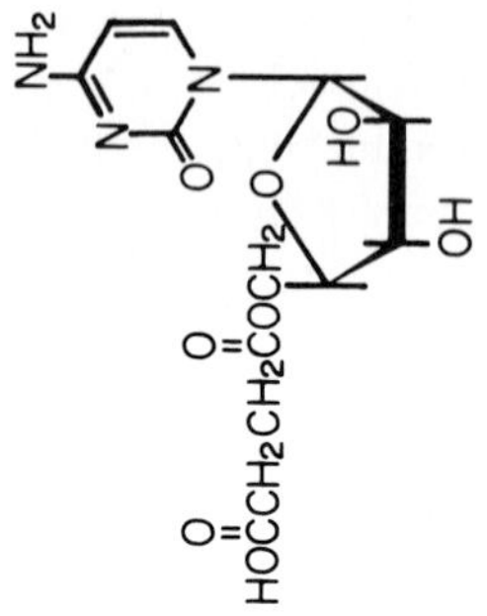

1-[5-*O*-(2,2,2-Trichloroethylsuccinyl)-β-D-arabinofuranosyl]cytosine (II)

A mixture of Ia (2.5 g, 10 mmol) and thionyl chloride (6.5 ml, 90 mmol) was heated at 65° for 30 min and then evaporated to dryness *in vacuo* and stored under high vacuum for 1 hr. The residue (**Ib**), which showed an intense IR absorption band at 1795 cm⁻¹ and no bands at 3000 or 1710 cm⁻¹, was dissolved together with Ara-C hydrochloride (2.79 g, 10 mmol) in anhydrous *N,N*-dimethylacetamide and stirred at 25° for 2 hr. Following evaporation of the solvent *in vacuo* at 65°, the residue was crystallized from 2-propanol giving 4.31 g (84%) of the hydrochloride of **II** with mp 158–161°.

Anal. Calcd. for $C_{15}H_{19}N_3Cl_4O_8$ (511.17): C, 35.25; H, 3.75; N, 8.22; Cl, 27.75. Found: C, 35.12; H, 3.70; N, 8.16; Cl, 27.58.

A portion of this material (2.0 g, 3.9 mmol) was dissolved in water (25 ml) and 0.5 M sodium carbonate (4 ml) was added. The resulting precipitate was collected, washed with water, dried, and crystallized from 2-propanol giving **II** as the free base (1.32 g, 71%) with mp 94–95°: λ_{max}(MeOH) 229 nm (sh, ϵ 7400), 273 (8500); NMR (DMSO-d_6) 2.71 ppm (m, 4, COCH_2), 3.94 (m, 3, C$_4$·H, C$_5$·H$_2$), 4.27 (m, 2, C$_2$·H, C$_3$·H), 4.88 (s, 2, CCl$_3$CH_2), 5.53 (m, 2, C$_2$·OH, C$_3$·OH), 5.66 (d, 1, $J_{5,6}$ = 7.5 Hz, C$_5$H), 6.07 (d, 1, $J_{1',2'}$ = 3.5 Hz, C$_1$·H), 7.05 (br s, 2, NH$_2$), 7.46 (d, 1, C$_6$H).

Anal. Calcd. for $C_{15}H_{18}N_3Cl_3O_8$ (474.70): N, 8.85; Cl, 22.41. Found: N, 8.77; Cl, 22.12.

1-(5-*O*-Succinyl-β-D-arabinofuranosyl)cytosine(S-ara-C, III)

Zinc dust (280 mg, 4 mmol) and glacial acetic acid (0.40 ml) were added to a solution of **II** (820 mg, 1.72 mmol) in dimethylformamide (10 ml) and the mixture was stirred at room temperature for 6 hr. It was then filtered and the solids were washed with water. The filtrates were evaporated *in vacuo* and an aqueous solution of the residue was passed through a column containing 25 ml of Dowex 50 (NH$_4$⁺) resin. Evaporation of the eluates and aqueous washings left a residue that was dissolved in water and passed through a column containing 20 g of Chromato Tokusei Shirasagi charcoal obtained from Takeda Pharmaceutical Industries (Osaka, Japan).[7] Following a water wash (1 liter), the product was eluted with 50% aqueous ethanol. Evaporation of the UV-absorbing eluates left 350 mg (59%) of **III** as a foam that was homogeneous by thin-layer chro-

---

[7] This particular brand of charcoal was selected because of the good recoveries of nucleotides obtained from it by elution with aqueous ethanol. Other brands of charcoal should be checked for such recovery before use in order to avoid reduced yields.

matography using 1-butanol:acetic acid:water (5:2:3) and thin layer electrophoresis in 0.05 M Na$_2$HPO$_4$, pH 8.9: λ_{max}(MeOH) 229 nm (sh, ϵ 7300),
273 (8400); [α]$_D^{23}$ 113.7° (c 0.4, MeOH); NMR (pyridine-D$_2$O-d$_5$) 2.33 ppm
(s, 4, COCH_2), 4.6–4.75 (m, 3, C$_3$,H, C$_4$,H, C$_{5'a}$H), 4.85 (dd, 1, J_{gem} =
11.5 Hz, $J_{4',5'b}$ = 4 Hz, C$_{5'b}$H), 5.02 (dd, 1, $J_{1',2'}$ = 3.5 Hz, $J_{2',3'}$ = 2 Hz,
C$_2$,H), 6.10 (d, 1, $J_{5,6}$ = 7.5 Hz, C$_5$H), 6.97 (d, 1, C$_1$,H), 8.10 (d, 1, C$_6$H).
    Anal. Calcd. for C$_{13}$H$_{17}$N$_3$O$_8$ (343.29): C, 45.48; H, 4.99; N, 12.24.
Found: C, 45.63; H, 5.19; N, 11.78.

Preparation of Antigen

    S-Ara-C is coupled to human serum albumin by the method of Oliver
et al.[8] Forty milligrams of human serum albumin is dissolved in 4.0 ml of
water followed by the gradual addition of 8.01 mg of S-Ara-C (23.33
μmol) and 20 mg of 1-ethyl-3-(3-dimethylaminopropyl)carbodiimide-HCl,
the pH being adjusted to 5.5 with 1 M sodium bicarbonate after each addition. The reaction mixture is incubated in the dark at 25° for 16 hr and
then dialyzed against 0.15 M NaCl:0.01 M sodium phosphate, pH 7.4, at
4° for 48 hr with eight changes of the dialyzing solution. The number of
hapten molecules conjugated to albumin can be estimated from the absorption spectrum of the conjugate and its albumin content determined by
the method of Lowry *et al.*[9] A conjugate which we used as the antigen
gave an absorption maximum at 275 nm at pH 7.4. On the basis of the
observed albumin content and of assumed molecular extinction coefficients of 6800 for S-Ara-C and 34,900 for albumin, the antigen contained
8.2 Ara-C residues per albumin molecule.

Immunization

    Rabbits are immunized with 1 mg of S-Ara-C albumin conjugate
(0.25 mg is injected into each foot pad) emulsified in Freund's complete
adjuvant, followed by booster injections totaling 0.5 mg injected at
monthly intervals. After 1 week, a sample of several milliliters of blood is
removed. Plasma is separated by centrifugation and adjusted to a final
concentration of 0.1% with sodium azide and 0.01% with sodium Merthiolate, then checked for antibody titer, affinity for [^{3}H]Ara-C, and specificity. These procedures revealed the presence of Ara-C antibody in all
seven rabbits immunized after three successive booster injections. Whole

[8] G. C. Oliver, B. M. Parker, D. L. Brasfield, and C. W. Parker, *J. Clin. Invest.* **47,** 1035
(1968).
[9] O. H. Lowry, N. J. Rosebrough, A. L. Farr, and R. J. Randall, *J. Biol. Chem.* **193,** 265
(1951).

blood was harvested after the fourth booster. In the presence of antiseptics, the antibodies can be stored without alteration of antibody titer or affinity for at least 2 years at 4°, or for a longer period after lyophilization and storage at $-20°$.

In the example presented below, experiments were performed using an antibody designated as Ab-3. This antibody was heated at 60° for 60 min before the addition of antiseptics. The heating is not always necessary, but is essential to inactivate a plasma esterase which hydrolyzes the 3'-O- or probably the 5'-O-ester linkage of acylated analogs of Ara-C. The heating did not appreciably alter the properties of the antibody.

Plasma Samples

Blood is collected from drug-treated animals in a heparinized tube and immediately centrifuged at 3000 g for 10 min at 4°. To one volume of the resulting plasma are added three volumes of absolute ethanol, and the mixture is stirred on a Vortex mixer and then kept in an ice bath for at least 10 min. The precipitate is removed by centrifugation and a calibrated volume of the supernatant is removed from the tube and evaporated to dryness under vacuum. The residue is dissolved in an appropriate volume of water (this is hereafter designated as the ethanol extract). Portions of the ethanol extract are immediately subjected to the assay or stored in a freezer. When stored at $-20°$, its Ara-C [and also 1-β-D-arabinofuranosyluracil (Ara-U)] content did not change appreciably for at least 3 months. The recovery by this procedure, as determined after adding known amounts (3 to 5 pmol) of Ara-C to 5 μl of plasma from a BD2F$_1$ mouse, was $99.4 \pm 2.2\%$ ($N = 18$).

Radioimmunoassay of Ara-C

Procedure

The assay is usually performed in duplicate or triplicate in small glass test tubes (0.8 × 8 cm). The standard assay mixture in a total volume of 100 μl includes 10 μl of 0.5 M sodium acetate buffer, pH 6.2, 10 μl of appropriately diluted antibody (diluted 1:33 in water in the case of Ab-3), 10 μl of 2% bovine serum albumin, 2 pmol of [³H]Ara-C (8.19 Ci/mmol, New England Nuclear), and various concentrations of Ara-C standard or unknown sample. The binding reaction is initiated by adding antibody, and the reaction tubes are incubated in an ice bath for 2 hr. Bound and free Ara-C are separated by filtration through a Millipore filter. A 75-μl sample of the assay mixture is placed with a micropipet on a Millipore filter (HA, 25 mm diameter, pore size 0.45 μm) presoaked with the acetate buffer and filtered gently under vacuum. The filter is washed four times using a medium or high vacuum with 2.5 ml of the buffer, placed in a counting vial, dissolved in 10 ml of a scintillation mixture (PPO 5 g, bis-MSB 0.5 g, naphthalene 120 g, made to 1 liter with dioxane) plus 0.5 ml H_2O, then the radioactivity is determined with a scintillation spectrophotometer.

Duplicate or triplicate reaction tubes in which the antibody is omitted are prepared and processed like the test specimens. The radioactivity retained on the filter, usually 20 to 30 cpm, is subtracted from those of other filters as the filter blank.

Standard Curve

Figure 2 gives a graphic representation of six replicate radioimmunoassays expressed as C_0/C_S versus S, where C_0 and C_S denote the radioactivity (cpm) of [³H]Ara-C retained on a Millipore filter (bound [³H]Ara-C) in the absence or presence of a particular amount (S pmol/tube) of unlabeled Ara-C, respectively. The curve is linear over a wide range of S. The sensitivity of this assay, defined as the lowest amount of Ara-C that can be significantly distinguished ($p < 0.05$) from a control with no Ara-C is

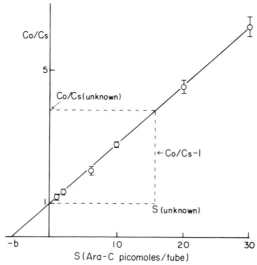

FIG. 2. Standard curve for Ara-C. Standard conditions were used and C_0/C_S was plotted against the amount of unlabeled Ara-C (S). Each point represents the mean ± SD of 6 replicates. C_0 was 3425 ± 169 cpm.

0.7 pmol. The assay precision, determined from the standard curve, is 0.10 at 1 pmol.

As illustrated in Fig. 2, the standard curve is a straight line which intersects the ordinate at $C_0/C_S = 1$ and the abscissa at $S = -b$. Thus, if an unknown sample gave bound radioactivity of $C_{S(unknown)}$ cpm, the Ara-C content in this sample, $S_{(unknown)}$, is equal to $(C_0/C_{S(unknown)} - 1) \times b$ pmol.

Applications

Determination of Plasma Ara-C Levels in BD2F$_1$ Mice after Injection of 1-(3-O-Octanoyl-β-D-arabinofuranosyl)cytosine (O-Ara-C)

O-Ara-C is a relatively water-insoluble derivative of Ara-C[10] which is susceptible to plasma esterase (see Fig. 4). Figure 3 shows that when an aqueous 2% suspension of O-Ara-C (particle size 0.8 to 1.2 μm) is administrated to mice im or sc at a dose of 200 mg/kg, a reasonably high plasma Ara-C level can be maintained for a fairly long period. In this experiment, 0.3 ml of blood was removed at the designated time from an eye vein.

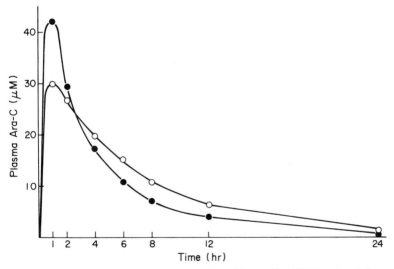

FIG. 3. Plasma Ara-C levels after administration of O-Ara-C to BD2F$_1$ mice. A fine-particle water suspension of O-Ara-C (200 mg/kg) was injected im (O——O) or sc (●——●).

[10] For synthesis, see E. K. Hamamura, M. Prystasz, J. P. H. Verheyden, J. G. Moffatt, K. Yamaguchi, N. Uchida, V. Sato, A. Nomura, O. Shiratori, S. Takase, and K. Katagiri, *J. Med. Chem.* **19**, 667 (1976).

Blood from three mice was pooled and processed as described above and portions of the ethanol extracts were subjected to the radioimmunoassay.

The validity of the assay data was checked using the remaining ethanol extracts and the following results were obtained. (1) Quantitative recoveries were obtained in experiments in which known amounts of Ara-C were added to portions of predetermined ethanol extracts and then reassayed for Ara-C. (2) When increasing volumes of an ethanol extract were assayed, there was a linear relationship between the volume and the values determined. (3) The values determined by the present method were practically identical to those determined by a deoxycytidine kinase method,[2] which we modified extensively to correct for interference by deoxycytidine and other substances.[11] The interassay coefficient of variation determined with a single ethanol extract five times on different days was 6.1%. The intraassay coefficient of variation determined with the same extract ($n = 10$) was 4.7%.

Portions of ethanol extracts were treated with 0.1 M NaOH at 40° for 10 min and after neutralization with 1 M HCl, the Ara-C content was determined again. The alkali treatment hydrolyzes O-Ara-C quantitatively to Ara-C, therefore the value determined after hydrolysis should be the sum of O-Ara-C and Ara-C. We could not detect O-Ara-C in any sample presented in the figure.

Relationship between Chain Length of the Acyl Group and Susceptibility to Plasma Esterase in 3'-O-Acylated Ara-C

The susceptibility of various 3'-O-acylated analogs of Ara-C[10] to esterase hydrolysis was compared by incubating the analogs in a total volume of 200 μl of 20 mM phosphate buffer (pH 7.4) containing 50 μl of BD2F$_1$ mouse plasma at 37° for 40 min. The incubation was terminated by adding 600 μl of ethanol. Ethanol extracts were prepared as described above, then Ara-C was determined by the radioimmunoassay. As no-plasma controls, samples of 3'-O-acylated Ara-C at a concentration of 5 μM were incubated at 37° without the plasma for 40 min and processed like the test specimens.

[11] M. Ide and T. Okabayashi, unpublished results (1975). The modification included the following three points. (1) The enzymatic reaction mixture contained [³H]Ara-C, in addition to the nonradioactive Ara-C substrate to correct for the recovery of the reaction product, [³²P]Ara-CMP, in subsequent steps. (2) The reaction product was separated first by paper chromatography on polyethyleneimine paper with 0.02 M potassium borate, and then by thin-layer chromatography on an Avicel SF plate using saturated $(NH_4)_2SO_4 : 1$ M K_2HPO_4 : isopropanol : H_2O — 15.5 : 1.25 : 1.25 : 2.80. (3) The separated [³²P]Ara-CMP, which was completely free of interfering substances, was counted for ³H and ³²P by double channel counting. Analysis of the same extracts as presented in Fig. 3 (sc) gave a curve that could be superimposed upon that indicated in the figure.

The results in Fig. 4 indicate that the effect of elongation of the 3'-*O*-acyl group is biphasic, and that O-Ara-C is the most susceptible to the esterase among the compounds tested. The figure also shows that values determined for the no-plasma controls expressed as Ara-C are not zero. This is due to the small, but recognizable, cross-reactivity of 3'-*O*-acylated Ara-C to the antibody. The values are, however, small enough to be neglected.

Factors Affecting the Radioimmunoassay

Specificity of the Antibody

The specificity of Ab-3 determined by measuring the displacement of bound [³H]Ara-C by bases, nucleosides, and nucleotides is given in Fig. 5. Deoxycytidine and cytidine, which have structures differing from Ara-C only at the 2'-position and are known to interfere with enzymatic[2] and microbiological[3] methods, show no appreciable cross-reactivity at concentrations 1000-fold greater than that of Ara-C. The affinity of Ara-U when determined at pH 6.2 for the antibody is only 0.1% of that of Ara-C. On

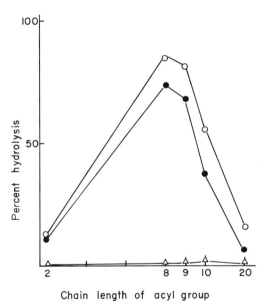

Chain length of acyl group

FIG. 4. Relationship between chain length of the acyl group in 3'-*O*-acyl-Ara-C and susceptibility to hydrolysis by plasma esterase. 3'-*O*-Acyl-Ara-C at two concentrations (0.5 μM, ●——●, and 5 μM, ○——○) was incubated with $BD2F_1$ mouse plasma and the resultant Ara-C was determined as described in the text. △——△, no-plasma control.

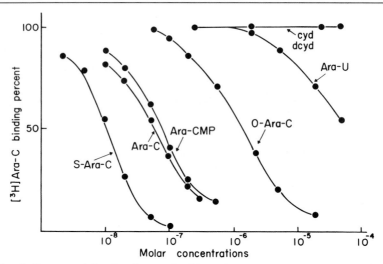

FIG. 5. Cross-reactivity of various nucleosides and the 5'-monophosphate of Ara-C (Ara-CMP) with Ab-3. The binding reaction was performed under the standard conditions except that compounds indicated in the figure were added to the assay tubes. [³H]Ara-C binding in the absence of inhibitors was taken as 100%.

the contrary, 5'-substituted analogs, the 5'-monophosphate of Ara-C and S-Ara-C, show affinity for the antibody equal to or greater than that of Ara-C. This is probably because we used as an antigen for immunization a conjugate of albumin with S-Ara-C. The affinity of 3'-acylated Ara-C for Ab-3, determined with O-Ara-C, is 5% of that shown by Ara-C. The following compounds show no cross-reactivity at concentrations of 20 μM: uridine, deoxyuridine, adenosine, guanosine, deoxyadenosine, deoxyguanosine, uracil, cytosine, thymine, thymidine, and dUMP.

Effects of Antibiotics and Antineoplastic Agents

Some antibiotics and antineoplastic agents are known to seriously interfere with a microbiological assay.[3] The following drugs tested in the present radioimmunoassay did not affect [³H]Ara-C binding at a concentration of 0.1 mM: penicillin V, ampicillin, erythromycin, cephalothin, tetracycline, chloramphenicol, lincomycin, nystatin, 6-mercaptopurine, 5-fluorouracil, cyclophosphamide, vinblastine, vincristine, mitomycin C, and actinomycin D.

Effect of pH

One important feature in the specificity of the antibody is that its affinity for Ara-U varies drastically depending upon the pH of the incubation mixture. Although the affinity for Ara-U determined at pH 6.2 is only 0.1% of that for Ara-C (Fig. 5), it increases progressively with an increase of the pH of the incubation mixture (Fig. 6). Since the plasma concentration of Ara-U in animals and human patients treated with Ara-C (or analogs of Ara-c) is usually higher than that of Ara-C, the pH of the incubation mixture is critical for the present radioimmunoassay.

Effect of Plasma

Plasma is known to affect enzymatic,[2] microbiological,[3] and spectrophotometric[4] assays. In the present example, the ethanol extract corresponding to up to 12 μl of original BD2F$_1$ mouse plasma did not affect the radioimmunoassay. This means that considering the sensitivity of the present assay, most, if not all, routine assays can be performed under conditions free from the effect of plasma.

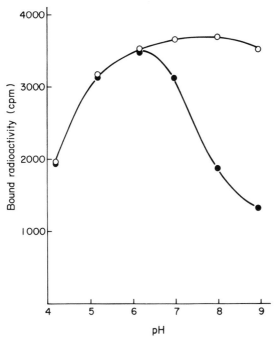

FIG. 6. Effect of pH on [³H]Ara-C binding in the absence (O——O) or presence (●——●) of unlabeled Ara-U (0.2 μM). The pH of the binding mixture was adjusted with the following buffers: sodium acetate, pH 4.2 to 6.2; Tris-maleate, pH 7.0; Tris–HCl, pH 8.0 to 9.0.

Radioimmunoassay of Ara-U

Ara-U is biologically inert, but is one of the most important metabolites of Ara-C because the highly active Ara-C deaminase is widely distributed in normal and neoplastic tissues of man and experimental animals.[12] The assay of low levels of Ara-U by conventional methods is difficult, but with the radioimmunoassay we can determine minute amounts of this nucleoside by procedures similar to those for Ara-C. The antibodies directed against Ara-U can be obtained by immunizing rabbits with a conjugate of 1-(5-*O*-succinyl-β-D-arabinofuranosyl)uracil as an antigen.[13] The procedures for the radioimmunoassay are similar to those for Ara-C, except that [³H]Ara-U is used as the labeled ligand.[13] Taking advantage of the high cross-reactivity of Ara-U with Ara-C-directed antibodies at alkaline pH values, a fairly reliable, but not very accurate, radioimmunoassay of Ara-U can also be performed by using [³H]Ara-C and the antibodies directed against Ara-C.[14]

Comments

The radioimmunoassay method described here seems to be applicable to many kinds of animal experiments and probably also for the determination of Ara-C levels in plasma of leukemic patients, but, of course, a variety of modifications and improvements can be made. One is that antibodies having higher titers may be obtained, as is now generally accepted, by immunizing animals with an antigen containing a higher molar content of hapten relative to albumin. Later, we found that such an antigen (17.2 Ara-C molecules per albumin) can be obtained by increasing the molar ratio of S-Ara-C to albumin (8.01 mg:20 mg) in the coupling reaction.

One unexpected, but important, observation is that all the Ara-C-directed antibodies that we obtained showed pH-dependent cross-reactivity to Ara-U. Another outstanding observation was that one out of four Ara-U-directed antibodies raised against 1-(5-*O*-succinyl-β-D-arabinofuranosyl)uracil showed a specificity closely resembling that of Ara-C-directed antibodies, while the others did not cross-react with Ara-C between pH 4 and 9.[13] We infer that all Ara-C-directed antibodies and one Ara-U-directed antibody recognized the enol form of Ara-U. If this is the case, it will be difficult to obtain an Ara-C-directed antibody which does not cross-react with Ara-U at neutral or alkaline pH values by the present

---

[12] D. H. W. Ho, *Cancer Res.* **33**, 2816 (1973).
[13] T. Okabayashi, S. Mihara, D. B. Repke, and J. G. Moffatt, *Cancer Res.* **37**, 3132 (1977).
[14] T. Okabayashi, S. Mihara, and J. G. Moffatt, *Cancer Res.* **37**, 625 (1977).

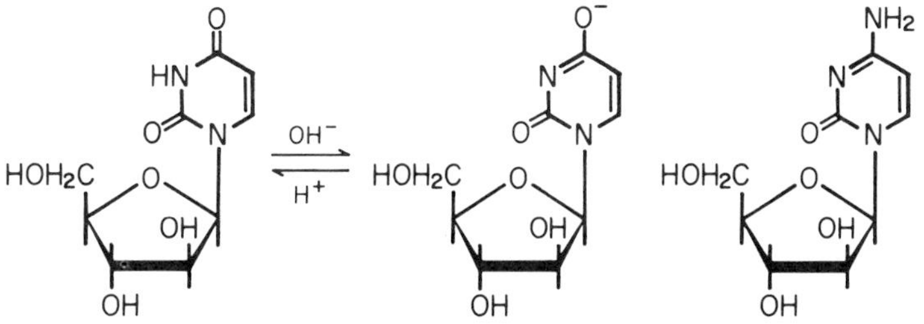

IV V VI

FIG. 7. Structures of Ara-U at neutral (**IV**) and alkaline (**V**) pH values and of Ara-C (**VI**).

immunization method because the gross structure of the enol form of Ara-U closely resembles that of Ara-C (Fig. 7).

Acknowledgments

The authors are grateful to Cancer Research Inc. for permission to reproduce part of this paper from *Cancer Research*.

[34] Radioimmunoassay of Neocarzinostatin, a Small Cytotoxic Protein Used in Cancer Chemotherapy

By VIC RASO

The F-41 variant of *Streptomyces carzinostaticus* produces a small protein antibiotic called neocarzinostatin (molecular weight 10,700). Isolated from the bacterial culture filtrate,[1] the protein has been purified to homogeneity.[2] Its primary structure has been described[3] and the molecule consists of a single polypeptide chain with 109 residues. Two intrachain disulfide bridges cross-link the structure which has an apparent compact configuration.[4]

Interest in neocarzinostatin (NCS) stems from its potent cytostatic and cytotoxic effects on eukaryotic cells, which coupled with its unique

[1] N. Ishida, K. Niyazaki, and M. Rikimoru, *J. Antibiot.* (*Tokyo*) *Ser. A* **18,** 68 (1965).
[2] T. S. A. Samy, J.-H. Hu, J. Meienhofer, H. Lazarus, and R. K. Johnson, *J. Natl. Cancer Inst.* **58,** 1765 (1977).
[3] J. Meienhofer, H. Maeda, C. B. Glaser, J. Czomkos, and K. Kuromizu, *Science* 178, 875 (1972).
[4] T. S. A. Samy, M. Atreyi, H. Maeda, and J. Meienhofer, *Biochemistry* **13,** 1007 (1974).

ISBN 0-12-181984-1

structural features has lead to its use as a novel agent for cancer chemotherapy.[5-7]

While the molecular basis of its cellular action is not adequately understood, NCS does effect DNA synthesis and replication since thymidine incorporation is inhibited[8] and single- or double-strand breaks in DNA are produced.[9] NCS also displays inhibitory effects on mitosis[8] and has been implicated in disrupting the microtubular systems which control membrane movements.[10] There is recent evidence that drug action is mediated by a dissociable chromophore moiety.[11]

The demonstrated steep dose-toxicity relationship for neocarzinostatin in animal studies made the need to reliably monitor drug levels during clinical trials compelling. A radioimmunoassay was developed[12] since the method is particularly suited to the task of rapidly measuring the large number of patient samples acquired in a pharmacologic study. The procedure requires no preparatory processing of biologic samples since the sensitivity and selectivity of antibody binding permits detection of low levels of NCS without nonspecific interference.[6,7,13]

Antigen Preparation and Immunization

A crude ammonium sulfate fraction of the bacterial culture filtrate served as a source for the purification of neocarzinostatin to homogeneity.[2] This pure protein, 2 mg in 1 ml PBS, was emulsified with an equal volume of complete Freund's adjuvant. The emulsion was injected weekly into New Zealand White rabbits at multiple intradermal and subcutaneous sites for three courses. Animals were bled from the ear and occassional iv booster injections were given to prolong antibody production.[12] The clinical preparation of neocarzinostatin is produced by Kayaku Antibiotics Co., Ltd., Tokyo, Japan and was obtained through

[5] S. S. Legha, D. D. Van Hoff, M. Rozencweig, D. Abraham, M. Slavik, and F. M. Muggia, *Oncology* **33**, 265 (1976).

[6] R. L. Comis, T. Griffin, V. Raso, and S. J. Ginsberg, *in* "Recent Results in Cancer Research" (S. K. Carter, H. Umezawa, J. Douros, and Y. Sakurai, eds.), Vol. 63, p. 261. Springer-Verlag, Berlin and New York, 1978.

[7] T. W. Griffin, R. L. Comis, J. J. Lokich, R. H. Blum, and G. P. Canellos, *Cancer Treat. Rep.* **62**, 2019 (1978).

[8] H. Sawada, K. Tatsumi, M. Sasada, S. Shirakawa, T. Nakamura, and G. Wakiska, *Cancer Res.* **34**, 3341 (1974).

[9] T. A. Beerman, and I. H. Goldberg, *Biochem. Biophys. Res. Commun.* **59**, 1254 (1974).

[10] T. Ebina, M. Satake, and N. Ishida, *Cancer Res.* **37**, 4423 (1977).

[11] L. S. Kappen, M. A. Napier, and I. H. Goldberg, *Proc. Natl. Acad. Sci. U.S.A.* **77**, 1970 (1980).

[12] T. S. A. Samy, and V. Raso, *Cancer Res.* **36**, 4378 (1976).

[13] R. L. Comis, T. W. Griffin, V. Raso, and S. J. Ginsberg *Cancer Res.* **39**, 757 (1979).

the National Cancer Institute, Bethesda, Maryland. This material is not completely pure but appears to be a satisfactory immunogen as well as an excellent source of drug for further purification.

An antiserum to rabbit antibody was elicited in a sheep by immunization with rabbit Fc fragments[14] in complete Freund's adjuvant. Comparable antibodies are also available commercially. This reagent may be advantageous for use in the "double antibody" technique since it is unlikely to interfere sterically with the binding of NCS to rabbit antibody sites.

Characterization of the Anti-NCS Serum

The specific antiserum reacted with both highly purified NCS and clinical preparations of the drug to produce a single line which showed identity on immunodiffusion plates either at optimal or elevated concentrations.[12] This indicated that the antibody was monospecific and not reactive with any impurities present in the clinical formulation. Furthermore, the antibody was able to precipitate the antibiotic activity of NCS, insuring that it was reactive with the biologically active form of NCS.[12]

Synthesis of ^{125}I-Labeled NCS

The single tyrosine residue at position 32 in native NCS is shielded within the tightly folded molecule.[4] Its pK value is high and it is not amenable to chemical modification unless the protein is first denatured. For this reason the molecule was radiolabeled using the ^{125}I-acylating agent introduced by Bolton and Hunter[15] instead of by direct iodination using chloramine-T or lactoperoxidase. In addition, the resulting 3-(4-hydroxyphenyl)propionamide derivative of NCS is both biologically active and reactive with anti-NCS antibody.[12,16] Typically 5 μg of the purified NCS in a 10 μl volume of 0.1 M borate buffer, pH 8.5, was reacted with 1 mCi of commercially prepared Bolton–Hunter reagent at 0° for 15–30 min. Instructions supplied with the product describe how to evaporate the solvent before using this reagent. The reaction mixture was separated on a 1 × 40 cm Sephadex G-50 column equilibrated and developed with PBS. The initial peak, which eluted sharply, contained radiolabeled NCS with an estimated specific activity of 10–40 Ci/mmol.[12,17] This ^{125}I-labeled derivative of NCS was very pure, of high specific activity and was completely precipitable with antiserum prepared against unmodified neocarzinostatin.

[14] R. B. Porter *Biochem. J.* **73,** 119 (1959).
[15] A. E. Bolton, and W. M. Hunter, *Biochem. J.* **133,** 529 (1973).
[16] T. S. A. Samy, *Biochemistry* **16,** 5573 (1977).
[17] I. S. Lowenthal, L. M. Parker, D. J. Greenblatt, B. L. Brown, and T. S. A. Samy, *Cancer Res.* **39,** 1547 (1979).

Radioimmunoassay Procedure

Materials and Reagents

Anti-NCS antiserum diluted 1/50 in normal rabbit serum

^{125}I-labeled NCS at 80,000 cpm/ml in a 1 : 1 mixture of PBS and 0.3 M EDTA, pH 8

Purified NCS ranging from 0.01 to 10 μg/ml to construct the standard displacement curve

Antiserum to rabbit Fc fragment ("second antibody")

Glass fiber filter discs, GF/C obtained from Arthur H. Thomas Co. were used on a 30 port multiple filtration manifold (Millipore Corp., Bedford, Mass.) for collection and washing immune precipitates

Protocol

1. The standard assay is carried out in duplicate using small 10 × 44-mm test tubes. Each reaction mixture receives 100 μl ^{125}I-labeled NCS, 100 μl of either (a) PBS alone, (b) PBS containing known amounts of unlabeled NCS for the standard displacement curve, or (c) an equivalent volume of any clinical specimen, and finally 5 μl of the appropriately diluted anti-NCS antiserum is added. Control tubes containing 5 μl of normal rabbit serum in place of specific antiserum are included to determine background "binding."

2. These reactants are mixed and incubated for 30 min at 37°.

3. Excess second antibody, 100 μl, (pretitered to ensure precipitation of all of the rabbit IgG in the mixture) is added, the contents are mixed and allowed to incubate at 37°for 2 hr.

4. Glass fiber filter discs arranged on the filtration manifold are premoistened with PBS and the reaction mixtures are passed through at a moderate rate. The assay tubes are rinsed out once and the precipitate trapped on each filter is washed with 8 ml of PBS to complete the separation of free ^{125}I-labeled NCS from antibody bound ^{125}I-labeled NCS.

5. Excess buffer is drawn off under maximum suction before the filters are removed for radioactivity measurements using a gamma counter.

Remarks. A preliminary precipitin curve had to be examined to ensure complete precipitation of the rabbit IgG in the assay. The combination of 100 μl of sheep anti-rabbit Fc plus 1–25 μl of normal rabbit serum produced results after 2 hr incubation at 37°. A volume of 5 μl rabbit serum gave slightly less than a maximal amount of precipitate (antibody excess) and was therefore, used as "carrier" throughout the assay. Anti-NCS antiserum was appropriately diluted into this normal rabbit serum so that its antibody content was sufficient to bind ~3000 cpm of ^{125}I-labeled NCS in the final precipitate. Background levels of only ca. 200 cpm adhered to precipitates when normal rabbit serum alone was used. A standard displacement curve was established by including known quantities of unlabeled NCS in the assay and relating these to the corresponding reduction in antibody-bound radioactivity (Fig. 1). The radioimmunoassay can reliably dectect as little as 1 ng of NCS.

Clinical Utilization of the Radioimmunoassay

Pursuant to pharmacologic use, clinical preparations of NCS should be evaluated using the assay. Clinical batches of the drug contained less immunoreactive material when compared on a protein basis to purified NCS (E_{277} = 9500 mol^{-1} cm^{-1})[4] in the radioimmunoassay.[12] Accordingly, during pharmacological studies it was discovered that 1000 units of the clinical formulation, which is designated to contain 1 mg total protein, had only 0.7 ± 0.04 mg NCS when measured by radioimmunoassay.[13] An equivalency of 0.7 mg/1000 units of NCS was therefore utilized for computations throughout that study.

NCS standards assayed in buffer, plasma, serum, bile, cerebral spinal fluid, and urine gave identical results, ensuring that these biologic specimens produced no nonspecific binding or inhibition in the assay.[6] A maximum of 100 μl of serum can be included in the assay and drug levels as low as 10 ng/ml are measurable. As a precaution against possible patient variations, 100 μl of serum sampled just prior to NCS administration is always assayed for comparison. All specimens should be stored at $-20°$ if they cannot be evaluated immediately.

The set of curves in Fig. 2 depict the time course for disappearance of NCS from serum of several patients who received drug via intravenous bolus. This data, which was obtained by radioimmunoassay, conformed to the sum of two exponential equations and corresponded to the rapid distribution plus slow elimination phases of NCS.[13] Radioimmunoassay was used to quantitate immunochemically intact NCS excreted in the urine, as well as to demonstrate its absence in bile and intracavitory effusions. Equivalent levels in simultaneously drawn hepatic arterial and venous blood samples established that NCS was not extracted upon passage through the liver.

Remarks. The high sensitivity of the radioimmunoassay is essential for the evaluation of serum drug levels during clinical studies of NCS. Toxicity considerations[6,7] dictate that this agent be administered at extremely low doses (2250 units/m^2 or ca. 0.05 mg/kg). Serum concentrations are therefore initially about 500 ng/ml and rapidly drop 10-fold within 2–3 hr (Fig. 2). These early low levels are crucial since the onset of NCS action may be very rapid (10 min) and irreversible[9] even though the culmination of its toxic effect, cell death, is often delayed for days[9] or weeks.[7]

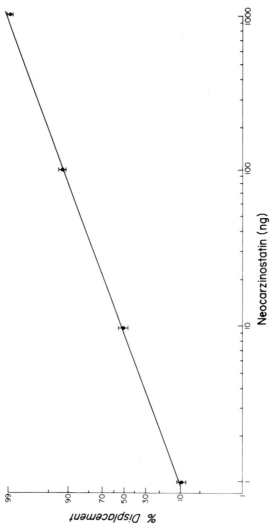

FIG. 1. Standard displacement calibration curve based upon the competition between NCS and its ^{125}I-labeled derivative for binding to antibody.

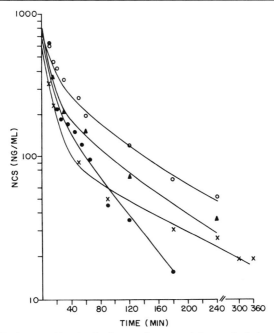

FIG. 2. NCS pharmacokinetics in four patients receiving an iv bolus of the drug. Reprinted, with permission, from Comis *et al.*[13]

Pharmacological studies which measure biologic activity or which utilize chemically modified, radiolabeled derivatives of NCS[17-19] are valuable but may not possess the sensitivity required for clinical studies. In animal experiments relatively large amounts of NCS or its derivatives have been administered (as high as 10 mg/kg) and pharmacokinetics as well as tissue levels have thus been followed by these methods.[17,19]

Measurement of biologically inactive species is a potential danger inherent in the utilization of radioimmunoassay to monitor drugs, particularly if the agent is a protein. Antibodies are directed to multiple determinant sites on NCS and fragments of the molecule may retain significant inhibitory activity in the assay. *In vitro* tests with various proteolytic enzymes showed NCS to be quite resistant to inactivation.[20] More recently, however, *in vitro* incubation at 37° with animal sera led to a time-dependent inactivation of NCS and also produced a slight shift of radiolabeled drug on a molecular sizing column.[21] NCS concentrations of 10^{-5} M were

[18] H. Fugita, M. Nakayama, T. Sawake, and K. Kimura, *Jpn. J. Antibiot.* **23,** 471 (1970).
[19] H. Maeda, N. Yamamoto, and A. Yamashita, *Eur. J. Cancer* **12,** 865 (1976).
[20] H. Maeda, K. Kumagai, and N. Ishida, *J. Antibiot. (Tokyo) Ser. A* **19,** 253 (1966).
[21] H. Maeda, and J. Takeshita, *Gann* **66,** 523 (1975).

used in this study so the significance of this inactivation to human pharmacology, where levels are ca. 10^{-8} M in blood, would depend upon the K_m of any enzymes involved. In similar experiments using an alternate radiolabeled derivative of NCS and incubation in human serum, no size changes were noted.[13]

NCS can be inactivated *in vitro* by several chemical and physical methods which do not involve peptide cleavage.[4,22,23] Indistinguishable inhibition profiles were obtained in the radioimmunoassay when native NCS, its partially active, monodeaminated derivative and an inactive dideaminated derivative[24] were compared. It is clear, therefore, that the biological activity of the NCS molecule can be affected without altering the structural features identified by antibody. Indeed NCS has recently been resolved into a biologically inactive apo-NCS form by methanol extraction of its chromophore moiety.[11] It is probable that most, if not all, of the immunoreactivity resides in this protein portion which serves to protect the labile chromophore from inactivation. Since, under physiological conditions this chromophore–protein complex is tightly associated, measurement by radioimmunoassay can be expected to reflect holo-NCS levels.

[22] M. Kono, I. Haneda, Y. Koyama and M. Kikuchi, *Jpn. J. Antibiot.* **27**, 707 (1974).
[23] H. Maeda and K. Kuromizu, *J. Biochem.* **81**, 25 (1977).
[24] K. Kumagai, H. Maeda, and N. Ishida, *Antimicrob. Agents Chemother.* 546 (1967).

[35] Radioimmunoassay of Benzodiazepines

By Ross Dixon

Introduction

Since the introduction of chlordiazepoxide hydrochloride in 1960, the benzodiazepines have had a major impact on modern psychopharmacology. Although benzodiazepine derivatives are the most commonly prescribed antianxiety drugs in clinical practice, certain members of the group are effective hypnotics and anticonvulsant agents.

The aim of this chapter is to describe a number of radioimmunoassay (RIA) procedures for different benzodiazepines, the reasons for their development and how the assays have been employed.

For a detailed account of the theory and basic principles of RIA, the reader is referred to two previous articles in this series.[1,2]

[1] D. Rodbard and H. A. Feldman, this series, Vol. 36, p. 3.
[2] G. D. Niswender, A. M. Akbar, and T. M. Nett, this series, Vol. 36, p. 16.

Diazepam

The antianxiety agent, diazepam (Fig. 1), which is the active ingredient in Valium, is the most widely prescribed centrally acting drug in the United States. A number of procedures, primarily employing electron-capture gas-chromatography (EC-GC),[3] have been reported for the determination of the drug and/or its metabolites in plasma and it is the first benzodiazepine for which an RIA was developed.[4] The original RIA procedure used a relatively low specific activity radioligand of [14C]diazepam and lacked sufficient sensitivity to measure low concentrations of diazepam in plasma without solvent extraction. This procedure has recently been modified using high specific activity [3H]diazepam and permits the determination of diazepam directly in micro samples of blood/plasma and saliva.[5] The procedure is ideally suited for routine screening of toxicology samples or in drug formulation bioequivalency studies where high sample throughput is required.

Synthesis of the Diazepam Hapten[6] (Fig. 1)

To a mixture of 1.2 ml (15 mmol) of concentrated hydrochloric acid and 5 g of ice is added 0.9 g (6 mmol) of p-aminoacetanilide followed by the dropwise addition of a solution of 0.45 g (6.2 mmol) of sodium nitrite in 5 ml of water. To a separate solution of 0.5 g (1.67 mmol) of 7-chloro-1,-3-dihydro-5-(4-hydroxyphenyl)-1-methyl-2H-1,4-benzodiazepin-2-one in 25 ml of tetrahydrofuran is added 12 ml (12 mmol) of 1 N NaOH and 20 g of ice. The diazonium salt solution is added rapidly with stirring to the basic solution and the mixture kept at 5° for 2.5 days. The precipitate is

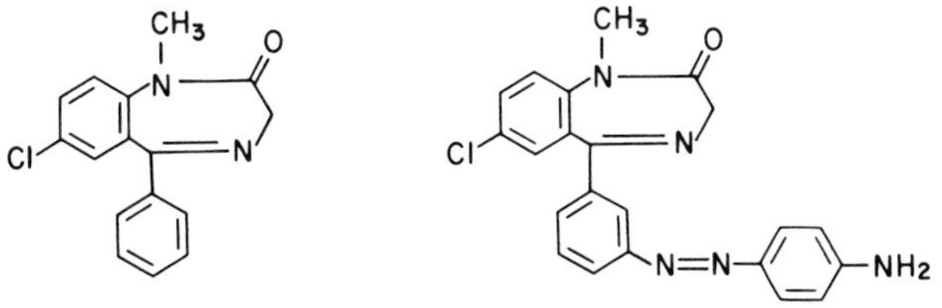

FIG. 1. Structure of diazepam and the hapten which was coupled to albumin following diazotization of the aromatic amine group.

[3] R. E. Weinfeld, H. N. Postmanter, K. C. Khoo, and C. V. Puglisi, *J. Chromatogr.* **143**, 581 (1977).
[4] B. Peskar and S. Spector, *J. Pharmacol. Exp. Ther.* **186**, 167 (1973).
[5] R. Dixon and T. Crews, *J. Anal. Toxicol.* **2**, 210 (1978).
[6] J. V. Earley, R. I. Fryer, and R. Y. Ning, *J. Pharm. Sci.* **68**, 845 (1979).

collected and recrystallized from dichloromethane:methanol to give about a 70% yield of 5-[3-4-acetaminophenylazo)-4-hydroxyphenyl]-7-chloro-1,3-dihydro-1-methyl-2H-1,4-benzodiazepin-2-one (A) as red prisms. Recrystallization from dichloromethane-petroleum ether gives yellow rods, mp 276–278°; IR (KBr) broad 1700 cm⁻¹ (2 C=O).

Anal. Calc. for $C_{24}H_{20}ClN_5O_3$: C, 62.44; H, 4.36; N, 15.16. Found: C, 62.24; H, 4.37; N, 14.97.

A solution of 0.6 g (1.3 mmol) of A in 50 ml of methanol and 10 ml of concentrated hydrochloric acid is heated for 10 min on a steam bath. After 2 hr at room temperature, the methanol is evaporated, the solution made basic with ammonium hydroxide, and extracted with 50 ml of dichloromethane. The organic layer is dried and evaporated. Crystallization of the residue from methanol and recrystallization from dichloromethane-methanol gives about 45% yield of the desired hapten (Fig. 1), 5-[3-(4-aminophenylazo)-4-hydroxyphenyl]-7-chloro-1,3-dihydro-1-methyl-2H-1,-4-benzodiazepin-2-one as brown prisms, mp 262–268°; IR (KBr) 3450, 3350 (NH₂), 1682 cm⁻¹(C=O).

Anal. Calc. for $C_{22}H_{18}ClN_5O_2$: C, 62.93; H, 4.32; N, 16.68. Found: C, 62.96; H, 4.49; N, 16.47.

Preparation of Diazepam Immunogen

Twenty milligrams (0.05 mmol) of the hapten (Fig. 1) as a solution in 2 ml of dimethylformamide and 0.6 ml of water is treated with 0.2 ml of 1 N hydrochloric acid and the dark brown solution cooled to 2°. Then 0.06 ml of a 1 M solution of sodium nitrite is added and the mixture stirred in the cold for 30 min. The excess nitrous acid is decomposed by the addition of 0.05 ml of a 1 M solution of ammonium sulfamate. The diazonium salt of the hapten is then added dropwise with stirring to a cold (2–4°) solution of 200 mg of bovine serum albumin in 5 ml of 0.16 M borate buffer (pH 9). Initially a dark blue solution is formed which after stirring for about 1 hr turns dark brown. It is important to keep the reaction mixture cold during the diazo coupling procedure and any drop in pH should be adjusted back to pH 9 by the addition of 1 N sodium hydroxide. The reaction mixture is then stored at 4° overnight followed by sequential dialysis against 2 × 2 liters of 0.05 M sodium bicarbonate and 2 × 3 liters of water to remove any uncoupled hapten, solvent, and salts. The dialyzed immunogen is then isolated by lyophilization as a dark brown fluffy powder.

Little, if any, uncoupled hapten is present after following the above coupling procedure which indicates that about 18 mol of the hapten are covalently coupled to 1 mol of albumin. This ratio is ideally suited for antibody production.

Immunization and Collection of Antiserum to Diazepam

Although there have been considerable advances made in the technique of RIA over the last decade, there is still no rational approach to immunization procedures. All aspects of the art such as choice of animal, immunization schedule, and the use of adjuvants are somewhat empirical. However, in the author's opinion, the widely used foot-pad immunization technique for rabbits is unnecessarily cruel and distressing to the animal in view of the inflammation and necrosis that can occur on the animal's feet and offers no advantages over other more humane procedures. Over the last 6 years we have successfully used a modification of the intradermal immunization technique of Vaitukaitis *et al.*[7] to produce antibodies to over 35 drug–albumin conjugates.

The immunogen is dissolved in 2 ml of 0.01 M phosphate-buffered saline (PBS) (pH 7.4) at a concentration of 2 mg/ml in a 10-ml glass beaker. An equal volume of Freund's complete adjuvant (Grand Island Biological Co., N.Y.) is added and the mixture emulsified using a Virtis 45 homogenizer. The plunger of a 5-ml plastic syringe is removed and the thick emulsion, which now has the consistency of whipped cream, is placed in the barrel using a wide spatula. The plunger is replaced holding the syringe upright and 0.1-ml aliquots are injected at 10 intradermal sites (0.5 mg of immunogen) on the shaved back of a New Zealand white rabbit. This immunization procedure is repeated twice at 2-week intervals using a fresh emulsion on each occasion. At about 2 months following the initial immunization, the immunogen is dissolved in PBS at a concentration of 1 mg/ml; 0.2 ml is injected intravenously via the marginal ear vein and 0.2 ml subcutaneously into each rabbit. Ten days later the rabbit is bled (20–30 ml of blood), the serum harvested and stored at $-20°$ or lyophilized in 1-ml aliquots.

Radioligand

[*methyl*-³H]Diazepam with a specific activity of around 40 Ci/mmol is now commercially available (NEN, Boston, MA) as a solution in ethanol and should be stored at $-20°$. Under these conditions the tracer is stable for up to 1 year but should be checked periodically for radiochemical purity. Thin-layer chromatography (TLC) on silica gel (Merck; silica gel 60, F-254) using chloroform:ethanol:concentrated ammonium hydroxide (80:10:1) is suitable for this purpose.

---

[7] J. Vaitukaitis, J. B. Robbins, E. Nieschlag, and G. T. Ross, *J. Clin. Endocrinol.* **33,** 988 (1971).

Diazepam RIA Procedure

The following reagents and standards are prepared:

Buffer. The assay buffer consists of 0.01 M phosphate-buffered saline (PBS) at pH 7.4 containing 0.1% bovine γ-globulin (Pentex, Fraction II) and 0.1% sodium azide. When not in use the buffer is stored at 4° and is stable for at least 1 month.

Radioligand Solution. An aliquot (0.1 ml) of the stock ethanolic solution of [³H]diazepam is added to 10 ml of assay buffer and the resulting solution diluted with buffer to give a concentration of 150,000 dpm/ml. The solution is stored at 4°.

Antiserum. Depending on the titer of antibodies to diazepam, antiserum is appropriately diluted with buffer so as to achieve 50–55% binding of the radioligand under the conditions of the assay[2] (vide infra). At 3–4 months following the initial immunization this dilution is in the region of 1:2000 to 1:3000. The diluted antiserum is stored at 4°.

Diazepam Standards

A 1 mg/ml solution of diazepam is prepared in methanol and stored in an amber glass vessel at 4°. A 20-μl aliquot of the methanolic solution is then added to 10 ml of control plasma to give a stock plasma standard containing 2000 ng/ml of diazepam. An aliquot (0.1–1 ml) of the stock standard is then sequentially diluted with an equal volume of control plasma to give a series of plasma standards containing 2000, 1000, 500, 250, 125, 62, and 31 ng/ml of diazepam. This method of preparing standards where the dilution is always one to one is less error-prone than making varying dilutions so as to arrive at preselected concentrations, e.g., 500, 200, 100, 50, and 20. The plasma standards are incubated at 37° for 30 min to allow the drug to equilibrate with plasma proteins and then frozen at −20° until use. In order to avoid any possible influence of repeated freezing and thawing, each individual standard may be aliquoted into 0.1 ml amounts and a new set of standards used for each assay run.

Saturated Ammonium Sulfate. A saturated solution of ammonium sulfate is prepared by dissolving ammonium sulfate in water until no more crystalline material will dissolve. The solution is decanted from the crystals and its pH adjusted to 7.4 by the addition of a few drops of 5 N sodium hydroxide. The solution is stored at room temperature and its pH checked prior to use.

Assay Method

Ten microliters of each standard and unknown sample are added to a 12 × 75-mm disposable glass tube and diluted with 1 ml of assay buffer.

The diluted samples are now either assayed immediately or can be stored at 4° for up to 1 week to await analysis.

To duplicate 12 × 75-mm tubes are now added 0.1 ml of each diluted standard or unknown sample (equivalent to 1 μl of plasma) followed by 0.2 ml of the [³H]diazepam solution (30,000 dpm) and 0.2 ml of diluted antiserum to give a total tube volume of 0.5 ml. Appropriate controls for the determination of nonspecific binding (absence of antiserum) and total binding (absence of unlabelled standard) are also included. The tube contents are mixed briefly on a vortex and allowed to stand at room temperature for 30 min. This incubation period is not critical and can be increased to several hours depending on the work schedule without any significant change in the overall assay performance. An equal volume (0.5 ml) of saturated ammonium sulfate is now added to each tube to precipitate the γ-globulins, the contents mixed and centrifuged at 3000 rpm for 20 min. The supernate is then decanted or removed by aspiration (taking care not to disturb the precipitate) and the precipitate washed once with 1 ml of 50% saturated ammonium sulfate and isolated as before. The washed pellet, containing the antibody bound [³H]diazepam, is dissolved in 0.3 ml of water, 3 ml of Aquasol (NEN, Boston, MA) added, the tube contents vortexed to obtain a clear solution, decanted into a Bio-Vial (Beckman Instruments Inc.), and counted in a liquid scintillation counter. The use of the counting vial may be avoided and the sample counted directly in the assay tube using a modified liquid scintillation counter.[8] The latter results in a considerable overall savings in the cost of the assay.

Saliva samples may be assayed in an identical fashion to plasma except that 0.1 ml of undiluted saliva must be used in order to measure the much lower concentrations of diazepam present in saliva.

The raw RIA data may be handled as outlined in detail by Rodbard[9] using a digital computer for iterative weighted linear regression analysis of logit B/B_0 versus log concentration of diazepam. However, satisfactory results can be more simply obtained by interpolation from a manual plot on logit–log graph paper (TEAM, Box 25, Tamworth, NH). A typical calibration curve is shown in Fig. 2.

Comments on Diazepam Assay Method

1. The RIA for diazepam is a very simple, stable and versatile procedure. As outlined above it is ideally suited for the routine determination of therapeutic levels of the drug in plasma and saliva. However, since the actual limit of sensitivity of the standard curve is about 30 pg/tube, the

[8] R. Dixon and E. Cohen, *Clin. Chem.* **22,** 1746 (1976).
[9] D. Rodbard, *Clin. Chem.* **20,** 1255 (1974).

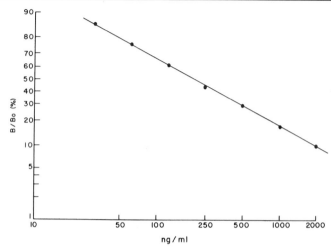

FIG. 2. Typical logit–log calibration curve for diazepam in plasma.

working limit of sensitivity i.e., pg/ml or ng/ml of plasma, can be readily adjusted depending on the amount of sample assayed.

2. Apart from dispensing each sample with a pipet, solutions of the radioligand, antiserum, and ammonium sulfate are best added from manually operated dispensers.

3. Ammonium sulfate is the reagent of choice for the separation of free and antibody-bound radioligand due to the time independent stability of the precipitate. The necessitity for low-temperature centrifugation in refrigerated centrifuges is also avoided.

4. It should be noted that all of the standards and unknowns are diluted and pipetted in an identical fashion at the same time. In this way inaccuracies and/or errors in the initial pipetting steps cancel each other out, i.e., it does not matter whether 12 μl of plasma instead of 10 μl are diluted with buffer or whether 87 μl instead of 100 μl of the diluted sample are added to the assay provided the error is consistent.

Diazepam RIA Characteristics. Using the previously described assay conditions the method has a limit of sensitivity, using 10 μl of plasma or 100 μl of saliva, of about 30 and 0.3 ng/ml of diazepam respectively (Fig. 2). Such sensitivity is ideally suited for routine screening procedures or pharmacokinetic studies. Plasma levels below 30 ng/ml are well below those required for any significant therapeutic or pharmacologic response in man.

The specificity of the diazepam antiserum in terms of its cross-reactivity with other benzodiazepines is shown in Table I. Only 3-hydroxydiazepam, which is a minor and mostly nondetectable metabolite of diazepam

TABLE I
SPECIFICITY OF DIAZEPAM ANTISERUM

Compound	% Cross-reactivity
Diazepam	100
N-Demethyldiazepam[a]	0.4
Oxazepam	0.3
3-Hydroxydiazepam	10
Chlordiazepoxide	<0.01
N-Demethylchlordiazepoxide	0.2
Demoxepam	0.1
Clonazepam	0.1
Lorazepam	<0.01
Prazepam	0.4
Flurazepam	<0.01
N¹-Dealkylflurazepam	0.2
Clorazepate	0.3
Medazepam	<0.1
Amitryptyline	<0.01

[a] Metabolite of diazepam present in plasma.

in plasma, shows any significant (10%) cross-reactivity. Specificity of the RIA has also been established by comparison with an EC-GC procedure.[3]

Intra- and interassay precision usually ranges from 7 to 10%, respectively, while recovery of diazepam from spiked control samples is quantitative.

Chlordiazepoxide

Chlordiazepoxide hydrochloride (Fig. 3), which was synthesized in the mid-1950s, was the first member of the 1,4-benzodiazepine class of compounds to be used clinically as an antianxiety agent and is still widely

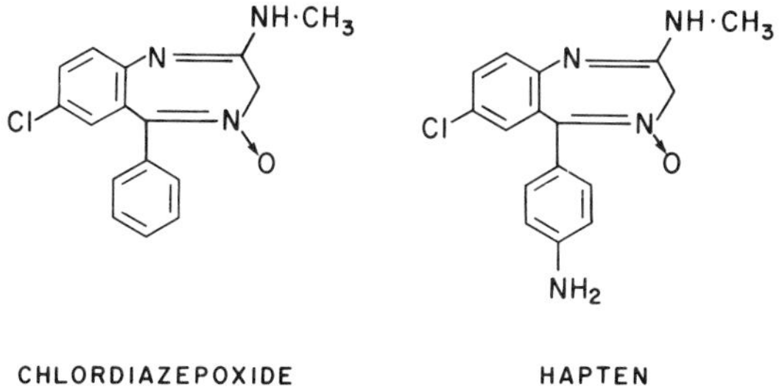

CHLORDIAZEPOXIDE	HAPTEN

FIG. 3. Structure of chlordiazepoxide and the hapten which was coupled to albumin after conversion to its reactive acyl azide with nitrous acid.

prescribed. This drug is the active ingredient in Librium and is now available in several generic formulations.

In 1975, Dixon *et al.*[10] reported on the development of the first RIA for chlordiazepoxide which used a rabbit antiserum and [^{14}C]chlordiazepoxide as the radioligand. More recently, this same group has described an improved procedure using a new antiserum and [^{3}H]chlordiazepoxide.[11] The RIA for chlordiazepoxide has been used extensively for studies involving the pharmacokinetic and biopharmaceutical evaluation of different formulations of the drug, either alone or in combination with amitriptyline.

The following account will describe the development and use of the more recent RIA procedure.

Synthesis of the Chlordiazepoxide Hapten[6] (*Fig. 3*)

To 100 ml of methanol saturated with methylamine in an ice bath is added 1.6 g (3.8 mmol) of 6-chloro-2-chloromethyl-4,4-(2,2,2-trifluoroacetamido)phenylquinazoline 3-oxide. After 18 hr at room temperature, the solution is evaporated to dryness. Crystallization of the residue from methanol followed by recrystallization from dichloromethane-methanol gives yellow prisms, mp 180–185° reset 288–290°. The filtrates are partitioned between 1 N hydrochloric acid and ether, the acid layer made basic with sodium hydroxide and filtered. The collected solids are crystallized from dichloromethane-methanol to give about 0.6 g (50%) of the desired hapten (Fig. 3), 5-(4-aminophenyl)-7-chloro-2-methylamino-3H-1,4-benzodiazepine 4-oxide. Elemental analysis may be problematic due to occlusion of solvents in the crystal; IR (KBr) 3250 (broad), 1625 cm^{-1}; mass spectrum m/e 314 (M^{+}).

Preparation of Immunogen and Production of Antiserum to
    Chlordiazepoxide

Thirty milligrams (0.1 mmol) of the hapten (Fig. 3) is suspended in 2 ml of water and treated with 0.36 ml of 1 N hydrochloric acid to give a yellow solution which is cooled to 2° on ice. The hapten is then diazotized by the addition of 0.09 ml (0.09 mmol) of a 1 M aqueous solution of sodium nitrite and kept cold for 30 min. (A nonequimolar amount of sodium nitrite is used to minimize side-reactions of the hapten with nitrous acid.) The diazonium salt of the hapten is then coupled to 72 mg of bovine serum albumin as described previously (see Diazepam) and the immunogen iso-

---

[10] R. Dixon, J. Earley, and E. Postma, *J. Pharm. Sci.* **64**, 937 (1975).
[11] R. Dixon, R. Dixon, R. Lucek, J. Earley, and C. Perry, *J. Pharm. Sci.* **68**, 261 (1979).

lated as a brick-red powder following exhaustive dialysis and lyophilization.

Rabbits are immunized, boosted, and bled in the usual manner.

Radioligand

[8-^{3}H]Chlordiazepoxide is synthesized according to the following general procedure:

2-Amino-4,5-dichlorobenzophenone, in methanol-trimethylamine, is selectively reduced wih tritium over Lindlar catalyst to yield 2-amino-5-chloro[4-^{3}H]benzophenone which is converted to the desired product as previously described.[12,13] The [8-^{3}H]chlorodiazepoxide is then crystallized as its hydrochloride salt from methanol-ether to yield material with a specific activity of about 18 Ci/mmol. The tracer is stored as a crystalline solid and its radiochemical purity checked by TLC as described previously for [methyl-^{3}H]diazepam.

RIA Procedure

The assay procedure is identical with that described for diazepam.

Chlordiazepoxide RIA Characteristics. The assay has approximately the same sensitivity and precision as the diazepam RIA and may be usefully employed for the determination of chlordiazepoxide in both plasma and saliva following therapeutic doses (10–30 mg) of the drug. Recent studies indicate that the concentration of chlordiazepoxide in saliva is about 3% of the concentration found in plasma.[11] This value closely approximates the amount of unbound chlordiazepoxide in plasma as determined by equilibrium dialysis. Thus measurement of the drug in saliva may provide an *in vivo* approach for determining the extent of protein binding of the drug in different individuals and the possible influence of concomitant medication or disease states on this parameter.

The specificity of the chlordiazepoxide antiserum is shown in Table II. It can be seen that it is highly specific for the parent drug with < 1% cross-reactivity with any of its known metabolites present in plasma and with other marketed benzodiazepines. Tricyclic antidepressants, which are commonly prescribed in combination with benzodiazepines, exhibit a similar lack of interference with the antiserum.

Clonazepam

Clonazepam (Fig. 4) is a benzodiazepine anticonvulsant agent which has been found to be clinically effective in controlling minor motor sei-

[12] G. Field and L. Sternbach, U.S. Patent No. 3,515,724 (1970).
[13] G. Field and L. Sternbach, U.S. Patent No. 3,398,139 (1968).

TABLE II
SPECIFICITY OF CHLORDIAZEPOXIDE ANTISERUM

Compound	% Cross-reactivity
Chlordiazepoxide	100
N-Demethylchlordiazepoxide[a]	<1
Demoxepam[a]	<1
N-Demethyldiazepam[a]	<1
Diazepam	<1
Clonazepam	<1
Medazepam	<1
Amitriptyline	<1

[a] Metabolite of chlordiazepoxide present in plasma.

zures (petit mal) in man.[14] The usefulness of determining blood concentrations of anticonvulsants in the clinical management of epileptic patients has been well documented. Knowledge of blood concentrations facilitates in dosage regulation and monitors patient compliance in taking the drug. In view of this situation an RIA for clonazepam was developed by Dixon et al.[15] which provided a simple procedure to relate therapeutic response to plasma concentrations of the drug in individual patients.

In the original procedure, a rabbit antiserum to clonazepam and [³H]clonazepam of relatively low specific activity was used. Separation of antibody bound and free radioligand was achieved using saturated ammonium sulfate. Since this method was first reported a number of improvements have been incorporated with the availability of high specific activity [9-³H]clonazepam and the use of polyethylene glycol-6000 for the separation of bound and free fractions. Furthermore, an RIA which

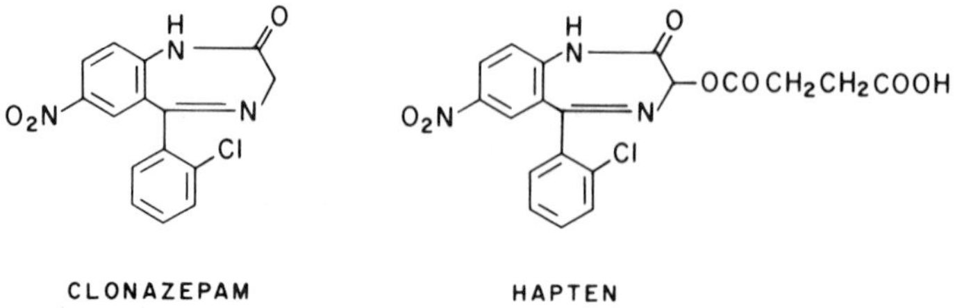

CLONAZEPAM   HAPTEN

FIG. 4. Structure of clonazepam and the hapten which was coupled to albumin through its carboxyl function as a mixed anhydride.

[14] R. Kruse, Epilepsia 12, 179 (1971).
[15] R. Dixon, R. L. Young, R. Ning, and A. Liebman, J. Pharm. Sci. 66, 235 (1977).

employs an ^{125}I-labeled radioligand has also been developed.[16] Both of these methods will be described below.

Synthesis of the Clonazepam Hapten[6] (Fig. 4)

To a suspension of 6.6 g of 5-(2-chlorophenyl)-1,3-dihydro-3-hydroxy-7-nitro-2H-1,4-benzodiazepin-2-one in 400 ml of dry tetrahydrofuran is added 1.2 g of a 50% suspension of sodium hydride in mineral oil (27 mmol hydride). After stirring for 0.5 hr under nitrogen, succinic anhydride (3.35 g; 33 mmol) is added to the clear solution in one portion and the stirring continued. After 1 hr, the tetrahydrofuran is evaporated *in vacuo,* the residue stirred with 200 ml of water, acidified with acetic acid, and extracted with 300 ml of dichloromethane. (The acidification and extraction must be performed rapidly so that the product will not crystallize out of the dichloromethane solution.) After drying the extract over anhydrous sodium sulfate, the solvent is concentrated to half its volume. Addition of hexane affords the desired hapten, 5-(2-chlorophenyl)-1,3-dihydro-3-hemisuccinyloxy-7-nitro-2H-1,4-benzodiazepin-2-one (Fig. 4). Recrystallization from tetrahydrofuran-dichloromethane-heptane yields colorless prisms, mp 172–180°; IR (KBr) 1770–1700 (strong unresolved), 1530 and 1350 cm^{-1}; UV$_{max}$ (2-propanol) 215 nm (32,800), 248 (16,500), and 308 (12,000).

Anal. Calc. for $C_{19}H_{14}ClN_3O_7$: C, 52.75; H, 3.08; N, 9.75. *Found:* C, 52.85; H, 3.27; N, 9.73.

Preparation of Immunogen and Production of Antiserum to Clonazepam

The mixed anhydride procedure of Erlanger *et al.*[17] is used to couple the hapten (Fig. 4) to bovine serum albumin.

The hapten (43 mg; 0.1 mmol) is dissolved in 2 ml of dioxane (dried and rendered peroxide-free by passage through a column of basic aluminum oxide, Woelm, Activity I), 0.16 ml of triethylamine:dioxane (1:10) added, and the solution cooled to 5° in an ice-water bath. Then 0.14 ml of isobutylchloroformate:dioxane (1:10) is added and the mixture kept at 5–10° for 20 min. (It is important to ensure that the reaction mixture does not solidify due to the dioxane freezing at low temperatures.) During this reaction period a precipitate of triethylamine hydrochloride will be observed. The mixed anhydride of the hapten is then added with rapid stirring of a solution of 70 mg of bovine serum albumin in 10 ml of water and

[16] R. Dixon and T. Crews, *Res. Commun. Chem. Pathol. Pharmacol.* **18,** 477 (1977).
[17] B. F. Erlanger, F. Borek, S. M. Beiser, and S. Lieberman, *J. Biol. Chem.* **234,** 1090 (1959).

8 ml of dioxane at pH 8.5 which has been cooled to 5°. [In order to avoid precipitation of the albumin by the dioxane, the 10-ml aqueous solution (pH 5) is adjusted to pH 8.5 with 0.1 N NaOH prior to addition of the 8 ml of dioxane with stirring.] Addition of the mixed anhydride reaction mixture to the albumin solution results in a gradual lowering of pH as the coupling proceeds and must be maintained at pH 8–9 by the addition of 0.1 N NaOH. The pale yellow reaction mixture is stirred in the cold for 2 hr and then dialyzed against 500 ml of dioxane : water (1 : 1) overnight. Dialysis is then continued against 2 liters of 0.05 M borate buffer (pH 9) followed by exhaustive dialysis against 2 liter volumes of water. The immunogen is isolated as a pale yellow powder by lyophilization.

The extent of coupling is estimated by comparing the absorbance at 365 nm of a 1 mg/ml solution of the immunogen in 0.1 N NaOH against a standard solution of the hapten (10 μg/ml) mixed with albumin (1 mg/ml). Using the described procedure about 35 mol of the hapten is covalently coupled to the lysine residues on 1 mol of albumin.

Rabbits are immunized intradermally and boosted intravenously and subcutaneously as described for diazepam. Satisfactory titers of antisera are obtained following the second booster immunization.

Radioligands

1. [9-^{3}H]Clonazepam is prepared according to the following general procedure: 2-amino-2'-chlorobenzophenone is brominated[18] and the resulting 2-amino-3,5-dibromo-2'-chlorobenzophenone acylated with bromacetyl bromide to give the amide, 2-bromoacetamido-3,5-dibromo-2'-chlorobenzophenone. Treatment of the amide with ammonia in methanol followed by refluxing in pyridine-toluene solution provides the benzodiazepine, 5-(2-chlorophenyl)-1,3-dihydro-7,9-dibromo-2H-1,4-benzodiazepin-2-one. Catalytic hydrogenolysis of the benzodiazepine with tritium gas in the presence of palladium-on-carbon catalyst selectively replaces both of the bromine atoms with tritium to yield, 5-(2-chlorophenyl)-1,3-dihydro-1,4-benzo-[7,9-^{3}H]diazepin-2(2H)-one, which is nitrated without isolation to give the desired radioligand [9-^{3}H]clonazepam. The final product is diluted with unlabeled clonazepam to reduce the specific activity to about 16 Ci/mmol. At this specific activity the [9-^{3}H]clonazepam is ideal for the RIA and stable when stored in toluene under argon at $-20°$.

2. The ^{125}I-labeled radioligand for the RIA of clonazepam is prepared as follows[16]: The acylating reagent, ^{125}I-labeled N-succinimidyl 3-(4-hydroxyphenyl)propionate, with a specific activity of approximately 1500

[18] L. H. Sternbach, R. I. Fryer, O. Keller, W. Metlesics, G. Sach, and N. Steiger, *J. Med. Chem.* **6,** 261 (1963).

Ci/mmol is commercially available (New England Nuclear, Boston, Mass.) and should be used within 1–2 days of arrival. One millicurie of the reagent as a solution in benzene is evaporated to dryness in its vial under a stream of dry nitrogen passing through a No. 22 needle inserted through the vial septum. The vapors are vented through another needle attached to the activated charcoal filter supplied with the reagent. Then 5 μg of 3-aminoclonazepam, as a solution in 0.1 ml of anhydrous tetrahydrofuran, is injected into the vial and the contents mixed. After allowing the clear solution to stand at room temperature overnight, the solvent is evaporated under nitrogen and the residue applied in dichloromethane as a single spot to the origin of a 5 × 20-cm silica gel plate (Merck; F-254). The plate is run 15 cm using the solvent system dichloromethane:ethanol:ammonia (90:10:0.5). Following TLC the plate is covered with Parafilm (American Can Co., Wis.) and scanned with a radiochromatogram scanner. The major radioactive peak ($R_f = 0.5$) is eluted with 2 ml of ethanol and stored at $-15°$. The excess unreacted 3-aminoclonazepam could be visualized under UV light (254 nm) at R_f 0.2.

Clonazepam RIA Procedures

The clonazepam RIA is carried out in an identical manner using both the ^{3}H and ^{125}I radioligands. The assay procedure is essentially the same as that previously described for diazepam.

A 10-μl sample of plasma is used for analysis with standards in the range of 5–200 ng/ml. The standards and unknown samples are added to 12 × 75-mm tubes followed by 0.3 ml of the radioligand (30,000 dpm of [9-^{3}H]clonazepam or 40,000 dpm of the ^{125}I derivative). After the addition of 0.2 ml of appropriately diluted antiserum (50–60% total binding) the tube contents (0.5 ml) are mixed and incubated at room temperature for 30 min. The antibody-bound fraction is precipitated by the addition of 1 ml of 25% (w/v) solution of polyethylene glycol-6000 (Carbowax-6000; Union Carbide) in water. Each tube is vortexed, centrifuged for 20 min, and the supernate aspirated. When using the ^{3}H-labeled radioligand the pellet is dissolved in 0.2 ml of water, 3 ml of Dioxal (Yorktown Research, Hackensack, N.J.) added, the contents mixed and counted as described in the diazepam assay method. When using the ^{125}I-labeled radioligand the assay tube containing the pellet is counted directly in a gamma counter.

The raw data are handled as previously described.

Comments on the Clonazepam RIA Procedures

1. Polyethylene glycol-6000 is the reagent of choice for the separation of antibody bound and free radioligand. Nonspecific binding is minimal

($<5\%$) compared to the use of saturated ammonium sulfate yet the separation procedure is still virtually time and temperature independent. The 0.1% γ-globulin present in the assay buffer provides sufficient mass for a discernible pellet using this reagent.

2. Liquid scintillation counting of the pellet obtained with polyethylene glycol-6000 can be problematic unless a dioxane-based scintillator is used. Cloudy solutions and inconsistent counting can result with other scintillators.

3. Although the ^{125}I procedure is somewhat faster and cheaper than the ^{3}H procedure, it cannot be recommended unless large numbers of samples are being continuously assayed on a routine clinical basis since the ^{125}I-labeled radioligand must be resynthesized about every 2 months in order to keep the assay parameters consistent.

4. It would appear that the commercially available ^{125}I-labeled acylating reagent may provide a convenient approach for the synthesis of certain ^{125}I-labeled radioligands and avoid some of the hazards associated with direct iodination procedures.

Clonazepam RIA Characteristics

Both the ^{3}H and ^{125}I procedures have the same limit of sensitivity of about 5 ng/ml of clonazepam which is quite adequate in the clinical situation. Intra- and interassay coefficients of variation are not greater than 4 and 9%, respectively.

The specificity of the clonazepam antiserum is shown in Table III. It can be seen that 7-amino and 7-acetylaminoclonazepam, the major metabolites of clonazepam present in plasma, exhibit a critical lack of cross-reactivity which allows the measurement of the parent drug directly in the sample. Although 3-hydroxyclonazepam cross-reacts 24%, its presence in

TABLE III
SPECIFICITY OF CLONAZEPAM ANTISERUM

Compound	% Cross-reactivity
Clonazepam	100
7-Aminoclonazepam[a]	<0.1
7-Acetylaminoclonazepam[a]	<0.1
3-Hydroxyclonazepam	24
Nitrazepam	3
Oxazepam	0.3
Diazepam	<0.01
N-Demethyldiazepam	<0.01
Non-benzodiazepine anticonvulsants	<0.01

[a] Metabolite of clonazepam present in plasma.

plasma has not been demonstrated. All other nonbenzodiazepine anticonvulsants do not cross-react with the antiserum. The specificity of the clonazepam RIA has also been established by comparison with an EC-GC procedure.

Nitrazepam

Nitrazepam (Fig. 5), a benzodiazepine structurally similar to clonazepam (Fig. 4), is a widely used hypnotic which has been on the market in the eastern hemisphere and Latin America for several years. In addition to being effective for the treatment of insomnia, the drug has also been used as an anticonvulsant particularly for the treatment of myclonic seizures in infants.[19] Apart from its clinical efficacy, much of its popularity may be attributed to its safety and low abuse potential.[20]

Recently an RIA for nitrazepam was reported.[21] The method has sufficient versatility to be used in studies which may require routine plasma level monitoring of the drug as an anticonvulsant or which are designed for pharmacokinetic evaluation. The small sample volume required (10 μl), which may be obtained by heel or finger-stick, is a particular attraction of the method when dealing with neonates or the elderly.

Synthesis of the Nitrazepam Hapten[6] (Fig. 5).

The desired hapten, 1,3-dihydro-3-hemisuccinyloxy-7-nitro-5-phenyl-2H-1,4-benzodiazepin-2-one (3-hemisuccinyloxynitrazepam), is prepared by succinylation of 3-hydroxy nitrazepam as in the previously described synthesis of the clonazepam hapten. Crystallization from dichloromethane-hexane yields needles, mp 187–190°, IR (KBr) unresolved carbonyl bands at 1690–1760 cm^{-1}; UV$_{max}$ (2-propanol) 219 nm (ϵ 24,500), 264

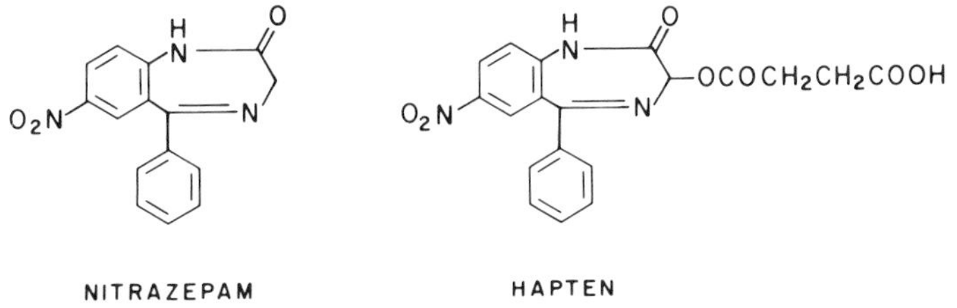

NITRAZEPAM   HAPTEN

FIG. 5. Structure of nitrazepam and the hapten which was coupled to albumin through its carboxyl function as a mixed anhydride.

[19] T. R. Browne and J. K. Penry, *Epilepsia* **14,** 277 (1973).
[20] D. J. Greenblatt and R. I. Shader, "Benzodiazepines in Clinical Practice." Raven, New York, 1974.
[21] R. Dixon, R. Lucek, R. Young, R. Ning, and A. Darragh, *Life Sci.* **25,** 311 (1979).

(17,100), and 305 (10,900); NMR (DMSO-d_6) δ 2.51–2.65 (m, 2, CH$_2$), 2.69–2.83 (m, 2, CH$_2$), 5.89 (s, 1, 3-H), 7.45–7.62 (m, 6, C$_6$H$_5$, and 9-H), 8.06 (d, J = 2.5 Hz, 1, 6-H), and 8.48 ppm (q, J = 2.5 and 9.0 Hz, 1, 8-H). *Anal.* Calc. for C$_{19}$H$_{15}$N$_3$O$_7$: C, 57.43; H, 3.81; N, 10.58. Found: C, 57.57; H, 3.87; N, 10.30.

Preparation of Immunogen and Production of Antiserum to Nitrazepam

The hapten is coupled to bovine serum albumin using the mixed anhydride procedure and rabbits immunized as previously described for clonazepam.

Radioligand

[9-^{3}H]Clonazepam cross-reacts considerably (30%) with the antiserum to nitrazepam and can thus be used as the radioligand in the RIA of the latter.

Nitrazepam RIA Procedure

The assay procedure is identical to the clonazepam RIA using polyethylene glycol-6000 for the separation of antibody bound and free radioligand.

Nitrazepam RIA Characteristics. Using a 10-μl sample of plasma the RIA has a limit of sensitivity of about 5 ng/ml of nitrazepam which is adequate for clinical samples.

The specificity of the nitrazepam antiserum is shown in Table IV. Its major metabolites, 7-amino- and 7-acetylaminonitrazepam, exhibit a simi-

TABLE IV
SPECIFICITY OF NITRAZEPAM ANTISERUM

Compound	% Cross-reactivity
Nitrazepam	100
7-Aminonitrazepam[a]	<0.1
7-Acetylaminonitrazepam[a]	<0.1
Clonazepam	30
Diazepam	0.3
N-Demethyldiazepam	14
Oxazepam	30
Chlordiazepoxide	2
Non-benzodiazepine anticonvulsants	<0.01

[a] Metabolite of nitrazepam present in plasma.

lar lack of cross-reactivity to that observed with the corresponding metabolites of clonazepam. However a number of other benzodiazepines, which include clonazepam, N-demethyldiazepam and oxazepam, show significant cross-reactivity and therefore the nitrazepam RIA cannot be used in patients who might be receiving other benzodiazepines concomitantly.

Flurazepam

Flurazepam (Fig. 6), which is the active ingredient in Dalmane, is a widely prescribed hypnotic for the treatment of insomnia. Studies on the metabolism of the drug in man have shown that it is extensively metabolized through N-dealkylation of the N-1 side-chain.[22] The blood level profile of flurazepam and its metabolites has also been studied in man following both single and chronic administration of therapeutic doses of the drug.[23] Although N^1-dealkylflurazepam could be readily detected in plasma using EC-GC,[24] intact flurazepam could not be measured and it was estimated that its maximum concentration in plasma at any time was $<3–4$ ng/ml. The latter situation presented an analytical challenge since it was of interest to ascertain whether any intact flurazepam reached the circulation and thereby contributed to the hypnotic activity of the drug or if the drug was subject to complete "first-pass" metabolism on oral administration. Although further attempts to measure the intact drug using EG-GC and GC-MS have thus far been unsuccessful,[24a] an RIA procedure

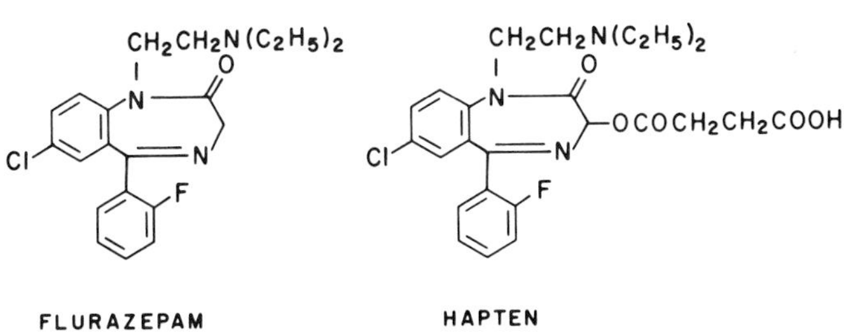

FIG. 6. Structure of flurazepam and the hapten which was coupled to albumin through its carboxyl function as a mixed anhydride.

[22] M. A. Schwartz and E. Postma, *J. Pharm. Sci.* **59**, 1800 (1970).

[23] S. A. Kaplan, J. A. F. de Silva, M. L. Jack, K. Alexander, N. Strojny, R. E. Weinfeld, C. V. Puglisi, and L. Weissman, *J. Pharm. Sci.* **62**, 1932 (1973).

[24] J. A. F. de Silva, C. V. Puglisi, M. A. Brooks, and M. R. Hackman, *J. Chromatogr.* **99**, 461 (1974).

[24a] A GC-MS procedure has recently been reported by B. Miwa, W. Garland, and P. Blumenthal, *Anal. Chem.* **53**, 793 (1981).

was recently reported which achieved this goal.[25] The following account describes this RIA procedure and illustrates the use of solvent extraction and chromatographic purification as a necessity to achieve sufficient sensitivity and specificity to measure some drugs in biological media.

Synthesis of Flurazepam Hapten (Fig. 6)

A solution of 17 g (42.1 mmol) of 3-hydroxyflurazepam and 6.1 g (60 mol) of succinic anhydride in 130 ml of dichloromethane is treated with 8 ml of triethylamine and the reaction mixture fefluxed for 30 min. After 1 hr at room temperature the mixture is partitioned between 500 ml of ether and 400 ml of dilute ammonium hydroxide. The basic layer is acidified with acetic acid and extracted with dichloromethane (2 × 200 ml) to yield the desired hapten, 3-hemisuccinyloxyflurazepam (Fig. 6) as an oil which is difficult to crystallize. The product may be crystallized, however, as its succinate salt by reaction of 3-hemisuccinyloxyflurazepam with one equivalent of succinic anhydride. The mixture is refluxed in tetrahydrofuran for 30 min, the solvent evaporated and the succinate salt crystallized from acetone ether to yield white prisms, mp 101–102°. *Analysis* Calc. for $C_{25}H_{27}$ Cl F $N_3O_5 \cdot C_4H_6O_4$: C, 56.00; H, 5.35; N, 6.76. Found: C, 57.18; H, 5.56; N, 7.38.

Preparation of Flurazepam Immunogen

The hapten, 3-hemisuccinyloxyflurazepam, is coupled to bovine serum albumin using the mixed anhydride procedure as described for clonazepam. Following the coupling procedure, the pale yellow reaction mixture is dialyzed against 0.05 M sodium bicarbonate followed by exhaustive dialysis against water and isolated by lyophilization. The extent of coupling is determined by recording the absorbance at 310 μm of a 1 mg/ml solution of the immunogen in 0.05 M Tris buffer (pH 8) against a standard solution of the hapten in buffer containing 1 mg/ml of albumin. About 18 mol of hapten are coupled to 1 mol of albumin under these conditions.

Production of Antisera to Flurazepam

Rabbits are immunized intradermally and boosted intravenously and subcutaneously as previously described. Satisfactory titers of antibodies are obtained following the second booster immunization.

Radioligand

A sample of 9-iodoflurazepam, 10.7 mg, in 1.5 ml of dry tetrahydrofuran is treated with 5 μl of triethylamine, 6.3 mg of 10% palladium on

[25] W. Glover, J. Earley, M. Delaney, and R. Dixon, *J. Pharm. Sci.* **69,** 601 (1980).

charcoal catalyst and 10 Ci of tritium gas. The reaction mixture is stirred magnetically at room temperature for 3 hr. After removing the unreacted tritium, the mixture is filtered and concentrated *in vacuo*. The residue is freeze dried several times from methanol to remove labile tritium and then chromatographed on a 3 g column of silica gel packed in ethyl acetate. The fractions containing [9-³H]flurazepam are pooled and concentrated to yield about 6 mg of radiochemically pure material with a specific activity of about 27 Ci/mmol. The radioligand is dissolved in toluene at a concentration of 1 mCi/ml and stored at −20°. For use in the RIA, a suitable aliquot of the toluene solution is evaporated to dryness under nitrogen, dissolved in a small volume of ethanol, and diluted with assay buffer (PBS).

Flurazepam RIA Procedures

Extraction and Chromatography. One milliliter of plasma is buffered to pH 9 with 1 ml of 1 *M* borate buffer and extracted twice with 5 ml of glass-distilled hexane (all solvents used in this RIA are obtained from Burdick and Jackson Laboratories Inc., Muskegon, Mich.) on a reciprocating shaker for 15 min. The combined hexane extracts are evaporated to dryness in a 15 ml conical tube at 40° under nitrogen and the inside wall of the tube rinsed down with 1 ml of hexane. Duplicate control plasma samples and internal standards containing 0.125, 0.5, and 1 ng/ml of flurazepam are processed along with the unknown samples.

Flurazepam is now isolated from the plasma extracts by chromatography on Sephadex LH-20. Prepacked columns containing 1 g of Sephadex LH-20 (Size QS-44; Isolab Inc., Akron, Ohio) are washed with methanol and then equilibrated with the solvent system hexane:benzene:methanol (95:5:4). Each plasma extract is dissolved in 0.2 ml of the solvent system, applied to the top of the column, and allowed to percolate into the column bed. The tube is then rinsed with 0.5 ml of solvent which is transferred to the column. The column is then developed and the first 3.5 ml of eluate discarded. The next 5 ml of eluate, which contains flurazepam but none of its known metabolites, are collected in a clean 20-ml glass scintillation vial and evaporated to dryness. The residue is dissolved in 0.5 ml of PBS, the vial capped, and incubated at 32° for 1 hr to obtain maximum dissolution of flurazepam prior to RIA analysis.

Assay of Flurazepam. A calibration curve is generated by adding [9-³H]flurazepam (15,000 dpm) in 0.1 ml of buffer to assay tubes containing 0.03 to 2 ng of flurazepam in 0.2 ml of buffer. Following the standards duplicate 0.2-ml aliquots of the reconstituted purified fractions from the LH-20 column are run. After incubation at room temperature for 1 hr the antibody bound and free fractions are separated using saturated ammonium sulfate as previously described. The concentration of flurazepam in each unknown sample is obtained by interpolation of its B/B_0 value from the external standard curve and correction for recovery from the internal standards.

Flurazepam RIA Characteristics

The logit–log plot of B/B_0 (%) verses concentration of flurazepam gives a linear response between 0.03 to 2 ng. The working limit of sensitivity is about 0.1 ng/ml of plasma using a 1 ml sample. Blank values for control plasma taken through the entire extraction and chromatographic-procedures are well below the limit of detection while recovery of internal standards over the range of 0.1–5 ng/ml of flurazepam averages about 83%. Intra- and interassay precision does not exceed 9%.

The specificity of the antiserum with regard to its cross-reactivity with the known metabolites of flurazepam present in plasma is shown in Table V. The extensive cross-reactivity with monodemethylflurazepam (17%) necessitates the chromatographic step since flurazepam cannot be selectively extracted from plasma without extensive carryover of its metabolites. Chromatography on LH-20, however, very effectively separates the parent drug from its metabolites; all of the latter compounds elute from the column considerably later than flurazepam. The flurazepam RIA procedure, as described, has sufficient sensitivity and specificity to measure plasma levels of the intact drug following administration of a single 30-mg oral dose of flurazepam hydrochloride.[25]

N-Demethyldiazepam

N-Demethyldiazepam (Fig. 7) is a benzodiazepine of particular clinical significance. It is a major pharmacologically active plasma metabolite of several widely prescribed benzodiazepines,[26] including diazepam, meda-

TABLE V
SPECIFICITY OF FLURAZEPAM ANTISERUM

Compound	% Cross-reactivity
Flurazepam	100
Monodeethylflurazepam[a]	17
Dideethylflurazepam[a]	4
N^1-Hydroxyethylflurazepam[a]	<1
N^1-Dealkylflurazepam[a]	<1

[a] Metabolite of flurazepam present in plasma.

[26] D. J. Greenblatt and R. I. Shader, *South. Med. J.* **71**, *Suppl.* **2**, 2 (1978).

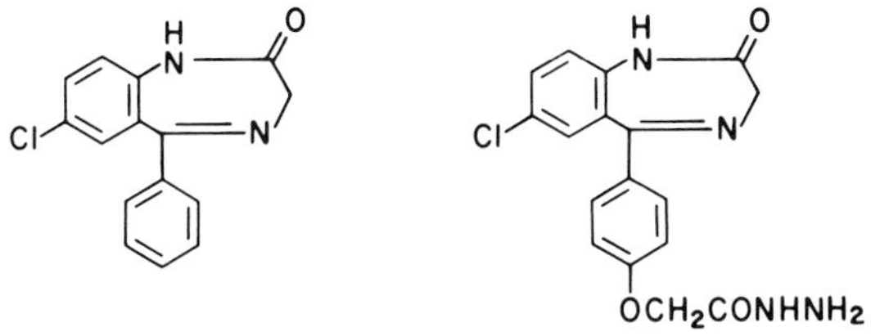

N-DESMETHYLDIAZEPAM   HAPTEN

FIG. 7. Structure of N-demethyldiazepam and the hapten which was coupled to albumin after conversion to its reactive acyl azide with nitrous acid.

zepam, prazepam, clorazepate, and chlordiazepoxide. In fact, prazepam and clorazepate are both "prodrugs" for N-demethyldiazepam with little, if any, of the parent drug reaching the circulation.[26] This is in contrast to results obtained with diazepam, for example, which showed that following its administration, intact diazepam is absorbed. In addition, pharmacologic activity following im and iv administration of diazepam has been demonstrated in the absence of any measurable levels of N-demethyldiazepam, indicating that the parent drug itself is pharmacologically active.

In view of these circumstances a specific RIA for N-demethyldiazepam has been developed which allows the investigator to follow the pharmacokinetic profile of this important metabolite following administration of any of the aforementioned marketed drugs.[27]

Synthesis of N-Demethyldiazepam Hapten[6] (Fig. 7)

To a solution of 0.5 g (1.3 mmol) of 7-chloro-1,3-dihydro-5-(4-methoxycarbonylmethoxyphenyl)-2H-1,4-benzodiazepin-2-one in a mixture of 25 ml of methanol and 30 ml of tetrahydrofuran is added 5 ml of hydrazine hydrate (85% in water). After 1 hr the solution is evaporated *in vacuo* and 25 ml of dichloromethane and 20 ml of water added. After standing overnight, the solution is filtered and the solids recrystallized from tetrahydrofuran-hexane to give about a 60% yield of the desired hapten, 7-chloro-1,3-dihydro-5-(4-hydrazinocarbonyl-methoxyphenyl)-2H-1,4-benzodiazepin-2-one, as colorless prisms, mp 180–185°, resets 210–220°; IR (KBr) 1700, 1680 cm^{-1} (C=O).

Anal. Calc. for $C_{17}H_{15}ClN_4O_3 \cdot 0.5 \, C_4H_8O$: C, 57.80; H, 4.88; N, 14.19. *Found:* C, 57.59; H, 4.79; N, 14.10.

[27] R. Dixon, W. Glover, J. Earley, and M. Delaney, *J. Pharm. Sci.* **68,** 1471 (1979).

Preparation of Immunogen and Production of Antiserum to
N-Demethyldiazepam

The hapten is coupled to bovine serum albumin via its reactive acyl azide which covalently couples to the ε-amino groups of the lysine residues on the albumin.

A solution of 48 mg (0.12 mmol) of the hapten in 1.5 ml of dimethylformamide is treated with 0.1 ml of 4.5 N HCl in dioxane to give a pale yellow solution which is cooled to $-25°$ in a dry-ice:isopropanol bath. The reactive acyl azide is then generated *in situ* by the addition of 0.16 ml of isoamyl nitrite:dioxane (1:10) and the reaction mixture kept at $-25°$ for 20 min. The excess nitrous acid is destroyed by the addition of 0.05 ml of 1 M ammonium sulfamate. The reaction mixture is then added dropwise with stirring to a cooled (2–4°) solution of 70 mg of bovine serum albumin in 3 ml of 0.16 M borate buffer (pH 8.5). The pH of the coupling reaction is kept at 8–8.5 by the addition of 1 N sodium hydride when necessary. After 4 hr in the cold the reaction mixture is dialyzed against 0.05 M sodium bicarbonate followed by water and the immunogen isolated by lyophilization.

Rabbits are immunized and boosted in the usual manner.

Radioligand

N-[9-^{3}H]Demethyldiazepam is used as the radioligand and synthesized in the following manner:

A sample of 9-iododemethyldiazepam, 11.5 mg, dissolved in 1 ml of absolute tetrahydrofuran is treated with 10.5 mg of 10% Pd/C, 5 μl of triethylamine, and 10 Ci of tritium gas. The reaction, carried out in a 3 ml round bottom flask is magnetically stirred at room temperature for 3 hr. Nonreacted tritium is first removed and then the mixture is filtered through celite. The filtrate is concentrated *in vacuo,* freeze dried from methanol several times to remove labile tritium, and finally the residue is chromatographed over a 3 g column of silica gel (E. Merck No. 7734) packed in ethyl acetate. The appropriate fractions, on evaporation, yield 4.1 mg of product having a specific activity of 98 mCi/mg (26.5 Ci/mmol).

N-Demethyldiazepam RIA Procedure

The RIA is carried out in diluted plasma exactly as described for diazepam.

Characteristics of the N-Demethyldiazepam RIA

The assay has a similar limit of sensitivity as the diazepam RIA; about 3 ng/ml of N-demethyldiazepam can be readily determined using 10 μl of undiluted plasma. Such sensitivity is however unnecessary in the analysis of clinical samples and 0.1 ml of a 1 : 100 dilution, equivalent to 1 μl of plasma, is a more appropriate aliquot. The specificity of the antiserum is shown in Table VI and it is apparent that it is highly specific for N-demethyldiazepam. Although clorazepate exhibits greater than 50% cross-reactivity, this is undoubtedly due, in part, to its decarboxylation to N-demethyldiazepam during the assay. The latter situation is not problematic, however, since clorazepate, as a "prodrug" of N-demethyldiazepam, undergoes almost complete decarboxylation prior to absorption. Thus, this RIA procedure may be usefully employed in clinical pharmacokinetic and/or biopharmaceutic studies for the determination of N-demethyldiazepam in plasma following administration of diazepam, chlordiazepoxide, clorazepate, prazepam, and medazepam.

Oxazepam

Oxazepam, which is the 3-hydroxy derivative of N-demethyldiazepam, is another benzodiazepine of particular interest since, not only is it marketed as a drug itself, but its glucuronide conjugate is a major urinary metabolite of diazepam, chlordiazepoxide, clorazepate, prazepam, and medazepam. Following oral administration of oxazepam, it is rapidly conjugated with glucuronic acid and plasma concentrations of both intact oxazepam and its pharmacologically inactive glucuronide metabolite can be readily detected.[20,26]

Although an RIA for oxazepam has not been reported a potentially

TABLE VI
SPECIFICITY OF N-DEMETHYLDIAZEPAM
ANTISERUM

Compound	% Cross-reactivity
N-Demethyldiazepam	100
Clorazepate[a]	>50
Diazepam[b]	0.6
Chlordiazepoxide[b]	0.2
N-Demethylchlordiazepoxide[b]	0.8
Demoxepam[b]	0.6
Oxazepam	5
Clonazepam	0.3
Lorazepam	1.4
Prazepam[a]	<0.1
Medazepam[b]	<0.01
Amitriptyline	<0.01

[a] Pro-drug of N-demethyldiazepam.
[b] Precursor of N-demethyldiazepam present in plasma.

useful procedure has been developed by the author using a previously described antiserum and radioligand. It will be noted in Table IV that the antiserum to nitrazepam crossreacts extensively (30%) with both clonazepam and oxazepam. Using clonazepam as the radioligand and the nitrazepam antiserum, satisfactory calibration curves can be generated for the determination of oxazepam directly in plasma with a sensitivity of about 5 ng/ml. The procedure has not been validated as yet by comparison with EC-GC methods and the extent of cross-radioactivity of oxazepam glucuronide, which would also be present in the plasma,[26] has not been evaluated. However, since the water-soluble glucuronide is the only known metabolite present in plasma, specificity for intact oxazepam could be achieved by its extraction into an organic solvent e.g., diethyl ether, leaving the oxazepam glucuronide in the aqueous phase.

The case of oxazepam serves to illustrate how advantage may be taken, in certain instances, of the nonspecificity of an antiserum to measure more than one drug using a single antiserum and radioligand.

Summary

In this chapter details of the development and routine application of specific radioimmunoassays for six important benzodiazepines have been outlined. There are a number of points, however, which should be emphasized:

1. As is well known in the art of developing an RIA, the specificity of the antiserum is dependent on the site through which the hapten is covalently coupled to the protein. In the case of the 1,4-benzodiazepines, it is apparent that considerable specificity can be achieved by suitable choice of attachment of haptens through either the 3-,7-,3'- or 4'-positions. Although these positions are by no means unique, they are open to considerable chemical manipulation by those familiar with the field of benzodiazepine chemistry. Although absolute specificity of an antiserum toward a single compound cannot be achieved, it is possible to obtain an antiserum against a benzodiazepine which can be used for a specific analytical task. Ideally, the specificity should be such that the drug can be quantitated directly in the biological sample without interference from its metabolites and/or other drugs which may be present in the sample. However, as in the case of the flurazepam RIA, such idealistic simplicity is not always achieved and some compromise must be made in the form of extraction, and possibly, chromatographic isolation of the compound to be determined. Selective extraction procedures, by suitable choice of pH and solvent, can be used to considerable advantage with certain drugs.[28,29] Re-

---

[28] R. Lucek and R. Dixon, *Res. Commun. Chem. Pathol. Pharmacol.* **18**, 125 (1977).

[29] R. Dixon, J. J. Carbone, E. Mohacsi, and C. Perry, *Res. Commun. Chem. Pathol. Pharmacol.* **22**, 243 (1978).

gardless, however, of the actual RIA procedure which is finally chosen for analysis of unknown biological samples, it is imperative that every effort is made to validate the specificity (or nonspecificity as often the case may be) of the RIA by comparison with a physico-chemical procedure if such a procedure is available. Total reliance on cross-reactivity data as evidence for assay specificity in the analysis of unknown samples can sometimes lead to erronous results. Such evidence is based on the assumption that one knows all of the metabolites and/or other substances (drugs) which are present in the sample to be analyzed. This assumption does not always hold true.

2. All of the previously described RIA procedures have used precipitation with either saturated ammonium sulfate or polyethylene glycol-6000 for the separation of antibody-bound and free radioligand. Although these reagents are by no means the only way to achieve such separations[2] they have been purposely employed in preference to the commonly used dextran-coated charcoal (DCC) reagent. In contrast to the use of DCC, the precipitation techniques are not influenced to any great extent by time and temperature and have proved to be more reliable when processing large numbers of samples.

3. It will be noted that all of the RIA procedures utilize tritium labelled radioligands. Tritium is used in preference to ^{125}I-labeled ligands for a number of reasons: (a) Compared to ^{125}I with a half-life of 60 days, ^{3}H has a half-life of 12 years. Since many of the assays described are used intermittently over a period of several months or even years, the ^{3}H-labeled ligands are much better suited to these purposes since the necessity to resynthesize ^{125}I-labeled ligands is avoided. (b) In assays which require extraction and/or chromatographic procedures, the ^{3}H-labeled ligands, since they behave similarly to the unlabeled drug, may be used to monitor recovery and chromatographic elution profiles. ^{125}I-labeled ligands cannot be used for these purposes. (c) The hazards associated with exposure to γ-radiation, particularly during iodination with ^{125}I, are avoided with the use of ^{3}H.

Radioimmunoassays for the hypnotics flunitrazepam[30] and triazolam[31] have also been reported.

Acknowledgments

The author is grateful for the outstanding cooperation he has received from Drs. Ian Fryer, Robert Ning, and Mr. James Earley for the synthesis of the haptens and from Drs. Arnold Liebman and Clark Perry for the synthesis of the radioligands. The secretarial assistance of Mrs. Peggy Althoff in the preparation of this manuscript is much appreciated.

[30] R. Dixon, W. Glover, and J. Earley, *J. Pharm. Sci.* **70**, 230 (1981).
[31] H. Ko, M. E. Royer, J. B. Hester, and K. T. Johnston, *Anal. Lett.* **10**, 1019 (1977).

[36] Radioimmunoassay of N-Substituted Phenylpiperidine Carboxylic Acid Esters and Dealkylated Metabolites

By DAVID S. FREEMAN and HILDA B. GJIKA

The development of radioimmunoassays (RIAs) for pharmaceutically important compounds can facilitate studies on their bioavailability, pharmacokinetic parameters, metabolism, and disposition. The N-substituted 4-phenylpiperidine-4-carboxylic acid esters (**I**) constitute an important series of chemical structures which include many useful therapeutic agents belonging to the myriad of synthetic narcotic analgesic compounds. These synthetic derivatives differ in their piperidine ring N-substituents and include the morphine surrogates meperidine (**Ia**), anileridine (**Ib**), and piminodine (**Ic**), and the antidiarrheal agent, diphenoxylate (**Id**).

In the development of RIAs useful for studying these phenylpiperidine drugs, structurally similar derivatives are used as haptens for antisera production. These relatively low-molecular-weight haptens must be covalently attached to a macromolecule for antigenic activity. Depending on the design for synthesis of these macromolecular conjugates, the antisera produced after animal immunization can have differing and predictable specificites. Thus, when the phenylpiperidine nucleus was conjugated to a macromolecule through its heterocyclic ring nitrogen, the exposed 4-phenylpiperidine-4-carboxylic acid ester portion of the molecule was the major antigenic determinant of the hapten.[1] The antiserum obtained after

[1] H. Van Vunakis, D. S. Freeman, and H. B. Gjika, *Res. Commun. Chem. Pathol. Pharmacol.* **12**, 379 (1975).

immunization with this conjugate was used in RIAs for many of the phenylpiperidine drugs with the 4-phenyl, 4-carboxylic acid ester substituents.[1] However, when the phenylpiperidine nucleus containing an unsubstituted ring nitrogen was conjugated to a macromolecule using the 4-carboxyl group, the major antigenic determinant of the hapten was now the unsubstituted phenylpiperidine ring.[2] The antiserum obtained with this conjugate was highly specific and therefore employed in RIAs for normeperidine (II), which is the N-dealkylated metabolite of the phenylpiperidine drugs.

The separating capabilities of high pressure liquid chromatography (HPLC) can be used to supplement and confirm the results of drug metabolism studies performed by RIA.[2-4] The collected HPLC fractions from a chromatographed biological extract can be analyzed by RIA both to detect the presence of the drug metabolite and to quantify its levels. The sensitivity and specificity of RIA and the separation capabilities of HPLC assure the identification and quantification of the metabolite.

Preparation of Macromolecular Conjugates for Immunization

Anileridine –Protein Conjugate

In order to synthesize an antigen capable of producing antiserum with specificity for the 4-phenylpiperidine-4-carboxylic acid ester fragment, advantage is taken of the N-(p-aminophenylethyl) substituent of commercially available anileridine (Ib). Anileridine (Ib) is conjugated to the protein through this substituent to form the antigen for immunization.

1. One milliliter of anileridine injection solution (Leritine, Merck, Sharp and Dohme, 25 mg/ml) is made alkaline with a few drops of 10 N NaOH and the free base extracted with three 5 ml portions of chloroform. The chloroform extract is dried with anhydrous Na_2SO_4 and the solvent then removed with a stream of N_2 at room temperature.

2. The free base is dissolved in 1.0 ml of anhydrous pyridine, 50 mg of succinic anhydride added, and the reaction mixture stirred overnight at room temperature. After cooling in ice, the white precipitate is collected by centrifugation and dried with a stream of N_2. Thin-layer chromatography (TLC) on silica gel eluted with ethyl acetate : methanol : acetic acid (90 : 10 : 1) shows anileridine (Ib), R_f = 0.85 and N-succinylanileridine, R_f = 0.11.

[2] D. S. Freeman, H. B. Gjika, and H. Van Vunakis, *J. Pharmacol. Exp. Ther.* **203**, 203 (1977).

[3] J. J. Langone and H. Van Vunakis, *Biochem. Med.* **12**, 283 (1975).

[4] L. J. Riceberg and H. Van Vunakis, *J. Pharmacol. Exp. Ther.* **206**, 158 (1978).

3. The N-succinylanileridine white solid is dissolved in 1.0 ml of dimethylformamide (DMF), and to this solution 21 mg of N-hydroxysuccinimide and 18 mg of N,N'-dicyclohexylcarbodiimide (DCCD) are added. The reaction mixture is stirred at room temperature for 30 min. The precipitate is removed by centrifugation and aliquots of the supernatant (containing the "activated" N-succinylanileridine) are used to prepare the protein conjugate for immunization and the tyramine derivative for ^{125}I labeling.

4. A 0.2-ml aliquot of the "activated" supernatant liquid is added to a 4° solution consisting of 60 mg of hemocyanin, 4.0 ml of 0.1 N NaHCO$_3$, and 2.0 ml of DMF. The reaction solution is stirred at 4° for 2 hr and the resulting hemocyanin conjugate separated using a Sephadex G-50M column (33 × 1.9 cm; elution buffer, 0.15 M NaCl, 0.005 M phosphate, pH 7.0). The fractions containing the anileridine–hemocyanin conjugate are pooled.

Normeperidinic Acid–Protein Conjugate

The synthesis of an antigen which can be used to obtain antiserum specific for the drug metabolite normeperidine (II) requires protein conjugation of the normeperidine hapten away from its unsubstituted heterocyclic nitrogen. Therefore, the 4-carboxylic acid ester substituent of normeperidine (II) is hydrolyzed and the resulting carboxylic acid is used for amide linkage to the protein.

1. The carboxylic acid derivative of normeperidine (II) is prepared by gently refluxing overnight 25 mg of normeperidine·HCl in 1.0 ml of 1 N HCl. The hydrochloride salt of normeperidinic acid is collected after lyophilization of the reaction solution, mp 250–255° dec.

2. A 15-mg sample of normeperidinic acid and 10 mg of bovine serum albumin (BSA) are dissolved in 2.0 ml of H$_2$O adjusted to pH 7.0. The conjugation reaction is started by the addition of 20 mg of 1-ethyl-3-(3-dimethylaminopropyl)-carbodiimide·HCl. The reaction solution is stirred overnight at room temperature.

3. Purification of the BSA–conjugate is accomplished by dialysis vs 0.15 M NaCl, 0.005 M phosphate buffer, pH 7.5, at 4° for 3 days with daily changes of buffer.

The type of macromolecule used for hapten conjugation is somewhat arbitrary. Although the use of hemocyanin and BSA provided macromolecular conjugates very effective for antisera production,[1,2] other macromolecules, such as human serum albumin or polylysine, may be equally as effective when used for hapten conjugation.

Immunization Procedure and Antisera Collection

Anileridine–Hemocyanin Conjugate

This conjugate is used for the immunization of a goat. A volume of the pooled, chromatographic fractions containing 5 mg of conjugate is emulsified with an equal volume of complete Freund's adjuvant. The goat is given an intramuscular (im) injection of this emulsion followed a month later by two consecutive daily im injections of the Freund's adjuvant emulsion each containing 10 mg of conjugate.

The goat is bled 3, 7, and 10 days after the final injection. The collected blood samples (40–60 ml) are allowed to clot and after centrifugation, the antisera are separated and stored frozen.

Normeperidinic Acid–BSA Conjugate

This conjugate is used for immunizing two rabbits. A 0.5-ml aliquot of conjugate dialyzate solution is emulsified with 2.0 ml of complete Freund's adjuvant. Each rabbit receives approximately 1 ml of this emulsion injected in small portions between eight of the rabbit's toe pads. The remaining emulsion is administered im into the rabbit's thigh muscles. All injection sites are carefully shaven and cleaned with alcohol before immunization. The injections are repeated two more times at 1-week intervals using the thigh muscles as the only site of administration.

Three weeks after the last injection, the rabbits are bled once on each succeeding week. During each bleeding a maximum of 40 ml of blood is collected from the rabbit's marginal ear vein. The blood samples are allowed to clot and centrifuged, and the separated antisera stored frozen.

The above procedures indicate that only a few animals are used for antiserum production. However, since the antigenic response of animals is unpredictable, several animals were immunized with each of the conjugates in order to increase the probability of obtaining antiserum with the best binding and specificity characteristics.[1,2] Immunization of the goat and rabbits provided the best antiserum samples for the phenylpiperidine drug and normeperidine RIAs, respectively.

[125]I Labeling of Hapten Derivatives

[125]I-Labeled N-Succinylanileridine Derivative

The radiolabeled ligand used in the RIA for phenylpiperidine drugs is produced by iodination of tyramyl-*N*-succinylanileridine. This tyramine derivative is synthesized utilizing the "activated" *N*-succinylanileridine used for protein conjugate formation.

1. A 0.2-ml aliquot of the "activated" *N*-succinylanileridine solution is added to a solution consisting of 8 mg of tyramine dissolved in 1 ml of DMF. The reaction solution is stirred for 2 hr at room temperature.

2. To verify formation of the tyramine derivative, TLC can be performed using silica gel and ethyl acetate : methanol : acetic acid (90 : 10 : 1). The tyramyl-*N*-succinylanileridine (R_f = 0.20) is colored upon treatment with β-nitrosonaphthol (presence of phenol moiety) and negative with ninhydrin treatment (absence of primary or secondary amino group).

3. The tyramine derivative is isolated and purified by preparative TLC on silica gel using the same solvent system. The product is eluted from the silica gel with ethanol and the concentration of tyramyl-*N*-succinylanileridine is determined by UV spectroscopy using the 280 nm peak of the phenolic moiety.

4. Five micrograms of this product is iodinated with ^{125}I by reaction with Na^{125}I and chloramine-T as described by Greenwood *et al.*[5]

5. The radiolabeled product is isolated and purified by applying the iodination reaction mixture to a Sephadex G-10 column (12 × 1.5 cm) and eluting with buffer (1.5 M NaCl, 0.05 M phosphate, pH 7.5, 0.2% gelatin) to remove unreacted radioactivity and other unwanted agents. The ^{125}I-labeled tyramyl-*N*-succinylanileridine is eluted and collected from the column using DMF:0.3 M NaCl (1 : 1).

^{125}I-Labeled Normeperidinic Acid Derivative

The radiolabeled ligand to be used in the RIA for the N-dealkylated metabolite of the phenylpiperidine drugs is produced by iodination of a tyramine derivative of normeperidinic acid. The procedure is very similar to that used for the radiolabeling of anileridine except that in this case tyramine is directly coupled by amide formation to the 4-carboxylate substituent of normeperidinic acid.

1. A 5-mg sample of normeperidinic acid dissolved in 0.2 ml of DMF is mixed with 3 mg of DCCD and 3.5 mg of *N*-hydroxysuccinimide. The reaction mixture is stirred for 30 min at room temperature and the precipitate which forms is removed by centrifugation.

2. The supernatant containing the "activated" normeperidinic acid is added to a solution of 5 mg of tyramine·HCl dissolved in 0.1 ml of 0.1 M NaHCO$_3$. This reaction solution is stirred well and then allowed to stand overnight at 4°.

3. To identify the tyramine derivative, TLC is performed using silica

[5] F. C. Greenwood, W. M. Hunter, and J. S. Glover, *Biochem. J.* **89**, 114 (1963).

gel and ethyl acetate:methanol:acetic acid (85:10:5). This tyramine amide of normeperidinic acid (R_f = 0.08) is positive upon spraying with β-nitrosonaphthol (presence of phenol moiety) and gives an orange color upon ninhydrin treatment (presence of heterocyclic secondary amine).

4. The tyramine derivative is isolated and purified by preparative TLC on silica gel using the same solvent system. The product is eluted from the silica gel with ethanol and the concentration is determined from the 275 nm peak in the UV spectrum due to the phenolic moiety.

5. Twelve micrograms of this product is iodinated with ^{125}I, isolated and purified in an identical fashion as described previously for tyramyl-N-succinylanileridine.

Radioimmunoassay Procedures

In order to select the best antisera for establishing an RIA, each sample of antiserum collected from the various bleedings needs to be tested for its binding of the appropriate radiolabeled hapten derivative and for its specificity.

Various dilutions of each antiserum sample are tested for maximum binding of its respective ^{125}I-labeled tyramyl compound by using no competitive binding compounds or standards in the RIA mixture. The samples which show the greatest binding are then used for the specificity studies. These studies involve a series of RIAs using varying concentrations of each of the possible compounds which may compete for the antibody binding sites. Standard curves are then constructed for each RIA competitor. The sample of antiserum having the best binding and specificity characteristics from the animal immunized with the anileridine-hemocyanin conjugate is used for the phenylpiperidine drug RIAs. The RIAs for the metabolite, normeperidine (II), use the best antiserum sample isolated from the animal immunized with the normeperidinic acid–BSA conjugate.

The general RIA procedure utilizing the double antibody technique to separate free and antibody-bound ^{125}I-labeled compound is outlined below:

1. All dilutions are made in Isogel–Tris buffer (0.15 M NaCl, 0.01 M Tris, 0.1% gelatin, pH 7.4).

2. The ^{125}I-labeled tyramyl derivative is diluted in buffer to give approximately 20,000 cpm/0.1 ml.

3. In order to provide a good competitive binding assay, an antiserum dilution that binds approximately 25–35% of the added ^{125}I-labeled tyramyl derivative is used in the RIA mixtures.

4. The RIA incubation mixtures are prepared in disposable glass cul-

ture tubes (10 × 75 mm) and all determinations are performed in duplicate.

5. The background or nonspecific binding RIA tubes each contain 0.1 ml of buffer and 0.1 ml of normal animal serum diluted in buffer to the same extent as the diluted RIA antiserum.

6. The total binding RIA tubes each contain 0.1 ml of buffer and 0.1 ml of diluted RIA antiserum.

7. To the other assay tubes are added 0.1 ml of serially diluted competitor standard (for constructing standard curves) or 0.1 ml of sample to be assayed (diluted plasma, enzyme digest or a chromatographed fraction) and then each assay tube receives 0.1 ml of diluted RIA antiserum.

8. A 0.1 ml portion of the diluted ^{125}I-labeled tyramyl derivative is added to all the tubes.

9. After mixing well, the RIA tubes are incubated for 1 hr at 37°.

10. If goat antiserum is used for the RIA, the immune complexes are precipitated by adding, to each assay tube, 0.1 ml of diluted normal goat serum (1 : 100) and 0.1 ml of a titered excess of rabbit anti-goat γ-globulin serum (double or second antibody). With the rabbit antiserum, 0.1 ml of diluted normal rabbit serum (1 : 50 or 1 : 100) and 0.1 ml of goat anti-rabbit γ-globulin serum is used for precipitation.

11. After mixing well, the RIA tubes are allowed to stand overnight at 4°.

12. The immune precipitates are collected by centrifugation, the supernatant solutions carefully decanted, and the inside walls of the RIA tubes wiped with folded strips of filter paper.

13. Radioactivity associated with each of the pellets is determined using a gamma counter.

14. The percentage inhibition of binding between the ^{125}I-labeled tyramyl derivative and the antiserum by competing nonlabeled compound is calculated for the standards and biological samples using the average cpm for each of the duplicates. In these calculations, the background cpm (BG) is substracted from all of the other RIA counts and 0% inhibition is defined as the total binding cpm (TB) minus BG.

$$\% \text{ Inhibition} = 100 - \left[\frac{(\text{cpm of RIA precipitate}) - \text{BG}}{\text{TB} - \text{BG}} \times 100 \right]$$

Using the calculated values for the standards, a standard curve is constructed and used for determining the amount of drug or metabolite contained in the biological samples.

If a volume of plasma larger than 0.01 ml per assay tube or an undiluted plasma sample needs to be analyzed, the RIA procedure is modified by adding 1.0 ml of the Isogel–Tris buffer to each of the assay tubes at the beginning of the RIA and by using a smaller dilution of antiserum (higher titer of antibody). This larger volume RIA procedure reduces the background cpm when larger volumes of plasma are analyzed and allows very low drug or metabolite levels to be quantified in plasma samples.

Applications of the Phenylpiperidine Drug and Normeperidine RIAs

Determination of Drug and Metabolite Plasma Levels

The utility of RIA for determining plasma levels of drugs or their metabolites is due to the high binding specificity of the RIA antiserum and the high specific radioactivity of the radiolabeled ligand. The specific binding of the antiserum allows analysis to be performed in the presence of impurities or weak drug associating molecules (e.g., albumin). The high specific activity of the radiolabeled competitor in the RIA enhances the sensitivity of the analysis. Thus, plasma samples can usually be directly analyzed by RIA avoiding extraction or other purification techniques.

Combining the phenylpiperidine drug RIA and normeperidine RIA allows both drug and a major metabolite of that drug to be quantified in plasma samples. Since the design of the antigen used for antiserum production for the phenylpiperidine drug RIA does not emphasize the type of substituent on the heterocyclic nitrogen of these compounds, this antiserum binds several drugs among this class of compounds [i.e., meperidine (**Ia**), anileridine (**Ib**), piminodine (**Ic**), and diphenoxylate (**Id**)].[1] Therefore, this RIA can be used for quantifying plasma drug levels for several of the phenylpiperidine agents as long as they are not administered in combination or succession of one another. This RIA has been used for determining levels of the analgesics meperidine (**Ia**) and anileridine (**Ib**) in plasma and red blood cell samples.[1] The antiserum showed very little cross-reactivity with other important narcotic analgesics such as morphine and methadone.

The antiserum used in the phenylpiperidine drug RIA binds the N-dealkylated metabolite, normeperidine (**II**). However, the best quantification of this metabolite is provided using the antiserum specifically produced to be most sensitive to the lack of an N-substituent on the phenylpiperidine moiety. This antiserum, when used in the normeperidine RIA, allows quantification of this N-dealkylated metabolite in plasma samples even though relatively large amounts of the parent phenylpiperidine drug is present.[2] The combination of the phenylpiperidine drug and normeperidine RIAs provided an accurate determination of drug and metabolite levels in plasma samples collected from mothers and newly born infants when meperidine (**Ia**) was administered as an obstetric analgesic.[2]

In Vitro Drug Metabolism

The high specificity of the normeperidine antiserum enables the RIA procedure to simplify drug metabolism studies on the N-dealkylation of several of the phenylpiperidine drugs. The normeperidine RIA is capable of quantifying levels of this metabolite in liver microsomal enzyme assays without purification steps or removal of protein. Aliquots of the enzymatic incubation mixtures need only to be diluted before analysis by RIA. Since the antiserum specifically binds with the common fragment from the enzymatic N-dealkylation of many of the N-substituted phenylpiperidine drugs, the normeperidine RIA can be used to study the metabolism of meperidine (**Ia**), anileridine (**Ib**), and other congeners among this chemical class of agents.[2]

The general enzymatic assay procedure is as follows:

1. In a 25-ml Erlenmeyer flask is placed 1.5 ml of 0.5 M Na^+/K^+ phosphate buffer, pH 7.4, containing 0.3 mg NADP, 1.5 mg glucose 6-phosphate, and 3 mg nicotinamide.

2. This incubation flask is placed in a shaking waterbath at 37°.

3. A 0.5 ml sample of phenylpiperidine drug diluted in 25 mM $MgCl_2$ [0.6–2.4 mM for meperidine (**Ia**), 0.06–1.0 mM for anileridine (**Ib**)] is added to the incubation flask.

4. After temperature equilibration, the metabolic reaction is started by the addition of 1.0 ml of liver homogenate (the 10,000 g supernatant from liver homogenized in 7 volumes of cold 1.15% KCl, 0.01 M Na^+/K^+ phosphate buffer, pH 7.4).

5. Aliquots of 50 μl are collected from the incubation mixture immediately after addition of the liver homogenate (zero time) and at various times thereafter.

6. Each aliquot is diluted with 2.0 ml of cold (0°) Isogel–Tris buffer immediately after collection and is maintained at 0° until analyzed.

7. The level of normeperidine (**II**) is directly determined in each aliquot using the normeperidine RIA.

The standard curve for the normeperidine RIA should be determined under the same conditions as used for the enzyme digest samples. Therefore, the RIA standard curves are determined by adding known amounts of normeperidine (**II**) to the aliquot collected at zero time which has been diluted to the same extent as for the enzyme digest samples.

Complementing RIA with HPLC

RIA may not unequivocally quantify levels of a drug or metabolite in a biological sample because of the possible formation of metabolites which cross-react with the antiserum. The specific compound being analyzed for by the RIA will then be measured along with these cross-reacting compounds. However, if an initial HPLC procedure is used for separation, the drug and each of the cross-reacting metabolites can be individually quantified with the RIA.[3,4] The RIA-HPLC technique enables very low amounts of drug or metabolite to be quantified in chromatographic fractions and is an important check for establishing whether all the serological reactivity detected by RIA in plasma or enzymatic digests is due only to the compound being measured rather than to unidentified metabolites or other cross-reactants.

A general description of the RIA-HPLC procedure, using the microsomal enzyme assay for normeperidine (II) formation as an example, is as follows[2]:

1. A 1.0-ml aliquot of the enzymatic incubation mixture is made alkaline (pH 11) with a NaOH solution and extracted 4 times with 1.0-ml portions of benzene.

2. Any emulsion formed during the extraction step is broken by saturating the aqueous phase with NaCl and centrifuging to separate the phases.

3. The solvent is removed from the combined benzene extract and the residue redissolved in 0.5 ml of methanol.

4. A 50-μl aliquot of the methanol solution is subjected to HPLC using an analytical diphenyl corasil column (2 feet × $\frac{1}{8}$ in.) and a mobile phase of acetonitrile–0.1% aqueous $(NH_4)_2CO_3$ (1:1) at a flow rate of 1 ml/min.

5. One milliliter fractions are collected immediately after injection and 0.1 ml of Isogel–Tris buffer added to each fraction. The Isogel–Tris buffer is used to minimize the loss of small amounts of metabolite by adsorption to the surface of the fraction collection tubes.

6. After mixing, the fractions are lyophilized to dryness and reconstituted with Isogel–Tris buffer.

7. The content of normeperidine (II) in each fraction is determined using the normeperidine RIA.

If the concentration of the parent drug or normeperidine (II) is great enough in the enzymatic incubation mixture, the HPLC elution pattern can be followed with the UV detector (254 nm). However, the normeperidine (II) present in the incubation mixtures at early time points is too low to be detected spectrophotometrically. The UV detector is useful in determining HPLC retention times with standards of the parent drug and normeperidine (II), but when extracts of enzymatic digests or plasma samples are chromatographed, RIA could quantify the low levels of metabolite eluted from the HPLC analytical column.

[37] Radioimmunoassay for Fentanyl

By ROBERT P. SCHLEIMER

Fentanyl is a synthetic narcotic analgesic with a potency about 150 times that of morphine. It is being used widely in the clinic since it has the advantage of a rapid onset and short duration of action. As a result of fentanyl's potency, very low doses are used (often less than 1 μg per person), consequently detection of the drug in body fluids and tissues is difficult. Radioimmunoassay provides a means by which the specificity and affinity of binding of antibodies with molecules can be exploited to enable the measurement of minute quantities of the molecules in fluids and tissues containing them. This report describes the development of a radioimmunoassay capable of detecting fentanyl at concentrations of less than 1 pmol/ml in samples of serum, urine, or cerebrospinal fluid.

Preparation of Immunogen

For purposes of the conjugation of fentanyl to BGG, carboxyfentanyl [N-phenyl-N-4-(1-(β-phenethyl)piperidine)succinimic acid] is prepared by the reaction scheme outlined in Fig. 1.[1] Five grams of 1-(β-phenethyl)-4-piperidone are dissolved in 50 ml of toluene, and mixed with 3.1 ml of freshly distilled aniline in a 100-ml round bottom flask fitted with a Dien-Stark trap. The solution is brought to reflux under an atmosphere of nitrogen. After 15 min, a catalytic amount of p-toluenesulfonic acid is added. The mixture is refluxed for 20 hr and then reduced in volume under a stream of nitrogen. The product, 1-(β-phenethyl)-4-piperidylidene aniline, is then dissolved in 50 ml of isopropanol and refluxed for 10 hr with 4 equivalents of sodium borohydride. The mixture is then cooled, and 30 ml of 7% NaHCO$_3$ is added. This mixture is then extracted with benzene, and purified. One gram of the resulting product [4-anilino-1-(β-phenethyl)piperidine] is dissolved in 50 ml of benzene and reacted with 0.36 g of succinic anhydride by refluxing for 40 hr under an atmosphere of nitrogen. The product, carboxyfentanyl, is crystallized from a benzene-hexane (3:1) system. ^{14}C-labeled carboxyfentanyl is prepared in the same manner using ^{14}C-labeled succinic anhydride. Conjugation of carboxyfentanyl to the carrier, bovine γ-globulin (BGG), is achieved by incubation of 50 mg carboxyfentanyl (containing a trace of ^{14}C internally labeled carboxyfentanyl) with 100 mg of BGG and 200 mg of 1-ethyl-3-(3-dimethylaminopro-

---

[1] G. L. Henderson, J. Frincke, C. Y. Leung, M. Torten, and E. Benjamini, *J. Pharmacol. Exp. Ther.* **192**, 489 (1975).

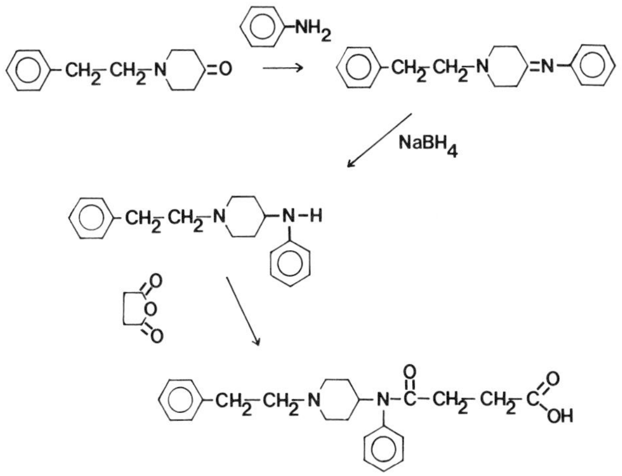

FIG. 1. Reaction scheme for the synthesis of carboxyfentanyl.

pyl)carbodiimide in 22 ml of phosphate-buffered saline, pH 7.4. Conjugate prepared by this procedure should contain approximately 12 mol of carboxyfentanyl per mole of BGG as determined by the radioactivity of the conjugate.

Preparation of Antiserum

Rabbits, or other suitable animals are immunized with 2 mg of the carboxyfentanyl–BGG conjugate emulsified in Freund's complete adjuvant. Injections are given in the footpad and subcutaneously in the back. A boosting injection is given 14 days later. Blood is collected 14 days after the second immunization, and serum is collected. Serum is kept frozen at $-20°$.

Radioimmunoassay

For the purpose of radioimmunoassay, radiolabeled fentanyl must be obtained. ^{3}H-labeled fentanyl with a specific activity of 10 Ci/mmol used in the procedure described herein was a gift of Janssen Pharmaceutica, Belgium. For maximal sensitivity in any radioimmunoassay, high specific activity of the radiolabeled ligand is essential. With a radiolabeled ligand of a given specific activity, maximal sensitivity of the radioimmunoassay will be achieved by using the lowest amount of ligand which will give binding sufficiently detectable to allow determination of binding inhibition by unlabeled ligand. A quantity of ^{3}H fentanyl which will yield about 2000 counts per minute is ideal. Measurement of the binding of ^{3}H-labeled fentanyl to anti-fentanyl is best carried out using the Farr assay (ammonium sulfate technique).[2] This method is based on the precipitation of antibodies with ammonium sulfate. Fentanyl is not precipitated by ammonium sulfate unless it is bound to the antibodies. The serum (0.1 ml), ^{3}H-labeled fentanyl (0.1 ml), test sample (0.3 ml), and BBS (0.1 ml) are added to a 10 × 75-mm test tube, and incubated at room temperature for 1 hr. Globulins and fentanyl bound to globulins are then precipitated by the addition of an equal volume (0.6 ml) of saturated ammonium sulfate. The precipitated proteins are then pelleted by centrifugation at 1300 g for at least 10 min. An aliquot (e.g., 0.6 ml) of each supernatant is then counted in PCS (or other) scintillation fluid in a scintillation counter. Specific binding of ^{3}H-labeled fentanyl to anti-fentanyl is calculated using Eq. (1):

Percentage specifically bound

$$= \left(1 - \frac{\text{cpm anti-fentanyl serum}}{\text{cpm control serum}}\right) \times 100 \qquad (1)$$

For highest sensitivity the radioimmunoassay is best carried out in a system where anti-fentanyl and ^{3}H-labeled fentanyl are proportioned such that approximately 50% of the ^{3}H-labeled fentanyl is specifically bound as determined by Eq. (1). In order to determine the amount of antiserum appropriate to bind 50% of the ^{3}H-labeled fentanyl (i.e., about 1000 of the added 2000 cpm), dilutions of the serum must be made and tested. The serum can be diluted in 20% normal sheep (or other) serum in borate-buffered saline (BBS—0.13 M NaCl, 0.113 M H$_3$BO$_3$, 0.013 M Na$_2$B$_4$O$_7$, pH 8). It is necessary to dilute the serum in 20% normal serum to maintain a protein concentration sufficiently high to allow satisfactory precipitation by the ammonium sulfate. Shown in Fig. 2 are data from an experiment in which dilutions of anti-fentanyl (expressed in μl) were tested for the ability to bind approximately 130 pg of ^{3}H-labeled fentanyl. As can be seen, a dilution of 1:5000 (0.02 μl) bound slightly greater than 60% of the ^{3}H-labeled fentanyl. This dilution of antiserum was chosen for radioimmunoassay.

Once the anti-fentanyl and ^{3}H-labeled fentanyl system is balanced (in this case a 1:5000 dilution of anti-fentanyl and approximately 2000 cpm of ^{3}H-labeled fentanyl), it can be used for radioimmunoassay and to test specificity. Both procedures are based on inhibition of the binding of ^{3}H-labeled fentanyl to anti-fentanyl.

For radioimmunoassay, or to test the specificity of the anti-fentanyl antibody, fentanyl or various compounds suspected of fentanyl activity

[2] R. S. Farr, *J. Infect. Dis.* **103**, 239 (1958).

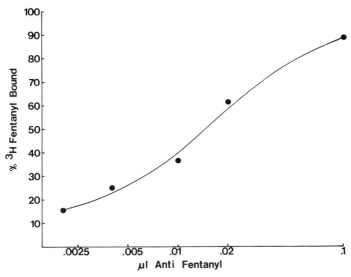

FIG. 2. Titration of anti-fentanyl serum.

are tested for their capacity to inhibit the reaction between ³H-labeled fentanyl and anti-fentanyl. For this purpose, the test compounds ("inhibitors") are added, at varied concentrations, to test tubes containing the anti-fentanyl and ³H-labeled fentanyl. The inhibitors are diluted in BBS, and the total volume of the reaction mixtures containing inhibitor is the same as those without, 0.6 ml. Inhibition of binding of ³H-labeled fentanyl and anti-fentanyl by the test compound indicates that the test compound is capable of binding with antifentanyl, and thus has immunochemical similarity to fentanyl. Inhibition of binding can be calculated by first determining the percentage binding in tubes containing inhibitor and tubes not containing inhibitor and then using Eq. (2):

Percentage inhibition
$$= \left(1 - \frac{\% \text{ bound in presence of inhibitor}}{\% \text{ bound in absence of inhibitor}} \right) \times 100 \qquad (2)$$

In Fig. 3 are data showing the inhibition of binding of ³H-labeled fentanyl to anti-fentanyl by a range of concentrations of unlabeled fentanyl (the "standard curve"). When plotted as percentage inhibition versus the log of the concentration of inhibitor (unlabeled fentanyl), the data appear as a straight line between 15 and 85% inhibition. The ability of fentanyl to inhibit the binding reaction between ³H-labeled fentanyl and anti-fentanyl is used to determine the content of fentanyl in a sample containing an unknown amount of fentanyl. The sample, or dilutions of the sample, is

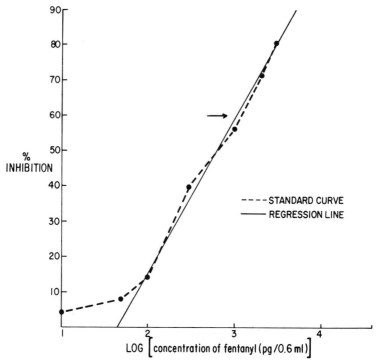

FıG. 3. Standard inhibition curve for radioimmunoassay of fentanyl.

added to tubes containing the ³H-labeled fentanyl and anti-fentanyl. If the sample does contain fentanyl, the binding reaction will be inhibited to an extent dependent on the quantity of fentanyl it contains. The fentanyl content of the sample is inferred by a comparison of the inhibition of binding by the sample to the inhibition by known quantities of fentanyl plotted in the standard curve. In order to facilitate this procedure, the standard curve is fitted with a regression line (Fig. 3) using only points in which inhibition is between about 15 and 85% inhibition. (A standard curve must be performed in each assay, although in any given assay, many different samples at various dilutions can be tested.) If, for example, in the assay in which Fig. 3 was the standard curve, a sample diluted 1 to 10 inhibited the binding reaction 60% (see arrow), it is inferred that the content of fentanyl in the diluted sample is 1000 pg/0.3 ml by finding the concentration coordinate of the point on the standard curve regression line that corresponds with an inhibition of 60%. The concentration of fentanyl in the sample is expressed per 0.3 ml since only 0.3 ml of the total reaction mixture is sample. The fentanyl content of the undiluted sample, of course, would be ten times that value (10 ng/0.3 ml) since the sample had been diluted 10-

fold. It is usually necessary to dilute samples (in BBS) in order to reduce the concentration of fentanyl they contain to a level which will inhibit the binding reaction less than 85%. The extent of dilution required to achieve this will vary, and experience with the particular type of sample (e.g., serum 2 hr after fentanyl injection) will aid in the selection of dilutions to be tested. Generally, 2-fold or 3-fold dilution series (e.g., 1:20, 1:40, 1:80, or 1:30, 1:90, 1:270, etc.) are the most successful in both reaching the proper dilution without over diluting, and conserving reagents. It is best to do all determinations (such as dilutions of samples, and points on the standard curve) in triplicate or at least duplicate. Variation among replicate determinations as a rule should not exceed 10%.

For a radioimmunoassay to have any usefulness, it must be specific for the test compound. It is necessary therefore to demonstrate that only fentanyl, and not a multitude of other compounds or drugs found in sample fluids, will inhibit the binding reaction between ^{3}H-labeled fentanyl and anti-fentanyl. Various compounds should be tested for their ability to inhibit the reaction when present at varied concentrations. In general, only compounds chemically related to the test compound will "cross react," and inhibit the binding reaction. The ability of compounds to inhibit the binding reaction is expressed relative to the ability of fentanyl to do so. This comparison is made using the concentrations necessary to achieve a 50% inhibition of binding. Thus, a compound which requires a concentration 100 times that of fentanyl to inhibit the reaction by 50% has a relative binding of 0.01. Shown in the table are such comparisons made with many

INHIBITION OF THE BINDING OF ^{3}H-LABELED
FENTANYL AND ANTI-FENTANYL BY VARIOUS
TEST COMPOUNDS

Compound	Relative activity[a]
Fentanyl	1.0
Carboxyfentanyl	1.0
4-Anilino-1-(β-phenethyl)piperidine	0.01
1-(β-Phenethyl)piperidone	0.001
Droperidol	10^{-6}
Morphine	0
Meperidine	0
Naloxone	0
Procaine	0
Diphenhydramine	0

[a] Relative ability to inhibit by 50% the reaction between ^{3}H-labeled fentanyl and anti-fentanyl. Values are relative to fentanyl which is valued at 1.0.

different compounds. It can be seen that carboxyfentanyl had a relative binding equal to one, and is therefore as effective as fentanyl itself as an inhibitor. This is not surprising since carboxyfentanyl was used as the original immunizing hapten. It is also apparent from the table that 4-anilino-1-(β-phenethyl)piperidine and 1(β-phenethyl)piperidone had relative binding activities of 0.01 and 0.001, respectively. These compounds, which are respectively a fentanyl metabolite in rodents, and a synthetic precursor of fentanyl, may be expected to give a small amount of interference in the radioimmunoassay if present in samples. It is possible, of course, that as yet unknown fentanyl metabolites may also interfere with radioimmunoassay. Finally, a host of compounds structurally unrelated to fentanyl show no ability to inhibit the binding reaction (table).

Concluding Remarks

The procedure outlined here is satisfactory for the development of a radioimmunoassay suitable for the detection of fentanyl at levels found in biological fluids of patients receiving fentanyl in the normal clinical situation. With ^{3}H-labeled fentanyl of a specific activity of 10 Ci/mmol, the limit of detection of this procedure is about 100 pg of fentanyl per milliliter of fluid. In patients receiving single intravenous doses of fentanyl as low as 3 μg/kg, fentanyl was still detectable in serum samples up to 6 hours after injection.[3]

[3] R. Schleimer, E. Benjamini, J. Eisele, and G. Henderson, *Clin. Pharmacol. Ther.* **23**, 188–194 (1978).

[38] Radioimmunoassay of Haloperidol

By ROBERT T. RUBIN, BARBARA B. TOWER, SALLY E. HAYS, and RUSSELL E. POLAND

The treatment of psychiatric patients with neuroleptic (antipsychotic) drugs has been largely empiric, because of the lack, until recently, of readily available techniques for determining the pharmacokinetics of these compounds. Large interindividual differences in therapeutic response are common with the usual clinical doses of neuroleptics, related to the wide range of serum concentrations among patients treated with identical regimens of these drugs.[1] This has led to the measurement of

[1] T. B. Cooper, *Clin. Pharmcokin.* **3**, 14 (1978).

METHODS IN ENZYMOLOGY, VOL. 84

blood levels of antipsychotic medications as a rational guide for achieving maximum therapeutic response.[2]

Haloperidol, 4-[4-(p-chlorophenyl)-4-hydroxypiperidino]-4-fluorobutyrophenone, is a butyrophenone neuroleptic widely used in the treatment of psychoses. While many other butyrophenones with antipsychotic properties have been synthesized, and several are in clinical use in Europe and Asia, haloperidol is presently the only one of its class licensed for marketing in the United States. It is an effective antipsychotic compound and is widely used throughout the world.

A specific and sensitive gas chromatographic (GC) method for the measurement of haloperidol concentrations in body fluids and tissues has been available for several years,[3] and large scale clinical studies utilizing GC determination of serum haloperidol have been accomplished.[4,5] However, this analytical technique is time-consuming and expensive compared to other potential methods, and it requires specialized apparatus not available in many laboratories. Radioimmunoassays (RIAs), traditionally applied to large, inherently immunogenic polypeptide hormones,[6] have been adapted to smaller molecules including neuroleptic drugs.[7-12] Two RIAs for haloperidol also have been developed recently, using hapten–protein conjugates at different sites on the haloperidol molecule.[13,14] The specificities and cross-reactivities of the resultant anti-haloperidol sera differ accordingly. This chapter describes the development of these two RIAs and compares them with GC, as an index of the accuracy of the radioassay techniques for haloperidol measurement in serum samples from chronically medicated patients.

[2] J. Koch-Weser, *New Engl. J. Med.* **287**, 227 (1972).

[3] A. Forsman, E. Mårtensson, G. Nyberg, and R. Öhman, *Arch. Pharmacol.* **286**, 113 (1974).

[4] A. Forsman and R. Öhman, *Curr. Ther. Res.* **21**, 396 (1977).

[5] S. E. Ericksen, S. W. Hurt, S. Chang, *et al.*, *Psychopharm. Bull.* **14**, 15 (1978).

[6] R. S. Yalow, *Science* **200**, 1236 (1978).

[7] K. Kawashima, R. Dixon, and S. Spector, *Eur. J. Pharmacol.* **32**, 195 (1975).

[8] L. J. M. Michiels, J. J. P. Heykants, A. G. Knaeps, and P. A. J. Janssen, *Life Sci.* **16**, 937 (1975).

[9] J. D. Robinson and D. Risby, *Clin. Chem.* **23**, 2085 (1977).

[10] D. H. Wiles and M. Franklin, *Br. J. Clin. Pharmacol.* **5**, 265 (1978).

[11] J. W. Hubbard, K. K. Midha, I. J. McGilveray, and J. K. Cooper, *J. Pharmaceut. Sci.* **67**, 1563 (1978).

[12] K. K. Midha, J. C. K. Loo, J. W. Hubbard, M. L. Rowe, and I. J. McGilveray, *Clin. Chem.* **25**, 166 (1979).

[13] M. Michiels, R. Hendricks, and J. Heykants, Preclinical Research Report R1625/2, Janssen Pharmaceutica, Beerse, Belgium, 1976.

[14] B. R. Clark, B. B. Tower, and R. T. Rubin, *Life Sci.* **20**, 319 (1977).

Preparation of Immunogens

Figure 1 illustrates the molecular structure of haloperidol and the configurations of its conjugation to bovine serum albumin (BSA) for antibody production in the two RIAs, designated in this chapter as RIA-I[14] and RIA-II.[13] RIA-I used haloperidol conjugated to BSA at the ketone group (Fig. 1A), and RIA-II used haloperidol conjugated to BSA via the tertiary alcohol (Fig. 1B).

For RIA-I,[14] fatty acid-free BSA was coupled to excess hydrazine hydrate in neutral aqueous solution using water-soluble 1-cyclohexyl-3-(2-morpholinoethyl)carbodiimide metho-*p*-toluene sulfonate. After 24 hr, the reaction mixture was dialyzed against 1000 volumes of water. Haloperidol, labeled with tritium at position 3 of the 4-chlorophenyl moiety to a specific activity of 10.5 Ci/mmole, was supplied by McNeil Laboratories, Fort Washington, PA. Equal volumes of dialyzed BSA-hydrazide solution and 1 M ^{3}H-labeled haloperidol (diluted to 100 μCi/mmol with unlabeled haloperidol) in methanol:acetic acid (5:1 v/v) were mixed and allowed to react in the dark at room temperature for 1 week. Ten volumes of ethyl acetate were added to the reaction mixture, and the resultant precipitate was centrifuged and washed repeatedly with 10 volumes of 0.1 M

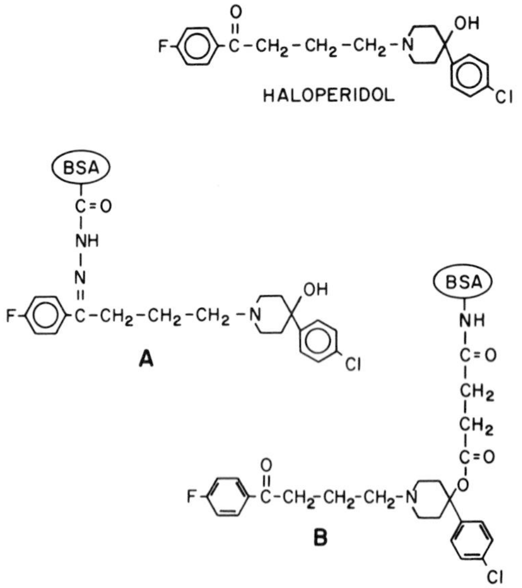

FIG. 1. Molecular structure of haloperidol and configurations of its conjugation to bovine serum albumin (BSA) for antibody production. Conjugation at the ketone group (compound A) was used to produce the RIA-I antibody, and conjugation via the tertiary alcohol (compound B) was used to produce the RIA-II antibody (Rubin *et al.*[18]).

ammonium hydroxide in ethyl acetate:methanol (5:1 v/v) until no radioactivity was detected in the supernatant. The precipitate was resuspended in water, lyophilized, and stored at −10°. From the specific activities of the ^{3}H-labeled haloperidol and the BSA–[3]H-labeled haloperidol conjugate, it was calculated that between 2 and 6 mmol of haloperidol were incorporated into each mmole of BSA.

For RIA-II,[13] haloperidol was converted to the 4-oxobutanoic acid to obtain a suitable hapten, which then was dissolved in dimethylacetamide and acidified with hydrochloric acid. This solution was added dropwise to 1-ethyl-3-(3-dimethylaminopropyl)-carbodiimide hydrochloride in distilled water. The pH was adjusted to 5.0 with sodium hydroxide, and BSA dissolved in phosphate buffer, pH 6.0, was added to this mixture. The solution was stirred overnight at 4°, after which the reaction mixture was dialyzed against distilled water for 12 hr at 4° and lyophilized. The molar incorporation of haloperidol hapten into BSA was not specified for this conjugate.

Production of Antibodies in Animals

For both RIAs, the haloperidol–BSA conjugate was dissolved in physiologic saline (phosphate-buffered saline, pH 7.4, for RIA-II) to a concentration of 1 mg/ml and then was emulsified with an equal volume of complete Freund's adjuvant. Female New Zealand albino rabbits were used for the production of both antibodies. For RIA-I, the rabbits were immunized with 0.5 ml doses of conjugate subcutaneously at multiple sites, and booster injections were given every 2 weeks for 2 months. Blood was collected from an ear vein, and the serum was separated and stored at −10°. For RIA-II, the rabbits were immunized with 1.0 ml doses of conjugate injected intradermally into multiple sites, and booster injections were given every 3 weeks for 3 months. Blood was collected from the central ear artery approximately 1 week after each booster injection and the serum tested for haloperidol antibodies. The rabbits were sacrificed by heart puncture 10 days after the last booster injection, and the sera from all the bleedings were pooled and stored at −20°. Thus, the method of immunization of the rabbits for the production of both antibodies was similar.

Radioimmunoassay Procedure

For the determination of the standard curve for RIA-I,[14] amounts of unlabeled haloperidol (Lot No. McN-JR-1625) ranging from 0.01 ng to 10 ng were added to glass assay tubes. ^{3}H-labeled haloperidol standard solution (0.025 ml) was added to each tube followed by addition of 0.1 ml diluted (1 : 400) antiserum and 50 μl normal human or dog serum. Gelatin buffer was added to give a final incubation mixture volume of 0.5 ml and a final antiserum dilution of 1 : 2000, at which 20–30% binding of ^{3}H-labeled haloperidol standard occurred. The antigen–antibody binding reaction reproducibly attained equilibrium after incubation for 2 hr at room temperature and a subsequent 30 min incubation at 4°, or after incubation for 18 hr at 4°. After incubation, antiserum-bound ^{3}H-labeled haloperidol was separated from free at 4° with 0.1 ml 0.75% dextran-coated charcoal and was quantitated by liquid scintillation counting of the supernatant suspended in Aquasol (New England Nuclear). The standard curve was calculated by the method of Rubin and Gouin.[15]

For the determination of the standard curve for RIA-II,[13] 0.2 ml of diluted (1 : 200) antiserum in 0.5 ml normal human plasma together with 0.2 ng ^{3}H-labeled haloperidol (10 Ci/mmol) was incubated with increasing amounts of unlabeled haloperidol, each contained in 0.05 ml 30% methanol-water. Antiserum-bound ^{3}H-labeled haloperidol was separated from free with 0.2 ml 2% dextran-coated charcoal and was quantitated by liquid scintillation counting of the supernatant suspended in Ria-fluor (New England Nuclear).

Antibody Specificity and Sensitivity

Neither antibody cross-reacted with compounds structurally dissimilar from the butyrophenone class of compounds, such as chlorpromazine or thioridazine, but they did differ in their profiles of cross-reactivity with metabolites of haloperidol and related butyrophenones (see the table). For RIA-I,[14] haloperidol conjugated to BSA at the ketone group produced an antibody that discriminated haloperidol from its acid metabolites and from certain related butyrophenones but that cross-reacted completely with reduced haloperidol, a major, relatively inactive metabolite present in the plasma of treated patients.[16] This antibody also cross-reacted about 8% with the piperidinyl moiety of haloperidol, a major metabolic fragment.[17] For RIA-II,[13] haloperidol conjugated to BSA via the tertiary alcohol produced an antibody that cross-reacted with other related butyrophenones but that discriminated haloperidol completely from its acid and piperidinyl metabolites and from reduced haloperidol.

For both RIAs the sensitivity has ranged between 0.05 and 0.3 ng ha-

[15] R. T. Rubin and P. R. Gouin, *Endocrinology* **97**, 1558 (1975).
[16] A. Forsman and M. Larsson, *Curr. Ther. Res.* **24**, 567 (1978).
[17] J. Heykants, Preclinical Research Report R1625/3, Janssen Pharmaceutica, Beerse, Belgium, 1977.

AMOUNTS OF VARIOUS BUTYROPHENONES AND HALOPERIDOL METABOLITES
REQUIRED TO REDUCE ANTIBODY BINDING OF ^{3}H-LABELED HALOPERIDOL BY 50%,
RELATIVE TO HALOPERIDOL ITSELF $(= 1.0$ ng$)^a$

Compound number	Reference number	Structure $F-\langle\bigcirc\rangle-\overset{O}{\overset{\|}{C}}-CH_2-CH_2-CH_2-\textcircled{R}$	Generic name	Antiserum RIA-I	Antiserum RIA-II
1	R 1625		Haloperidol	1.0 ng	1.0 ng
2	R 4749		Droperidol	>500 ng	16 ng
3	R 5147		Spiperone	>500 ng	38 ng
4	R 11 333		Bromperidol	1.2 ng	1.4 ng
5	R 1658		Moperone	1.3 ng	1.7 ng
6	R 2498		Trifluperidol	8.5 ng	2.9 ng
7	R 11 302		Acid Metabolite	>500 ng	>5000 ng
8	X 680		idem	>500 ng	>5000 ng
9	T 473		idem	385 ng	>5000 ng
10	R 1515		Metabolite (Piperidinyl Moiety)	12 ng	>5000 ng
11	R 2572		Reduced Haloperidol	0.6 ng	380 ng

^{a} From Rubin et al.[18]

loperidol/ml, and the intra- and interassay coefficients of variation have averaged 10%. Figure 2 illustrates a standard curve for RIA-I (curve A) and indicates the parallel displacements of ³H-labeled haloperidol from antiserum by increasing volumes of sera from psychiatric patients treated with different doses of haloperidol.

Comparison of Radioimmunoassay with Gas Chromatographic Determination of Haloperidol

For the comparison of the radioimmunoassay and gas chromatographic methods of haloperidol measurement, 21 serum samples collected from 11 schizophrenic patients chronically medicated with haloperidol were analyzed by gas chromatography and by the two RIAs.[18] RIA-I was used to assay the serum samples in our laboratory in California, and RIA-II was used to assay the samples, under somewhat different conditions, both in our laboratory and at the Janssen laboratories in Belgium. For the assay of haloperidol in Belgium, a commercial RIA kit containing the RIA-II antiserum was used (I.R.E., Brussels, Belgium). The sensitivities and intraassay coefficients of variation were as stated earlier in this chap-

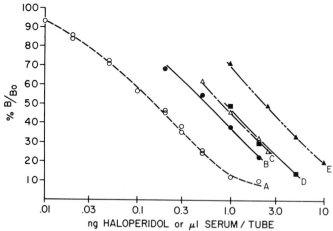

FIG. 2. Curve A: Radioimmunoassay standard curve for ³H-labeled haloperidol displacement from antiserum (expressed as percentage of ³H-labeled haloperidol bound in the absence of unlabeled haloperidol) by increasing amounts of unlabeled haloperidol. Curves B–E: Displacement of ³H-labeled haloperidol from antiserum by increasing volumes of sera from psychiatric patients receiving oral haloperidol (Clark *et al.*[14]).

[18] R. T. Rubin, A. Forsman, J. Heykants, R. Öhman, B. B. Tower, and M. Michiels, *Arch. Gen. Psychiat.* **37**, 1069 (1980).

ter. GC measurements of haloperidol were performed by Dr. Anders
Forsman in Sweden.

Figure 3 presents a comparison of total haloperidol concentrations determined by the three RIA methods (RIA-I in California and RIA-II both
in California and in Belgium) with total haloperidol concentrations determined by GC. Figure 3A shows RIA values related to GC values between
0 and 10 ng/ml, and Fig. 3B shows similar data on a compressed scale for
GC values between 10 and 70 ng/ml. These data indicate a consistent, linear overestimation of GC values (average of 54%) by the RIA-II procedure performed in California, a still larger overestimation of GC values
(average of 91%) by the commercial RIA-II kit used in Belgium, and a
severalfold nonlinear overestimation of the GC values (average of 223%)
by the RIA-I procedure in California. Since the GC analysis is specific and
the extraction is consistent over a wide range of serum haloperidol concentrations,[3] and because the GC values matched fairly closely those
from one of the RIA procedures, it can be assumed that the GC values are
accurate. The large nonlinear overestimation of values by the RIA-I
method is most likely due to the relative nonspecificity of this antibody

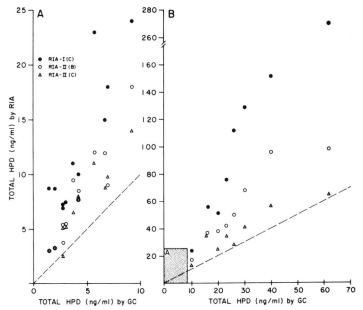

FIG. 3. Comparison of total serum haloperidol determined by three RIA systems [one
using the RIA-I antibody in California (C) and two using the RIA-II antibody, in California
(C) and in Belgium (B)] with total haloperidol determined by GC. Twenty-one serum samples from 11 chronically medicated schizophrenic patients were analyzed. Figure 3A represents a magnification of shaded area in Fig. 3B (Rubin *et al.*[18]).

which, as indicated earlier, binds not only the parent compound but also some metabolites such as the piperidinyl moiety and reduced haloperidol (table).

Liquid Scintillation Sample Quenching

When relatively large serum samples (e.g., 500 μl) are needed for the quantitation of low concentrations of haloperidol, the potential problems of color quenching from even minor hemolysis and/or photoluminescence from other compounds in serum become important. Bleaching of the solubilized RIA mixture in the scintillation vials is an effective remedy and may be accomplished by the addition to each vial, with mixing, of 0.25 ml 30% hydrogen peroxide followed by a 12 h incubation at room temperature. The peroxides formed then are removed by the addition, with mixing, of 0.25 ml 5 N hydrochloric acid followed by 0.5 ml 10% ascorbic acid. The scintillation vials should be held in a refrigerated scintillation counter, in complete darkness, for at least 24 hr before counting.

Discussion

The assessment of blood levels of neuroleptic (antipsychotic) drugs requires analytic techniques that are specific, sensitive, and precise. Some compounds, such as the phenothiazines, may have important active metabolites. As assay technique such as GC, that specifically measures the parent compound, may not assess the combined therapeutic blood level of all active metabolites. On the other hand, a receptor assay such as the striatal dopamine receptor assay for neuroleptic drugs does assess both a particular compound and its active metabolites,[19] but it cannot distinguish diverse neuroleptic drugs simultaneously present in patients' sera.

RIA also offers a number of practical advantages over GC for the measurement of haloperidol, including its application to small quantities (microliters) of unextracted plasma or serum, its cheaper cost, and its ease of applicability to hundreds of samples per day. Both of the haloperidol RIA methods are more sensitive than the GC method. The RIA-II antibody has the requisite specificity for application to serum samples from patients chronically treated with haloperidol, in whom there may be a significant buildup of metabolites. The RIA-I antibody has shown utility in measuring serum haloperidol concentrations acutely over a several hour period in subjects given single parenteral doses of drug as low as 0.5 mg, as indicated in Fig. 4.[20] RIA-I also may be useful in assessing the buildup of the reduced haloperidol and piperidinyl metabolites over time, by a subtrac-

---

[19] I. Creese and S. H. Snyder, *Nature* (*London*) **270**, 180 (1977).
[20] S. E. Hays and R. T. Rubin, *Psychopharmacology* **61**, 17 (1979).

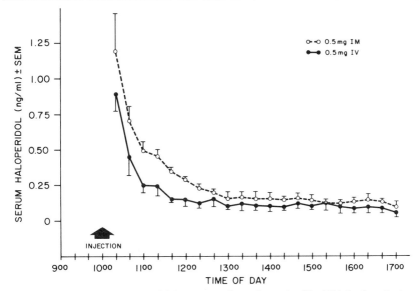

FIG. 4. Average serum haloperidol concentrations, determined by RIA-I, after the intramuscular (IM) and intravenous (IV) injection of 0.5 mg haloperidol in normal men. Each point represents the mean of seven subjects ± SEM (Hays and Rubin[20]).

tion technique utilizing the RIA-I antiserum for haloperidol and its metabolites and the RIA-II antiserum for haloperidol alone.[20a] This possibility, which has been applied to other systems,[21,22] will require systematic verification in future studies. The cross-reactivity of both antisera with several other butyrophenones suggests the applicability of this RIA to other clinical circumstances, such as the determination of the pharmacokinetics of bromperidol as a neuroleptic or droperidol as an adjunct to anesthesia.

Antibodies against haloperidol that appear to have the requisite specificity for application to sera from chronically medicated patients have been developed by other investigators as well (Drs. Michael Shostak and David Fenimore, personal communications). However, it should be cautioned that all presently available RIAs for haloperidol require further systematic validation; for example, the overestimation of haloperidol values by RIA compared to GC is an important methodologic issue requiring further refinement of this technique. In fact, using our own RIA-II antiserum we have found that a simple organic extraction of serum yields

[20a] R. T. Rubin and R. E. Poland, in "Clinical Pharmacology in Psychiatry: Neuroleptic and Antidepressant Research" (E. Usdin, S. Dahl, L. Gram, and O. Lingjaerde, eds.), p. 217. Macmillan, London, 1981.

[21] A. A. Rosenbloom and D. A. Fisher, Endocrinology 95, 1726 (1974).

[22] W. A. Ratcliffe, S. M. Fletcher, A. C. Moffat, J. G. Ratcliffe, W. A. Harland, and T. E. Levitt, Clin. Chem. 23, 169 (1977).

statistically equivalent RIA and GC haloperidol values.[23] The availability of a convenient RIA method for haloperidol measurement should stimulate further large scale clinical studies of this drug and, with refinement and validation, should prove to be an important tool for assessing therapeutic blood levels of haloperidol in individual patients.

Acknowledgments

Drs. Theodore S. Roosevelt and Mark A. Goldberg provided valuable consultation on liquid scintillation counting techniques. This research was supported by NIMH grant MH 29491 and by McNeil Laboratories Inc., Fort Washington, PA. R.T.R. is the recipient of NIMH Research Scientist Development Award MH 47363.

[23] R. E. Poland and R. T. Rubin, *Life Sci.* **29**, 1837 (1981).

[39] Radioimmunoassay of Pimozide

By M. MICHIELS, R. HENDRIKS, and J. HEYKANTS

Although the neuroleptics belonging to the butyrophenone and diphenylbutylamine series have already been used for many years in psychiatry, research on the pharmacokinetic profile of the prototype drugs of these series, i.e., haloperidol and pimozide, started only in 1974. The reason for this delay was the lack of a suitable analytical method accurate enough to measure therapeutic plasma levels of these drugs in man. At that time, a sensitive radioimmunoassay was developed for pimozide[1] which is the only assay available so far to measure the very low plasma levels of the neuroleptic. The assay has been adapted for the analysis of large numbers of samples, required for pharmacokinetic studies and drug monitoring of pimozide in man.

Materials and Methods

Specifically tritium-labeled pimozide (specific activity 12 Ci/mmol) was synthesized at I.R.E. (Fleurus, Belgium) according to the procedure developed in our laboratory. The radiochemical purity of ³H-labeled pimozide was determined by radio-HPLC (Waters Associates HPLC equipment with on-line radiodetector Berthold 5026 HP) and was found to be greater than 98%. It was stored as an ethanolic solution at −20°. Dextran-Radioimmunoassay Grade was purchased from Schwarz/Mann, Orange-

[1] M. Michiels, R. Hendriks, and J. Heykants, *Life Sci.* **16**, 937 (1975).

METHODS IN ENZYMOLOGY, VOL. 84

burg, N.Y., bovine serum albumin (Cohn fraction V) from Sigma, St. Louis, Mo., Norit A-supra from C.M.I.-Codepa, Belgium, and Pico-fluor 30 from Packard Instrument. All drugs and test compounds mentioned were originally synthesized and analyzed in the Janssen Research Laboratories, Beerse, Belgium.

Preparation of the Hapten

Pimozide was converted to the more reactive 3-{1-[4, 4-bis(4-fluorophenyl)butyl]-4-piperidinyl}-2,3-dihydro-2-oxo-1*H*-benzimidazole-1-acetic acid as shown in Fig. 1. Therefore equimolar amounts of pimozide and sodium amide (0.05 mmol) were dissolved in toluene, with application of heat to cause gentle refluxing for 24 hr. A mixture of equimolar amounts of chloroacetic acid and sodium amide (0.05 mmol), dissolved in toluene, was added slowly to the former and the mixture was allowed to react overnight at reflux. After cooling, the reaction mixture was acidified to pH 5 with diluted acetic acid and extracted with chloroform. The hapten, obtained in 62% yield, melted at 252.5° and gave the following elemental

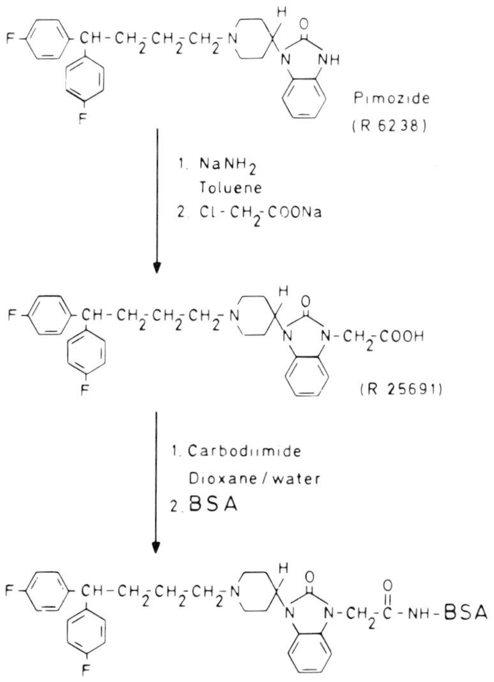

FIG. 1. The synthesis of the pimozide-hapten and its covalent attachment to bovine serum albumin.

analysis: calculated for $C_{30}H_{31}F_2N_3O_3 \cdot C_3H_8O$: N, 7.25%; C, 68.37%; H, 6.78%; found: N, 7.28%; C, 68.05%, H, 6.65%.

Preparation of the Pimozide Immunogen

An amount of 100 mg of the pimozide derivative was dissolved in 5 ml of boiling dioxane and added dropwise to 400 mg of 1-ethyl-3-(3-dimethyl-aminopropyl)carbodiimide hydrochloride, dissolved in 5 ml of distilled water. To this mixture 200 mg of bovine serum albumin (BSA) in 20 ml of 0.05 M phosphate buffer (pH 6.0) was added dropwise and the solution was stirred overnight at 4° (Fig. 2). The resulting mixture was dialyzed extensively against 0.005 M phosphate buffer of pH 7.4 for about 12 hr at 4° and freeze-dried afterward. The extent of covalent binding of the hapten to the protein carrier was determined previously.[1] Using ^{3}H-labeled pimozide as the marker, the degree of substitution was calculated from the specific activity of the hapten–BSA conjugate and was estimated to be 2 to 3 mol of pimozide per mole of albumin.

Immunization

The protein conjugate was dissolved in phosphate-buffered saline of pH 7.4 at a concentration of 1 mg/ml and emulsified on a Vortex mixer with an equal volume of complete Freund's adjuvant. Two female New Zealand albino rabbits, weighing about 3.0 kg, were injected with 1 ml of the emulsion into 2 to 4 intramuscular and into multiple intradermal sites around the neck and in the thigh. Subsequent booster injections were given similarly at intervals of about 3 weeks. To follow the immunization process, the animals were test-bled by ear vein puncture 7 to 10 days after each booster injection and serum was tested for antibodies. The rabbits were exsanguinated 10 days after the last booster and the collected serum was pooled and stored at −20°.

Assay of Pimozide

Antiserum titers were determined by incubating 0.2 ml of serially diluted rabbit serum with 0.5 ng of ^{3}H-labeled pimozide (28,000 dpm, contained in 0.05 ml of 30% methanol/water, v/v) in 0.5 ml of 2% (w/v) BSA –phosphate buffer (0.05 M, pH 7.4). Incubations were carried out in 1.3 ml plastic tubes (type Eppendorf) by continuous rotation (25 rpm) for 2 hr at room temperature. Thereafter, bound and free pimozide were separated by selective adsorption on dextran-coated charcoal: aliquots of 0.2 ml of a suspension containing 2 g activated charcoal and 200 mg dextran per 100 ml BSA–phosphate buffer (0.05 M, pH 7.4) were added to the incubation mixture and allowed to equilibrate at room temperature for 1 hr with continuous rotation. After centrifuging at 8000 g for 10 min (Micro-fuge, Heraeus-Christ) the supernatant was pipetted into a minivial containing 3 ml of Pico-fluor and the radioactivity was determined in a liquid scintillation spectrometer (Packard-Prias). By external standard channel ratios and a counting efficiency curve, cpm were converted by programmed analysis to dpm (Wang PCS 2200 II 4-2, Wang Labs. Inc., Mass.). Appropriate controls were included to determine nonspecific binding, for which undiluted normal rabbit serum was substituted for the antiserum.

Standard curves of pimozide were obtained by incubating increasing amounts of the unlabeled drug together with a fixed 0.5-ng amount of ^{3}H-labeled pimozide in the presence of 0.2 ml of an antiserum dilution which bound almost approximately 40–45% of the tracer, as found by previous titration. Incubations were carried out in BSA–buffer by measuring the inhibition of the antibody–pimozide complex formation produced by increasing amounts of up to 1000 ng of either the known metabolites of pimozide or of various structurally related drugs, incubated with 0.5 ng of ^{3}H-labeled pimozide and 0.2 ml of the appropriate antiserum dilution.

The inter- and intraassay variability and the accuracy of the assay were tested either by repetitive analysis over a 3-week period of pimozide standard added to control human plasma or by assaying increasing volumes of the same sample added with known amounts of standard. All samples were assayed in duplicate and the data were calculated as outlined by Rodbard[2] using a Wang desk-top computer system for iterative weighed linear regression analysis of logit B/B_0 versus log dose.

Procedure in Man

Plasma samples were obtained at various time intervals from a group of nine patients after oral administration of a daily dose of 6 mg for 4 consecutive days or, 2 weeks later, of a single 24-mg dose,[3] and from a group of 10 male patients after a single oral dose of 4 mg. Analysis of the unknown samples was carried out on 0.05- to 0.5-ml aliquots of plasma as described for the standard curves. Drug concentrations in the plasma samples were calculated from the degree to which unknown amounts of pimozide caused inhibition of binding of ^{3}H-labeled pimozide as compared to standard curves obtained simultaneously.

[2] D. Rodbard, *Clin. Chem.* **20**, 1255 (1974).
[3] Plasma samples were obtained through the courtesy of Dr. R. G. McCready, Gartnovel Royal Hospital, Glasgow. R. G. McCready, J. J. P. Heykants, A. Chalmers, and A. M. Anderson, *Br. J. Clin. Pharmacol.* **7**, 533 (1979).

Results and Discussion

Sensitivity of Assay

Antiserum titrations demonstrated that the final pooled rabbit sera contained antibodies capable of binding ³H-labeled pimozide. Nearly 45% of the added tracer was bound specifically to antibodies at 1/10,000 antiserum dilution (0.2 ml) under the assay conditions described. Nonspecific binding to control rabbit serum was less than 1.5%. Following logit transformation, a standard curve for unlabeled pimozide gave a linear response from 0.05 to 6 ng in both BSA–buffer (Fig. 2) and control human plasma if the limit of significance is restrained to 10% inhibition. When applied to 0.5 ml of plasma, a concentration of 0.1 ng/ml of pimozide could be readily detected.

Specificity of the Antiserum

The specificity of the pimozide antiserum against various structural congeners and the major metabolites of pimozide[4] is illustrated in the

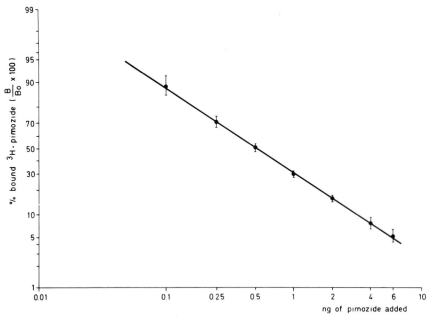

FIG. 2. Logit–log plot of standard curve: displacement of ³H-labeled pimozide effected by various amounts of unlabeled pimozide, added to 0.2 ml of 1/10,000 antiserum dilution in 2% (w/v) albumin–buffer (pH 7.4). Mean ± SD of 8 independently obtained curves.

[4] G. E. Abraham, *J. Clin. Endocrinol.* **29,** 866 (1972).

table. The results are expressed as percentage of cross-reactivity according to Abraham.[4] Only the hapten N-carboxymethylpimozide and the long-acting neuroleptic drug clopimozide (R 29 764) were bound to the same degree as pimozide itself. Fluspirilene (R 6218) and penfluridol (R 16 341), both neuroleptic congeners belonging to the diphenylbutylamine series also displaced ^{3}H-labeled pimozide from binding sites, although to a lesser extent. The cardiovascular drug lidoflazine, having the diphenylbutylamine moiety of the molecule in common with pimozide, showed also less but considerable cross-reaction. The affinity of the antibody for fluspirilene enabled the antiserum to be used also for radioimmunoassay of this neuroleptic,[5] whereas the high degree of cross-reaction of the cardiovascular drug lidoflazine made this assay unsuited to be applied to measuring plasma pimozide levels in patients treated with both the neuroleptic and cardiovascular drug. On the contrary, neither 4, 4-bis(4-fluorophenyl)butanoic acid nor 1,3-dihydro-I-(4-piperidinyl)-2H-benzimidazol-2-one, both reported to be the major metabolites of pimozide,[6] inhibited antibody-binding of labeled pimozide to any extent even at 1000-fold molar excess. This ensures that these metabolites, even in large amounts, are not able to interfere with the assay of the parent drug, which does away with laborious extraction procedures. Previous results with analogous antisera[1] already demonstrated the strict dependence of antibody specificity on the structure of the pimozide-hapten only, by the lack of cross-reactivity of drugs having a butyrophenone or cyanodiphenylpropyl group substituted for the diphenylbutyl moiety of pimozide, such as the neuroleptics benperidol and droperidol or the analgesic bezitramide. These results suggest that the antigenic determinant is contained in nearly the intact pimozide molecule, which seems obvious, since the length of the alkyl chain between the hapten and the protein carrier and the site of attachment of the hapten to this carrier are considered to be determinant for the ability of the antibody to differentiate between the parent drug and related structures such as its metabolites. This also implies that the design of the hapten is of the utmost importance and must be carried out taking full account of the metabolic pathways of the drug, which is facilitated for pimozide by its rather simple degradation pattern.[6] In this view it seems reasonable that antibody specificity is predictable to some degree.

Accuracy of the Assay

The intra- and interassay coefficients of variation did not exceed 6.3 and 8.2%, respectively, over a range of 0.25 to 5 ng per incubation tube.

[5] J. Vranckx-Haenen, *Acta Psychiatr. Belg.* **79,** 459 (1979).
[6] W. Soudijn and I. Van Wijngaarden, *Life Sci.* **8,** 291 (1969).

SPECIFICITY OF PIMOZIDE ANTISERUM

Structure and reference number	Non proprietary name (Therapeutic class)	% Cross-reaction
$R = F-\langle\bigcirc\rangle-CH-CH_2CH_2CH_2$ (with second F-phenyl)		
(R 6238)	Pimozide (Neuroleptic)	100
(R 25691)	Hapten (Inactive)	100
(R 29764)	Clopimozide (Neuroleptic)	100
(R 6218)	Fluspirilene (Neuroleptic)	42
(R 7904)	Lidoflazine (Cardiovascular drug)	21
(R 16341)	Penfluridol (Neuroleptic)	11
(R 21506)	Metabolite (Inactive)	< 0.1
(R 30507)	Metabolite (Inactive)	< 0.1

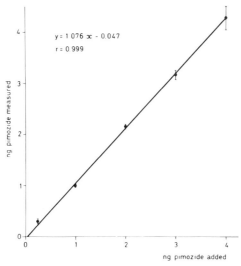

FIG. 3. Correlation of pimozide measured and pimozide added to human plasma (r = 0.999). The quantitative recovery assures the accuracy of the assay, independently of the amounts to be measured.

The angular coefficient of the regression line of Fig. 3 calculated for the correlation of the amount of pimozide added versus the amounts measured, assures an almost quantitative recovery of the pimozide added, independently of the amounts to be measured within the range indicated (r = 0.999).

The linear regression for the correlation of the sample volume versus the estimated amount (Fig. 4) confirms that the estimates are independent

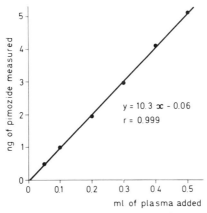

FIG. 4. Correlation of pimozide measured and volume of the same plasma sample added (r = 0.999), confirming that the assay is independent of the composition of the incubation mixture.

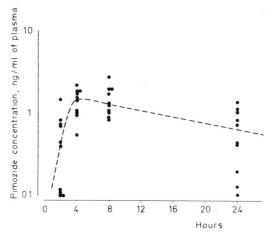

FIG. 5. Pimozide plasma levels after a single oral dose of 4 mg in patients ($n = 10$).

of both concentrations and sample dilution with respect to composition of the incubation mixture, thus also of interferences which may be present in the sample ($r = 0.999$). Due to the lack of an alternative analytical procedure for pimozide, further validation of the method by comparing results with those of independent assay systems could not be performed.

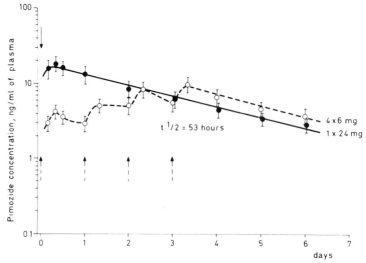

FIG. 6. Pimozide plasma levels after a single oral dose of 24 mg (●——●) and after administration of 6 mg once daily for 4 consecutive days (○——○) (mean ± SD, $n = 9$).

Serum Levels of Pimozide in Man

The radioimmunoassay procedure described above was used to measure pimozide in human plasma, taken at various time intervals after a single oral dose of 4 mg (n = 10) (Fig. 5), and of 24 mg or 4 subsequent daily doses of 6 mg (Fig. 6). The plasma level data reveal a gradual absorption of pimozide lasting up to 8 hr after administration, followed by a slow elimination from plasma. Depending on the dose regimen, plasma levels declined with a terminal half-life of nearly 53 hr.[3] The correlation between the curves demonstrates that the systemic availability, even after chronic administration, can be extrapolated from single dose estimates. These results, however, primarily demonstrate that the radioimmunoassay reported is sufficiently sensitive and specific for study of the pharmacokinetics of pimozide in man.

[40] Radioimmunoassay for Morphine

By SYDNEY SPECTOR

The radioimmunoassay has been used to quantitate both hormones[1-23] and small molecular weight compounds such as drugs[24-37]. The number of radioimmunoassays for drugs is ever increasing. Drugs being used thera-

[1] S. A. Berson and R. S. Yalow, *J. Clin. Invest.* **38,** 1966 (1959).
[2] S. A. Berson and R. S. Yalow, *Harvey Lect.* **62,** 107 (1968).
[3] R. S. Yalow and S. A. Berson, *J. Clin. Invest.* **39,** 1157 (1960).
[4] G. M. Grodsky, T. Hayashida, C. T. Peng, and I. Geschwind, *Proc. Soc. Exp. Biol. Med.* **107,** 491 (1961).
[5] A. M. Lawrence, *Proc. Natl. Acad. Sci. U.S.A.* **55,** 316 (1966).
[6] R. H. Unger, A. M. Eisentrant, M. S. McCall, and L. L. Madison, *J. Clin. Invest.* **40,** 1280 (1961).
[7] J. B. Bauman, J. Girard, and M. Vest, *Immunochemistry* **6,** 699 (1970).
[8] W. R. Butt and S. S. Lynch, *Clin. Chim. Acta* **22,** 79 (1968).
[9] A. R. Midgley, Jr., *J. Clin. Endocrinol.* **27,** 295 (1967).
[10] W. D. Odell, A. F. Parlow, C. M. Cargille, and G. T. Ross, *J. Clin. Invest.* **47,** 2551 (1968).
[11] K. Abe, W. E. Nicholson, G. W. Liddle, D. P. Island, and D. N. Orth, *J. Clin. Invest.* **46,** 1609 (1967).
[12] M. B. Clark, G. W. Boyd, P. G. H. Byfield, and G. V. Foster, *Lancet* **2,** 74 (1969).
[13] A. H. Tashjiian, Jr., *Endocrinology* **34,** 140 (1969).
[14] W. Schopman, W. H. C. Hackeng, and R. M. Lequin, *Acta Endocrinol.* **63,** 643 (1970).
[15] R. D. Utiger, *J. Clin. Invest.* **44,** 1277 (1965).
[16] M. Miller and A. M. Moses, *Endocrinology* **84,** 557 (1969).
[17] M. B. Vallotton, L. B. Page, and E. Haber, *Nature, (London)* **215,** 741 (1967).
[18] T. Chard, N. R. H. Boyd, M. L. Forsling, A. S. McNeilly, and J. Landon, *J. Endocrinol.* **48,** 223 (1970).

METHODS IN ENZYMOLOGY, VOL. 84

peutically are monitored in biological fluids or tissues in order to (a) assist in establishing a dose-therapeutic effect relationship, (b) investigate pharmacokinetic and metabolic studies [those drugs which are being used for recreational purposes and are being abused, their drug concentrations are determined for reasons cited in (a) and (b)], but also to (c) establish a clinical diagnosis, (d) medicolegal reasons, and (e) ascertain the prevalence of drug abuse for epidemiological studies.

The morphine molecule (Fig. 1) has a molecular weight of 285, and as such is not able to stimulate antibody formation. The hapten requires chemical conjugation to a larger molecule in order for it to become immunogenic. There are a number of sites which are amenable for coupling to a carrier protein. The phenolic OH at C_3 can be used, as can the alcoholic OH at C_6. It is also possible to diazotize p-aminobenzoic acid and then the phenolic OH of morphine activates the ring so that the diazo compound is directed ortho or para to the phenolic OH. The COOH group of benzoic acid can then be used as a bridge between the amino groups of the carrier and the derivatized morphine. Finally, one can prepare a conjugated protein of morphine through the nitrogen atom. The specificity of the antibodies produced is influenced by the site on the morphine molecule at which the protein is conjugated. It is possible therefore to generate antibodies that can recognize various portions of the molecule as determinant groups. Conjugation of the carrier protein to C_2, C_3, or C_6 of the

[19] T. Chard, J. M. Kitau, and J. Landon, *J. Endocrinol.* **46**, 269 (1970).
[20] J. Hansky, M. G. Korman, C. Soveny, and D. J. B. St. John, *Gut* **12**, 97 (1971).
[21] M. G. Korman, C. Soveney, and J. Hansky, *Scand. J. Gastroenterol.* **6**, 71 (1971).
[22] J. E. McGuigan, *Gastroenterology* **56**, 429 (1969).
[23] J. F. Stremple and R. C. Meade, *Surgery* **64**, 165 (1968).
[24] B. F. Erlanger, F. Borek, S. M. Beiser, and S. Lieberman, *J. Biol. Chem.* **228**, 713 (1957).
[25] B. F. Erlanger, F. Borek, S. M. Beiser, and S. Lieberman, *J. Biol. Chem.* **234**, 1090 (1957).
[26] L. Goodfriend and A. H. Sehon, *Nature (London)* **185**, 764 (1960).
[27] L. Goodfriend and A. H. Sehon, *Can. J. Biochem. Physiol.* **39**, 941 (1961).
[28] S. J. Gross, D. H. Campbell, and H. H. Weetall, *Immunochemistry* **5**, 55 (1969).
[29] V. P. Butler, *Lancet* **1**, 186 (1971).
[30] T. W. Smith, V. P. Butler, and E. Haber, *N. Engl. J. Med.* **281**, 1212 (1969).
[31] T. W. Smith, V. P. Butler, and E. Haber, *Biochemistry* **9**, 331 (1970).
[32] A. L. Steiner, D. M. Kipnis, R. Utigen, and C. W. Parker, *Proc. Natl. Acad. Sci. U.S.A.* **64**, 367 (1969).
[33] A. L. Steiner, C. W. Parker, and D. M. Kipnis, *Advan. Biochem. Psycopharmacol.* **3** (1969).
[34] S. Spector, *J. Pharmacol. Exp. Ther.* **178**, 253 (1971).
[35] S. Spector and C. W. Parker, *Science* **168**, 1347 (1970).
[36] E. J. Flynn and S. Spector, *J. Pharmacol. Exp. Ther.* **181**, 547 (1972).
[37] V. P. Butler, *Pharmacol. Rev.* **29**, 103 (1978).

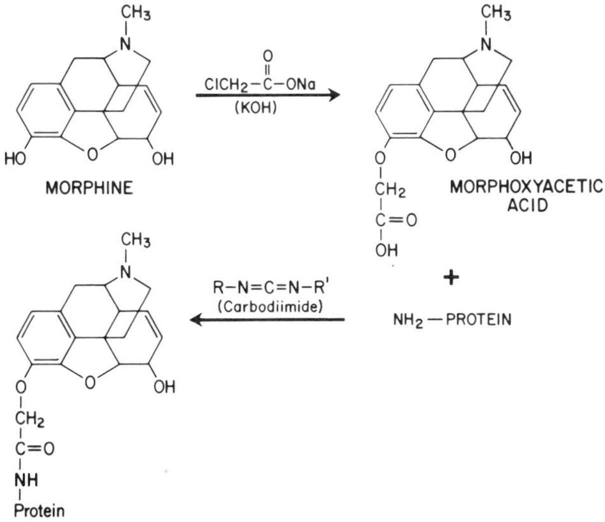

FIG. 1. The morphine molecule.

morphine molecule generates antibodies that can recognize alterations on the nitrogen atom of the molecule. However, when the carrier protein was conjugated to the nitrogen atom of the opiate alkaloid then the phenolic hydroxy group on C_3 and the alcoholic ground on C_6 became determinant groups so that codeine and heroin can be differentiated from morphine.

Preparation of Immunogens. (a) 3-*O*-Carboxymethylmorphine. Morphine-free base (0.02 mol) is dissolved in 20 ml KOH solution containing (0.02 mol) KOH. Evaporate the solution at 60° at 25 mm Hg pressure and add 100 ml of benzene-azeotropically distilled water and distill until the benzene is evaporated off. To the solids add 120 ml absolute ethanol plus 0.025 *M* chloroacetic acid sodium salt and then reflux for 3 hr. Cool to room temperature and filter off the solids. The solids are then washed with 100 ml distilled water. Adjust pH to 7 with 1 *N* NOH. Evaporate to dryness and then add 100 ml absolute ethanol to the solids. Reflux for 3 hr. Filter while solution still hot.

Carboxymethylmorphine (8 mg) was dissolved in 2 ml distilled water containing 10 mg of BSA. The pH was adjusted to 5.5 and 8 mg of 1-ethyl-3(3-dimethylaminopropyl)carbodiimide was added (Fig. 1). The mixture was incubated overnight at room temperature, and then dialyzed for 7 days against distilled water, with four to five changes per day.

(b) *N*-Carboxymethylnormorphine—Normorphine (0.03 mol) is added to 2.5 g sodium bicarbonate and dimethylformamide (60 ml) at room temperature following which ethylbromacetate (0.03 mol) is added. The mixture is refluxed for 4 hr. The residue is suspended in water (60 ml) and then extracted with chloroform (1 : 1). The extract was dried and re-crystallized from acetone-ether. The dried material was heated and re-fluxed for 2 hr in 100 ml 2 N HCl. The acid was evaporated and the resi-due dissolved in hot water (80 ml) and some charcoal (Norite A) was added. The hot suspension was filtered and the filtrate allowed to stand at room temperature for 72 hr.

The N-carboxymethylnormorphine (10 mg) was dissolved in 2 ml of 0.01 N HCl which was then heated at 65° for 10 min. To the solution, 10 mg of BSA was added, the pH was adjusted to 5.5, and immediately followed by the addition of 10 mg of 1-ethyl-3(3-dimethylaminopro-pyl)carbodiimide. The solution was left for 24 hr at room temperature and then dialyzed against distilled water for 2 days with four to five changes per day.

Immunization. The immunogens (100 mg) were dissolved in 0.5 ml of saline and emulsified with an equal volume of Freund's complete adju-vant. The emulsion, 0.25 ml, was injected into each footpad of the rabbit. Booster injections were given weekly for 2 weeks and monthly thereafter.

Bleedings were taken from the central ear artery 1 to 2 weeks after booster injections, the first one as a control before immunization. The blood was allowed to clot overnight at 4° and then centrifuged to separate the serum. Aera was stored frozen until used.

Radioimmunoassay. To 0.1 ml of a 1 : 10 dilution of normal rabbit serum were added 0.1 ml of a diluted antiserum and 0.01 ml of [³H]dihy-dromorphine, 100 pmol. Unlabeled morphine at concentrations from 10 pg to 1 ng was used to generate a standard curve and was added in vol-umes ranging from 10 to 100 μl. These volumes were also used when ana-lyzing unknown samples in urine, plasma, or acid-precipitated tissue. The volume was adjusted to 0.5 ml with phosphate-buffered saline, pH 7.4. The mixture was incubated from 1 hr to overnight at 4° and then free [³H]dihydromorphine and antibody-bound [³H]dihydromorphine were separated by the Farr[38] technique. After two washings with 50% saturated ammonium sulfate, the precipitate was dissolved in 0.5 ml of NCS solubi-lizer, and the radioactivity was counted in a liquid scintillation spectrome-ter. Appropriate blanks containing normal rabbit serum instead of anti-serum were included in each assay.

Antibody-bound counts in the presence of cold ligand were calculated as a percentage of total counts bound in the absence of cold ligand. A standard curve was constructed as described by Rodbard *et al.*[39] and the concentration of morphine in the samples determined by interpolation.

[38] A. S. Farr, *J. Infect. Dis.* **103**, 239 (1958).
[39] D. Rodbard, W. Bridson, and P. L. Rayford, *J. Lab. Clin. Med.* **74**, 770 (1969).

[41] Radioimmunoassay for Acetylcholine

By Sydney Spector

There currently exist methods to measure acetylcholine which are very sensitive and specific, namely mass spectrometric and radioenzymatic procedures as well as the bioassay technique. The production of antibodies to acetylcholine affords the neurobiologist a simple method to measure the levels of the neurotransmitter in various tissues and following a variety of substances and physiological conditions. The molecular weight of acetylcholine necessitates it being conjugated to a larger molecule; also it does not have any reactive groups for coupling. As a consequence the acetylcholine molecule was built into another compound.

Synthesis of Immunogen. The amino group of 6-aminohexanoic acid was initially blocked to form N-benzyloxycarbonylhexanoic acid. A solution of 6-aminohexanoic acid (0.1 mol) in 40 ml of 2.5 N NaOH was cooled to 0° and 0.11 mol benzyloxycarbonylcholoride in ether (25 ml) together with 50 ml of 2.5 N NaOH were added. The solution was stirred vigorously for 1 hr. The reaction mixture was then extracted with ether, cooled, and acidified with 6 N HCl. Having blocked the amino group, the next step was to build the choline moiety onto the molecule. The trimethylaminoethyl ester was prepared in two stages. A solution of N-benzyloxycarbonylhexanoic acid (0.01 mol) in 60 ml ethyl acetate was then treated with triethylamine (0.02 mol) and 0.025 mol of β-dimethylaminoethylchloride and refluxed for 16 hr, following which it was filtered and extracted with saturated NaCl, 5% $NaHCO_3$, and again saturated NaCl. It was dried over $MgSO_4$, filtered, and the solvent evaporated *in vacuo*. The evaporated material was dissolved in 10 ml ethyl acetate and treated with methyl p-toluenesulfonate (4.46 mmol). After standing at 25° for 3 hr the product was precipitated and was recrystallized from ethanol-ether to give a white crystalline product (mp 105–106.5°).

The benzyloxycarbonyl group was removed from the above compound by dissolving it in trifluoracetic acid (25 ml) and passing hydrogen bromide gas through the solution, which was then evaporated to dryness. The residue was crystallized from ethanol-ether and recrystallized from dimethylformamide ether to yield a white crystalline product (mp 139–143°). This crystalline product was conjugated to bovine serum albumin (BSA) using the carbodiimide procedure. The 6-aminohexanylcholine (0.2 mmol) and BSA (25 mg) were dissolved in 3 ml H_2O, and the pH was adjusted to 5.5 by adding 0.2 mmol 1-ethyl-3-(3-dimethylaminopropyl)-carbodiimide (Fig. 1). The reaction proceeded for 2 days at 25° and then was

METHODS IN ENZYMOLOGY, VOL. 84

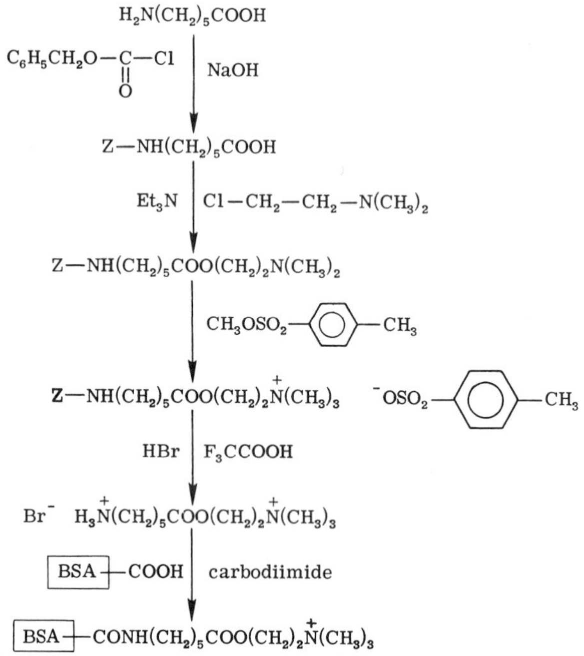

H$_2$N(CH$_2$)$_5$COOH

C$_6$H$_5$CH$_2$O—C—Cl   NaOH
            ‖
            O

Z—NH(CH$_2$)$_5$COOH

Et$_3$N | Cl—CH$_2$—CH$_2$—N(CH$_3$)$_2$

Z—NH(CH$_2$)$_5$COO(CH$_2$)$_2$N(CH$_3$)$_2$

CH$_3$OSO$_2$—⟨○⟩—CH$_3$

Z—NH(CH$_2$)$_5$COO(CH$_2$)$_2$N$^+$(CH$_3$)$_3$ $^-$OSO$_2$—⟨○⟩—CH$_3$

HBr | F$_3$CCOOH

Br$^-$ H$_3$N$^+$(CH$_2$)$_5$COO(CH$_2$)$_2$N$^+$(CH$_3$)$_3$

[BSA—COOH] carbodiimide

[BSA—CONH(CH$_2$)$_5$COO(CH$_2$)$_2$N$^+$(CH$_3$)$_3$]

FIG. 1. Synthesis of an acetylcholine immunogen.

dialyzed against water for 3 days with frequent changes per day. The conjugated protein was then freeze dried.

The cross-reactivity of various ligands for the acetylcholine antibody was determined by preparing several dilutions of all ligands in the radioimmunoassay system and determining the amount which inhibited binding of the 10 pmol label by 50% (IC$_{50}$). Choline showed only 0.1% of the activity of acetylcholine in antibody binding, while no significant interference was shown by other normal tissue constituents examined, such as acetate, choline, carnitine, and phosphorylcholines (Figs. 2 and 3). Appreciable cross reactivity was shown by a number of drugs, including decamethonium, dimethylphenylpiperazinum (DMPP), succinylcholine, and carbachol.

Radioimmunoassay Technique. Stock solutions of ACh standard and labeled ACh were prepared in 0.42 M NaH$_2$PO$_4$ (pH 4) and stored at 4°. Working solutions were made up in PBS just before use. Serial dilutions of the standard (1–137 pmol) or aliquots of the samples (10–50 μl) were incubated with 10 pmol of labeled ACh and antiserum (final dilution 1:1000) in PBS containing 0.025° bovine γ-globulin and 2 × 10^{-4} M

Compound	Structure	ID_{50} (nmol)
Acetylcholine	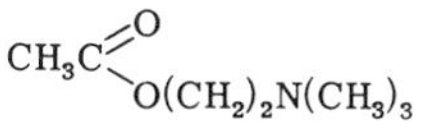 $CH_3C\overset{O}{\underset{O(CH_2)_2N(CH_3)_3}{<}}$	0.022
Choline	$HO(CH_2)_2N(CH_3)_3$	24
Butyrylcholine	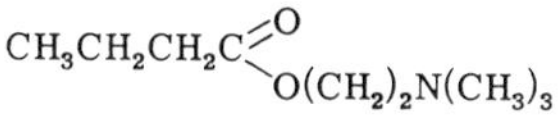 $CH_3CH_2CH_2C\overset{O}{\underset{O(CH_2)_2N(CH_3)_3}{<}}$	0.0055
Succinylcholine	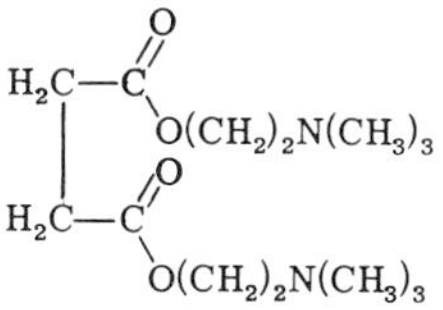 $\begin{array}{l} H_2C-C\overset{O}{\underset{O(CH_2)_2N(CH_3)_3}{<}} \\ \quad\mid \\ H_2C-C\overset{O}{\underset{O(CH_2)_2N(CH_3)_3}{<}} \end{array}$	0.012
Carbachol	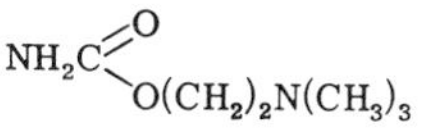 $NH_2C\overset{O}{\underset{O(CH_2)_2N(CH_3)_3}{<}}$	0.015

FIG. 2. Cross-reactivity of various compounds with acetylcholine antibody.

eserine in a final volume of 500 μl. Assays could be performed using a 1 or 16 hr incubation at 4°. Antibody-bound label was precipitated by addition of an equal volume of saturated ammonium sulfate.[1] Following centrifugation, the pellet was dissolved in distilled water (500 μl), and an aliquot (400 μl) was taken for liquid scintillation counting in 10 ml of Aquasol (New England Nuclear, Boston, MA). A blank was included in which the sample was reassayed after alkaline hydrolysis (pH 11, 60° for 20 min), and this value was subtracted from the determination of the original homogenate.

Compound	ID_{50} (nmol)
Nicotine	4.5
Atropine	15
Curare	70
Acetyl-β-methylcholine	0.3
Neostigmine	0.4
L-α-Lecithin	50 μg
L-α-Lysophosphatidylcholine	65 μg
Sphingomyelin	> 10 μg

FIG. 3. Cross-reactivity of various compounds with acetylcholine antibody.

[1] A. S. Farr, J. Infect. Dis. 103, 239 (1958).

Antibody-bound counts in the presence of cold ligand were calculated as a percentage of total counts bound in the absence of cold ligand. A standard curve was constructed as described by Rodbard et al.[2] and the concentration of ACh in the samples determined by interpolation.

The previous paper[3] describes radioimmunoassays for morphine and includes references to radioimmunoassays for other compounds.

[2] D. Rodbard, W. Bridson, and P. L. Rayford, *J. Lab. Clin. Med.* **74,** 770 (1969).
[3] S. Spector, this volume [40].

[42] Immunoassay of Digoxin and Other Cardiac Glycosides

By VINCENT P. BUTLER, JR. and DORIS TSE-ENG

Cardiac glycosides obtained from various plant sources, notably digitalis leaf and strophanthus seed, are widely used in the treatment of congestive heart failure and of various disturbances of cardiac rhythm. The use of cardiac glycosides in patients with congestive heart failure is based on their ability to increase the force of myocardial contraction, thereby increasing the output of the failing heart. Cardiac glycosides also exert significant electrophysiological effects on the heart, notably on atrioventricular conduction, which cause a beneficial slowing of the ventricular rate in patients with certain tachycardias of supraventricular origin. Toxic amounts of cardiac glycosides exert undesirable and potentially lethal electrophysiological effects, notably atrial tachycardia, complete atrioventricular block, atrioventricular nodal tachycardia, ventricular extrasystoles, ventricular tachycardia, and ventricular fibrillation.[1] By virtue of their ability to inhibit Na^+,K^+-ATPase-mediated influx of potassium into cells, toxic concentrations of cardiac glycosides may cause severe, life-threatening hyperkalemia.[2] The clinical use of these valuable, but potentially dangerous, therapeutic agents is complicated by the fact that individual patients vary considerably both in the dosage required to produce a beneficial therapeutic response and in their sensitivity to the toxic effects of these drugs. Studies with radiolabeled cardiac glycosides and with cardiac glycoside immunoassays have provided evidence that inadequate body stores of these drugs, as reflected clinically in relatively low

[1] B. F. Hoffman and J. T. Bigger, Jr., *in* "The Pharmacological Basis of Therapeutics" (A. G. Gilman, L. S. Goodman, and A. Gilman, eds.), 6th ed., p. 729. Macmillan, New York, 1980.
[2] C. Bismuth, M. Gaultier, F. Conso, and M. L. Efthymiou, *Clin. Toxicol.* **6,** 153 (1973).

serum concentrations, are often a major factor in a poor clinical response to this form of therapy; conversely, excessive accumulation of glycosides, as reflected clinically in relatively high serum levels, is often a major factor in the appearance of serious cardiac glycoside intoxication. Accordingly, radioimmunoassays for cardiac glycosides are now widely used clinically as aids in the determination of appropriate dosage schedules for patients receiving these drugs.[3] Radioimmunoassays have also been employed extensively in studies of the bioavailability, intestinal absorption, pharmacokinetics, metabolism, and excretion of this group of drugs.[4]

Digoxin, derived from the leaves of *Digitalis lanata,* is the cardiac glycoside prescribed most frequently in English-speaking countries. Accordingly, the most extensive experience with cardiac glycoside immunoassays has been obtained with digoxin radioimmunoassay procedures. Thus, the digoxin radioimmunoassay, as performed in our laboratory, will be described in detail but other cardiac glycoside immunoassay procedures will be referred to, when appropriate.

Synthesis of Digoxin–Protein Conjugates

Digoxin, with a molecular weight of 780.92, is too small a molecule to be antigenic by itself and thus, in order to elicit digoxin-specific antibodies in experimental animals, it was necessary to conjugate digoxin covalently as a hapten to antigenic protein carriers, such as bovine serum albumin (BSA) or human serum albumin (HSA). This was accomplished by a modification of the periodate oxidation method of Erlanger and Beiser.[5] As shown in Fig. 1, digoxin consists of the pharmacologically active aglycone, digoxigenin (containing a steroid nucleus and a lactone ring), linked at the C-3 position to three pharmacologically inactive glycosidic digitoxose residues. The terminal digitoxose residue is oxidized by sodium metaperiodate to form the dialdehyde derivative shown at the upper right of Fig. 1. When this dialdehyde derivative is added to a protein solution at a mildly alkaline pH, it reacts with free amino groups (ϵ-amino groups of lysine and the α-amino group of the NH_2-terminal amino acid) to form an unstable Schiff-base type compound which, after reduction with sodium borohydride, yields the stable conjugate depicted at the lower right of Fig. 1. After two dialyses and a precipitation step to remove unconjugated gly-

---

[3] V. P. Butler, Jr. and J. Lindenbaum, *Am. J. Med.* **58,** 460 (1975).
[4] G. Bodem and H. J. Dengler (eds.), "Cardiac Glycosides." Springer-Verlag, Berlin and New York, 1978.
[5] B. F. Erlanger and S. M. Beiser, *Proc. Natl. Acad. Sci. U.S.A.* **52,** 68 (1964).

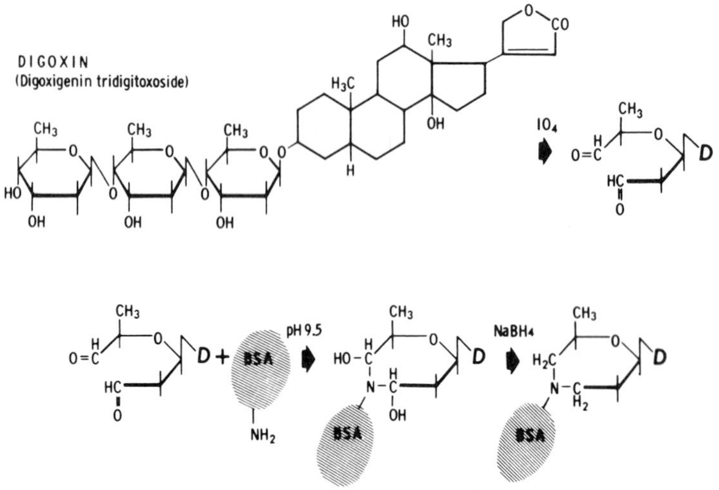

FIG. 1. Schematic representation of conjugation of digoxin to a representative protein carrier, bovine serum albumin (BSA). D, Digoxigenin didigitoxoside. See text for details. From V. P. Butler, Jr., D. H. Schmidt, J. F. Watson, and J. D. Gardner, *Ann. N.Y. Acad. Sci.* **242**, 717 (1974).

coside and other reaction products, the conjugate is lyophilized and stored.[6]

The periodate oxidation method has also been used to conjugate ouabain,[7] a cardiac glycoside obtained from *Strophanthus gratus* seed, and proscillaridin,[8] a glycoside derived from the bulb of squill (sea onion), to the amino groups of protein carriers via their glycosidic rhamnose residues. Succinyl derivatives of aglycones of two digitalis glycosides, digitoxigenin[9] and gitaloxigenin,[10] have been conjugated to the amino groups of protein carriers by the carbodiimide method and, in the case of 3-*O*-succinyldigitoxigenin, also by the mixed anhydride coupling method.[9]

The periodate oxidation method, as used to conjugate digoxin to bovine (or human) serum albumin, is described in detail below.

Materials Needed

Digoxin powder, Burroughs Wellcome Company, Research Triangle Park NC

95% ethanol

[6] V. P. Butler, Jr. and J. P. Chen, *Proc. Natl. Acad. Sci. U.S.A.* **57**, 71 (1967).

[7] T. W. Smith, *J. Clin. Invest.* **51**, 1583 (1972).

[8] G. G. Belz, W. J. Brech, U. R. Kleeberg, G. Rudofsy, and G. Belz, *Naunyn Schmiedebergs Arch. Pharmacol.* **279**, 105 (1973).

[9] G. C. Oliver, Jr., B. M. Parker, D. L. Brasfield, and C. W. Parker, *J. Clin. Invest.* **47**, 1035 (1968).

[10] M. Lesne and R. Dolphen, *J. Pharmacol. (Paris)* **7**, 619 (1976).

0.1 M sodium metaperiodate (ordinarily prepared on day of use; may be kept for up to 1 week at 4° in dark container)

Bovine serum albumin (BSA), obtained as Fraction V powder, Miles Laboratories, Kankakee IL; human serum albumin (HSA) has also been used as a digoxin carrier in our laboratory with equally good results

5% potassium carbonate

1 M ethylene glycol

Sodium borohydride

1 M formic acid

0.1 N hydrochloric acid

10% trichloroacetic acid

0.15 M NaHCO$_3$

Dialysis tubing, 20 mm and 27 mm (flat width)

Procedure

1. Suspend 438 mg digoxin (0.56 mmol) in 20 ml 95% ethanol. Insert glass-covered or Teflon-coated magnet and stir on magnetic stirrer (digoxin will not dissolve completely).

2. To the digoxin suspension, add 20 ml freshly prepared 0.1 M NaIO$_4$ dropwise over a 5-min period, with continual magnetic stirring. Keep stirring for 30 min, during which time the suspension will become somewhat clearer. During this 30-min period, complete Step 3.

3. Dissolve 560 mg BSA in 20 ml deionized water, using a 50-ml beaker. Place a small glass-covered or Teflon-coated magnet in the beaker and insert a small pH meter electrode. Determine pH and initiate magnetic stirring. Add 5% K$_2$CO$_3$ dropwise until pH is approximately 9.5 (five or six drops will usually be required). Discontinue stirring until start of Step 5.

4. After 30 min of interaction between digoxin and NaIO$_4$, add 0.6 ml 1 M ethylene glycol (to inactivate excess periodate) and continue stirring for 5 additional min.

5. Resume stirring of albumin solution and add the periodate-oxidized digoxin mixture dropwise, with continual magnetic stirring, to the albumin, maintaining the pH in the 9.3–9.5 range by the simultaneous dropwise addition of 5% K$_2$CO$_3$ (usually 2 to 3 ml is required). Midway during this procedure (after approximately 20 ml of the periodate-oxidized digoxin solution has been added), it will be necessary to transfer the contents of the reaction mixture (including the magnet) to a 100-ml beaker; after this transfer, continue to add the periodate-oxidized digoxin and 5% K$_2$CO$_3$ with magnetic stirring and pH monitoring as before. CAUTION: *Do not mouth-pipet the digoxin-containing solution! Ingestion of as little as 10 mg may be lethal.*

6. After the addition of periodate-oxidized digoxin to albumin has

been completed, continue stirring for 45–60 min or until the pH has been stable for 20–30 min. Use 5% K_2CO_3 to maintain a pH of 9.3–9.5; approximately 0.5–1.0 ml is usually required.

7. After the pH has stabilized in the 9.3–9.5 range, add 300 mg $NaBH_4$ *freshly* dissolved in 20 ml deionized water. Note pH and transfer the reaction mixture to a 125-ml Erlenmeyer flask; cover *loosely* with parafilm (do not use rubber stopper; release of H_2 could cause explosion). Set flask aside for 4 hr (or overnight) at room temperature.

8. After 4–18 hr, transfer reaction mixture to beaker. Note pH and, with continual pH monitoring and magnetic stirring, add 1 M formic acid dropwise until pH is approximately 6.5 (7–8 ml is usually required). Allow to stand 1 hr at room temperature.

9. After 1 hr, add 1 M NH_4OH dropwise with stirring to pH 8.5 (approximately 1.5–2 ml will be required).

10. Transfer entire reaction mixture to 27-mm dialysis tubing (allow sufficient air-free dead space to permit a 100% increase in volume during dialysis). Affix glass stopper as anchor to one end of tubing, place in a 2000-ml cylinder or other large vessel and dialyze overnight against cold running tap water (may dialyze for up to 72 hr, if overnight dialysis is inconvenient).

11. Transfer contents of dialysis bag to a 250-ml beaker and add 0.1 N HCl dropwise with continual magnetic stirring and pH monitoring until pH is 4.5 (or until precipitation is maximal).

12. Resuspend the precipitate and transfer the suspension to a 250-ml round-bottomed centrifuge bottle. After 1 hr of standing at room temperature and 3 hr at 4°, centrifuge for 1 hr at 4° and 1000 g.

13. Decant supernatant into any vessel. Add 1 ml of supernatant to 9 ml 10% trichloroacetic acid; lack of a visible precipitate signifies absence of protein in the supernatant, which can then be discarded.

14. Add 0.15 M $NaHCO_3$ to centrifuge bottle slowly until precipitate has redissolved. Usually, no more than 10 ml is required.

15. Transfer to 20-mm dialysis tubing and dialyze for 36–120 hr against cold running tap water, as described above in Step 10.

16. Transfer contents of dialysis bag to a lyophilization flask or bottle, and lyophilize to dryness. Weigh powder to determine yield and store immediately in well-stoppered container at 4°. Conjugate will remain stable indefinitely in the lyophilized state at this temperature.

Characterization of Digoxin–Protein Conjugates

Prior to immunization of animals with a newly prepared digoxin–protein conjugate, it is useful to determine the extent of incorporation of digoxin into the conjugate. This is most conveniently done by a simple colorimetric procedure which takes advantage of the fact that, when dissolved in concentrated sulfuric acid, digoxin exhibits absorption maxima at 388 and at 466 nm.[11] The optical density of albumin solutions at these wavelengths is negligible. Hence, any absorption by a digoxin–albumin conjugate is caused by its digoxin moiety and, if the molar extinction coefficient of digoxin at these wavelengths is simultaneously determined, the extent of digoxin incorporation into the conjugate can readily be calculated, as described below.

In our experience, various conjugates containing between three and ten digoxin residues per molecule of albumin carrier have been excellent immunogens. We have not determined the minimum (or maximum) number of molecules of digoxin per molecule of carrier which will elicit satisfactory antibody production.

Materials Needed

Digoxin
Unconjugated carrier protein
Digoxin–protein conjugate
Concentrated sulfuric acid (analytical reagent grade)

Procedure

1. On an analytical balance which is accurate to 0.1 mg or less, weigh out the following in separate 25-ml glass-stoppered Erlenmeyer flasks: 5 mg of the digoxin–protein conjugate, 5 mg digoxin, 20 mg of the same lot of protein to which the digoxin was conjugated.

2. Add 4 ml deionized water to each flask. When the conjugate and protein have dissolved (digoxin is insoluble in water), carefully add 20 ml concentrated sulfuric acid. Allow to stand at room temperature for several hours or overnight. A reddish-brown color will appear in the flask containing digoxin or conjugate. (Since the coloration appears to increase with time, it is important to prepare and analyze all three solutions simultaneously.)

3. When all three solutions are clear, determine the optical density (OD) of each at 388 and at 466 nm.

4. Determine the molar extinction coefficient, E_{Molar}, of each compound at both wave-lengths, according to the equation: $E_{Molar} = OD/Molarity$. The molecular weight of digoxin is 780.92. Since the molecular weight of the conjugate represents that of the protein carrier plus that of conjugated digoxin, one must initially use an estimated molecular weight; in the case of a BSA carrier, we use an initial estimate of five digoxin residues per molecule of albumin carrier and correct this figure after our initial calculations.

---

[11] B. T. Brown and S. E. Wright, *J. Am. Pharm. Assoc. Sci. Ed.* **49**, 777 (1960).

5. To calculate n, the number of digoxin residues conjugated to each molecule of protein carrier, use the equation:

$$n = \frac{E_{\text{Molar}}(\text{Conjugate}) - E_{\text{Molar}}(\text{Protein carrier})}{E_{\text{Molar}}(\text{Digoxin})}$$

Immunization

Antibodies of high titer, specificity, and affinity have been obtained by immunization of rabbits and sheep with digoxin–protein conjugates in complete Freund's adjuvant mixture. It should be noted, however, that antibodies obtained after a few weeks of immunization may be of high titer but their specificity and affinity may be considerably less than the specificity and affinity attained after several months of immunization.[12]

Materials Needed

Digoxin–protein conjugate

0.15 M NaCl or isotonic phosphate-buffered NaCl, pH 7.4

Arlacel A, Hill Top Research, Miamiville OH

Light mineral oil (Marcol 52, Exxon Company, Houston TX, is a very satisfactory preparation)

Dried, killed tubercle bacilli, Lederle Laboratories, Pearl River NY

Procedure

1. Prepare complete Freund's adjuvant mixture by suspending 85 mg dried, killed tubercle bacilli in a mixture of 15 ml Arlacel A and 85 ml mineral oil (stable for several months at room temperature or at 4°). A preparation of complete Freund's adjuvant mixture is available from Difco Laboratories, Detroit MI, but we have not had experience with its use.

2. Dissolve digoxin–albumin conjugate at a concentration of 2 mg/ml in 0.15 M NaCl or in isotonic phosphate-buffered saline, pH 7.4.

3. Mix exactly equal volumes of Freund's complete adjuvant mixture and of the digoxin–albumin solution and emulsify. Emulsification can be achieved by the use of two glass syringes connected by a double-hub needle or by polyethylene tubing, as described in detail elsewhere in this series.[13] When adequately emulsified, a single drop should maintain a spherical shape after being placed on a water surface. Digoxin–protein conjugates are stable in Freund's complete adjuvant mixture for at least 4–6 weeks at 4°C; however, separation into two phases often occurs and reemulsification should be carried out immediately prior to all animal injections.

4A. Rabbits. Select 2.5–4 kg New Zealand white rabbits. Obtain, if

---

[12] V. P. Butler, Jr., *Pharmacol. Rev.* **29**, 103 (1977).

[13] B. A. L. Hurn and S. M. Chantler, this series, vol. 70 [5].

desired, a preimmunization blood specimen and cut away hair covering toepads, using curved scissors. Using a small syringe with a 25-gauge needle, inject 0.1 ml of antigen emulsion into one toepad on each front foot and 0.2 ml into one toepad on each rear foot; repeat twice at weekly intervals. Then, inject 0.4 ml of antigen emulsion intramuscularly in divided doses of 0.2 ml at each of two sites once weekly for 5 additional weeks (intramuscular injections may be given in the haunches or paraspinal muscles); continue intramuscular injections at fortnightly intervals for 2–4 months, and at monthly intervals thereafter.

4B. Sheep. Using a syringe and 20-gauge needle, inject 1.0 ml of antigen emulsion intramuscularly in divided doses of 0.5 ml at each of two sites in the buttocks. Repeat at fortnightly intervals for 3–4 months, and at monthly intervals thereafter.

5. Animals may be bled twice monthly, preferably at least 5 days after the last previous antigen injection. From rabbits, 40–50 ml blood can be obtained. The volume of blood which can be obtained from sheep varies with body size. As much as 500 ml may be obtained fortnightly from 60 to 70 kg sheep; if this volume is removed regularly, 5 ml of iron dextran (Merrell-National Laboratories, Cincinnati OH) should be injected intramuscularly after each bleeding and periodic determinations of hematocrit (or hemoglobin concentration) should be performed. Sheep in an animal care facility will tolerate a hematocrit of 30% without apparent adverse effects but, if the hematocrit falls below this level, smaller or less frequent bleedings would be advisable.

Serum should be separated promptly and stored at 4° or in the frozen state. If the serum has not been obtained aseptically, the addition of thimerosal (Merthiolate, Eli Lilly and Company, Indianapolis IN; 0.1 ml of a 1% solution per 10 ml of serum) as a preservative is advisable.

Antibodies can ordinarily be detected in serum within 1 month after the initiation of immunization. However, these antibodies are often of relatively low affinity and specificity. Thus, it is our practice to initiate bleeding after 3 months of immunization.

Comment. Antiserum may be purchased from commercial sources but affinity, specificity, and titer are not uniformly satisfactory. It is advisable, before such antiserum is purchased, to obtain data from the vendor concerning affinity, titer, and reactivity with steroid hormones and drugs (as described in later sections of this chapter).

Selection of a Labeled Digitalis Preparation

The most commonly used methods for the detection of antibodies to cardiac glycosides involve the demonstration of the binding of ³H-labeled digitalis preparations or of ^{125}I-labeled digitalis derivatives; enzyme-labeled digoxin derivatives have been used in the detection of digoxin-specific antibodies.[12] For the detection of digitalis-specific antibodies, ^{125}I-labeled derivatives have major advantages over tritiated digitalis preparations in view of the greater convenience of determining γ-radiation in comparison with measuring β-radiation in liquid scintillation counting fluid. The measurement of ^{3}H-labeled antibody-bound haptens is further complicated by the fact that specific quenching of radioactivity may occur in the presence of an excess of antibody. Such quenching is not detected by the ordinary methods for the determination of counting efficiency in liquid scintillation counters; this effect can be eliminated by heating the ^{3}H-labeled digitalis complex in liquid scintillation fluid (presumably due to denaturation of the antibody in the presence of an organic solvent), but this extra step makes the use of ^{3}H-labeled digitalis preparations somewhat cumbersome in the detection of digitalis-binding antibodies.

Despite their clearcut advantages in the detection of antibodies to digitalis, some ^{125}I-labeled preparations have limitations in the radioimmunoassay of clinical and other biological specimens. Variations in values obtained from individuals who have not received digitalis have been reported.[12] Accordingly, before an ^{125}I-labeled digitalis derivative is employed in a radioimmunoassay procedure, it is advisable to compare the results obtained with the results observed with a tritium radioimmunoassay method or other digitalis assay procedure.[14,15]

Significant amounts of radiochemical impurities may be present in radiolabeled digitalis preparations. These may be present upon delivery from the manufacturer or they may appear due to chemical degradation of the radiolabeled compound during storage in the laboratory. Tritium-labeled compounds may remain stable for a year or more when stored at $-20°$, but ^{125}I-labeled digitalis derivatives may deteriorate within 6–8 weeks of preparation. The appearance of radiochemical impurities may be detected by chromatographic analysis, but a more convenient method is routine determination, with each assay performed, of the percentage of radioactivity bound by an excess of digitalis antibody. In the case of ^{3}H-labeled digitalis preparations, 90–95% of the radioactivity should be bound and, in our experience, a decrease in binding is accompanied by an increase in the percentage of radiochemical impurities detected by thin-layer chromatography.

[14] F. I. Marcus, J. N. Ryan, and M. G. Stafford, *J. Lab. Clin. Med.* **85**, 610 (1975).
[15] V. P. Butler, Jr., *in* "Cardiac Glycosides" (G. Bodem and H. J. Dengler, eds.), p. 1. Springer-Verlag, Berlin and New York, 1978.

Detection of Antidigoxin Antibodies

If a radiolabeled digitalis preparation is used for the detection of antibodies, a method for the separation of antibody-bound ("bound") from unbound ("free") radiolabel must be employed. Equilibrium dialysis was originally employed for this purpose,[6] but this method is too cumbersome for routine use. Numerous other methods have subsequently been used to separate bound from free label; these methods include the dextran-coated charcoal separation method, ammonium sulfate precipitation, polyethylene glycol precipitation, Somogyi precipitation, double antibody precipitation, membrane filtration, gel equilibration, and a number of solid-phase methods.[12]

In our experience, the dextran-coated charcoal method of Herbert *et al.*[16] has proved to be reliable and to be the most convenient method for the detection of antidigoxin antibodies.[17,18] Accordingly, this method as currently employed in our laboratory[19] will be described in detail. This method is based on the fact that free, non-antibody-bound radiolabeled digitalis (or digitalis derivative) is rapidly and almost completely adsorbed to the surface of charcoal whereas, in the presence of high concentrations of dextran and albumin, antibodies are not adsorbed to a significant extent; accordingly, after addition of dextran-coated charcoal followed by immediate centrifugation, free tracer will be present in the charcoal precipitate while antibody-bound radioactivity will be present in the supernatant where, after aspiration or decanting, it can be readily counted. In our experience, this method is more reliable and reproducible than precipitation of antibody-bound radioactivity by the ammonium sulfate (Farr), polyethylene glycol, or Somogyi methods and it is more convenient for antibody detection than double antibody or solid-phase antibody methods.

The principal limitation of the dextran-coated charcoal method is the fact that any bound tracer which dissociates from antibody during exposure to charcoal may be adsorbed to the charcoal. With antibodies of high affinity and with brief incubation periods with charcoal, such adsorption should be minimal; however, with prolonged exposure of low-affinity antibodies to charcoal, it can be appreciable.[20,21] This limitation must be borne in mind in using this method but, if centrifugation is carried out im-

[16] V. Herbert, K.-S. Lau, C. W. Gottlieb, and S. J. Bleicher, *J. Clin. Endocrinol. Metab.* **25,** 1375 (1965).

[17] T. W. Smith, V. P. Butler, Jr., and E. Haber, *New Engl. J. Med.* **281,** 1212 (1969).

[18] T. W. Smith, V. P. Butler, Jr., and E. Haber, *Biochemistry* **9,** 331 (1970).

[19] C. J. Hayes, V. P. Butler, Jr., and W. M. Gersony, *Pediatrics* **52,** 561 (1973).

[20] R. C. Meade and T. J. Kleist, *J. Lab. Clin. Med.* **80,** 748 (1972).

[21] T. W. Smith and E. Haber, *Pharmacol. Rev.* **25,** 219 (1973).

mediately after the addition of coated charcoal, it should not constitute a significant problem when one uses this method for antibody detection.

Materials Needed

Antidigoxin serum

Control serum from a nonimmunized animal of the same species

Radiolabeled digitalis or digitalis derivative

a. ³H-labeled digitalis preparations. Tritium-labeled digoxin, digitoxin, and ouabain (specific activity > 10 Ci/mmol) are available from the New England Nuclear Corp., Billerica MA, and from Amersham Corporation, Arlington Heights IL. Radiochemical purity may be assessed by thin-layer chromatography. In the case of ³H-labeled digoxin, this is conveniently performed on silica gel G plates (Analtech, Newark DE) in a solvent system consisting of cyclohexane : acetone : acetic acid in a ratio of 49:49:2. After developing, 0.5-cm segments of the gel are transferred to vials containing liquid scintillation counting solution and counted in a liquid scintillation spectrometer. More than 95% of the recovered radioactivity should migrate with an R_f identical to that of a simultaneously analyzed nonradioactive digoxin standard (which may be detected by spraying with an aqueous solution containing (w/v) 3% chloramine-T and 25% trichloroacetic acid.[22]

b. ¹²⁵I-labeled digitalis derivatives. Such derivatives may be synthesized by the method of Oliver *et al.*[9] or may be purchased from a commercial source such as Wellcome Reagents Division, Burroughs Wellcome Company, Research Triangle Park NC; Clinical Assays Inc., Cambridge MA; New England Nuclear Corp., Billerica MA; or, Schwarz/Mann Division, Becton Dickinson and Company, Orangeburg NY.

Tris base (Sigma Chemical Company, St. Louis, MO)

Human serum albumin, HSA (Albumisol, 5% HSA, Merck Sharpe & Dohme Co, West Point PA, is a highly satisfactory preparation)

Dextran T-70 (Pharmacia Fine Chemicals, Piscataway NJ)

Activated charcoal (Norit, neutral, Fisher Scientific Company, Fair Lawn NJ)

Liquid scintillation counting solution (only if ³H-labeled tracer is used). Any solution which can be used with aqueous samples of at least 0.5 ml volume should be satisfactory. Solutions which we have found to be useful include Bray's solution,[23] a toluene-Triton

[22] J. F. Watson and V. P. Butler, Jr., *J. Clin. Invest.* **51**, 638 (1972).
[23] G. A. Bray, *Anal. Biochem.* **1**, 279 (1960).

X-100 scintillant,[24] Biofluor (New England Nuclear, Billerica MA) and Liquiscint (National Diagnostics, Parsippany NJ).

Preparation of Reagents

Tris-buffered saline (TBS; 0.14 M NaCl, 0.01 M Tris–Cl, pH 7.4)

Dissolve 81.6 g NaCl and 12.1 g Tris base in water, adjust pH to 7.5 ± 0.05 with 2 N HCl (40–42 ml usually required) and add water to a final volume of 2 liters (5× strength); dilute one volume with four volumes water before use (pH falls about 0.1 unit with dilution). Stable for up to 1 year at room temperature.

Isotonic phosphate buffered saline, pH 7.4, may also be used.

Tris-buffered albumin

Dilute 7 ml Albumisol to 100 ml with TBS. Stable for up to 7–10 days at 4°; discard if turbidity develops.

Radiolabeled digitalis or digitalis derivative

If supplied in an organic solvent, the solvent should be evaporated with air or nitrogen and the tracer dissolved in Tris-buffered albumin.

Dilute to a concentration of 0.1–0.2 μCi/ml with Tris-buffered albumin, divide into aliquots, and store at −20°. Avoid repetitive freezing and thawing.

Dextran-coated charcoal

Thoroughly mix equal volumes of dextran T-70 (0.5% solution in TBS) and of activated charcoal (5% suspension in TBS) and store in glass container (charcoal adheres to Teflon and other plastics). Stable for up to 1 month at 4°. Shake well to resuspend before use.

Antiserum and control sera

a. Prepare 1:50 dilutions of antiserum and of control sera in Tris-buffered albumin. Store in small aliquots in the frozen state (avoid frequent thawing and refreezing which causes antibody denaturation).

b. Thaw, mix well, and dilute 1:10 with Tris-buffered albumin when ready for use. Dilute antiserum only in 5 to 9 further serial twofold steps.

Procedure

1. To 0.5 ml Tris-buffered albumin, add 50 μl tracer and 50-μl aliquots of all antiserum and control serum dilutions or of Tris-buffered albumin (to be used for the determination of standard and blank values); prepare additional tubes containing 0.6 ml of Tris-buffered albumin for background determination. All determinations should be performed in duplicate. Mix well on a vortex mixer. Incubate 1 hr at room temperature or at 37°.

[24] M. S. Patterson and R. C. Greene, *Anal. Chem.* **37**, 854 (1965).

2. Resuspend dextran-coated charcoal by vigorous shaking. Maintain a uniform suspension by continuous magnetic stirring. Using a 1-ml Cornwall syringe pipet and 4-in. 14-gauge cannula (Arthur H. Thomas Co., Philadelphia), add 0.25 ml dextran-coated charcoal to all tubes, except to the duplicate tubes to be used for the determination of standard values ("total counts"); add 0.25 ml of Tris-buffered albumin to this latter pair of tubes. Subject all tubes to immediate vortex mixing.

3. Centrifuge immediately at 4° for 15 min at 2000 g (or for 30 min at 1000 g).

4. Promptly transfer supernatant to tube or vial to be used for radioactivity determinations. Supernatant may be decanted into counting tube or vial or a measured aliquot (usually 0.5 ml) may be aspirated and transferred. The supernatant radioactivity is then counted.

5. The amount of supernatant radioactivity present in blank specimens ("dextran-coated charcoal control" specimens which contain no antibody) is subtracted from the supernatant radioactivity present in all test and standard ("total count") tubes. The bound radioactivity is then expressed as the percentage of supernatant radioactivity present in each tube when compared with that present in standards.

6. It is convenient to express the titer of each antiserum as the greatest dilution of that antiserum capable of binding 50% of a standard amount of the radiolabeled digitalis preparation under the specific conditions employed. It is apparent that the titer of a given antiserum may vary under different experimental conditions. Moreover, since titer reflects both antibody affinity and content, it cannot be used as an absolute measurement of antibody content. Nevertheless, serial titers are useful in assessing the responses of individual animals to immunization; titers are also useful in estimating the number of assay determinations which can be performed per ml of antiserum.

Special Procedure for Liquid Scintillation Counting of Tritiated Digitalis Preparations

1. After selection of a liquid scintillation counting solution, one should determine whether that scintillant permits accurate detection of tracer in the presence of antibody as follows: Add 50 μl tracer (0.1–0.2 μCi/ml) to 0.5–1.0 ml buffered albumin and to an equal volume of several dilutions (e.g., 1:20, 1:200, 1:2000, 1:20,000) of an antiserum of known high potency. Mix with scintillant and count (if it is necessary to use an internal standard to determine counting efficiency, it is essential that a nondigitalis tracer be employed). If >95% of the added radioactivity is detected in all tubes, the scintillant may be employed without a heating

step. If detection of radioactivity is suboptimal, one should determine appropriate conditions of heating to ensure complete recovery. Heating for 10 min at 60° has proved useful for counting in Bray's solution,[17] while a 30 min incubation at 80° has been necessary when using a toluene-Triton X-100 scintillant.[19]

2. Because of problems inherent in counting in liquid scintillation spectrometers, occasional spurious counts may be recorded. Furthermore, chemiluminescence may be present in biological specimens being assayed (this may be accentuated by the heating step, if this is required). Accordingly, all specimens should be counted at least twice to ensure that the recorded counts are accurate and stable.

3. If the liquid scintillation spectrometer being employed does not express its results in disintegrations per minute (dpm), counting efficiency must be determined in all vials (by an automatic external standardization device, by a channels ratio method, or by the addition of an internal standard) and quenching corrections must be made before results can be calculated.

Determination of Antibody Specificity

Because digoxin is conjugated to protein carriers via its glycosidic portion, animals immunized with digoxin–protein conjugates form antibodies with specificity for the digoxigenin portion of the molecule. Antidigoxin sera cannot effectively distinguish between digoxin and other digoxigenin-containing glycosides such as deslanoside, lanatoside C, methyldigoxin, and acetyldigoxin. This cross-reactivity presents no problems, if it is known with certainty which glycoside a patient has been receiving; as a matter of convenience, the digoxin RIA procedure may be used to measure any of these glycosides.[12] Antidigoxin sera also cannot distinguish effectively between digoxin and certain of its metabolites, specifically digoxigenin and its mono- and bisdigitoxosides; since these metabolites are pharmacologically active, their contribution to the serum immunoreactive digoxin concentration creates no major clinical problem since it reflects cardioactivity as well.[12,25] All antidigoxin sera cross-react to some extent with cardiac glycosides other than digoxin, notably digitoxin, gitoxin, ouabain, etc.; again, this cross-reactivity creates no problems if it is known what glycoside a patient is receiving, but it must be remembered that other cardiac glycosides (notably digitoxin) will give a spurious and misleading value when serum from patients receiving these drugs is subjected to a serum digoxin RIA procedure.[25] Antidigoxin sera also cross-react to some extent with most steroid hormones, prednisone, spironolac-

[25] V. P. Butler, Jr., *Prog. Cardiovasc. Dis.* **14**, 571 (1972).

tone, and cardioinactive digoxin reduction products, notably dihydrodigoxigenin and its digitoxosides. These latter cross-reactions can create serious problems in the clinical interpretation of results and it is therefore important that antisera employed in clinical assays do not cross-react extensively with these compounds; as with antibody affinity, specificity is generally greater when one examines antisera obtained after several months of immunization than when earlier bleedings are examined.[12] Accordingly, it is essential that the cross-reactivity of an antidigoxin serum with these compounds be assessed before that antiserum is employed in clinical assays. In our experience, testosterone and progesterone have been the most reactive of the endogenous steroid hormones.[18] We therefore routinely assess specificity by comparing the relative abilities of these two steroid hormones, spironolactone, prednisone, and digoxin to inhibit the binding, by antidigoxin serum, of radiolabeled digoxin or a radiolabeled digoxin derivative.

Reagents

Digoxin

Prepare 3 mg/ml solution in 95% ethanol

Dilute with Tris-buffered albumin, pH 7.4, to the 0.5–25 pmol/ml range.

Testosterone, progesterone, spironolactone, prednisone (Sigma Chemical Co., St. Louis, MO)

Dissolve at highest possible concentration in 95% ethanol

Dilute 1:20 in Tris-buffered albumin, and then serially dilute twofold to a final dilution of 1 pmol/ml

Antidigoxin serum

Diluted in Tris-buffered albumin to a concentration which will bind 40–50% of immunoreactive tracer

All other reagents are as described above for antibody detection.

Procedure

1. To duplicate 500-μl aliquots of all concentrations of digoxin and the other inhibitors, add 50 μl radiolabeled digoxin or digoxin derivative; mix vigorously with vortex mixer. Include buffer control tubes to determine antibody binding in the absence of inhibitor; also include other control tubes (standard, blank, background) as described above for antibody detection.

2. Add 50 μl of an appropriate dilution of antiserum, followed by vortex mixing.

3. Incubate 1 hr at room temperature or at 37°.

4–7. Separate free from bound radioactivity by the coated charcoal method, count radioactivity, and calculate results as described in Steps 2–5 for antibody detection.

8. Percentage of tracer (or of immunoreactive tracer) bound is plotted on the ordinate of semilogarithmic graph paper against the $\log_{10}$ of the concentration of digoxin and of each of the other inhibitors. The concentration of each of these other inhibitors required to produce 50% inhibition of binding under these conditions (I_{50}) is determined and compared with the I_{50} of digoxin in the same assay.

Comment. The ratio of the I_{50} of each inhibitor to the I_{50} of digoxin under identical assay conditions should be at least 100-fold greater than the ratio of the highest anticipated physiological or pharmacological concentration of that inhibitor (expressed in pmol/ml) to a low therapeutic concentration of digoxin (1 pmol/ml). With satisfactory antisera, ratios 1000–10,000 times as great may be anticipated.

Antibody Affinity

With immunization for 3 months or longer, average intrinsic association constants of antidigoxin sera regularly exceed 10^9 and sometimes are as great as 10^{10}. Although the association constants of antidigoxin sera can readily be determined by equilibrium dialysis,[18] it is rarely necessary to determine these values directly. The adequacy of the association constant can be inferred indirectly from the sensitivity of the assay procedure employing a given antiserum and from the steepness of the slope of the inhibition curve with digoxin. From a practical point of view, the dissociation rate constant, if too great, can create problems in the coated charcoal assay procedure, as discussed above in the antibody detection section. It is therefore important to determine that significant dissociation of bound antibody does not occur during the charcoal separation step.[20,21] This can readily be assessed by determining the decrease in the percentage of tracer bound to antibody as a function of time after the addition of charcoal; if there is more than a 10% decrease in bound tracer within 10–15 min, an antiserum may give variable and unpredictable results, particularly when an assay is a lengthy one or when unpredictable delays are encountered during the charcoal separation step.

Radioimmunoassay Procedure

The method described below is a reliable one which has yielded precise and reproducible results in our hands. Numerous other reliable methods have been described, while other methods are less precise and reproducible for various reasons. Assay procedures which are useful for large-scale, daily routine laboratory use and which are susceptible to automation sometimes are less precise because: they employ ^{125}I-labeled tracers rather than ^{3}H-labeled glycosides; they employ solid-phase anti-body procedures in which antibody binding to a solid surface is noncova-lent and somewhat variable; the volume of serum assayed is too small; or, the antibody employed has suboptimal specificity or affinity. Although somewhat less precise than the method described below, such assay pro-cedures often provide results which are quite satisfactory for their pri-mary clinical use, i.e., the determination of an appropriate digoxin dosage schedule for each patient whose serum is assayed.

The serum radioimmunoassay procedure described below may also be used for the measurement of digoxin in heparinized plasma; however, when heparinized plasma is employed, clots often form during the assay procedure (perhaps because of the presence of traces of thrombin in the diluted antiserum) and hence the assay of serum is preferable to the assay of heparinized plasma. In our experience, the use of citrate or oxalate as anticoagulants often leads to spurious results and thus blood processed with these anticoagulants should not be employed in the digoxin RIA pro-cedure.

Reagents

Digoxin standards (0, 0.2, 0.5, 1.0, 1.5, 2.0, 3.0, 5.0 pmol/ml in Tris-buffered albumin)

Prepared as described in above section on antibody specificity

Include additional concentrations if desired

Pooled normal human serum (from subjects not treated with digoxin)

Antidigoxin serum

Stored in aliquots at a 1:50 dilution (avoid frequent freezing and thawing) and diluted appropriately on the day of assay, as de-scribed above in section on antibody specificity

All other reagents are as described above in the section on antibody detection.

Procedure

1. To duplicate 250-μl aliquots of each standard digoxin dilution, add 250 μl normal serum.

2. To duplicate 250-μl aliquots of serum from each patient, add 250 μl Tris-buffered albumin, pH 7.4.

3. Include control tubes (standard, blank, background) as described above in section on antibody detection.

4. Add 50 μl tracer, followed by vigorous vortex mixing.

5. Add 50 μl appropriately diluted antiserum; mix well and incubate for at least 15 min at room temperature or at 37°.

6-9. Separate free from bound tracer by the coated charcoal method, determine radioactivity, and calculate the percentage of radioactivity bound as described in Steps 2–5 in the section on antibody detection.

10. Plot percentage of tracer (or of immunoreactive tracer) bound in standard specimens on the ordinate of semilogarithmic graph paper against the $\log_{10}$ of the digoxin concentration in each pair of standards. Use the standard curve generated in this manner to determine digoxin concentrations in serum from patients.

Comments. This procedure may be modified for the measurement of digoxin in urine, other body fluids, and tissues.[12] In the case of urine, specimens are ordinarily diluted 1:20 in Tris-buffered albumin to permit restoration of isotonicity, a physiological pH, and an appropriate protein concentration (spurious RIA results may be observed at protein concentration < 0.2%); normal urine from subjects not receiving digoxin, diluted in a similar manner, is employed in place of serum in tubes containing digoxin standards. Concentrations of digoxin are often quite high, and further dilution of urine from digoxin-treated subjects is often required to obtain accurate measurements in such instances. Other body fluids and tissue extracts may be assayed in a similar manner, but it is important to remember that the protein concentration in all assay and standard tubes should be no less than 0.35%.

Occasionally, it is necessary to measure digoxin concentrations in the serum and urine of a patient who has been treated with sheep digoxin-specific Fab fragments to reverse severe digoxin intoxication. The Fab fragments in the serum and urine of such patients bind radiolabeled digoxin or digoxin derivatives, thus spuriously altering the result of the digoxin RIA procedure; in the case of the coated charcoal method, the percentage of bound digoxin will be high and the apparent digoxin concentration low while, in the case of solid-phase RIAs, the reverse is true. To eliminate this interference, serum (and digoxin standards in serum) may be diluted 1:5 in phosphate-buffered saline, pH 7.4, placed in a loosely stoppered tube in a boiling water bath for 10 min, centrifuged to sediment denatured proteins, including the Fab fragments, and assayed as described above. To facilitate precipitation of denatured Fab fragments from urine, urine (and urine digoxin standards) should be diluted in the presence of normal human serum (1:5 in phosphate-buffered saline, pH 7.4) prior to the heating step and, following centrifugation, the supernatant may be assayed as described in the preceding paragraph.[26]

If a patient has recently received a radioisotope for diagnostic or therapeutic purposes, the presence of tracer in serum may, if appropriate control tubes are not included, cause sufficient interference with the assay to yield inaccurate results; to detect such interference, it is advisable to

[26] T. W. Smith, E. Haber, L. Yeatman, and V. P. Butler, Jr., *New Engl. J. Med.* **294,** 797 (1976).

use an additional channel in one's scintillation spectrometer to detect isotopes other than ^3H (or, in the case of radioiodinated digoxin derivatives, ^{125}I).[25] Chemiluminescence may temporarily cause falsely elevated liquid scintillation counting rates, particularly when urine or serum from uremic patients is assayed; this can readily be detected by a progressive decrease in counts over a period of several hours, and hence it is always advisable to count each assay specimen two (or, preferably, three) times when the assay procedure involves liquid scintillation counting.[25]

When assays are performed regularly over a period of time, a slow decrease in the binding of tracer may be noted. This may be due either to the deterioration of tracer or of antibody. In the case of tracer deterioration, a new working solution should be prepared from one's stock solution; in the case of antibody deterioration, use of slightly less diluted antiserum often will correct the problem.

Interpretation of Results

All cardiac glycosides studied to date cross-react with antidigoxin antibodies and thus their presence in serum may cause apparent digoxin immunoreactivity; this has been a problem with digitoxin, in particular, because therapeutic serum concentrations of this glycoside which are about 10-fold greater than digoxin yield appreciable apparent serum values when subjected to the digoxin RIA procedure. It is therefore essential that it be known with assurance that each patient is receiving the specific glycoside which is being measured in his or her serum. Serum glycoside measurements are significant only when measured in the so-called "steady state" after absorption and tissue equilibration are complete. In the case of digoxin, serum should not be obtained for assay until 6–8 hr after the last dose; serum obtained earlier may yield spuriously high results.[25]

The finding of a low serum digoxin concentration (<0.6 ng/ml) often may provide an explanation for a poor clinical response to the drug. A low serum digoxin concentration may reflect inadequate dosage, patient noncompliance, use of a digoxin preparation of low bioavailability, malabsorption (rare), interference with absorption caused by other drugs, excessive conversion by gut flora to cardioinactive digoxin reduction products, increased metabolic degradation (perhaps drug-induced), or hyperthyroidism. The finding of a low serum level in a patient with a poor clinical response may enable the identification and correction of the cause of the low level but, even when this is not possible, one may carefully adjust the digoxin dosage upward in a rational and controlled manner.[3]

The finding of a high serum digoxin concentration (>2.5 ng/ml) does not necessarily indicate the presence of digoxin toxicity. Some patients may require a relatively high level to achieve a desirable therapeutic effect; this is clearly true in some patients with digitalis-responsive supraventricular arrhythmias and may also be the case in some infants with congenital heart disease. However, high serum levels put a patient at increased risk for the development of digitalis toxicity and, if possible, the dosage should be adjusted accordingly in a downward direction. Elevated blood levels generally reflect excessive dosage and/or impaired renal excretion due to renal disease or to aging.[12] The institution of therapy with quinidine, an antiarrhythmic agent, also may cause a potentially dangerous increase in the serum digoxin concentration.[27]

The presence of a serum digoxin concentration in the so-called "therapeutic range" (0.6–2 ng/ml) does not exclude the possibility of digitalis intoxication. It should be remembered that factors other than digoxin may contribute to the development of arrhythmias in patients without striking elevations in serum digoxin; such factors may include underlying myocardial disease (the most important such factor), altered autonomic tone, other drugs (e.g., aminophylline, adrenergic agents), electrolyte disturbances (e.g., hypokalemia), and hypoxia.[28]

Acknowledgments

Supported by USPHS research grant HL 10608.

[27] E. B. Leahey, Jr., *Ann. Intern. Med.* **93**, 775 (1980).
[28] T. W. Smith, *Am. J. Med.* **58**, 470 (1975).

[43] Radioimmunoassay of *n*-Butylbiguanide

By F. ÖTTING

Introduction

Sensitive gas chromatographic methodology for the determination of biguanides in biological fluids is available.[1,2] However, the ease and simplicity of immunological techniques for the determining of *n*-butylbiguanide especially when utilizing an iodinated tracer may be preferable to the somewhat more tedious extraction and derivatization procedures necessary for gas chromatographic determinations.

[1] M. Mottale and Ch. J. Stewart, *J. Chromatogr.* **106**, 263 (1975).
[2] S. B. Matin, J. H. Karam, and P. H. Forsham, *Anal. Chem.* **47**, 545 (1975).

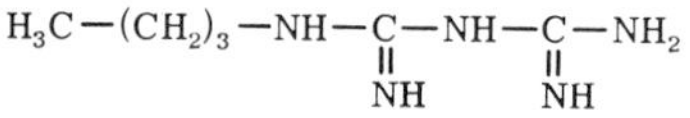

$$H_3C-(CH_2)_3-NH-\underset{\underset{NH}{\|}}{C}-NH-\underset{\underset{NH}{\|}}{C}-NH_2$$

n-Butylbiguanide

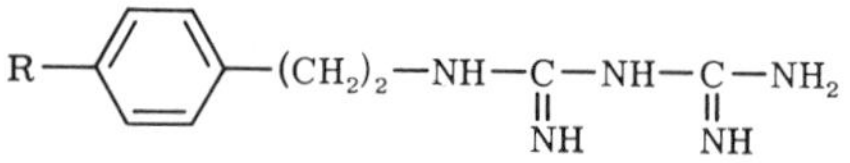

$$R-\langle\!\!\!\!\!\!\bigcirc\!\!\!\!\!\!\rangle-(CH_2)_2-NH-\underset{\underset{NH}{\|}}{C}-NH-\underset{\underset{NH}{\|}}{C}-NH_2$$

R = H;   Phenethylbiguanide
R = OH;   p-Hydroxyphenethylbiguanide
R = HOOC;   p-Carboxylphenethylbiguanide

Since the n-butylbiguanide molecule cannot be iodinated, the phenolic metabolite of phenethylbiguanide is used instead.[3] For this heterogeneous RIA an antiserum of high affinity toward both the radioactive ligand and n-butylbiguanide is necessary to give a sensitive drug assay.

Reagents

Buffer pH 7.6; 70 mM NaH_2PO_4, 137 mM NaCl, 1 mM EDTA, 1% bovine serum albumin (w/v)
Chloramine-T (Fluka)
1-Ethyl-3-(3-dimethylaminopropyl)carbodiimide hydrochloride (EDAC) (Bio-Rad)
Bovine serum albumin (BSA) (Serva)
Norit-A (Serva)
Complete Freund's adjuvant (Difco)
$Na^{125}I$ (Cat. No NEZ-033 H) (NEN Chemicals)
Precoated TLC-sheets:
Polyethyleneimine-coated cellulose
(CEL 300 PEI/UV_{254}, Macherey-Nagel)
Cellulose F_{254} (Merck)

Phenethylbiguanide Hydrochloride[4]

2-Phenethylamine hydrochloride, 100 mmol (15.8 g) and 100 mmol (8.4 g) of dicyandiamide were heated gradually in an oil bath to 150° and kept at that temperature for ca. 1 hr. After cooling the product was recrystallized twice from isopropanol (mp 175–178°).

p-Carboxylphenethylbiguanide Sulfate

A solution of 10 mmol (1.66 g) p-carboxyl-2-phenethylamine and 10 mmol (1.55 g) O-methylamidinoisourea in 10 ml 1 N sodium hydroxide was kept for 1 week at room temperature. The precipitate which is formed

[3] F. Ötting, *Arzneim. Forsch.* (*Drug Res.*) **25**, 524 (1975).
[4] S. L. Shapiro, V. A. Parrino, and L. Freedman, *J. Am. Chem. Soc.* **81**, 2220 (1959).

is collected and dissolved in a few milliliters of diluted ammonia. After addition of 10 mmol copper sulfate in water the biguanide copper complex precipitates. The product is collected, washed with water, and an aqueous suspension of it is treated with hydrogen sulfide to cleave the complex. The copper sulfide is filtered off at pH 4.0 and the solution of the biguanide is concentrated until crystals begin to appear. After addition of 2 *N* sulfuric acid the biguanide sulfate precipitates. It is collected and washed with water and ethanol. For further purification the biguanide sulfate is dissolved in boiling 50% acetic acid, treated with charcoal, and recrystallized with the addition of isopropanol (mp 250–252°).

p-Hydroxyphenethylbiguanide Hydrochloride

The compound is synthetised exactly like the *p*-carboxyl-derivative of phenethylbiguanide described above except that copper chloride is used to form the copper complex which is reprecipitated by solution in diluted aqueous hydrogen chloride and treatment with ammonia to pH 5.0. Crystallization was performed in ethanol-ether (mp 184–185°).

O-Methylamidinoisourea

Dicyandiamide, 0.21 mol (17.6 g) and 0.1 mol (17.3 g) copper(II)-chloride in 80 ml methanol were boiled under reflux for 2 hr. The copper complex of the *O*-methylamidinoisourea is collected, washed with ethanol, and suspended in 200 ml of water. Treatment with hydrogen sulfide cleaves the complex and precipitates copper sulfide which is filtered off. The solution of the *O*-methylamidinoisourea is evaporated to dryness and recrystallized twice from ethanol-ether (mp 154–156°).

Preparation of the Immunogen

n-Butylbiguanide is nonantigenic due to its small molecular size and must therefore be covalently linked to a carrier protein to render it a suitable immunogen. For this purpose the *p*-carboxyl-derivative of phenethylbiguanide was bound to bovine serum albumin via a carbodiimide-promoted reaction according to Gharib *et al.*[5]

The coupling procedure was as follows (room temperature): to 50 mg of protein, dissolved in 25 ml doubly distilled water, 30 mg of EDAC was added followed by 3.6 mg (10 μmol) of *p*-carboxylphenethylbiguanide in 1 ml water. The reaction mixture was stirred on a magnetic stirrer, the pH checked with a glass electrode and constantly adjusted to 5.0 with dilute

[5] H. Gharib, R. J. Ryan, W. E. Mayberry, and T. Hockert, *J. Clin. Endocrinol.* **33**, 509 (1971).

NaOH. After 10 min an additional 10 mg of solid EDAC was added. The reaction was allowed to continue for 24 hr and then was exhaustively dialyzed against doubly distilled water and freeze-dried. To determine the number of moles biguanide bound per mole of bovine serum albumin a biguanide determination was performed according to Bell[6] or alternatively indirectly on the dialysis water according to Shapiro *et al.*[4] utilizing the decrease in absorbance at 233 nm when the pH of aqueous biguanide solutions is changed from 7.0 to 1.0. In this latter procedure varying concentrations of phenethylbiguanide were used as standards. In general a molar ratio of between 10 to 20 mol biguanide per mole protein was observed.

Immunization

One milligram of biguanide–albumin conjugate per 0.5 ml of sterile physiological sodium chloride solution was emulsified with an equal volume of complete Freund's adjuvant and administered intradermally to New Zealand white rabbits at 10 different sites on their back. Each animal was injected with 1 mg of conjugate. At 4-week intervals the rabbits were reimmunized, over a period of 5 to 8 months. Every 2 weeks the animals were bled from their ear arteries, their sera separated, and screened for presence of antibody in the following manner: 0.1 ml buffer containing the radioactive ligand (usually 64 pg), 0.1 ml human serum free of biguanides, and 0.1 ml buffer containing serial dilutions of the potential antiserum were incubated in 1.5 ml Eppendorf reaction vials at room temperature. The vials were slowly rotated in a suitable device. After 16 hr the samples were transferred to the cold room and 0.5 ml of a freshly prepared ice-cold charcoal suspension (400 mg Norit-A per 100 ml doubly distilled water) was added and the samples rotated for another 15 min followed by centrifugation. The radioactivity of an aliquot of the supernatant, representing the bound fraction was determined in a gamma counter with 70% counting efficiency for ^{125}I. The data were plotted as percentage ligand bound versus antibody dilution. A typical antibody titer gave 50% binding of tracer at a dilution of 1:1500 after 5 months of immunization.

Sera with antibodies suitable for the assay were dialyzed against doubly distilled water, freeze-dried, and stored at room temperature.

Preparation of the Radioactive Ligand

The iodination procedure followed the original chloramine-T method of Hunter and Greenwood.[7] Our routine was as follows (room temperature): 2 mCi of carrier-free ^{125}I is used for 2 μg *p*-hydroxyphenethylbi-

[6] E. A. Bell, *Biochem. J.* **75,** 618 (1960).

[7] W. M. Hunter and F. C. Greenwood, *Nature (London)* **194,** 495 (1962).

guanide in 20 μl 125 mM phosphate buffer pH 7.65. Chloramine-T (250 μg) in 30 μl water is admixed and the mixture agitated for 60 sec. The reaction is stopped with 3 mg sodium metabisulfite in 0.1 ml water and the solution applied to the origin of a polyethyleneimine-coated cellulose thin layer sheet and developed with 0.05 M citrate pH 2.5. The separation is evaluated by means of a Berthold scanner. The scan of a typical chromatogram is shown in Fig. 1. The peak zone of the iodinated biguanide (R_f = 0.67) is scraped off and eluted with ca. 50 μl water. Rechromatography in the same system proves the product to be at least 95% pure. The tracer is appropriately diluted with buffer and stored at 4°. Since separation of the iodinated *p*-hydroxyphenethylbiguanide from its precursor is not achieved in this system the calculation of the specific radioactivity is based on the amount of radioactivity found in the particular peak of the chromatogram by comparing the area of this peak of the scanner diagram to the total area of all peaks. In general an iodination yield of 50–70% has been obtained, and specific radioactivity being 0.5–0.7 mCi/μg. The structure of the iodinated compound has not been elucidated. It is assumed, however, that monoiodination occurs because of the large molar excress of the phenolic precursor. For routine assays the tracer was used with this specific radioactivity for 2 weeks without repurification.

It poses no problem to separate the labeled *p*-hydroxyphenethylbi-

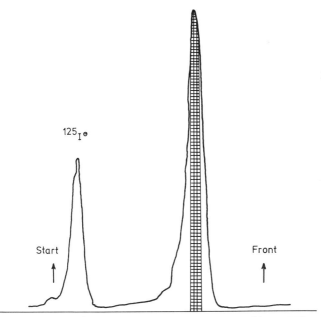

^{125}I●

Start

Front

FIG. 1. Purification of ^{125}I-labeled *p*-hydroxyphenethylbiguanide on PEI-coated cellulose. For details see text. Hatched area, elution zone.

guanide from its precursor by a second preparative TLC [precoated cellulose sheet, amylalcohol–pyridine–water 7:7:6 (v/v); R_f p-hydroxyphen-
ethylbiguanide $= 0.50$; R_f ^{125}I-labeled p-hydroxyphenethylbiguanide $= 0.67$] thus improving the specific radioactivity to its theoretical value of
ca. 6.3 mCi/μg depending on the isotopic purity of the label.

Specificity of the Antibody

Several structural analogs were checked for cross-reactivity by competitive assays with ^{125}I-labeled p-hydroxyphenethylbiguanide for the antibody in the standard procedure. For details see assay procedure. The
results (Fig. 2 and Table I) show that n-butylbiguanide has a considerably
lower affinity toward the antibody as compared with the aromatic biguanides. Physiological substances such as urea, guanidine, and L-tyrosine
gave no significant cross-reactions in the concentration range tested, although guanidine gave an indication of some interference at the highest
level tested (3.3 mg/ml). Therefore false positive results in serum determinations of n-butylbiguanide are not to be expected in the presence of
physiological serum values of L-tyrosine, urea, and guanidine (10 μg/ml,
0.2–0.3 mg/ml, and 1.5–2 μg/ml, respectively.[8,9]

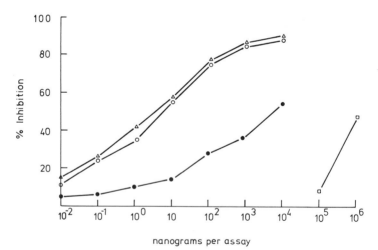

nanograms per assay

FIG. 2. Competitive binding studies with phenethylbiguanide (○——○), p-hydroxyphenethylbiguanide (△——△), n-butylbiguanide (●——●), guanidine (□——□). Each reaction
mixture: 64 pg ^{125}I-labeled p-hydroxyphenethylbiguanide and sufficient antiserum to bind
ca. 50% of the tracer with no addition of unlabeled compounds. Data are from F. Ötting[3]
with permission of the publisher.

[8] D. Cascio, N.Y. State J. Med. **49,** 1685 (1949).
[9] W. Stein and S. Moore, J. Biol. Chem. **211,** 915 (1954).

TABLE I

INHIBITION OF BINDING OF [125]-LABELED
p-HYDROXYPHENETHYLBIGUANIDE TO
THE ANTIBODY[a]

	Nanograms required for 50% inhibition
Urea	$>10^6$
Guanidine	$\sim 10^6$
L-Tyrosine	$>10^6$
Phenethylbiguanide	~ 10
p-Hydroxyphenethylbiguanide	~ 10
n-Butylbiguanide	$\sim 10^4$

[a] Data are from F. Ötting[3] with permission of the publisher.

Assay Procedure

The assay of *n*-butylbiguanide in serum was the same as described in the section screening for antibodies except that the 100 μl serum sample contained varying amounts of *n*-butylbiguanide for the standard curve: 15 to 500 ng/ml serum. The amount of antibody per assay was adjusted such as to bind approximately 50% of the radioligand unter working conditions. A flow sheet is given in Table II. The assay was run in triplicate. For calculation of results the bound radioactivity is plotted as a function of the logarithm of unlabeled *n*-butylbiguanide added. For larger sample quantities however the data were processed utilizing computer programs, the logit–log transformation or spline function being most satisfactory.

Characterisation of the Radioimmunoassay

A typical standard curve is shown in Fig. 3. The sensitivity of each assay was defined as three times the standard deviation of the blank determination. In 26 separate assays a mean value of 8.7 ng/ml (1.4–40.6 ng/ml) has been obtained. Intraassay variation, assessed by measuring identical samples in triplicate, was found to be 2 to 8%.

Interassay variation was evaluated by determining different control sera in the range 50, 250, and 500 ng/ml in a number of separate assays. As can be seen from Table III, a mean coefficient of variation of 12% (7–27%) and a mean recovery rate of 102.6% is observed.

It is advisable to include similar control sera into each assay in order to control the reliability of the RIA. According to our standards deviations of $\geq 15\%$ from the expected value for one of the control sera led to rejection of the assay.

TABLE II
Assay Method

Rotate in disposable 1.5-ml Eppendorf vials at room temperature
  100 μl tracer[a]
  100 μl antibody[b]
  100 μl serum[c]
After 16 hr cool to 4° and add 500 μl charcoal suspension[d]
Rotate for 15 min
Centrifuge at 3000 rpm for 2 min
Count 500-μl aliquot in a gamma counter

[a] Depending on titration curves of the antibody; i.e., 64 pg, 50,000 dpm in buffer.
[b] Depending on titration curves; sufficient quantity to bind 50% of tracer.
[c] Normal human serum with added amounts of unlabeled *n*-butylbiguanide (15–250 ng/ml serum) for standard curve; unknowns; control sera.
[d] Norit-A 0.4% (w/v) in doubly distilled water at 4° kept on a magnetic stirrer.

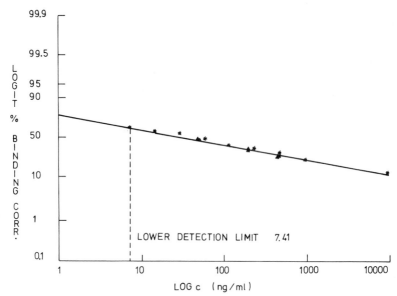

Fig. 3. Computer plot of a typical standard curve utilizing a logit–log transformation; (▲——▲) control sera.

TABLE III
QUALITY CONTROL DATA

Control serum (ng/ml)	n^a	Measured $\bar{x} \pm$ SD (ng/ml)	VC^b (%)	Recovery (%)
47.0	10	52.4 ± 4.4	8.4	111.5
52.1	8	54.4 ± 14.8	27	104.4
242.0	14	246.0 ± 17.0	7	101.6
260.7	8	267.5 ± 27.2	10.2	102.8
484.0	14	488.0 ± 40.2	8.2	100.8
521.3	7	493.6 ± 43.0	8.8	94.7

^{a} Number of separate determinations.
^{b} Variation coefficient.

Occasionally serum samples show unusually high values. These sera were diluted with biguanide-free normal human serum and the measurement repeated. The cause of these effects is not known.

Acknowledgment

The biguanides were synthesized and kindly provided by Dr. Uragg of the Organic Chemistry Department of Grünenthal GmbH.

[44] Radioimmunoassay of Metyrapone and Reduced Metyrapone

By A. WAYNE MEIKLE

Introduction

Metyrapone is a potent inhibitor of 11β-hydroxylase of the adrenal cortex.[1-3] This action has been exploited in the development of a test of the functional integrity of the pituitary–adrenal axis.[1,4] The performance of the metyrapone test is the major use of the drug clinically, but it is also used to inhibit cortisol synthesis in some patients with hypercortisolemia

[1] G. W. Liddle, D. Island, E. M. Lance, and A. P. Harris, J. Clin. Endocrinol. Metab. 18, 906 (1958).
[2] J. S. Jenkins, J. W. Meakin, D. H. Nelson, and G. W. Thorn, Science 128, 478 (1958).
[3] J. G. Sprunt, M. C. K. Browning, and D. M. Hannah, Mem. Soc. Endocrinol. 17, 193 (1967).
[4] A. W. Meikle, S. C. West, J. A. Weed, and F. H. Tyler, J. Clin. Endocrinol. Metab. 40, 290 (1975).

and Cushing's syndrome.[5] Metyrapone appears to be a less potent inhibitor of the conversion of cholesterol to pregnenolone,[6] 18,19-[7,8] and 21-hydroxylases,[9] and 17,20-lyase[10] than of the 11β-hydroxylase, but in sufficient concentration it may impair steroidogenesis by blocking all of these enzymes. It apparently competes with substrates for the cytochrome P-450 binding site(s) and thereby blocks enzyme reactions.[11-13] In addition to its modification of steroidogenesis, investigators have been intrigued by a wide range of actions in mammalian tissues: acceleration of the clearance of cortisol from plasma,[14,15] modification of catecholamine synthesis,[16,17] inhibition of glucose utilization by brain and muscle,[18] and hydrocortisone uptake and binding in brain,[19] suppression of insulin release from the pancreas[18] and metabolism of drugs by hepatic tissue,[20,21] and stimulation of the exchange of cortisol through the mesentry.[22]

In vivo the ketone group of metyrapone (Fig. 1) is reduced to form reduced metyrapone which is almost as active as metyrapone in the *in vitro* inhibition of adrenal 11β-hydroxylase as shown in Fig. 2.[2,23,24] Radioimmunoassays (RIA) were developed to investigate the metabolism of the inhibitors in humans.[4] It was postulated that a carboxymethyloxime derivative of metyrapone or a hemisuccinate of reduced metyrapone could then be conjugated to a large protein. The hapten–protein complex would

[5] D. N. Orth, *Ann. Intern. Med.* **89**, 128 (1978).
[6] A. Carballeira, L. M. Fishman, and J. D. Jacobi, *J. Clin. Endocrinol. Metab.* **42**, 687 (1976).
[7] G. F. Kahl and S. Simon, *Naunyn Schmiedebergs, Arch. Parmakol.* **268**, 1 (1971).
[8] T. Bledsoe, D. P. Island, A. M. Riondel, and G. W. Liddle, *J. Clin. Endocrinol. Metab.* **24**, 740 (1964).
[9] D. C. Sharma and R. I. Dorfman, *Biochemistry* **3**, 1093 (1964).
[10] G. Betz and D. Michels, *Biochem. Biophys. Res. Commun.* **50**, 134 (1973).
[11] A. G. Hildebrandt, K. C. Leibman, and R. W. Estabrook, *Biochem. Biophys. Res. Commun.* **37**, 477 (1969).
[12] G. F. Kahl and K. J. Netter, *Biochem. Pharmacol.* **19**, 27 (1970).
[13] H. D. Colby and A. C. Brownie, *Arch. Biochem. Biophys.* **138**, 632, (1970).
[14] J. Levin, B. Zumoff, and D. K. Fukushima, *J. Clin. Endocrinol. Metab.* **47**, 845 (1978).
[15] M. Blickert-Toft, K. Folke, and M. L. Nielsen, *J. Clin. Endocrinol. Metab.* **35**, 59 (1972).
[16] H. Parvez, S. Parvez, and J. Roffi, *Arch. Int. Pharmacodyn. Ther.* **198**, 187 (1972).
[17] H. Parvez and S. Parvez, *Horm. Metab. Res.* **4**, 398 (1972).
[18] O. D. Bruno, P. Metzger, and W. J. Malaisse, *Acta Endocrinol. (Copenhagen)* **70**, 710 (1972).
[19] R. D. Stith, R. J. Person, and R. C. Dana, *Neuroendocrinology* **22**, 183 (1976).
[20] K. C. Leibman, *Mol. Pharmacol.* **5**, 1 (1969).
[21] K. C. Leibman and E. Ortiz, *Drug Metab. Disp.* **1**, 184 (1973).
[22] E. Couturier, O. D. Bruno, P. Metzger, R. Leclercq, and G. Copinschi, *Acta Endocrinol. (Copenhagen) Suppl.* **155**, 145 (1971).
[23] I. Kraulis, H. Traikov, M. P. Li, C. P. Lantos, and M. K. Birmingham, *Can. J. Biochem.* **46**, (1968).
[24] W. Dünges and G. F. Kahl, *Naunym Schmiedebergs Arch. Pharmakol.* **267**, 293 (1970).

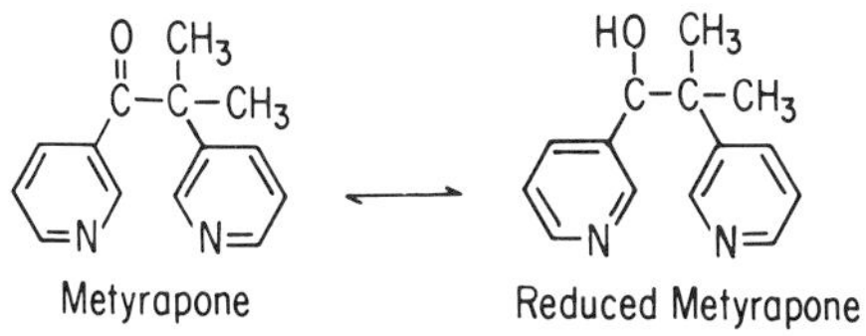

Metyrapone   Reduced Metyrapone

FIG. 1. The structures of metyrapone and reduced metyrapone are shown.

then produce antisera that would have high binding affinity for both metyrapone and reduced metyrapone. The presumption was that the reaction site used in formation of the oxime or the hemisuccinate would then have high binding affinity of both compounds. The oxime derivative of metyrapone was arbitrarily selected.

Previous Methods

Bioassay

Sprunt *et al.*[3] developed an *in vitro* bioassay based on inhibiton of 11β-hydroxylase of the adrenal cortex. This technique gave substantially higher values than those obtained with the fluorometric assay for metyrapone developed in 1969 by Meikle *et al.*[25] This suggested that metabolites of metyrapone, possibly reduced metyrapone, may have been detected with the bioassay but not with the fluorometric assay.

Fluorometric

With the fluorometric method, metyrapone was extracted from plasma chromatographed on miniflorisil columns, reacted with cyanogen bromide and then *p*-aminoacetophenone which produced a fluorescent derivative. It had a sensitivity of much less than 1 μg.

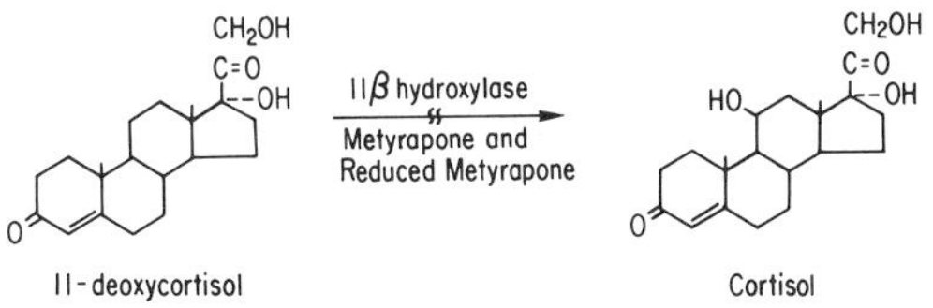

11-deoxycortisol   Cortisol

FIG. 2. Inhibition of 11β-hydroxylase by metyrapone and reduced metyrapone and the structure of cortisol and 11-deoxycortisol are shown.

[25] A. W. Meikle, W. Jubiz, C. D. West, and F. H. Tyler, *J. Lab. Clin. Med.* **74**, 515 (1969).

Spectrophotometric

Szeberenyi and associates[26,27] also in 1969 reported a spectrophotometric technique for assay of both metyrapone and reduced metyrapone with a sensitivity of about 1 μg. It was based on extraction of plasma, followed by thin-layer or paper chromatography and spectrophotometric determination after reaction with cyanogen bromide and *p*-aminosalicylic acid.

Methods of Radioimmunoassay of Metyrapone and Reduced Metyrapone and Source of Metyrapone and Reduced Metyrapone

Crystalline compounds were generous gifts of the Ciba Pharmaceutical Company and kept in a desiccator at 20°. Metyrapone was tritiated by New England Nuclear Corp. It had a specific activity of 5 curies/millimole and labeled and unlabeled compounds were over 95% pure as determined from paper chromatography with the Bush B-3 system (petroleum ether:benzene:methanol:water, 33:17:40:10).

Stock solutions of metyrapone and reduced metyrapone were prepared by dissolving 100 mg of substances in 100 ml of redistilled methanol and diluted serially.

Preparation of Oxime Derivative

One and one-half millimoles of metyrapone (MW 226), 12.5 μCi of tritiated metyrapone, 1.6 mmol of *o*-carboxymethyloxime hemihydrochloride (Aldrich Chemical Co.) (MW 218) in 14 ml of ethanol, and 1.95 ml of 2 *N* KOH were reflexed for 3 hr (Fig. 3).[28–30] The volume of the reaction mixture was reduced by vacuum. Ten milliliters of water was added, and the pH adjusted to 10.5 with 2 *N* KOH and extracted twice with 2 volumes of ethyl acetate which was discarded. The aqueous phase was ad-

PREPARATION OF METYRAPONE OXIME DERIVATIVE

METYRAPONE + O-CARBOXYMETHYLOXIME (OXIME) —→ METYRAPONE-OXIME

PREPARATION OF HAPTEN PROTEIN COMPLEX

METYRAPONE-OXIME + BSA + CARBODIIMIDE —→ METYRAPONE-OXIME - BSA

FIG. 3. The reactions involved in preparation of metyrapone-oxime and its conjugation to bovine serum albumin (BSA) are presented.

[26] S. Szeberenyi, M. T. Tacconi, and S. Garattini, *Endocrinology* **85,** 575 (1969).
[27] S. Szeberenyi, K. S. Szalay, and M. T. Tacconi, *J. Chromatogr.* **40,** 417 (1969).
[28] B. F. Erlanger, F. Borek, S. M. Beiser, and S. Lieberman, *J. Biol. Chem.* **228,** 713 (1957).
[29] D. K. Mahajan, J. D. Wahlen, F. H. Tyler, and C. D. West, *Steroids* **20,** 609 (1972).
[30] A. W. Meikle, L. G. Lagerquist, and F. H. Tyler, *Steroids* **22,** 193 (1973).

justed to pH 2.0 with HCl and placed in a refrigerator at 4° for 24 hr. The precipitate was filtered and washed with cold water. The yield was 50% as determined from recovery of radioisotope.

Conjugation to Bovine Serum Albumin

Forty-five milligrams of metyrapone oxime was dissolved in 9 ml of water by adjusting the pH 8.5 with 0.1 N NaOH while stirring. After it was in solution the pH was adjusted to pH 7.0 with 0.1 N HCl. Sixty milligrams of bovine serum albumin (BSA, Pentex) was dissolved, 1.5 ml of an aqueous solution 60 mg of 1-ethyl-3-(3-dimethylaminopropyl)carbodiimide-HCl (carbodiimide, Ott Chemical Co.) were added "dropwise," and the reaction mixture was then dialyzed overnight at room temperature (Fig. 3).[28-30] The mixture was dialyzed against three to four daily changes of 2 liters of physiologic saline, pH 7.2 for 3 days. The number of metyrapone oxime residues conjugated per mole of BSA was calculated according to the method of Gross et al.[31] from the measurement of the optical densities at 265 and 280 nm. Five residues of the oxime derivative were conjugated per mole of BSA.[31]

Immunization of Rabbits

Four rabbits were injected subcutaneously in four sites and two foot pads with 1 mg of metyrapone oxime–BSA conjugate in complete Freund's adjuvant (Difco). Boosters were injected in four sites in the back at monthly intervals thereafter. Ten days after the first two and subsequent injections they were bled from the ear vein. The blood was chilled and the serum collected. Sodium azide was added (1 mg/ml) to the serum, the titer checked, and it was stored at −20° until use.

Characterization of Antisera

The buffer was prepared: 0.01 M sodium phosphate, adjusted to pH 7.8 with 2 N NaOH; 0.14 M sodium chloride: 0.01 M EDTA, sodium azide, 1 mg/ml, and rabbit γ-globulin (Pentex) 0.25 mg/ml. Tritiated metyrapone was dried under a jet of nitrogen and the buffer added to give about 10,000 dpm/0.1 ml. The antiserum was diluted 1:100 and diluted serially with the antisera buffer, and incubated overnight at 4°. Two hundred microliters of the buffer was added to test tubes and 20 μl of goat anti-rabbit γ-globulin second antibody (grade P-3, Antibody, Inc., Calif.) was added. The mixture was incubated for 1 hr at 4°, and to ensure the precipitation of the antigen–antibody, the tubes were centrifuged in a re-

---

[31] S. J. Gross, D. H. Campbell, and A. H. Weetall, *Immunochemistry* **5**, 55 (1968).

frigerated centrifuge at 2500 rpm for 10 min. One hundred microliters of the supernatant was pipetted with a capillary pipet and added to a radio-isotope counting vial, and 15 ml of radioisotope counting solution cocktail (Bio-Solv, Beckman; Liquifluor, New England Nuclear) and toluene were added. One hundred microliters of the original antibody solution before addition of second antibody was also used for radioisotope counting. The titer of M-2 was 1:750 and M-4 1:2000 as shown in Figure 4.

Affinity Constant

The bound to free ratio was plotted as a function of the picomoles bound and the affinity constant was determined according to Scatchard.[32]

Radioimmunoassay

The antisera were diluted to give approximately 50% bound [total counts per minute (cpm) per 100 μl of immunoassay buffer-free CPM of supernatant after precipitation metyrapone antibodies with second anti-body]. Standards with 0, 50, 100, 200, 500, 750, and 1000 pg of metyra-pone in 0.1 ml of methanol and 0, 25, 50, 100, 200, and 500 pg of reduced metyrapone in 0.1 ml of methanol were pipetted into assay tubes (glass, 1 × 10 cm) in triplicate. Plasma extracts dissolved in methanol (0.02,

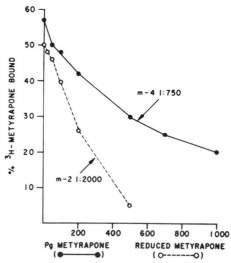

FIG. 4. Standard curves for metyrapone (solid lines, M-4) and reduced metyrapone (dashed lines, M-2) are shown [from Meikle et al., J. Clin. Endocrinol. Metab. 40, 290 (1975)].

[32] G. Scatchard, Ann. N.Y. Acad. Sci. 51, 660 (1949).

0.05, and 0.1 ml) were also added to assay tubes in triplicate. The volume of methanol was adjusted to 0.1 ml, the tubes were evaporated to dryness in a vacuum oven (Lab-Line, Melrose Park, Ill.) at 55° and cooled to 4° followed by addition of 0.2 ml of radiolabeled antisera solution. The tubes were covered with parafilm and incubated at 4° overnight. The bound and free were separated by addition of the second antibody as described above. All values were corrected for procedural loss. The picograms from the standard curve times the dilution factors and procedural losses gave the concentration of compounds in the extracts of plasma.

Plasma Extraction

One-half milliliter of plasma containing either standards of metyrapone, or reduced metyrapone used for assessment of procedural losses or aliquots of plasma samples with unknown levels of the haptens was extracted with 10 ml of dichloromethane. Column chromatography was performed on minisilica gel columns. They were prepared by adding a silica gel (2.0 g) to a glass wool stoppered disposable pipet (Pasteur pipet, 0.6 × 14.6 cm) and were washed with methanol and then dichloromethane before use. Two milliliters of the dichloromethane extract was added to the column and an additional 2 ml of the solvent was added and discarded. Two milliliters of 8% methanol in dichloromethane eluted more than 95% of the metyrapone and less than 4% of the reduced metyrapone. Elution with an additional 2 ml of 40% methanol in dichloromethane recovered more than 80% of the reduced metyrapone. The eluates were evaporated to dryness under a jet of air, dissolved in methanol, and were used for radioimmunoassay.

Paper Chromatography

Whatman No. 1 paper was prewashed by development several times with methanol. Two-centimeter lanes were cut and methanol-containing metyrapone or reduced metyrapone or extracts of plasma was chromatographed with the Bush B-3 system. Reference standards or tritiated metyrapone were also developed on separate strips. Metyrapone had an R_f (relative to solvent front) of 0.55 and reduced metyrapone an R_f of 0.12. Fifty to 100 micrograms of the compounds can be identified by ultraviolet scanning and the tritiated metyrapone by a radiochromatogram scanner (Packard).

Specificity

The cross-reaction of the antisera with metyrapone and reduced metyrapone was determined. The ratio of the mass of metyrapone to reduced metyrapone or reduced metyrapone to metyrapone required to displace 50% of the bound tritiated metyrapone in the RIA was used to calculate the cross-reaction of the antisera. M-2 was more specific for reduced metyrapone with a cross-reaction of 35% for metyrapone and M-4 for metyrapone with a cross-reaction of 50% for reduced metyrapone.

Plasma levels observed from patients not receiving metyrapone (plasma blanks) had values for both substances of less than 1 μg/dl. A high correlation ($r = 0.91$, $p < 0.001$) was observed between values determined with both the fluorometric and RIA. However, values averaged 25% higher with the RIA as compared to the fluorometric method. Plasma from a patient treated with metyrapone had a concentration of both metyrapone and reduced metyrapone of over 100 μg/dl. They were extracted and chromatographed with the Bush B-3 system. The metyrapone and reduced metyrapone sections were eluted and assays performed for the compounds. Dilution of the respective metyrapone and reduced metyrapone sections had ultraviolet absorption spectra similar to authentic compounds and gave RIA curves parallel with the standards. Other areas of the strip were also cut and eluted, but their cross-reactions were negligible with the antisera. These assays were made to determine if plasma contained other metabolites of metyrapone which might cause interference with the RIA. None was observed.

Recovery and Accuracy

Known quantities of metyrapone and reduced metyrapone were added to a plasma pool at concentrations of 25 to 100 μg/dl. The average recovery for the samples following the extraction and column chromatography procedures was 100% for metyrapone and 60% for reduced metyrapone. The accuracies determined by correcting the recoveries for procedural loss were 98 ± 20% ($n = 10$) for metyrapone and 95 ± 14% ($n = 10$) for reduced metyrapone.

Precision

The intraassay coefficient of variation (standard error/mean) for 10 analyses of the plasma pool with the same group of assays was less than 20% for both compounds. The interassay coefficient of variation for assays on 10 occasions was less than 15% for metyrapone and less than 10% for its reduced metabolite.

Plasma Content of Metyrapone and Reduced Metyrapone

In Fig. 5, the plasma values of metyrapone and reduced metyrapone and 11-deoxycortisol and cortisol are shown after a 5.0 g infusion of me-

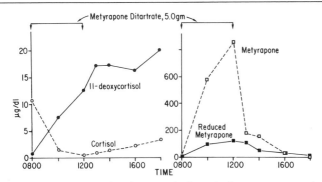

FIG. 5. The plasma contents of 11-deoxycortisol, cortisol, metyrapone, and reduced metyrapone in a normal subject during and following a 5.0 g infusion of metyrapone ditartrate from 0800 to 1200 hr are presented.

tyrapone ditartrate (about 40% metyrapone) from 0800 to 1200 hr. During the infusion of metyrapone, its content in plasma exceeds that of its metabolite by approximately 6 to 7 times.

Eight hours after a single oral midnight dose of metyrapone (30 mg/kg body weight), the mean content of reduced metyrapone was about 1.5 times that of metyrapone (100 ± 111 μg/dl, $n = 34$). As shown in Figs. 6 and 7, the plasma levels of metyrapone and the content of both metyrapone and its metabolite were highly variable from one individual to another (metyrapone range 8.4 to 473 μg/dl, and reduced metyrapone 15–540 μg/dl).

Correlation between the Ratio of 11-Deoxycortisol to Cortisol and the Metyrapone Compounds

A highly significant correlation (Fig. 6) was observed between metyrapone and the ratio of the plasma concentration of 11-deoxycortisol (the precursor of cortisol) to cortisol which reflects the effect of the compounds on the adrenals 11β-hydroxylase activity in persons with normal adrenal–pituitary reserve. A similar high correlation was observed between the total levels of the inhibitors and the ratio of steroids (Fig. 7). Marked adrenal inhibition of cortisol synthesis as indicated by a plasma cortisol level of less than 2 μg/dl was observed when the total plasma content of the inhibitors exceeded 400 μg/dl.

Comments

The radioimmunoassays are simple, specific, accurate, and their reproducibility is satisfactory. The sensitivity of the RIA is approximately

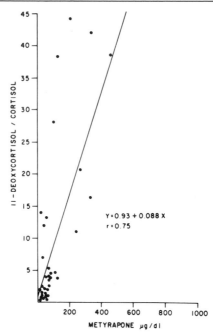

FIG. 6. The comparisons of the ratio of the 0800 hr plasma concentration of 11-deoxycortisol to cortisol and metyrapone 8 hr after a single oral dose of metyrapone at midnight to normal subjects are shown [from Meikle *et al.*, *J. Clin. Endocrinol. Metab.* **40**, 290 (1975)].

50 pg, as compared to several micrograms for the bioassay,[3] about 1 μg for the spectrophotometric technique,[26-27] and nanogram quantities with the fluorometric assay.[25] High specificity of the RIA was achieved by selecting antisera which were more specific for respective compounds combined with a simple column chromatography procedure which separated the substances. Nonspecific materials of plasma and other possible metabolites of metyrapone if present on a paper chromatogram of plasma extracts failed to cross-react with the antisera. The RIA could no doubt be adapted to tissue, and it has the advantages of sensitivity and specificity over previous methods.

With the RIA techniques for metyrapone and reduced metyrapone, it was documented that in humans, metyrapone is rapidly reduced to an active metabolite, reduced metyrapone. Both compounds appear to be about equally effective in inhibition of 11β-hydroxylase of the adrenal.[3] Their plasma concentration correlated with inhibition of the enzyme as documented by comparing the ratio of 11-deoxycortisol (the precursor of cortisol) to cortisol and the plasma content of metyrapone or total plasma concentration of both inhibitors. It was assumed, therefore, that their

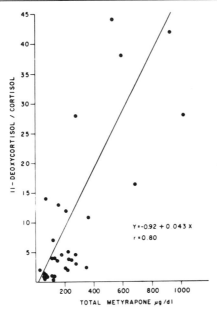

FIG. 7. The comparisons of the ratio of the 0800 hr plasma levels of 11-deoxycortisol to cortisol and the total content of both metyrapone and reduced metyrapone are presented [from Meikle *et al.*, *J. Clin. Endocrinol. Metab.* **40**, 290 (1975)].

plasma levels reflect their content in the adrenal cortex. This assumption is based on the fact that the adrenal is less active than liver and kidney in reducing metyrapone to form the active metabolite, and that their lipid solubilities are similar.[23]

It was concluded from assay of metyrapone and reduced metyrapone in plasma of humans following administration of the former that metyrapone was the major inhibitor of 11β-hydroxylase during an infusion of metyrapone ditartrate. In contrast, 8 hr after oral metyrapone, reduced metyrapone was the major inhibitor as its plasma concentration exceeded the plasma content of metyrapone by 1.5 times.

Measurement of plasma levels of metyrapone and its metabolite have been useful clinically.[4,25,33-35] Patients receiving drugs that induce hepatic drug-metabolizing enzymes fail to exhibit normal response of the pitui-

---

[33] A. W. Meikle, W. Jubiz, S. Matsukura, G. Harada, C. D. West, and F. H. Tyler, *J. Clin. Endocrinol. Metab.* **30**, 259 (1970).

[34] A. W. Meikle, W. Jubiz, S. Matsukura, C. D. West, and F. H. Tyler, *J. Clin. Endocrinol. Metab.* **29**, 1553 (1969).

[35] W. Jubiz, S. Matsukura, A. W. Meikle, G. Harada, C. D. West, and F. H. Tyler, *Arch. Intern. Med.* **125**, 468 (1970).

tary–adrenal axis to metrapone.[34] These drugs accelerate the clearance of metyrapone and reduced metyrapone from blood and their plasma levels are often inadequate to produce substantial inhibition of 11β-hydroxylase. Further, patients who appear to metabolize the metyrapone slowly may have very high levels of the drug after a single oral dose (30 mg/kg at midnight). The high levels might markedly impair all steroidogenesis by the adrenal apparently by inhibiting side chain cleavage of cholesterol to form pregnenolone, the early step in steroidogenesis.[6,36]

There are no doubt many practical applications of the RIA of metyrapone and reduced metyrapone in investigation of their numerous biological actions besides blockage of adrenal synthesis of cortisol at the 11β-hydroxylase step. Other drugs that either increase or decrease the activity of hepatic drug-metabolizing enzymes may be identified by observing alterations of the plasma content of the inhibitors following administration of metyrapone. The content of metyrapone and reduced metyrapone in various tissues after administration of the metyrapone is unknown. It would, however, reflect the equilibrium between uptake and release and their interconversion and degradation within the tissues.

Acknowledgments

This was supported by the Veterans Administration Hospital Research funds and CA 25031 from N.I.H.

[36] A. Carballeira, W. U. Cheng, and L. M. Fishman, *Acta Endocrinol. (Copenhagen)* **76,** 689 (1974).

[45] Radioimmunoassay of Pyrazolone Derivatives

By TAKEHIKO TAKATORI

Some drugs and other chemicals are well known as inducing a hypersensitive state in the host.[1,2] Since pyrazolone derivatives can elicit drug allergies, they are among those substances that have antigenic properties.[2] However, in order to obtain antibodies against drug-haptens in experimental animals it is necessary in general to enhance antigenicity by conjugating the hapten to proteins or to natural or synthetic polymers which act as carriers. In this article, methods for the preparation of antipyrine-conjugated bovine serum albumin are presented and details are

[1] C. W. Parker, J. Shapiro, M. Kern, and H. N. Eisen, *J. Exp. Med.* **115,** 821 (1962).
[2] B. N. Halpern, A. Hotzer, P. Liacopoulos, and J. Meyer, *J. Allergy* **29,** 1 (1958).

given for the production of specific antibodies against pyrazolone derivatives and for their radioimmunoassay.

Preparation of the Immunogens

4-Azoantipyrine-Conjugated Bovine Serum Albumin[3]

The azoprotein was prepared by the method of Tabachnik and Sobotka.[4] To 4-aminoantipyrine (0.2 mmol) dissolved in 6 ml of an aqueous solution containing 0.5 mmol hydrochloric acid and 0.04 mmol sodium bromide, was added sodium nitrite (0.2 mmol) with continuous stirring for 10 min in an ice-bath. The mixture was then stirred for 1 hr in a cold bath and the volume adjusted to 25 ml by the addition of cold water. The solution of diazotized antipyrine was added dropwise to 30 ml of 1% bovine serum albumin (BSA) solution in 0.01 M borate buffer (pH 9.3) at 0° over a period of 20 min, the pH value being maintained at 9.0–9.5 with 0.5 N NaOH during the addition. The reaction mixture was stirred for 4 more hours, dialyzed against distilled water for 6 days, and then against 0.01 M phosphate-buffered saline (pH 7.6) for 1 day in the cold. The 4-azoantipyrine conjugated to BSA (4-AAP-BSA) was kept at −20° until used for immunization.

For determination of the amount of antipyrine coupled to the protein, an aliquot of the dialysate containing 4-AAP-BSA was hydrolyzed to amino acids with 6 N HCl at 110° for 24 hr. The amino acids were analyzed with an amino acid analyzer. Based on the decrease in lysine residues, approximately 21.3 mol of hapten were bound to 1 mol of BSA (molecular weight 68,000). The protein content was determined by the method of Lowry et al. using BSA as standard.[5]

4-Succinamidoantipyrine-Conjugated Bovine Serum Albumin[6]

The method of synthesis of 4-succinamidoantipyrine-conjugated bovine serum albumin (4-SAAP-BSA) is illustrated in Fig. 1. 4-Succinamidoantipyrine was prepared as follows: 100 mg of succinic anhydride was suspended in 10 ml of benzene and the solution boiled. To the boiling solution was added 2 ml of benzene containing 200 mg of 4-aminoantipyrine and the mixture was boiled for several minutes until the crystalline compound was produced. The precipitate was then filtered and washed 3

[3] T. Takatori and A. Yamaoka, Forensic Sci. Int. 12, 151 (1978).
[4] M. Tabachnik and H. Sobotka, J. Biol. Chem. 235, 1051 (1960).
[5] O. H. Lowry, N. J. Rosebrough, A. L. Farr, and R. J. Randall, J. Biol. Chem. 193, 265 (1951).
[6] T. Takatori and A. Yamaoka, J. Immunol. Methods 35, 147 (1980).

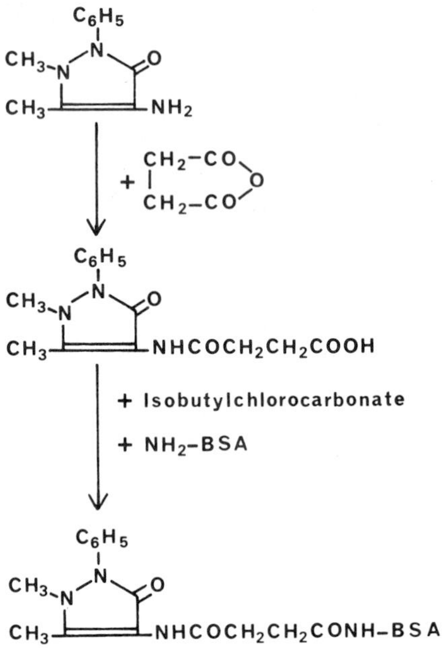

FIG. 1. Synthesis of the 4-succinamidoantipyrine-conjugated bovine serum albumin. Data from T. Takatori and A. Yamaoka, *J. Immunol. Methods* **35**, 147 (1980).

times with boiling benzene. The product was detected by thin-layer chromatography on plates precoated with silica gel G using a solvent system of acetone–water (4:1, v/v). The R_f value of 4-succinamidoantipyrine (4-SAAP) was 0.71 and its melting point was 183.5–184.5°. Analysis for $C_{15}H_{17}O_4N_3$: calculated: C, 59.4; H, 5.6; N, 13.9; found: C, 59.6; H, 5.7; N, 13.9%.

The antipyrine hapten, 4-SAAP, was coupled to BSA according to the method of Erlanger *et al.*[7] with a slight modification. 4-SAAP (32 mg) and 55 μl of tri-*n*-butylamine were dissolved in 2 ml of dry pyridine. After the solution was cooled to 4°, 30 μl of isobutylchlorocarbonate was added to form the mixed anhydride and the mixture was then stirred for 30 min at 4°. This solution was added dropwise to a cooled solution of BSA (140 mg) dissolved in 3.8 ml of distilled water, 2 ml of pyridine, and 0.145 ml of 1 N NaOH. During the addition, the pH was maintained between 7.5 and 9.0 with 1 N NaOH. The reaction mixture was stirred overnight at 4°, and dialyzed against distilled water for 5 days. Some precipitate in the dialysate was dissolved by addition of 8% aqueous $NaHCO_3$. For the determination of the amount of antipyrine coupled to BSA,

---

[7] B. F. Erlanger, F. Borek, S. M. Beiser, and S. Lieberman, *J. Biol. Chem.* **228**, 713 (1957).

2.0 μCi of [1,4-^{14}C]succinic anhydride was added to the unlabeled 4-aminoantipyrine, and the above procedure was followed. By measuring the radioactivity in the dialysate, the degree of conjugation was calculated to be 29.6 mol of hapten per mole of BSA. The dialysate was kept at $-20°$ until used for immunization.

Immunization and Antisera

A 0.5 ml volume of either 4-AAP–BSA or 4-SAAP–BSA conjugate solution in saline, which contains 1 mg protein, was emulsified with an equal volume of complete Freund's adjuvant. Domestic male albino rabbits, weighing 2.5–3.0 kg, were immunized subcutaneously in six different sites at the back and in two thighs. Three rabbits received 1 ml of the emulsion of 4-AAP–BSA conjugate twice the first month and then once a month as a booster injection. One milliliter of the emulsion of 4-SAAP–BSA conjugate was used to immunize 3 rabbits 3 times during the first month and then once every 3 weeks.

Blood samples were collected from the ear vein 7–10 days after each booster injection and allowed to clot at 4°. Serum was separated by centrifugation and served as the source of antibody. All animals produced detectable antibodies after 2 months, and the titers increased by repeated booster injections. The maximum titers against both immunogens were produced by 6 months after immunization and then leveled off. Anti-antipyrine antibodies obtained with the 4-AAP–BSA conjugate and the 4-SAAP–BSA conjugates were used in all experiments. Nonimmunized rabbit serum did not bind antipyrine.

Radioimmunoassay Procedures[6,8]

The radioimmunoassay (RIA) is based on competitive binding of unlabeled and ^{3}H-labeled antipyrine to the specific anti-antipyrine antibodies present in rabbit antisera. Incubation and subsequent separation of the free from the bound labeled hapten were carried out at room temperature.

A. RIA using the antibody against 4-AAP–BSA conjugate (Antibody-I) was carried out as follows: 0.01 M phosphate-buffered saline (pH 7.6) was used to dilute the reagents. Each assay tube (0.25 ml) contained the following components; 0.1 ml of [^{3}H]antipyrine (specific activity 0.199 Ci/mmol, approx. 5000 dpm, 2.13 ng), 0.1 ml of diluted antiserum (1:60), and 0.05 ml of either unlabeled antipyrine or other cold ligand solutions. The incubation mixture was allowed to stand for 2 hr to form the hapten–antibody complex. After 0.25 ml of saturated ammonium sulfate solution

[8] T. Takatori, A. Yamaoka, and K. Terazawa, *J. Immunol. Methods* **29**, 185 (1979).

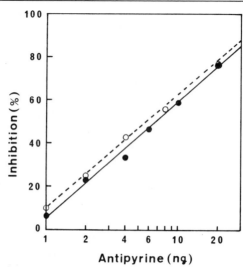

FIG. 2. Inhibition by nonradioactive antipyrine of [³H]antipyrine by Antibody-I
(O------O) and Antibody-II (●——●).

was added to the mixture, the precipitate containing antipyrine bound to
antibody was collected by centrifugation for 15 min at 3000 rpm.

B. RIA using the antibody against the 4-SAAP–BSA conjugate (Antibody-II) was carried out as follows: antiserum, normal rabbit serum,
[³H]antipyrine, and other cold ligands were diluted with 0.1 M phosphate
buffer (pH 7.6). Each assay tube (0.5 ml) contained 0.1 ml of [³H]anti-
pyrine (approx. 5000 dpm), 0.1 ml of diluted antiserum (1:300), 0.1 ml of
20-fold diluted normal rabbit serum added to obtain sufficient ammonium
sulfate precipitate, 0.1 ml of either unlabeled antipyrine or other cold lig-
and solution, and 0.1 ml of phosphate buffer. The mixed solution was incubated for 2 hr. After the addition of 0.5 ml of saturated ammonium sulfate solution to the mixture, the precipitate was collected as described
above.

Both precipitates were washed twice with 0.5 ml of 50% saturated ammonium sulfate solution and then dissolved in 1 ml of distilled water;
0.5 ml of each solution was transferred to 10 ml of dioxane scintillation
fluid containing 7 g of DPO, 0.3 g of POPOP, and 100 g of naphthalene per
1000 ml of dioxane. Radioactivity was measured for 10 min in a liquid
scintillation spectrometer.

Comments

A constant amount of [³H]antipyrine (5000 dpm, 2.13 ng) was added to
a test tube containing 0.1 ml of antiserum diluted by serial dilution. In the

absence of nonradioactive antipyrine, Antibody-I and -II diluted 1:60 and 1:300, respectively, bound approximately 50% of the radioactivity. The calibration curves of antipyrine in the RIAs are shown in Fig. 2. When increasing amounts of unlabeled antipyrine were added to the incubation mixture, the formation of [³H]antipyrine–Antibody-I and -II complex was inhibited competitively and linearly. The sensitivity of both RIAs shows that as little as 1 ng of antipyrine is detectable (Fig. 2). The 2 hr-incubation at room temperature was established to be sufficient for the assay without reduction in sensitivity.

The inhibition of binding by various pyrazolone derivatives using Antibody-I and -II is shown in Table I. Addition of 5.8 and 6.8 ng of unlabeled antipyrine gave 50% inhibition of immune binding with these two antisera. The antibodies showed different binding affinities for the ligands tested, depending on the similarity of the chemical structure of the inhibi-

TABLE I

INHIBITION BY ANTIPYRINE LIGANDS OF BINDING OF [³H]ANTIPYRINE BY ANTI-ANTIPYRINE ANTISERA[a]

Ligands	R_1	R_2	50% Inhibition (ng) Antibody-I	Antibody-II
Antipyrine	CH_3-	$H-$	5.8	6.8
4-Aminoantipyrine	CH_3-	NH_2-	6.5	6.4
4-Succinamidoantipyrine	CH_3-	$HOOC(CH_2)_2CONH-$		6.7
Isopropylantipyrine	CH_3-	$(H_3C)_2CH-$	6.4	1320
Sulpyrine	CH_3-	$\begin{matrix}NaSO_3H_2C\\H_3C\end{matrix}N-$	6.6	35.5
Aminopyrine	CH_3-	$(H_3C)_2N-$	15.0	2820
4-Acetaminoantipyrine	CH_3-	CH_3CONH-	66.0	8.3
Aminopropylon	CH_3-	$\begin{matrix}H_3C\\H_3C\end{matrix}NCHCONH-$ (CH_3)	70.0	8.5
1-Phenyl-3-methyl-5-pyrazolone	$H-$	$H-$	1150	6100

[a] Data from T. Takatori, A. Yamaoka, and K. Terazawa, *J. Immunol. Methods* **29**, 185 (1979); and T. Takatori and A. Yamaoka, *J. Immunol. Methods* **35**, 147 (1980).

TABLE II
INHIBITION BY PYRAZOLIDINE DERIVATIVE LIGANDS OF BINDING OF
[³H]ANTIPYRINE BY ANTI-ANTIPYRINE ANTISERA[a]

Ligands	R_1	R_2	X	50% Inhibition (ng) Antibody-I	Antibody-II
1-Phenyl-3-pyrazolidone	H—	H—	$H_2=$	4,200	9,700
Sulfinpyrazone	(phenyl)—	(phenyl)—S(CH₂)₂— (with ↓O)	O=	>20,000	15,900
Phenylbutazone	(phenyl)—	$n\text{-}C_4H_9$—	O=	>20,000	19,500
Oxyphenbutazone	HO—(phenyl)—	$n\text{-}C_4H_9$—	O=	>20,000	29,000
Ketophenylbutazone	(phenyl)—	$CH_3CO(CH_2)_2$—	O=	>20,000	33,500

[a] Data from T. Takatori, A. Yamaoka, and K. Terazawa, *J. Immunol. Methods* **29,** 185 (1979), and T. Takatori and A. Yamaoka, *J. Immunol. Methods* **35,** 147 (1980).

tor to that of antipyrine. Alteration of substituents on carbon-4 of the pyrazolone ring generally decreased binding to the antibodies (Table I[6,8]). The binding of 4-acetaminoantipyrine (an aminopyrine metabolite[9]) and aminopropylon to Antibody-I, was approximately 10 times weaker than that of antipyrine, suggesting that the chain length of substituents on carbon-4 of the pyrazolone ring is discriminated by Antibody-I. However Antibody-I does not discriminate among individual pyrazolone derivatives. On the other hand, the binding ability of 4-acetaminoantipyrine and aminopropylon to Antibody-II was almost equal to that of antipyrine, revealing that the straight chain containing amido bond at carbon-4 of the pyrazolone ring is not recognized by this antibody. Although aminopyrine is structurally quite similar to sulpyrine, the binding affinity of the former drug for Antibody-I is weaker than that of the latter and Antibody-II does not cross-react significantly with aminopyrine. These differences may be due to steric hindrance of binding to the antibodies. Both Antibody-I and

[9] F. Pechtold, *Arzneim.-Forsch.* **14,** 972 (1964).

-II showed no significant affinity for 1-phenyl-3-methyl-5-pyrazolone which does not have a methyl group at the carbon-2 position of the pyrazolone ring, indicating that this methyl group in the pyrazolone skeleton is immunodominant. In addition, both antibodies gave no cross-reaction with any pyrazolidine derivatives (Table II[6,8]).

Compared to Antibody-I produced by immunization with the 4-AAP–BSA conjugate, Antibody-II obtained against the 4-SAAP–BSA conjugate showed different specificities to individual pyrazolone derivatives; the degree of conjugation of 4-SAAP to the carrier protein was higher than that of 4-AAP, and anti-antipyrine antiserum of higher titer was produced against the former compound compared to that against the latter compound. These differences in specificities and titer may be attributable to the presence of the succinyl spacer group on the carbon-4 position of the pyrazolone ring.[10] Similar observations have been reported by Chang et al.[11] using N-(4-antipyrinyl)succinamic acid BSA as antigen. However the titer of their antibodies was lower than our Antibody-II, which may be due to the lower degree of hapten conjugation. The presence of a hapten-carrier spacer as well as degree of conjugation of the hapten to the carrier and selection of the hapten-conjugate site seem to be important in obtaining a suitable antibody against the hapten. These observations are consistent with the effect of different distances from the gel matrix backbone in affinity chromatography reported by Cuatrecasas.[12]

[10] L. T. Cheng, S. Y. Kim, A. Chung, and A. Castro, FEBS Lett. 36, 339 (1973).
[11] R. L. Chang, A. W. Wood, W. R. Dixon, A. H. Conney, K. E. Anderson, J. Eisen, and A. P. Alvares, Clin. Pharmacol. Ther. 20, 219 (1976).
[12] P. Cuatrecasas, J. Biol. Chem. 245, 3059 (1970).

Section VIII

Environmental Agents

[46] Radioimmunoassay for 2-Acetylaminofluorene–DNA Adducts

By Miriam C. Poirier and Robert J. Connor

Introduction

The experimental chemical carcinogen 2-acetylaminofluorene (2-AAF) serves as a useful model for investigating the role of DNA modification by carcinogens. The interaction of 2-AAF and its target cells during the process of carcinogenesis *in vivo* results in 3 DNA adducts.[1-5] While the identity of each adduct is now well established, the relative or absolute quantification of each from target tissue has been difficult and required the use of radioactively labeled carcinogen. Recently, we have demonstrated that these adducts can be distinguished by a sensitive and specific RIA. Furthermore quantification of the two major DNA binding products can be obtained utilizing this assay at the low binding levels generally observed *in vivo*.[6,7]

When DNA *in vitro* is reacted with N-acetoxy-2-acetylaminofluorene (N-Ac-AAF) approximately 85% of the bound products are in the form of N-(8)-deoxyguanosinylacetylaminofluorene (dG-8-AAF) and the other 15% are 3-deoxyguanosin-(N²)-ylacetylaminofluorene (dG-N²-AAF),[5,8,9] (structures are shown in Fig. 1). The latter compound is present as only about 5% of the total bound products in DNA when 2-AAF, N-hydroxy-AAF, or N-Ac-AAF are administered either to whole animals or cultured cells.[5,10,11] The amount of dG-8-AAF found *in vivo* varies with the enzy-

[1] E. Kriek, J. A. Miller, U. Juhl, and E. C. Miller, *Biochemistry* **6**, 177 (1967).

[2] E. Kriek, *Chem. Biol. Interact.* **1**, 3 (1969).

[3] E. Kriek, *Cancer Res.* **32**, 2042 (1970).

[4] E. Kriek, *Biochim. Biophys. Acta* **355**, 177 (1974).

[5] J. G. Westra, E. Kriek, and H. Hittenhausen, *Chem. Biol. Interact.* **15**, 149 (1976).

[6] M. C. Poirier, S. H. Yuspa, I. B. Weinstein, and S. Blobstein, *Nature (London)* **270**, 186 (1977).

[7] M. C. Porier, M. A. Dubin, and S. H. Yuspa, *Cancer Res.* **39**, 1377 (1979).

[8] H. Yamasaki, P. Pulkrabek, D. Grunberger, and I. B. Weinstein, *Cancer Res.* **37**, 3756 (1977).

[9] R. P. P. Fuchs, *Analyt. Biochem.* **91**, 663 (1978).

[10] P. A. Cerutti, in "DNA Repair Mechanisms: ICN-UCLA Symposia on Molecular and Cellular Biology" (P. C. Hanawalt, E. C. Friedberg, and C. F. Fox, eds.), Vol. 9, p. 1. Academic Press, New York, 1978.

[11] F. A. Beland, K. L. Dooley, F. E. Evans, and C. D. Jackson, *Proc. Am. Assoc. Cancer Res.* **19**, 128 (Abstr. 518) (1979).

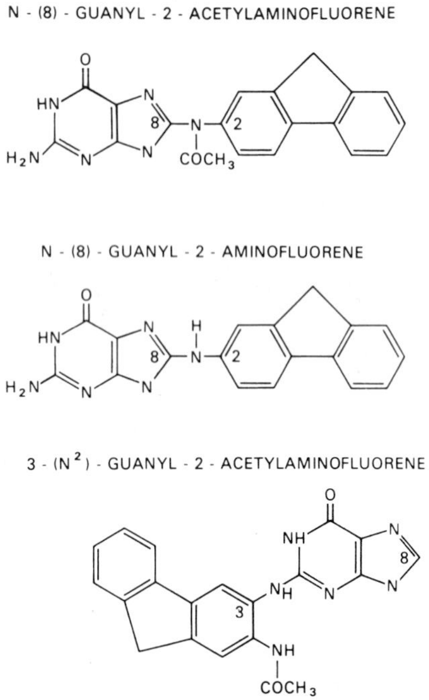

N - (8) - GUANYL - 2 - ACETYLAMINOFLUORENE

N - (8) - GUANYL - 2 - AMINOFLUORENE

3 - (N^2) - GUANYL - 2 - ACETYLAMINOFLUORENE

FIG. 1. Structures of the three adducts formed upon interaction of the carcinogen 2-AAF with DNA *in vivo*.

mological capabilities of the target cell, but frequently the deacetylated C-8 adduct, N-(8)-deoxyguanosinylaminofluorene (dG-8-AF) is the major adduct formed[2,3,7,10,11] (see Fig. 1).

This report will describe methods for the synthesis of G-8-AAF, its coupling to bovine serum albumin and the immunization of rabbits. Procedures for the establishment of an RIA for both dG-8-AAF and dG-8-AF will also be outlined, and examples of the application of this assay to quantify DNA-bound carcinogen in hydrolyzed DNA from different types of cultured cells exposed to N-Ac-AAF will be presented. The quantification of acetylated and deacetylated C-8 adducts described here depends on the specificity of this particular antiserum.[6,7] Similar specificity has been demonstrated for antibodies raised in an identical fashion in another laboratory,[12] however one must assume that each new antiserum will have unique characteristics and should be thoroughly tested.

[12] M. Guigues and M. Leng, *Nucleic Acids Res.* **6**, 733–744 (1979).

Developmental Procedures

G-8-AAF Synthesis, Coupling, and Immunization

Ten milligrams of guanosine (Sigma, St. Louis, MO.) in 7 ml 0.01 *M* Tris–HCl pH 7.3 is mixed with 10 mg of *N*-Ac-AAF (supplied by the National Cancer Institute Standard Chemical Carcinogen Reference Repository, Division of Cancer Cause and Prevention, Bethesda, MD), in 3 ml of 95% ethanol and incubated for 18–24 hr at 37° in the dark. After the ethanol is removed by nitrogen evaporation, the G-8-AAF can be purified by Sephadex LH-20 chromatography. Because Sephadex LH-20 (Pharmacia, Piscataway, N.J.) has affinity for aromatic compounds, the adduct will bind to the column in aqueous medium and elute in 40% methanol/water. Any unreacted guanosine will be removed in the initial water wash, and unreacted 2-AAF degradation products will remain on the column unless eluted with 100% methanol. The 40% methanol/water material is further purified by additional LH-20 chromatography until the ratio of the UV absorbances at 302 nm/278 nm is approximately 0.50 (see UV spectra in Ref. 1). The yield of purified compound is about 50%.

Reagents and Procedure for Coupling G-8-AAF to Bovine
Serum Albumin (BSA)

12 mg G-8-AAF is added to 1.2 ml of 0.05 *M* sodium metaperiodate
Incubate 45 min, room temperature in the dark
0.025 ml of 1 ethylene glycol is added for 5 min
35 mg of BSA in 1.25 ml water is added and 5% K_2CO_3 introduced dropwise until the pH reaches 9
Stir 45 min, room temperature in the dark
19 mg $NaBH_4$ in 1.25 ml water is added
Stir 1 hr, 4° in the dark
Dialyze overnight against 3 changes of deionized water and store frozen.

Due to a small portion of UV absorbance in the range of 310–330 nm[12] in the final G-8-AAF-BSA it has been suggested that the conjugate contains a small portion of deacetylated adduct formed at pH 9.

Three New Zealand white rabbits were given intramuscular injections of G-8-AAF-BSA in complete Freund's adjuvant in the hindquarters (1 ml) on seven occasions. Each rabbit received 0.3 mg of adduct per injection and the highest titer antisera were obtained from all three rabbits at 4–5 months.

Radioimmunoassay

Two radiolabeled compounds have been utilized as trace in the RIA, [³H]G-8-AAF and [³H]G-8-AF. The former compound is synthesized from

[5'-^{3}H]guanosine, (24 Ci/mmol, New England Nuclear, Boston, Mass.) and N-Ac-AAF according to the procedure previously outlined. Deacetylation can be accomplished under alkaline conditions (25% methanol, 15% acetone, pH 11 and 72° for 5 hr) to yield [^{3}H]G-8-AF with a specific activity similar to that of the original precursor. Isolation of the deacetylated adduct from the reaction mixture is accomplished by Sephadex LH-20 chromatography, with the G-8-AF eluting at about 75% methanol. A similar synthesis can be performed using nonradioactive dG-8-AAF to yield purified dG-8-AF for use as inhibitor in the RIA.[1]

RIA Reagents and Procedure

0.1 ml specific antiserum diluted 1:600

0.1 ml of nonradioactive standard G-8-AAF or G-8-AF (0.2–20 pmol) or 2–35 μg of hydrolyzed unknown DNA sample as inhibitor

0.1 ml of [^{3}H]G-8-AAF (20,000 cpm) or [^{3}H]G-8-AF (10,000 cpm)

0.010–0.10 ml (2–35 μg) of hydrolyzed control DNA (added to standard curve tubes in quantity comparable to unknown DNA)

0.050–0.100 ml Tris–HCl pH 7.3 to equalize volume in each tube

0.1 ml of goat-anti-rabbit IgG (0.5 mg) (Miles, Elkhart, Ind.)

The RIA is performed in a volume of approximately 0.5 ml with all components diluted in 0.01 M Tris–HCl pH 7.3.

The first 5 components are incubated together for 1.25 hr at 37° and the second antibody added for 1 hr at 22°. By utilizing nonequilibrium conditions, in which labeled adduct is added for 5 min at the end of the 37° incubation, the sensitivity of the assay is increased 2- to 3-fold. Immunoprecipitates are sedimented by centrifugation, pellets are dissolved in 0.2 ml 0.1 N NaOH, and counted in 13 ml of Aquasol-2 (New England Nuclear, Boston, Mass.) at an efficieny of 50%.

Antibody Specificity

Antibodies raised against G-8-AAF-BSA recognize ribo- and deoxyribo-C-8 adducts almost equally since the 2' and 3' OH groups were bound to BSA during immunization. Therefore, radioactive adduct synthesized from [^{3}H]guanosine reacts in a fashion identical to that synthesized from [^{3}H]deoxyguanosine in the RIA. The former compound is generally used as precursor because the latter is more expensive and unstable.

A standard curve at equilibrium is shown in Fig. 2 (O——O). In this assay the trace is [^{3}H]G-8-AAF and nonradioactive dG-8-AAF competes as inhibitor. The sensitivity is such that there is 50% inhibition at about 2.5 pmol (5 ng). Figure 2 also shows that the following related molecules do not inhibit in this assay within the same dose magnitude: deoxyguanosine, N-Ac-AAF, or dG-N^2-AAF (supplied by Dr. J. G. Westra of the

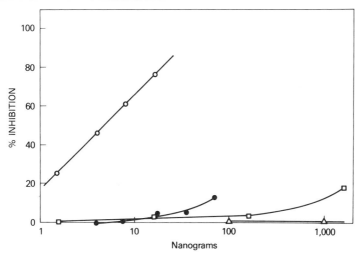

FIG. 2. RIA standard curve (equilibrium conditions) in which [³H]G-8-AAF as trace is competed against dG-8-AAF as inhibitor (O——O). No significant competition is observed in the same concentration range with dG-N^2-AAF (O——O), N-Ac-AAF (□——□), or deoxyguanosine (△——△) in the same assay.

Netherlands Cancer Institute, Amsterdam, The Netherlands). In addition, up to 80 μg of hydrolyzed unmodified DNA will not inhibit binding, and up to 35 μg of the same DNA can be added to the standard curve tubes without changing the profile of the curve (data not shown).

Figure 3 (O------O) shows [³H]G-8-AAF in competition with dG-8-AAF under nonequilibrium conditions, resulting in a 3-fold increase in sensitivity (50% inhibition at 0.9 pmol). In the same assay, competition with dG-8-AF (O——O) results in saturation at about 40% inhibition with concentrations above 3 pmol of inhibitor. This saturation is not due to dG-8-AF insolubility because when assayed with [³H]G-8-AF as trace a complete inhibition curve is obtained (Fig. 3, ●——●). Thus, it seems possible that a smaller portion of cross-reacting antibody is able to recognize both dG-8-AAF and dG-8-AF. When unsaturating mixtures of dG-8-AAF and dG-8-AF are assayed against [³H]G-8-AAF the percentage inhibition produced by the mixture will be the sum of the percentage inhibitions observed in individual assays.[7]

Determination of the Deacetylated Adduct Utilizing
Synthetic Standards

Upon observing the plateau or saturation at 40% inhibition when dG-8-AF was assayed against [³H]G-8-AAF (Figs. 3 and 4A, O——O) we considered the possibility that small known amounts of dG-8-AAF added to

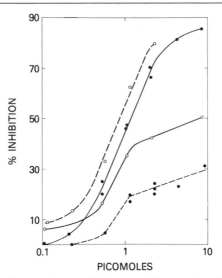

FIG. 3. RIA standard curves (nonequilibrium conditions) using either [³H]G-8-AAF (○) or [³H]G-8-AF (●) as trace and inhibiting with increasing picomole concentrations of either dG-8-AAF (------) or dG-8-AF (———).

saturating concentrations of dG-8-AF would increase the percentage inhibition, and that the increase observed with such mixtures might be quantifiable. This possibility was of considerable interest since we anticipated differences in the proportion of acetylated and deacetylated adduct in different cell types[7,10,11] and were searching for a procedure to distinguish between the two adducts in a mixture. The feasibility of performing relative measurements became apparent when the total amount of C-8 adduct was maintained at 6 pmol, and relative proportions of the acetylated adduct were increased to about 10% dG-8-AAF : 90% dG-8-AF. A marked increase in the observed % inhibition (Fig. 4A, ▲——▲ and △——△) over that of 100% of dG-8-AF (○——○) was observed.

Ten experiments were performed, each using newly diluted antiserum and standard dG-8-AAF mixtures of 0, 5, and 8.7%. The results showed a near linear increase in percentage inhibition as percentage dG-8-AAF increased averaging 40.0 at 0, 51.5 at 5, and 61.5 at 8.7 (see Fig. 4B). An analysis of covariance[13] indicated that a regression line for each experiment gave a significantly better fit ($p < 0.001$) than did a single regression line for the 10 experiments combined. The regression lines have a common slope, estimated from the 10 data sets, but distinct intercepts, each

[13] G. H. Snedecor and W. G. Cochran, in "Statistical Methods," p. 419. Iowa State Univ. Press, Ames, Iowa, 1967.

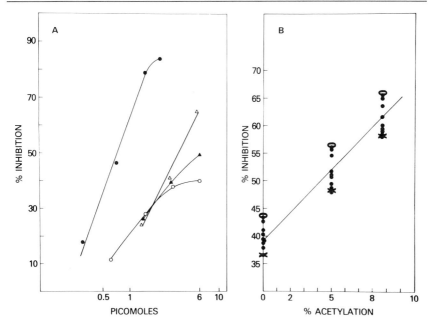

Fig. 4. (A) RIA standard curves (nonequilibrium conditions) in a competition assay utilizing [³H]G-8-AAF and a series of standards or standard mixtures of nonradioactive dG-8-AAF and/or dG-8-AF. The curves are as follows: dG-8-AAF (●——●), dG-8-AF (○——○), 8.7% dG-8-AAF:91.3% dG-8-AF (△——△), and 5% dG-8-AAF:95% dG-8-AF (▲——▲). (B) Percentage inhibition observed in the RIA at 6 pmol of total adduct (ordinate) against the percentage acetylation in standard mixtures of 0, 5, and 8.7% dG-8-AAF (abscissa). The regression line is for the average over the 10 experiments of the percentage inhibitions against the standard mixtures. The points with the open circles are from the experiment with the highest observed percentage inhibitions. The points with the ×s are from the experiment with the lowest observed percentage inhibitions.

estimated using the experiments average percentage inhibition. Thus, we have

$$Y_{ij} = \bar{Y}_i + b(X_j - \bar{X}) \tag{1}$$

where

    Y_{ij} is the percentage inhibition at X_j in the ith assay
    $\bar{Y}_i$ is the average percentage inhibition in the ith assay
    X_j is the percentage dG-8-AAF in the jth mixture
    $\bar{X}$ is the average percentage dG-8-AAF of the standard mixtures, 0, 5, and 8.7%, used in the assays, 4.56
    b is the common slope, 2.46

Note that this result is consistent with the percentage inhibition of each of the standard mixtures increasing or decreasing in tandem due to small variations in antibody dilution which occur between assays.

More importantly this result allows the percentage dG-8-AAF of an unknown mixture to be estimated and confidence limits given. The procedure uses the regression line in reverse (linear calibration[14]). To estimate the percentage dG-8-AAF the unknown mixture's percentage inhibition is compared with the average percentage inhibition for the three standard mixtures assayed simultaneously. Formally we estimate the unknown percentage dG-8-AAF (acetylation) using

$$Z = \bar{X} + (Y - \bar{Y})/b \qquad (2)$$

where

Z is the estimate of the percentage dG-8-AAF in the unknown mixture

Y is the percentage inhibition for the unknown mixture at 6 pmol total modification

$\bar{Y}$ is the average percentage inhibition for 6 pmol of the three standard mixtures assayed simultaneously with the unknown mixture

$\bar{X}$ and b are as defined above

Confidence limits depend on the reproducibility of the percentage inhibition at 6 pmol modification and improve with increasing numbers of assays.

RIA of Acetylated and Deacetylated C-8 Adducts in DNA

Synthesis of DNA Modified with 2-AAF in Vitro

Single-stranded, or double-stranded DNA is modified *in vitro* at neutral pH by reaction with N-Ac-AAF for 24 hr at 37° in 30% ethanol. If the amount of DNA is kept constant the level of modification is directly proportional to the amount of N-Ac-AAF employed. For example 5 mg of DNA reacted with less than 0.5 mg N-Ac-AAF will yield a level of modification about 1–2%, whereas 5 mg or more N-Ac-AAF will yield levels of modification as high as 27%.[6,15] At least 85% of the total bound material will be dG-8-AAF under these circumstances and the amount of dG-N^2-AAF formed will be variable.[5,8,9] The reaction mixture should be extracted 6 times with ether, to remove unreacted carcinogen, and reprecipitated with ethanol at least twice or until the ratio of absorbance at 278 nm/302 nm is a constant. If the level of modification is above 10% there will be a visible shoulder at A_{302} and the DNA peak absorbance normally

[14] G. H. Snedecor and W. G. Cochran, *in* "Statistical Methods," p. 159. Iowa State Univ. Press, Ames, Iowa, 1967.
[15] R. Fuchs and M. Daune, *Biochemistry* **11**, 2659 (1972).

at A_{260} will be shifted to A_{270}.[16] Since DNA does not usually absorb at A_{302}, the amount of modification can be calculated from an ϵ of 15,000 at that wavelength, but the amount of DNA should be determined by some method other than UV spectrum (for example, diphenylamine).[6,16,17]

Assay of Highly Modified M. luteus DNA

Initial validation of the RIA was performed with an *M. luteus* DNA denatured and modified to 27% (27 out of 100 nucleotides adducted). The level of modification was determined by UV spectrum and diphenylamine assay before analysis by RIA.[6] RIA data indicated modification values of 1.5% for intact material, 5.5% for heat denatured material, and 27% for enzyme hydrolyzed material. Thus, enzymatic hydrolysis with DNase, alkaline phosphatase, and venom phosphodiesterase was necessary in order to determine all of the C-8 adduct by RIA.[6] Subsequent experiments have demonstrated that similar results are obtained by S_1 nuclease hydrolysis after denaturation.[7]

Reagents and Procedure

To 300 g DNA in about 1 ml water add
0.05 ml of 1.5 N NaCl
0.025 ml of 1 N Na acetate pH 4.6
0.01 ml of 0.4 N Zn SO_4
Heat to 100° for 5–10 min and cool rapidly and add 5 × 10^3 units of S_1 nuclease (Sigma Chemical Co., St. Louis, MO)
Incubate 3 hr at 37°
Adjust pH to 6 with 0.1 N Tris–HCl pH 9 and store frozen at $-20°$.

Assay of AAF-Modified (2%) Calf Thymus DNA

While the highly modified G-C rich *M. luteus* DNA gave *different* apparent modification values when assayed as intact or denatured DNA, the calf thymus DNA modified to 2% with AAF (CT-AAF-DNA) gave the *same* RIA profile with either native or denatured DNA (data not shown). However, both modified DNAs were assayed with greater sensitivity and accurate quantification only after denaturation and S_1 nuclease hydrolysis. A similar phenomenon has been reported by Guigues and Leng.[12] The RIA profile for increasing concentrations of hydrolyzed CT-AAF-DNA is shown in Fig. 5B and is identical to the dG8-AAF standard curve for the same experiment, shown in Fig. 5A (●——●). When concentration of adduct in the DNA sample shown in Fig. 5B is plotted with the standard curve in Fig. 5A, the resulting profile (○——○) is superimposable on the

[16] E. C. Miller, U. Juhl, and J. A. Miller, *Science* **153**, 1125 (1966).
[17] E. Sage, R. P. P. Fuchs, and M. Leng, *Biochemistry* **18**, 1328 (1979).

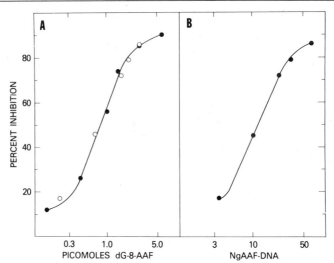

FIG. 5. (A) RIA standard curve (nonequilibrium conditions) in which [³H]G-8-AAF is utilized as trace and dG-8-AAF as inhibitor (●——●). The open circles (○——○) are values
for picomoles of dG-8-AAF adduct from a calf-thymus DNA modified *in vitro* with *N*-Ac-
AAF (see B). (B) RIA profile (●——●) showing percentage inhibition for increasing nanogram concentrations of a calf thymus DNA modified *in vitro* with *N*-Ac-AAF. When picomoles of adduct of this DNA are plotted simultaneously with a dG-8-AAF standard curve
(Fig. 5A, ○——○) both curves are superimposable.

standard curve. Thus, the adducts measured in hydrolyzed, *in vitro* modified DNA give RIA profiles identical to standard profiles of dG-8-AAF
competition against [³H]G-8-AAF. Previous reports have shown that
DNAs modified *in vitro* with *N*-Ac-AAF contain all of the C-8 modification in the form of dG-8-AAF.[5,8,9]

Assay of DNA from Cells Exposed to *N*-Ac-AAF *in Vivo*

Exposure of Cultured Cells and Preparation of DNA

Semiconfluent cells in medium containing serum are exposed to *N*-Ac-
AAF (10^{-5} *M*) in 0.4% DMSO for 1 hr, a time at which maximum binding
to DNA by this direct-acting carcinogen occurs.[18,19] For DNA repair experiments a portion of the cultures are harvested at 1 hr while parallel
dishes are washed 3× with physiological saline to remove any unbound
carcinogen and replated with fresh medium for an additional incubation

[18] D. E. Amacher and M. W. Lieberman, *Biochem. Biophys. Res. Commun.* **74,** 285 (1977).
[19] D. E. Amacher, J. A. Elliott, and M. W. Lieberman, *Proc. Natl. Acad. Sci. U.S.A.* **74,**
1553 (1977).

period. DNA is prepared on CsCl gradients, dialyzed against water to remove CsCl, quantified by optical density, and concentrated to about 300
μg/ml before heat denaturation and S_1 nuclease hydrolysis. Up to 35 μg
of hydrolyzed DNA from unexposed cells can be added to the standard
curve samples with no loss in sensitivity, and the amount chosen to add
should parallel that assayed in the unknown DNA samples.

*Quantification of C-8 Adducts and Determination of the Proportion of
Acetylated and Deacetylated Adducts*

Accurate determination of binding levels depends upon appropriate
choice of tracer and standard curve. As shown in Fig. 3, acetylated and
deacetylated adducts can be assayed with similar sensitivities using
[^{3}H]G-8-AAF or [^{3}H]G-8-AF, respectively. These four curves also show
that when the unlabeled competitor is not the same as the trace (i.e., dG-
8-AF in a [^{3}H]G-8-AAF assay) there is a decrease in sensitivity and a saturation of the curve below the usual 80–90% inhibition. When cells are regularly exposed to N-Ac-AAF under the same conditions of confluency,
serum concentration, and type of medium the proportion of acetylated
and deacetylated C-8 adducts will remain constant between experiments
although the binding levels will vary. Different cell types, however, will
form different proportions of acetylated and deacetylated adducts. To determine which standard curve and/or trace is most appropriate for routine
assay of modified DNA from a particular cell type experiments should be
performed as follows.

Step 1. Assay samples of an unknown modified DNA simultaneously
with [^{3}H]G-8-AAF and [^{3}H]G-8-AF. If most of the C-8 adduct is acetylated a higher percentage inhibition will be observed with [^{3}H]G-8-AAF.
Likewise, if most of the C-8 adduct is deacetylated, there will be a higher
percentage inhibition in the [^{3}H]G-8-AF assay. The same amount of DNA
should be used in both assays.[20]

Step 2. Increasing concentrations of the unknown modified DNA
should be assayed with [^{3}H]G-8-AAF and the profile compared to those
shown in Fig. 4A. As a point of reference the percentage inhibition of dG-
8-AF saturation should be determined simultaneously by assaying 6 pmol
dG-8-AF plus control DNA at each concentration of DNA.

If the profile of modified DNA assayed against [^{3}H]G-8-AAF (Step 2)
seems to be linear beyond 50% inhibition (similar to Fig. 4A, ●——●)
and if a lower percentage inhibition is observed (for the same amount of
DNA) in the [^{3}H]G-8-AF assay than the [^{3}H]G-8-AAF assay (Step 1), the

[20] Poirier, M. C., Williams, G. M., and Yuspa, S. H. *Mol. Pharmacol.* **18,** 581 (1980).

DNA in question has more acetylated than deacetylated C-8 adduct and Fig. 4A (●——●) would be the appropriate standard curve.[20]

If, however, the profile of modified DNA assayed in Step 2 is similar to Fig. 4A (○——○, △——△, or ▲——▲) and the same amount of DNA gives a higher percentage inhibition in the [³H]G-8-AF assay than the [³H]G-8-AAF assay (Step 1), there is much more deacetylated than acetylated adduct. The proportion of each, up to about 10% acetylated:90% deacetylated can be determined by comparing the percentage inhibition of the unknown DNA at 6 pmol of modification with 6 pmol values of the standard 0, 5, and 8.7% acetylated mixtures and utilizing Eq. (1) (see Fig. 4B). Since standard curves for the [³H]G-8-AAF assay are essentially identical below saturation (about 40% inhibition) for mixtures between 0 and 10% dG-8-AAF (Fig. 4A), a 5% dG-8-AAF:95% dG-8-AF standard curve will be appropriate for any sample < 10% acetylated, when the level of modification is calculated from values of unknown sample giving 20–40% inhibition. Alternatively, a DNA sample containing 10% dG-8-AAF can be assayed with [³H]G-8-AF and a dG-8-AF standard curve since a small portion of dG-8-AAF does not shift that standard curve. Once the appropriate standard curve has been chosen one can assume that the same type of cells treated under the same conditions will have a stable proportion of acetylated and deacetylated C-8 adducts[7] and utilize the chosen assay routinely.

Conclusion

The assays presented here for determination of acetylated and deacetylated C-8 adducts of 2-AAF with deoxyguanosine in DNA are sensitive to approximately 10 fmol/μg DNA (one adduct in 3 × 10⁵ bases). Since the DNAs assayed in this laboaratory have been either ≥95% deacetylated (mouse epidermal cells and human dermal fibroblasts) or >80% acetylated (rat hepatocytes), emphasis has been placed on characterizing these mixtures. Adequate procedures are still being established for DNAs between 20 and 80% acetylated since in these ranges saturation occurs at 90% inhibition and a shifting of the standard curve toward greater sensitivity is observed with increasing dG-8-AAF. We anticipate that this sensitive and versatile assay will become a prototype for a wide variety of similar immunoassays to be used as probes for the interaction of chemical carcinogens with cellular macromolecules.

[47] Radioimmunoassay of Chlorinated Dibenzo-*p*-dioxins

By PHILLIP W. ALBRO, MICHAEL I. LUSTER, KUN CHAE, GEORGE CLARK, and JAMES D. MCKINNEY

The polychlorinated dioxins are a series of tricyclic aromatic compounds which have no known commercial uses themselves but occur as associated by-products and impurities in the production of chlorinated phenols and their conversion products such as the phenoxy acid herbicides. Dioxins are not generally considered to have ubiquitous distribution in the environment although this has recently been challenged by the "chemistry of fire" proposal of Dow Chemical Company.[1] The number of chlorine atoms can vary between one and eight and the number of possible positional cognates is quite large. The tetra series is of particular analytical interest since it contains the most toxic isomer (2,3,7,8-TCDD) known to date.[2] The hexa and hepta series are also of interest since some of these isomers are similarly toxic. A number of these isomers are found associated with Herbicide Orange (2,4,5-trichlorophenoxyacetic acid) and pentachlorophenol (a wood preservative).

Because of the high toxicity of 2,3,7,8-TCDD, it has been necessary to develop analytical methods with extreme specificity and sensitivity to achieve a detection limit in the low part per trillion (ppt) range. It is essential that any analytical method include a prior clean-up procedure to separate the desired analyte from most of the potential interfering components of the sample if such high sensitivity and specificity are to be achieved. The methods presently in use generally consist of some form of low or high resolution chromatography (usually GC) and high or low resolution mass spectrometry and are currently not unequivocal for specific isomers analysis.[3]

Any assay procedure must also deal with the substantial problems of cross-reactivity to related isomers of the dioxins and other compounds such as the chlorinated dibenzofurans. In addition, it is often difficult to distinguish interference from specific binding, therefore it is essential to include well matched control samples (blanks) for simultaneous determination.

[1] Report by Rebecca L. Rawls, *Chem. Eng. News* Feb. 12 (1979). Dow Finds Support, Doubt for Dioxin Ideas.

[2] E. E. McConnell, J. A. Moore, J. K. Haseman, and M. W. Harris, *Toxicol. Appl. Pharmacol.* **44**, 335 (1978).

[3] J. D. McKinney, *Ecol. Bull.* (*Stockholm*) **27**, 55 (1978).

METHODS IN ENZYMOLOGY, VOL. 84   ISBN 0-12-181984-1

RIA has several advantages which make it particularly attractive for use in dioxin analysis. Since the method is relatively (to mass spectrometric methods) simple and cheap to perform, it is possible to run large numbers of samples simultaneously permitting better quality control of the method as well as screening of a larger population of a given sample type. Furthermore, the potential for interferences is limited not only by antibody specificity but also by the limited ability of the detergent additive to solubilize potential interferences. The availability of RIA methods for dioxins should permit preliminary screening of a variety of sample types to lower the use of the more inconvenient, time-consuming, and costly instrumental methods of analysis. In some cases, it should also be useful as a confirmatory method for the findings from instrumental methods of analysis.

Preparation of Antigens

The starting point for the preparation of the hapten–protein conjugates used as antigens (immunogens) is the amino derivative of the halogenated aromatic compound (hapten moiety). In general, the amino compound should have one fewer chlorine atoms than the parent hapten for greatest specificity; thus we use 1-amino-3,7,8-trichlorodibenzo-p-dioxin in the production of an antiserum that binds 2,3,7,8-tetrachlorodibenzo-p-dioxin. The amino compounds are made by reduction of the corresponding nitro derivatives, which are either synthesized *de novo,* or obtained by direct nitration of the parent compound. These syntheses have been described in detail elsewhere.[4-6]

Preparation of Adipamide Derivative

An adipamide side chain contributes to the effectiveness of these conjugates by holding the desired determinant structures out from the matrix of the protein carrier. Shorter side chains lead to the formation of cyclic by-products and are not effective. The general procedure will be illustrated with the dioxin derivative, but the same process will apply for all substituted aromatic amines.

One millimole (302.5 mg) of 1-amino-3,7,8-trichlorodibenzo-p-dioxin is dissolved in 10 ml of dry pyridine (freshly distilled from barium oxide) in a 50-ml round bottom flask and cooled in ice. Methyl adipoyl chloride

[4] K. Chae, L. K. Cho, and J. D. McKinney, *J. Agric. Food Chem.* **25,** 1207 (1977).
[5] A. Norstrom, S. K. Chaudhary, P. W. Albro, and J. D. McKinney, *Chemosphere* No. 6, 331 (1979).
[6] S. K. Chaudhary and P. W. Albro, *Org. Prep. Proc. Intern.* **10,** 46 (1978).

(made from the monomethyl ester of adipic acid and thionyl chloride), 1.5 mmol (0.27 ml), is added dropwise with constant stirring. A precipitate of pyridine hydrochloride should appear. A calcium chloride drying tube is added and stirring continued for 1 hr. The reaction mixture is stored at 4° overnight, suspended in 25 ml of diethyl ether, and filtered through glass wool. The filtrate is washed twice with 25 ml of 1 N HCl, once with 10 ml of saturated aqueous KCl, once with 0.4 M K_2CO_3, and once with saturated aqueous KCl, dried over anhydrous sodium sulfate and filtered through glass wool. Solvent is removed using a rotary evaporator at ≤ 40°.

The product (no residual amine is detectable by gas chromatography) is suspended in 15 ml of 95% ethanol in a 100-ml round bottom flask containing a few carborundum boiling chips. Seven milliliters of 0.3 N aqueous NaOH is added and the mixture refluxed for 90 min. The yield of the adipic acid amide of 1-amino-3,7,8-trichlorodibenzo-p-dioxin is essentially quantitative. The basic solution, after cooling, is acidified with 30 ml of 1 N HCl and extracted three times with 25 ml portions of diethyl ether. The combined ether extract is washed once with 30 ml of water, dried over anhydrous sodium sulfate, and filtered through glass wool. Solvent is removed on a rotary evaporator; yield = 425 mg; ϵ = 2.50 × 10⁴ liters cm⁻¹ mol⁻¹ in 0.1 N NaOH, λ_{max} = 238 nm. At 302 nm, ϵ = 3.83 × 10³ liters cm⁻¹ mol⁻¹.

Preparation of Conjugates

Although a wide variety of coupling procedures have been tested (carbodiimide, Schiff base formation, N-hydroxysuccimide activated ester, etc.), the most reproducible procedure involves the formation of a mixed anhydride, as follows.

Dioxane is passed through basic alumina to remove peroxides and moisture. The adipic acid amide derivative (0.06 mmol, 25.9 mg) is dissolved in 2.4 ml of freshly purified dioxane with the addition of 15 μl of tri-n-butylamine. The solution is stirred for 5 min at room temperature (magnetic stirrer), after which 8 μl of fresh isobutyl chloroformate are added. Stirring is continued for 20 min yielding the mixed anhydride.

Protein, either thyroglobulin or albumin (200 mg in either case), is dissolved in 5.25 ml of water plus 0.2 ml of 1 N NaOH. Freshly purified dioxane, 4.25 ml, is stirred in while the protein solution is cooled in an ice bath. The whole of the mixed anhydride solution is added dropwise with stirring to the protein solution, and stirring is continued for 1 hr. The initially cloudy suspension should clear after about 15 min. Another 0.1 ml of 1 N NaOH is added and stirring at ice bath temperature is continued for an additional 3 hr. The solution is then dialyzed for 24 hr against two 2L changes of 0.1 N ammonium formate at 4°. The dialysin is lyophilized.

Being careful to avoid moisture pickup, the dry, conjugated protein is extracted with 100 ml of dry chloroform in a small blender, filtered onto Whatman No. 541 paper, and stored in a vacuum desiccator over Carbowax 20 M. Failure to protect the antigen from moisture during the extraction will result in an isoluble product. The overall recovery of antigen should exceed 150 mg.

Lyophilization by freeze-drying removes all of the ammonium formate from the dialyzed antigen, so the final weight is valid. The salt is necessary, however, to keep the protein from being denatured during dialysis. Removal of absorbed hapten is required since dioxin derivatives in free form may induce immune tolerance to the hapten.

Characterization of Antigens

Stock solutions of antigen and original parent protein are made by dissolving the dry material in 0.1 N NaOH. The solutions are adjusted to contain 4 mg of protein per ml based on the biuret assay.[7] Aliquots are diluted with 0.1 N NaOH to 0.4 mg/ml and the UV difference spectrum recorded in the range 340–220 nm. The total amount of bound dioxin is estimated from the ΔOD at 302 nm. For example, the antigen made as described above with bovine thyroglobulin as parent protein gave $\Delta OD_{1\,cm}^{302} = 0.23$. Thyroglobulin has a reported molecular weight of 670,000, so 0.4 mg = 0.597 nmol. Then $0.23/(3.83 \times 10^3) = 6.0 \times 10^{-5}\,M$, and 60 nmol $\div$ 0.597 nmol = 100.5 mol of dioxin per mole of protein. ΔOD at 302 nm was suitable for the dioxin hapten, but appropriate wavelengths for other haptens must be chosen from their UV spectra.

The antigen is assayed for free amino groups in comparison with the parent protein by the method of Harding and MacLean,[8] using ethanolamine to generate a standard curve. Briefly, one combines 0.5 ml of stock (4 mg/ml) protein solution in 0.1 N NaOH with 0.5 ml of 0.1 N HCl, 1 ml of 10% aqueous pyridine, and 1 ml of 2% aqueous ninhydrin. Heat 20 min in a boiling water bath, cool, dilute to 50 ml with water, and read ΔA against a reagent blank at 570 nm. Procedures based on trinitrobenzenesulfonate (e.g.,[9]) can be used. In the example being considered, thyroglobulin appeared to contain 188 free (lysine) amino groups per molecule while the antigen contained only 111. Presumably, then, 77 hapten molecules bound to thyroglobulin at the amino group of lysine (possibly plus the N-terminal amino group).

The antigen is assayed for phenolic hydroxyl (tyrosine) by a modified Folin–Ciocalteau procedure. To 2 mg of protein in 1 ml of 0.1 N NaOH is

[7] A. G. Gornall, C. J. Bardawill, and M. M. David, *J. Biol. Chem.* **177**, 751 (1949).
[8] V. J. Harding and R. M. MacLean, *J. Biol. Chem.* **24**, 503 (1916).
[9] A. F. S. A. Habeeb, *Anal. Biochem.* **14**, 328 (1966).

added 5 ml of 20% aqueous sodium carbonate (wt%). Folin–Ciocalteau phenol reagent is diluted to 0.111 N acidity. Nine milliliters of diluted phenol reagent is added rapidly to the protein solution with constant mixing (Vortex). The resulting solution is heated for 5 min at 40°, cooled, and left to stand 30 min at room temperature. The OD is read at 750 nm against a procedure blank, using p-hydroxyphenylacetic acid to generate a standard curve. In the present example thyroglobulin appeared to contain 114 free tyrosine –OH moieties per molecule, while the antigen contained only 94, indicating that 20 hapten molecules were bound to tyrosine. In this concentration of carbonate, only tyrosine reacts. Thus, both ribonuclease, which contains no tryptophan, and Avidin, which contains much more tryptophan than tyrosine, gave the correct amounts of apparent tyrosine in this assay. Should the protein be insoluble in sodium carbonate solution, one can substitute 0.2 M potassium tetraborate.

The sum of the ninhydrin and phenol reagent results accounted for 97 mol of bound hapten, in excellent agreement with the 100 mol approximated from the UV difference spectrum. Spectrophotometric assay for intact tryptophan[10] and free sulfhydryl groups[11] showed no changes in going from thyroglobulin (or albumin) to antigen by this procedure. Both the total amount of dioxin derivative bound (approximately one molecule per 6000–7000 of protein MW) and ratio of amino-linked to tyrosine-linked dioxin were reproducible over the approximately eight batches that have been prepared using albumin and three batches using bovine thyroglobulin.

Production and Screening of Antiserum

Since the antigen is water soluble, immunizations should be performed with an adjuvant or by a route that encourages slow absorption. Evaluation of a variety of immunization techniques and schedules indicated that relatively high antibody titers can be obtained in New Zealand rabbits following weekly or biweekly injections of 0.5 mg of the thyroglobulin-conjugated antigen.[12] For the primary injection, 1 mg of antigen is dissolved in 0.4 ml of phosphate-buffered saline (PBS; 20 mM potassium phosphate, 140 mM NaCl, and 0.2% sodium azide; pH 7.3), emulsified with 0.6 ml of complete Freund's adjuvant and 0.5 ml of the emulsion is injected intradermally into multiple sites of the back and hind foot pad. Usually 0.25 ml is injected into the footpads and 50 μl portions into each of 5 sites on the back. Booster injections are given using similar volumes,

[10] M. K. Gaitonde and T. Dovey, *Biochem. J.* **117,** 907 (1970).

[11] H. M. Klouwen, *Arch. Biochem. Biophys.* **99,** 116 (1962).

[12] P. W. Albro, M. I. Luster, K. Chae, S. K. Chaudhary, G. Clark, L. D. Lawson, J. T. Corbett, and J. D. McKinney, *Toxicol. Appl. Pharmacol.* **50,** 137 (1979).

by intramuscular injection (hind legs), with incomplete Freund's adjuvant. Approximately 12 rabbits are injected with each antigen. Serum samples are drawn weekly and stored frozen at −70°. Antiserum titer slowly increases over a 4-month period, reaching an eventual plateau. At this time, booster injections are temporarily discontinued. After a rest period of 2–3 months, a secondary response (yielding an antiserum of greater specificity than the primary response) is elicited with another booster injection as described above. The animal may be bled 10 days after this secondary boost.

Initially, antibody production is detected by double diffusion analysis in agarose gel as described by Ouchterlony.[13] The gel consists of 1% agarose in barbital buffer (103 mM NaCl, 47 mM sodium barbitol, and 1% thimerosal, pH 7.6). Each antiserum is tested against the dibenzo-p-dioxin derivative conjugated to a heterologous protein, such as bovine serum albumin (BSA) or rabbit serum albumin (RSA) at concentrations of 0.5, 1.0, and 2.0 mg/ml in PBS. Specificity controls routinely used include wells containing BSA, BSA with the adipamide side chain linked to aniline, and BSA linked to the adipamide of 2-methoxy-3-chloroaniline (2M3CA). Antisera revealing precipitin reactions when tested against BSA-trichlorodibenzo-p-dioxin but not BSA, BSA-aniline adipamide, or BSA-2M3CA are further evaluated by RIA.

Principle of RIAs for Dibenzo-p-dioxin

For data analysis, if the percentage of radiolabeled dioxin derivative bound to antibody (after equilibrium is reached) is given by %B, a plot of logit %B, where logit %B = ln[(%B)/(100 %B)], against the logarithm of the concentration of tetrachlorodibenzo-p-dioxin should be a straight line. This is most conveniently accomplished by plotting (%B)/(100 − %B) vs [TCDD] on log–log paper.

A straight line is not always required or desired. A plot of B/B_0, where B is the amount of bound radioactivity in the presence of test material and B_0 the amount of bound radioactivity in complete reagent blanks, versus log concentration of test material (nonradioactive dioxin), provides a smooth, usable (though not quite linear) calibration curve that is automatically normalized relative to the affinity of a given preparation of antiserum and the specific radioactivity of the labeled dioxin derivative.

Central in the performance of RIAs for small molecules is a means of separating free hapten from antibody-bound hapten. The double-antibody procedure used in these studies was chosen because it is compatible with the detergent used to solubilize the dioxins. In this procedure all rabbit γ-globulin present in the assay (hapten bound or unbound) is precipitated

---

[13] O. Ouchterlony, in "Progress in Allergy" (P. Kallos, ed.), p. 1. Karger, Basel, 1958.

with a second antibody produced in goats, and directed against rabbit γ-globulins.

A unique feature of the assay is the use of nonionic detergents (Cutscum or Triton X-305) to solubilize the extremely hydrophobic dibenzo-p-dioxins in a manner permitting their binding by antibodies. Triton X-305 is used when the samples are expected to contain low levels of dibenzo-p-dioxin while Cutscum is used to quantify wider ranges of the chemical.

As with other RIAs, it is necessary to optimize a variety of conditions and reagents in order to approach maximum sensitivity. Optimization of incubation times and conditions of second antibody (dilution and type) may vary slightly from one laboratory to another and should be individually standardized. Optimization of these parameters in the dibenzo-p-dioxin assay presents no unique problems and details of these procedures have been described in an earlier publication.[12] The sensitivity of the various antisera produced in rabbits needs to be carefully compared in order to obtain the most sensitive antisera. Although time consuming, this may accurately be performed by a series of antibody dilution curves as described by Hunter.[14] Briefly, the RIA is performed by testing serial dilutions of antisera against various concentrations of radiolabeled hapten. The antiserum revealing maximum slope of the dilution curve at the lowest concentrations of radiolabeled hapten added is presumably the most sensitive antiserum. Finally, it is necessary to examine the specificity of the antisera by determining potential cross-reactivity of the antiserum to structurally similar compounds.

Tissue Extraction and Clean-up

Tissues (liver, adipose, serum) are processed as recommended previously.[15] Soil samples should be extracted with acetone in a Soxhlet extraction apparatus[16] and the extracts purified as for tissue extracts. The final residues are stored frozen in small volumes of benzene. Detailed procedures for extraction and clean-up of samples for assay vary widely according to the particular sample matrix involved, and are beyond the scope of the present paper.

Preparation of Radiolabeled Compounds

The radiolabeling procedure involves synthesis of the N-(5-bromovaleramide) derivative of the hapten to be assayed, replacement of the bro-

---

[14] W. M. Hunter, *in* "Handbook of Experimental Immunology" (D. M. Weir, ed.), p. 171. Blackwell, Oxford, 1973.

[15] P. W. Albro and J. T. Corbett, *Chemosphere* **6**, 381 (1977).

[16] G. Seidl and K. Ballschmiter, *Chemosphere* **5**, 373 (1976).

mine with iodine, and finally exchange of the unlabeled iodo group with Na^{125}I. For example, 1-amino-3,7,8-trichlorodibenzo-p-dioxin (15 mg, 45 μmol) is dissolved in 2 ml of dry pyridine. The 5-bromovaleryl chloride, prepared from 5-bromovaleric acid (20 mg, 110 μmol) and thionyl chloride, is added to the above solution in an ice-bath with magnetic stirring. The mixture is left in a refrigerator for 16 hr. The product is extracted with 100 ml of ether and the etheral solution washed with 50 ml each of 10% HCl, saturated Na$_2$CO$_3$ solution, and water, then dried over anhydrous sodium sulfate. After evaporation of the solvent, 1-N-(5-bromovaleramido)-3,7,8-trichlorodibenzo-p-dioxin (methane chemical ionization mass spectrum: m/e 492 = M + 29, 464 = M + 1, 384 = M − 79) is obtained in nearly quantitative yield.

The bromo derivative (15 mg) is treated with 5 mg of NaI in 2 ml of acetone at 50° for 20 hr. The product is extracted with 50 ml of ether, washed with 10 ml each of 10% sodium thiosulfate and water, and dried over anhydrous sodium sulfate. After evaporation of the solvent, the crude product (17 mg) is chromatographed on 7 g of activated silica gel (Brinkmann). After a benzene elution (100 ml), the iodovaleramide derivative is eluted with 100 ml of 10% chloroform in benzene. The product should give a single spot (R_f 0.6) on a silica gel TLC plate with benzene as solvent. The methane supported chemical ionization spectrum shows m/e 540 = M + 29, 512 = M + 1, 384 = M − 127.

Unlabeled   1-N-(5-iodovaleramido)-3,7,8-trichlorodibenzo-p-dioxin (10 μg, 20 nmol) is dissolved in 100 μl of dry acetone and injected into a vial containing 5 mCi of carrier-free Na^{125}I (low pH, New England Nuclear). The mixture is heated at 50° for 60 hr in a sand bath. After cooling to room temperature, the product is extracted with 10 ml of chloroform, washed with 2 ml each of 10% sodium thiosulfate and water, and dried over anhydrous sodium sulfate. The chloroform is evaporated, the residue dissolved in 1 ml of hexane and applied on a silica gel column (6 × 170 mm) made in a glass pipet. Two fractions are collected: (1) 10 ml of hexane, (2) 10 ml of hexane–chloroform (1:1); the second fraction contains the desired product. Radiopurity may be determined by silica gel TLC using benzene as solvent. The final product, 1-N-[5-^{125}I]iodovaleramido-3,7,8-trichlorodibenzo-p-dioxin, should be greater than 98% radiopure with a specific activity of ∼78 Ci/mmol.

Assay Procedure

Dilutions of samples (standards, unknowns, etc.) are added in triplicate in 0.2 ml of benzene to 12 × 75-mm disposable glass tubes (B-D No. 7313). The tubes are dried under a nitrogen stream and 0.2 ml of 1% Cutscum or Triton X-305 in absolute methanol is added. The solvent is again blown off with nitrogen and 0.2 ml of PBS is added. The tubes are placed in an Ultramet II sonic cleaner for 45 min, cooled, and 0.2 ml of diluted antiserum is added. The antiserum diluent is PBS containing 0.1% bovine γ-globulin and 0.02% rabbit γ-globulin as carrier. A dilution of antiserum that binds about 40% of the radioactive hapten is used. After incubation for 30 min at 37° approximately 7000 cpm (14,000 dpm) of radiolabeled hapten in PBS with 1% detergent, prepared as described above for sample extracts, usually 14,000 dpm/25 μl, is added. This is followed by an additional 30 min incubation at 37° and a further 68–72 hr at 4°. Each tube then receives 200 μl of cold goat anti-rabbit γ-globulin appropriately diluted in PBS containing 0.05 mM ethylenediamine-tetraacetic acid (EDTA). The preparation is then incubated for an additional 6 hr at 4°. All tubes are centrifuged (4°, 500 g, 30 min), the supernatant is discarded, and the radioactivity in the precipitate measured in a gamma counter. The first two triplicates are used to determine the background and receive radiotracer, 0.2 ml of PBS with 1% detergent, 0.2 ml of antibody diluent (without antibody) and anti-rabbit γ-globulin. The third and last triplicates serve as control tubes, receiving all reagents but no unlabeled standards or unknowns. All the remaining triplicates receive the above reagents as well as either working standards, unknown samples, or compound being examined for cross-reactivity. The standard curve is prepared by including amounts of unlabeled TCDD between 10 pg and 100 ng. Tubes containing just radiotracer are randomly inserted into the assay to detect possible pipetting errors.

Calibration Curve

All counts are corrected by subtracting the cpm in the precipitate of the background tubes from all values and the means determined for each triplicate. The corrected cpm in tubes lacking unlabeled dibenzodioxin (control tubes) are designated B_0. The mean cpm in a triplicate containing unlabeled dibenzo-p-dioxin (calibration standard or test sample) is designated B. The ratio $(B/B_0) \times 100$ is calculated for each level of standard. A calibration curve is prepared by plotting $(B/B_0 \times 100)$ for the standards vs ng TCDD in each triplicate using linear-log graph paper with ng TCDD on the log axis. The amount of unknown dibenzo-p-dioxin is then read from the linear region of this graph with use of the measured $B/B_0 \times 100$ value.

Sensitivity and Specificity

Both the antibody affinity and avidity, which determine the limits of specificity and sensitivity of the assay, will vary from animal to animal as well as between primary and secondary response. One should be able with little difficulty to obtain antisera permitting as little as 30 pg of 2,3,7,-8-tetrachlorodibenzo-*p*-dioxin to be detected. Other dioxin isomers and some chlorinated dibenzofurans cross-react,[12] but classes of compounds other than these should not interfere. At the present time, the lower detection limit of the assay is limited by the specific radioactivity of the ^{125}I-labeled dioxin derivative.

[48] Radioimmunoassay of Nicotine, Cotinine, and γ-(3-Pyridyl)-γ-oxo-*N*-methylbutyramide

By JOHN J. LANGONE and HELEN VAN VUNAKIS

Nicotine (Fig. 1) is the major alkaloid present in tobacco. Up to 1–2 mg may be found in the average cigarette, although special processing can reduce the level markedly in "low tar–low nicotine" tobacco. In addition to smoking tobacco, smokeless products (e.g., snuff and chewing tobacco) are sources of nicotine. The metabolism of nicotine is complex and can lead to the formation of several biotransformation products.[1-4] In several species, including humans, cotinine is a major metabolite. The minor product, γ-(3-pyridyl)-γ-oxo-*N*-methylbutyramide (oxoamide) is derived from cotinine.

General Considerations

Since 1973,[5] several radioimmunoassays (RIA) for nicotine, cotinine, and the oxoamide (Table I) have been developed and in some cases used to determine levels of these compounds in physiological fluids and other

[1] H. McKennis, Jr., *in* "Tobacco Alkaloids and Related Compounds" (U. S. von Euler, ed.), p. 53. Pergamon, Oxford, England, 1965.
[2] J. Gorrod and P. Jenner, *Essays Toxicol.* **6**, 35 (1975).
[3] A. Pilotti, *Acta Physiol. Scand. Suppl.* **479**, 13 (1980).
[4] M. A. H. Russell and C. Feyerabend, *Drug Metab. Rev.* **8**, 29 (1978).
[5] J. J. Langone, H. B. Gjika, and H. Van Vunakis, *Biochemistry* **12**, 5025 (1973).
[6] J. J. Langone, J. Franke, and H. Van Vunakis, *Arch. Biochem. Biophys.* **164**, 536 (1974).
[7] C. F. Haines, Jr., D. K. Mahajan, D. Miljkovic, M. Miljkovic, and E. S. Vesell, *Clin. Pharmacol. Ther.* **16**, 1083 (1974).
[8] H. Matsushita, M. Noguchi, and E. Tamkai, *Biochem. Biophys. Res. Commun.* **57**, 1006 (1974).

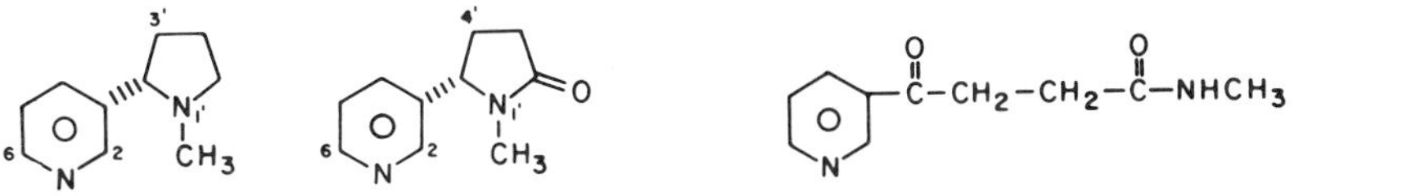

(−)-Nicotine   (−)-Cotinine γ-(3-Pyridyl)-γ-Oxo-N-Methylbutyramide

FIG. 1.

experimental samples.[5−20] The RIAs for nicotine, cotinine, and the oxo-amide that were developed in our laboratory are described in detail.

Preparation of Hapten Derivatives

General

Since nicotine and cotinine do not have functional groups suitable for coupling to protein carriers, carboxylic acid derivatives were prepared according to the reaction sequence shown in Fig. 2. Two relatively simple syntheses yield *trans*-4′-carboxycotinine.[5,21] Modification of the carboxyl group over three steps gives succinylated 3′-hydroxy-methylnicotine.[5]

---

[9] S. Matsukura, N. Sakamoto, H. Imura, H. Matsuyama, T. Tamada, T. Ishiguro, and H. Muranaka, *Biochem. Biophys. Res. Commun.* **64**, 574 (1975).

[10] A. Castro and I. Prieto, *Biochem. Biophys. Res. Commun.* **67**, 583 (1975).

[11] A. Castro, N. Monji, H. Malkus, W. Eisenhart, H. McKennis, Jr., and E. R. Bowman, *Clin. Chim. Acta* **95**, 473 (1979).

[12] A. Castro, N. Monji, H. Ali, J. M. Yi, E. R. Bowman, and H. McKennis, Jr., *Eur. J. Biochem.* **104**, 331 (1980).

[13] H. Van Vunakis, J. J. Langone, and A. Milunsky, *Am. J. Obstet. Gynecol.* **120**, 64 (1974).

[14] J. J. Langone, H. Van Vunakis, and P. Hill, *Res. Commun. Chem. Pathol. Pharmacol.* **10**, 21 (1975).

[15] P. Zeidenberg, J. H. Jaffe, M. Kanzler, M. D. Levitt, J. J. Langone, and H. Van Vunakis, *Compr. Psych.* **18**, 93 (1977).

[16] S. Matsukura, N. Sakamoto, Y. Seino, T. Tamada, H. Matsuyama, and H. Muranaka, *Clin. Pharmacol. Ther.* **25**, 555 (1979).

[17] E. R. Gritz, V. Baer-Weiss, N. L. Benowitz, H. Van Vunakis, and M. E. Jarvik, *Clin. Pharmacol. Ther.* **30**, 201 (1981).

[18] N. J. Wald, M. Idle, J. Boreham, A. Bailey, and H. Van Vunakis, *Lancet* Oct. 10, **775** (1981); Jan. 2, **40** (1982).

[19] C. L. Williams, A. Eng, G. J. Boivin, P. Hill, and E. L. Wynder, *Am. J. Public Health* **69**, 1272 (1979).

[20] E. W. Weiler and M. H. Zenk, this series, Vol. 73 [27].

[21] M. Cushman and N. Castagnoli, Jr., *J. Org. Chem.* **37**, 1268 (1972).

TABLE I
RADIOIMMUNOASSAYS FOR NICOTINE, COTININE, AND γ-(3-PYRIDYL)-γ-OXO-N-
METHYLBUTYRAMIDE (OXOAMIDE)

Hapten derivative	Labeled hapten	Reference
O-Succinyl-3′-hydroxymethylnicotine	[³H]Nicotine	5
N-Succinyl-6-aminonicotine	[³H]Nicotine	7
6-(p-Aminobenzamido)nicotine	[³H]Nicotine	8, 9
N-Succinyl-6-aminonicotine or 6-(ε-aminocapramido)nicotine	¹²⁵I-labeled tyrosine methyl ester derivative of 6-aminonicotine	10, 11
trans-4′-Carboxycotinine	[³H]Cotinine or ¹²⁵I-labeled tyramine derivative of 4′-carboxycotinine	5
l-1(β-Aminoethyl)cotinine	¹²⁵I-labeled-1-β-(p-hydroxybenzamidol)-ethylcotinine betaine	16
γ-(3-Pyridyl)-γ-oxobutyric acid (POBA)	¹²⁵I-labeled tyramine derivative of POBA	6

Cotinine Derivative

N-3-Pyridylidenemethylamine[21]

Reagents

11.0 g methylamine solution prepared by bubbling the gas into 150 ml benzene at 20° with stirring in a fume hood

35 g pyridine-3-carboxaldehyde (Aldrich Chemical Co.)

25 g molecular sieve drying agent

Procedure. In the first step N-3-pyridylidenemethylamine is prepared by adding a solution of pyridine-3-carboxyaldehyde in 50 ml benzene to the benzene solution of methylamine and molecular sieve pellets. The mixture is stirred at 20° in a fume hood overnight in a flask protected from moisture by a drying tube containing anhydrous sodium sulfate. The solution is filtered through two layers of Whatman number 2 filter paper and evaporated under reduced pressure (rotary evaporator) to give the imine as a yellow oil (30–35 g). The crude product is suitable for the second step.

trans-4-Carboxycotinine[21]

Reagents

25 g imine prepared above

30 g succinic anhydride

Procedure. To prepare the acid, the imine and succinic anhydride are refluxed for 24 hr in 100 ml xylene. After the mixture cools, the top

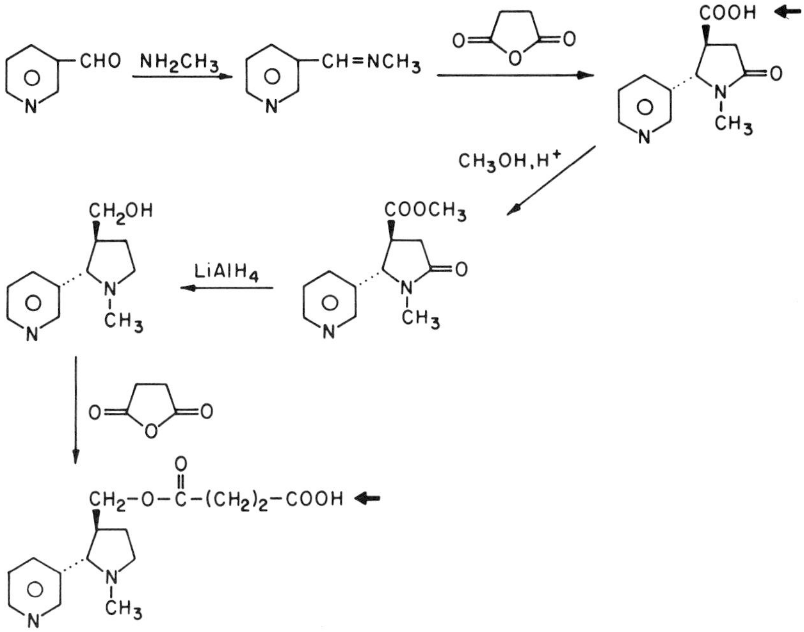

FIG. 2. Preparation of hapten derivatives for conjugation to macromolecules; two products are formed.

layer is decanted and discarded. The residual brown oil is dissolved in 5% sodium bicarbonate solution (approximately 300 ml), washed with two 250-ml portions of chloroform, and decolorized by swirling with 2 g activated charcoal. The suspension is filtered through several layers of Whatman number 2 filter paper and the yellow filtrate heated on a steam bath to remove traces of chloroform. The pH is carefully adjusted to 4.7 with phosphoric acid to precipitate the product. If it does not precipitate spontaneously, the walls of the beaker should be scratched vigorously with a glass rod to initiate the process. The crude carboxylic acid (15 g) is collected by filtration and recrystallized from a minimum volume of boiling ethanol to give 10–12 g of white or cream colored crystals, mp 193–195°. This product is used to prepare the immunogen for production of antibodies to cotinine.

Nicotine Derivative

Methyl Ester of trans-4'-Carboxycotinine[21]

Reagents
10 g 4'-carboxycotinine
100 ml 2 *N* methanolic sulfuric acid
5 g molecular sieve drying agent

Procedure. Add the carboxycotinine to the methanolic sulfuric acid and molecular sieve. Allow the mixture to stir at 20° overnight. Filter the mixture through two layers of Whatman number 2 filter paper. Add the filtrate carefully and in portions to an 8% solution of sodium bicarbonate (150 ml). Extract the solution with four 100-ml portions of chloroform. The combined chloroform washes are dried over anhydrous magnesium sulfate and concentrated on a rotary evaporator to give the ester as a yellow oil. Crystallization from ether–acetone (3:1 v/v) gave 8.0 g of needles (mp 82–84°).

trans-3'-Hydroxymethylnicotine

Reagents
6.0 g of methyl ester
3.6 g lithium aluminum hydride (LAH)

Procedure. Extreme caution must be exercised when using lithium aluminum hydride. All glassware and solvents must be absolutely dry since LAH reacts vigorously with water (and other solvents containing reactive hydrogen) to produce hydrogen gas which may ignite from the heat of reaction.

Add the LAH *slowly* to 200 ml *dry* diethyl ether in a flask protected from moisture by a drying tube containing anhydrous sodium sulfate. Add a solution of the ester in 25 ml ether, and allow the solution to stir at 20° overnight. Cool the solution in an ice bath and, while stirring, destroy excess LAH by addition of cold water (3.0 ml added dropwise), 15% sodium hydroxide (3 ml), and then water (10 ml). Filter the suspension through 2 layers of Whatman number 2 filter paper, dry the filtrate over anhydrous magnesium sulfate, and remove the ether by evaporation in a rotary evaporator. The crude hydroxymethylnicotine is a pale yellow oil which can be used directly in the subsequent reaction. Alternatively, it can be purified by distillation through a short-path apparatus, bp 85–87° (10 μm).

Succinylated trans-3'-Hydroxymethylnicotine

Reagents
300 mg hydroxymethylnicotine
157 mg succinic anhydride

Procedure. A solution of hydroxymethylnicotine and succinic anhydride in 30 ml dry benzene is allowed to stir at 60° overnight. The solution is concentrated on a rotary evaporator to give the succinylated product as a semisolid which can be used to prepare the immunogen for production of antibodies to nicotine.

γ-3-Pyridyl-γ-oxobutyric acid[22,23]

Reagents
  28 g sodium ethoxide
  69 g diethylsuccinate
  33 g ethylnicotinate

Procedure. Diethylsuccinate and ethylnicotinate are added to the stirred suspension of sodium ethoxide in 150 ml dry benzene. The mixture is heated at reflux for 2 hr and allowed to cool. Concentrated hydrochloric acid (34 ml) in 200 ml water is added, followed by sufficient sodium bicarbonate (solid) to saturate the system. The mixture is extracted with three 100-ml portions of 5% hydrochloric acid and the aqueous phase neutralized with 10% sodium bicarbonate solution and extracted with three 200-ml portions of ether. The combined ether layers are dried over anhydrous sodium sulfate and concentrated on a rotary evaporator. The oily product is distilled under reduced pressure to give two fractions. The first is unreacted ethylnicotinate (bp 75–80°; 0.75 mm). The second fraction is 12–15 g of the desired product diethyl α-nicotinoylsuccinate, bp 155–165° (0.75 μ).

γ-(3-Pyridyl)-γ-oxobutyric acid is prepared by hydrolysis with concomitant decarboxylation of the above ester. Ten grams of ester is heated under reflux with 50 ml N sulfuric acid. The flask is cooled in an ice bath and the solution adjusted to pH 4.3 by addition of concentrated aqueous ammonia. The oxoacid is isolated by filtration as a white crystalline powder, mp 160–163°, and is sufficiently pure for preparation of the immunogen. The resulting antibodies are specific for the N-methylamide derivative (see below), presumably since it mimics the amide linkage joining the acid to the carrier protein, and this linkage is an immunodominant structural feature.

Production of Antibodies and Tracers[5,6]

Preparation of Immunogens

Reagents
  10 mg succinylated hydroxymethylnicotine, carboxycotinine, or γ-3-pyridyl-γ-oxobutyric acid
  10 mg keyhole limpet hemocyanin or human serum albumin
  10 mg 1-ethyl-3-(3-dimethylamino)propyl carbodiimide hydrochloride (EDAC)

[22] H. McKennis, Jr., L. B. Turnbull, H. N. Wingfield, Jr., and L. J. Dewey, *J. Am. Chem. Soc.* **80,** 1634 (1958).
[23] R. N. Castle and A. Burger, *J. Am. Pharm. Assoc. Sci. Ed.* **43,** 163 (1954).

Procedure. EDAC is added to a solution of hapten derivative and protein in a total volume of 1.0 ml water at pH 6.8–7.0. Adjust the pH by addition of 0.01 N sodium hydroxide and allow the solution to stir at room temperature overnight. The immunogen is separated from low-molecular-weight components by chromatography on a column (1.9 × 35 cm) of Sephadex G-50 wet-packed and eluted with saline–0.05 M phosphate buffer, pH 7.2. Fractions (1.5–2.5 ml) are collected and the tubes corresponding to blue dextran are pooled. Based on the difference in the absorption spectrum at 260 nm, the incorporation is 6–18 mol hapten per mole protein.

Immunization. New Zealand albino rabbits were injected in the toe pads and im with 1 mg conjugate (1 ml) emulsified in 1 ml complete Freund's adjuvant. Three weeks later they were bled from the ear vein. Every 6 weeks, they were given booster injections and bled 1 week later.

Radiolabeled Tracers

L(−)[^{3}H]Nicotine (sp. act. 2.4 Ci/mmol) was prepared commercially by random catalytic tritium exchange on the pyridine ring and purified by chromatographic procedures.

L(−)[^{3}H]Cotinine was prepared from the L(−)[^{3}H]nicotine using the supernatant fluid from a 10,000 g rabbit liver homogenate as the source of the enzymes.

Procedure. L(−)[^{3}H]Nicotine (6 × 10^{10} cpm) in methanol was dried under N$_2$ to remove the organic solvent, then incubated with 1.0 ml liver extract (25 mg protein in phosphate–KCl buffer, 0.1 M, pH 7.5) containing 4 mg NADPH for 3 hr at 37° with shaking. At the end of the first 1.5 hr, another portion of NADPH was added and pH adjusted to 7.5. Water (0.5 ml) was added and the tube heated for 3 min in boiling water bath. After cooling, the reaction mixture was centrifuged to remove denatured proteins. The pellet was washed once with 0.5 ml H$_2$O, centrifuged, and the fluid combined with the supernatant fraction. The [^{3}H]cotinine was extracted into 3 ml of CH$_2$Cl$_2$, and the organic phase centrifuged to remove debris and break the emulsion. The organic extract was transferred to a vial and dried with sodium sulfate. The liquid was carefully transferred to a clean vial and the solvent removed by evaporation under N$_2$. TLC was used to isolate and purify the labeled cotinine [silica gel plates, solvent EtOAc, MeOH, glacial HAc (17:2:1); R_f for nicotine N'-oxide = 0.01, nicotine = 0.17, cotinine = 0.5]. The [^{3}H]cotinine was removed from the plate by extraction with methanol, and dried under N$_2$. The chromatography was repeated when necessary. It was stored in MeOH at −20°.

Tyramine derivatives of carboxycotinine and the oxoacid were prepared by the following procedure and labeled with [125]I.

N-(p-Hydroxyphenethyl-*trans*-cotinine carboxyamide

Reagents
  20 mg *trans*-4'-carboxycotinine or γ-(3-pyridyl)-γ-oxobutyric acid
  15 mg oxalyl chloride
  14 mg triethylamine
  14 mg tyramine

Procedure. Oxalyl chloride is added to a stirred solution of carboxylic acid in 1.0 ml dimethylformamide. After 30 min at 20°, tyramine is added and the solution allowed to stir for another 18 hr. The product is isolated by preparative thin-layer chromatography using Brinkmann glass plates (20 × 20 cm) coated with 2-mm-thick silica gel G containing a fluorescent indicator. Bands are visualized by exposure to a UV light source at 254 nm wave length.

The reaction mixture is applied as a narrow band at the origin and developed in a mixture of ethylacetate–methanol–ammonium hydroxide, 65:35:11 by volume. From the carboxycotinine reaction three bands are observed with R_f values 0.15–0.30 (carboxycotinine). 0.41–0.52 (tyramine), and 0.56–0.61 (product). Similarly, the tyramine derivative of the oxoacid has $R_f = 0.67$–0.81. The tyramine derivative is collected by scraping the band from the plate with a spatula, breaking up the pieces in a 100-ml beaker, and eluting wih 40 ml ethanol. The silica suspension is stirred vigorously at 20° for 15 min, then filtered. This procedure is repeated 2 more times. The combined ethanol filtrates are concentrated to dryness on a rotary evaporator and the residue dissolved again in 3 ml ethanol and filtered to remove particles of silica gel. The concentration (generally 0.3–0.5 mg/ml) is determined by comparing the UV absorption spectrum to that for a standard solution of tyramine at pH 12. At this pH, the phenolate group shows a well-defined maximum at 293 nm.

Labeling Procedure. The tyramine derivatives (20–25 μg) are labeled with ^{125}I by the chloramine-T procedure as described[5,6,24] and separated by column chromatography on Sephadex G-10 (1.5 × 10 cm). After free ^{125}I plus cold iodide carrier has been eluted with buffer (1.5 M sodium chloride, 0.05 M phosphate, pH 7.5 containing 0.2% gelatin), the labeled product is eluted with 50% dimethylformamide–0.3 M sodium chloride. The specific activities of the cotinine and oxoacid derivatives are 32 and 5 Ci/mmol, respectively.

---

[24] P. McConahey and F. Dixon, this series, Vol. 70 [11].

Radioimmunoassay Procedure

The double-antibody technique was used to separate free labeled antigen from antibody bound antigen.[25] The buffer was 0.15 M sodium chloride, 0.01 M Tris, pH 7.4 containing 0.1% gelatin. Plastic tubes (from SARSTEDT, W. Germany, No. 55.535, 3.5 ml. 55 × 12 mm) were used with tritiated ligands. Typically, 0.1-ml portions of labeled tracer (8000–20,000 cpm), appropriately diluted antiserum and buffer were incubated at 37° for 1 hr. For inhibition experiments, 0.1 ml-aliquots containing known amounts of standard or dilutions of test sample were added in place of buffer. Goat anti-rabbit γ-globulin (0.1 ml previously calibrated to be in antibody excess with respect to rabbit γ-globulin) was added and the mixture incubated at 4° overnight. To control for nonspecific binding, normal rabbit serum was used in place of the immune serum. The precipitate was collected by centrifugation at 1000 g for 30 min at 4°, the supernatant poured off, and the walls of the tubes wiped dry. To count [³H]nicotine or [³H]cotinine, the precipitate was dissolved in 0.2 ml 0.1 N sodium before adding 2.5 ml scintillation fluid (5 g, 2,5-diphyloxazole/liter and 0.33 g 1,4-bis-[2-(5-phenyloxazolyl)]benzene/liter). When ¹²⁵I-labeled tracers were used, the assay was carried out in 12 × 75-mm borosilicate disposable culture tubes. After completion of the assay, the precipitates were counted directly in a gamma counter.

Table II shows part of a typical protocol that is used when assaying physiological fluids (serum, plasma, saliva, cerebrospinal fluid, urine, tissue extracts, enzymic reaction mixtures, etc.). The larger volume employed in this assay minimizes nonspecific effects that may be encountered with some samples (i.e., differences in pH and ionic strength of urine). We routinely dilute all urine and saliva samples in buffer before assay. Depending upon the amount of compound present in the experimental sample, additional dilutions may be required to obtain inhibition values on the proper portion of the standard curve. It is possible to assay up to 0.1 ml of serum (or plasma if EDTA is incorporated into the buffer) without prior extraction of the sample. When analyzing plasma or serum, equivalent volumes of nicotine and cotinine-free fluids should be added instead of buffer to construct the standard curve.

The standard curve is obtained by estimating the percentage tracer bound in the presence of known amounts of test compound. The amount of compound in physiological fluids is calculated by comparing its inhibition with that obtained in the standard inhibition curve.[26,27]

[25] A. R. Midgley, Jr., and M. R. Hepburn, this series, Vol. 70 [16].
[26] D. Rodbard and H. A. Feldman, this series, Vol. XXXVI [1].
[27] C. J. Halfman, this series, Vol. 74 [32].

TABLE II

INHIBITION OF [³H]COTININE ANTICOTININE BINDING BY COTININE

	1	2	3	4	5	6
M1 Isogel Tris buffer	0.8	0.8	0.7	0.7	0.7	0.7
M1 inhibitor, cotinine, serially diluted	—	—	0.1	0.1	0.1	0.1
Ng inhibitor, cotinine, added	—	—	2.5	1.25	0.63	0.31
M1[³H]cotinine, 8000 cpm/0.1 ml	0.1	0.1	0.1	0.1	0.1	0.1
M1 normal rabbit serum (1:200)	0.1	—	—	—	—	—
M1 rabbit anticotinine (1:200)	—	0.1	0.1	0.1	0.1	0.1
Incubate at 37° for 60 min						
M1 normal rabbit serum (1:25)	0.1	0.1	0.1	0.1	0.1	0.1
M1 goat anti-rabbit γ-globulin (1:2)	0.1	0.1	0.1	0.1	0.1	0.1
Incubate at 4° overnight						
cpm in precipitate	128	3688	1190	1688	2220	2693
	117	3514	1113	1666	2227	2781
Percentage Inhibition	—	—	70	55	40	25

Sensitivity and Specificity

The extent to which nicotine, cotinine, and the oxoamide inhibit their respective antigen–antibody reactions is shown in Table III. In the homologous system, 0.018 nmol (3.2 ng) nicotine gives 50% inhibition. The lower limit of detection is 350 pg. Cotinine, the major metabolite, is a poor inhibitor and gives less than 20% inhibition at the 50 nmol level. With anticotinine, 1.2 ng cotinine gives 50% inhibition. Nicotine is 2000 times less effective. The oxoamide has an I_{50} of 5 ng in the homologous RIA. The geometry of the five-membered ring may be a factor in determining the specificity of these antisera. Thus, antibody combining sites may recognize the "envelope" conformation of the nicotine N-methylpyrrolidine ring or the more planar conformation of the cotinine N-methyl-2-pyrrolidone ring. The opening of the pyrrolidine ring during the metabolic formation of the oxoamide yields a compound of greatly altered structure.

Over 50 compounds (nicotine metabolites and structurally related molecules) have been tested as inhibitors (Table III).[5] Nicotine (Fig. 1) contains two distinct ring systems, i.e., the pyridine and N-methylpyrrolidine rings. The inhibition data indicate that the specificity of the antisera is directed toward both rings of the structure. The pyridine or pyrrolidine derivatives which possess only one of the rings show essentially no inhibition at the 50 nmol levels. The specificity of the anticotinine is also directed toward both ring systems, i.e., the pyridine and N-methyl-2-pyrrolidone rings.

TABLE III
INHIBITION OF NICOTINE, COTININE, AND γ-(3-PYRIDYL)-γ-OXO-N-
METHYLBUTYRAMIDE ANTIGEN–ANTIBODY REACTIONS

Compound	[³H]Nicotine–antinicotine	[³H]Cotinine–anticotinine	¹²⁵I-labeled oxamide–antioxamide
(−)-Nicotine	100[a] (0.018 nmol)[b]	0.05	0.02
(−)-Cotinine	<0.01	100[a] (0.0058 nmol)[b]	0.01
γ-3(-Pyridyl)-γ-oxo-N-methylbutyramide	<0.01	<0.01	100[a] (0.026 nmol)[b]
Nicotine N'-oxide	0.5	0.01	0.05
(−)-Nornicotine	0.9	0.01	0.1
N'-Nitrosonornicotine	0.04	0.2	—
(−)-Cotinine N-oxide	<0.01	0.1	<0.01
DL-Desmethylcotinine	<0.01	0.3	0.3
γ-(3-Pyridyl)-γ-oxobutyric acid	<0.01	<0.01	0.8
(−)-Anabasine	0.3	<0.01	<0.01
DL-2-Aminonicotine	2.4	0.01	0.01
DL-6-Aminonicotine	0.6	<0.01	<0.01
6-Hydroxynicotine	0.02	<0.01	<0.01
N'-Formylnornicotine[c]	<0.01	—	—
N'-Acetylnornicotine[c]	<0.01	—	—
N'-Hexanoylnornicotine[c]	<0.01	<0.01	—
N'-Carbomethoxynornicotine[c]	<0.01	—	—
N-Methylpyrrolidine	<0.01	<0.01	<0.01
N-Methyl-2-pyrrolidone	<0.01	<0.01	0.01
Pyridine	<0.01	<0.01	<0.01
Nicotinic acid	<0.01	<0.01	<0.01
Nicotinamide	<0.01	<0.01	<0.01
2-Aminopyridine	<0.01	<0.01	<0.01
2-Acetamidopyridine	<0.01	<0.01	<0.01
(+)-Nicotine	<0.01	<0.01	<0.01
(+)-Cotinine	<0.01	<0.01	<0.01

[a] Percentage cross-reactivity.
[b] The quantity of compound required to produce 50% inhibition in the respective assay.
[c] Compounds synthesized and assayed by Dr. Andre Castonguay.

Antibodies that were produced in this laboratory by immunizing rabbits with conjugates in which 2- or 6-aminonicotine was linked to macromolecules by their amino groups showed strong specificity for the N-methylpyrrolidine ring. Other laboratories[7,11] have been successful in producing more specific antibodies by immunizing animals with 6-amino nicotine that had been succinylated before linkage to protein.

As shown in Table I, differences exist in the structures of the hapten derivatives used for immunization and in the procedural details of the assays (e.g., use of tritiated or iodinated ligands, the methods of separating free-labeled ligand from that bound to antibody, etc.). However, in specificity and sensitivity the RIAs described in this paper for L(−)-nicotine and cotinine equal and, in most cases, surpass the other assays referred to in Table I. With the availability of D(+)-[^{3}H]nicotine (sp. act. 60 Ci/mmol) we can detect 0.5 ng D(+)-nicotine/ml serum when 0.1 ml of the sample is analyzed.

Applications

The RIAs have proved to be useful analytical tools to quantify levels of nicotine and cotinine in physiological fluids and tissues of smokers and others who come into contact with the parent alkaloid.

Duplicate aliquots of serum samples obtained from experimental subjects were analyzed for nicotine and cotinine by gas chromatography (GLC) using nitrogen-phosphorous detection[28] and by RIA. The results were in general agreement[17] with the RIA values being, in general, somewhat higher. Although GLC is a more time-consuming and expensive assay, it can detect 1–2 ng nicotine/ml serum. The RIA can detect 6–8 ng nicotine/ml serum. Neither this nor any other direct RIA (Table I) is sufficiently sensitive to measure nicotine levels in certain subjects, e.g., nonsmokers who may have taken in nicotine by passive means. In such cases, larger volumes of fluid would have to be extracted in order to obtain sufficient compound for analysis.[7]

The distributional half-life of nicotine is relatively short (about 10 min) and is followed by an elimination half-life of 80–110 min depending on urinary pH.[29] Nicotine can be present in the plasma of active smokers at concentrations as high as 50–100 ng/ml but such peak values last only for a few minutes. Cotinine has a relatively long half-life ($\cong 30$ hr)[5] and its level tends to remain fairly constant in a subject who smokes about the same number of cigarettes/day.[30] Cotinine levels average about 300 ng/ml in cigarette smokers,[18] and a few subjects have had levels as high as 900 ng/ml serum.

The cotinine assay is remarkably specific; analysis of serum or plasma samples from approximately 500 nonsmokers has yielded no false positives. The relatively high levels of cotinine present in the physiological fluids of cigarette, pipe, and cigar smokers[18] and in habitual users of

---

[28] P. Jacob, M. Wilson, and N. L. Benowitz, *J. Chromatogr.* **222**, (1981).

[29] J. Rosenberg, N. L. Benowitz, P. Jacob, and K. M. Wilson, *Clin. Pharmacol. Ther.* **28**, 517 (1980).

[30] J. J. Langone, H. Van Vunakis, and L. Levine, *Accounts Chem. Res.* **8**, 335 (1975).

smokeless tobacco products[17] makes cotinine the ideal component to monitor when it is necessary to confirm the oral testimony given by individuals regarding their status as smokers or abstainers.[15,19] Since the RIA can be carried out on saliva and urine samples, it is possible to use noninvasive procedures to obtain specimens for analysis. The reliance on cotinine analysis rather than on oral testimony gives greater credence to epidemiological data. As far as is known,[1-3] nicotine is the only source of cotinine. COHb and thiocyanate are also compounds that appear in the bloodstream as a result of smoking but both can also originate from other sources, i.e., smoking of nonnicotine-containing material for COHb and eating certain vegetables for thiocyanate. Nicotine is not an illicit drug yet a significant number of individuals partake of its pleasures and when questioned continue to deny its use.

Cotinine can also be a reliable indicator to determine absorption of the parent alkaloid through the skin. Based on RIA results,[31] nicotine is now considered to be the probable cause of green tobacco sickness, an occupational illness of young tobacco workers who are not smokers and who come into contact with the nicotine-containing dew on the tobacco leaves. Such subjects show elevated levels of cotinine in their urine. (Cotinine does not occur in tobacco or its dew.) When harvesters were outfitted with waterproof coats, they showed up to a 78% reduction in cotinine excretion compared to harvesters who did not wear protective clothing.[32]

Levels of nicotine and/or cotinine also have been estimated by RIA in spinal fluid and amniotic fluid of smokers.[13] We also have used the oxoamide RIA to determine levels of this metabolite in the urine of cigarette smokers. Although the levels were low relative to cotinine (approximately 0.3 μg/ml), oxoamide activity was characterized after extraction and isolation by high-pressure liquid chromatography.[5] In addition to studying the fate of nicotine in animals and man, the RIA for cotinine has been used to obtain kinetic values for the conversion of nicotine to cotinine in *in vitro* systems containing microsomal cytochrome P-450, aldehyde oxidase, and the necessary cofactors. Since the $I_{50\,product}:I_{50\,substrate}$ ratio is greater than 10^3, it is possible to detect the formation of less than 1% cotinine in incubation mixtures containing large amounts of nicotine. The RIAs are carried out on suitably diluted aliquots of the incubation mixtures; there is no need to extract or process the samples before analysis.

[31] S. H. Gehlbach, L. D. Perry, W. A. Williams, J. I. Freeman, J. J. Langone, L. V. Peta, and H. Van Vunakis, *Lancet* 1, 479 (1975).
[32] S. H. Gehlbach, W. A. Williams, and J. I. Freeman, *Arch. Environ. Health* 111 (1979).

[49] Radioimmunoassays for *N'*-Nitrosonornicotine and *N'*-Acylnornicotine Analogs

By ANDRE CASTONGUAY and HELEN VAN VUNAKIS

Nornicotine **1** and several analogs with an N-substituted pyrrolidine ring are present in tobacco and tobacco smoke[1-7] (Fig. 1). Nornicotine has also been isolated from the leaves of the Australian shrub *Duboisia hopwoodii*.[8] In tobacco, nitrosation of nornicotine (and nicotine) yields *N'*-nitrosonornicotine (NNN **3**) during curing and smoking.[9] The other nicotinoids (**4–14**) originate very likely from nornicotine and the lower fatty acids which are present in tobacco plants.[10]

Some of the biological and pharmacological effects of nornicotine[11-13] and N-acetylnornicotine[14-16] are known but in general the *N'*-acylnornicotine derivatives have not been studied extensively. Of special interest is NNN **3**. This tobacco-specific nitrosamine induces lung adenomas in mice,[9,17] esophageal papillomas and nasal cavity carcinomas in rats,[18,19] and tracheal papillomas in Syrian golden hamsters.[20,21]

[1] I. Schmeltz and D. Hoffmann, *Chem. Rev.* **77**, 295 (1977).

[2] H. Matsushita, Y. Tsufino, D. Yoshida, A. Saito, T. Kisaki, K. Kato, and M. Noguchi, *Agric. Biol. Chem.* **43**, 193 (1979).

[3] A. Kowai, Y. Mikami, H. Matsushita, and T. Kisaki, *Agr. Biol. Chem.* **43**, 1421 (1979).

[4] M. Miyano, H. Matsushita, N. Yasumatsu, and K. Nishida, *Agr. Biol. Chem.* **43**, 2205 (1979).

[5] J. J. Piade and D. Hoffmann, *J. Liq. Chromatogr.* **3**, 1505 (1980).

[6] M. Miyano, N. Yasumats, H. Matsuski, and K. Nishida, *Agr. Biol. Chem.* **45**, 1029 (1981).

[7] E. Leete, *Phytochemistry* **20**, 1037 (1981).

[8] W. Bottomley, B. A. Nottle, and D. E. White, *Aust. J. Sci.* **8**, 18 (1945).

[9] S. S. Hecht, C. B. Chen, N. Hirota, R. M. Ornaf, T. C. Tso, and D. Hoffmann, *J. Natl. Cancer Inst.* **60**, 819 (1978).

[10] T. C. Tso, *in* "Physiology and Biochemistry of Tobacco Plants", p. 272. Dowden, Hutchinson and Ross, Stroudsburg, Pennsylvania, 1972.

[11] C. S. Hicks and H. Le Messurier, *Austr. J. Exp. Biol.* 175 (1935).

[12] S. Kakuo, A. Kihei, T. Marushige, S. Nakajima, and Z. Fukai, *Folia Pharmacol. Jpn.* **65**, 76 (1969).

[13] T. Besslo, *Nippon Yakurogaku Z.* **56**, 1223 (1960).

[14] A. H. Rees and H. Schnieden, *Br. J. Pharmacol.* **18**, 299 (1962).

[15] M. Mattila, *Ann. Med. Exp. Fenn.* **41**, Suppl. No. 3 (1963).

[16] M. Mattila, Academic Dissertation, University of Helsinki, 1963.

[17] E. Boyland, F. J. C. Roe, and J. W. Gorrod, *Nature (London)* **202**, 1126 (1964).

[18] D. Hoffmann, R. Raineri, S. S. Hecht, R. Maronpot, and E. L. Wynder, *J. Natl. Cancer Inst.* **55**, 977 (1975).

[19] S. S. Hecht, C. B. Chen, T. Ohmori, and D. Hoffmann, *Cancer Res.* **40**, 298 (1980).

[20] J. Hilfrich, S. S. Hecht, and D. Hoffmann, *Cancer Lett. (Amsterdam)* **2**, 169 (1977).

[21] D. Hoffmann, A. Castonguay, A. Rivenson, and S. S. Hecht, *Cancer Res.* **41**, 2386 (1981).

METHODS IN ENZYMOLOGY, VOL. 84

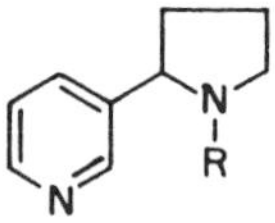

1 R = H
2 R = $CH(CH_3)_2$
3 R = NO
4 R = CHO
5 R = CO_2CH_3
6 R = $CO_2CH_2CH_3$
7 R = $COCH_3$
8 R = $COCH_2CH_3$
9 R = $CO(CH_2)_2CH_3$
10 R = $CO(CH_2)_4CH_3$
11 R = $CO(CH_2)_6CH_3$
12 R = $CO(CH_2)_4CHOHCH_2CH_3$
13 R = $CO(CH_2)_5CHOHCH_3$
14 R = $CO(CH_2)_3N(CH_3)_2$
15 R =

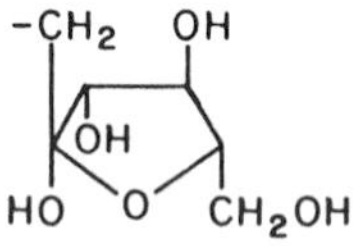

FIG. 1. Structure of nornicotine and nornicotine analogs isolated from tobacco and tobacco smoke.

Radioimmunoassays for nicotine and several of its metabolites are described in this volume.[22,23] The production of antibodies and the development of radioimmunoassays for NNN, *N'*-formylnornicotine, and *N'*-acetylnornicotine are the subject of this paper.[24–26]

Synthesis of Derivatives

Since the nornicotine analogs lack a functional group that can be used to form a covalent bond to proteins, the corresponding 5'-carboxynornicotine derivatives were synthesized (Fig. 2). Although several syntheses of nornicotine **1** have been published,[27–32] only the procedure of Hellmann

[22] J. J. Langone and H. Van Vunakis, this volume [48].
[23] W.-C. Shen, this volume [50].
[24] A. Castonguay and H. Van Vunakis, *Anal. Biochem.* **95,** 387 (1979).
[25] A. Castonguay and H. Van Vunakis, *J. Org. Chem.* **44,** 4332 (1979).
[26] A. Castonguay and H. Van Vunakis, *Toxicol. Lett.* **4,** 475 (1979).
[27] N. Polonovski and M. Polonovski, *Bull. Soc. Chim.* 1190 (1927).
[28] J. C. Craig, N. Y. Mary, N. L. Goldman, and L. Wolf, *J. Am. Chem. Soc.* **86,** 3866 (1964).
[29] B. P. Mundy, B. R. Larson, L. F. McKenzie, and G. Broden, *J. Org. Chem.* **37,** 1635 (1972).
[30] E. Leete, M. R. Chedekel, and G. B. Boden, *J. Org. Chem.* **37,** 4465 (1972).
[31] M. W. Hu, W. E. Bondinell, and D. Hoffmann, *J. Labelled Compd.* **10,** 79 (1974).
[32] W. B. Edwards III, D. F. Glenn, F. Greene, and R. H. Newman, *J. Labelled Compd. Radiopharm.* **14,** 255 (1978).

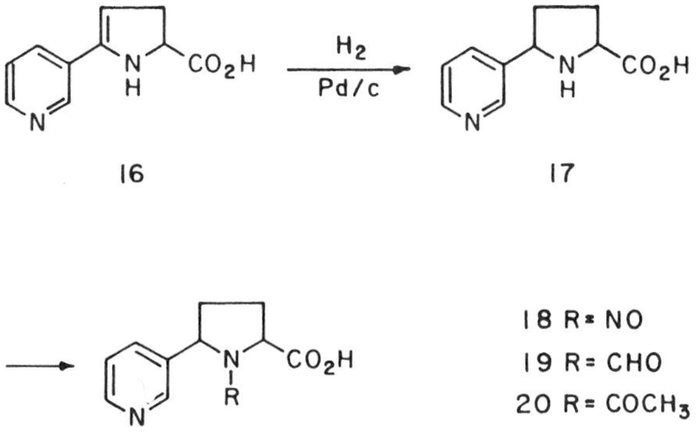

FIG. 2. Syntheses of N'-substituted 5-carboxynornicotines.

and Dietrich[33] has led to the preparation of 2',3'-dehydro-5'-carboxynornicotine **16**, which can be reduced to yield the nornicotine hapten **17**. Substitution on its pyrrolidine nitrogen yield the functionalized derivative of NNN **18**, *N'*-formylnornicotine **19**, and *N'*-acetylnornicotine **20**. The possible cis and trans isomers generated during the reduction were not separated prior to synthesis of the conjugate. Haptens, radiolabeled ligands, and inhibitors were racemic mixtures unless otherwise specified. Thin-layer chromatography (TLC) on silica gel plates was used to isolate the synthetic products.

5'-Carboxynornicotine **17**

Reagents
2',3'-dehydro-5'-carboxynornicotine, 100 mg
10% Palladium; charcoal, 100 mg
Hydrogen gas
Procedure. A solution of the starting material in 5 ml of 90% ethanol was stirred with the catalyst under hydrogen (1 atmosphere) for 1.5 hr. The catalyst was collected by filtration on celite and the filtrate concentrated *in vacuo*. The residue was separated by TLC (methanol, three migrations) and the band at R_f 0.28–0.40 was extracted with methanol. A colorless oil (80 mg, 79% yield) was obtained after evaporation of the solvent *in vacuo*.

N'-Nitroso-5'-carboxynornicotine **18**

Reagents
5'-Carboxynornicotine, 150 mg
Sodium nitrite (215 mg) in 50% acetic acid (2 ml)

---

[33] H. Hellmann and D. Dietrich, *Justus Liebigs Ann. Chem.* **672,** 97 (1964).

Procedure. The starting material was added to a cooled solution (0–4°) of sodium nitrite in acetic acid. The mixture was stirred at 4° for 24 hr then lyophilized. The residue was extracted with methanol and after TLC (ethyl acetate:methanol 1:1, three migrations), the band at R_f 0.63 was extracted with methanol. Evaporation of the organic solvent left a semi-solid product (130 mg, 75% yield).

N'-Formyl-5'-carboxynornicotine **19**

Reagents
5'-Carboxynornicotine, 80 mg
Formic acid, 4 ml
Formic acetic anhydride, 1 ml

Procedure. The starting material was stirred overnight with formic acid and formic acetic anhydride[34] at 4°. Excess reagent was removed by codistillation with cyclohexane and the residue chromatographed (methanol:ethyl acetate, 1:1) Extraction of the band at R_f 0.28–0.40 with methanol and evaporation of the solvent *in vacuo* gave a colorless oil (61 mg, 66% yield).

N'-Acetyl-5'-carboxynornicotine **20**

Reagents
5'-Carboxynornicotine
Acetic anhydride, 2 ml

Procedure. The starting material and acetic anhydride were stirred overnight at 4°. After addition of 2 ml of water, the mixture was concentrated under vacuum and the residue purified by TLC (methanol:ethyl acetate, 1:1). Extraction of the band at R_f 0.34 with methanol gave a pale yellow oil (24 mg, 49% yield).

Syntheses of Tritiated Nornicotine Analogs

[2'-³H]Nornicotine

Reagents
Myosmine, 16 mg
10% Palladium: charcoal
Tritium gas

Procedure. A solution of myosmine[31,35] in methanol (1.0 ml) was

[34] L. Maramatsu, M. Murakami, T. Yoneda, and A. Hagitani, *Bull. Chem. Soc. Jpn.* **38,** 244 (1965).
[35] S. Brandänge and L. Lindblom, *Acta Chem. Scand. B* **30,** 93 (1976).

stirred with the catalyst and tritium gas for 4.5 hr. The solvent was distilled under vacuum and the residue dissolved in methanol. The product was isolated by TLC (methanol, two migrations). Extraction of the band at R_f 0.3 yielded [^{3}H]nornicotine with a specific activity of 8 Ci/mmol.

[$2'$-^{3}H]N'-Nitrosonornicotine

Reagents
[$2'$-^{3}H]Nornicotine, 5.22 mCi
Sodium nitrite, 57 mg
Acetic acid:water, 1:1, 1 ml

Procedure. Sodium nitrite was added in portions to a cooled solution of starting material in 50% acetic acid. The mixture was stirred at 0–4° overnight, brought to pH 12 with concentrated NaOH, saturated with NaCl, and extracted with chloroform (4 × 1.5 ml). TLC (ethyl acetate: methanol, 7:3) of the organic phase showed only one spot with an R_f of 0.48 identical to unlabeled N'-nitrosonornicotine. The band was extracted with chloroform and dried with anhydrous sodium sulfate. The solvent was evaporated under nitrogen and the residue (4.13 mCi, 79% yield) was stored in ethanol at −50°.

[$2'$-^{3}H]N'-Acylnornicotine Derivatives

Reagents
[$2'$-^{3}H]Nornicotine
Formic acid, or acetic anhydride

Procedure. [^{3}H]N'-Formyl and N'-acetylnornicotine were obtained by treatment of [^{3}H]nornicotine with either formic acid or acetic anhydride. [$2'$-^{3}H]Nornicotine was stirred with 1 ml of the appropriate acid or anhydride at 70° for 1 hr. The pH was adjusted to ≃10 with concentrated NaOH or Na$_2$CO$_3$ solution and the reaction product extracted with methylene chloride. TLC showed only one radioactive component comigrating with the corresponding unlabeled nornicotine derivative. After removal of the organic solvent, the radiolabeled ligands were stored in Isogel Tris buffer at −50°.

Preparation of Hapten–Protein Conjugates[36]

Reagents
N,N'-Dicyclohexylcarbodiimide (CDI), 6.7 mg
N-Hydroxysuccinimide (recrystallized), 7.5 mg
Human serum albumin, 20 mg

---

[36] S. Bauminger and M. Wilchek, this series, Vol. LXX [7].

Procedure. A solution of ≃9 mg of the hapten in 0.5 ml dimethylformamide was reacted with CDI and *N*-hydroxysuccinimide. The mixture was stirred for 30 min at 25°, then centrifuged. The supernatant was added to a cold (0–4°) solution of serum albumin in 0.5 ml of 0.1 *M* sodium bicarbonate. After standing overnight at 4°, the reaction mixture was exhaustively dialyzed against phosphate buffer (0.005 *M* sodium phosphate pH 7.5, 0.15 *M* NaCl).

Immunization Procedure

New Zealand albino adult rabbits were immunized with complete Freund's adjuvant. The conjugate (2.6 mg in 0.4 ml of buffer) was emulsified with 4.6 ml of adjuvant in a Sorvall omnimixer for 4 min. The material was administered to rabbits by injection into the leg muscles at three weekly intervals. Each animal was bled weekly for 3 weeks, then given booster injections for 3 more weeks. The rabbits were bled, rested, and the booster injections repeated as necessary.

Radioimmunoassay

A double-antibody technique similar to that described[22] was used to separate free labeled antigen from antibody-bound labeled antigen. When 8500 cpm of the labeled ligand was used, approximately 125 cpm were precipitated nonspecifically and 3400 cpm precipitated with the NNN antibody used at a 1/200 dilution. Approximately the same number of counts were bound when the homologous ^{3}H-labeled ligand was used with the *N'*-formylnornicotine and *N'*-acetylnornicotine antibodies at dilutions of 1/100 and 1/500, respectively.

Sensitivity and Specificity of the Assays

The ability of NNN and various structurally related compounds to inhibit the binding of [2'-^{3}H]NNN to the homologous antibody is shown in Table I. Fifty percent inhibition (I_{50}) was achieved with 3.5 ng of NNN. NNN contains two distinct heterocyclic rings, i.e., the pyridine and pyrrolidine rings. Both of these rings and the *N'*-nitroso group are necessary for effective binding of a compound to the antibody. As an inhibitor, nornicotine is only 0.07% as effective as NNN. The immunodominance of the nitroso group was also observed in another radioimmunoassay[37] in which N^6-(methylnitroso)adenosine and N^6-methyladenosine had I_{50}s of 9.6 pmol and 200 nmol, respectively.

[37] R. L. Taylor, H. B. Gjika, and H. Van Vunakis, *Biochem. Biophys. Res. Commun.* **80,** 213 (1978).

TABLE I

INHIBITION OF THE N'-NITROSONORNICOTINE
ANTIGEN–ANTIBODY REACTION[24]

Compound	mmol required for 50% inhibition
N'-Nitrosonornicotine[a]	0.02
(−)-N'-Nitrosoanabasine[a]	0.05
N'-Formylnornicotine[a]	0.6
Nornicotine	27.0
2-Methylnornicotine	105.0
Nicotine	20.3
2-Methylnicotine	260.0
(−)-Cotinine	35.2
Norcotinine	24.6
2′,3′-Dehydronornicotine	9.6
Myosmine	10.9
(−)-Anabasine	57.9
Bridged nicotine[b]	170.0
Ethylnicotinate	132.3
Pyridine	126.4 (26%)[c]
Nicotinamide	81.9 (21%)[c]
Nicotinic acid	81.2 (11%)[c]
(−)-Cotinine N-oxide	52.0 (15%)[c]
Pyrrolidine	d
N-Nitrosopyrrolidine	d
N-Methyl-2-pyrrolidone	d
N-Nitroso-3-hydroxypyrrolidine[a]	d
2-Pyrrolidone	d
(−)-Proline	d
(−)-N-Nitrosoproline[a]	d
N-Nitrosodimethylamine	d

[a] Mixture of Z(-syn) and E(-anti) conformers.
[b] 1,2,3,5,6,10b-Hexahydropyrido[2,3-g]indolizine.
[c] Percentage inhibition achieved at the level tested.
[d] Inhibited less than 10% at the 80 nmol level.

N'-Formylnornicotine cross-reacts with the antibody (3.3%; I_{50} = 0.6 nmol). The fact that the N-nitroso and N-formyl groups are isoelectronic, have similar dipole moments, and three-dimensional geometry[38,39] might explain the affinity that N'-formylnornicotine has for the antibody.

[38] A. L. McClellan, "Tables of Experimental Dipole Moments." Freeman, San Francisco, 1963.
[39] A. L. Friedman, F. M. Muklamethshin, and S. S. Novikov, *Russ. Chem. Rev.* **40**, 34 (1971).

The antibody recognizes the six-membered piperidine ring in N'-nitrosoanabasine (I_{50} = 0.05 nmol compared to 0.02 nmol for NNN). Since N'-nitrosoanabasine has not been detected in tobacco, this cross-reactivity should not pose a problem in tobacco-related studies. Other N-nitroso compounds found in tobacco, e.g., N-nitrosodimethylamine and N-nitrosopyrrolidine, cross-react to an insignificant extent (Table I).

N'-Nitroso-5'-carboxynornicotine, the derivative prepared for synthesis of the conjugate, is 20% as effective an inhibitor as NNN. Among the reasons for this difference is the fact that the derivative, in contrast to NNN, has a carboxyl group that is negatively charged at the pH of the immunoassay.

Approximately 0.01 nmol (1.7 ng) of N'-formylnornicotine is required for 50% inhibition of the $[^3H]N'$-formylnornicotine anti-N'-formylnornicotine system (Table II). The structural similarities of NNN and N'-formylnornicotine were discussed in relation to the cross-reactivity observed with N'-formylnornicotine and the NNN antibody. As expected, NNN cross-reacts with the N'-formylnornicotine antibodies (Table II) and to almost the same extent as N'-formylnornicotine did in the heterologous reaction, i.e., 4.2% (Table I).

TABLE II

INHIBITION OF THE N'-FORMYLNORNICOTINE AND N'-ACETYLNORNICOTINE
ANTIGEN–ANTIBODY REACTION

Compound	nmol required for 50% inhibition	
	$[^3H]N'$-Formylnornicotine anti-N'-formylnornicotine	$[^3H]N'$-Acetylnornicotine anti-N'-acetylnornicotine
N'-Formylnornicotine	0.0096	0.03
N'-Acetylnornicotine	>50.0[a]	0.0064
N'-Hexanoylnornicotine	—	0.05
N'-Carbomethoxynornicotine	>50.0	0.14
N'-Nitrosonornicotine	0.23	0.14
Nornicotine	80.0	25.0
(−)-Nicotine	6.4	20.0
(−)-Cotinine	24.0	120.0
Myosomine	350.0	140.0
N'-Nitrosoanabasine	0.31	—
trans-3'-Hydroxymethyl-N'-acetylnornicotine	—	47.0
Ethyl nicotinate	66.0 (36%)	—
Pyridine	>50.0[a]	—
N'-Nitrosopyrrolidine	>50.0[a]	—

[a] Inhibited less than 10% at the 50 nmol level.

Nicotine and its major metabolite, cotinine, are poor inhibitors of the antigen–antibody reaction, and compounds with substituents larger than the formyl group on the pyrrolidine nitrogen of nornicotine (e.g., N'-acetylnornicotine and N'-carbomethoxynornicotine) are even less effective (Table II).

Approximately 6.4 pmol (1.2 ng) of N'-acetylnornicotine inhibits 50% of the binding of the [³H]N'-acetylnornicotine to its homologous antibodies. The antibody does not recognize differences in the size of the substituents on the pyrrolidine nitrogen; N'-hexanoylnornicotine, despite its size, is an effective inhibitor. Compared to N'-acetylnornicotine, NNN is 20 times less effective. While the N'-acetylnornicotine antibodies crossreact with derivatives having various N'-substituents, 3'-hydroxymethyl-N'-acetylnornicotine, a derivative with a substituent on a position further from the 5' position used as the locus for conjugation, exhibits a 7300-fold decrease in binding effectiveness relative to the homologous hapten.

General Comments

The radioimmunoassay for NNN has been used to monitor nitrosation of nornicotine and anabasine under chemical conditions and to determine the rate constants for the reactions.[24] Since the I_{50} (substrate):I_{50} (product) ratio is greater than 10^3 (Table I) it was possible to detect small amounts of nitrosated product in a reaction mixture that contains relatively large amounts of substrate. Appropriately diluted samples were assayed directly; there was no need for extraction or isolation of the products. The *in vitro* transnitrosation of nornicotine by N-nitrosamines, C-nitrocompounds, and N-nitrosamide and the inhibitory effects of some antioxidants on this reaction were studied in a similar way.[26] Some attempts at using this assay in the analysis of human plasma samples have been made.[24]

The assay for N'-formylnornicotine and N'-acetylnornicotine should prove useful in biosynthetic studies. The I_{50} substrate (i.e., nornicotine):I_{50} product ratios are sufficiently large ($>4 \times 10^3$, Table II) to permit direct assay of enzymatic reaction mixtures.

The fact that the N'-acetylnornicotine antibodies crossreact with several N'-acylnornicotine analogs would preclude their use in radioimmunoassays carried out directly on tobacco samples. However, antisera specific for a class of related compounds can be used to advantage to quantify several compounds after they have been separated by, for example, high pressure liquid chromatography.[40,41]

---

[40] J. G. Loeber and J. Verhoef, this series, Vol. 73 [18].
[41] I. Alam and L. Levine, this series, Vol. 73 [19].

Acknowledgments

The authors are indebted to Dr. D. Dietrich of the Bayer AG, 5090 Leverkusen, Bayer-werk, Federal Republic of Germany, and Dr. H. Hellmann, Member of the Board of Chemische Werke Hüls AG, Federal Republic of Germany for supplying us with a generous sample of 2',3'-dehydro-5'-carboxynornicotine and to Dr. E. Leete of the University of Minnesota for his gifts of 2-methylnornicotine, 2-methylnicotine, and the "bridged" nicotine.

[50] Radioimmunoassay of Nicotinamide Nucleotide Analogs of Nicotine and Cotinine

By WEI-CHIANG SHEN

In the presence of NADase (NAD glycohydrolase, E.C. 3.2.2.5), nicotine and its major metabolite, cotinine, can displace the nicotinamide moiety in NAD or NADP to form the corresponding dinucleotide analogs, i.e. (nicotine)AD,[1] (nicotine)ADP,[2] (cotinine)AD,[3] and (cotinine)ADP.[3] These dinucleotide analogs can be further metabolized in liver homogenates to mononucleotide analogs, i.e., (nicotine)RP and (cotinine)RP.[3] In order to investigate their role in diseases related to cigarette smoking, radioimmunoassays for these nucleotide analogs were developed to measure their formation *in vivo,* and to detect their presence in biological samples.[4]

Materials and Methods

Preparation of (Nicotine)AD and (Cotinine)AD

To a solution of NAD (20 mM) and nicotine or cotinine (100 mM) in 50 ml sodium phosphate buffer, pH 7.4 (50 mM), 9 units of pig brain NADase (Sigma Chemical Co., sp. act. 0.007 units/mg) is added. The reaction mixture is incubated in a shaking water bath at 37°. After 16 hr, the mixture is transferred to a boiling water bath and heated for 3 min to stop the reaction. After cooling, the protein precipitate is removed by centrifugation. Cold acetone (200 ml) is added slowly with stirring, to the supernanant solution. The resulting precipitate is collected, dried *in vacuo,* redissolved in 5 ml H$_2$O, and desalted by passing through a Sephadex G-10 column (2 × 30 cm) equilibrated with 50 mM NH$_4$HCO$_3$. The crude product is lyophilized, redissolved in 30 ml of 50 mM sodium phos-

[1] W.-C. Shen and H. VanVunakis, *Res. Commun. Chem. Pathol. Pharmacol.* **9,** 405 (1974).
[2] W.-C. Shen and H. VanVunakis, *Biochemistry* **13,** 5362 (1974).
[3] W.-C. Shen, J. Franke, and H. VanVunakis, *Biochem. Pharmacol.* **26,** 1835 (1977).
[4] W.-C. Shen, K. M. Greene, and H. VanVunakis, *Biochem. Pharmacol.* **26,** 1841 (1977).

phate buffer, pH 7.4, and incubated 16 hr in a 37° water bath, with 2.5 units of NADase from *N. crassa* (Sigma Chemical Co., sp. act. 3 units/mg) to hydrolyzed unreacted NAD.[5] The incubated solution is lyophilized, desalted, and redissolved in 20 ml H_2O. After adjusting to pH 8 by NH_4OH, the final solution is loaded into a column (1.5×12 cm) packed with Dowex-1 ion-exchange resin in the formate form. (Nicotine)AD and (cotinine)AD can be eluted from the column by 0.1 and 0.2 *M* ammonium formate buffer, pH 3.6 (equal molar of ammonium formate and formic acid), respectively. Absorption at 260 nm can be used to monitor the fractions. Molar absorption at 260 nm of NAD and these analogs are almost identical, i.e., 1.8×10^4.

Preparation of (Nicotine)RP and (Cotinine)RP

(Nicotine)AD or (cotinine)AD (150 mg) is dissolved in 5 ml Tris buffer (50 m*M* Tris-acetate and 5 m*M* $MgCl_2$, pH 7.4). Snake venom nucleotide pyrophosphatase (25 units, Sigma Chemical Co.) is added to the solution. After 18 hr incubation at 37°, the solution is diluted to 100 ml with H_2O, and passed through a Dowex-1 formate column (1.5×12 cm). (Nicotine)RP does not absorb to the resin and can be collected as effluent fluids, while (cotinine)RP can be eluted from the column by 0.08 *M* ammonium formate buffer, pH 3.6. After lyophilization, both products need to be further purified by a Sephadex G-10 column eluted with 50 m*M* NH_4HCO_3.

Preparation of Mononucleotide–BSA Conjugates and the
    Immunization

1-Ethyl-3-(3-dimethylaminopropyl)carbodiimide (10 mg) is added to a solution containing 5 mg bovine serum albumin (BSA) and 10 mg of (nicotine)RP or (cotinine)RP in 0.1 ml H_2O. The reaction mixture is kept in the dark at 25° overnight. It is then diluted to 0.3 ml with buffer-saline (0.14 *M* NaCl–0.01 *M* Tris–Cl, pH 7.2) and loaded on a Sephadex G-50 column (1.2×30 cm). The column is eluted with the same buffer-saline, and fractions at the void volume with a high absorbance at 280 nm are pooled (total approximately 3 ml). From spectral measurements, the concentration of protein can be determined. When small amounts of ^{3}H-labeled nucleotides were used in a typical preparation of the conjugate, the degree of coupling was estimated by radioactivity measurements, and was found to be approximately 4 mol hapten/mol of BSA.

The mononucleotide–BSA conjugate solution collected from the Sephadex G-50 column is mixed with an equal volume of complete

[5] N. O. Kaplan, this series, Vol. 2 [114].

Freund's adjuvant. The emulsion mixture (about 1 mg BSA/ml; 1.25 ml) is injected into the toe pads and leg muscles of four New Zealand albino rabbits. After 6 weeks, the animals are given booster injections and bled 1 week later. Antibodies that bind ([³H]nicotine)AD or ([³H]cotinine)AD can be detected in the sera of all animals. Rabbit sera with good specificity and sensitivity are chosen to develop radioimmunoassays for nicotine nucleotides and cotinine nucleotides.

Preparation of Labeled Haptens

The source and preparation of tritiated alkaloids is described in this volume.[6] [³H]Nicotine or [³H]cotinine (1 mCi, sp. act. 2.4 Ci/mmol) in 0.5 ml of 50 mM sodium phosphate buffer, pH 7.4, is incubated at 37° with 3 mg NAD and 8 mg of NADase (pig brain, sp. act. 0.007 units/mg). After 1 hr, the reaction mixture is heated in a boiling water bath for 3 min. The protein precipitate is removed by centrifugation. The supernatant solution is diluted to 15 ml with H_2O and passed through a column packed with Dowex-1 (formate) resin. The column is washed with an excess of H_2O to remove free radioactivity and then eluted with ammonium formate buffer, pH 3.6 (0.1 M for [³H]nicotine and 0.2 M for [³H]cotinine products). Fractions that contain (nicotine)AD or (cotinine)AD are pooled and lyophilized. The residue from lyophilization is dissolved in a small amount of H_2O and stored at $-20°$. These radioactive NAD analogs undergo large dilutions prior to use in the radioimmunoassay procedure and, therefore, can be used without further purification. Since the antisera produced from rabbits immunized with the mononucleotide analog–BSA conjugate bind the homologous NAD analog and mononucleotide equally well, these radioactive NAD analogs can be used as the labeled antigens in the radioimmunoassays without further hydrolysis to the mononucleotide analogs.

Radioimmunoassay Procedures

The double antibody technique, described in this volume,[6] can be used to separate free labeled antigen from antibody–antigen complexes. Antisera for rabbits immunized with mononucleotide–BSA conjugates are diluted to 1/1000 for anticotinine nucleotide and 1/200 for antinicotine nucleotide. When 15,000 cpm of labeled antigens was incubated with these diluted antisera, approximately 2500 cpm could be precipitated by the goat anti-rabbit γ-globulin. Only 200 cpm was found in the precipitate when normal rabbit serum was used to substitute the antisera.[4] The table shows the 50% inhibition of binding of labeled NAD analogs to the anti-

---

[6] J. J. Langone and H. Van Vunakis, this Volume [48].

INHIBITION OF THE NICOTINE NUCLEOTIDE AND COTININE
NUCLEOTIDE ANTIGEN–ANTIBODY REACTIONS[a]

	Amount required for 50% inhibition (nanomoles)	
	---	---
Compound	([³H]Nicotine)AD-(antinicotine nucleotide)	([³H]Cotinine)AD-(anticotinine nucleotide)
(Nicotine)AD	0.010	10.0 (45.8%)[b]
(Nicotine)ADP	0.013	
(Nicotine)RP	0.017	10.0 (38.4%)[b]
Nicotine-ribose	0.320	
Nicotine	10.0 (14.7%)[b]	
(Cotinine)AD	10.0 (45.2%)[b]	0.008
(Cotinine)ADP		0.008
(Cotinine)RP	10.0 (42.8%)[b]	0.010
Cotinine-ribose		0.450
Cotinine		10.0 (19%)[b]
NAD	10.0 (3.6%)[b]	10.0 (0%)[b]
NMN	10.0 (4.6%)[b]	10.0 (0%)[b]
Nicotinamide	10.0 (0%)[b]	10.0 (0%)[b]

[a] From Shen et al.[4]
[b] Percentage inhibition at 10.0 nmol.

sera by various nicotine and cotinine derivatives. In both cases, the homologous mononucleotide, NAD and NADP analogs are the most effective inhibitors; only 8 to 13 pmol is required to produce 50% inhibition of the binding. The ribosides, i.e., nicotine-ribose and cotinine-ribose, are 30- to 40-fold less sensitive than their nucleotide derivatives, while the free bases are essentially inactive in the binding assays.

Examples

This RIA procedure has been used to measure the formation of nicotine and cotinine nucleotide analogs in rabbits injected with either nicotine or cotinine.[4] In these studies, New Zealand albino rabbits were injected (iv) daily with 5 μmol nicotine or 50 μmol cotinine for 23 days. Tissue extracts from the treated rabbits were partially purified by Dowex-1 ion-exchange column and desalted by Sephadex G-10 column prior to the assay procedure. In both nicotine and cotinine treatments, cotinine mononucleotide was the major analog formed in vivo. It was found primarily in liver (up to 48 pmol/g wet tissue), but was also found with appreciable amounts in lung and kidney.[4] Since the pyridine–ribose bond in the nucleotide analogs is labile in alkaline conditions,[7] the presence of

[7] N. O. Kaplan, S. P. Colowick, and C. C. Barnes, J. Biol. Chem. **191,** 461 (1951).

nicotine or cotinine nucleotides can be confirmed by measuring the free alkaloid released during mild NaOH hydrolysis. (Nicotine and cotinine can be measured with RIA methods described in this volume.[6]) For example, the amount of free cotinine present in the hydrolyzate of a cotinine-treated rabbit liver extract was found to be equal to that of the cotinine nucleotide present before hydrolysis.[4]

Most recently, the presence of these nucleotide analogs has also been found in human autopsy tissues of cigarette smokers, but not of nonsmokers, with a procedure similar to that described above.[8]

Discussion

This method of antibody preparation for the development of RIA of nucleotide analogs is similar to that used by Halloran and Parker for the preparation of antibodies against mono-, oligo-, and polynucleotides,[9,10] and by Humayun and Jacob for antideoxyadenylate antibodies.[11] Such a direct linking of the phosphate group of a nucleotide analog to the amino group of a protein to form an immunogenic conjugate has several advantages: (a) It requires a minimum of synthetic procedure for conjugate preparation. (b) The antibody produced from this conjugate is highly specific to the whole nucleotide structure. Maximal binding occurs only when the base, ribose and phosphate are all present in the hapten. (c) The nucleoside moiety is the most important determinant for the antibody binding. Antinucleotide analog antibodies recognize the homologous nucleosides much better than nucleotides consisting of other base groups (table). This specificity makes it possible to measure each individual nucleotide analog in the presence of other structurally similar nucleotides, especially of the naturally abundant NAD, NADP, and NMN.

For nucleotides other than pyridine nucleotide analogs, this carbodiimide conjugation method has been reviewed in this series.[12]

[8] W. C. Shen, R. L. Taylor, M. S. Dunnill, and H. Van Vunakis (unpublished result).
[9] M. J. Halloran and C. W. Parker, *J. Immunol.* **96,** 373 (1966).
[10] M. J. Halloran and C. W. Parker, *J. Immunol.* **96,** 379 (1966).
[11] M. Z. Humayun and T. M. Jacob, *Biochim. Biophys. Acta* **331,** 41 (1973).
[12] B. D. Stollar, this series, Vol. 70 [3].

Section IX

Summary

[51] Previously Published Articles from Methods in Enzymology Related to Immunochemical Techniques

By HELEN VAN VUNAKIS and JOHN J. LANGONE

I. Immunoassays for Various Compounds

A. Proteins and Peptides

1. Enzymes

Vol. X [106]. Preparation and Use of Antisera to Respiratory Chain Components. S. D. Davis, T. D. Mehl, R. J. Wedgewood, and B. Mackler.

Vol. X [107]. Antibody Against F_1. J. Fressenden and E. Racker.

Vol. XLIII [6]. Immunological Techniques for Studying β-Lactamase. M. H. Richmond and V. Betina.

Vol. LVII [6]. Determination of Creatine Kinase Isoenzymes in Human Serum by an Immunological Method Using Purified Firefly Luciferase. A. Lundin.

Vol. 66 [102]. Preparation of an Antiserum to Sheep Liver Dihydropteridine Reductase. S. Milstein and S. Kaufman.

Vol. 70 [31]. Quantitative Micro Complement Fixation: Serologic Properties of Pig Liver Carboxylesterase. L. Levine, A. Baer, and W. P. Jencks.

Vol. 73 [37]. Use of Antibodies and the Primary Enzyme Immunoassay (PEIA) to Study Enzymes: The Arylsulfatase A–Anti-Arylsulfatase A System. E. A. Neuwelt.

Vol. 74 [12]. Radioimmunoassay of Creatine Kinase Isoenzymes. R. Roberts and C. W. Parker.

Vol. 74 [13]. Specific Radioimmunoassays for Rabbit Liver Fructose-Biphosphatase, Pyruvate Kinase, and Glycerol-3-Phosphate Dehydrogenase. C. M. Veneziale, J. C. Donofrio, J. B. Hansen, M. L. Johnson, and M. Y. Mazzotta.

METHODS IN ENZYMOLOGY, VOL. 84

Vol. 74 [14]. Specific Radioimmunoassays for Rabbit Skeletal and Cardiac Muscle-6-Phosphofructokinase and Pyruvate Kinase. C. M. Veneziale, J. B. Hansen, and M. L. Johnson.

Vol. 74 [15]. Preparation of Site-Specific Anti-Cytochrome *c* Antibodies and Their Application. R. Jemmerson and E. Margoliash.

Vol. 74 [16]. Production and Application of Antibodies to Rat Liver Cytochrome *P*-450. L. S. Kaminsky, M. J. Fasco, and F. P. Guengerich.

Vol. 74 [17]. Radioimmunoassay Determination of Circulating Pancreatic Endopeptidases. C. Largman, J. W. Brodrick, and M. C. Geokas.

Vol. 74 [18]. Radioimmunoassay of Human Pancreatic Amylase. M. Ogawa, Y. Takatsuka, T. Kitahara, K. Matsuura, and G. Kosaki.

Vol. 74 [19]. Radioimmunoassay of the Regulatory Subunit of Type 1 cAMP-Dependent Protein Kinase. C. L. Kapoor, J. A. Beavo, and A. L. Steiner.

Vol. 74 [20]. Radioimmunoassay of Bovine Type II cAMP-Dependent Protein Kinase. N. Fleischer, D. Sarkar, C. Rubin, and J. Erlichman.

Vol. 74 [21]. Immunotitration of 3-Hydroxy-3-Methyl-Glutaryl-CoA Reductase. T. J. Scallen. J. E. Hardgrave, and R. A. Heller.

Vol. 74 [22]. Production of Antibodies to Catalase and Their Effect on Enzyme Activity. R. N. Feinstein and B. N. Jaraslow.

Vol. 74 [23]. Radioimmunoassay of Human Adenosine Deaminase. P. E. Daddona, M. A. Frohman, and W. N. Kelley.

Vol. 74 [24]. Quantitation of Human Cuprozinc Superoxide Dismutase (SOD-1) by Radioimmunoassay and Its Possible Significance in Disease. B. C. Del Villano and J. A. Tischfield.

Vol. 74 [25]. Radioimmunoassay and Immunotitration of Human Serum Dopamine-Hydroxylase. J. Dunnette and R. Weinshilboum.

2. Hormones and Endorphins

Vol. XXXVII [2]. General Considerations for Radioimmunoassay of Peptide Hormones. D. N. Orth.

Vol. XXXVII [3]. Development and Application of Sequence-Specific Radioimmunoassays for Analysis of the Metabolism of Parathyroid Hormone. G. V. Segre, G. W. Tregear, and J. T. Potts, Jr.

Vol. XXXVII [16]. Methods for Assessing Immunologic and Biologic Properties of Iodinated Peptide Hormones. J. Roth.

Vol. XXXVII [28]. Methods for the Assessment of Peptide Precursors. Studies on Insulin Biosynthesis. H. S. Tager, A. H. Rubenstein, and D. F. Steiner.

Vol. XXXVII [29]. Technique for the Identification of a Biosynthetic Precursor to Parathyroid Hormone. J. F. Habener and J. T. Potts, Jr.

Vol. 70 [23]. The Talc–Resin–Trichloroacetic Acid Test for Screening Radioiodinated Polypeptide Hormones. B. B. Tower, M. B. Sigel, R. E. Poland, W. P. VanderLaan, and R. T. Rubin.

Vol. 73 [18]. High-Pressure Liquid Chromatography and Radioimmunoassay for the Specific and Quantitative Determination of Endorphins and Related Peptides. J. G. Loeber and J. Verhoef.

3. Immunoglobulins (Including Monoclonal Antibodies) and Immune Complexes

Vol. XLVI [53]. Affinity Labeling of Antibody Combining Sites as Illustrated by Anti-Dinitrophenyl Antibodies. D. Givol and M. Wilchek.

Vol. XLVI [54]. *p*-Azobenzenearsonate Antibody. M. J. Ricardo and J. J. Cebra.

Vol. XLVI [55]. Affinity Cross-Linking of Heavy and Light Chains. M. Wilchek and D. Givol.

Vol. XLVI [56]. Bivalent Affinity Labeling Haptens in the Formation of Model Immune Complexes. P. H. Plotz.

Vol. XLVI [58]. Labeling of Antilactose Antibody. P. V. Gopalakrishnan, U. J. Zimmerman, and F. Karush.

Vol. 70 [6]. Preparation of Fab Fragments from IgGs of Different Animal Species. M. G. Mage.

Vol. 70 [24]. Labeled Antibodies and Their Use in the Immunoradiometric Assay. C. N. Hales and J. S. Woodhead.

Vol. 70 [25]. ^{125}I-Labeled Protein A: Reactivity with IgG and Use as a Tracer in Radioimmunoassay. J. J. Langone.

Vol. 70 [26]. The RAST Principle and the Use of Mixed-Allergen RAST as a Screening Test for IgE-Mediated Allergies. T. G. Merrett and J. Merrett.

Vol. 73 [1]. Preparation of Monoclonal Antibodies: Strategies and Procedures. G. Galfrè and C. Milstein.

Vol. 73 [33]. Magnetic Solid-Phase Enzyme Immunoassay for the Quantitation of Antigens and Antibodies: Application to Human Immunoglobulin E. J. L. Guesdon and S. Avrameas.

Vol. 73 [34]. The Amplified ELISA: Principles of and Applications for the Comparative Quantitation of Class and Subclass Antibodies and the Distribution of Antibodies and Antigens in Biochemical Separates. J. E. Butler.

Vol. 73 [38]. Immunonephelometric Assay for Immunoglobulins Released by Cultured Lymphocytes. J. Muñoz, G. Virella, and H. H. Fudenberg.

Vol. 73 [39]. Solid-Phase Radioimmunoassay for Immunoglobulins and Influenza Antibodies. H. Daugharty.

Vol. 73 [40]. A Luminol-Assisted Competitive Binding Immunoassay of Human Immunoglobulin G. L. S. Hersh, W. P. Vann, and S. A. Wilhelm.

Vol. 73 [41]. Preparation and Radioimmunoassay of IgM Domains. A. Shimizu and S. Watanabe.

Vol. 73 [42]. Quantitation of Secretory Protein Levels by Radioimmunoassay. J. L. Klein and J. R. Dawson.

Vol. 73 [43]. Double-Antibody Radioimmunoassay for IgE. S. L. Dunnette and G. J. Gleich.

Vol. 73 [44]. Radioimmunoassay for IgE Using Paper Disks. M. Ceska.

Vol. 73 [45]. Enzyme-Linked Immunosorbent Assays (ELILSA) for Immunoglobulin E and Blocking Antibodies. D. R. Hoffman.

Vol. 73 [46]. Development and Clinical Application of Radioimmunoassay Techniques for Measuring Low Levels of Immunoglobulin Classes G, A, M, D, and E in Cerebrospinal and Other Body Fluids. S. T. Nerenberg and R. Prasad.

Vol. 74 [34]. The Raji, Conglutinin, and Anti-C3 Assays for the Detection of Complement-Fixing Immune Complexes. A. N. Theofilopoulos.

Vol. 74 [35]. The [^{125}I]Clq Binding Assay for the Detection of Soluble Immune Complexes. R. H. Zubler, N. Carpentier, and P.-H. Lambert.

Vol. 74 [36]. Quantitation of Circulating Immune Complexes by Combined PEG Precipitation and Immunoglobulin-Specific Radioimmunoassay (PICRIA). H. C. Siersted, I. Brandslund, S.-E. Svehag, and J. C. Jensenius.

Vol. 74 [37]. Detection and Quantitation of Immune Complexes with a Rapid Polyethylene Glycol Precipitation Complement Consumption Method (PEG-CC). I. Brandslund, H. C. Siersted, J. C. Jensenius, and S.-E. Svehag.

Vol. 74 [38]. Detection and Quantitation of Circulation Immune Complexes by the Clq–Protein A Binding Assay (Clq-PABA). G. Glikmann and S.-E. Svehag.

4. Interferons and the Immune Response

B. Nucleic Acids and Related Compounds

Vol. VI [8]. Preparation of Lamb Brain Phosphodiesterase. J. W. Healy, D. Stollar, and L. Levine.

Vol. XIIB [173].[a] Purine- and Pyrimidine–Protein Conjugates. S. M. Beiser, S. W. Tatenbaum, and B. F. Erlanger.

Vol. XIIB [174]. Preparation and Assay of Nucleic Acids as Antigens. O. J. Plescia.

Vol. XIIB [175].[a] Preparation of Nucleoside-Specific Synethic Antigens. M. Sela and H. Ungar-Waron.

Vol. XIIB [176]. Immunological Detection of Ribonucleic Acids by Agar Diffusion. F. Lacour.

Vol. XXXVIII [13]. Assay of Cyclic Nucleotides by Radioimmunoassay Methods. A. L. Steiner.

Vol. XL [22]. Use of Antibodies to Nucleosides and Nucleotides in Studies of Nucleic Acids in Cells. B. F. Erlanger, W. J. Klein, Jr., V. G. Dev, R. R. Shreck, and O. J. Miller.

Vol. 70 [3]. The Experimental Induction of Antibodies to Nucleic Acids. B. D. Stollar.

Vol. 70 [14]. *In Vitro* Iodination of Nucleic Acids. S. L. Commerford.

Vol. 79 [28]. Radioimmune and Radiobinding Assays for A2'p5'A2'p5'A, pppA2'p5'A2'p5'A, and Related Oligonucleotides. M. Knight, D. H. Wreschner, R. H. Silverman, and I. M. Kerr.

C. Carbohydrates—Simple and Complex

Vol. VIII [5]. Immunological Methods for Characterizing Polysaccharides. G. Schiffman.

Vol. XXVIII [16].[a] Carbohydrate Antigens: Coupling of Carbohydrates to Proteins by Deazonium and Phenylisothiocyanate Reactions. C. R. McBroom, C. H. Samanen, and I. J. Goldstein.

Vol. XXVIII [17].[a] Carbohydrate Antigens: Coupling of Carbohydrates to Protein by a Mixed Anhydride Reaction. G. Ashwell.

Vol. XXVIII [18].[a] Carbohydrate Antigens: Coupling of Carbohydrates to Proteins by Diazotizing Aminophenylflavazole Derivatives. K. Himmelspach and G. Kleinhammer.

Vol. XXVIII [19]. Preparation of Antisera Against Glycolipids. S. I. Hakomori.

Vol. L [5]. Direct Identification of Specific Glyco-Proteins and Antigens in Sodium Dodecyl Sulfate Gels. K. Burridge.

Vol. L [12].[a] Antibodies to Carbohydrates: Preparation of Antigens by Coupling Carbohydrates to Proteins by Reductive Amination with Cyanoborohydride. G. R. Gray.

Vol. L [13].[a] Carbohydrate Antigens: Coupling Melibionic Acid to Bovine Serum Albumin Using Water-Soluble Carbodiimide. J. Lönngren and I. J. Goldstein.

Vol. L [14]. Carbohydrate Antigens: Coupling of Oligosaccharide-Phenethylamine Derivatives to Edestin by Diazotization and Characterization of Antibody Specificity by Radioimmunoassay. D. A. Zopf, C.-M. Tsai, and V. Ginsburg.

Vol. L [15].[a] Carbohydrate Antigens: Coupling of Oligosaccharide Phenethylamine-Isothiocyanate Derivatives to Bovine Serum. D. F. Smith, D. A. Zopf, and V. Ginsburg.

Vol. L [16]. Affinity Purification of Antibodies Using Oligosaccharide-Phenethylamine Derivatives Coupled to Sepharose. D. A. Zopf, D. F. Smith, Z. Drzeniek, C-M Tsai, and V. Ginsburg.

Vol. 70 [1]. Basic Principles of Antigen–Antibody Reaction. E. A. Kabat.

Vol. 83 [13]. Immunological Characterization of Cartilage Proteoglycans. J. R. Baker, B. Caterson, and J. E. Christner.

[a] Only the preparation of hapten–protein conjugates is described.

D. Lipids, Prostaglandins, and Steroids

Vol. XXXV [35]. Immunology of Prostaglandins. R. M. Gutierrez-Cernosek, L. Levine, and H. Gjika.

Vol. XXXVI [2]. Use of Specific Antibodies for Quantification of Steroid Hormones. G. D. Niswender, A. M. Akbar, and T. M. Nett.

Vol. XXXVI [4a]. Assays of Cellular Steroid Receptors Using Steroid Antibodies. E. Castañeda and S. Liao.

Vol. 73 [12]. Effect of Catalytic Hydrogenation of Cellular Lipid and Fatty Acid on the Susceptibility of Tumor Cells to Humoral Immune Killing. S. I. Schlager.

Vol. 73 [19]. Qualitative and Quantitative Analyses of Arachiodonic Acid Metabolites by Combined High-Performance Liquid Chromatography and Radioimmunoassay. I. Alam and L. Levine.

E. Receptors, Membranes, and Cell Surface Antigens

Vol. XXXII [6]. Use of Antibodies for Localization of Components on Membranes. W. C. Davis.

Vol. XXXIV [88]. Thyrotropin Receptors and Antibody. R. L. Tate, R. J. Winand, and L. D. Kohn.

Vol. XXXVI [4a]. Assays of Cellular Steroid Receptors Using Steroid Antibodies. E. Castañeda and S. Liao.

Vol. LVI [21]. Use of Antibodies for Studying the Sidedness of Membrane Components. S. H. P. Chan and G. Schatz.

Vol. 69 [47]. Antibody Approach to Membrane Architecture. R. J. Berzborn.

Vol. 69 [65]. Inhibitors in Electron Flow: Tools for the Functional and Structural Localization of Carriers and Energy Conservation Sites. A. Trebst.

Vol. 70 [15]. Radioiodination of Cell Surface Lipids and Proteins for Use in Immunological Studies. S. I. Schlager.

Vol. 73 [12]. Effect of Catalytic Hydrogenation of Cellular Lipid and Fatty Acid on the Susceptibility of Tumor Cells to Humoral Immune Killing. S. I. Schlager.

Vol. 73 [28]. Methods for Binding Cells to Plastic: Application to Solid Phase Immunoassays for Cell-Surface Antigens. C. H. Heusser, J. W. Stocker, and R. H. Gisler.

Vol. 73 [30]. The Use of [^{125}I]Clq Subcomponent for the Measurement of Complement-Binding Antibodies on Cell Surfaces. P. S. Shepherd and C. J. Dean.

Vol. 74 [26]. Antibodies to Prolactin Receptors and Growth Hormone Receptors. R. G. Drake and H. G. Friesen.

Vol. 74 [27]. Measurement of Thyrotropin Receptor Antibodies. B. R. Smith and R. Hall.

Vol. 74 [28]. Quantitation of Estradiol Receptors by Radioimmunoassay. A. Floridi.

Vol. 74 [29]. Production and Assay of Antibodies to Acetylcholine Receptors. J. Lindstrom, B. Einarson, and S. Tzartos.

Vol. 74 [30]. Production and Application of an Antibody Specific for the Cardiac-Adrenergic Receptor. S. M. Wrenn, Jr.

Vol. 74 [31]. Insulin Receptor Antibodies. S. Jacobs and P. Cuatrecasas.

F. Viruses, Bacteria, and Toxins

Vol. LVIII [35]. Purification and Assay of Murine Leukemia Viruses. C. J. Sherr and G. J. Todaro.

Vol. LX [69]. Toxin Inhibitors of Protein Synthesis: Production, Purification, and Assay of *Pseudomonas aeruginosa* Toxin A. B. H. Iglewiski and J. C. Sadoff.

Vol. 78 [48]. Staphylococcal Enterotoxin A (SEA). L. Spero and J. F. Metzger.

G. Other Low-Molecular-Weight Compounds (Including Vitamins, Hormones, Drugs, and Environmental Agents)

Vol. XVIIIA [93]. Preparation and Properties of Antigenic Vitamin and Coenzyme Derivatives. J-C. Jaton and H. Ungar-Waron.

Vol. XVIIIB [189]. Preparation and Properties of Antigenic Vitamin and Coenzyme Derivatives. J-C. Jaton and H. Ungar-Waron.

Vol. 62 [57]. Antibodies That Bind Biotin and Inhibit Biotin-Containing Enzymes. M. Berger.

Vol. 73 [5]. ^{125}Iodinated Tracers for Hapten-Specific Radioimmunoassays. J. E. T. Corrie and W. M. Hunter.

Vol. 73 [26]. Ultrasensitive Enzymic Radioimmunoassay. I. C. Hsu, R. H. Yolken, and C. C. Harris.

Vol. 74 [7]. Spin Immunoassay. G. C. Yang and E. S. Copeland.

Vol. 74 [8]. Membrane Immunoassay: A Spin Membrane Immunoassay for Thyroxine. C. T. Tan, S. W. Chan, and J. C. Hsia.

II. General Immunoassays and Immunological Techniques: Principles and Procedures

Vol. VI [119]. Immunodiffusion Two-Dimensional. D. Stollar and L. Levine.

Vol. XI [91]. Immunological Techniques (General). R. K. Brown.

Vol. XI [92].[b] Micro Complement Fixation. L. Levine and H. Van Vunakis.

Vol. XXII [31]. Affinity Chromatography. P. Cuatrecasas and C. B. Anfinsen.

Vol. XXXIV [90]. Immunoadsorbents. J. B. Robbins and R. Schneerson.

Vol. XXXIV [91]. Immunoaffinity Chromatography of Proteins. D. M. Livingston.

Vol. XLIV [48].[b] Immunoenzymatic Techniques for Biomedical Analysis. S. Avrameas.

Vol. 70 [1]. Basic Principles of Antigen–Antibody Reaction. E. A. Kabat.

Vol. 70 [4].[a,b] The Preparation of Antigenic Hapten-Carrier Conjugates: A Survey. B. F. Erlanger.

Vol. 70 [5]. Production of Reagent Antibodies. B. A. L. Hurn and S. M. Chantler.

Vol. 70 [7]. The Use of Carbodiimides in the Preparation of Immunizing Conjugates. S. Bauminger and M. Wilchek.

Vol. 70 [9]. Immunochemical Analysis by Antigen–Antibody Precipitation in Gels. J. Oudin.

Vol. 70 [10]. Radioimmunoassays: An Overview. H. Van Vunakis.

Vol. 70 [13].[b] Radioiodination by Use of the Bolton–Hunter and Related Reagents. J. J. Langone.

Vol. 70 [16]. Use of the Double-Antibody Method to Separate Antibody Bound from Free Ligand in Radioimmunoassay. A. Rees Midgley, Jr., and M. R. Hepburn.

Vol. 70 [17]. Use of Charcoal to Separate Antibody Complexes from Free Ligand in Radioimmunoassay. W. D. Odell.

---

[b] An extensive list of references to original papers describing immunoassays for various compounds is included.

Vol. 70 [18]. Ammonium Sulfate and Polyethylene Glycol as Reagents to Separate Antigen from Antigen–Antibody Complexes. T. Chard.

Vol. 70 [19]. Use of Hydroxyapatite in Radioimmunoassay. D. J. H. Trafford and H. L. J. Makin.

Vol. 70 [21]. Microfiltration as a Means of Separating Free Antigen from Antigen–Antibody Complexes in Immunoassay. S. R. Chalkley and A. Renshaw.

Vol. 70 [22]. Gel Centrifugation: Separation of Free Labeled Antigen from Antibody-Bound Labeled Antigen. J. Larsen and W. F. Rasmussen.

Vol. 70 [24]. Labeled Antibodies and Their Use in the Immunoradiometric Assay. C. N. Hales and J. S. Woodhead.

Vol. 70 [25].[b] ^{125}I-Labeled Protein A: Reactivity with IgG and Use as a Tracer in Radioimmunoassay. J. J. Langone.

Vol. 70 [27]. Semiautomation of Immunoassays by Use of Magnetic Transfer Devices. K. O. Smith and W. D. Gehle.

Vol. 70 [28].[b] Enzyme Immunoassay ELISA and EMIT. E. Engvall.

Vol. 70 [29]. Electrode-Based Enzyme Immunoassays Using Urease Conjugates. M. E. Meyerhoff and G. A. Rechnitz.

Vol. 70 [30]. Passive Hemagglutination and Hemolysis for Estimation of Antigens and Antibodies. F. L. Adler and L. T. Adler.

Vol. 73 [1]. Preparation of Monoclonal Antibodies: Strategies and Procedures. G. Galfre and C. Milstein.

Vol. 73 [2]. Production of Antisera with Small Doses of Immunogen: Multiple Intradermal Injections. J. L. Vaitukaitis.

Vol. 73 [3]. Production of Specific Antisera by Immunization with Precipitin Lines. J. Krøll.

Vol. 73 [4]. Polymers for the Sustained Release of Macromolecules: Their Use in a Single-Step Method of Immunization. R. Langer.

Vol. 73 [5].[b] [125]Iodinated Tracers for Hapten-Specific Radioimmunoassays. J. E. T. Corrie and W. M. Hunter.

Vol. 73 [6].[b] Radioiodination by Use of the Bolton–Hunter and Related Reagents. J. J. Langone.

Vol. 73 [7]. Use of ^{125}I-Labeled Anti-2,4-Dinitrophenyl (DNP) Antibodies as a General Tracer in Solid-Phase Radioimmunoassays. A. R. Neurath.

Vol. 73 [8]. Tritiation of Proteins to High Specific Activity: Application to Radioimmunoassay. B. F. Tack and R. L. Wilder.

Vol. 73 [9].[b] Methods for the Preparation of Enzyme–Antibody Conjugates for Use in Enzyme Immunoassay. M. J. O'Sullivan and V. Marks.

Vol. 73 [10]. Purification of Peroxidase-Conjugated Antibody for Enzyme Immunoassay by Affinity Chromatography on Concanavalin A. J. Arends.

Vol. 73 [11]. Preparation of Protein A-Enzyme Monoconjugate and Its Use as a Reagent in Enzyme Immunoassays. A. Surolia and D. Pain.

Vol. 73 [13]. Use of Particulate Immunosorbents in Radioimmunoassay. L. Wide.

Vol. 73 [14]. Antibody-Coated Plastic Tubes in Radioimmunoassay. G. H. Parson, Jr.

Vol. 73 [15]. Quantitation of Antibodies Immobilized on Plastics. J. E. Herrmann.

Vol. 73 [16]. Radioimmunoassay of Peptide Hormones Using Killed *Staphylococcus aureus* as a Separating Agent. S. Y. Ying.

Vol. 73 [17]. Microencapsulation of Antibody for Use in Radioimmunoassay. F. Lim and R. S. Buehler.

Vol. 73 [20]. Single Radial Immunodiffusion. J. P. Vaerman.

Vol. 73 [21]. The Two-Cross Immunodiffusion Technique for Determining Diffusion Coefficients and Precipitating Titers of Antigen and Antibody. B. Pokrić and Z. Pučar.

Vol. 73 [22]. Antigen Quantitation by a Reverse Hemolytic Assay. G. A. Molinaro, W. C. Eby, and C. A. Molinaro.

Author Index

Numbers in parentheses are reference numbers and indicate that an author's work is referred to although the name is not cited in the text.

M

Subject Index